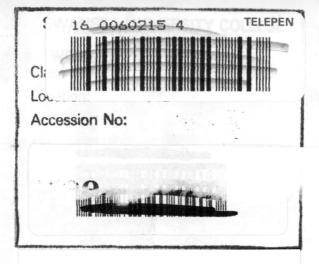

This book must be returned immediately
it is asked for by the Librarian, and in
any case by the last date stamped below.

Wave Dynamics and Radio Probing of the Ocean Surface

WAVE DYNAMICS AND RADIO PROBING OF THE OCEAN SURFACE

Edited by

O. M. PHILLIPS

The Johns Hopkins University
Baltimore, Maryland

and

KLAUS HASSELMANN

Max Planck Institute for Meteorology
Hamburg, Federal Republic of Germany

PLENUM PRESS • NEW YORK AND LONDON

Library of Congress Cataloging in Publication Data

Wave dynamics and radio probing of the ocean surface.

"Proceedings of a symposium . . . held May 13–20, 1981, in Miami, Florida, sponsored by the Inter-Union Commission on Radio Meteorology"—T.p. verso.
 Bibliography: p.
 Includes index.
 1. Ocean waves—Congresses. 2. Oceanography—Remote sensing—Congresses. 3. Microwave devices—Congresses. I. Phillips, O. M. (Owen M.), 1930–   . II. Hasselmann, Klaus, 1931–   . III. Inter-Union Commission on Radio Meteorology.
GC206.W38   1986   551.47'022   86-8106
ISBN 0-306-41992-0

Proceedings of a symposium on Wave Dynamics and Radio Probing of the Ocean Surface, held May 13–20, 1981, in Miami, Florida, sponsored by the Inter-Union Commission on Radio Meteorology

© 1986 Plenum Press, New York
A Division of Plenum Publishing Corporation
233 Spring Street, New York, N.Y. 10013

Printed in the United States of America

PREFACE

In 1960, Dr. George Deacon of the National Institute of Oceanography in England organized a meeting in Easton, Maryland that summarized the state of our understanding at that time of ocean wave statistics and dynamics. It was a pivotal occasion: spectral techniques for wave measurement were beginning to be used, wave–wave interactions had just been discovered, and simple models for the growth of waves by wind were being developed. The meeting laid the foundation for much work that was to follow, but one could hardly have imagined the extent to which new techniques of measurement, particularly by remote sensing, new methods of calculation and computation, and new theoretical and laboratory results would, in the following twenty years, build on this base. When Gaspar Valenzuela of the U.S. Naval Research Laboratory perceived that the time was right for a second such meeting, it was natural that Sir George Deacon would be invited to serve as honorary chairman for the meeting, and the entire waves community was delighted at his acceptance.

The present volume contains reviewed and edited papers given at this second meeting, held this time in Miami, Florida, May 13–20, 1981, with the generous support of the Office of Naval Research, the National Aeronautics and Space Administration, and the National Oceanic and Atmospheric Administration. The range of contributions illustrates clearly the much more powerful measurement techniques that have become available since 1960 and the firmer and more extensive observational base that has been built up as a result. They show the greatly enhanced level of dynamical sophistication with which we view ocean wave characteristics and structure. Many important questions remain but it is to be hoped that the advances described in this book will serve as a timely summary of the state of the art that will be of interest to oceanographers, ocean engineers, and meteorologists and will, at the same time, reset the stage for future directions of research.

Klaus Hasselmann
O. M. Phillips

CONTENTS

Keynote Addresses

Sir George Deacon, Honorary Chairman 1
Dr. William Rainey, NASA 2
Mr. Herbert Rabin, Deputy Assistant Secretary of the Navy 3

I. Ocean Wave Spectra

1. The Equilibrium Ranges in Wind–Wave Spectra: Physical Arguments and Experimental Evidence for and against Their Existence 9
   S. A. Kitaigorodskii

2. Nonlinear Energy Transfer between Random Gravity Waves: Some Computational Results and Their Interpretation. 41
   Akira Masuda

3. The Interaction between Long and Short Wind-Generated Waves 59
   M. T. Landahl, J. A. Smith, and S. E. Widnall

4. Energy Distribution of Waves above 1 Hz on Long Wind Waves 75
   Karl Richter and Wolfgang Rosenthal

5. The Effects of Surfactant on Certain Air–Sea Interaction Phenomena 95
   H. Mitsuyasu and T. Honda

6. Experimental Study of Elementary Processes in Wind-Waves Using Wind over Regular Waves 117
   Yoshiaki Toba, Mitsuhiko Hatori, Yutaka Imai, and Masayuki Tokuda

7. An Experimental Study of the Statistical Properties of Wind-Generated Gravity Waves 129
   Norden E. Huang, Steven R. Long, and Larry F. Bliven

8. On Finite-Depth Wind-Wave Generation and Dissipation. 145
   C. E. Knowles

9. The Equilibrium Range for Waves in Water of Finite Depth 161
   Dinorah C. Esteva

II. Wave Propagation

10. The 1978 Ocean Wave Dynamics Experiment: Optical and *in Situ* Measurement of the Phase Velocity of Wind Waves 165
    G. B. Irani, B. L. Gotwols, and A. W. Bjerkaas

11. Transformation of Statistical Properties of Shallow-Water Waves   181
    B. Le Méhauté, C. C. Lu, and E. W. Ulmer
12. Aspects of the Velocity Field and Dispersion Relation in Surface Wind Waves.   193
    V. V. Yefimov and B. A. Nelepo

III. Wave Instabilities and Breaking

13. Advances in Breaking-Wave Dynamics   209
    M. S. Longuet-Higgins
14. Experimental Studies of Strong Nonlinear Interactions of Deep-Water Gravity
    Waves   231
    Ming-Yang Su and Albert W. Green
15. The Instability and Breaking of a Deep-Water Wave Train   255
    W. K. Melville
16. Measurements of Breaking Waves: Implications for Wind-Stress and Wave
    Generation   257
    M. S. Longuet-Higgins and N. D. Smith
17. Statistical Characteristics of Breaking Waves   265
    Yeli Yuan, C. C. Tung, and Norden E. Huang
18. On Microwave Scattering by Breaking Waves   273
    Lewis Wetzel
19. Observation of Breaking Ocean Waves with Coherent Microwave Radar . .   285
    W. C. Keller, W. J. Plant, and G. R. Valenzuela
20. An Estimate of the Influence of Breaking Waves on the Dynamics of the Upper
    Ocean   295
    Norden E. Huang
21. Stability of Nonlinear Capillary Waves   315
    S. J. Hogan

IV. Air Flow over Waves

22. A Comparison of the Wave-Induced Momentum Flux to Breaking and
    Nonbreaking Waves   321
    M. L. Banner
23. Observations and Measurements of Air Flow over Water Waves   335
    M. A. Weissman
24. Measurements of Wave-Induced Pressure over Surface Gravity Waves . . .   353
    D. Hasselmann, J. Bösenberg, M. Dunckel, K. Richter, M. Grünewald, and H. Carlson

V. Methods of Remote Sensing

25. The SAR Image of Short Gravity Waves on a Long Gravity Wave.   371
    Robert O. Harger
26. The Response of Synthetic Aperture Radar to Ocean Surface Waves   393
    Klaus Hasselmann and Werner Alpers

27. On the Ability of Synthetic Aperture Radar to Measure Ocean Waves . . .   403
J. F. Vesecky, R. H. Stewart, R. A. Shuchman, H. M. Assal, E. S. Kasischke, and
J. D. Lyden

28. Limitations of the SEASAT SAR in High Sea States   423
F. M. Monaldo and R. C. Beal

29. Microwave Scattering from Short Gravity Waves: Deterministic, Coherent, Dual-
Polarized Study of the Relationship between Backscatter and Water Wave
Properties   443
Daniel S. W. Kwoh and Bruce M. Lake

30. Remote Sensing of Directional Wave Spectra Using the Surface Contour Radar   449
E. J. Walsh, D. W. Hancock, III, D. E. Hines, and J. E. Kenney

31. The Visibility of rms Slope Variations on the Sea Surface   465
R. D. Chapman

VI. Sea Surface Measurements

32. Southern Ocean Waves and Winds Derived from SEASAT Altimeter Measurements   479
Nelly M. Mognard, William J. Campbell, Robert E. Cheney, James G. Marsh, and
Duncan V Ross

33. Marineland Aircraft Observations of L-Band Radar Backscatter Dependence upon
Wind Direction   491
T. W. Thompson, D. E. Weissman, and W. T. Liu

34. Microwave Measurements over the North Sea   505
G. P. De Loor, P. Hoogeboom, R. Spanhoff, and J. Bruinsma

35. Some Skywave Radar Measurements of Wind Vectors and Wave Spectra: Comparison with Conventional Data for JASIN 1978   517
P. E. Dexter and S. Theodoridis

36. Study of the Modulation by Correlation in the Time and Frequency Domains of
Wave Height and Microwave Signal: Preliminary Results   529
Danielle de Staerke and André Fontanel

37. HF Radar Measurements of Wave Spectral Development   541
Dennis B. Trizna

38. Passive Microwave Probing of Roughened Sea   555
A. M. Shutko

VII. Wave Modeling

39. Inverse Modeling in Ocean Wave Studies   571
Robert Bryan Long

40. Comparisons of Hurricane Fico Winds and Waves from Numerical Models with
Observations from SEASAT-A   595
Duncan Ross, Linda M. Lawson, and William McLeish

41. Modeling Wind-Driven Sea in Shallow Water   615
J. W. Sanders and J. Bruinsma

42. An Evaluation of Operational Wave Forecasts on Shallow Water   639
    E. Bouws, G. J. Komen, R. A. van Moerkerken, H. H. Peeck, and M. J. M. Saraber

43. Anomalous Dispersion in Numerical Models of Wave Spectra   661
    William Carlisle Thacker

44. Some Problems in the Development of the National Coastal Waves Program .   671
    L. Baer, D. Esteva, L. Huff, W. Iseley, R. Ribe, and M. Earle

45. Models for the Hurricane Wave Field   677
    D. Lee Harris

Participants   683

Index   689

KEYNOTE ADDRESSES

SIR GEORGE DEACON, HONORARY CHAIRMAN

Twenty years ago, Walter Munk welcomed the wave conference at Easton as a great concentration of talent marking the change of wave studies from an art to a science.

Since then we have seen a great awakening by industry and governments to the commercial and safety value of reliable information. They sometimes complain about apparent deficiencies, but considering the late start and primary dependence on individual enthusiasms, it is remarkable how much progress has been made. It is rather surprising how little attention was paid to waves in the early days. M. F. Maury does not say much about them, nor did they receive much attention in the Challenger Reports. The voluminous Sailing Directions, much used by seamen in the late 19th and early 20th centuries, are content with thorough reporting of wind patterns and do not say much about waves. Krümmel (1911), Thorade (1931), and Cornish (1934) are great sources of information, but the real growth of the subject, and the essential cooperation of theory and observation, did not begin till its problems became urgent during the Second World War.

The 1961 meeting highlighted key problems and a few of them, like the need to record air pressures close to the surface with sufficient accuracy for comparison with wave profiles, and the need for more measurements of directional spectra and the attenuation of swell over long distances, remain hard-core problems. But good progress has been made in many others, in improvement of measuring techniques, in understanding the nonlinear processes involved in wave interaction and breaking of waves, and in collecting long time-series of observations. It is encouraging to see how long-term studies of the physical processes, essential for ultimate satisfaction, are surviving the increasing demands for immediate figures. The scatter between individual methods of forecasting and hindcasting is much less than it was, though there are more uncertainties in the methods of predicting extreme conditions. One of them is an apparent change, usually a decrease in height, as waves pass from the deep ocean to coastal waters. There are still many problems requiring specialists and active cooperation between theoretical physicists, instrumentalists, and engineers.

The rapid development of radio measuring techniques calls for another discussion of all aspects of the subject. Every now and then in oceanography, and, I suppose, in other applied sciences, a new method of approach promises final solutions to many of our problems, and only afterwards do we learn some details that limit its usefulness in some directions and make it fully effective in others. In such a period of active growth we have to be as clear as possible about what measurements are most needed and how they can best be made.

I suspect that my invitation to Miami is largely a recollection of the happy, friendly conditions which helped to bring success to the meeting 20 years ago. The organizers and sponsors of this new opportunity have set up equally promising conditions, in which, I am sure, much good work will be done.

SIR GEORGE DEACON   •   Institute of Oceanographic Sciences, Wormley, England.

1

WILLIAM RAINEY

On behalf of Tony Calio, the Associate Administrator for NASA's Office of Space and Terrestrial Applications, I am delighted to address this symposium, and I wish to thank the Inter-Union Commission on Radio Meteorology, the American Meteorological Society, NOAA, and ONR for their joint sponsorship.

In preparing for this meeting, I went back to two books to familiarize myself with some past history in the field of ocean waves. The first article I looked at was on "Wave Generation by the Wind" by Fritz Ursell, in the G. I. Taylor birthday book. Twenty-five years ago, he wrote:

> Wind blowing over a water surface generates waves in the water by physical processes which cannot be regarded as known.... What part is played by the air boundary layer and the air turbulence level? At what distance (if any) do the wave motions at different points cease to interact significantly? What part (if any) is played by minute roughnesses on the sea surface?

These comments and questions are still applicable today, 25 years later.

The other book to which I turned was *Ocean Wave Spectra*, which was the product of a symposium, much like this one, held in Easton, Maryland, 20 years ago this month, to discuss the measurement of ocean wave spectra. Sir George Deacon, the honorary chairman of this present symposium, reviewed the status of wave research at that symposium. Let me quote briefly from his review:

> The question of how waves are generated must be regarded as one of overriding importance.... It seems superfluous to insist that the greatest need is still for closer integration of theory and experiment. Observations of one sort and another are urgently required for a variety of purposes, but a lot of effort will be saved when we have a better understanding of the physical processes involved. A systematic approach is not only more satisfying, it is forced on us by the scope and difficulty of the problems that have still to be solved. We must not be afraid of venturing outside the circle of past experience....

These comments also are still applicable today, 20 years later. Much progress has been made in the past two decades, of course, although the specific topics to which these generalized comments apply now are certainly different. For example, of the 30 papers presented at Easton, only one was directed toward remote sensing—a radar paper by Frank McDonald. Of the 75 papers at this symposium, 26 are on remote sensing. The goal of most of these papers, however, is a common one: the understanding and measurement of ocean wave spectra.

One of the goals of NASA's Oceanic Processes Branch is to provide the capability for measuring and making available to users (Navy, NOAA, commercial, and academic research) the two-dimensional ocean wave height spectrum. Let me give you a brief summary of some of our accomplishments in this field, and then tell you of our future plans.

NASA workers and NASA-supported researchers have participated in numerous field experiments involving the remote sensing of waves, such as Marineland, West Coast Experiment, Joint Air–Sea Interaction Study (JASIN), Marine Remote Sensing Experiment (MARSEN), Gulf of Alaska Surface Experiment (GOASEX), Storm Response Experiment (STREX), and Atlantic Remote Sensing Land Ocean Experiment (ARSLOE). The SEASAT mission provided two instruments for wave measurements: the synthetic aperture radar (SAR) and the altimeter.

Other remote sensing instruments developed and flown by NASA investigators are:

1. Dual-frequency scatterometer for measuring the 2-D spectrum from aircraft
2. Short-pulse radar, also for measuring the 2-D spectrum from aircraft
3. Surface contour radar, which measures the 2-D wave height field (and thence spectra through Fourier transformation)
4. Airborne oceanic lidar (AOL), which can determine a 1-D spectrum
5. SAR, both SEASAT and aircraft. The SAR can be used to determine wave length and direction, and recently some quantitative wave height information has been extracted.

You will hear at this symposium the latest research efforts of the scientists who have made use of these five instruments. (There are 18 scientists listed in the program who have been directly or indirectly funded by NASA.) The ocean wave height directional spectrum was not available in detail at the time of the Easton meeting, although schemes for determining it through the use of buoys were presented. Now we can routinely make available, through the use of buoys, a reasonable approximation of this spectrum. Further, aircraft measurements of this spectrum are now available, and with relatively high degrees of precision, as you will hear later. This represents real progress in remote sensing.

The future missions presently under consideration by NASA's Oceans Branch are TOPEX, FIREX, and some form of scatterometer for wind field determination. TOPEX (Topography Experiment) is an altimeter mission to determine the ocean surface topography with sufficient precision to be able to determine the ocean circulation. FIREX (Free-flying Imaging Radar Experiment) is a bilateral study being presently conducted by NASA and Canada. In particular, the requirements for ocean surface waves that can be met by the use of SAR are now being determined. Finally, in order to determine the wind stress field necessary to complement the TOPEX altimeter measurements for the study of ocean circulation, we are initiating some studies to determine the feasibility of incorporating a scatterometer on one of the next series of NOAA satellites (H, I, or J).

In closing, let me say that it would be presumptuous to assume that we can do everything by remote sensing. We at NASA are certainly aware of the limitations of remotely sensed data. We need concerned scientists, such as you here at this symposium, to provide us with help and advice in order that NASA can assist you with the development and refinement of the oceanic remote sensing techniques that you may require. We are grateful for the help you have given us in the past. We look forward to continued close collaboration with you and the other agencies in the future. I wish you every success in your discussions at this symposium, and hope your deliberations will result in an increased wave research effort and a more complete understanding of ocean waves.

WILLIAM RAINEY ● Deputy Administrator for Science and Applications, NASA Headquarters, Washington, D.C. 20546.

HERBERT RABIN

On behalf of the Secretary of the Navy, I take pleasure in welcoming you to this, the 8th Inter-Union Commission Symposium on "Wave Dynamics and Radio Probing of the Ocean Surface." For myself, I am privileged to be here to provide a few opening remarks before this distinguished gathering.

It is noteworthy to have so many of the scientists present who have made the pioneering and insightful discoveries in the fields of wave dynamics and radar probing of the ocean surface. I would like to extend a special welcome to all our guests who have come from scientific institutions beyond the borders of the United States. We are particularly pleased that you were able to attend this meeting. I hope you will find your visit here to be rewarding both technically and socially.

This is the third such meeting that the U.S. Navy has cosponsored through its Office of Naval Research in collaboration with NASA and NOAA which I believe gives some indication of our continuing interest in this forum to address research in the ocean sciences. We are pleased that a number of papers to be presented here will be given by investigators sponsored directly or indirectly by the Navy. The U.S. Navy is no newcomer to this field. The year 1981 marks the 150th anniversary of the U.S. Naval Oceanographic Office, and to that organization and those of its members who may be in this audience, I would like to take this opportunity to extend my congratulations in this anniversary year for their many important contributions to the understanding of ocean sciences.

Although the world's navies and merchant fleets have long since abandoned sails, there is still a strong requirement for precise knowledge of the ocean environment, the atmosphere, and the interrelationships between them. Of all the environmental constraints, those generated by ocean waves are by for the most visible, and potentially the most dangerous. Although waves may pose a threat to the vessel itself, there are many disturbances on the sea which can interfere with or prevent effective ship utilization. Severe sea state reduces the ability to launch and recover aircraft, conduct replenishment operations, and use small craft. For all users of the oceans, it is desirable to operate ships at their maximum efficiency. With the enormous increase in the cost of fuel, reduction in ship transiting time equates to significant cost savings. For these and other reasons there is increasing rationale to have accurate and up-to-date knowledge of global wave conditions.

Oceanographic prediction, like weather forecasting, is a child of modern science. The advent of rapid worldwide communications, the high-speed digital computers, and the ability to sense remotely parameters related to ocean behavior are relatively recent developments which have given strong impetus to the field. It was only in the 1950s that the first synoptic wave charts became available in the U.S., and in the 1960s that the fleet Numerical Oceanography Center began providing computer-generated ocean wave analyses and forecasts to a variety of users. The first known operational ocean wave spectral forecasts for large areas—the Mediterranean Sea— were available at the Center in 1972, and two years later were expanded to cover the northern hemisphere. Of course, also during the late 1960s and 1970s, spaceborne cameras were developed and the whole technology of satellite remote sensing came into being to stimulate an already expanding field.

But there was a more subtle advance in the field which is tacitly recognized in the coupling of the two disciplines in the title of this symposium. The work of Crombie and Wright in the United States and the independent efforts of Bass, Fuks, and their colleagues in the Soviet Union established the intimate relationship that exists between the ocean wave spectrum and the radio energy scattered from it. The ability to probe the ocean remotely with radar energy, and investigate the behavior of the Bragg-resonant wave components, has produced great insight into dynamics of wave growth and decay, wind-wave interactions, and a variety of other phenomena hitherto inaccessible to the investigator. These insights have provided the basis for a powerful interaction between the theoretical and experimental communities, and have established the basis for joint symposia such as the one to begin shortly.

Let me look to the future for a moment and discuss a few specifics of a practical nature that I see as fundamental to continued advance in the area of wave prediction and analysis. Global problems require global solutions and I believe that the future of the environmental sciences lies in space-based surveys. I say this despite the Navy's recent decision not to participate in the National Oceanic Satellite System, a reluctant decision based on affordability and competition with other Naval priorities. We recognize that our need for oceanic data still exists, and we will continue to explore alternative options to satisfy them.

Returning to specifics, I believe we first and foremost need a good grasp of the phenomenology of the observables sensed. The signals received by remote sensors are, in general, complex functions of viewing geometry, the propagation medium, the noise background, and the parameter of interest. The parameter must often then be related to a quantity of oceanographic interest. Good calibration of the measuring instrument is obviously a necessity, and for spaceborne sensors this is a nontrivial task, but we must also understand the phenomena. The false alarm is not an affliction of search radar alone. Confusing one phenomenon with another, for example, an oil slick may be mistaken for local calm on a radar image. Precise knowledge of both phenomena can resolve this kind of issue.

Further, efficient sensor designs must be evolved. I note in the Symposium Program several papers involving radars operating in space. Radar systems are complex instruments, requiring power above the average, large apertures compared to, say, optical instruments, and extensive processing. These characteristics, in general, determine satellite weight and satellite weight correlates well with costs. Moreover, complexity also relates to the mean time to failure. Developing affordable and reliable instrumentation which can meet a reasonable mission life is a challenge that must be met if the promise of remote ocean wave prediction and analysis is to be realized.

Space systems offer, as I have said previously, the possibility of rapid surveys and repetitive surveys on a global scale. The field therefore needs large throughput data processing. High-resolution sensors operating over the ocean areas repetitively deliver unbelievable quantities of data which must be calibrated, cataloged, and interpreted if they are to be of value. It has been several years since the short-duration SEASAT mission, and investigators are still mulling over interpretation of the data. We must learn to cope with this problem.

Remote sensing will require careful plans for ground truth to interpret and validate measurements. Ground truth, of course, may be greatly assisted by autonomous *in situ* oceanographic sensors which report through satellite relay as well as by ocean surveys in areas of interest of oceanographic vessels. Again, because of the high cost of the space segment of a remote sensing program, there is a tendency to constrain the ground truth portion. This economy may be expeditious, but it is not sensible in the long run. We will need continuing discipline to resist this tendency.

Remote sensing provides essentially a two-dimensional view of a three-dimensional body, the ocean. Our ability to connect the remotely sensed surface to 3-D features will be dependent on new theoretical understanding of the ocean surface boundary and the water column beneath. Of course, to the extent we are able to extend the 2-D observation to 3-D processes, the more powerful and useful will remote sensors become.

I will stop here, secure in the knowledge that the community interested in the ocean environment has enough to concern itself with in the future to keep it fully occupied and out of trouble.

In conclusion, I hope I have given you some of Navy's orientation toward the subject of this symposium. We encourage your efforts, and will continue to do so. We share a common interest in the sensors, the processing algorithms, and the phenomenology related to ocean waves and ocean wave spectra. Your continued efforts will bring more insight and understanding to this challenging research area. I wish you an interesting and productive symposium.

HERBERT RABIN ● Deputy Assistant Secretary of the Navy, Research, Application and Space Technology, Pentagon, Washington, D.C. 20350.

I

Ocean Wave Spectra

1

THE EQUILIBRIUM RANGES IN WIND–WAVE SPECTRA

Physical Arguments and Experimental Evidence for and against Their Existence

S. A. KITAIGORODSKII

ABSTRACT.   A review is given of the present ideas on the structure of the equilibrium range in the spectrum of wind-generated gravity waves in deep and shoaling water. It is also shown that an exact analog of Kolmogoroff's spectrum in a random field of weakly nonlinear surface gravity waves gives a spectral form for frequency spectra $S(\omega) \sim \omega^{-4}$ in close agreement with the results of recent observational studies. A suggestion is made for the description of a "transitional" range of wavenumbers (frequencies), where the deviation from Kolmogoroff's equilibrium is due to gravitational instability (wavebreaking). Because of this it is suggested that one of the possible equilibrium forms for the spectrum of wind generated waves has two asymptotic regimes: Kolmogoroff's and Phillips' type of equilibrium with a relatively rapid transition from the first to the second at high frequencies.

1. INTRODUCTION

Since Phillips's fundamental contribution to the concept of the equilibrium range in the spectrum of wind-generated waves (Phillips, 1958), numerous experimental studies have demonstrated quite clearly that a substantial portion of the wind-wave spectrum above the frequency of the spectral peak is saturated and in an equilibrium with the local wind. However, questions concerning the dynamical processes in the wind-wave field that are responsible for such equilibrium, and the precise form of the spectrum in equilibrium conditions are still the subject of serious controversies. Probably the first interesting experimental fact about striking deviations from Phillips's frequency spectrum $S(\omega) = \beta g^2 \omega^{-5}$ (β is Phillips's constant, g is gravity) was reported in Kitaigorodskii et al. (1975) in connection with the description of the equilibrium range in wind-generated wave spectra in *shoaling water depths*. In such conditions, according to the data of Kitaigorodskii et al. (1975), the high-frequency regions of $S(\omega)$ above the spectral peak roughly obey an ω^{-3} law (instead of ω^{-5}). Their explanation of this feature was based on purely kinematic considerations. Noting that Phillips's (1958) similarity hypothesis for spatial and temporal (frequency) spectra are far from being of equivalent validity, it was suggested that the existence of Phillips's equilibrium range can be postulated *only* for spatial statistical character-

S. A. KITAIGORODSKII ● Department of Earth and Planetary Sciences, The Johns Hopkins University, Baltimore, Maryland 21218.

istics of wind waves, for example, for wavenumber spectrum $F(\mathbf{k})$, which according to Phillips (1958) must be equal to

$$F(\mathbf{k}) = Bk^{-4}\phi(\vartheta) \tag{1}$$

where $\mathbf{k} = (k\cos\vartheta, k\sin\vartheta)$ and $\varphi(\vartheta)$ is a certain universal function describing the angular energy distribution and satisfying the standard normalization condition

$$\int_{-\pi}^{+\pi} \phi(\vartheta)d\vartheta = 1 \tag{2}$$

The nondimensional coefficient B in (1) must, according to Phillips's similarity arguments, be an *absolute constant*. If (1)–(2) is *chosen* to represent an equilibrium form of the high-wavenumber portion of $F(\mathbf{k})$, then by using (1) and the relationships

$$S(\omega) = \int_{-\pi}^{+\pi} \left[F(k,\vartheta)\frac{k}{G(k,\vartheta)} \right]_{k=k(\omega,\vartheta)} d\vartheta \tag{3}$$

where

$$G(k,\vartheta) = \partial\omega/\partial k \tag{4}$$

$$\omega = \omega(k,\vartheta) = [gk\,\mathrm{th}\,(kH)]^{1/2} + kU\cos(\vartheta - \gamma_0) \tag{5}$$

(H is depth, γ_0 is the angle between the direction of mean current $|\mathbf{U}| = U$ and mean wind, say), we can obtain much richer information about the corresponding frequency spectra $S(\omega)$, than those obtained by using only dimensional analysis. In particular, it was found by Kitaigorodskii *et al.* (1975) that for wind-generated waves in *shoaling water depths* ($kH \lesssim 1$), the asymptotic form of frequency spectra $S(\omega)$ will be

$$S(\omega) = BgH\omega^{-3} \tag{6}$$

instead of Phillips's *deep-water* law

$$S(\omega) = 2Bg^2\omega^{-5} \tag{7}$$

One of the advantages of this method of determination of $S(\omega)$ is related to the possibility of finding experimentally the values of Phillips's constant $\beta = 2B$ by two independent ways [(6),(7)]. Since 1975, a number of observational studies of wind waves in *shallow waters* have demonstrated that the rear face of the frequency spectrum indeed has a slope much closer to ω^{-3} than to ω^{-5} (Thornton, 1977; Gadzhiyev and Krasitskii, 1978; Gadzhiyev, et al., 1978; Vincent, 1982). Therefore, the approach based on the transformation of *one universal wavenumber spectrum* (1) into frequency spectra of deep (7) and shallow (6) water waves has certainly received some support. However, it cannot be considered as a universal tool for the explanation of *any kind of deviations* in *observed* spectral form from $S(\omega) = \beta g^2\omega^{-5}$ ($\beta = 2B$) by *purely kinematic effects*, i.e., by differences in the form of the dispersion relationship $\omega = \omega(k, H, U, \vartheta)$ for surface gravity waves. This approach can be constructive only if we know the universal form of the wavenumber spectrum in the *equilibrium range*. However, at present there are very serious indications that the Phillips spectrum (1) cannot serve as a *universal basis* for the description of the *observed* variability of energy contained *high-frequency regions* of ocean wind-wave spectra. These indications can be seen first of all in the well-known collections of empirical data illustrating the variability of Phillips's constant β (Hasselmann et al., 1973). Such variability usually is described in the form

$$\beta = \beta(\tilde{X}), \qquad \tilde{X} = gX/U_a^2 \tag{8}$$

where $\tilde{X}$ is nondimensional fetch and U_a is wind speed.

Figure 1, reproduced from the widely known JONSWAP paper (Hasselmann *et al.*, 1973), is a typical example of such a presentation. The same caption appears here as in the original paper, to suggest that, according to Fig. 1, there is more sense to call β Phillips's *variable*, but not constant. To explain the *fetch dependence of β without abolishing* the universal wavenumber Phillips's spectrum (1), it was suggested by Zaslavskii and Kitaigorodskii (1971, 1972) that one should take into account the modulation of short-wave components by longer ones, which leads to the appearance of a random gravity acceleration $\hat{g} = g + \hat{a}$ ($\hat{a} = \partial^2 \hat{\zeta}/\partial t^2$; $\hat{\zeta}$ is the surface displacement associated with long waves) and random orbital velocities $\hat{U}$ associated with long waves, instead of fixed g and u, in the dispersion relation (5). The generalization of Phillips's equilibrium theory for frequency spectra, based on (1) and a *random dispersion relation*

$$\omega = \omega(k, \vartheta) = [(g + \hat{a})k]^{1/2} + k\hat{U}\cos(\vartheta - \gamma^\circ) \tag{9}$$

where γ° is a *fixed* angle between the propagation direction of the long-wave component and the

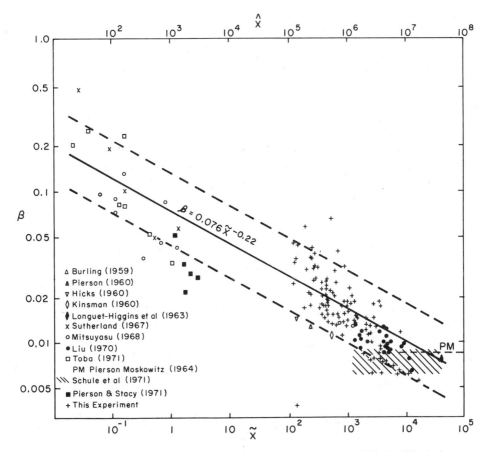

FIGURE 1. Phillips's constant $\beta = 2B$ vs fetch scaled according to Kitaigorodskii. Small-fetch data were obtained from wind-wave tanks (capillary-wave data were excluded where possible). Measurements by Sutherland (1967) and Toba (1971) were taken from Pierson and Stacy (1973). The coordinate $\tilde{X} = gX/u_*^2$, where u_* is the friction velocity in the atomospheric surface layer.

mean wind, say, was discussed in detail in Kitaigorodskii *et al.* (1975). The general conclusion of the theoretical analysis in these papers was that the range of variability of $\beta = S(\omega)/g^2\omega^{-5}$ shown in Fig. 1 *cannot be* attributed to the kinematic effects of long waves (Doppler shift due to orbital velocities of long waves and additional vertical acceleration, imposed on short waves moving on long waves). Recently, Kitaigorodskii (1981) attempted to explain the variability of β by considering the *intermittency* of short waves in the presence of long waves. Such intermittency was assumed to have a dynamical origin (either due to dynamical interaction between the short and the long waves, or due to dynamics of the short-wave generation in the presence of long waves). However, the results also were far from being relevant to explain the features of Fig. 1. It is important to notice that the theories mentioned above were essentially based on a *two-scale model* of surface displacement and precisely because of this, their general applicability to explain the variability of β is very questionable. The reason for this latter statement is that "observed" values of β usually characterize the rear face of the spectrum $S(\omega)$ *in the energy-containing range* (where spectral levels are greater than say 1% of the peak frequency ω_m), whereas in the theories based on two-scale models, it is assumed that equilibrium spectra (1) characterize the range of wavenumbers (and frequencies) *far above* $k_m = \omega_n^2/g$. Therefore, we can conclude that the parameterization of the rear face of the frequency spectrum $S(\omega)$ by Phillips's formula (7) *with variable values of B* (according, let us say, to Fig. 1) *cannot be based* on a universal form of wavenumber Phillips's spectrum (1), whatever modifications of dispersion relations we take into account.

But probably the most important indication of systematic deviations from Phillips's equilibrium form of wind-wave spectra [(1),(7)] has come very recently, with three independent observational studies (Forristall, 1981; Kahma, 1981a, b; Donelan *et al.*, 1982) demonstrating very convincingly that the frequency spectra in the range of frequencies just above the peak frequency ω_m have a *wind-dependent* form $S(\omega) = \alpha_n g^{5-n} U_a^{5-n} \omega^{-n}$ with values of n close to 4.0, and values of the nondimensional coefficient α_n practically fetch independent. The value $n = 4.0$ leads to a universal equilibrium of the form

$$S(\omega) = \alpha_u g U_a \omega^{-4} \tag{10}$$

This form of equilibrium in wind-wave spectra was *first suggested by* Kitaigorodskii (1962) as a possible consequence of Kolmogoroff's type of energy cascade from low to high frequencies, and was derived later on the basis of a calculation of the dynamical resonant interaction between weakly nonlinear surface-gravity deep-water waves by Zakharov and Filonenko (1966). The very fact that there can exist an exact analog of Kolmogoroff's spectra in a random wave field which gives a spectral form in qualitative agreement with observations, indicates the *importance* of the weak nonlinear interactions between wave components in the formation of steady-state wave spectra (or the equilibrium interval in it). Therefore, it seems that the concept of an equilibrium range based *solely* on a consideration of the process of wave breaking [(1),(7)] due to gravitational instability (Phillips, 1958) certainly needs either some revision or generalization. Such a generalization can be made (Kitaigorodskii, 1983) by assuming that the growth of shorter waves on the rear face of the spectrum (limited by breaking) is not due to direct energy input from wind, but rather to energy flux from the lower wavenumbers. This suggestion, made very recently by this author, led to an important and far-reaching conclusion about the possible *simultaneous* existence of two different types of statistical equilibrium in the spectra of wind-generated waves (Kolmogoroff's and Phillips's equilibrium subranges). One of the main purposes of this review chapter is to convince the reader that experimental data favor such an interpretation.

2. THE SPECTRAL DESCRIPTION OF THE RANDOM WIND-WAVE FIELD IN THE OCEAN

It is instructive for further discussions to underline here the basic differences in the spectral description of random turbulent and wave fields. For a weakly nonlinear wave field, the Fourier series representation of the surface $\zeta(\mathbf{x}, t)$

$$\zeta(\mathbf{x},\ t) = \int\int \exp\{i(\mathbf{k}\mathbf{x} - \omega t)\} dZ_\zeta(\mathbf{k},\ \omega) = \int d\mathbf{k} \int d\omega \zeta_{\mathbf{k}\omega} \exp\{i(\mathbf{k}\mathbf{x} - \omega t)\}$$

[$\mathbf{k} = (k_1, k_2)$ is the wavenumber vector, and ω is the frequency] is a more natural tool for theoretical analysis than in studies of random turbulent fields. The reason is simple: the linear free surface waves are an acceptable basic approximation in the description of wind waves because weak nonlinearities as well as weak forcing can produce only *slow* variations of the statistical characteristics of *linear free waves*. The simplest model of a random wave field as a superposition of free linear waves provides the possibility for a useful specialization of the spectral theory of random fields by introducing the deterministic relationship between frequency ω and wavenumber vector $\mathbf{k}$. To demonstrate this, let us first list below the general results of the spectral theory of the random field $\zeta(\mathbf{x}, t)$ without specifying its wave "nature." If waves are statistically homogeneous (SH) and stationary (SS), the covariance of the surface displacement

$$B(\mathbf{x}_1 = \mathbf{x} + \mathbf{r},\ t_1 = t + \tau,\ \mathbf{x}_2 = \mathbf{x},\ t_2 = t) = \langle \zeta(\mathbf{x} + r,\ t + \tau)\zeta(\mathbf{x}, t) \rangle \tag{12}$$

is

$$B(\mathbf{x} + \mathbf{r},\ t + \tau,\ \mathbf{x},\ t) = B(\mathbf{r}, \tau) \tag{13}$$

By changing variables in (13), $\mathbf{x} \to \mathbf{x} - \mathbf{r}, t \to t - \tau$, we will get the symmetry conditions for $B(\mathbf{r}, \tau)$, i.e.,

$$B(\mathbf{r},\ \tau) = B(-\mathbf{r},\ -\tau) \tag{14}$$

$$B(-\mathbf{r},\ \tau) = B(\mathbf{r},\ -\tau) \tag{15}$$

but

$$B(\mathbf{r},\ \tau) \neq B(\mathbf{r},\ -\tau) = B(-\mathbf{r},\ \tau) \tag{16}$$

The wave spectrum $E(k, \omega)$, defined as

$$E(\mathbf{k},\ \omega)\delta(\mathbf{k} - \mathbf{k}')\delta(\omega - \omega') = \langle \zeta_{\mathbf{k}\omega} \zeta^*_{\mathbf{k}'\omega'} \rangle = \langle \zeta_{\mathbf{k}\omega} \zeta_{-\mathbf{k}'-\omega'} \rangle \tag{17}$$

is the Fourier transform of the covariance $B(\mathbf{r}, \tau)$:

$$E(\mathbf{k},\ \omega) = (2\pi)^{-3} \int d\mathbf{r} \int d\tau B(\mathbf{r},\ \tau) \exp\{-i(\mathbf{k}\mathbf{r} - \omega\tau)\} \tag{18}$$

with the normalization condition

$$\langle \zeta^2 \rangle = B(0,\ 0) = \int d\mathbf{k} \int d\omega E(\mathbf{k},\ \omega) \tag{19}$$

The symmetry conditions (14)–(16) for $B(\mathbf{r}, \tau)$ leads to the following symmetry conditions for $E(\mathbf{k}, \omega)$:

$$E(\mathbf{k},\ \omega) = E(-\mathbf{k},\ -\omega) \tag{20}$$

$$E(-\mathbf{k},\ \omega) = E(\mathbf{k},\ -\omega) \tag{21}$$

but

$$E(\mathbf{k}, \omega) \neq E(\mathbf{k}, -\omega) = E(-\mathbf{k}, \omega) \tag{22}$$

Reduced spectra can be obtained from $E(\mathbf{k}, \omega)$ by integration over ω and over $\mathbf{k}$:

$$\psi(\mathbf{k}) = \int d\omega E(\mathbf{k}, \omega) \tag{23}$$

$$S(\omega) = \int d\mathbf{k} E(\mathbf{k}, \omega) \tag{24}$$

where $\psi(\mathbf{k})$ is wave number and $S(\omega)$ are frequency spectra. Formulas (23) and (24) are *unable to establish the correspondence between wavenumber and frequency spectra.* In the theory of turbulent random fields, this produces certain difficulties, because the theoretical derivations are often performed in terms of $\psi(\mathbf{k})$, whereas in experiments the data are usually available concerning $S(\omega)$. Therefore, the relationship between $\psi(\mathbf{k})$ and $S(\omega)$ is found in turbulence theory only via *additional* information of the type of a "frozen turbulence" model, or as it is usually called, Taylor's hypothesis. For a *random wave* field, such difficulties can be avoided by assuming that in (11)

$$\zeta_{\mathbf{k}\omega} = \eta_{\mathbf{k}}^{+} \delta(\omega - \sigma) + \eta_{\mathbf{k}}^{-} \delta(\omega + \sigma) \tag{25}$$

where the random coefficients $\eta_{\mathbf{k}}^{+}, \eta_{\mathbf{k}}^{-}$ are the amplitudes of free linear surface waves propagating in the positive and negative directions of the vector $\mathbf{k}$, and σ satisfies the dispersion relationship for capillary–gravity waves in finite depth H:

$$\sigma = \sigma(k) = [(gk + \gamma k^3) \operatorname{th} kH]^{1/2} \tag{26}$$

where γ is the coefficient of surface tension divided by water density. To fulfill the conditions of statistical homogeneity and stationarity, i.e., to have (13), the coefficients $\eta_{\mathbf{k}}^{+}, \eta_{\mathbf{k}}^{-}$ must satisfy the following conditions:

$$\langle \eta_{\mathbf{k}}^{+} (\eta_{\mathbf{k}'}^{+})^* \rangle = \tfrac{1}{2} F(\mathbf{k})\delta(\mathbf{k} - \mathbf{k}') \tag{27}$$

$$\langle \eta_{\mathbf{k}}^{-} (\eta_{\mathbf{k}'}^{-})^* \rangle = \langle \eta_{-\mathbf{k}'}^{+} (\eta_{-\mathbf{k}}^{+})^* \rangle = \tfrac{1}{2} F(-\mathbf{k})\delta(\mathbf{k} - \mathbf{k}') \tag{28}$$

$$\langle \eta_{\mathbf{k}}^{+} (\eta_{\mathbf{k}'}^{-}) \rangle = \langle \eta_{\mathbf{k}}^{+} \eta_{-\mathbf{k}'}^{+} \rangle = 0 \qquad \text{for all } \mathbf{k} \neq \mathbf{k}' \tag{29}$$

$$\langle \eta_{\mathbf{k}}^{-} (\eta_{\mathbf{k}'}^{+})^* \rangle = \langle \eta_{\mathbf{k}}^{-} \eta_{-\mathbf{k}'}^{-} \rangle = 0 \qquad \text{for all } \mathbf{k} \neq \mathbf{k}' \tag{30}$$

where $F(\mathbf{k})$ can be called the *nonsymmetrical* wavenumber spectrum to distinguish it from *symmetrical* spectra $E(\mathbf{k}, \omega)$, $\psi(\mathbf{k})$, and $S(\omega)$. From (27)–(30) it follows that a contribution to $F(\mathbf{k})$ for given $\mathbf{k}$ comes only from waves propagating in the positive direction of vector $\mathbf{k}$ and that $F(\mathbf{k}) \neq F(-\mathbf{k})$ (and thereby the term nonsymmetrical spectra). The relationship between $E(\mathbf{k}, \omega)$, $\psi(k)$ and $F(\mathbf{k})$ are given by the formulas

$$E(\mathbf{k}, \omega) = \tfrac{1}{2} [F(\mathbf{k})\delta(\omega - \sigma) + F(-\mathbf{k})\delta(\omega + \sigma)] \tag{31}$$

$$\psi(\mathbf{k}) = \tfrac{1}{2} [F(\mathbf{k}) + F(-\mathbf{k})] \tag{32}$$

From (31) it follows that

$$F(\mathbf{k}) = 2 \int_{0}^{\infty} E(\mathbf{k}, \omega) d\omega \tag{33}$$

Finally, substituting (31) in (24) and using the well-known formula for δ-function,

$$\delta[f(x)] = \sum_{n=1}^{N} \left(\frac{\partial f}{\partial x}\right)_{x=x_n}^{-1} \delta(x - x_n) \tag{34}$$

where $f(x)$ is a given function and x_n are the roots of the equation $f(x) = 0$, we can get the expression for $S(\omega)$:

$$S(\omega) = \int d\mathbf{k}\, E(\mathbf{k}, \omega) = \tfrac{1}{2} \int d\mathbf{k}\,[F(\mathbf{k})\delta(\omega - \delta) + F(-\mathbf{k})\delta(\omega + \sigma)]$$

$$= \frac{1}{2} \int_{-\pi}^{+\pi} d\vartheta \int_{0}^{\infty} k\,dk\,\{F(k\cos\vartheta, k\sin\vartheta)\delta(\omega - \sigma(k))$$

$$+ F(-k\cos\vartheta, -k\sin\vartheta)\delta[\omega + \sigma(k)]\}$$

$$= \frac{1}{2} \int_{-\pi}^{+\pi} d\vartheta \sum_{n=1}^{N} \left[\frac{k}{\left|\dfrac{\partial\sigma}{\partial k}\right|} F(k\cos\vartheta, k\sin\vartheta)\right]_{k=\kappa_n^{+}(\omega)} +$$

$$+ \frac{1}{2} \int_{-\pi}^{+\pi} d\vartheta \sum_{m=1}^{M} \left[\frac{k}{\left|\dfrac{\partial\sigma}{\partial k}\right|} F(-k\cos\vartheta, -k\sin\vartheta)\right]_{k=\kappa_m^{-}(\omega)} \tag{35}$$

where κ_n^{+} the roots of the equation $\omega = \sigma(k)$ and κ_m^{-} are the roots of the equation $\omega = -\sigma(k)$. Formulas (31), (32), and (35) express the usual symmetrical spectra $E(\mathbf{k}, \omega)$, $\psi(\mathbf{k})$, and $S(\omega)$ through $F(\mathbf{k})$, but the reverse is true only for $E(\mathbf{k}, \omega)$ and even in this case the relationship between $F(\mathbf{k})$ and $E(\mathbf{k}, \omega)$ has, according to (33), an integral character. Therefore, for practical determination of $F(\mathbf{k})$, we still need to know *the wavenumber frequency spectrum* $E(\mathbf{k}, \omega)$. This, generally speaking, shows that the advantages in the description of a random wave field in terms of $F(\mathbf{k})$ compared for example, with $\psi(\mathbf{k})$, are not too evident. However, in some important applications, in particular in the theory of *wind-generated* waves, it is reasonable *to assume* that

$$F(\mathbf{k}) = F(k\cos\vartheta, k\sin\vartheta) = \begin{cases} \neq 0 & \text{for } \vartheta\in(\text{I, IV}) \\ = 0 & \text{for } \vartheta\in(\text{II, III}) \end{cases} \tag{36}$$

where I, II, III, IV are quadrants in the plane (k, ϑ). If (36) is fulfilled, then

$$F(-\mathbf{k}) = F(-k\cos\vartheta, -k\sin\vartheta) = \begin{cases} = 0 & \text{for } \vartheta\in(\text{I, IV}) \\ \neq 0 & \text{for } \vartheta\in(\text{II, III}) \end{cases} \tag{37}$$

and

$$F(-k\cos\vartheta, -k\sin\vartheta) = F[k\cos(\vartheta\pm\pi), k\sin(\vartheta\pm\pi)] \tag{38}$$

In this case according to (32) there is a correspondence between $F(\mathbf{k})$ and $\psi(\mathbf{k})$ described as

$$\psi(k\cos\vartheta, k\sin\vartheta) = \begin{cases} \tfrac{1}{2}F(k\cos\vartheta, k\sin\vartheta) & \text{for } \vartheta\in(\text{I, IV}) \\ \tfrac{1}{2}F(-k\cos\vartheta, -k\sin\vartheta) & \text{for } \vartheta\in(\text{II, III}) \end{cases} \tag{39}$$

which permits us to express the right-hand side of (35) for $S(\omega)$ through *symmetrical* (measurable) wavenumber spectra $\psi(\mathbf{k})$:

$$S(\omega) = \int_{(I,IV)} d\vartheta \sum_{n=1}^{N} \left[\frac{k}{\frac{\partial \sigma}{\partial k}} \psi(k \cos \vartheta, k \sin \vartheta) \right]_{k = \kappa_n^+(\omega, \vartheta)}$$

$$+ \int_{(II,III)} d\vartheta \sum_{m=1}^{M} \left[\frac{k}{\frac{\partial \sigma}{\partial k}} \psi(k \cos \vartheta, k \sin \vartheta) \right]_{k = \kappa_m^-(\omega, \vartheta)} \tag{40}$$

This solves the problem of finding the relationship between symmetrical frequency and wavenumber spectra for *wind*-generated waves. In the simplest case of an isotropic *dispersion* relationship for surface gravity waves

$$\sigma = \sigma(k) = (gk \, \text{th} \, kH)^{1/2} \tag{41}$$

we have $n = 1$, $m = 0$ in (40), and (40) reduces to

$$S(\omega) = \int_{-\pi}^{+\pi} \left[\psi(k, \vartheta) \frac{k}{G(k)} \right]_{k = k(\omega, H)} d\vartheta = \frac{x(k)}{\left. \frac{\partial \sigma}{\partial k} \right|_{k = (\omega^2/g)Z(\omega_H)}} \tag{42}$$

where $G = \partial \sigma / \partial k$ and $Z(\omega_H = \omega H^{1/2}/g^{1/2})$ satisfy the equation

$$Z[\text{th} \, (\omega_H^2 Z)] = 1 \tag{43}$$

In (42), $x(k)$ is the spectrum of wavenumber modulus, characterizing energy distribution over k regardless of the direction of wave component propagation, i.e.,

$$x(k) = \int_{|\mathbf{k}| = k} \psi(\mathbf{k}) d\mathbf{k} = \int_{-\pi}^{+\pi} \psi(k, \vartheta) k \, d\vartheta \tag{44}$$

The relationship (42) which together with (41) determines the correspondence between symmetrical wavenumber and frequency spectra of a random dispersive wave field, is significantly different from those widely used in the description of a turbulent field.

If $\psi_{ij}(k)$ is the spectral tensor of isotropic 3d turbulent velocity field $\mathbf{u}^t$, defined as

$$\psi_{ij}(k) = \frac{E(k)}{4\pi k^2} \left(\delta_{ij} - \frac{k_i k_j}{k^2} \right) \tag{45}$$

where $E(k)$ is the energy spectrum ($\int_0^\infty E(k) dk = \frac{1}{2} \langle u_j^t u_j^t \rangle$), then the *frequency spectra* $E_{ij}(\omega)$ of the fluctuating velocity field $\mathbf{u}^t$ *at fixed x* can be expressed through the 1d wavenumber longitudinal and transverse spectra

$$E_1(k_1) = \int_{k_1}^{\infty} \left(1 - \frac{k_1^2}{k^2} \right) \frac{E(k)}{k} dk \quad \text{and} \quad E_2(k_1) = \frac{1}{2} \int_{k_1}^{\infty} \left(1 + \frac{k_1^2}{k^2} \right) \frac{E(k)}{k} dk$$

of an isotropic field $\mathbf{u}^t(\mathbf{x}, t)$ for *fixed t*, by using the model of "frozen" turbulence:

$$E_1(\omega) = \frac{1}{U_a} E_1 \left(\frac{\omega}{U_a} \right), \qquad E_2(\omega) = \frac{1}{U_a} E_2 \left(\frac{\omega}{U_a} \right) \tag{46}$$

where U_a is a constant transport velocity (wind speed for atmospheric turbulence). Therefore, the shape of the function $E(\omega)$ will coincide with the shape of $E(k)$, whereas in case of a *dispersive* wave field [(41),(42)], it will not.

After this short discussion about the *differences* between the methods of spectral description of *wave* and *turbulent* fields, we will, in the next section, examine the possible *similarities* in the dynamical regimes in a statistical ensemble of weakly nonlinear surface gravity waves and locally isotropic 3d turbulence.

3. THE KOLMOGOROFF TYPE OF EQUILIBRIUM IN THE SPECTRA OF WEAKLY NONLINEAR WIND-GENERATED SURFACE GRAVITY WAVES

A very brief description of Kolmogoroff's treatment (1941, 1962) of the *equilibrium* regime for locally isotropic 3d turbulence is needed here to make more clear the kind of analogy between the dynamics of "strong" Kolmogoroff's turbulence and "weak turbulence" in the field of interacting weakly nonlinear surface waves.

In a turbulent flow the energy dissipation

$$\varepsilon = \frac{v}{2}\sum_{i,j}\left(\frac{\partial u_i^t}{\partial x_j} + \frac{\partial u_j^t}{\partial x_i}\right)^2 \tag{47}$$

is a random function of the coordinate $x_i (i = 1, 2, 3)$ and time t, fluctuating together with the velocity field. If the *mean* value of ε, i.e., $\bar{\varepsilon} = (v/2)\langle\sum_{i,j}(\partial u_i^t/\partial x_j + \partial u_j^t/\partial x_i)^2\rangle$, is equal to the *energy transfer rate* from large- to small-scale motion, then according to Kolmogoroff's similarity hypothesis, $\bar{\varepsilon}$ and the kinematic viscosity of the fluid v can be considered as the only parameters which determine the spectral energy density $E(k)$. Therefore,

$$E(k) = \bar{\varepsilon}^{2/3} k^{-5/3} \psi(k\lambda_v), \qquad \lambda_v = (v^3/\bar{\varepsilon})^{1/4} \tag{48}$$

where λ_v is Kolmogoroff's internal scale, which determines the scale of eddy motions which are *directly* influenced by viscous effects. If we consider the range of scales $k\lambda_v \ll 1$ and also $k\lambda_v \gg \lambda_v/L_0$ (L_0 is the scale of turbulence influenced by mean flow characteristics, i.e., where there is *direct energy* supply of turbulence from mean motion), then according to the second Kolmogoroff similarity hypothesis

$$\text{for } \lambda_v/L_0 \ll k\lambda_v \ll 1, \qquad \psi(k\lambda_v) = \text{const} = A_t \tag{49}$$

and

$$E(k) = A_t \bar{\varepsilon}^{2/3} k^{-5/3} \tag{50}$$

Here A_t is an *absolute number*, for which experimental data give a value close to 1. Thus, the Kolmogoroff *inertial* subrange (50) can exist if, in wavenumber space, the regions of "generation" and "dissipation" of turbulence are *separated by a sufficiently broad transparent region* $L_0^{-1} \ll k \ll \lambda_v^{-1}$. Because energy transfer among different scales of turbulent motions cannot be successfully treated in terms of eigen-oscillations (waves), the Kolmogoroff results (48)–(50) are based solely on similarity arguments and dimensional analysis. Physically, they are equivalent to the assumption that the energy of a mode with scale k^{-1} can be transferred only to the modes of nearly the same scale (local interactions in wavenumber space). Thus, a portion of energy handed down from larger to smaller scales must pass through the entire range of scales (*cascade*) almost to λ_v, where kinetic energy starts to dissipate into heat. Therefore, (50) means that for a given wavenumber k, the mean value $\bar{\varepsilon}$ is determined only by the level of fluctuations at this scale, i.e., by $E(k)$. In other words, $\bar{\varepsilon}$ must be expressible in terms of k and $E(k)$ only. This of course leads to the famous "$-5/3$" Kolmogoroff–Obuchoff law for $E(k)$, (50). To find an analogy with

Kolmogoroff's spectrum (50) for a field of wind-generated surface gravity waves, we also *must assume* that the regions (in Fourier space) of wave "generation" and "dissipation" are separated. However, since for weakly nonlinear dispersive waves we have an equation for energy density, *derived* from first principles, it is very interesting and important to determine if it will be possible for nonlinear interaction between different wave modes to form an equilibrium spectrum of Kolmogoroff's type. In other words, the question is, can wave energy spectra in certain regions be influenced only by energy flux from low to high wavenumbers and if it can, what form must the wind-wave spectra have in such a case? To get an answer, let us consider first of all the *free surface dispersive waves as a potential motion*. Then the general system of equations with boundary conditions at the free surface $\zeta(x, y, t)$ and bottom $z = -H = \text{const}$, will be

$$(\nabla_x^2 + \partial^2/\partial z^2)\phi = 0 \tag{51}$$

$$(\partial\phi/\partial z)_{-H} = 0 \tag{52}$$

$$\frac{\partial\zeta}{\partial t} = \left(\frac{\partial\phi}{\partial z}\right)_\zeta - (\nabla_x\phi)_\zeta\nabla_x\zeta \tag{53}$$

$$\left(\frac{\partial\phi}{\partial t}\right)_\zeta + g\zeta + \frac{1}{2}\left[(\nabla_x\phi)^2 + \left(\frac{\partial\phi}{\partial z}\right)^2\right]_\zeta - \gamma F(\zeta) = -\frac{\rho_\zeta^a}{\rho_w} \tag{54}$$

where ρ_w is the water density; for free surface waves the atmospheric pressure ρ_ζ^a will be considered as a constant and

$$F(\zeta) = \frac{\nabla_x^2\zeta}{[(1 + \nabla_x\zeta)^2]^{3/2}} \approx \nabla_x^2\zeta \tag{55}$$

An important consequence of the irrotational character of wave motion is the dependence of total *mean wave energy* $\bar{E}^w$

$$\bar{E}^w = \frac{1}{2}\rho_w\overline{\int_{-H}^\zeta\left[(\nabla_x\phi)^2 + \left(\frac{\partial\phi}{\partial z}\right)^2\right]dz} + \frac{1}{2}\rho_w g\overline{\zeta^2} + \rho_w\gamma\{[1 + \overline{(\nabla_x\zeta)^2}]^{1/2} - 1\} \tag{56}$$

only on *two functions* of **x** and *t*, the surface displacement $\zeta(\mathbf{x}, t)$ and the velocity potential at the free surface $\psi = \phi_\zeta$. Indeed, after some transformations in (56) [integration of the z-derivative term by parts and recalling (51)] it is possible to show that

$$E^w = \frac{1}{2}\rho_w\overline{\left[\psi\left(\frac{\partial\phi}{\partial z} - \nabla_x\phi\cdot\nabla_x\zeta\right)_\zeta\right]} + \frac{1}{2}\rho_w g\overline{\zeta^2}$$
$$+ \rho_w\gamma\{[1 + \overline{(\nabla_x\zeta)^2}]^{1/2} - 1\} \tag{57}$$

or, with use of the kinematic boundary condition (53), alternatively

$$\bar{E}^w = \frac{1}{2}\rho_w\overline{\psi\frac{\partial\zeta}{\partial t}} + \frac{1}{2}\rho_w g\overline{\zeta^2} + \rho_w\gamma\{[1 + \overline{(\nabla_x\zeta)^2}]^{1/2} - 1\} \tag{58}$$

In (56)–(58) the overbar denotes the average over the horizontal plane.

Conservation of total energy for free irrotational waves implies that the equation of motion is given by variational equations

$$\frac{\delta\mathcal{H}}{\delta\psi} = \frac{\partial\zeta}{\partial t}, \quad \frac{\delta\mathcal{H}}{\delta\zeta} = -\frac{\partial\psi}{\partial t} \tag{59}$$

where $\delta/\delta\psi$, $\delta/\delta\zeta$ are variational derivatives of the total Hamiltonian $\mathscr{H} = \bar{E}^w/\rho_w$. To determine the form of Hamiltonian function $\mathscr{H}$ in (59), the solution of the boundary problem for potential [(51)–(54)] is needed. With the differential relation

$$(\nabla_x\phi)_\zeta = \nabla_x\psi - \nabla_x\zeta\mathscr{W}, \qquad \mathscr{W} = (\partial\phi/\partial z)_\zeta \tag{60}$$

(53) and (54) can now be written in terms of canonical variables ψ and ζ:

$$\frac{\partial\zeta}{\partial t} - \mathscr{W} = -\nabla_x\zeta\cdot\nabla_x\psi + \mathscr{W}(\nabla_x\zeta)^2 \tag{61}$$

$$\frac{\partial\psi}{\partial t} + g\zeta = -\tfrac{1}{2}(\nabla_x t)^2 + \frac{\mathscr{W}^2}{2} + \frac{\mathscr{W}^2}{2}(\nabla_x\zeta)^2 \tag{62}$$

The solution of Laplace's equation (51) in the fluid interior can be written as

$$\phi(\mathbf{x}, z, t) = \int_{-\infty}^{+\infty} \phi_\mathbf{k}(t)\frac{\cosh[k(z+H)]}{\cosh kH}e^{i\mathbf{k}\mathbf{x}}d\mathbf{k} \tag{63}$$

Here the Fourier transformation for ϕ is written in the form which automatically satisfies (51) and the boundary condition (52). The Fourier series for ψ, ζ, and $\mathscr{W}$ can be written as

$$\begin{Bmatrix} \psi(\mathbf{x}, t) \\ \zeta(\mathbf{x}, t) \\ \mathscr{W}(\mathbf{x}, t) \end{Bmatrix} = \int_{-\infty}^{+\infty} \begin{Bmatrix} \psi_\mathbf{k} \\ \zeta_\mathbf{k} \\ \mathscr{W}_\mathbf{k} \end{Bmatrix} e^{i\mathbf{k}\mathbf{x}}d\mathbf{k} \tag{64}$$

Now the velocity potential at the free surface ψ can be written as the Taylor series expansion about the $z = 0$ plane:

$$\psi(\mathbf{x}, t) = \sum_{n=0}^{\infty} \frac{\zeta^n}{n!}\frac{\partial^n\phi}{\partial z^n}\bigg|_{z=0} \tag{65}$$

In a similar manner, we can write the vertical fluid velocity at the surface as a Taylor series with an additional vertical derivative:

$$\mathscr{W}(\mathbf{x}, t) = \sum_{n=0}^{\infty} \frac{\zeta^n}{n!}\frac{\partial^{n+1}\phi}{\partial z^{n+1}}\bigg|_{z=0} \tag{66}$$

We can express using (63), (65), and (66) $\mathscr{W}_\mathbf{k}$ in terms of $\psi_\mathbf{k}$ and $\zeta_\mathbf{k}$ only. With $n < 3$, i.e., with accuracy of terms cubic in ζ in Taylor's series expansions (65) and (66), one can get (Zakharov and Filonenko, 1966)

$$\mathscr{W}_\mathbf{k} = |\mathbf{k}|\psi_\mathbf{k} + \int |\mathbf{k}_1|(|\mathbf{k}_1| - |\mathbf{k}|)\psi_{\mathbf{k}_1}\zeta_{\mathbf{k}_2}\delta(\mathbf{k} - \mathbf{k}_1 - \mathbf{k}_2)d\mathbf{k}_1 d\mathbf{k}_2$$

$$+ \int |\mathbf{k}_1|\left[\frac{|\mathbf{k}_1|(|\mathbf{k}_1| - |\mathbf{k}|)}{2} + |\mathbf{k} - \mathbf{k}_2|(|\mathbf{k}| - |\mathbf{k} - \mathbf{k}_2|)\right]\psi_{\mathbf{k}_1}\zeta_{\mathbf{k}_2}\zeta_{\mathbf{k}_3}$$

$$\cdot\delta(\mathbf{k} - \mathbf{k}_1 - \mathbf{k}_2 - \mathbf{k}_3)d\mathbf{k}_1 d\mathbf{k}_2 d\mathbf{k}_3 \tag{67}$$

Then substituting (64) and (67) in the Fourier transform of (61) and (62), we can derive the equation for $\zeta_\mathbf{k}$. However, it is more convenient at this stage to introduce another set of canonical variables according to the formulas

$$\zeta_{\mathbf{k}} = \frac{1}{(2V_{\mathbf{k}})^{1/2}}(a_{\mathbf{k}} + a_{\mathbf{k}}^*), \qquad \psi_{\mathbf{k}} = -i\left(\frac{V_{\mathbf{k}}}{2}\right)^{1/2}(a_{\mathbf{k}} - a_{\mathbf{k}}^*) \tag{68}$$

where $V_{\mathbf{k}}$ weights each mode by the phase speed of the small-amplitude gravity wave $[V_{\mathbf{k}} = \sigma(k)/k$, and $\sigma(k)$ is given by (41)]. Then the Hamiltonian equation (59) with new variables $a_{\mathbf{k}}, a_{\mathbf{k}}^*$ will take the form

$$\frac{\partial a_{\mathbf{k}}}{\partial t} = -i\frac{\delta \mathcal{H}}{\delta a_{\mathbf{k}}^*}, \qquad \frac{\partial a_{\mathbf{k}}^*}{\partial t} = i\frac{\delta \mathcal{H}}{\delta a_{\mathbf{k}}} \tag{69}$$

Because $\zeta_{\mathbf{k}} = \zeta_{-\mathbf{k}}^*$, $\psi_{\mathbf{k}} = \psi_{-\mathbf{k}}^*$ (reality condition), the second of (69) is complex-conjugable to the first and therefore can be omitted. Thus, the hydrodynamic equations reduce to one equation:

$$\frac{\partial a_{\mathbf{k}}}{\partial t} + i\frac{\delta \mathcal{H}}{\delta a_{\mathbf{k}}^*} = 0 \tag{70}$$

and the Hamiltonian $\mathcal{H}$ can be calculated by substituting in (58) the Fourier transform of (61) and (62) together with (67) and (68). This procedure corresponds to a perturbation series for the Hamiltonian $\mathcal{H}$

$$\mathcal{H} = \mathcal{H}_1 + \mathcal{H}_2 + \mathcal{H}_3 + \cdots \tag{71}$$

where the subscripts indicate the order of products of the variables $\zeta_{\mathbf{k}}$ and $\psi_{\mathbf{k}}$ on the right-hand side of (67). The Hamiltonian representation (70) and (71) clearly specifies the wave dynamics as an ordering of the nonlinear interactions in powers of the surface slope. Each term in (71) is independent of the vertical coordinate and the explicit expressions for $\mathcal{H}_1, \mathcal{H}_2, \mathcal{H}_3$ can easily be found. Substituting in (70) a linear part of the Hamiltonian $\mathcal{H}_1$, equal to

$$\mathcal{H}_1 = \int \sigma a_{\mathbf{k}} a_{\mathbf{k}}^* d\mathbf{k} \tag{72}$$

corresponds to

$$\partial a_{\mathbf{k}}/\partial t + i\sigma a_{\mathbf{k}} = 0 \tag{73}$$

with the solution

$$a_{\mathbf{k}} = A_{\mathbf{k}} e^{-i\sigma t}, \qquad A_{\mathbf{k}} = a_{\mathbf{k}}(0) \tag{74}$$

Therefore, $a_{\mathbf{k}}$ are usually called normal variables in the statistical description of a random field of weakly nonlinear surface waves; it is easy to see that amplitudes $\eta_{\mathbf{k}}^{\mathrm{s}}(\mathrm{s} = \pm$ is the sign index) introduced in Section 2 are uniquely related to $A_{\mathbf{k}}(a_{\mathbf{k}})$. In the case of $interacting$ waves, the amplitudes $A_{\mathbf{k}}$ are not constant and their evolutions are described by the Hamiltonian equation (70), which in terms of $A_{\mathbf{k}}$ with cubic accuracy in $\zeta_{\mathbf{k}}$ will have the form

$$\begin{aligned}
\frac{\partial A_{\mathbf{k}}}{\partial t} = &-i\int d\mathbf{k}_1 \int d\mathbf{k}_2 \left[\!\left[U_{-\mathbf{k}2,\mathbf{k}1,\mathbf{k}2} A_{\mathbf{k}1} A_{\mathbf{k}2} \right.\right. \\
&\left. \cdot \exp\{-i[\sigma_{\mathbf{k}_1} + \sigma_{\mathbf{k}_2} - \sigma_{\mathbf{k}}]\} + \}\delta(-\mathbf{k} + \mathbf{k}_1 + \mathbf{k}_2) + \cdots \right]\!\right] \\
&-i\int d\mathbf{k}_1 \int d\mathbf{k}_2 \int d\mathbf{k}_3 \left[\!\left[Q_{-\mathbf{k},-\mathbf{k}1,\mathbf{k}2,\mathbf{k}3} A_{\mathbf{k}1}^* A_{\mathbf{k}2} A_{\mathbf{k}3} \right.\right. \\
&\left. \cdot \exp\{-i[\sigma(\mathbf{k}_2) + \sigma(\mathbf{k}_3) - \sigma(\mathbf{k}) - \sigma(\mathbf{k}_1)]t\}\delta(-\mathbf{k} - \mathbf{k}_1 + \mathbf{k}_2 + \mathbf{k}_3) + \cdots \right]\!\right]
\end{aligned} \tag{75}$$

Here the terms with other combinations of wave amplitudes A_k, A_{k_1}, A_{k_2}, ... are omitted and the explicit expressions for coefficients U and Q can be found in Hasselmann (1968), Zakharov (1968), or West (1981). The solution of (75) depends on the form of the dispersion relation $\sigma = \sigma(k)$. It is useful to divide the dispersion relations into two groups—decay and nondecay relations, according to whether decay of one wave into two is or is not permitted, or in other words whether *triad resonant* interaction is or is not possible. It was discovered by Phillips (1960, 1961) that for *deep-water gravity* waves $[\sigma = (gk)^{1/2}]$ to obtain a nonvanishing interaction, "four-wave" processes, which are cubic in A_k, must be considered. Later, Hasselmann (1962, 1963) made the *first* theoretical calculations of the energy transfer due to weak nonlinear interactions in the random field of surface gravity wind-generated waves, using the so-called wave kinetic equation for spectral density of wave action per unit mass $N(k)$, defined as

$$\langle a_k a_k^* \rangle = \langle A_k A_k^* \rangle \tag{76}$$

$N(k)$ is related to nonsymmetrical spectra $F(k)$ as

$$N(k) = gF(k)/\sigma \tag{77}$$

The equation for $N(k)$ can be derived for example from (75) by multiplying it by $A_{k'}^*$, adding to it its complex-conjugate counterpart, averaging and integrating over k'. This equation will then have the following form:

$$
\frac{\partial N(k)}{\partial t} = \iiint U_{k,k_1,k_2,k_3} \{N(k_1)N(k_2)N(k_3) + N(k)N(k_2)N(k_3)
$$

$$
- N(k)N(k_1)N(k_2) - N(k)N(k_1)N(k_3)\}
$$

$$
\cdot \delta(k + k_1 - k_2 - k_3)\delta(\sigma_k + \sigma_{k1} - \sigma_{k2} - \sigma_{k3})dk_1 dk_2 dk_3 \tag{78}
$$

(where U_{k,k_1,k_2,k_3} is a complicated homogeneous function of wavenumbers k, k_1, k_2, k_3 of order $|k|^6$ [see its determination in Hasselmann (1968), Zakharov and Filonenko (1966).

The wave kinetic equation (78) describes the so-called "weak turbulence" phenomena, when the influence of nonlinear interactions in the random wave field on the dispersion properties of waves is negligible. The integrals on the right-hand side of the integro-differential equation (78) are called collision integrals, denoted usually as $I[N(k)]$, so that (78) can be written as

$$\frac{\partial N(k)}{\partial t} = I[N(k)] \tag{79}$$

In spite of the complexity of the wave kinetic equation (78), it does not give a complete description of weak turbulence in the field of surface gravity waves, because in its form (79), it does not include sources of generation and damping (dissipation). The generation of waves by wind is usually described in terms of $N(k)$ as $\alpha(k) + \beta(k)N(k)$ (Phillips–Miles theory). Usually, the experimental data on $\beta(k)$ are presented in the form $\beta(k) = \beta(k^{1/2}U/g)$, and in the range $k^{1/2}U_a/g = U_a/C(k) \approx 1.0$–$2.0$, they do not show a strong variation in values of $\beta(k)$ with k. Atmospheric turbulent pressure fluctuations, responsible for resonant initial linear growth of $N(k)$ with time, can be represented by $\alpha(k) \sim k^{-7/3}$ (Kitaigorodskii and Lumley, 1983). Therefore, if dissipative effects are strong only at very high wavenumbers [as for example viscous damping, described as $\gamma(k)N(k)$, $\gamma(k) = \nu k^2$, ν being the kinematic viscosity of water], it is possible to introduce a *physical hypothesis that the sources and sinks of energy for wind-generated waves are separated in wavenumber space*. In such a case, a "weak turbulence" can have some properties of "strong turbulence," in particular the presence of a cascade of energy from larger to smaller scales. Therefore, the intriguing and important question in the analysis of (78)–(79) is: does the stationary distribution corresponding to Kolmogoroff's inertial subrange, i.e., the conditions for predominantly *local* (in wavenumber space) energy transfer, exist for dispersive nonlinear surface gravity waves? The necessary (but not sufficient) condition for this is *localness* of the collision integral I, when the coupling coefficient U_{k,k_1,k_2,k_3} rapidly diminishes for $k \gg k_1, k_2, k_3$ or $k \ll k_1, k_2, k_3$. In such a case, let us, following Kolmogoroff's idea, introduce the energy transfer rate through the energy spectrum $gF(\mathbf{k})$ and assume that there exists a certain range of wavenumbers k in the wind-wave spectrum where this energy flux is constant (inertial subrange). It is convenient for further discussion to deal with average (over all directions of wave component propagation ϑ) values $F(\mathbf{k})$ and $N(\mathbf{k})$ defined as

$$F_k = \int_{\vartheta} F(\mathbf{k})d\vartheta = \int_{-\pi}^{+\pi} F(k, \vartheta)d\vartheta \tag{80}$$

$$N_k = \int_{\vartheta} N(\mathbf{k})d\vartheta = \int_{-\pi}^{+\pi} N(k, \vartheta)d\vartheta$$

For the isotropic dispersion relationship $\sigma(\mathbf{k}) = \sigma(k)$,

$$N_k = gF_k/\sigma(k) \tag{81}$$

Now F_k and N_k characterize the energy distribution as a function of wavenumber modulus $|\mathbf{k}| = k$. The condition of constant energy flux $\varepsilon(k)$ through the spectra gF_k can be written as

$$\varepsilon(k) = gF_k k^2 \tau_k^{-1} = N_k \sigma_k k^2 \tau_k^{-1} = \text{const} = \varepsilon_0 \tag{82}$$

where τ_k is a characteristic time of nonlinear interactions in a narrow interval of wavenumbers around $|\mathbf{k}| = k$. From (78) we can easily find an estimate of this quantity in terms of the collision integral I [right-hand side of (78)]:

$$\tau_k^{-1} = \frac{1}{N_k} \frac{\partial N_k}{\partial t} = \frac{\int_{\vartheta} I d\vartheta}{N_k} = \frac{I_k}{N_k} \tag{83}$$

For deep-water waves $[\sigma(k) = (gk)^{1/2}]$ the contribution from δ-functions in I_k can be estimated and the scaling of I_k for $U_{kkkk} \sim k^6$ will give us

$$I_k \simeq U_{kkkk} N_k^3 k^{7/2} g^{-1/2} \sim \frac{N_k^3 k^{9.5}}{g^{1/2}} \tag{84}$$

which in turn will lead to the following estimate of τ_k^{-1}.

$$\tau_k^{-1} = \frac{k^{9.5} N_k^2}{g^{1/2}} = g^{1/2} F_k^2 k^{8.5} \tag{85}$$

Substituting (85) in (82), we obtain the analog of Kolmogoroff's energy spectra for deep-water surface waves:

$$gF_k \sim \varepsilon_0^{1/3} g^{1/2} k^{-3.5} \tag{86}$$

or

$$F_k \sim \varepsilon_0^{1/3} g^{-1/2} k^{-3.5} \tag{87}$$

The corresponding frequency spectra $S(\omega)$ can be immediately found for the isotropic dispersion relation $\sigma = (gk)^{1/2}$ from (3), (4), and (87):

$$S(\omega) = \left. \frac{kF_k}{\dfrac{\partial \sigma}{\partial k}} \right|_{k = \omega^2/g} \sim \varepsilon_0^{1/3} g \omega^{-4} \qquad (88)$$

It was shown by Zakharov and Filonenko (1966) that frequency spectra $S(\omega) \sim \omega^{-4}$ for an isotropic field of weakly nonlinear waves really are an exact stationary solution of wave kinetic equation (78), corresponding to $I = I(N_k) = 0$. The collision integrals $I(N_k)$ [(78)] converge both for $\omega \to \infty$ and for $\omega \to 0$ [this solution was overlooked by Hasselmann in 1962 (Hasselmann, 1962, 1963) when *he first* raised the question about the existence of Kolmogoroff's energy cascade due to nonlinear wave–wave interactions in the spectra of wind-generated deep-water gravity waves].

We now can consider (86)–(88) as an *exact analog of Kolmogoroff's spectra* for so-called "weak turbulence" in a random field of weakly nonlinear surface gravity waves. The question as to the properties of the angular energy distribution, i.e., the anisotropy in the wavenumber spectrum $F(\mathbf{k})$ in such formulation, remains open, however. Also, the low (in wavenumber and frequency) boundary of Kolmogoroff's interval in spectra of wind waves are to be determined empirically. They certainly must lie above the peak frequency of $S(\omega)$, at which supposedly the direct atmospheric forcing must be important. There have been practically no attempts *to identify the appearance of Kolmogoroff's* subrange in the observed wind-wave spectra. The reason for this is more or less clear—the parameterization of the observed spectra was based in most cases on the concept of equilibrium, suggested by Phillips (1958). Only very recently, three independent observational studies (Forristall, 1981; Kahma, 1981a,b; Donelan *et al.*, 1982) have demonstrated very convincingly that *in the energy-containing region of the frequency spectra*

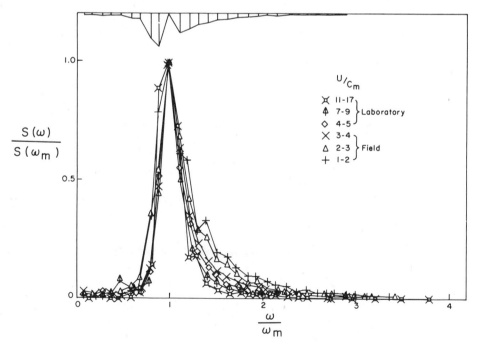

FIGURE 2. The normalized frequency spectra $S(\omega)/S(\omega_m)$ of wind-generated gravity waves for different values of the nondimensional parameter u_a/c_m ($c_m = g/\omega_m$) (according to Donelan *et al.*, 1982). At the top of the figure, the 95% confidence limits in each band of width $\omega/10\omega_m$ are indicated. The position of the vertical line indicates the average value of ω/ω_m in each band.

$S(\omega)$ [i.e., where spectral levels are greater than say 1% of the peak], the rear face of the spectrum is well described by an ω^{-4} power law. In Fig. 2 we present the frequency spectra of Donelan *et al.* (1982) on axes normalized by the magnitude of the peak spectral density and of the peak frequency. The spectra have been grouped into classes by the parameter U_a/C_m ($C_m = g/\omega_m$, ω_m is the peak frequency). The 95% confidence limits in each band of width $\omega/10\omega_m$ are indicated at the top of Fig. 2. The position of the vertical line indicates the average value of ω/ω_m in each band. To illustrate the slope of the rear face, these spectra have been multiplied by ω^4 and normalized by the average level of the spectral estimates multiplied by ω^4 in the frequency region $\omega > 1.5\,\omega_m$ (see Fig. 3). It is clear that an ω^{-4} power law is a good description of the rear face of the spectrum in the energy-containing region. Although harmonic peaks are clearly evident in the largest U_a/C_m spectral group, the mean spectral level is in good agreement with an ω^{-4} line. Both laboratory and field data support an ω^{-4} description of the rear face of the spectrum in the frequency region of the wind-generated gravity waves. That, of course, *can be considered as an indication of the existence of Kolmogoroff's subrange* [(87),(88)] in the observed statistical characteristics of wind-wave fields. Other examples of noticeable deviations from an ω^{-5} law for energy-containing components were given recently by Forristall (1981) and Kahma (1981a, b). In Fig. 4, the ensemble average of the measured spectra was compared with the JONSWAP spectrum. It is clear that above 0.2 Hz the normalized spectral density $S(\omega)\omega^5/g^2$ continues *to increase with* ω. This increase can be seen more *clearly as linear* in Fig. 5 in the region of the abscissa between 1.0 and 3.0 which in terms of nondimensional frequency $\tilde{\omega} = \omega U_a/g$ with $u_* \sim (1/30)U_a$ gives $5.5 > \omega U_a/g > 1.8$. It was pointed out by Forristall (1981) that above a transitional frequency $\omega_{tr} = 0.0275 \cdot 2\pi g/U_*$ ($\omega U_a/g > 5.1$) the ω^{-4} approximation for the near face is not good enough—the spectral densities begin to decrease more rapidly and as a possible parameterization of the region $\omega > \omega_{tr}$ he suggested Phillips's ω^{-5} power law. The data presented an Fig. 5 [numerous similar examples can be found in Forristall (1981) and Kahma (1981a, b)] indicate the possible limitations in applying Kolmogoroff's subrange theory [(87), (88)] to the higher frequencies than transitional. The proper question to ask now is, what can be the reasons for the *deviations from* Kolmogoroff's type of equilibrium at high frequencies? The concept of weak turbulence, when nonlinear wave–wave interactions in a random wave field are so weak that they

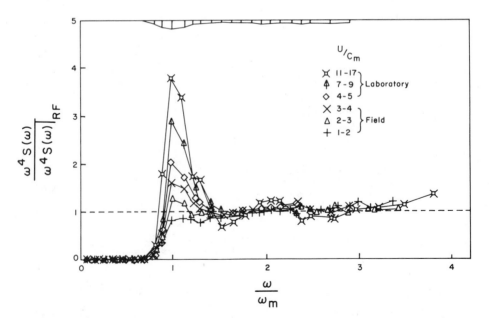

FIGURE 3. The spectra of Fig. 2 multiplied by ω^4 and normalized by the *average* level of $S(\omega)\omega^4$ (in the frequency range $\omega > 1.5\omega_m$) (according to Donelan *et al.*, 1982). $\omega^4 S(\omega)|_{RF}$ is the average value of $S(\omega)\omega^4$ in the frequency range $\omega > 1.5\omega_m$.

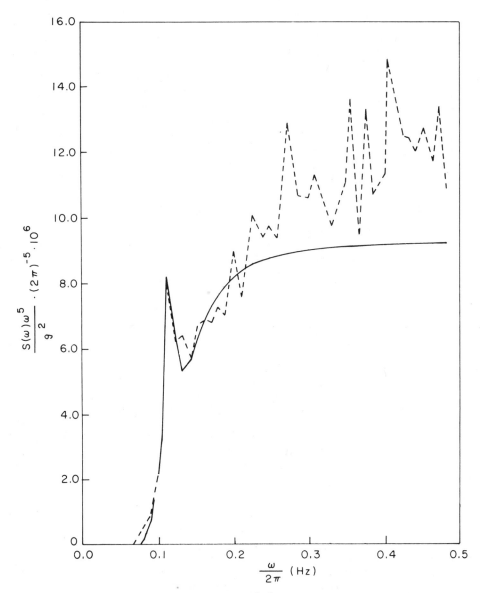

FIGURE 4. The frequency spectra $S(\omega)$ multiplied by ω^5/g^2 (according to Forristall, 1981). The dashed line is the ensemble average of individual spectra during hurricane Eloise for winds $U_a \approx 22.4$ to 24.6 m/s (U_a corresponds to 19.5 m wind). The solid line is the JONSWAP parameterization.

do not influence the dispersive properties of waves, is fundamental in the derivation of the wave kinetic equation (78). However, it is valid only when the ratio of characteristic time of nonlinear interactions τ_k to wave period $\tau \sim \sigma_k^{-1}$ is much larger than 1, or the quantity $\mu = (\sigma_k \tau_k)^{-1} \ll 1$. We can now estimate μ in Kolmogoroff's subrange using (85)–(87). We find for deep-water gravity waves $[\sigma_k = (gk)^{1/2}]$

$$\mu = (\sigma_k \tau_k)^{-1} = \frac{N_k^2 k^{9.0}}{g} \sim \frac{\varepsilon_0^{2/3} k}{g} \tag{89}$$

It is now clear from (89) that there must exist the wavenumber $k = k_\mu$ for which interactions are so

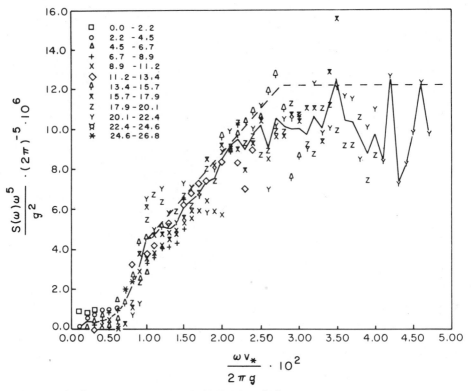

FIGURE 5. The frequency spectra $S(\omega)$ multiplied by ω^5/g^2 from the Pacesetter data (according to Forristall, 1981). The symbols are for various wind speed in m/s. The nondimensional frequency ω_{u*}/g was scaled by the friction velocity V_* in the atmospheric surface layer.

fast that $\mu \gtrsim 1$ and Kolmogoroff's concept, *based on scaling, (83)–(85) cannot be valid anymore.* The criteria for the applicability of the weak turbulence model [(78),(83)–(85)] for a description of Kolmogoroff's subrange based on critical values of $\mu = \mathrm{const} < 1$ [(89)] give a characteristic cutoff wavenumber k_μ for Kolmogoroff's subrange:

$$k_\mu \sim g/\varepsilon_0^{2/3} \tag{90}$$

This can be interpreted as the wavenumber at which the energy flux ε_0 will create *strong* nonlinearities in the Fourier component of the wave field with $k \gg k_\mu$ which ultimately will lead *to breaking.* Therefore, the slope of wave components with $k > k_\mu$ must also be achieving a critical value. The dynamics of components with $k \gg k_\mu$ cannot be described theoretically on the basis of the wave kinetic equation (78) based solely on resonance four-wave interactions. Therefore, the similarity arguments can be of some help here, and it was Phillips (1958) who suggested that there can exist an equilibrium range of scales where the statistical characteristics of wind waves are determined solely by the process *of wave breaking.* To proceed further with the construction of a general theory of the equilibrium range we must discuss the basic foundation and conclusions of Phillips's theory (Phillips, 1958).

4. THE PHILLIPS TYPE OF EQUILIBRIUM IN THE SPECTRA OF STRONGLY NONLINEAR WIND-GENERATED SURFACE GRAVITY WAVES

According to Phillips (1958, 1977), space and time scale ranges can exist in the spectra of wind-generated surface gravity waves within which the energy distribution must depend only on physical parameters characterizing the formation of sharp crests in *breaking waves*. If we consider that region of space and time scales in which viscosity and surface tension have no influence on the wave motion, the only significant parameter, according to Phillips (1958), *is the acceleration of gravity*. He deduced from this hypothesis the following well-known expressions for wind-wave spectra:

$$F(\mathbf{k}) = Bk^{-4}\phi(\vartheta) \tag{91}$$

$$S(\omega) = \beta g^2 \omega^{-5} \tag{92}$$

For a long time, (91) and (92) were used as basic and in fact the only constructive inferences about the shape of wind-generated gravity wave spectra at sufficiently large k and ω. Phillips (1958) derived (91) and (92) independently and from identical assumptions, although it is obvious in advance that similarity hypotheses for spatial and temporal spectra are *far from being of equivalent validity*. Therefore, it was suggested by Kitaigorodskii *et al.* (1975) that the expression for $S(\omega)$ should be derived on the basis of Phillips's equilibrium wavenumber spectra (91) and the formula (42), using the dispersion relationship for free surface waves. For an *isotropic* dispersion relationship (41), this gives the following result for frequency spectra of wind-generated waves in a finite-depth sea (Kitaigorodskii *et al.*, 1975):

$$S(\omega) = \beta g^2 \omega^{-5} \Phi(\omega_H) \tag{93}$$

where $\Phi(\omega_H)$ is a universal nondimensional function of nondimensional frequency $\omega_H = \omega H^{1/2}/g^{1/2}$ equal to

$$\Phi(\omega_H) = Z^{-2}\left\{1 + \frac{2\omega_H^2 Z}{\mathrm{sh}(2\omega_H^2 Z)}\right\}^{-1} \tag{94}$$

and $Z(\omega_H)$ satisfies (43). It is easily verified that when $\omega_H \to \infty$, then $\Phi(\omega_H) \to 1$, leading to Phillips's spectra (92) with

$$\beta = 2B \tag{95}$$

In another extreme case at $\omega_H \to 0$, the function $\Phi(\omega_H) \to \omega_H^2/2$ and (93) leads to another universal power law:

$$S(\omega) = BgH\omega^{-3} \tag{96}$$

We will not discuss here the numerous data about the "variability" of Phillips's constant β [(92)] (see last section) but prefer to deal with a more general form of frequency spectra [(93)]. Some measurements of frequency spectra of wind waves in the Caspian Sea (Gadzhiyev and Krasitskii, 1978) were used to estimate the universal function $\Phi(\omega_H)$ in (93). To do this, the average experimental value of $\beta = 6 \times 10^{-3}$ was used and then $\Phi(\omega_H)$ was constructed using the frequency spectra measured at different wind speeds ($U_a \approx 10\text{--}12$ m/s) and *different depths* [$H = 40$ m (deep sea), 12 m, 6 m]. The results of such determinations of $\Phi(\omega_H)$ in (93) are shown in Fig. 6, together with the theoretical curve (94) (Fig. 7). The *qualitative* agreement between theory and observation is noticeable, indicating that at frequencies $\omega_H \leqslant 1.0$ (see Fig. 7), the observed spectra have a slope closer to ω^{-3} than to ω^{-5}. We have already mentioned in the Introduction some other data on frequency spectra *of shoaling wind waves*, which also give support to an ω^{-3} rather than ω^{-5} law over a certain range of frequencies. However, the universal equilibrium wavenumber spectra (91) according to Phillips (1958) must characterize the process of wave breaking, when downward acceleration of particles can reach (or exceed) g. The typical acceleration $a_k = \{[\int_k^{\zeta} d\mathbf{k}]^2\}^{1/2}$ *associated only with the energy transfer rate* for given wavenumber

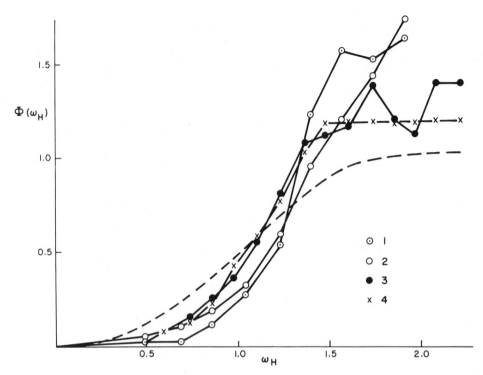

FIGURE 6. The empirical estimates of the universal function $\Phi(\omega_H)$ in (93) (according to Gadzhiyev and Krasitskii, 1978). $\odot$ 1, $\bigcirc$ 2: the experimental data corresponding to depth $H = 12$ m. $\odot$ 1: based on observed values of spectral density $S(\omega)$ in the range of frequencies $0.5\,\text{Hz} > \omega/2\pi > 0.1\,\text{Hz}$ and wind speed $U_a = 12$ m/s. $\bigcirc$ 2: based on observed "average" values of spectral densities $S(\omega)$ in the range of frequencies $0.5\,\text{Hz} < \omega/2\pi > 0.1\,\text{Hz}$ and wind speeds $U_a = 14.5, 13.5, 12, 6.5$ m/s. N3, X4: the experimental data corresponding to depth $H = 6$ m. $\bullet$ 3: based on observed values of spectral density $S(\omega)$ in the range of frequencies $0.5\,\text{Hz} > \omega/2\pi > 0.1$ Hz for wind speed $U_a = 11.7$ m/s. X4: based on observed "average" values of spectral densities $S(\omega)$ for wind speeds $U_a = 10, 10.5, 11.7, 12.3$ m/s. The dashed line is the theoretical estimates of $\Phi(\omega_H)$ according to (43) and (94) (see also Fig. 7).

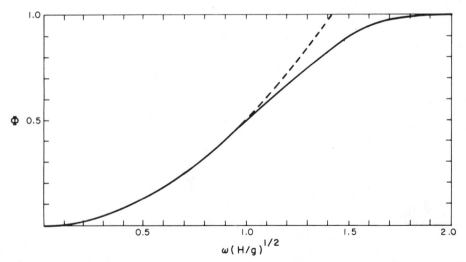

FIGURE 7. The universal dimensionless function $\Phi(\omega_H)$ from (93) and (94) (solid line) and the function $\omega_H^2/2$ (dashed curve).

k can be estimated as

$$a_k = a_k(k, \varepsilon_0) \sim \varepsilon_0^{2/3} k \tag{97}$$

and therefore Phillips's (1958) "condition for breaking" can be written as

$$a_k/g \sim \varepsilon_0^{2/3} k/g \sim \text{const} \sim 1 \tag{98}$$

It follows from (97) and (98) that *cutoff wavenumber k_g for Kolmogoroff's subrange* in the wave field *which can be considered also as the lowest wavenumber* for applicability of Phillips's law (91) is

$$k_g \sim g/\varepsilon_0^{2/3} \tag{99}$$

The corresponding cutoff frequency for *deep-water gravity waves* is

$$\omega_g = (gk_g)^{1/2} \sim g/\varepsilon_0^{1/3} \tag{100}$$

Naturally, the definition of the cutoff wavenumber k_g for Kolmogoroff's subrange coincides with the criteria of applicability of the weak turbulence model (78) for the description of Kolmogoroff's subrange based on critical value of μ [(89), (90)]. Therefore, if we assume that the growth of shorter waves on the rear face of the spectrum (limited by breaking) *is not due to direct* energy input from wind, but rather *to energy flux from lower wavenumbers*, we come to the conclusion that Phillips's equilibrium can exist in the region $k \gg k_g$, i.e., at wavenumbers *higher than those where to Kolmogoroff's subrange can develop*. It follows from this that in the *finite-depth sea* the applicability of Phillips's equilibrium wavenumber spectra (91) at least must be limited to the range of wavenumbers satisfying the condition $k > k_g$. We will not attempt to describe here the form of spectra for *shoaling* wind waves, but instead we will try to incorporate the possibility of growth of shorter waves due to energy flux *into the general theory of equilibrium range*, using an approach similar to the theory of locally isotropic 3d turbulence; namely by *postulating* the existence of Kolmogoroff's type of equilibrium in spatial statistical characteristics of a random wave field and then *determining* the "internal scale" for surface gravity waves, at which deviations from the energy cascade mechanism can occur due to gravitational instability (wave breaking).

5. THE GENERAL SIMILARITY HYPOTHESIS FOR THE EQUILIBRIUM RANGE IN SPATIAL STATISTICAL CHARACTERISTICS OF A WIND-WAVE FIELD

In this section we will mostly follow the theory of the equilibrium range, proposed recently by Kitaigorodskii (1983). To start with, it is important to notice that among three statistical characteristics of the wave field, which are conserved in the process of nonlinear wave–wave interaction, i.e., energy, momentum, and action density, only the latter underlines the conservation of energy in *dispersive wave* motion. Therefore, the similarity arguments must be applied first of all to action density $N(\mathbf{k})$ or N_k. According to the discussion in the previous section, we can now formulate the first similarity hypothesis as follows: if in wavenumber space we restrict our attention to wavenumbers well below those associated with capillary ripples and those directly influenced by viscosity, so that

$$k \ll k_T = (\rho_w g T^{-1})^{1/2}, \qquad k \ll k_v = g^{1/4}/v^{1/2} \tag{101}$$

where T is the surface tension, ρ_w the density, v the kinematic viscosity, and well above those associated with strong direct energy input from wind ($k \sim k_m$) so that

$$k \gg k_m = \omega_m^2/g \tag{102}$$

where k_m is associated with the peak frequency of the spectrum $S(\omega)$, then for sufficiently large fetch and duration of the wind the values of N_k are determined only by the values of the parameters ε_0, g, and k.

It follows from this hypothesis that

$$N_k = \varepsilon_0^{1/3} k^{-4} \psi(k\lambda_g), \qquad \lambda_g = \varepsilon_0^{2/3}/g \tag{103}$$

where ψ is a nondimensional function of nondimensional wavenumber $\tilde{k} = k\lambda_g = k\varepsilon_0^{2/3}/g$. The second similarity hypothesis can be formulated for the range of wavenumbers satisfying the inequality $k\lambda_g \ll 1$. In this range of scales, if it exists and is broad enough, gravitational instability is not important because wave components do not reach critical amplitudes to produce breaking, and N_x depends only on ε_0 and k so that

$$\text{for } k\lambda_g \ll 1, \qquad \psi(k\lambda_g) = \text{const} = A \tag{104}$$

$$N_k = A\varepsilon_0^{1/3} k^{-4} \tag{105}$$

where A is an *absolute constant*. By analogy with Kolmogoroff's theory of locally isotropic turbulence (see Section 3), *we can expect here that constant A must be close to unity*. It is easy to see that according to (81) and (88) this corresponds to the following expressions for wave spectra for deep-water gravity waves:

$$F_k = A\varepsilon_0^{1/3} g^{-1/2} k^{-3.5} \tag{106}$$

$$S(\omega) = \alpha\varepsilon_0^{1/3} g\omega^{-4}, \qquad \alpha = 2A \tag{107}$$

Finally, the third similarity hypothesis can be formulated following Phillips's theory (Phillips, 1958), according to which we can assume that for the range of wavenumbers $k\lambda_g \gg 1$, the governing parameters are those that determine the continuity of the wave surface, and therefore asymptotically N_k become independent of ε_0 and will depend only on g and k, so that

$$\psi(k\lambda_g) = B(k\lambda_g)^{-1/2} \tag{108}$$

$$\text{for } \left.\begin{matrix} k_T\lambda_g \\ k_v\lambda_g \end{matrix}\right\} \gg k\lambda_g \gg 1$$

$$N_k = Bg^{1/2} k^{-9/2} \tag{109}$$

where B is an *absolute constant*. It follows from (81) and (88) that this corresponds to the expressions for wave spectra, first suggested by Phillips (1958),

$$F_k = Bk^{-4} \tag{110}$$

$$S(\omega) = \beta g^2 \omega^{-5}, \qquad \beta = 2B \tag{111}$$

where β is Phillips's constant. There are a few remarks that should be made at this point. The first is that independence of N_k, F_k, and $S(\omega)$ from ε_0 means that the geometry of the limiting wave configuration near the sharp crests is determined by the condition (98). An increase in ε_0 would have the effect of increasing the rate at which wave crests are passing through the transient limiting configuration, but should not influence the geometry of sharp crests itself. According to (99), it simply means that the increase in ε_0 will lead to growth of *longer* wave components up to this limiting configurations, so that in the wavenumber spectra the boundary of Phillips's equilibrium range will move with increase of ε_0 toward a lower wavenumber.

The other remark is that according to our third similarity hypothesis, we consider the *asymptotic* situation, corresponding to indefinitely large values of $k\lambda_g$ (indefinitely large values of

k or indefinitely large values of ε_0), where the statistical characteristics of the wave field are determined solely by the process of wave breaking.

Therefore, the magnitude of the spectrum in Phillips's *subrange* represents an upper limit of $N_k[F_k, S(\omega)]$, dictated by the requirement of crest attachment. Generally speaking, we cannot disregard the possibility (because of the very nature of asymptotic arguments) that for $\varepsilon_0 \to \infty$ ($k\lambda_g \to \infty$) the values of N_k continue to depend, no matter how slightly, on ε_0, so that instead of (108) and (109) we will have

$$N_k = A\varepsilon_0^{1/3} k^{-4} (k\lambda_g)^{-p} \tag{112}$$

where $0 < p < \frac{1}{2}$ is some small power exponent. However, it seems that assumption of independence of N_k from ε_0 ($p = \frac{1}{2}$) which leads to Phillips's spectra (110) and (111) is the most elegant and attractive approximation to describe the asymptotic form of wind-wave spectra in the range of scales where breaking is primarily important in limiting the growth of wave components. If we accept (103), (104), and (108), then the *simplest* interpolation formula for universal function (103) in the equilibrium interval will be

$$\psi(k\lambda_g) = A + B(k\lambda_g)^{-1/2} \tag{113}$$

so that

$$N_k = A\varepsilon_0^{1/3} k^{-4} + B k^{-4.5} g^{1/2} \tag{114}$$

This lead to a definition of transitional wavenumber k_g as

$$A\varepsilon_0^{1/3} k_g^{-4} \approx B k_g^{-4.5} g^{1/2} \tag{115}$$

or

$$k_g^{1/2} \approx (B/A)\varepsilon_0^{-1/3} g^{1/3} \tag{116}$$

For deep-water gravity waves the "transitional" frequency will be

$$\omega_g = (gk_g)^{1/2} \approx (B/A) g \varepsilon_0^{-1/3} \tag{117}$$

As mentioned earlier, the transition from inertial Kolmogoroff's subrange to Phillips's asymptotic regime, as well as the structure of Phillips's subrange by itself, cannot be described theoretically on the basis of the wave kinetic equation (78) based solely on *weak resonance four-wave interactions*. Therefore, to discover the existence of the transition between Kolmogoroff's and Phillips's subranges, we must examine experimental data. Fortunately, recent observational studies by Forristall (1981) and Kahma (1981a, b) give enough material to make a *search* for such a transition well justified. We will deal with this in the next section.

6. EXPERIMENTAL EVIDENCE FOR TRANSITION FROM KOLMOGOROFF'S TO PHILLIPS'S EQUILIBRIUM IN THE SPECTRA OF WIND-GENERATED DEEP-WATER GRAVITY WAVES

The universal form of equilibrium wave spectra for deep-water gravity waves according to (103) can be written as

$$F_k = \varepsilon_0^{1/3} g^{-1/2} k^{-3.5} \psi(k\lambda_g) \tag{118}$$

$$S(\omega) = 2\varepsilon_0^{1/3} g^{-1} \{k^{-2.0} \psi(k\lambda_g)\}_{k=\omega^2/g} \tag{119}$$

If we use the simple linear interpolation formula for $\psi(k\lambda_g)$, which *satisfies the asymptotic behaviour* of $\psi(k\lambda_g)$ for small and large $k\lambda_g$ (for Kolmogoroff's and Phillips's regimes), then (118) and (119) will be

$$F_k = A\varepsilon_0^{1/3} g^{-1/2} k^{-3.5} + Bk^{-4} \tag{120}$$

$$S(\omega) = 2A\varepsilon_0^{1/3} g\omega^{-4} + 2Bg^2\omega^{-5} \tag{121}$$

It has been mentioned that Forristall (1981) showed that the high-frequency portion of *all* observed spectra can be parameterized successfully by ω^{-4} and ω^{-5} laws *below* and above a *transitional* frequency ω_t, which according to his empirical determination was

$$\omega_t = 0.173 \frac{u_*}{g} \tag{122}$$

where u_* is the friction velocity in the atmospheric surface layer, which he calculated by using the logarithmic law

$$U_a/u_* = (1/\kappa) \ln(Z_a/Z_0) \tag{123}$$

and roughness length Z_0 according to Garrat (1977):

$$Z_0 = 0.0144 \, u_*^2/g \tag{124}$$

Similar values of the transitional frequency (122) were also mentioned by Kahma (1981a), who used mean wind speed U_a (instead of u_*) in describing the variability of wave spectra and their characteristics. If we want to use such data for a determination of the universal function $\psi(k\lambda_g)$ in (118) and (119), or of the universal constants A and B in (120) and (121), we must determine the energy flux ε_0, or relate it in some way to such measurable external parameters as wind speed u_a, fetch, and duration of wind. If the range of frequencies from which energy is extracted from the wind is distinct from the range over which energy is lost, and if the internal (viscous) energy dissipation is negligible in the wave field, then ε_0 represents the difference between the rate of gain of energy from wind W_a and the rate of increase of energy of those components of the wave system that are still developing and have not yet attained a statistical equilibrium [(118),(119)]. In the absence of direct measurements of the atmospheric input, we can use only *general energy* balance to estimate ε_0, i.e., assume that

$$\rho_w \varepsilon_0 \approx W_a \tag{125}$$

where W_a characterizes the energy flux per unit area from wind to waves. Because W_a is *proportional* to $\rho_a U_a^3$, we can also write that

$$\varepsilon_0 = m \, (\rho_a/\rho_w) \, U_a^3 \tag{126}$$

where m is an *a priori* unknown nondimensional coefficient in which the ratio $W_a/\rho_a U_a^3$ is also incorporated. It follows from (126) and (118)–(121) that the Kolmogoroff type of equilibrium in the spectrum of wind-generated waves corresponds to a *wind-dependent* saturation of the form

$$F_k = A_u g^{-1/2} U_a k^{-3.5} \tag{127}$$

$$S(\omega) = \alpha_u g U_a \omega^{-4}, \qquad \alpha_u = 2A_u \tag{128}$$

where A_u, α_u are nondimensional coefficients in which the absolute constant A, coefficient m, and constant ratio ρ_a/ρ_w are incorporated according to the expression

$$\alpha_u = 2A\,(m\rho_a/\rho_w)^{1/3} \tag{129}$$

If in accordance with the similarity hypothesis (Kitaigorodskii, 1962, 1981) the spectra of wind waves in various stages of their development are assumed to have a similar form, the expression for $S(\omega)$ can be written in the form

$$\tilde{S}(\omega) = F(\tilde{\omega}, \tilde{x}, \tilde{t}) \tag{130}$$

where

$$\tilde{S}(\omega) = S(\omega)g^3/U_a^5, \qquad \tilde{\omega} = \omega U_a/g, \qquad \tilde{x} = gx/U_a^2, \qquad \tilde{t} = gt/U_a \tag{131}$$

Within the framework of similarity theory [(130),(131)], the existence of an equilibrium range in the frequency spectra of wind-generated waves means that for sufficiently large fetch and duration, there are ranges of frequencies $\tilde{\omega} > \tilde{\omega}_m(\tilde{x}, \tilde{t})$ where the universal function F of nondimensional frequency $\tilde{\omega}$ must satisfy the following conditions:

$$\tilde{\omega}_m < \tilde{\omega} \ll \tilde{\omega}_g \qquad F = \psi_1(\tilde{\omega}) = \alpha_u \tilde{\omega}^{-4} \tag{132}$$

$$\tilde{\omega} \gg \tilde{\omega}_g \qquad F = \psi_2(\tilde{\omega}) = \beta \tilde{\omega}^{-5} \tag{133}$$

Here β is a universal constant and α_u according to (129) can depend on $\tilde{X}$ (or $\tilde{t}$) only because $\rho_w\varepsilon_0/\rho_a U_a^3 = m$ can be variable. It is important to underline the fact that according to (133), Phillips' constant β can be accurately determined *only with the data in the range of frequencies* $\tilde{\omega} > \tilde{\omega}_g$[(100), (117)]. If the transition in frequency spectra between Kolmogoroff's [(132)] and Phillips' [(133)] equilibrium is very sharp in the frequency domain, we can determine the transition frequency $\tilde{\omega}_g$ [(117)] using (132) and (133), i.e., assuming

$$\psi_1(\tilde{\omega}_g) \approx \psi_2(\tilde{\omega}_g) \tag{134}$$

which leads to

$$\tilde{\omega}_g = \beta/\alpha_u \tag{135}$$

which is another form of (117).

To illustrate the existence of two asymptotic regimes [(132),(133)] and sharp transition between them [(134), (135)], we analyzed the data collected by Kahma (1981a, b). These data were based on the Waverider buoy measurements in Bothnia Bay and include the values of $S(\omega)$ in the narrow and fixed low-frequency interval between 0.35 and 0.5 Hz. To avoid the bias due to the overshooting effect at the peak frequency, only spectra $S(\omega)$ in which peak frequency ω_m was *below* 0.25 Hz were accepted. The range of variation of 10-min average wind speed U_a was for those data from 3 to 18 m/s. This new set of data was combined with previously observed spectra $S(\omega)$ at exceptionally steady wind conditions (Kahma, 1981a, b). Figure 8 shows the *experimentally determined* values of universal function $F(\tilde{\omega}, \tilde{x}, \tilde{t})$ [(130)], presented in the form $F\tilde{\omega}^5 = \tilde{S}(\omega)\tilde{\omega}^5$ versus $\tilde{\omega}$ and wind speed. The *linear* variation of $\tilde{S}(\omega)\tilde{\omega}^5$ with $\tilde{\omega}$ (or with wind speed U_a for fixed (ω) is evident up to values of $\tilde{\omega} \sim 4.0$, indicating that for $\tilde{\omega} < 4.0$, $F(\omega) = \psi_1(\tilde{\omega}) \sim \tilde{\omega}^{-4}$. When the *average* value of $F(\tilde{\omega})\tilde{\omega}^5$ was calculated, the corresponding curve in Fig. 9 very closely follows the straight line, predicted by (132) up to a wind speed of 14 m/s, or in terms of nondimensional frequency $\tilde{\omega}$ up to about 4.0. The important feature of Figs. 8 and 9 is the noticeable leveling of the function $F(\tilde{\omega})\tilde{\omega}^5$ at frequencies $\tilde{\omega} > 4.0$. It demonstrates that the behavior of $F(\tilde{\omega})$ here satisfies (133) when

$$F(\tilde{\omega})\tilde{\omega}^5 = \text{const} = \beta \tag{136}$$

It is important to point out that it follows from Figs. 8 and 9 that the constant value of $F(\tilde{\omega})\tilde{\omega}^5$ $= \beta$ for $\tilde{\omega} > 4.0$ is close to 1.5×10^{-2}. Such a value of Phillips's constant is in agreement with the well-known determination of β, based on experiments by Burling (1959) and Mitsuyasu (1977). One of the reasons for the variability of β, reported by many authors (see, e.g., our Fig. 1 from Hasselmann *et al.*, 1973), is certainly due to the fact that the ω^{-5} law was applied to the region were $S(\omega) \sim gU_a\omega^{-4}$, i.e., for the range of frequencies $\omega < \omega_g$, where saturation is at least *wind*

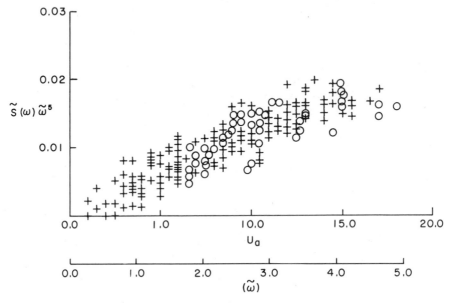

FIGURE 8. The nondimensional function $F(\tilde{\omega})\tilde{\omega}^5 = \tilde{S}(\omega)\tilde{\omega}^5$ [(130), (131)] according to the observations of Kahma (1981a, b). The crosses show the values of $\tilde{S}(\omega)\tilde{\omega}^5$ determined from a fixed frequency interval $0.35 - 0.5\,\text{Hz}$ as a function of wind speed U_a. The scale below shows the dimensionless angular frequency $\tilde{\omega} = \omega U_a/g$, which corresponds to the frequency $0.425\,\text{Hz}$ in the middle of this interval. The open circles correspond to the data of Kahma (1981a).

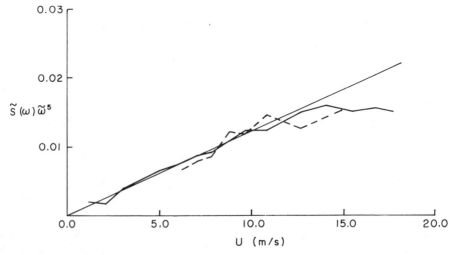

FIGURE 9. The nondimensional function $F(\tilde{\omega})\tilde{\omega}^5 = \tilde{S}(\omega)\tilde{\omega}^5$. The continuous curve is the average of the observations in Fig. 8. The straight solid line is calculated from (132) with $\alpha_u = 4.5 \times 10^{-3}$. The dashed line represents observations under exceptionally steady-state wind conditions, reported by Kahma (1981a).

dependent [can also be, in principle, fetch dependent through $m(\tilde{x})$]. Therefore, among all the reported values of β, only a few from our point of view can be trusted as really *adequate* for the description of the asymptotic Phillips's law (133) in the range of frequencies $\tilde{\omega} > \tilde{\omega}_g$. One of the consequences of approximation of the *whole* equilibrium range [(132),(133)] by the ω^{-5} law is the *necessity* to describe high-frequency portions of the spectra $S(\omega)$ with a *variable* β, as was first suggested in the JONSWAP Experiment (Hasselmann *et al.*, 1973) in the form $\beta = \beta(\tilde{x})$. However, it is clear that if in Kolmogoroff's subrange $\omega_m < \omega < \omega_g$ we determine the value of β, it will vary with wind $U_a(\beta \sim U_a)$ but not necessarily show the variations with fetch. Such "artificial" variability of β with wind speed U_a was very often incorporated in an empirical relationship $\beta(\tilde{x}) \sim \tilde{x}^{-n} \sim U_a^n$. To demonstrate this, one can calculate the values $\tilde{S}(\omega)\tilde{\omega}^5$ by using the data of $S(\omega)$ in the region $\omega_m < \omega < \omega_g$ (in the inertial subrange) at a *fixed* frequency and in a *very narrow* range of wind speed. Then if the relationship $\beta = \beta(\tilde{x})$ is correct, we must expect the ratio $S(\omega_i)/g^2\omega_i^{-5} = \beta_i$ to be fetch dependent (for constant wind!). The experimentally derived values β_i for a fixed frequency in the range of frequencies $\tilde{\omega} < 4.0$ for the data in Fig. 8, are shown in Fig. 10 for a very narrow range of wind speeds $U_a \approx 7$–9 m/s. There is no variation of β with fetch x; even so, the values of $\beta \approx 1 \times 10^{-2}$ itself are different from those derived from the range of frequencies $\tilde{\omega} > \tilde{\omega}_g$ (where $\beta \approx 1.5 \times 10^{-2}$).

According to Figs. 8 and 9, the value of α_u in Kolmogoroff's spectra (128) was found to be 4.4×10^{-3} which is practically equal to the values of α_u reported by Kahma (1981) in his previous experiments and also very close to the value of α_u determined by Forristall (1981), where α_u was 4.5×10^{-3}. The results reported in Kahma (1981a, b) and Forristall (1981), as well as the data presented in Figs. 8 and 9 give, from our point of view, serious support to the idea that the variability of high-frequency parts of the wave spectrum can be explained *by equilibrium range theory*, which includes existence *both* of an *inertial subrange* where $S(\omega) = \alpha_u\tilde{\omega}^{-4}$ and *Phillips's subrange* where $S(\omega) = \beta\tilde{\omega}^{-5}$, with the relatively sharp transition between the two at the "transitional" frequency $\tilde{\omega}_g \approx 4.0$–$5.0$. The determinations of the nondimensional coefficient α_u according to these studies lead to values of $\alpha_u \approx (4.3$–$4.5) \times 10^{-3}$ which are *practically fetch independent*. Therefore, we can consider the nondimensional coefficient α_u in the wind-dependent spectrum (128) as an *empirical constant*. This permits us to make attempts to determine (very crudely) the value of the *universal constant* $A(\alpha)$ in the inertial (Kolmogoroff) subrange [(106),(107)]. To do this, let us write the expression form

$$W_a = \gamma_a \tau_a \bar{c} = \gamma_a \rho_a C_f U_a^2 \bar{c} \tag{137}$$

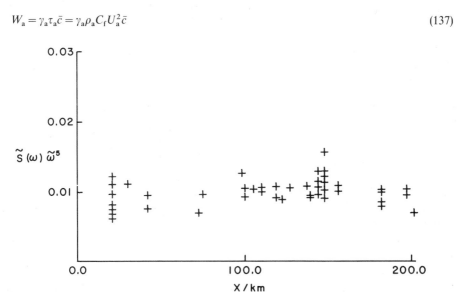

FIGURE 10. The dependence of the nondimensional function $\tilde{S}(\omega)\tilde{\omega}^5$ on the fetch X for fixed frequency $\tilde{\omega}_i < 4.0$ in the narrow range of wind speed $U_a = 7 - 9$ m/s.

where $\bar{c} = g/\sigma$ is the mean phase speed of the waves and $\bar{\sigma}$ is defined as

$$\bar{\sigma}^2 \int_0^\infty S(\omega)d\omega = \int_0^\infty \omega^2 S(\omega)d\omega \tag{138}$$

The product $\gamma_a \tau_a$ in (137) characterizes the fraction of the momentum flux in atmosphere τ_a which goes to waves and C_f in (137) is the drag coefficient. Comparing (137) with (125) and (126), we can find an expression for the nondimensional coefficient m in (126) as

$$m = \gamma_a C_f (\bar{c}/U_a) \tag{139}$$

In general, we must take into account the fact that both γ_a and C_f can vary with $\bar{c}/U_a$ (see, e.g., Benilov et al., 1978a, b), so that

$$m = \gamma_a(\bar{c}/U_a)C_f(\bar{c}/U_a)\bar{c}/U_a \tag{140}$$

Taking $\gamma_a = 0.1-1.0$, $C_f \simeq (1.0-1.3) \times 10^{-3}$, and $\bar{c}/U_a = 0.7-0.9$, we can reasonably well estimate the range of variations of m in (140). We have

$$m = (0.07-1.17) \times 10^{-3} \tag{141}$$

From (129) with $\rho_a/\rho_w = 1.2 \times 10^{-3}$, we have

$$A = 5\alpha_u/m^{1/3} \tag{142}$$

For experimentally derived values of $\alpha_u = 4.5 \times 10^{-3}$ (Kahma, 1981a, b; Forristall, 1981) and values of m [(141)], we obtain

$$A = 0.55-0.22 \tag{143}$$

Therefore, we can expect the universal constant A to be of order 1 under any circumstances. This gives additional confirmation that in our similarity theory in Section 5 the choice of governing parameters was made correctly.

7. CONCLUSIONS

In the last two sections I have made an attempt to present a *unifying* theory of equilibrium range in the spectra of wind-generated waves, as well as some observations whose interpretation at least does not contradict the proposed theory. The theory suggests of course only *asymptotic forms* for energy-containing high-wavenumber and high-frequency parts of wave spectra. However, it can be, as I hope, successfully used in the future in the parameterization of observed statistical characteristics of wind waves. For example, the theory probably gives us a chance to determine the *variability* of *energy flux* ε_0 from the *observed spectra* in the *inertial* (Kolmogoroff) *subrange* exactly in the same way as was very successfully done in studies of atmospheric turbulence. Once the relationship

$$\varepsilon_0/U_a^3 = \phi(\tilde{x}) \tag{144}$$

has been experimentally established and the values of the universal constants A and B in (106) and (110) determined by accurate procedures similar to those described in Section 6, we can really hope to have a proper method to describe the behavior of the rear face of wave spectra under different wind-wave generation conditions. Special attention must, however, be given to the determination of a universal form of the spectra in the equilibrium range:

$$F_k = A\varepsilon_0^{1/3} g^{-1/2} k^{-3.5} f_1(k\varepsilon_0^{2/3}/g) \tag{145}$$

$$S(\omega) = SA\varepsilon_0^{1/3} g\omega^{-4} f_2(\omega\varepsilon_0^{1/3}/g) \tag{146}$$

This means that the universal functions $f_{1,2}$ in (145) and (146) must be constructed and their asymptotic behavior studied on the basis of reliable experimental data. This will permit more accurate descriptions of the transitional region between inertial subrange (where $f_1 = f_2 = 1$) and subrange of "dissipation" due to breaking [where supposedly $f_1 \approx (B/A)(k\varepsilon_0^{2/3}/g)^{-1/2}$ and $f_2 = (B/A)(\omega\varepsilon_0^{1/3}/g)^{-1}$]. The numerical calculations of wave–wave interactions in the spectra of wind waves, pioneered by Hasselmann et al. (1973), must be based on a parameterization close to (145) and (146). In such cases, the calculation of collision integrals can be used for determination of the energy transfer rate $\varepsilon(k)$ and its deviations from a constant value ε_0, characterizing the inertial subrange. It will certainly help to estimate *under what conditions the asymptotic theory of equilibrium range* can be really adequate in the description of the observed wave characteristics.

The other question which deserves serious attention is, in what way can the direct energy input from wind *modify* the equilibrium form of the spectra (145) and (146)?

Theoretically, of course, it is possible (however, experimental studies do not show) that at very short fetches the direct energy input from wind will not give a possibility for development of a broad inertial subrange. Under such conditions, wave spectra will approach an upper limit, based on the *breaking* criteria (Phillips, 1958) without having a Kolmogoroff cascade at *lower wavenumbers*. This can be relevant for example for the range of wavenumbers (frequencies) much *higher than those studied in this chapter*. However, in general, it seems to us that there are very serious indications that for wave components of a deep-water wind-generated wave field, whose phase velocities $c(k)$ are in the range $\frac{1}{5} \lesssim c(k)/U_a < 1$, the direct energy input from wind is negligible compared to energy flux from the region of the peak, and the concept of the equilibrium range in wind-wave spectra, suggested in the last two sections, is really applicable.

ACKNOWLEDGMENTS. A part of the work presented in this chapter was supported by NASA–Ames Research Center under Grant MSG-2382.

REFERENCES

Benilov, A. Y., M. M. Zaslavskii, and S. A. Kitaigorodskii (1978a): Construction of small parametric models of wind-wave generation. Oceanology **18**, 387–390.

Benilov, A. Y., A. I. Gumbatov, M. M. Zaslavskii, and S. A. Kitaigorodskii (1978b): A nonstationary model of development of the turbulent boundary layer over the sea with generation of surface waves. *Izv. Akad. Nauk SSSR Fiz. Atmos. Okeana* **14**, 830–836.

Bouws, E., Günther, H., Rosenthal, W., and Vincent, C. L. (1986a): Similarity of the wind-wave spectrum in finite depth water. I. Spectral form, *J. Geophys. Res.* (in press).

Bouws, E., Günther, H., Rosenthal, W., Vincent, C. L. (1986b): Similarity of the wind-wave spectrum in finite depth water. II. Statistical relations between shape and growth parameters, *J. Geophys. Res.* (in Press).

Burling, R. W. (1959): The spectrum of waves at short fetches. *Dtsch. Hydrogr. Z.* **12**, 45–64, 96–117.

Donelan, M. A., J. Hamilton, and W. H. Hui (1982): Directional spectra of wind-generated waves. Unpublished manuscript.

Forristall, Z., (1981): Measurements of a saturated range in ocean wave spectra. *J. Geophys. Res.* **86**, 8075–8084.

Gadzhiyev, J. Z., and V. P. Krasitskii (1978): On equilibrium range in the frequency spectra of wind waves in the finite depth sea. *Izv. Akad. Nauk SSSR Fiz. Atmos. Okeana* **14**, 335–339.

Gadzhiyev, J. Z., S. A. Kitaigorodskii, and V. P. Krasitskii (1978): On a high-frequency region in the spectra of wind-driven waves in the presence of currents in a shallow sea. *Oceanology* **18**, 411–416.

Garrat, T. R. (1977): Review of drag coefficients over oceans and continents. *Mon. Weather Rev.* **105**, 915–922.

Hasselmann, K. (1962): On the nonlinear energy transfer in a gravity-wave spectrum. Part 1. General theory. *J. Fluid Mech.* **12**, 481–500.

Hasselmann, K. (1963): On the nonlinear energy transfer in a gravity-wave spectrum. Part 2. Conservation theorems, wave–particle correspondence, irreversibility. *J. Fluid Mech.* **15**, 273–381.

Hasselmann, K. (1968): Weak interaction theory of ocean waves. *Basic Developments in Fluid Dynamics* (M. Holt, ed.), Academic Press, New York, **2**, 117.

Hasselmann, K., T. P. Barnett, E. Bouws, H. Carlson, D. E. Cartwright, K. Enke, J. A. Ewing, H. Gienapp, D. E. Hasselmann, P. Kruseman, A. Meerburg, P. Müller, D. J. Olbers, K. Richter, W. Sell, and H. Walden (1973): Measurements of wind-wave growth and swell decay during the Joint North Sea Wave Project (JONSWAP). *Dtsch. Hydrogr. Z. Suppl. A* **8**(12).

Kahma, K. K. (1981a): A study of the growth of the wave spectrum with fetch. *J. Phys. Oceanogr.* **11**, 1503–1515.

Kahma, K. K. (1981b): On the wind speed dependence of the saturation range of the wave spectrum. Paper presented at Geofgslikan paivat, Helsinki.

Kitaigorodskii, S. A. (1962): Applications of the theory of similarity to the analysis of wind-generated wave motion as a stochastic process. *Izv. Akad. Nauk SSSR Ser. Geofiz.* **1**, 105–117.

Kitaigorodskii, S. A. (1981): The statistical characteristics of wind-generated short gravity waves. *Spaceborne Synthetic Aperture Radar for Oceanography* (R. C. Beal, P. S. DeLeonibus and I. Katz, eds.), Johns Hopkins Press, Baltimore.

Kitaigorodskii, S. A. (1983). On the theory of equilibrium range in the spectrum of wind-generated gravity waves. *J. Phys. Oceanogr.* **13**(N5), 816–827.

Kitaigorodskii, S. A. and Hansen, C. (1986): The general explanation of the quasi-universal form of the rear face of the spectra of wind-generated waves at different stages of their development, *J. Phys. Oceanogr.* (submitted).

Kitaigorodskii, S. A., V. P. Krasitskii, and M. M. Zaslavskii (1975): On Phillips' theory of equilibrium range in the spectra of wind-generated gravity waves. *J. Phys. Oceanogr.* **5**, 410–420.

Kolmogoroff, A. N. (1941): The local structure of turbulence in an incompressible viscous fluid for very large Reynolds numbers. *C.R. Acad. Sci. USSR* **30**, 301.

Kolmogoroff, A. N. (1962): A refinement of previous hypothesis concerning the local structure of turbulence in a viscous incompressible fluid at high Reynolds numbers. *J. Fluid Mech.* **13**, 82–85.

Mitsuyasu, H. (1977): Measurement of the high frequency spectrum of ocean surface waves. *J. Phys. Oceanogr.* **7**, 882–891.

Phillips, O. M. (1958): The equilibrium range in the spectrum of wind-generated waves. *J. Fluid Mech.* **4**, 426–434.

Phillips, O. M. (1977): *The Dynamics of the Upper Ocean*, 2nd ed., Cambridge University Press, London.

Phillips, O. M. (1985): Spectral and statistical properties of the equilibrium range in wind generated gravity waves, *J. Fluid Mech.* **156**: 505–531.

Vincent, C. L. (1982): Shallow water wave modelling. *Proceedings, 1st Conference on Air–Sea Interaction in Coastal Zone*.

West, B. T. (1981): On the simpler aspects of nonlinear fluctuating deep water gravity waves. (Weak interaction theory.) Center for Studies of Nonlinear Dynamics, La Jolla Institute.

Zakharov, V. E. (1974): Hamiltonian formulation for nonlinear dispersive waves. *Izv. Yvzov Radiofz.* **17**, N4.

Zakharov, V. E. and Zaslavskiy, M. M. (1982): The kinetic equation and Kolmogorov spectra in the weak turbulence theory of wind-waves, *Izv. Atmos. Ocean. Phys.* **18**:747–752 (English translation).

Zakharov, V. E. and Zaslavskiy, M. M. (1983): Shape of the spectrum of energy carrying components of a water surface in the weak-turbulence theory of wind-wave, *Izv. Atmos. Ocean. Phys.* **19**:207–212 (English translation).

Zakharov, V. E., and N. N. Filonenko (1966): The energy spectrum for random surface waves. *Dokl. Akad. Sci. SSSR* **170**, 1291–1295.

Zaslavskii, M. M., and S. A. Kitaigorodskii (1971): On equilibrium range in the spectrum of wind-generated surface gravity waves. *Izv. Akad. Nauk SSSR Fiz. Atmos. Okeana* **7**, 565–570.

Zaslavskii, M. M., and S. A. Kitaigorodskii (1972): Intermittent pattern effects in the equilibrium range of growing wind-generated waves. *Izv. Akad. Nauk SSSR Fix. Atmos. Okeana* **5**, 1230–1234.

DISCUSSION

TRIZNA: We have a third piece of evidence for ω^{-4} spectra in addition to those you showed, with additional information which perhaps can offer some insight into the problem. We normally measure wave spectra with a Waverider buoy while simultaneously measuring directional spreading for several wave frequencies with a scanning narrow-beam HF radar using first-order Bragg scatter. During one test period, although we observed ω^{-5} spectra on several occasions under very high wind conditions, when the radar was not operable, on more frequent occasions we saw the ω^{-4} slope. On some of these latter occasions we measured extremely narrow angular spreading with the radar, between $\cos^{32}\theta$ and $\cos^{64}\theta$, along the local wind direction [Trizna et al. (1980): J. Geophys. Res. **85**, 4946]. This is in agreement with your statement that the spectrum is in equilibrium with the wind while exhibiting the ω^{-4} slope. Although wave–wave interactions which allow interchange of energy along and opposite the wind direction are possible during this time by implication, such interactions are not in effect to any great degree which transfer energy to other directions because of the observed angular narrowness. It can follow that such interactions which cause wave spreading in angle may also be connected with transition from ω^{-4} to ω^{-5} slopes of the spectrum.

KITAIGORODSKII: Thank you, Dr. Trizna.

HUANG: Laboratory data showed the growth of the spectral components did reach a saturated level after initial overshoot. Is it possible that -5 law is the saturation upper limit while the -4 wind-dependent law only represents the unsaturated state after overshoot which is still growing back to the saturation state?

KITAIGORODSKII: It is quite possible that -5 law is the saturation *upper limit*, while the -4 wind-dependent law represents the state *after overshoot*, which can be considered as another type of equilibrium spectrum (energy density doesn't change with fetch or duration).

PIERSON: A large part of the difference between various families of published spectra that attempt to demonstrate f^{-4}, f^{-5}, $U_* f^{-4}$ and others, if any, spectral forms and representations may be due to differences in calibration and frequency response of the multitude of different wave recording systems that have been used. Do you know of any examples when conclusions from a particular experiment may be questioned because of calibration problems of the wave recording system?

KITAIGORODSKII: I know that the calibration problems are a serious obstacle for accurate measuring of the high-frequency part of wave spectrum. However, the data which I have shown here (M. Donelan and K. Kahma) to demonstrate the possibility of a wind-dependent saturation regime, were for the range of frequencies less than 1 Hz, and I guess for different instrumentation, which their people used, the calibration problems were not big sources of errors.

DONELAN: Our observations (Donelan et al., 1982) of an "equilibrium range" slope of ω^{-4} are in the energy-containing region at frequencies for which the directional distribution is far from isotropic, being contained by at most $\pm\pi/2$ about the wind direction. Therefore, they cannot be taken as support for Zakharov's theory. If one avoids the difficulties pointed out by Professor Hasselmann by examining the frequency dependence of the directional spectrum in the wind direction only $F(\omega,\bar{\theta})$, then the dimensional arguments advanced by Professor Phillips (1958) apply and, indeed, one finds that $F(\omega,\bar{\theta})$ is roughly proportional to ω^{-5}. Since the spread of $F(\omega,\theta)$ increases with frequency for any spectrum, it follows that $S(\omega) = F(\omega,\theta)$ must decrease with frequency less rapidly than ω^{-5}. Our data suggest that a convenient description of the rear face of the spectrum is $S(\omega) \propto \omega^{-4}$, $1.5 < \omega_p < 3$. We believe that the dominant mechanism acting to limit the spectrum on the rear face is in fact wave breaking, but the angular distribution of breaking broadens with increasing frequency for any spectrum. We (Donelan et al.) have described the balance on the rear face using ideas like this applied to the directional spectrum $F(\omega,\theta)$. The resulting frequency spectrum $S(\omega)$ is consistent with the ω^{-4} observed dependence.

K. HASSELMANN: While it is formally correct that one can apply Phillips's dimensional argument to the two-dimensional spectrum in a very narrow band of directions parallel to the wind, the argument is then very strongly dependent on the locality of the whitecapping process in the wavenumber domain. This is difficult to reconcile with the observed locality of process in physical space. Rather than trying to explain the observed frequency and directional distribution in terms of a thin directional band of saturated whitecapping and a broad directional distribution around this band which is nonsaturated, it may be easier to consider the complete energy balance. The main features can then be explained quite naturally without the concept of a saturation spectrum using Zakharov's concepts. Although I agree that Zakharov's theory is not applicable in the strict form in which it was originally proposed (as it would also indeed require an isotropic spectrum), our nonlinear transfer calculations show that for an f^{-5} or f^{-4}

spectrum and a slowly increasing angular spread with frequency, the divergence of the nonlinear flux in the wavenumber plane is small. This can be understood physically in terms of the fourth-order diffusion approximation of the nonlinear transfer, which is valid in this region of the spectrum.

KITAIGORODSKII: I agree with both M. Donelan and K. Hasselmann that experimental evidence on the existence of a saturation wind-dependent regime of the type $S(\omega) \propto u_* g \omega^{-4}$ cannot be considered as an application of Kolmogoroff's type of cascade equilibrium in wind-wave spectrum (Zakharov's theory). However, I feel that a model which includes the balance between atmospheric forcing and weak nonlinear interaction between waves will not be very sensitive to the assumptions about the form of angular energy distribution (isotropy in Zakharov's theory), and can give results consistent with wind-dependent saturation regime $S(\omega)$ or ω^{-4}. The only alternative to the nonlinear transfer of energy in such an energy balance, which can produce a similar spectrum, is the wave dissipation due to wave–turbulence interactions. But it remains to be seen how important they can be.

NOTE ADDED IN PROOF

Since this review was prepared many interesting publications have appeared in oceanographic literature. I would like to mention a few of them, which can be considered as new concepts, or as a direct continuation or detailization of the ideas described in this chapter (Zakharov and Zaslavskiy, 1982, 1983; Bouws et al., 1986a, b; Phillips, 1985: Kitaigorodskii and Hansen, 1985).

2

Nonlinear Energy Transfer between Random Gravity Waves

Some Computational Results and Their Interpretation

Akira Masuda

ABSTRACT. A computational scheme for calculating transfer functions is applied to a few typical spectra. The results clarify the change of the transfer function with the increase in the spectral sharpness. That is, they show how the transfer function obtained by Webb for a broad spectrum of Pierson–Moskowitz on the basis of Hasselmann's model changes into those by Fox or Dungey and Hui for the narrow spectra of JONSWAP on the basis of Longuet-Higgins's model or its improved version. Furthermore, an example is presented which shows that for a spectrum expressed as a sum of two spectra with peak frequencies and peak spectral densities different from each other, energy flows so as to smooth out the spectral peak at the higher frequency much more intensely than expected from a simple superposition. The basic features of transfer functions thus obtained are explained well from the two properties: (1) that for most cases of resonant four waves, energy flows from the pair of intermediate frequencies toward the pair of outer (higher or lower) frequencies and (2) that the coupling coefficient depends strongly on the mean frequency of the resonant four waves and the configuration of their wavenumber vectors.

1. INTRODUCTION

Since Phillips (1960) and Hasselmann (1962, 1963a, b) established weakly nonlinear resonance theories, much attention has been paid to the nonlinear energy transfer among wind waves. Because of computational difficulties, however, transfer functions had rarely been calculated until Longuet-Higgins (1976) proposed a simpler model, which is an approximation to Hasselmann's under the assumption of narrow bandwidth, and which was used by Fox (1976) to compute the transfer function for the JONSWAP spectrum. His results differed considerably from those obtained by Sell and Hasselmann (1972) for the same spectrum. This discrepancy seems to have caused some confusion and, at the same time, provided a motivation for recent studies.

Phillips (1977) suggested that Fox's results were preferable, judging partly from numerical stability and accuracy and partly from the intuition that energy should be transferred from high-energy components toward low-energy ones. Webb (1978) made a detailed calculation of the transfer function for the Pierson–Moskowitz spectrum. He confirmed that, for this spectrum with a broad bandwidth, the peak gains energy rather than loses it, at the expense of high-

AKIRA MASUDA ● Research Institute for Applied Mechanics, Kyushu University, Kasuga, 816, Japan.

frequency waves. Dungey and Hui (1979) improved the Longuet-Higgins model and showed that the inadequacy of Fox's scheme (1976) originates in the constant value of the coupling coefficient. The same conclusion was reached independently by the author (Masuda, 1980) without the assumption of narrow bandwidth; he developed a computational scheme with rather general applicability on the basis of Hasselmann's model and obtained reliable transfer functions with numerical stability by treating singularities analytically. In short, recent progress has been focused on how to carry out precisely and conveniently the integration of Hasselmann's equation.

Thus, we now have definite answers to the question: what are transfer functions like for typical spectra of wind waves? They are: (1) Webb (1978) for the broad spectrum of Pierson–Moskowitz; (2) Dungey and Hui (1979) for the narrow spectrum of JONSWAP; and (3) Masuda (1980), which connects the two extreme cases by covering the whole range of spectral bandwidth. In this chapter, we briefly describe the results of Masuda (1980), and then turn to another important question: why are transfer functions such as computed? The answer is not automatically provided by the results obtained through a black box of "numerical integration." Therefore many studies have been concerned with this problem of interpretation. For instance, Hasselmann (1963a) derived three conservation laws, introduced the entropy, and proved a proposition analogous to Boltzmann's H-theorem. Webb (1978) explained the creation of order near the spectral peak as the by-product of the large creation of disorder at high frequencies, in terms of "diffusion" and "pumping" of energy. Masuda (1980) suggested a method of interpretation, which consists of decomposing the transfer function into contributions from individual combination of resonance and tracing out all the combinations.

In spite of the above efforts, however, our understanding still remains unsatisfactory. Therefore, an approach is proposed to argue the nature of nonlinear energy transfer. According to this, a discussion is made on the energy flow among four interacting waves and an interpretation is given with the use of a simple metaphor, which is devised for the convenience of explanation. The model succeeds in reproducing the basic features of the transfer function, which are explained very well from the two properties of nonlinear energy transfer. The same interpretation applies almost directly to the real case of Hasselmann's model as well.

2. TRANSFER FUNCTIONS FOR TYPICAL SPECTRA

This section only outlines the contents of Masuda (1980), which is to be referred to for details. Deep water is assumed throughout and all quantities are made nondimensional so that $\rho = g = \omega_m = 1$, where ρ is the density of water, g the gravitational acceleration, and ω_m the spectral peak (angular) frequency. The basic equation is (1.12) of Hasselmann (1963a):

$$\frac{\partial n_4}{\partial t} = \iiint d\mathbf{k}_1 d\mathbf{k}_2 d\mathbf{k}_3 \, G(\mathbf{k}_1, \mathbf{k}_2, \mathbf{k}_3, \mathbf{k}_4) \delta(\mathbf{k}_1 + \mathbf{k}_2 - \mathbf{k}_3 - \mathbf{k}_4)$$

$$\cdot \delta(\omega_1 + \omega_2 - \omega_3 - \omega_4)[n_1 n_2 (n_3 + n_4) - n_3 n_4 (n_1 + n_2)] \tag{1}$$

where $n_j = n(\mathbf{k}_j), j = 1, 2, 3, 4$, denotes the action density for the wavenumber $\mathbf{k}_j, \omega_j = |\mathbf{k}_j|^{1/2}$ the corresponding frequency, t the time, δ the delta function, and $G(\mathbf{k}_1, \mathbf{k}_2, \mathbf{k}_3, \mathbf{k}_4)$ is the coupling coefficient.

Without showing the details of the computational scheme, we present numerical results. Here the directional distribution is always assumed as proportional to $\cos^2 \theta$, irrespective of ω. That is,

$$\phi(\omega, \theta) = \begin{cases} \psi(\omega)\left(\dfrac{2}{\pi}\right)\cos^2 \theta & 0 \leqslant |\theta| \leqslant \pi/2 \\ \\ 0 & \pi/2 \leqslant |\theta| \leqslant \pi \end{cases}$$

where $\psi(\omega)$ is the frequency spectrum normalized so that $\psi(1) = 1$. Dimensional values are obtained if we multiply the nondimensional results by $\psi^3_m \omega^{11}_m / g^4$, where ψ_m is the spectral density at the peak frequency.

Two-dimensional and one-dimensional transfer functions are defined by

$$T_2(\omega, \theta) = \frac{\partial \phi(\omega, \theta)}{\partial t}$$

$$T_1(\omega) = \frac{\partial \psi(\omega)}{\partial t} = \int \frac{\partial \phi(\omega, \theta)}{\partial t} d\theta$$

respectively.

The solid curve in Fig. 1 shows the present calculation of $T_1(\omega)$ for the JONSWAP spectrum. We find that the present result supports Sell and Hasselmann (1972) and Dungey and Hui (1979) rather than Fox (1976). Also, it is evident that the present scheme gives more stable $T_1(\omega)$ than that in Sell and Hasselmann.

The dashed curve in Fig. 1 is calculated by setting $G = G_0 = 4\pi$ for comparison with Fox's results. Note that Fox's scheme differs from the present one in four ways: (1) $G = G_0 = $ constant, (2) resonance conditions are simpler, (3) integration is over the wavenumber rather than the (ω, θ) space, and (4) the spectrum is approximated by a sum of normal distributions. It appears that the

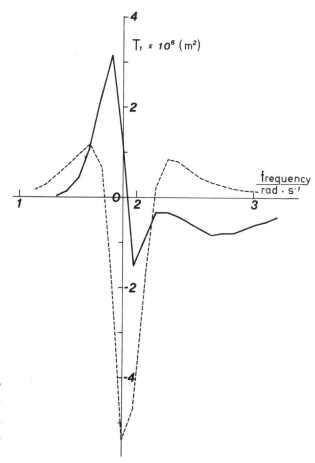

FIGURE 1. $T_1(\omega)$ for the JONSWAP spectrum obtained in this chapter. The solid curve was determined rigorously, while the dashed one by setting $G = G_0 = 4\pi$ for the purpose of comparison with Fox's results.

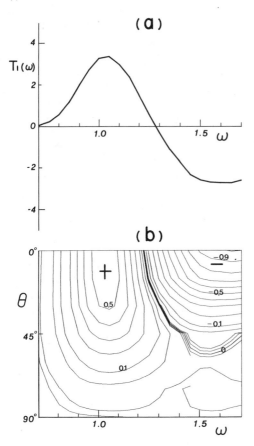

FIGURE 2. (a) $T_1(\omega)$ and (b) $T_2(\omega, \theta)$ for the Pierson–Moskowitz spectrum. The contours are $\pm(0.99, 0.9, 0.8, 0.7, 0.6, 0.5, 0.4, 0.3, 0.2, 0.1, 0.05, 0.01, 0.005, 0.0)$ in units of $\max|T_2(\omega, \theta)|$. The darker curves denote zero values. There are some spurious unclosed contour lines due to the rough spacing of data points.

approximation $G = $ constant is the most serious of them. The same conclusion was obtained independently by Dungey and Hui (1979), though on the basis of an improved model of Longuet-Higgins's one. Figure 1 shows (1) that the discrepancy between Fox and Sell and Hasselmann is essentially due to the approximation $G = G_0$ used in Fox and (2) that the present calculations are reliable enough because Fox's results are very precise within the framework of the Longuet-Higgins model.

We now concentrate our attention on the dependence of T_1 (or T_2) on the sharpness of the spectral form. We examine the ocean type of spectra:

$$\psi_{\text{ocean}}(\omega; \gamma) \sim \omega^{-5} \exp\left(-\frac{5}{4}\omega^{-4}\right) \cdot \gamma^{\exp\left[-\frac{(\omega-1)^2}{2\sigma^2}\right]}$$

where $\sigma = 0.07$ for $\omega \leqslant 1$ and 0.09 for $\omega > 1$ type.

Figures 2 and 3 display (a) $T_1(\omega)$ and (b) contour maps of $T_2(\omega, \theta)$ for the two spectra—the Pierson–Moskowitz spectrum ($\gamma = 1$) and the JONSWAP spectrum. ($\gamma = 3.3$) Contour lines are those for $\pm(0.99, 0.9, 0.8, 0.7, 0.6, 0.5, 0.4, 0.3, 0.2, 0.1, 0.05, 0.02, 0.005, 0.0) \times \max|T_2(\omega, \theta)|$ in each case. To see in detail the dependence of $T_1(\omega)$ on the sharpness of the spectral form, Fig. 4 is presented, in which $T_1(\omega)$ for ocean-wave types of spectra are shown for various values of γ. An apparent feature is the change of $T_1(\omega)$ both in magnitude and in pattern. The variation of the pattern of the energy flow is one of the most important results, and requires careful interpretation. When $\gamma = 1$, the maximum $T_1(\omega)$ is found near the spectral peak frequency with negative $T_1(\omega)$ at higher frequencies. This is no other than the "creation of order near the peak" (Webb, 1978). The behavior of $T_1(\omega)$ with respect to ω is smooth and slow in this case. As γ

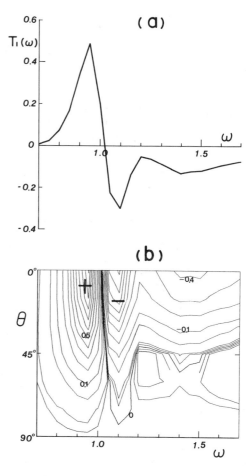

FIGURE 3. (a) $T_1(\omega)$ and (b) $T_2(\omega,\theta)$ for the JONSWAP spectrum. For details, see Fig. 2 caption.

increases, the magnitude decreases rapidly and the first maximum shifts slightly toward lower frequencies ($\omega \approx 0.95$). At the same time the minimum (< 0) at $\omega \approx 1.1$ and the second maximum appear. They break the smooth pattern of $T_1(\omega)$. Eventually, the contrast between the minimum and the second maximum becomes more and more evident. Finally, $T_1(\omega)$ has a resemblance to that in Dungey and Hui (1979). For large γ, notable energy transfer seems to be confined within a comparatively narrow frequency range ($0.75 \leqslant \omega \leqslant 1.35$).

When two spectra are superposed, for example, as

$$\psi(\omega) \sim \psi_{\text{labo}}(\omega; \delta = 0.37) + 0.3 \times \psi_{\text{ocean}}(\omega/1.3; \gamma = 3.3) \tag{2}$$

where

$$\psi_{\text{labo}}(\omega, \delta) \sim \omega^p (1 - \delta\omega^{[(1/\delta)-1]p})$$

p being 11 for $\omega \leqslant 1$ and -10 for $\omega > 1$, $T_1(\omega)$ turns out to be remarkably different from a mere superposition of individual $T_1(\omega)$ for each basic spectrum. Figure 5 shows $T_1(\omega)$ for (1) the superposed spectrum (2); (2) $\psi_{\text{labo}}(\omega; \delta = 0.37)$; and (3) $0.3 \times \psi_{\text{ocean}}(\omega/1.3; \gamma = 3.3)$. This seems to suggest that a rugged distribution of the spectral density induces an intense energy flow so as to smooth out the ruggedness and increase the spectral density near the peak frequency. This phenomenon might be very important in the following problems: (1) how the interaction occurs between two coexisting spectra of different origins such as mechanically generated waves and wind waves in a laboratory or swell and wind waves in the ocean; (2) why the second spectral peak at $\omega \approx 1.3$ is never observed in the actual spectrum in spite of the presence of positive $T_1(\omega)$ at the corresponding frequency for sharp spectra. The latter might have a close relation to the "self-stabilization of the spectral form" due to nonlinear energy transfer.

3. ENERGY FLOW AMONG FOUR INTERACTING WAVES

The previous section presents many interesting results. Especially, Fig. 1 shows that the pattern of the transfer function is much affected by the functional form of the coupling coefficient G. Figure 4 demonstrates the change of transfer functions both in pattern and in magnitude with the increase of the sharpness of the spectral form. Figure 5 illustrates that the transfer function for a spectrum composed of two different spectra differs remarkably from a mere superposition of individual transfer function for each basic spectrum. The purpose of this and the following sections is to interpret the above results, which are never easy to predict beforehand or to understand even after the computation.

We adopt the following approach as suggested in Masuda (1980). The transfer function is a sum over all the resonance configurations as shown by Longuet-Higgins's interaction chart, in each of which there are four combinations a wavenumber $\mathbf{k}$ is concerned with, since $\mathbf{k}$ may be any of $\mathbf{k}_1, \mathbf{k}_2, \mathbf{k}_3$, and $\mathbf{k}_4$. For a fixed resonance configuration, the energy gain of a component is a sum over those for the four combinations. The total energy income of a component is obtained by summation over the whole configurations. Conceptually, the above recipe is the most elementary and complete, though actually it cannot be carried out in a rigorous manner because of continuous distribution of resonance configurations.

We begin with the investigation of energy flow in a resonance combination. First, we define

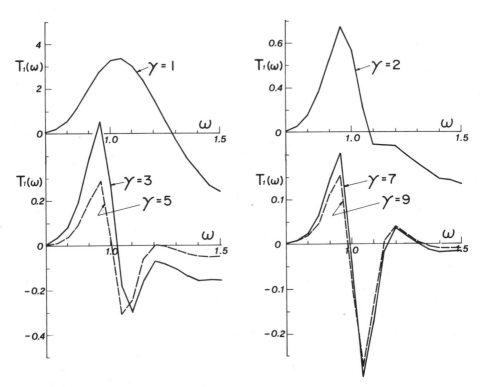

FIGURE 4. $T_1(\omega)$ for various ocean-wave-type spectra ($\gamma = 1, 2, 3, 5, 7$, and 9). Large γ corresponds to a sharp spectrum.

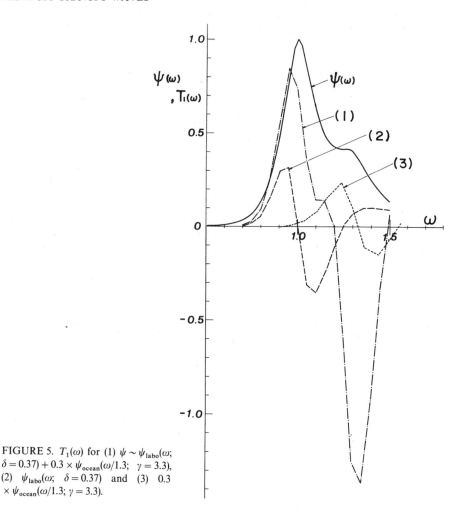

FIGURE 5. $T_1(\omega)$ for (1) $\psi \sim \psi_{labo}(\omega;$ $\delta = 0.37) + 0.3 \times \psi_{ocean}(\omega/1.3; \quad \gamma = 3.3)$, (2) $\psi_{labo}(\omega; \quad \delta = 0.37)$ and (3) 0.3 $\times \psi_{ocean}(\omega/1.3; \gamma = 3.3)$.

the term "acceptance" by the inverse of the action density:

$$a(\mathbf{k}) = 1/n(\mathbf{k}) \tag{3}$$

The distribution of $a(\mathbf{k})$ is determined immediately from that of $n(\mathbf{k})$. On the basis of the graph of acceptance, we can conveniently argue the energy flow in the wavenumber space. Although this concept does not provide any new information, it gives a fine perspective of the energy flow as shown below.

For a fixed combination of interaction, the sign of $n_1 n_2 (n_3 + n_4) - n_3 n_4 (n_1 + n_2)$ determines the direction of the energy flow. This quantity is rewritten as

$$n_1 n_2 (n_3 + n_4) - n_3 n_4 (n_1 + n_2) = n_1 n_2 n_3 n_4 [(a_3 + a_4) - (a_1 + a_2)] \tag{4}$$

Since the coupling coefficient and action density are nonnegative, energy is transferred from the pair of lower acceptance to that of higher acceptance. In particular, if n_4 is negligibly small (or zero) as compared with the other three, its acceptance $a_4 = n_4^{-1}$ is far larger than those of the other three, and the pair of n_3 and n_4 receives the energy from that of n_1 and n_2. Equations (1) and (4) show that the following three conditions are necessary in order for a combination of interaction to cause a large magnitude of energy transfer: (1) high action densities of the components which participate in the interaction, (2) large difference of acceptance between the interacting pairs, and (3) large coupling coefficient.

Next let us consider what will happen to the energy flow for a small increment of the action density, e.g., n_1. The variation of $n_1 n_2 (n_3 + n_4) - n_1 n_2 (n_3 + n_4)$ is expressed as

$$\delta[n_1 n_2 (n_3 + n_4) - n_3 n_4 (n_1 + n_2)] = n_1 n_2 n_3 n_4 \left(\frac{\delta n_1}{n_1}\right)[(a_3 + a_4) - a_2] \tag{5}$$

where δ denotes a small increment. We find that small increase in the action density n_1 causes energy flow from the pair to which n_1 belongs toward the counterpair (n_3 and n_4), so long as the acceptance of the partner a_2 is less than that of the counterpair. In other words, energy flows so as to reduce the change of n_1, i.e., in a stabilizing manner; this case is easy to understand. However, if the acceptance of the partner a_2 is higher than that of the counterpair ($a_3 + a_4$), n_1 and n_2 gain energy more and more from the counterpair. Although this phenomenon might seem contradictory to intuition at first sight, it is explained as follows: when the acceptance of n_2 is so high that energy tends to flow from n_3 and n_4 toward n_1 and n_2, this reaction is enhanced by the presence of n_1 as a catalyst.

Now we turn to the discussion of a general tendency of energy transfer, which proves of central importance to interpretation. Consider an interaction among wavenumbers $\mathbf{k}_1, \mathbf{k}_2, \mathbf{k}_3$, and $\mathbf{k}_4$, where $\mathbf{k}_1 + \mathbf{k}_2 = \mathbf{k}_3 + \mathbf{k}_4$ and $\omega_1 + \omega_2 = \omega_3 + \omega_4$. Let us assume $\omega_1 \leqslant \omega_3 \leqslant \omega_4 \leqslant \omega_2$ or $|\mathbf{k}_1| \leqslant |\mathbf{k}_3| \leqslant |\mathbf{k}_4| \leqslant |\mathbf{k}_2|$ and call the pair (ω_3, ω_4) of the intermediate frequencies as the inner pair and the other pair (ω_1, ω_2) as the outer pair. If the graph of $a(\mathbf{k})$ is moderately convex in the region including the four wavenumbers, we will obtain

$$a(\mathbf{k}_1) + a(\mathbf{k}_2) \geqslant a(\mathbf{k}_3) + a(\mathbf{k}_4) \tag{6}$$

According to (4), energy flows from the inner pair toward the outer. In such a case as studied by Fox (1976), where $n(x, \xi) \propto \exp[-(P/2)x^2 - Q\xi^2]$ (see Fox, 1976, for notations), we can prove the outward energy flow for each combination of resonance. Of course, this direction of energy flow is not a strict rule; the opposite direction of energy flow is possible in the case of double-peaked spectra, for instance. However, for most combinations of interaction in smooth single-peaked spectra (especially for those related with components of high energy), one will find the general tendency of the energy flow from the inner pair to the outer.

Another notable fact is that the coupling coefficient G strongly depends on the characteristic frequency $\bar{\omega}$ of resonant four waves and the configuration of their wavenumber vectors. In particular, G is proportional to $\bar{\omega}^{12}$ for a fixed configuration of resonance and consequently intensifies the interaction at high frequencies. These two properties make most important keys to the understanding of nonlinear energy transfer.

4. METAPHOR AND INTERPRETATION

The preparation in the previous section is sufficient for the discussion of the results in Section 2. Instead of interpreting them directly, however, we give a full explanation in a simple metaphor (model) and let it indicate the interpretation in the real case of Hasselmann's model. The reason is as follows.

As is well recognized, solving the full equations does not necessarily lead to the understanding of the phenomenon. Rather, the degree of understanding is measured by the degree to which we can simplify the problem, omitting digressive factors and retaining the essence. If our interpretation is to the point, the model thus simplified must reproduce basic features of the original model; in this sense the metaphor plays the role of a touchstone. Therefore, it is interesting and desirable to investigate the results of such simplification. Moreover, a clear-cut explanation is obtained much more conveniently in the purest framework.

Before proceeding further, we point out three characteristics which seem essential for the interpretation of transfer functions: (1) the term $n_1 n_2(n_3 + n_4) - n_3 n_4(n_1 + n_2)$ in the integrand of Eq. (1), (2) fine symmetry of the coupling coefficient G and delta functions, and (3) the strong dependence of G on the representative frequency and the resonance configuration. Details of G are probably insignificant for a qualitative argument. Note that if the first two points are retained, we can equally argue conservation laws or the increase in the entropy (disorder).

Now we construct a model, which is never an approximation to but an artificial metaphor of Hasselmann's model. First, we simplify (1) into a one-dimensional form:

$$\frac{\partial n(\omega_4)}{\partial t} = \int \int \int d\omega_1 d\omega_2 d\omega_3 \, G(\omega_1, \omega_2, \omega_3, \omega_4) \times h(\omega_1, \omega_2, \omega_3, \omega_4)$$

$$\times [n_1 n_2(n_3 + n_4) - n_3 n_4(n_1 + n_2)] \tag{7}$$

where h denotes the interaction condition. We require that G is symmetric with respect to its arguments and that h is invariable with the exchange of ω_1 with ω_2, ω_3 with ω_4, and (ω_1, ω_2) with (ω_3, ω_4). As the interaction condition, h should include $\delta(\omega_1 + \omega_2 - \omega_3 - \omega_4)$. Further, we assume the frequency ratio to be $j:j+1:j+2:j+3$ for interaction to occur. This choice corresponds to the extraction of one specified configuration of resonance. With the property of G in Hasselmann's model in mind, we put

$$G(\omega_1, \omega_2, \omega_3, \omega_4) = G(\bar{\omega}) = \bar{\omega}^p \tag{8}$$

where $\bar{\omega} = (\omega_1 + \omega_2 + \omega_3 + \omega_4)/4$ and p is a constant. This equation implies that the strength of interaction is proportional to the p powers of the mean frequency. If $p = 0$, the strength of coupling is constant, but for large p, it increases with $\bar{\omega}$.

There are four interaction combinations with which the frequency ω is concerned and they are labeled by $k = 1, 2, 3, 4$; ω is located at the highest frequency when $k = 1$, the second highest when $k = 2$, and so on. This interaction condition is illustrated in Fig. 6. The final equation becomes

$$\frac{\partial n(\omega)}{\partial t} = \sum_{k=1}^{4} A_k G(B_k \omega) \cdot \{ n(P_k \omega) n(Q_k \omega)[n(R_k \omega) + n(\omega)] $$

$$- n(R_k \omega) n(\omega)[n(P_k \omega) + n(Q_k \omega)] \} \tag{9}$$

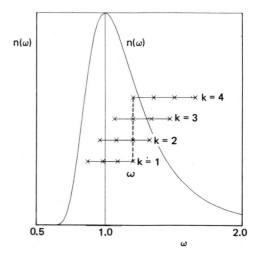

FIGURE 6. Schematic graph of the interaction condition. For a given frequency component, there are four combinations of interaction with which the component is related.

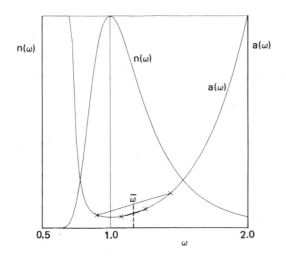

FIGURE 7. Schematic distributions of action $n(\omega)$ and acceptance $a(\omega)$. The symbols × denote a combination of interaction (resonance). In this case, energy flows from the inner pair toward the outer, since the acceptance of the latter is higher than that of the former.

though details of manipulation and the coefficients A_k etc. are omitted. Figure 7 illustrates the tendency to the outward energy flow discussed in the previous section.

We put $j = 5$ henceforth and examine the following form of spectra:

$$n(\omega) = f(\omega; \gamma) = \omega^{-6} \exp\left\{-\tfrac{6}{5}(\omega^{-5} - 1)\right\}\gamma\exp\left[-\frac{(\omega-1)^2}{2\sigma^2}\right] - 1\}$$ (10)

where $\sigma = 0.08$ and γ is an index of the sharpness of the spectral form. We call also $\partial n/\partial t$ as the transfer function, for simplicity. We investigate the change of transfer functions when two parameters are varied: (1) γ, the sharpness of the spectral form, and (2) p, the exponent of the coupling coefficient G [see (8)].

Figures 8a–d show transfer functions and action spectra for $\gamma = 1, 2, 5$, and 15, respectively, when $p = 0$, and Figs. 9a–d the same except that $p = 12$. Also, $T_k(\omega)$, $k = 1, 2, 3, 4$, is exhibited, where T_k denotes the contribution from the combination labeled k. Therefore, $T_1(\omega)[T_4(\omega)]$ means the action flows toward ω from the inner pair of lower (higher) frequencies. On the other hand, $T_2(\omega)$ and $T_3(\omega)$ denote the action inflow from the outer pair. As is observed from the figures, action always flows from the inner pair to the outer, since T_1, $T_4 \geqslant 0$ and T_2, $T_3 \leqslant 0$.

First we investigate the former case ($p = 0$). Since G is constant, the first two conditions for large energy flow (see Section 3) determine the tendency of energy transfer. Thus, as is apparent, action flows from around the peak toward higher and lower frequencies because of the largest outward energy flow near $\omega = 1$. Another remarkable fact is the rapid decrease in magnitude of transfer functions as γ increases. Note that scales TM differ from one figure to another. This is attributed to the decrease in action density as a whole; since we adopt such a normalization that $n(1) = 1$, large γ implies the decrease of $n(\omega)$ except at $\omega = 1$. Because the energy transfer is proportional to the product of three action densities, the decrease of $n(\omega)$ inevitably reduces the magnitude of transfer functions. It is worthwhile referring to an important conclusion of Dungey and Hui (1979) that large angular spread reduces the magnitude of one-dimensional transfer functions. This phenomenon is interpreted by a similar argument, as follows. Wide angular dispersion thins down the action density, though the total action is conserved. Consequently, individual interaction much weakens on account of the resonance character mentioned above. This decrease in "intensity" of interaction is so efficient that it reduces the magnitude of one-dimensional transfer functions, though somewhat is compensated by the increase in the "number" of interaction due to wide angular distribution.

On the other hand, large p yields quite different features. Figure 9 reminds us of the change in pattern and magnitude of transfer functions found in Fig. 4. When $\gamma = 1$, the maximum value

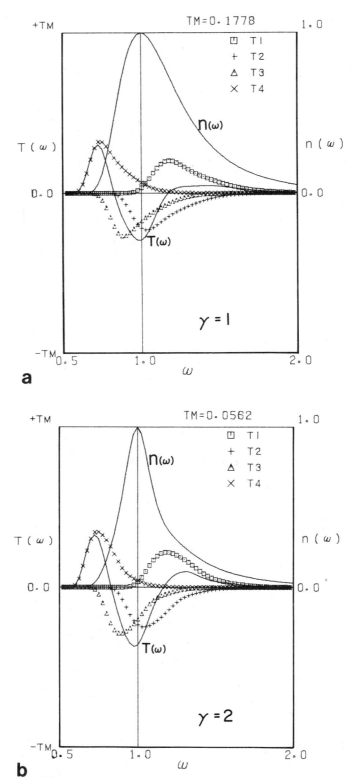

FIGURE 8. (Action) transfer functions $T(\omega)$ according to the present model when $p = 0$; (a) $\gamma = 1$, (b) $\gamma = 2$, (c) $\gamma = 5$, and (d) $\gamma = 15$. Here, $T_k(\omega)$ ($k = 1, 2, 3,$ and 4) shows the action transfer into the frequency ω due to each of four combinations of interaction with which the frequency is concerned. The scale TM of the transfer function is shown at the shoulder of each figure.

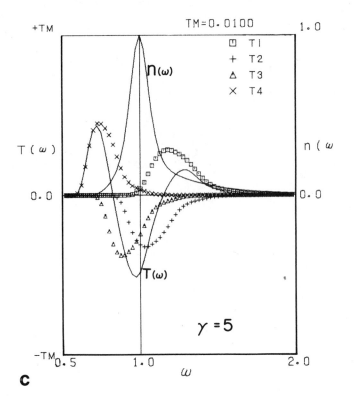

c

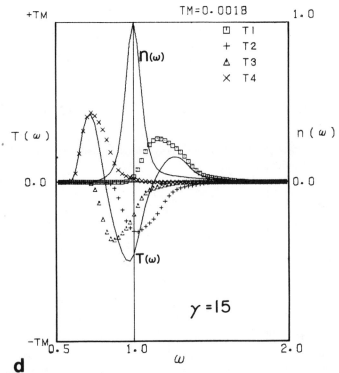

d

FIGURE 8. (*Continued*)

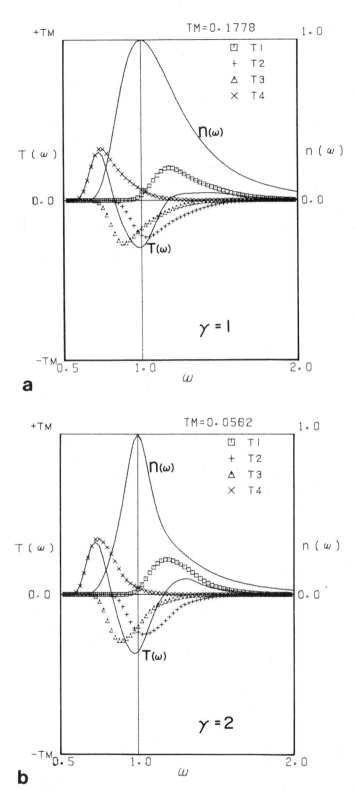

FIGURE 8. (Action) transfer functions $T(\omega)$ according to the present model when $p = 0$; (a) $\gamma = 1$, (b) $\gamma = 2$, (c) $\gamma = 5$, and (d) $\gamma = 15$. Here, $T_k(\omega)$ ($k = 1, 2, 3,$ and 4) shows the action transfer into the frequency ω due to each of four combinations of interaction with which the frequency is concerned. The scale TM of the transfer function is shown at the shoulder of each figure.

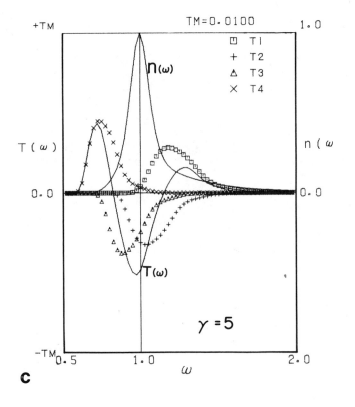

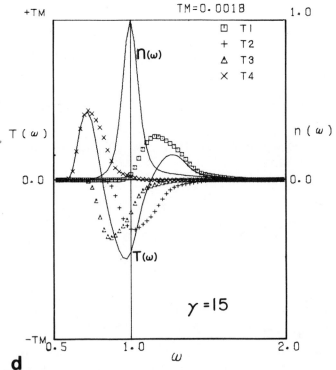

FIGURE 8. (*Continued*)

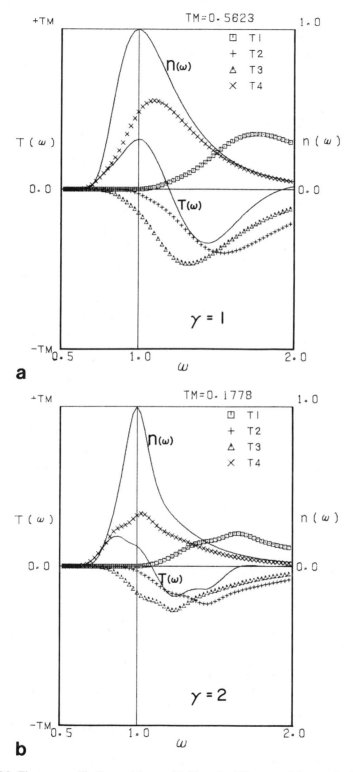

FIGURE 9. The same as Fig. 8 except for $p = 12$. Note the differences of the transfer functions both in pattern and in magnitude with those in the case of $p = 0$ shown in Fig. 8.

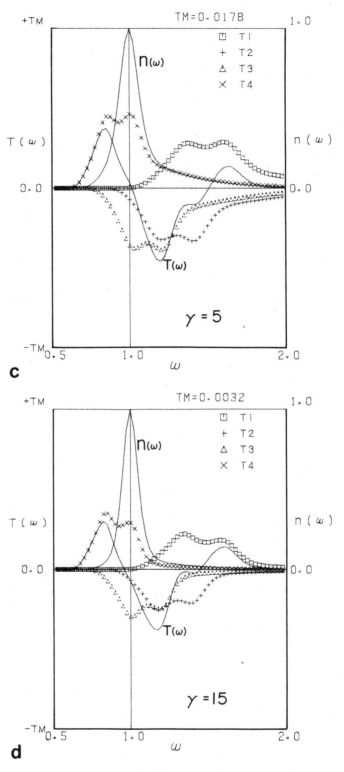

FIGURE 9. (*Continued*)

of T appears at around $\omega = 1$ and the minimum near $\omega = 1.5$. The spectral peak gathers more and more energy at the expense of higher frequencies. This seemingly peculiar phenomenon may be described by the "diffusion" and "pumping" mechanism of Webb (1978), but here it is explained simply as follows. For each combination of interaction, action certainly flows from the inner pair to the outer. However, large p implies that interaction is intensified at high frequencies. Accordingly, the center of the outward energy flow shifts to the high-frequency side. Note that the heavy weight of G makes the magnitude of energy transfer very large as compared with the case $p = 0$. Thus, the peak frequency accepts more action from higher frequencies than it gives to higher and lower frequencies. As γ increases, action density decreases except near $\omega = 1$ and energy transfer reduces its intensity in general. Since the decrease of the action inflow from high frequencies to the peak is more rapid than that of the action outflow from the peak, Fig. 8d ($\gamma = 15$) again shows negative values of transfer functions at the peak.

Similar results are obtained when another frequency ratio for interaction is assumed. Therefore, we can conclude that for the real model, which is considered as a sum over an infinite number of resonance configurations, the synthetic interpretation remains almost the same as for a single resonance configuration.

Closing this section, we examine two cases of double-peaked spectra

$$n(\omega) \sim f(\omega; 4) + 0.2f(0.8\omega; 4) \tag{11}$$

$$n(\omega) \sim f(\omega; 4) + 0.2f(1.4\omega; 4) \tag{12}$$

Figure 10 (11) shows T for (1) the first spectrum of (11), (12), (2) the second spectrum, and (3) the superposed spectrum, where $p = 12$. It is apparent that the result of Fig. 5 is reproduced in Fig. 10. This enhanced outward energy flow from the second peak is caused by the increase in

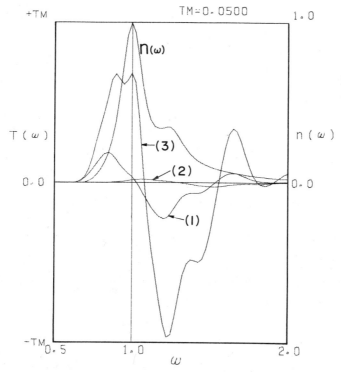

FIGURE 10. Transfer function for a doublepeaked spectrum; $T(\omega)$ for (1) $n \sim f(\omega; 4)$, (2) $n \sim 0.2 \times f(0.8\omega; 4)$, and (3) $n \sim f(\omega; 4) + 0.2f(0.8\omega; 4)$. See the text for the definition of $f(\omega; \gamma)$.

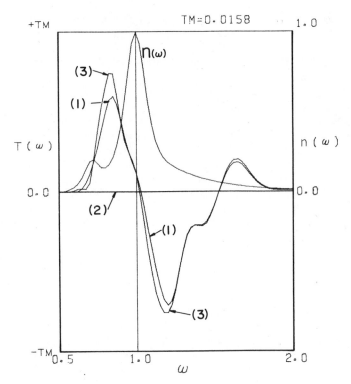

FIGURE 11. The same as Fig. 10; $T(\omega)$ for (1) $n \sim f(\omega; 4)$, (2) $n \sim 0.2 \times f(1.4\omega; 4)$, and (3) $n \sim f(\omega; 4)$ $+ 0.2f(1.4\omega; 4)$.

action densities near the second peak as action sources, from the viewpoint that the second spectrum is superposed on the first spectrum. Likewise, it is interpreted to be due to the presence of large action densities near the first peak as catalysts, from the viewpoint that the first spectrum is added to the second spectrum. On the other hand, when a small irregularity is located at the lower-frequency side as in (12), the transfer function near the peak frequency is not so influenced as when it is located at the opposite side. This difference arises obviously from the strong dependence of the coupling coefficient on the mean frequency ($p = 12$).

5. SUMMARY

Application of a computational scheme to typical spectra clarifies the change of the transfer function with the increase in the spectral bandwidth; that is, they connect the two extreme cases of Webb for the broad spectrum of Pierson–Moskowitz and Fox or Dungey and Hui for the narrow spectra of JONSWAP. Moreover, an example is presented which shows that for a spectrum composed of two different spectra, energy flows so as to smooth out the spectral peak at the high-frequency side much more intensely than expected from simple superposition.

An interpretation is obtained as follows. Hasselmann's equation (1) is represented by an integration over all the resonance configurations. If the energy flow is understood for each configuration, synthesis yields the total transfer function. For that purpose, the energy flow among four resonant waves is discussed first and a convincing interpretation is given to the behavior of the transfer function with the use of a simple metaphor, which is devised for the convenience of explanation and corresponds to a system with a single interaction condition. Note that the metaphor reproduces basic features of the obtained transfer functions. This fact supports that the metaphor retains the essence of the problem. The same interpretation applies to the Hasselmann model as well, because it is considered as a synthesized system of a variety of resonance configurations and each single configuration produces a similar transfer function.

In conclusion, the basic features of one-dimensional transfer functions are explained well from two properties: (1) that for most cases of resonant four waves, energy flows from the pair of intermediate frequencies toward the pair of the outer (higher or lower) frequencies and (2) that the coupling coefficient depends strongly on the mean frequency of the resonant four waves and on the configuration of their wavenumber vectors.

ACKNOWLEDGMENTS. I thank Professors H. Mitsuyasu, M. Takematsu, and S. Mizuno for valuable comments. Comments of an anonymous referee were helpful in revising the manuscript. Thanks are due to Mr. K. Marubayashi for drawing the figures and Miss M. Hojo for typing the manuscript.

REFERENCES

Dungey, J. C., and W. H. Hui (1979): Nonlinear energy transfer in a narrow gravity-wave spectrum. II. *Proc. R. Soc. London Ser. A* **368**, 239–265.

Fox, M. J. (1976): On the nonlinear transfer of energy in the peak of a gravity-wave spectrum. *Proc. R. Soc. London Ser. A* **348**, 467–483.

Hasselmann, K. (1962): On the non-linear energy transfer in a gravity-wave spectrum. Part 1. General theory. *J. Fluid Mech.* **12**, 481–500.

Hasselmann, K. (1963a): On the non-linear energy transfer in a gravity-wave spectrum. Part 2. Conservation theorems; wave-particle analogy; irreversibility. *J. Fluid Mech.* **15**, 273–281.

Hasselmann, K. (1963b): On the non-linear energy transfer in a gravity-wave spectrum. Part 3. Evaluation of the energy flux and swell-sea interaction for a Neumann spectrum. *J. Fluid Mech.* **15**, 385–398.

Longuet-Higgins, M. S. (1976): On the nonlinear transfer of energy in the peak of a gravity-wave spectrum: A simplified model. *Proc. R. Soc. London Ser. A* **348**, 311–328.

Masuda, A. (1980): Nonlinear energy transfer between wind waves. *J. Phys. Oceanogr.* **10**, 2082–2093.

Phillips, O. M. (1960): On the dynamics of unsteady gravity waves of finite amplitudes. Part 1. The elementary interactions. *J. Fluid Mech.* **9**, 193–217.

Phillips, O. M. (1977): *The Dynamics of the Upper Ocean*, 2nd ed., Cambridge University Press, London.

Sell, W., and K. Hasselmann (1972): Computations of nonlinear energy transfer for JONSWAP and empirical wind-wave spectra. Institute of Geophysics, University of Hamburg.

Webb, D. J. (1978): Non-linear transfers between sea waves. *Deep-Sea Res.* **25**, 279–298.

3

THE INTERACTION BETWEEN LONG AND SHORT WIND-GENERATED WAVES

M. T. LANDAHL, J. A. SMITH, AND S. E. WIDNALL

ABSTRACT. The interaction of wind-induced short (capillary) and long (gravity) surface waves of disparate scales is analyzed with the use of an expansion procedure accounting for the lowest-order effects of finite short-wave amplitude on the growth of a small-amplitude long wave. Interactions in both water and air are considered. In the water there is transfer of momentum from the short to the long waves arising from the modulation of the short-wave growth rate by the long wave; this may be interpreted as the effect of the short-wave surface-stress modulations in phase with the long-wave surface elevation and is consistent with earlier treatments employing energy considerations. The interaction in the air is treated with an inviscid model. This shows that the modulation of the short-wave Reynolds stresses by the long wave can give rise to a phase shift between long-wave pressure and surface slope which may significantly increase the rate of transfer of momentum from the wind to the long wave.

1. INTRODUCTION

The mechanism whereby wind generates water waves has long proven a difficult and challenging problem in fluid mechanics, and is still incompletely understood. The simple linear mechanisms of forcing by pressure fluctuations (Phillips, 1957) and by instability induced by the mean wind-shear field (Miles, 1957, 1959, 1962) have been found inadequate to account for the high values of energy transfer from wind to waves observed for the longer waves, both in the laboratory and on the open sea. For short waves, in the capillary regime, laboratory experiments (Larson and Wright, 1975) have given good agreement between observed growth rates and Miles's instability theory, particularly when the surface drift velocity in the water is taken into account (Valenzuela, 1976). For waves in the gravity range, however, experiments by Plant and Wright (1977) give growth rates much in excess of that predicted by instability theory, with the discrepancy beginning at a wavelength of about 10 cm and increasing with wavelength. Open ocean measurements have also produced energy-transfer rates for gravity waves which are much in excess of the values according to Miles.

In view of the failure of linear theory, one is forced to look for significant nonlinear mechanisms for energy transfer to longer waves. One possible mechanism by which the growth rate of the longer waves could be augmented is weak resonant wave–wave interactions (see Hasselmann, 1962, 1963). Another is the modulation of the turbulent stresses in the air by the waves, which could increase the pressure component in phase with the surface slope, an effect that

M. T. LANDAHL, J. A. SMITH, AND S. E. WIDNALL ● Department of Aeronautics and Astronautics, Massachusetts Institute of Technology, Cambridge, Massachusetts 02139.

has been studied by many investigators, employing a variety of turbulence models (e.g., Manton, 1972; Davis, 1972; Townsend, 1972, 1980). These show that such a mechanism could indeed be important.

An interesting possibility for transfer of energy from the wind to the longer gravity waves is through nonresonant interaction with the short waves. According to linear theory, the rate at which short waves can draw energy from the wind is much higher than for the longer waves because the momentum transfer rate is proportional to the square of the surface slope, which is much higher for the capillary waves. Also, the phase angle between pressure and surface elevation should be larger for the shorter waves.

The interaction between short and long waves has been the subject of a number of investigations reported in the literature. Longuet-Higgins (1969a) suggested that the radiation stress set up by the short-wave train could transfer momentum to the long waves. This "maser" mechanism was reexamined by Hasselmann (1971), who demonstrated that, in the absence of modulation of the short-wave growth, the net transfer would be negligible, due to cancellation of the transfers of kinetic and potential energy. Valenzuela and Laing (1972) developed a theory for capillary–gravity wave interaction, and Plant and Wright (1977) suggest that part of the measured excess growth rate in the gravity-wave regime could be due to such interactions at resonance. The capillary–gravity wave interaction has also been considered by Benney (1976).

None of the above analyses considered the interaction between the long and short waves in the air. In their analysis of the interaction between long and short gravity waves, Garrett and Smith (1976) allowed for the possibility that the growth rates of the short waves could be modulated by the long waves. Assuming the short-wave field to be stationary with respect to the long-wave phase, and using different assumed models for the wind stress, they found that the effect of the modulation of the short-wave radiation stress on the growth of the long waves would be negligible compared to the effect of the modulation of the wind stress acting on the short waves. Also, they found that only a fraction of the total wind stress, proportional to long-wave slope, could go into long-wave momentum by this mechanism.

Valenzuela and Wright (1976) also compared the contribution to long-wave growth from short-wave radiation stress with that due to modulation of the input from the wind to the short waves and found that the latter would dominate. Their comparisons with the Joint North Sea Wave Project (JONSWAP) data indicated that the sum of energy flux due to the modulated wind stress and the direct transfer according to Miles's mechanism is still inadequate to explain the observed growth rate of the long waves.

In an earlier paper (Landahl et al., 1979), the modulation of the input to short waves by the long waves in the wind was investigated. A two-scale analysis was employed, leading to an inhomogeneous Orr–Sommerfeld equation which was solved numerically. It was found that the modulation of the short-wave growth rate in phase with the long-wave deflection gives rise to an added momentum input to the long wave, proportional to the mean-square slope of the short waves, and inversely proportional to the difference between the long-wave phase velocity and the short-wave group velocity. It was concluded that the mechanism investigated could give a contribution to the long-wave growth rate which could conceivably be of the same order of magnitude as that provided by Miles's mechanism.

In this chapter the interaction between long and short waves is reexamined with additional effects of modulation of the air flow taken into account. A theory which includes some terms of third order in the wave slopes (second order in the short-wave slope times first order in the long-wave slope) is developed, by use of Hasselmann's (1971) approach. The cubic terms retained are those accounting for the direct interaction, whereas those giving indirect interactions through higher harmonics are ignored. Also, it is assumed that the ratio of short to long wavelengths is small, so that only terms of first order in this ratio are retained. The effects of short-wave dissipation are not treated here, but are presently being studied. The transfer of momentum from short to long waves in the water due to short-wave growth rate modulation by the long waves may be directly related to the modulation of the surface shear stress exerted by the short waves on the long waves and can then be seen to be consistent with previous work (e.g., Garrett and Smith, 1976). Possibly more important, the effect of varying short-wave Reynolds stress on the boundary layer in the air is estimated. Using the long-wave limit for the long waves (comparing wavelength to boundary-layer thickness), and evaluating the short-wave Reynolds stresses according to inviscid theory, we find that the variation of stress on the short waves with the long-wave phase may induce a substantial increase in the correlation of pressure with long-wave slope, and hence in the long-wave growth rate. Although inviscid theory is certainly not appropriate to short waves (which may be confined to the viscous sublayer), it serves to demonstrate the possibly overwhelming importance of this effect.

2. BASIC EQUATIONS

We consider interacting short and long two-dimensional (x, z) waves in air and water for which the longer wavelength, $\lambda^l = 2\pi/k^l$, is much greater than the shorter, $\lambda^s = 2\pi/k^s$ (see Fig. 1). Thus

$$\varepsilon \equiv \lambda^s/\lambda^l \equiv k^l/k^s \ll 1 \tag{1}$$

In the following analysis, only terms of lowest order in ε are considered. Also, the long waves are assumed to have a small slope, σ_l, so that terms nonlinear in σ_l may be neglected in the long-wave dynamics. The short-wave slope, σ_s, is allowed to be much larger, so interaction effects of order $\sigma_s^2 \sigma_l$ need be retained.

We start from the momentum equations for a constant-density flow,

$$\rho\frac{\partial U_i}{\partial t} + \frac{\partial}{\partial x_j}(\rho U_i U_j - \tau_{ij}) + \frac{\partial}{\partial x_i}(p + \rho g x_3) = 0 \tag{2}$$

where

$$\tau_{ij} = \mu\left(\frac{\partial U_i}{\partial x_j} + \frac{\partial U_j}{\partial x_i}\right) \tag{3}$$

is the viscous stress tensor. For incompressible flow, the equation for continuity is

$$\partial U_i/\partial x_i = 0 \tag{4}$$

which holds both in the air and in the water. Whenever needed for clarity, we will distinguish quantities in the air and water by subscripts a or w, respectively.

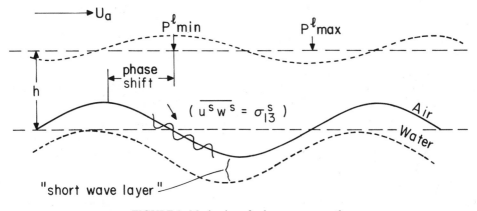

FIGURE 1. Mechanisms for long-wave growth.

The velocity field is divided into mean and fluctuating parts

$$U_i = U_i^m + u_i \tag{5}$$

and the fluctuating field in turn into long-wave, short-wave, and random components

$$u_i = u_i^l + u_i^s + u_i' \tag{6}$$

The pressure, p, and the surface displacement, ζ, are similarly divided:

$$p = p^m + p^l + p^s + p', \qquad \zeta = \zeta^m + \zeta^l + \zeta^s + \zeta' \tag{7}$$

This subdivision of the dependent variables is accomplished by application of three distinct averaging procedures. The first is that used in the ordinary ensemble average, here denoted by $\langle \rangle$, in which the average is formed at a fixed point over an ensemble of long and short waves of given wavenumber and amplitude but with random phases. This gives the mean velocity

$$\langle U_i \rangle = U_i^m \tag{8}$$

Next, the long-wave average is formed over an ensemble of waves of the same given properties, but with the long-wave phase fixed while the short-wave phase is random. This long-wave average is denoted by an overbar; thus,

$$\bar{u}_i = \bar{U}_i - \langle U_i \rangle \equiv u_i^l \tag{9}$$

Third, the short-wave average, denoted by a tilde, is taken over an ensemble in which both long- and short-wave phases are fixed, giving the short-wave velocity field

$$u_i^s = \tilde{u}_i - u_i^l \tag{10}$$

The turbulent velocity field is found as the remainder; in this work we shall not consider the effects of turbulence explicitly, and thus all the turbulent quantities are treated as if negligible. Also, we assume that the mean velocity in the water is negligible and thus in essence we consider the flow in the water to be inviscid and irrotational. It is clear from the calculations by Valenzuela (1976) that the surface drift current in the water is of importance for the growth rate of the capillary waves, but for the sake of simplicity we ignore the effect here.

For the air we consider a parallel mean flow, i.e.,

$$U_i^m = U(z)\delta_{1i} \tag{11}$$

With $U^m = 0$ and $u_3 = w$, integration of the equations for the mean flow under the assumption of horizontal homogeneity gives, for the water

$$p^m = -\rho g z - \rho \langle w^2 \rangle \tag{12}$$

The water depth and the height of the atmosphere are assumed infinite, so the disturbances must vanish far from the interface. At the interface we have the kinematic boundary condition

$$w = \zeta_t + (U + u)\zeta_x \qquad \text{at } z = \zeta \tag{13}$$

with $u_1 = u$, in both air and water. It is useful to transform this condition to the long-wave mean surface (see Hasselmann, 1971). Integration of the continuity equation for the fluctuating velocities

$$\partial u/\partial x + \partial w/\partial z = 0 \tag{14}$$

from $z = \zeta^1$ to $z = \zeta$ gives

$$w(\zeta) - w(\zeta^1) = -\frac{\partial}{\partial x} \int_{\zeta^1}^{\zeta} (U + u)dz + \zeta_x(U + u)_{z=\zeta} - \zeta_x^1(U + u)_{z=\zeta^1} \tag{15}$$

Substituting (13) into this and taking the long-wave average we obtain the boundary condition for the long wave,

$$w^1 = \zeta_t^1 + (U + u^1)\zeta_x^1 + \bar{S}_x \tag{16}$$

where all the flow quantities are to be evaluated at the long-wave interface, $z = \zeta^1$, and where

$$\bar{S} = \overline{\int_{\zeta^1}^{\zeta} (U + u)dz} \tag{17}$$

is the net Stokes drift induced by the short waves. Substitution of (6) into (17) and expansion in powers of ζ^s yields

$$\bar{S} = S^s + \tfrac{1}{2}(U_z + u_z^1)\overline{(\zeta^s)^2} \tag{18}$$

where S^s is the long-wave averaged Stokes drift due to the short waves alone, in the absence of mean flow and long waves,

$$S^s = \overline{\int_{\zeta^1}^{\zeta} u^s dz} \tag{19}$$

and where the z derivatives in (18) are to be evaluated at $z = \zeta^1$. By subtracting (16) from (13) we find the corresponding interface condition for the short waves alone,

$$w^s = \zeta_t^s + u^1\zeta_x^s + u^s\zeta_x^1 + u_x^1\zeta^s \tag{20}$$

Terms representing differences between the square of short-wave quantities and their averages have been neglected since they merely give rise to higher harmonics.

The second kinematic interface condition is that, with $U^m = 0$ for the water,

$$(U + u)_a = u_w \qquad \text{at } z = \zeta \tag{21}$$

This may also be expanded to $z = \zeta^1$, but this will not be needed.

The dynamical boundary conditions at the interface are that the tangential stress is continuous and that the difference in normal stress is balanced by surface tension. The tangential stress condition is mainly balanced by a thin near-surface shear layer in the water. Variations in this stress contribute to long-wave momentum a fraction on the order $\langle u^1\tau_{13}\rangle/c^1$ (see Longuet-Higgins, 1969b); this effect is ignored temporarily compared to that due to variation in momentum transfer to the short waves. The normal stress condition, to second order in wave slope, is

$$(p - \tau_{33})_w = (p - \tau_{33})_a - T\zeta_{xx} \tag{22}$$

where T is the surface tension. The viscous normal stresses are small compared to the pressures, and are neglected. This condition is transformed to the long-wave surface $z = \zeta^1$ through expansion in ζ^s and use of the z component of the momentum equation (2) to determine $\partial p/\partial z$ in the water; to second order, this yields

$$p_w = p_a - T\zeta_{xx} + \rho_w\zeta^s(g + w_t) \qquad . \tag{23}$$

Taking the long-wave average of this, and subtracting the mean pressure according to (12), we find

$$p_w^l = p_a^l - T\zeta_{xx}^l + \rho_w(g\zeta^l + \overline{\zeta^s w_t^s} + \langle w^2 \rangle) \tag{24}$$

3. DYNAMICS OF THE WATER WAVES

Because of the large ratio of water to air density, the dynamics of the water and air may be treated as approximately independent, with the propagation velocity of the waves set by the water dynamics. The main coupling between air and water then manifests itself through the component of wind stress in phase with the wave slope, to which the wave growth rate will be approximately proportional. It should be noted that this approximation is poor for cases when the wave velocity is close to that for the Tollmien–Schlichting wave velocity in the air (see Miles, 1962).

For the water waves we roughly follow the procedure employed by Hasselmann (1971). Except for the near-surface shear mentioned above, the flow in the water may be treated as potential flow. The x component of the momentum equation (2) then becomes

$$\rho\left[\frac{\partial u}{\partial t} + \frac{\partial}{\partial x}(u^2) + \frac{\partial}{\partial z}(uw)\right] + \frac{\partial}{\partial x}(p - \tau_{11} + \rho gz) = 0 \tag{25}$$

The viscous normal stress component τ_{11} is ignored hereafter. Integrating from $z = -\infty$ to ζ^l, and after some manipulations using the interface conditions (16) and (24) at $z = \zeta^l$, we obtain

$$\frac{\partial}{\partial t}\int_{-\infty}^{\zeta^l} \rho u\,dz + \frac{\partial}{\partial x}\int_{-\infty}^{\zeta^l} (p + \rho gz + \rho u^2)dz = \zeta_x^l(p_a - T\zeta_{xx} + \rho gz + \overline{\rho\zeta^s w_t^s})$$
$$+ \rho u(w^s - u^s\zeta_x^l + \bar{S}_x) \tag{26}$$

All the flow quantities outside the integrals are to be evaluated at $z = \zeta^l$. If desired, the part of the viscous shear stress at the surface which, according to Longuet-Higgins (1969b), acts to increase the long-wave momentum may now be reinserted: namely, $\langle u^l\tau_{13}\rangle/c^l$. The corresponding input from the shear stress to the short waves should be included in the evaluation of S_x^s, rather than here. The pressure inside the integral may be approximated with the aid of Bernoulli's equation

$$p + \frac{\rho}{2}(u^2 + w^2) + \rho gz + \rho\phi_t = 0 \tag{27}$$

where ϕ is the velocity potential. Thus,

$$p + \rho u^2 + \rho gz = -\rho\phi_t + \frac{\rho}{2}(u^2 - w^2) \tag{28}$$

The overall mean $\langle\rangle$ of (26) then yields

$$M_t^l + \frac{\partial}{\partial x}\left\langle \rho\int_{-\infty}^{\zeta^l} [-\phi_t + \frac{\rho}{2}(u^2 - w^2)]dz \right\rangle = \langle \zeta_x^l p_a^l\rangle - \langle \rho\zeta_x^l(\overline{\zeta^s w_t^s} + uu^s)\rangle$$
$$+ \frac{1}{c^l}\langle u^l\tau_{13}\rangle + \rho\langle u(w^s + S_x^s)\rangle - \frac{\partial}{\partial x}\langle \tfrac{1}{2}T(\zeta_x^l)^2 + \rho g(\zeta^l)^2\rangle \tag{29}$$

where M^l is the momentum of the long waves, defined by

$$M^l = \left\langle \rho \int_0^{\zeta^l} u^l dz \right\rangle \simeq \rho \langle u^l \zeta^l \rangle \tag{30}$$

For travelling waves

$$\phi_t = -c\phi_x \tag{31}$$

where c is the phase velocity. For linear water waves in a stationary medium, $c = v(k)$, where $v(k)$ is the dispersion relation. In deep water

$$v(k) = (g/k + Tk/\rho)^{1/2} \tag{32}$$

Setting $\langle (\zeta_x^l)^2 \rangle = \langle (k^l \zeta^l)^2 \rangle$, and noting that for deep-water waves $\langle u^2 \rangle = \langle w^2 \rangle$, then, we may rewrite the last term of (29) as

$$\frac{1}{2} \frac{\partial}{\partial x} \langle (k^{l2} T + \rho g)(\zeta^l)^2 \rangle = \frac{\partial}{\partial x} [(c_g^l - c^l) M^l] \tag{33}$$

where c_g^l is the group velocity of the long waves,

$$c_g^l = (\partial/\partial k^l)[k^l v(k^l)] \tag{34}$$

Introducing also the momentum for the short waves

$$M^s = \int_{\zeta^l}^{\zeta} \rho u^s dz = \rho S^s \simeq \overline{\rho u^s \zeta^s} \tag{35}$$

we obtain the following equation for the evolution of the long waves:

$$M_t^l + \frac{\partial}{\partial x}(c_g^l M^l) = \langle p_a^l \zeta_x^l \rangle + \langle u^l \tau_{13} \rangle / c^l - \langle u^l M_x^s \rangle \tag{36}$$

Because the mean background field U^m is assumed negligible, c_g^l is constant along the long-wave characteristics

$$x - c_g^l t = \text{const} \tag{37}$$

If U^m also varies, an additional term arises due to the radiation stress; see the analysis for the short waves, below. The physical interpretation of (36) is that the long wave gains (or loses) momentum from the air pressure component in phase with the long-wave slope (the first term on the right-hand side), from the variation of the surface shear stress in phase with u^l (as described by Longuet-Higgins, 1969b), and also from the interaction with the short waves (the third term). Note that exchanges of kinetic and potential energy arising from short-wave dissipation are not included in (36); Hasselmann (1971) showed that these virtually cancel, so the omission is justified. In the above, then, M_x^s can be interpreted as just those changes of short-wave momentum found without considering dissipation.

The pressure term may be expressed in terms of M^l by setting (in the fashion of Miles, 1957)

$$p_a^l = \rho_a U_a^2 (\alpha^l k^l \zeta^l + \beta^l \zeta_x^l) \tag{38}$$

where U_a is a reference air velocity, and α^l and β^l are aerodynamic coefficients to be found. For linear long waves,

$$u^l = c^l k^l \zeta^l \tag{39}$$

so that (36) takes the form

$$dM^l/dt = sU_a^2 \beta^l k^l M^l/c^l + \langle u^l \tau_{13} \rangle/c^l - \langle u^l M_x^s \rangle \tag{40}$$

where

$$\frac{d}{dt} = \frac{\partial}{\partial t} + c_g^l \frac{\partial}{\partial x} \tag{41}$$

denotes time rate of change following a long-wave ray, and $s = \rho_a/\rho_w$ is the air-to-water density ratio. For the case of *no* interaction with the short waves, the long waves grow exponentially (following the wave group),

$$M^l = M_0^l \exp(2\Omega_{iM}^l t) \tag{42}$$

where M_0^l is the initial momentum and Ω_{iM}^l is Miles growth rate

$$\Omega_{iM}^l = sU_a^2 \beta^l k^l/2c^l \tag{43}$$

Again, the interaction term $-\langle u^l M_x^s \rangle$ involves just those changes of the short-wave momentum induced by the air and by straining; dissipation should not be included. The exchanges in the air will also have a direct effect on the long-wave disturbances in the air, and thereby on β^l, as will be investigated in the next section.

   To derive the corresponding equations for the short-wave momentum, we integrate (25) from $z = \zeta^l$ to ζ, and take the long-wave average (overbar) instead. For the sake of simplicity, we continue to neglect short-wave dissipation; this is the subject of further study. After manipulations analogous to the above, and again neglecting the shear stresses, we find

$$M_t^s + (\partial/\partial x)(c_g^s M^s) = \overline{p_a^s \zeta_x^s} - u_x^l M^s \tag{44}$$

where c_g^s is the group velocity for the short waves

$$c_g^s = (\partial/\partial k^s)[k^s v(k^s)] + u^l \tag{45}$$

This should be calculated with the effective gravity when accelerating with the long-wave surface, $g' = g + w_t^l$, but this modification is unimportant for small long-wave steepness. For the short waves, the wavenumber k^s and phase velocity c^s will vary with x and t because of the modulation by the long waves. From kinematic wave theory,

$$dk^s/dt = -k^s u_x^l \tag{46}$$

where d/dt now denotes the rate of change along short-wave ray

$$dx/dt = c_g^s \tag{47}$$

As for the long waves, we describe the induced pressure in the air as

$$p_a^s = \rho_a U_a^2 (\alpha^s k^s \zeta^s + \beta^s \zeta_x^s) \tag{48}$$

but we also allow these coefficients to be modulated by the long waves. Consequently, we set

$$\alpha^s = \alpha_0^s + k^l \zeta^l \alpha_1^s + \zeta_x^l \alpha_2^s \tag{49}$$

$$\beta^s = \beta_0^s + k^l \zeta^l \beta_1^s + \zeta_x^l \beta_2^s \tag{50}$$

Thus,

$$\overline{\zeta_x^s p_a^s} = sU_a^2 \left[\alpha^s \frac{k^s}{2} \frac{\partial}{\partial x} \left(\frac{M^s}{k^s c^s} \right) + \beta^s k^s / c^s \right] \tag{51}$$

The first term is of order k^l/k^s times the second and is hence negligible. Inserting (51) into (44), we obtain

$$M_t^s + \frac{\partial}{\partial x}(c_g^s M^s) = sU_a^2 \beta^s k^s M^s / c^s - u_x^l M^s \tag{52}$$

All the coefficients in this equation are functions of the long-wave phase

$$\xi = x - c^l t \tag{53}$$

and the amplitude of the long wave. If the growth rate of the long wave is small, one may consider these coefficients to be quasi-steady. The characteristics of (52) defined by (47), with $x = x_0$ for $t = 0$, can then be obtained approximately from

$$t = \int_{x_0}^x \frac{d\xi'}{c_g^s - c^l} \tag{54}$$

and the solution of (52) may be written

$$M^s = M_0^s \exp \left\{ \int_{x_0}^{\xi} \frac{[sU_a^2 \beta^s k^s / c^s - U_\xi^l - c_{g\xi}^s]}{c_g^s - c^l} d\xi' \right\} \tag{55}$$

where M_0^s is the value of M^s at time $t = 0$. This yields the x derivative required in (41). Assuming M_0^s to be independent of x, we find

$$M_x^s = \frac{M^s}{c_g^s - c^l}[sk^s U_a^2(\beta^s - \beta_0^s)/c^s - u_\xi^l - c_{g\xi}^s] \tag{56}$$

Substitution into (40) then gives the growth of the long waves,

$$\frac{1}{M^l} \frac{dM^l}{dt} = sk^l U_a^2 \left[\frac{\beta^l}{c^l} - \frac{\beta_1^s \sigma_s^2}{c_g^s - c^l} \right] \tag{57}$$

where $\sigma_s^2 = \langle (\sigma_x^s)^2 \rangle$ is the mean-square slope of the short waves. It should be noted that a term like u_ξ^l in (56) gives no contribution to the average $\langle u^l M_x^s \rangle$. The second term, which gives the growth due to the interaction, is apparently quite large near the condition of resonance, $c_g^s = c^l$, as would be expected from physical reasoning. However, the singularity is not real, and the simple theory presented here does not hold in the neighborhood of this resonance.

Both of the aerodynamic coefficients β^l and β_1^s of (57) depend on the interaction between the long and short waves. In the absence of short waves, β^l could be obtained from Miles's theory; in the presence of short waves, the modulation of short-wave Reynolds stress in the air will give a contribution to the growth of the long wave which is proportional to the square of the short-wave amplitude as will be shown in the following section.

4. DYNAMICS OF THE WIND WAVES

The presence of strong shear and turbulence in the air flow precludes use of potential theory. Instead, interactions with the shear and turbulence may be expected to dominate the dynamics.

One avenue of approach is to model the effects of wave-induced straining on the turbulence. Townsend (1972) assumed constant stress ratios (but not isotropic), and found growth rates similar to those of Miles's calculations (1957, 1959, 1962), where variations in the turbulence were neglected. Gent and Taylor (1976) assumed rough flow and isotropic turbulence to gage the effect of varying roughness. Constant roughness yields results in agreement with Townsend (1972), in spite of the isotropic assumption; large variations in roughness are found to alter greatly the net transfer to the waves. Others (e.g., Davis, 1972; Townsend, 1980) have found marked changes in the results with different turbulent closure schemes.

The variation in input to the short waves induced by the long waves was considered in an earlier work (Landahl *et al.*, 1979). There, a two-scaling asymptotic method was used, leading to a modified Orr–Sommerfeld equation. This was solved numerically; some results from this calculation are presented in the next section.

In this section, we present a simplistic analysis of the effect of the short waves on the long-wave pressure, and hence on Ω^{li}. As in Miles's work (1957, 1959), the effect of turbulence is ignored in favor of the wavelike forced motion in the air. The boundary-layer thickness h is treated as small compared to the long wavelength, and some results from inviscid theory are employed in the examination of the wave-induced stresses.

For the flow in the air, we ignore the effect of gravity, and again neglect the viscous stresses. The long-wave average of the momentum equation (2) gives, to lowest order in long-wave quantities,

$$\frac{\partial p^l}{\partial x_i} = -\rho \left[\frac{\partial u_i^l}{\partial t} + U_j^m \frac{\partial u_i^l}{\partial x_j} + u_j^l \frac{\partial U_i^m}{\partial x_j} \right] + \frac{\partial}{\partial x_j} \sigma_{ij}^s \tag{58}$$

where

$$\sigma_{ij}^s = -\rho(\overline{u_i^s u_j^s} - \langle u_i^s u_j^s \rangle) \tag{59}$$

is the part of the short-wave Reynolds stress correlated with long-wave phase. The $\langle \rangle$ term arises from subtraction of the momentum equation for the mean flow. To calculate the effect of modulation of the short-wave Reynolds stress on the long-wave pressure field, we start from the x component of (58). Neglecting terms nonlinear in u^l,

$$\partial p^l / \partial x_i = -\rho(u_t^l + U u_x^l + w^l U_z) + \frac{\partial}{\partial z} \sigma_{13}^s + \frac{\partial}{\partial x} \sigma_{11}^s \tag{60}$$

In the long-wave approximation, in which the wavelength is assumed much greater than the thickness, h, of the atmospheric boundary layer, the pressure is assumed to be constant through the layer. The pressure may then be obtained by considering potential flow over the wavy surface presented by the topmost streamline of the boundary layer. Let the latter be defined by

$$z = h + d_1(x, t) \equiv h + \hat{d}_1 e^{ik^l(x - c^l t)} \tag{61}$$

From potential theory then,

$$p^l = -\rho_a k^l (U_a - c^l)^2 d_1 \tag{62}$$

Of the short-wave Reynolds stress terms, only the shear stress component σ^s_{13} is important in the long-wave limit. To estimate the short-wave Reynolds stresses, we use inviscid theory (Lin, 1955), according to which the stresses exhibit a discontinuous jump at the critical level $z = z^s_c$, where the wave phase velocity matches the mean velocity. The momentum transmitted through the critical layer from the mean shear flow is transferred to the wave by the pressure component in phase with the wave slope (see Fig. 1). Hence, in the inviscid case,

$$\sigma^{sc}_{13} = \rho_a U^2_a \beta^s \zeta^2_x \tag{63}$$

where σ^{sc}_{13} is the value inside the critical layer. Since, for the short waves,

$$\beta^s = \beta^s_0 k^1 \zeta^1 \beta^s_1 + \zeta^1_x \beta^s_2 \tag{50}$$

we obtain the modulated portion of the Reynolds stress directly

$$(\partial/\partial z)\sigma^s_{13} = -\rho_a U^2_a \sigma^2_s (\beta^s_1 k^1 \zeta^1 + \beta^s_2 \zeta^1_x)\delta(z - z^s_c) \tag{64}$$

where $\delta(z - z^s_c)$ is Dirac's delta function. By substituting (62) and (64) into (60), we find

$$k^1(U_a - c^1)^2 d_{1x} = u^1_t + U u^1_x + w^1 U_z + \sigma^2_s U^2_a(\beta^s_1 k^1 \zeta^1 + \beta^s_2 \zeta^1_x)\delta(z - z^s_c) \tag{65}$$

Now we introduce the streamline displacement

$$d(x, z, t) = \hat{d}(z)e^{ik^1(x - c^1 t)} \tag{66}$$

which, with $u^1 = \hat{u}^1 \exp[ik^1(x - c^1 t)]$ (etc.), gives

$$\hat{w}^1 = ik^1 \hat{d}(U - c) \tag{67}$$

$$\hat{u}^1 = -\frac{\partial}{\partial z}[\hat{d}(U - c^1)] \tag{68}$$

and after substitution into (65),

$$(U - c^1)\hat{d}_z = -k^1(U_a - c^1)\hat{d}_1 - i\sigma^2_s U^2_a(\beta^s_1 + i\beta^s_2)\zeta^1\delta(z - z^s_c) \tag{69}$$

which may be integrated to relate $d(0) = d_0$ to d_1. From the boundary condition (17) and (67), the linear approximation gives

$$\hat{d}_0 = \zeta^1 - (S^s_a)_x/ik^1 c^1 \tag{70}$$

The second term arises from the displacement due to the short-wave Stokes transport in the air. This involves an additional factor s and is henceforth neglected. Integrating (69) between $z = 0$ and h, and using (70) for d_1 in (62), we find the complex pressure

$$\hat{p}^1 = \frac{-\rho_a(U_a - c^1)^2 k^1 \zeta^1}{1 - k^1(U_a - c^1)I}\left[1 - i\sigma^2_s U^2_a \frac{\beta^s_1 + i\beta^s_2}{(c^s - c^1)^2}\right] \tag{71}$$

where

$$I = I_R + iI_I = -\int^h_0 \frac{dz}{(U - c^1)^2} \tag{72}$$

The singularity in this integral is handled in the usual way for inviscid parallel-flow stability theory, i.e., c^1 is assumed to have a small positive imaginary part. Thus,

$$I_1 = \pi [U_{zz}/(U_z)^3] c^1 \tag{73}$$

The imaginary part of the complex pressure gives the desired component in phase with the long-wave surface slope, and from this we calculate the coefficient β^1. With

$$[1 - k^1 (U_a - c^1)^2 I]^{-1} \simeq 1 + k^1 (U_a - c^1)^2 I + \cdots \tag{74}$$

one may write the result as

$$\beta^1 = \beta_M^1 + \beta_R^1 \tag{75}$$

where

$$\beta_M^1 = \pi k^1 (1 - c^1/U_a)^4 U_a^2 |U_{zz}/(U_z)^1| c^1 \tag{76}$$

$$\beta_R^1 = \sigma_s^2 \beta_1^s \frac{(U_a - c^1)^2}{(c^s - c^1)^2} \tag{77}$$

are the contributions due to the mean shear without wave–wave interactions (Miles's term in this long wave approximation), and due to the long-wave modulation of the short-wave Reynolds stresses, respectively.

5. RESULTS AND DISCUSSION

It is convenient to express the long-wave growth rate in the form

$$\frac{1}{M^1} \frac{dM^1}{dt} = 2\Omega_i^1 \tag{78}$$

where

$$\Omega_i^1 = \Omega_{iM}^1 + \sigma_s^2 [K_m + K_R] \tag{79}$$

and where $\Omega_{iM}^1 = s U_a^2 \beta^1 k^1 / 2c^1$ is the growth rate due to interaction with the mean shear in the absence of wave–wave interactions (Miles's term) and K_m and K_R are interaction coefficients representing the contributions due to the interaction of the long waves with the short waves in the water, and to the modulation of the short-wave Reynolds stresses in the air, respectively. From (57) and (77),

$$K_m = \frac{s k^1 U_a^2 \beta_1}{2(c^1 - c_g^s)} \tag{80}$$

$$K_R = \frac{s k^1 \beta_1^s U_a^2 (U_a - c^1)^2}{c^1 (c^1 - c^s)^2} \tag{81}$$

The term in (80) arising from the modulation of the short waves by the long waves was considered previously by Landahl et al. (1979). In that work the modified Orr–Sommerfeld equation arising from a two-scale analysis was solved numerically. The mean wind velocity was modeled as a turbulent flat-plate boundary-layer profile, including the viscous sublayer, following Reichardt

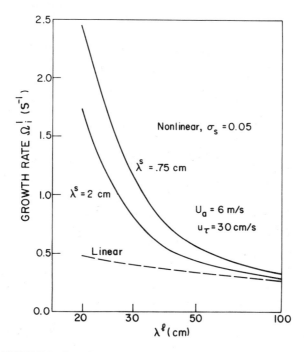

FIGURE 2. Growth rates of long waves in presence of short waves.

(1951). Calculations were carried out for a friction velocity u_τ of 30 cm/s; conditions were chosen so that the ratio u_τ/U_a is about 0.05, typical of wind-tunnel experiments. Interactions between "long" waves of length 100, 75, 36, 20, and 16.5 cm with "short" waves of 2, 1, 0.75, and 0.6 cm were studied. The linear temporal growth rate of unmodulated wind waves, Ω_{iM}, is a direct output of these calculations. Comparison with values from Miles's (1962) viscous calculation shows satisfactory agreement for the shorter wavelengths, but for the longer waves, in the short gravity-wave regime, our calculations give somewhat higher values than Miles's, the differences probably arising because of our use of a different mean wind profile; Miles employed a logarithmic profile extending to infinity.

The results for the growth rate for the long waves due to the short wave–long wave interaction, for 0.75 and 2-cm short waves with an rms steepness, σ_s, of 0.05, are shown in Fig. 2 for k^l ranging from 20 cm to 100 cm. For such waves the interaction in the air, due to modulation of the short-wave Reynolds stresses by the longer waves, was found to be the dominant effect and larger by two orders of magnitude than the contribution due to interaction in the water. That this would generally be expected to be the case for short gravity waves subjected to strong winds can be seen by forming the ratio of K_R to K_m. From (80) and (81) it follows that

$$\frac{K_R}{K_m} = \frac{(U_a - c^l)^2 (c^l - c_g^s)}{c^l (c^l - c^s)^2} \tag{82}$$

For large ratios of wind to wave velocity, this behaves roughly as $(U_a/c^l - 1)^2$ which would be of the order 100 for the cases considered. Thus, from this simplified model one would conclude that the effect of Reynolds stress modulations in the air would be by far the dominant effect. However, use of the inviscid results for the short waves is clearly not justified when the short-wave critical level is in the viscous wall layer; thus, these results for the air interaction are at best only qualitative.

6. CONCLUSIONS

The analysis shows that the interaction between short and long surface waves may substantially increase the transfer of momentum from the wind to waves. The effect of the short waves is found to be proportional to the mean-square slope of the short waves, and the constant of proportionality may be large. Of the effects considered, that due to the modulation of the short-wave Reynolds stress in the air by the longer waves appears, from the simplified analysis above, to be the strongest and is found to give contributions to the long-wave growth rate of comparable magnitude to Miles's term. Since the long-wave inviscid theory used is clearly inadequate in most geophysical applications, however, the results must be viewed as merely indicative. In principle, the effect could be analyzed on the basis of quasi-parallel flow theory with the effects of viscosity included, but this rather laborious numerical task has not yet been attempted.

The variation in stress on the short waves is in some ways analogous to the variation in the effective surface roughness in turbulent flow over a wavy boundary, as considered by Gent and Taylor (1976); they too found that this could be quite important.

From the simple analysis of the short-wave Reynolds stress modulation, one might draw the tentative conclusion that the effects of the modulation of the turbulent Reynolds stresses by the long waves should also be important. However, how best to model the turbulent field in such circumstances is likely to remain a subject of research for some time to come.

ACKNOWLEDGMENT. This work was supported by the National Science Foundation under Grant CME-79-12132.

REFERENCES

Benney, D. J. (1976): Significant interaction between small and large scale surface waves. *Stud. Appl. Math.* **55**, 93.

Davis, R. E. (1972): On prediction of the turbulent flow over a wavy boundary. *J. Fluid Mech.* **52**, 287.

Garrett, C., and J. Smith (1976): On the interaction between long and short surface waves. *J. Phys. Oceanogr.* **6**, 926.

Gent, P. R., and P. A. Taylor (1976): A numerical model of the air flow above water waves. *J. Fluid Mech.* **77**, 105.

Hasselmann, K. (1962): On the non-linear energy transfer in a gravity wave spectrum. Part 1. *J. Fluid Mech.* **12**, 481.

Hasselmann, K. (1963): On the non-linear energy transfer in a gravity wave spectrum. Part 2. *J. Fluid Mech.* **15**, 273; Part 3 **15**, 385.

Hasselmann, K. (1971): On the mass and momentum transfer between short gravity waves and larger-scale motions. *J. Fluid Mech.* **50**, 189.

Landahl, M. T., S. E. Widnall, and L. Hultgren (1979): An interaction mechanism between large and small scales for wind-generated water waves. *Proceedings, Twelfth Symposium on Naval Hydrodynamics*, National Academy of Sciences, 541.

Larson, T. R., and J. Wright (1975): Wind generated gravity–capillary waves: Laboratory measurements of temporal growth rates using microwave backscatter. *J. Fluid Mech.* **70**, 417.

Lin, C. C. (1955): *The Theory of Hydrodynamic Stability*, Cambridge University Press, London.

Longuet-Higgins, M. S. (1969a): A non-linear mechanism for the generation of sea waves. *Proc. Soc. London Ser. A* **311**, 371.

Longuet-Higgins, M. S. (1969b): Action of a variable stress at the surface of water waves. *Phys. Fluids* **12**, 737.

Manton, M. J. (1972): On the generation of sea waves by a turbulent wind. *Boundary-Layer Meteorol.* **2**, 348.

Miles, J. W. (1957): On the generation of waves by shear flow. *J. Fluid Mech.* **3**, 185.

Miles, J. W. (1959): On the generation of waves by shear flows. Part 2. *J. Fluid Mech.* **6**, 568.

Miles, J. W. (1962): On the generation of waves by shear flows. Part 4. *J. Fluid Mech.* **13**, 433.

Phillips, O. M. (1957): On the generation of waves by turbulent wind. *J. Fluid Mech.* **2**, 417.

Plant, W. J., and J. W. Wright (1977): Growth and equilibrium of short gravity waves in a wind wave tank. *J. Fluid Mech.* **82**, 767.

Reichardt, H. (1951): Vollständige Darstellung der turbulenten Geschwindigkeitsverteilung in glatten Leitungen. *Z. Angew. Math. Mech.* **31**, 208.

Townsend, A. A. (1972): Flow in a deep turbulent boundary layer over a surface distorted by water waves. *J. Fluid Mech.* **55**, 719.

Townsend, A. A. (1980): The response of sheared turbulence to additional distortion. *J. Fluid Mech.* **81**, 171.

Valenzuela, G. R. (1976): The growth of gravity–capillary waves in a coupled shear flow. *J. Fluid Mech.* **76**, 229.

Valenzuela, G. R., and M. B. Laing (1972): Nonlinear energy transfer in gravity–capillary wave spectra, with application. *J. Fluid Mech.* **54**, 507.

Valenzuela, G. R., and J. W. Wright (1976): Growth of waves by modulated wind stress. *J. Geophys. Res.* **81**, 5795.

DISCUSSION

MOLLO-CHRISTENSEN: This analysis may provide an explanation for the observed effect of an oil slick on the phase angle between Reynolds stresses and sea surface slope, as found by Ruggles *et al.* (1969). It also seems to be supported by recent laboratory observations of long wave–short wind wave interactions by Ramamonjiarisoa and me.

PLANT: I think that the phenomenon of augmental growth of long waves due to short-wave modulation probably is important. However, as you show, it probably can only increase long-wave growth by a factor of two or less over Miles's expression. I think that this cannot account for the large temporal growth rates we observed for 36-cm waves. These were more than five times Miles's values. I think such high rates are probably explained by nonlinear transfer from near the spectral peak of higher-frequency waves which have already grown to equilibrium.

LANDAHL: Clearly, spectral transfer between a continuum of wavenumbers may give an important contribution to the growth of the long waves. The simple theory does give quite large effects, however; the shown increase by a factor of two or so obtained with a short-wave steepness of only 0.05. The inviscid model for the short-wave Reynolds stresses would probably exaggerate the effect of Reynolds stress modulations.

M. WEISSMAN: I question the use of linear theory for the air flow. Even for small-amplitude waves, it is more appropriate to use nonlinear critical layer theory. However, even this theory may not be valid in many cases in which there is flow separation. In any case, perhaps your theory could be applied for any *given* air flow field.

LANDAHL: The question of whether the nonlinear structure of the critical layer has an important effect on the momentum transfer rate is still an unresolved one. However, linear stability theory does seem to give growth rates in reasonable agreement with observations, at least for amplitudes below which higher instability modes appear.

LONGUET-HIGGINS: Is the principal effect that you found [due to mechanism (3)] strongly dependent on the amplitude of the longer waves, so that the steeper (long) waves would receive a disproportionately large amount of energy from the wind?

LANDAHL: The mechanism (3) of momentum transfer due to modulation of the short-wave Reynolds stresses is, in the simple model considered by us, proportional to long-wave amplitude. One would expect it to be fairly local and not crucially dependent on the assumption of an infinite uniform wave train for the long waves.

ALLENDER: Can you compare your present results with the results of Garrett and Smith (1976), where they found the net transfer to be limited by long-wave steepness to 5% or so of the total stress?

SMITH: The term $\langle u' m_x^s \rangle$ here is essentially the same as the term $\langle u' \tau^s \rangle$ from the paper by Garrett and myself (1976) so the same arguments give an upper bound on this term. The modification of the pressure field acting on the long wave, however, takes place in the *air*, where there is additional energy and momentum available to tap for long-wave growth, and is not so limited.

VALENZUELA: At the Naval Research Laboratory we also have investigated the straining of short gravity–capillary waves by long waves and find the process is basically a perturbation of the short waves and then relaxation back to equilibrium by resonant nonlinear interactions. The relaxation rate of the short waves is equal to their initial temporal growth rates. Also, the nonlinear resonant interactions do remove the singularity in the extra growth rate of long waves by this mechanism as Professor Landahl has indicated.

LANDAHL: Here we have considered just pairs of monochromatic waves and have avoided the singularity near $c_g^s = c^l$; to consider a spectrum of short waves, some treatment like yours is required to remove this singularity.

SMITH: Relaxation of the short waves to "equilibrium" has to involve some assumption about dissipation as well. We have avoided discussion of dissipation here but some work is under way to model it, and we will try to compare the results with your work at the Naval Research Lab.

4

Energy Distribution of Waves above 1 Hz on Long Wind Waves

Karl Richter and Wolfgang Rosenthal

ABSTRACT. A new technique is developed to study the spatial distribution of wind-generated ripples on long wind waves in the ocean. For that purpose the deformation of ripple spectra by currents originating from the long waves has to be removed. The data taken during the JONSWAP 75 Experiment indicate that the distribution of short surface waves (ripples) on long waves has a maximum in the trough of the long waves, so far as the contribution coherent with the long waves is concerned. If the raw time series are considered, the maximum shifts to the forward front of the long waves due to the effect of orbital motion of the long waves.

1. INTRODUCTION

During JONSWAP 75, measurements have been made with a capillary wave staff to study the distribution of short waves on long waves. This problem is important both for the remote sensing microwave techniques and for the generation (or dissipation) of long waves and currents. The basic equations for the latter problem can be found in Hasselmann (1971), Garrett and Smith (1976), and in a simpler fashion already in Longuet-Higgins's (1969) paper on the maser-type generation of waves. With these equations it can be shown that, depending on the distribution of the short wave source terms, long waves can grow on the expense of ripple energy and momentum.

We use the term "short waves" or "ripples" for waves in the capillary-gravity transition range from $\simeq 1$ Hz to $\simeq 15$ Hz and connected wavelengths smaller than 1.5 m. "Long waves" have wave frequencies less than $\simeq 0.5$ Hz and wavelengths longer than $\simeq 6$ m.

In the literature we find two kinds of measurements of ripples:

1. Averages over long records of the order of 20 min have been performed by several groups (e.g., Mitsuyasu and Honda, 1975; Stolte, 1979). They obtained the dependence of the high-frequency tail of the spectrum on the average wind velocity and on properties of the low-frequency spectrum. A theory for this kind of spectrum has been given by Kitaigorodskii et al. (1975) and Kitaigorodskii (1981).
2. Statistical distribution of ripples relative to the phase of long waves. This is important for wave generation and remote sensing. Measurements for this distribution are done mostly by radar either in a tank (Keller and Wright, 1975) or in the ocean (Alpers and Jones, 1978; Wright et al., 1980).

KARL RICHTER ● Deutsches Hydrographisches Institut, Hamburg, West Germany.
WOLFGANG ROSENTHAL ● Royal Netherlands Meteorological Institute, De Bilt, The Netherlands.

High-frequency time series sampled conditionally with respect to the phase of long waves have only recently been studied by Reece (1978) and by Wu (1979) in a wave tank and by Evans and Shemdin (1980) in the ocean.

In the wave measurements with a fixed probe, ripples are advected with the orbital velocity of the underlying long waves. Therefore, the wave probe measures a frequency spectrum of "encounter." The main problem is to convert it to the "true" ripple spectrum which should be measured in a frame of reference moving with the orbital motion of long waves. This problem is addressed in Section 4. In Section 2 we describe the experimental setup and the main points in our data analysis. In Section 3 we give a description of the measured spectra of encounter relative to the long wave phases and Section 5 gives our results of short wave distribution on long waves in terms of the demodulated or "true" ripple spectra. Section 6 is a summary of the work done, some comparison with other work, and the conclusions that could be drawn.

2. EXPERIMENTAL SETUP

Figure 1 shows the experimental setup. As a contribution to the JONSWAP 75 Experiment, two wave staffs were mounted together with a radar scatterometer (Alpers and Jones, 1978) and a Stilwell camera (Monaldo and Kasevich, 1980).

In this section, we concentrate on the measurements of one of the wave staffs, which has a high-pass and a low-pass output.

The output of the resistance wire was high and low pass filtered, so that waves lower than 0.7 Hz (long waves) and higher than 1 Hz (high pass) can be processed separately.

The sampling rate for high-frequency data was 80 Hz, but due to the filter used to cut off the background noise, we can only analyze data up to about 15 Hz. The low-frequency data were sampled at 2 Hz.

The low-frequency output of the wave staff gives the surface wave height $\zeta(t)$, from which we calculated the "instantaneous" orbital velocity as follows. From the Fourier decomposition of

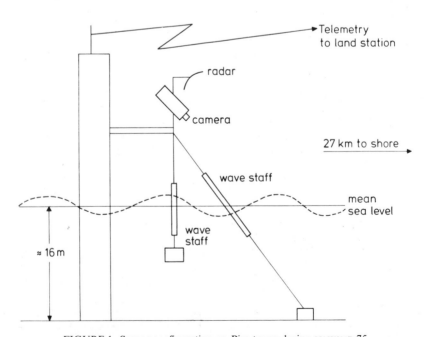

FIGURE 1. Sensor configuration on Pisa tower during JONSWAP 75.

$\zeta(t)$,

$$\zeta(t) = \sum_{f=0}^{0.7\,\mathrm{Hz}} \zeta(2\pi f)e^{i2\pi ft} \qquad\qquad (1)$$

the orbital velocity is given by

$$v(t) = \sum_{f=0}^{0.7\,\mathrm{Hz}} 2\pi f\,\zeta(2\pi f)e^{i2\pi ft} \qquad\qquad (2)$$

if we assume linear unidirectional long waves.

It should be mentioned that after the transform into frequency space, we corrected the undesirable phase shift in $\zeta(t)$, which is introduced by the low-pass filter. The phase shift in the high-pass filtered data can be neglected because it has no influence on the final results of power spectra.

In Fig. 2 we sketch the procedure for further data analysis. From the high-frequency time series we obtained power spectra using time intervals of about 1.5 s (exactly 128/80 s). (Since we began a new time series every 0.5 s, the time series overlap each other.) We receive by this procedure a time series of power spectra. Each power spectrum has a frequency spacing of 80/128 Hz $= 0.625$ Hz and a Nyquist frequency of 40 Hz.

We will use the term "variance spectrum of encounter" for this kind of power spectrum, and use the symbol $E(\omega/2\pi, t)$. The argument $\omega/2\pi$ is measured in hertz, t is the discrete time at intervals of 0.5 s. There are two degrees of freedom only for each frequency component. An improved statistical significance can be reached by fitting the spectrum between 2.5 Hz and 15 Hz to the two-parameter function:

$$E(\omega/2\pi, t) = \alpha(t)[g^2/(2\pi)^4](\omega/2\pi)^{n(t)}$$

$$\log E(\omega/2\pi, t) = \log g^2/(2\pi)^4 + \log \alpha(t) + n(t)\log \omega/2\pi \qquad\qquad (3)$$

The factor $g^2/(2\pi)^4$ is chosen to give the Phillips dimensionless parameter $\alpha_0 \approx 10^{-2}$ for $n = -5$. It may be mentioned that α in (3) is not dimensionless unless $n = -5$. $\alpha(t)$ and $n(t)$ are now time

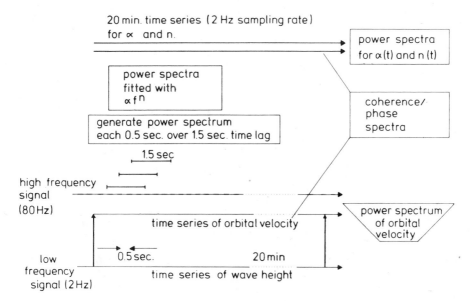

FIGURE 2. Structure of correlation analysis.

series. At each time step t the temporary configuration of the long waves (considered to remain unchanged over the considered time interval of $128/80$ s) determines the sampling variability of $\alpha(t)$ and $n(t)$. The ensemble average at time step t is therefore not identical with a time average over time intervals that are not short compared to the period of the long waves.

To get an estimate of the sampling variability of $\alpha(t)$ and $n(t)$ at a fixed time step t, let us define

$$\delta\alpha(t) = \alpha(t) - \langle\alpha(t)\rangle$$
$$\delta n(t) = n(t) - \langle n(t)\rangle$$
$$\delta E(\omega/2\pi, t) = E(\omega/2\pi, t) - \langle E(\omega/2\pi, t)\rangle \tag{4}$$

where angle brackets denote the ensemble average for the time step t. The statistic of δE obeys a χ_2^2 distribution [we follow the notation of Jenkins and Watts (1968)]. The standard deviation $\sigma(\delta E)$ of δE, according to this distribution, is connected with its mean value by

$$\sigma(\delta E) = (\langle\delta E^2\rangle)^{0.5} = \langle\delta E\rangle$$

The standard deviation of log $E(\omega/2\pi, t)$ is therefore in a linear approximation

$$\sigma[\log\ E(\omega/2\pi, t)] = \frac{\sigma(E)}{\langle E\rangle}\cdot\log e = \log e = 0.43$$

The sampling statistics of the regression line parameters log $\alpha(t)$ and $n(t)$ follow from (3) as described in standard textbooks (Jenkins and Watts, 1968). We get for the numerical values of the standard deviations at any time step t

$$\sigma[\log\alpha(t)] = 0.1, \qquad \sigma[n(t)] = 0.05$$

This random scatter of the time series log $\alpha(t)$ and $n(t)$ transform into a white background noise in the connected power spectra, where the integrals over the noise spectrum are given by the square of $\sigma[\log\alpha(t)]$ and $\sigma[n(t)]$. The power spectra are derived from time series of 1280 s by averaging over variance spectra of 10 partial time series of 128 s.

According to Jenkins and Watts (1968), the variance of the power spectra for $n(t)$, $\alpha(t)$ are equivalent to 30 degrees of freedom, the same as the velocity power spectra (20-min time series are transferred to spectra with spectral resolution of $1/128$ Hz by a Bartlett procedure).

It is convenient to define the deviations of variables from their temporal means, denoted by subscript "0":

$$\Delta\alpha(t) = \alpha(t) - \alpha_0$$
$$\Delta\alpha(t) = n(t) - n_0$$
$$\Delta E(\omega/2\pi, t) = E(\omega/2\pi, t) - E_0(\omega/2\pi) \tag{5}$$

For the temporal means α_0, n_0 we got

$$n_0 = 4.1, \qquad \alpha_0 = 2.0 \times 10^{-2}$$

These temporal means correspond approximately with an empirical formula from Stolte (1979) derived in a similar environment and defining the wind dependence of n_0 and α_0 if U_{10} is measured in m/s:

$$\log\alpha_0 = -2.87 + 2.73\cdot10^{-1}\cdot U_{10} - 1.53\cdot10^{-2}\cdot U_{10}{}^2$$
$$n_0 = -5 + 1.22\cdot10^{-1}\cdot U_{10} + 3.91\cdot10^{-3}\cdot U_{10}{}^2 \tag{6}$$

To avoid confusion between our two scales of frequencies, we use ω for the frequency of encounter, and f for the low frequencies ($f \lesssim 1$ Hz) corresponding to the long waves.

As regards the quantities $\alpha(t)$, log $\alpha(t)$, $n(t)$, $\Delta\alpha(t)$, $\Delta n(t)$, $\zeta(t)$, and $V_{orb}(t)$, we use the same symbols for their Fourier transforms. We therefore write explicitly the argument t or f, if it is necessary to distinguish between the time series of their Fourier transforms.

3. PROPERTIES OF RIPPLE SPECTRA OF ENCOUNTER

During the experiment we obtained wave data for about one month. The resulting spectra had a variety of properties probably because of the complicated influence of tidal currents and orbital motion of long waves. To illustrate the general features we describe two cases where homogeneous and stationary wind blew in the same direction with those of tidal currents and

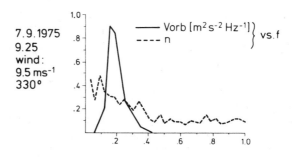

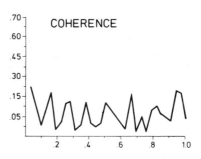

FIGURE 3. Low-frequency spectra of parameters characterizing the high-frequency spectrum of encounter, case I. (Top) Solid curve: autocovariance spectrum for orbital velocity; dashed curve: autocovariance spectrum for exponent n. (Middle) Coherence spectrum between n and orbital velocity. (Bottom) Phase spectrum between n and orbital velocity (positive values mean n is leading the orbital velocity).

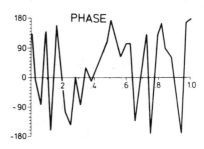

long waves. These very special conditions were encountered only four times during the experiment.

Case 1: Windsea Case

The wind and the waves came from 330° relative to north and the wind speed was 9.5 m/s. Using these data, the peak frequency of the orbital velocity spectrum was higher than that of the Pierson–Moskowitz spectrum, where the latter is estimated from the local wind. Therefore, this case corresponds to the growing windsea.

Figures 3 and 4 summarize the results. The power spectrum of $n(t)$ shows no pronounced peak and low coherence with the orbital velocity. The rms of the time series $\Delta n(t)$ is 0.37, so that

$$[(\Delta n)^2]^{1/2} \simeq 0.1 n_0$$

7.9.1975
9.25
wind:
9.5 ms^{-1}
330°

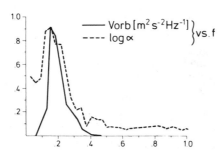

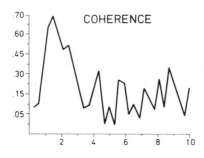

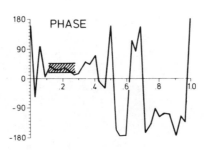

FIGURE 4. Low-frequency spectra of parameters characterizing the high-frequency spectrum of encounter, case I. (Top) Solid curve: autocovariance spectrum for orbital velocity; dashed curve: autocovariance spectra for $\log \alpha$. (Middle) Coherence spectrum between $\log \alpha$ and orbital velocity. (Bottom) Phase spectrum between $\log \alpha$ and orbital velocity (positive values mean $\log \alpha$ is leading the orbital velocity). The hatched area is the 90% confidence interval.

The power spectrum of $\log \alpha(t) \simeq \log e \cdot \Delta\alpha(t)/\alpha_0$ shows a peak at the peak of the velocity spectrum and a coherence of 0.7 at the peak. The phase of $\log \alpha$ is 30° relative to the crest of long waves for frequencies of high coherence. This means $\alpha(t)$ is greatest on the forward side of the crest. The rms of $\Delta\alpha(t)/\alpha_0$ is 1.1.

The behavior of α and n could be interpreted in terms of the spectral densities $E(\omega/2\pi, t)$. Because from Eq. (3), $n(t)$ determines the slope and $\alpha(t)$ the average level of $\log E(\omega/2\pi, t)$ versus $\log \omega$, the high coherence of $\alpha(t)$ and the low coherence of $n(t)$ with the orbital velocity means that $\log E(\omega/2\pi, t)$ varies in the same way over the whole part of the $\log \omega$ axis used for the regression line in (3).

Remarkable is the similar spectral shape of $|\log \alpha(\omega/2\pi, f)|^2$ and $|V_{orb}(f)|^2$.

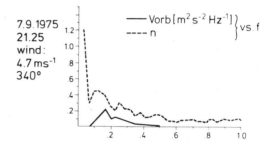

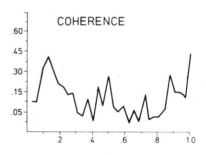

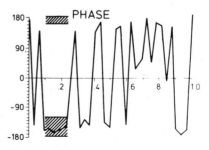

FIGURE 5. Low-frequency spectra of parameters characterizing the high-frequency spectra of encounter, case II. The meaning of the different curves is the same as in Fig. 3.

Case 2: Swell Case

In this case, the waves and the wind came from 340°. The local wind speed was 4.7 m/s. The results are given in Figs. 5 and 6.

In contrast to case 1, the coherence of n with V_{orb} is significantly larger ($\simeq 0.4$) at the peak. The phase is 180° (in the vicinity of high coherence), indicating that n is greatest in the trough of long waves. The rms of Δn is 0.4. The coherence between $V_{orb}(t)$ and log $\alpha \simeq (\Delta\alpha/\alpha_0)$ (log e) is 0.4 near the peak frequency of V_{orb}. The phase is 20° in the vicinity of the peak frequency, similar to case 1. The rms of $(\Delta\alpha/\alpha_0)$ is 1.0.

As in case 1, a qualitative consideration of the spectral behavior of ripples can be made. Since in this case not only α but also n is correlated with the orbital velocity, the high-frequency waves (near the upper spectral limit of 15 Hz) are more strongly enhanced and attenuated than waves at the lower frequencies (near our lower spectral limit of 2.5 Hz).

7.9.1975
21.25
wind:
4.7 ms⁻¹
340°

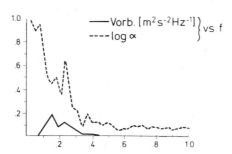

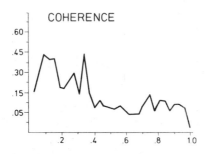

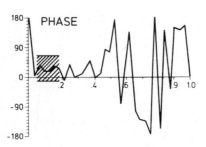

FIGURE 6. Low-frequency spectra of parameters characterizing the high-frequency spectra of encounter, case II. The meaning of the different curves is the same as in Fig. 4.

Though many other data were analyzed, no simple definite conclusions could be derived about the encounter spectra of ripples on long waves.

4. RIPPLE SPECTRA IN A FRAME MOVING WITH THE ORBITAL MOTION OF LONG WAVES

To deduce ripple behavior from our analysis done so far is difficult, because we derived "spectra of encounter" instead of "true spectra," since the wave staff is moving relative to the water column because of tidal currents and long wave orbital motion. The high-frequency waves are Doppler shifted by the orbital velocity of the long waves by an amount

$$\frac{\Delta\omega}{2\pi} = \frac{1}{2\pi}k\cdot V_{\text{orb}} \simeq 50 \text{ Hz} \qquad \text{for } 2\pi/k = 2 \text{ cm and } V_{\text{orb}} = 1 \text{ m/s} \tag{7}$$

The main purpose of our data analysis is to correct the Doppler shift due to tidal currents and orbital velocity.

In addition to f and ω defined previously, we introduce the third notation σ for frequency, where σ denotes the intrinsic frequency for all wavenumber $\mathbf{k}$, which is expressed via the deep-water dispersion relation, by

$$\sigma^2 = g\cdot|\mathbf{k}| \tag{8}$$

where g is the gravitational acceleration. The intrinsic frequency σ would be measured for a wavenumber vector $\mathbf{k}$ in a frame of reference fixed to a water column large compared to the ripple

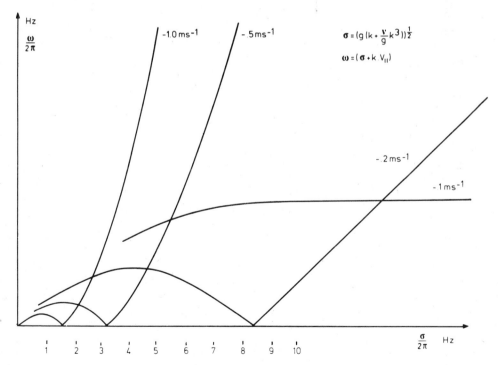

FIGURE 7. Relationship between frequency of encounter $\omega/2\pi$ and true frequency $\sigma/2\pi$ for different negative current velocities.

wavelength $2\pi/|\mathbf{k}|$ but small compared to the long waves. It should be small enough to neglect the spatial variation of the orbital velocity of long waves.

This water column is moved with the orbital velocity of the long waves V_{orb} and the mean current $\mathbf{V}$. The frequency $\omega(k)$ observed in the fixed frame of reference is related to the intrinsic frequency $\sigma(\mathbf{k})$ by

$$\omega(\mathbf{k}) = \sigma(\mathbf{k}) + \mathbf{k} \cdot (V_{\mathrm{orb}} + \mathbf{V}) \tag{9}$$

In (7) we gave estimates of possible shifts between $\sigma(\mathbf{k})$ and $\omega(\mathbf{k})$ for $2\pi/|\mathbf{k}| = 2$ cm and frequently occurring velocities. Figures 7 and 8 show relation (9) for different velocities.

There is another effect masking the "true" spectra (i.e., spectra in the σ frame). Since the interval $\Delta\sigma$ is mapped into the interval $\Delta\omega$ through (9), energy density $E(\omega)$ in the ω frame is related to $\hat{E}(\sigma)$ by

$$E(\omega)\Delta\omega = \hat{E}(\sigma)\Delta\sigma \tag{10}$$

or, for infinitesimally small $\Delta\sigma$,

$$E(\omega)\frac{d\omega}{d\sigma} = \hat{E}(\sigma) \tag{11}$$

Thus, the Jacobian between σ and ω gives rise to an additional difference of the energy densities

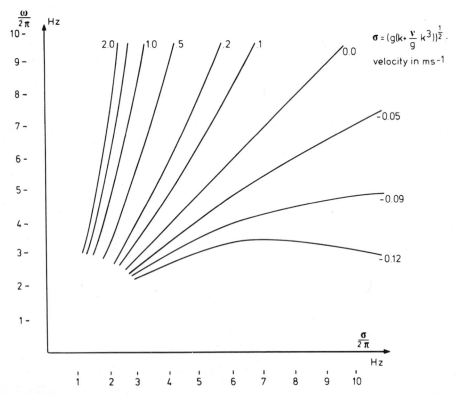

FIGURE 8. Relationship between frequency of encounter $\omega/2\pi$ and true frequency $\sigma/2\pi$ for various positive and slightly negative current velocities.

$\hat{E}(\sigma)$ and $E(\omega)$. The explicit relation for the Jacobian is

$$\frac{d\omega}{d\sigma} = \left[1 + (\mathbf{V}_{orb} + \mathbf{V}) \cdot \frac{d\mathbf{k}(\sigma)}{d\sigma} \right]$$

$$= \left[1 + \frac{(\mathbf{V}_{orb} + \mathbf{V})}{V_G(\mathbf{k})} \cdot \frac{k}{|\mathbf{k}|} \right] \tag{12}$$

where $V_G(\mathbf{k})$ is the group velocity for $\mathbf{k}$ in the frame of reference moving with the water column. We get a numerical value of $d\omega/d\sigma = 6$ for a water velocity of 1 m/s and a group velocity of 20 cm/s. This shows that we cannot neglect this effect for field conditions.

Under certain assumptions it is possible to derive the true frequency spectrum $\hat{E}(\sigma)$ from the encounter spectrum $E(\omega)$. We assume the ripples to have the same direction with the water velocity $\mathbf{V}_{orb} + \mathbf{V}$. Figure 8 shows that relation (9) provides a one-to-one mapping between σ and ω. Our measured spectrum has a frequency step $\Delta\omega = 0.625$ Hz $\times 2\pi$ as explained in Section 2. We convert the center frequency ω_i $(i = 1, \cdots 20)$ of each step to the corresponding intrinsic frequency $\sigma_i = \sigma(\omega_i)$ via relation (9). The measured spectral variance density $E(\omega_i)$ was translated via relation (11) to $E(\sigma_i)$. While the ω_i remains the same, the σ_i varies according to the variation of the orbital velocity. For each time step, however, it is possible to deduce the functional relationship $\hat{E} = \hat{E}\,(\sigma/2\pi)$ if we assume that this relationship is always given by a relation

$$\hat{E}\left(\frac{\sigma}{2\pi}, t\right) = \alpha(\hat{t})\left(\frac{\sigma}{2\pi}\right)^{\hat{n}(t)} \cdot \frac{g^2}{(2\pi)^4} \tag{13}$$

with this assumption, we fitted each half second a regression line

$$\log \hat{E}\left(\frac{\sigma_i}{2\pi}, t\right) = \log \hat{\alpha}(t) + \hat{n}(t) \log\left(\frac{\sigma_i}{2\pi}\right) + \log\left(\frac{g^2}{(2\pi)^4}\right), \qquad i = 1, \cdots 20 \tag{14}$$

to our 20 points of the true spectrum. The two time series $\hat{\alpha}(t)$, $\hat{n}(t)$ now contain according to (13) the information on the true spectrum. It is thus obvious that we can correlate these two time series with the orbital velocity instead of calculating from (13) the energy density and do corrleation analysis for $\hat{E}(\sigma, t)$ and the orbital velocity.

To analyze the correlation of $\hat{E}$, $\hat{n}$, $\hat{\alpha}$ with V_{orb}, we may divide $\hat{E}$, $\hat{\alpha}$, $\hat{n}$ into a constant and a time-dependent part:

$$\hat{E}(\sigma, t) = \hat{E}_0(\sigma) + \Delta\hat{E}(\sigma, t)$$

$$\hat{\alpha}(t) = \hat{\alpha}_0 + \Delta\hat{\alpha}(t) \tag{15}$$

$$\hat{n}(t) = \hat{n}_0 + \Delta\hat{n}(t)$$

Inserting (15) into (14), linearizing, and assuming zero coherence between $\Delta\alpha(t)/\alpha_0$ and $\Delta\hat{n}(t)$, we have

$$|\Delta\hat{E}(\sigma, f)/E_0(\hat{\sigma})|^2 = |\Delta\hat{\alpha}(f)/\alpha_0|^2 + |\Delta\hat{n}(f)|^2 \ln^2(\sigma/2\pi) \tag{16}$$

The cross-correlation spectrum CSP with the long wave orbital velocity is

$$\text{CSP}\left[\frac{\Delta\hat{E}(\sigma, f)}{E_0(\sigma)}, V_{orb}(f)\right] = \text{CSP}\left[\frac{\Delta\hat{\alpha}(f)}{\alpha_0}, V_{orb}(f)\right] + \text{CSP}\left[\Delta\hat{n}(f), V_{orb}(f)\right] \ln\left(\frac{\sigma}{2\pi}\right) \tag{17}$$

It may be suitable at this point to comment on a different technique used by Evans and

Shemdin (1980) for the West Coast Experiment data to account for the orbital velocity of the long waves. A wave record of short waves

$$z(t) = \sum_{\mathbf{k}} a_{\mathbf{k}} \cos \omega(k) \cdot t$$

with

$$\omega(k) = \sigma(k) + \mathbf{k} \cdot \mathbf{u}$$

is transferred to a time variable τ defined by

$$\sigma(\mathbf{k}_0) \cdot \tau = \omega(\mathbf{k}_0) \cdot t$$

for a fixed wavenumber $\mathbf{k}_0$.

If the sum over $\mathbf{k}$ degenerates to one term $\mathbf{k} = \mathbf{k}_0$, the Fourier coefficient could easily be determined by a Fourier transform with regard to $\exp[i\sigma(k_0)\tau]$. In cases, however, where other components appear under the sum over $\mathbf{k}$, a Fourier transform would also include contributions from these other terms and we felt unable to estimate the magnitude of this error. We therefore did not apply this method to our data.

7.9.1975
9.25
wind:
9.5 ms⁻¹
330°

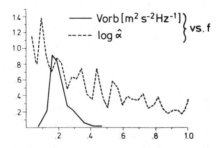

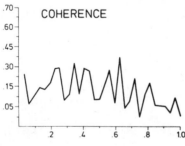

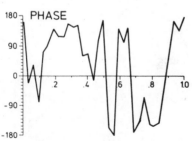

FIGURE 9. Low-frequency spectra of parameters characterizing the high-frequency spectra in true frequency space $\sigma/2\pi$, case I. (Top) Solid curve: autocovariance spectrum for orbital velocity; dashed curve; autocovariance spectrum for $\log \alpha$. (Middle) Coherence spectrum between orbital velocity and $\log \hat{\alpha}$. (Bottom) phase spectrum between orbital velocity and $\log \hat{\alpha}$ (positive values mean $\log \hat{\alpha}$ is leading the orbital velocity).

5. DEMODULATED SPECTRA

In Section 4 we described our inverse technique for positive water velocities only. In principle, it could be applied to negative velocities also, but the relation between σ and ω becomes multivalued and changes strongly for small changes of the velocity. This is clearly shown in Fig. 7 for curves with -0.1 m/s and -0.2 m/s. The inverse technique is therefore very sensitive to small errors in the velocity and our tests to apply the inverse technique to cases with negative velocities were not successful. We therefore selected carefully those cases where we were sure that tidal currents, surface waves, and ripples (wind) run in the same direction and no negative currents occur during the 20-min measurement period.

From the four selected cases we show for brevity two examples (Figs. 9–12). These cases are the same from which the analysis of the uncorrected spectra is given in Section 3.

Case 1 (Windsea Case) in Figs. 9 and 10

The "log $\hat{\alpha}$" has low coherence with the orbital velocity and therefore the phase is strongly scattered. The rms variance can be expressed by

$$(|\Delta\hat{\alpha}/\hat{\alpha}_0|^2)^{0.5} = 1.59$$

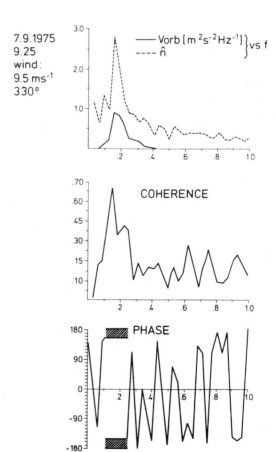

7.9.1975
9.25
wind:
9.5 ms^{-1}
330°

FIGURE 10. Low-frequency spectra of parameters characterizing the high-frequency spectra in true frequency space $\sigma/2\pi$, case I. (Top) Solid curve: autocovariance spectrum for orbital velocity; dashed curve: autocovariance spectrum for $\hat{n}$. (Middle) Coherence spectrum between orbital velocity and $\hat{n}$. (Bottom) Phase spectrum between orbital velocity and $\hat{n}$ (positive values mean $\hat{n}$ is leading the orbital velocity). The hatched area is the 90% confidence intervall.

In the vicinity of the peak f_p of the orbital velocity power spectrum, we have

$$|\log \hat{\alpha}(f_p)|^2 = 0.8 \ \text{Hz}^{-1}$$

which implies

$$|\Delta\alpha(f_p)/\hat{\alpha}|^2 = 4.24 \ \text{Hz}^{-1}$$

The $\hat{n}$ spectrum follows the frequency dependence of the orbital velocity spectrum very well and in the vicinity of f_p is modeled by

$$|\Delta\hat{n}(f)|^2 = 3.6 \cdot |V_{orb}(f)|^2 \qquad (18)$$

when $|V_{orb}(f)|^2$ is given in m^2/s^2 per Hz. The coherence around f_p is 0.5 and the surprising result is that the phase is 180° in the high-coherency region, a result found earlier by Reece for some of his wave-tank measurements. The rms variance of $\hat{n}$ is 0.76.

Case 2 (Swell Case) in Figs. 11 and 12

The "$\log \hat{\alpha}(f)$" has again low coherence with the orbital velocity, but the phase is still around 30°. The total variance is smaller than that obtained from the spectra of encounter. The

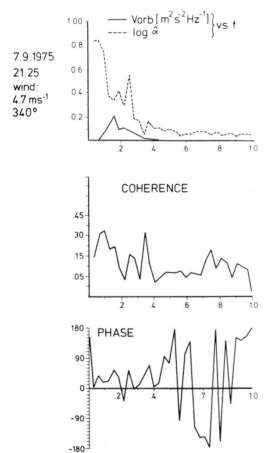

FIGURE 11. Low-frequency spectra of parameters characterizing the high-frequency spectra in true frequency space $\sigma/2\pi$, case II. The meaning of the different curves is the same as in Fig. 9.

7 9.1975

21.25

wind:

4.7 m s^{-1}

340°

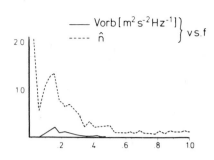

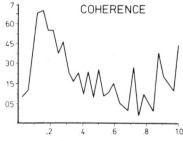

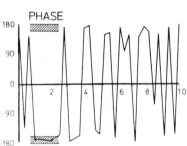

FIGURE 12. Low-frequency spectra of parameters characterizing the high-frequency spectra in true frequency space. The meaning of the different curves is the same as in Fig. 10.

numerical value is about

$$(\overline{|\Delta\hat{\alpha}/\hat{\alpha}_0|^2})^{0.5} = 0.88$$

In the vicinity of f_p we get

$$|\log\hat{\alpha}(f_p)|^2 = 0.35$$

$$|\Delta\hat{\alpha}(f_p)/\hat{\alpha}_0|^2 = 1.85$$

The coherence of $\hat{n}$ with the orbital velocity is 0.5 near the peak frequency f_P. In the neighbourhood of f_P, the relation between $|\Delta n(f)|^2$ and $|V_{orb}(f)|^2$ is given by

$$|\Delta\hat{n}(f)|^2 = 5.2\cdot|V_{orb}(f)|^2 \tag{19}$$

if $|V_{orb}(f)|^2$ is measured in m^2/s^2 per Hz. The phase for frequencies of high coherency lies near 180°. The total variance is 0.59.

6. CONCLUSIONS

The present analysis shows a maximum exponent n for the parameterized spectral shape (13) in the trough of the long waves. We show now that $E(\sigma, t)$ and $n(t)$ have the same spatial behavior, which means that also $E(\sigma,t)$ is maximum in the trough of the long waves. The similar behavior of n and $E(\sigma)$ is easily deduced from (17), assuming zero coherence between $\Delta\hat{\alpha}$ and V_{orb}. We get

$$\text{CSP}\left[\frac{\Delta\hat{E}(\sigma,f)}{\hat{E}_0(\sigma)}, V_{orb}(f)\right] = \text{CSP}\left[\Delta\hat{n}(f), V_{orb}(f)\ln\left(\frac{\sigma}{2\pi}\right)\right] \tag{20}$$

This means that phase and coherence of $\Delta\hat{n}$ relative to the long wave orbital velocity is the same as phase and coherence of $\Delta E(\sigma,f)$. It also shows that the energy density variation increases logarithmically with σ.

The behavior of the phases of $\Delta\hat{E}(\sigma,f)$ is different from what could be expected from radar measurements.

Let us consider the parameter

$$m(\sigma,f) = \frac{c(f)\cdot\text{CSP}[\Delta\hat{E}(\sigma,f), V_{orb}(f)]}{\hat{E}_0(\sigma)|V_{orb}(f)|^2} \tag{21}$$

$[c(f) = \text{phase speed for frequency } f]$, which is similar to the modulation transfer function of Wright et al. (1980). We replaced the power received by the radar, occurring in their expression, by the spectral energy density of the ripples. In a linear approximation of the radar signal from the sea surface, both quantities are proportional. The proportionality constant drops out, because m is normalized with the mean radar signal or the mean energy density, respectively.

The choice of the parameter $m(\sigma,f)$ can be motivated by assuming that the time series $\Delta\hat{E}(\sigma,t)$ is a linear functional of the history of the large-scale velocity

$$\frac{\Delta\hat{E}(\sigma,t)}{\hat{E}_0(\sigma)} = \int_0^{-\infty} \tilde{m}(t-\tau)V_{orb}(\tau)d\tau \tag{22}$$

which gives after Fourier transformation

$$\frac{\Delta\hat{E}(\sigma,f)}{\hat{E}_0(\sigma)} = \tilde{m}(\sigma,f)V_{orb}(f) \tag{23}$$

where $\tilde{m}(\sigma,f)$ is the Fourier transform of the response function $m(\sigma,t)$. It is convenient to define

$$m(\sigma,f) = c(f)\cdot\tilde{m}(\sigma,f)$$

for nondimensional representation. Multiplying (23) with $V_{orb}(f)$ and ensemble averaging gives formula (21).

The cross-spectrum between $\hat{E}(\sigma)$ and V_{orb} has been expressed by the cross-spectrum of V_{orb} with $\Delta\alpha/\alpha_0$ and $\Delta\hat{n}$ respectively in (17). Since the data show no significant coherence between $\Delta\hat{\alpha}$ and V_{orb}, we have within the statistical error

$$\text{CSP}\left[\frac{\Delta\hat{E}(\sigma,f)}{\hat{E}_0(\sigma)}, V_{orb}(f)\right] = \text{CSP}[\Delta\hat{n}(f), V_{orb}(f)]\ln\frac{\sigma}{2\pi}$$

$$= \{[|\Delta\hat{n}(f)|^2]^{0.5}[|V_{orb}(f)|^2]^{0.5}\text{coh}[\Delta\hat{n}(f), V_{orb}(f)]$$

$$\cdot\exp(i(\text{phase}[\Delta\hat{n}(f), V_{orb}(f)]))\}\ln\frac{\sigma}{2\pi} \tag{24}$$

where the coherence spectrum "coh" and the phase spectrum "phase" are defined as usual in terms of the cross- and power spectra:

$$\coh[\Delta\hat{n}(f), V_{orb}(f)] = \left\{ \frac{|CSP[\Delta\hat{n}(f), V_{orb}(f)]|^2}{|\Delta\hat{n}(f)|^2 |V_{orb}(f)|^2} \right\}^{0.5}$$

$$tg\,(\text{phase}\,[\Delta\hat{n}(f), V_{orb}(f)]) = \frac{\text{Im}(CSP[\Delta\hat{n}(f), V_{orb}(f)])}{\text{Re}(CSP[\Delta\hat{n}(f), V_{orb}(f)])} \tag{25}$$

Im(A) and Re(A) denote the imaginary and real part of the argument A, respectively. Thus, (21) becomes

$$m(\sigma,f) = c(f)\cdot\ln\frac{\sigma}{2\pi}\frac{|\Delta\hat{n}(f)|\coh\,[\Delta\hat{n}(f), V_{orb}(f)]}{|V_{orb}(f)|}$$

$$\cdot\exp\{i(\text{phase}\,[\Delta\hat{n}(f), V_{orb}(f)])\}$$

Using the relationship between $|\Delta\hat{n}(f)|$ and $V_{orb}(f)$ and the coherence and phase values in the previous section, we get for the modulation transfer function

$$m(\sigma,f) = -0.9\cdot c(f)\cdot\ln(\sigma/2\pi) \qquad \text{for case 1} \tag{26}$$

$$m(\sigma,f) = -1.1\cdot c(f)\cdot\ln(\sigma/2\pi) \qquad \text{for case 2} \tag{27}$$

where $c(f)$ has units of m/s.

Together with our approximation of vanishing coherence between $\hat{a}$ and V_{orb}, we estimate the statistical uncertainty to less than 50% of rms error for the above-derived values of $m(\sigma,f)$. It may be recalled that $\sigma/2\pi$ in our investigation lies within an interval of 2.5 Hz and 15 Hz. At the low border, neglecting the modulation of $\hat{a}$ will cause the largest error. This can be easily seen from (16), in which $\Delta\hat{E}(\sigma,f)$ for $\sigma/2\pi = 1$ Hz would be totally determined by $\Delta\hat{a}(f)$ since the second term vanishes. The above-derived values of $m(\sigma,f)$ give the same order of magnitude as the measurements of Wright *et al.* (1980) for the modulus of $m(f)$. The measurements cited in Section 1 give for this modulus of $m(f)$ also the same order of magnitude. The main difference occurs for the phase of m. It scatters strongly for most of the cited measurements, having a tendency to be less than 90°. Only Reece showed measurements which have, just like our transfer function, 180° phase differences. From his error bars, however, he gave for the phase of m a range of angle from 45° to 180°.

An explanation for the observed phases near the troughs of the long waves might be a local balance for the oscillating amplitude of the short wave components. Let $\mathbf{k}_s$ be the wavenumber vector of a short wave component with momentum density $M(\mathbf{k}_s, \mathbf{x}_s)$. With $\mathbf{x}_s$ we denote the position in a frame of reference with the orbital velocity of the long waves. In this frame of reference, $\mathbf{x}_s$, $\mathbf{k}_s$ vary according to the canonical equations

$$\frac{d\mathbf{k}_s}{dt} = -\frac{\partial\sigma(\mathbf{k}_s, \mathbf{x})}{\partial x}\bigg|_{\mathbf{x}=\mathbf{x}_a}$$

$$\frac{d\mathbf{x}_s}{dt} = \frac{\partial\sigma(\mathbf{k}, \mathbf{x}_s)}{\partial k}\bigg|_{\mathbf{k}=\mathbf{k}_s} \tag{28}$$

The first of these equations describes the straining of the short waves by the long waves. [The modulation in frequency space caused by straining alone cannot account for the large modulation measured, as has been pointed out by Wright *et al.* (1980).]

For the source function $S(\mathbf{k}_s)$ we assume a Phillips-type generation term $g(\mathbf{k}_s)$, a relaxation-type dissipation with relaxation time τ and a term oscillating with the orbital motion, being proportional to $M(\mathbf{k}_s)$:

$$S(\mathbf{k}_s) = g(\mathbf{k}_s) - \frac{M(\mathbf{k}_s)}{\tau} + M(\mathbf{k}_s)\left\{\sum_f h(\mathbf{k}_s,f)U(f)\cos\left[2\pi ft + \phi(\mathbf{k}_s,f)\right]\right\} \tag{29}$$

where $\phi(\mathbf{k}_s,f)$ is the phase angle and $h(\mathbf{k}_s,f)$ is the interaction parameter, being positive. Assuming the local equilibrium for $M(\mathbf{k}_s)$, we have

$$S(\mathbf{k}_s) = 0 \tag{30}$$

It follows directly from (29) that

$$M(\mathbf{k}_s,t) = g(\mathbf{k}_s)\cdot\left\{\tau^{-1} - \sum_f h(\mathbf{k}_s,f)U(f)\cos\left[2\pi ft + \phi(\mathbf{k}_s,f)\right]\right\}^{-1} \tag{31}$$

To the linear approximation, (31) reduces to

$$M(\mathbf{k}_s,t) = g(\mathbf{k}_s)\cdot\tau\cdot\left\{1 + \tau\cdot\sum_f h(\mathbf{k}_s,f)U(f)\cos\left[2\pi ft + \phi(\mathbf{k}_s,f)\right]\right\} \tag{32}$$

For different source mechanisms it is now possible to deduce the influence on the phase of $M(\mathbf{k}_s,t)$ from (32). In the case that microscale wave breaking, suggested by Phillips and Banner (1974), is the dominant negative source term, this would cause a phase of 180° in the undulating part of (29) because it should be the most negative at the crest of long waves. From (32) this results in a maximum of $M(\mathbf{k}_s)$ near the troughs of the long waves.

7. SUMMARY

We investigated the variation of the ripple spectrum on long waves. For that purpose we generated the ripple spectrum on time steps 0.5 s apart over time series of 20 min. The ripple spectrum was corrected for Doppler shift and for the Jacobian $d\sigma/d\omega$ introduced by the relation between the intrinsic frequency and the encounter frequency. At each time step a Phillips-type spectral shape was fitted to the ripple spectrum. From the variance spectra of the parameters of the fitted curve and their covariance spectra with the orbital velocity of the long waves, we could retrieve the variance and covariance spectra of the ripple (between 1 Hz and 15 Hz) with the orbital motion of the long waves.

The transfer function that relates the intrinsic spectra with the spectra of orbital motion of long waves, has a modulus of the same order of magnitude as the transfer function determined from scatterometer measurements. The phase of the transfer function has a 90% confidence interval between 160 and 200° relative to the crest of long waves.

REFERENCES

Alpers, W., and L. Jones (1978): The modulation of the radar backscattering cross section by long ocean waves. *12th International Conference on Remote Sensing of Environment* Manila.

Evans, O. D., and O. H. Shemdin (1980): An investigation of the modulation of capillary and short gravity waves in the open ocean. *J. Geophys. Res.* **85**, 5019–5024.

Garrett, C., and J. Smith (1976): On the interaction between long and short surface waves. *J. Phys. Oceanogr.* **6**, 925–930.

Hasselmann, K. (1971): On the mass and momentum transfer between short gravity waves and larger-scale motion. *J. Fluid Mech.* **50**, 189–205.

Jenkins, G. M., and D. G. Watts (1968): *Spectral Analysis and Its Application*, Holden–Day, San Francisco.

Keller, W. C., and J. W. Wright (1975): Microwave scattering and the straining of wind generated gravity waves. *Radio Sci.* **10**, 139–147.

Kitaigorodskii, S. A. (1981): The statistical characteristics of wind generated short gravity waves. *Spaceborne Synthetic Aperture Radar for Oceanography* (R. C. Beal, P. S. DeLeonibus, and I. Katz, eds.), Johns Hopkins Press, Baltimore.

Kitaigorodskii, S. A., V. P. Krasitskii, and M. M. Zaslavskii (1975): On Phillips' theory of equilibrium range in the spectra of wind-generated gravity waves. *J. Phys. Oceanogr.* **5**, 410–420.

Longuet-Higgins, M. S. (1969): A nonlinear mechanism for the generation of sea waves. *Proc. Roy. Soc. London Ser. A* **311**, 371–389.

Mitsuyasu, H., and I. Honda (1975): The high frequency spectrum of wind generated waves. *Rep. Res. Inst. Appl. Mech. Kyushu Univ.* **22**, No. 71.

Monaldo, F. M., and R. S. Kasevich (1980): Wave–wave interaction study using fine time series optical spectra. Applied Physics Laboratory, Johns Hopkins University, SIR 80U-016.

Phillips, O. M., and M. L. Banner (1974): Wave breaking in the presence of wind drift and swell. *J. Fluid Mech.* **66**, 625–640.

Reece, A. (1978): Modulation of short waves by long waves. *Boundary-Layer Meteorol.* **13**, 203–214.

Stolte, S. (1979): Ein Modell des kurzwelligen Seegangs im spektralen Frequenzbereich von 0.8 Hz bis 5.0 Hz. *Forschungsanstalt der Bundeswehr für Wasserchall- und Geophysik, Bericht.*

Wright, J. W., W. J. Plant, W. C. Keller, and W. L. Jones (1980): Ocean wave-radar modulation transfer functions from the West Coast Experiment. *J. Geophys. Res.* **85**, 4957–4966.

Wu, J. (1979): Distribution and steepness of ripples on carrier waves. *J. Phys. Oceanogr.* **9**, 1014–1021.

DISCUSSION

PLANT: Let me be sure I understand you. For your short-wave spectral form $\gamma/f^n - n$ is larger in the troughs of long waves. That is, short waves are of larger amplitude in long wave troughs?

ROSENTHAL: Yes. We find average values of n of 4.5 with values of 6.3 near the crest and 3.3 near the trough.

5

THE EFFECTS OF SURFACTANT ON CERTAIN AIR–SEA INTERACTION PHENOMENA

H. MITSUYASU AND T. HONDA

ABSTRACT. A comprehensive laboratory study has been made on the following typical phenomena at the air–sea interface both for ordinary tap water and for water containing a soluble surfactant ($NaC_{12}H_{25}SO_4$): (1) generation of wind waves; (2) wind shear stress and wind setup; and (3) growth of regular waves by the wind. The addition of the surfactant to the water shows a large suppression of the wind-generated waves, and its effect increases with increasing concentration of the surfactant within the range of our experiment. When the wind waves attenuate partially, the spectral density near the dominant frequency region shows similarity. For the maximum concentration used ($2.6 \times 10^{-2}\%$), wind waves are almost completely suppressed up to a wind speed $U_{10} \cong 15$ m/s, where U_{10} is the wind speed at height $z = 10$ m. For $U_{10} > 19$ m/s, wind waves are generated which are similar to those on tap water. A large decrease of the wind shear stress is observed when the wind waves are suppressed almost completely by the surfactant. An empirical relation for the drag coefficient has been obtained for this case, which is slightly different from that by Van Dorn (1953). An empirical relation for the drag coefficient covering a very wide range of wind speed (U_{10}:8–35 m/s) has also been obtained for ordinary tap water. The surface slope has been measured and related to the friction velocity of the wind. It is shown that the same relation holds for tap water and for water containing surfactant, if the friction velocities measured for the respective waters are used in the relation. The measured growth rate of the fundamental frequency component of regular waves on tap water is greater, by a factor of 2, than Miles' (1959) growth rate, and a little greater than Snyder and colleagues' (1981) growth rate. The growth rate of the regular waves is greatly reduced by the addition of surfactant. However, the relation between the growth rate and the friction velocity of the wind is little affected by the surfactant, because the friction velocity is diminished by the presence of surfactant.

1. INTRODUCTION

Many studies have been published on the effects of surfactants on air–sea interaction phenomena. The problems treated in these studies may be classified as follows: (1) changes of wind shear stress, water surface roughness, and wind setup (Keulegan, 1951; Van Dorn, 1953; Fitzgerald, 1963; Barger et al., 1970), (2) damping of short water waves (Lamb, 1932; Levich, 1962; Dorrestein, 1951; Goodrich, 1962; Davies and Vose, 1965; Lombardini, 1978), (3)

H. MITSUYASU AND T. HONDA ● Research Institute for Applied Mechanics, Kyushu University, Kasuga, 816, Japan.

95

suppression of wind-generated waves (Miles, 1967; Gottifredi and Jameson, 1968; Craik, 1968; Hino et al., 1969; Smith and Craik, 1971; Scott, 1972; Hühnerfuss et al., 1981a, b). However, these problems are not independent but mutually related. For example, the addition of surfactant to the water greatly suppresses the generation of wind waves, the wind structure changes corresponding to the reduction of form drag of the waves, and wind setup changes accordingly.

One purpose of the present study was to clarify the effects of surfactant on the following phenomena studied in parallel: (1) generation of wind waves, (2) wind shear stress on the water surface, and (3) wind setup.

Another purpose was to clarify the growth mechanism of water waves by the wind. An experimental study of the growth of monochromatic water waves is a useful approach to clarify one aspect of the growth mechanism of wind waves, because we are free of the difficult problem of nonlinear wave–wave interactions among spectral components (Hasselmann, 1968). Even in this approach, the problem is not as simple as we expect, since when the wind blows over the mechanically generated waves, short wind waves are generated superimposed on the mechanically generated waves. This introduces complicated problems of the interaction between long and short water waves (Phillips, 1963; Mitsuyasu, 1966; Longuet-Higgins, 1969; Hasselmann, 1971; Phillips and Banner, 1974; Garrett and Smith, 1976; Valenzuela and Wright, 1976). The use of surfactant simplifies the phenomena, because wind waves do not develop on the regular waves under the action of the wind. In the present study, the growth of mechanically generated waves by the wind is measured for water containing surfactant as well as for clean water (ordinary tap water).

2. EXPERIMENTS

2.1. Equipment and Procedure

Our design consists of three similar but independent experiments for studying (1) the generation of wind waves, (2) the growth of regular waves by the wind, and (3) the wind shear stress and wind setup.

All experiments were made in a wind-wave flume 0.8 m high, 0.6 m wide, and with a usual test section 15 m long.

In the first experiment, a beach for absorbing wave energy and a centrifugal fan for sucking air through the flume were situated downwind of the test section (Fig. 1, right-hand side). In the second experiment, the arrangements were the same as for the first one, except that a flap-type wave generator at the upwind side was used for generating regular oscillatory waves. The period and height of the waves were varied. Water depth in the flume was kept at 0.335 m in this experiment. The elevation of the water surface $\eta(t)$ was measured simultaneously at 11 stations by using resistance-type wave gages. Four mean wind speeds in the flume—U_r: 5, 7.5, 10, 12.5 m/s— were used, and vertical wind profiles over the water surface were measured with a Pitot-static tube at fetches X of 1.85, 3.03, 4.26, 5.49, 6.72, 9.15, 10.62, 12.09 m. In the second experiment, particularly in the experiment using a surfactant, mean wind speed $U_r = 11.4$ m/s was used

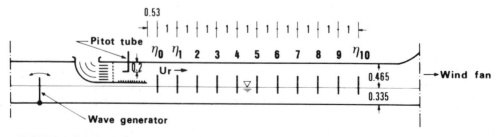

FIGURE 1. Schematic diagram of wind-wave flume arranged for the first and second experiments (units in meters).

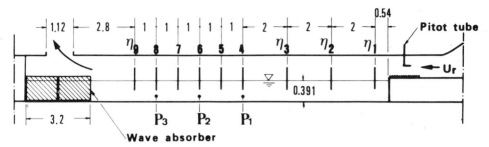

FIGURE 2. Schematic diagram of wind-wave flume arranged for the third experiment (units in meters).

instead of 12.5 m/s, because the water surface was covered by a large amount of foam of the surfactant at $U_r = 12.5$ m/s.

Wind shear stress and wind setup were measured in the first experiment, but the maximum wind speed in this arrangement (Fig. 1) was limited to $U_r = 12.5$ m/s due to the effects of turning vanes, honeycombs, and screen. In order to obtain data at higher wind speeds, the third experiment was designed especially for measuring wind shear stress and wind setup. In this experiment, water depth in the flume is kept at 0.391 m, and the centrifugal fan generates the wind from the right to the left in the flume (Fig. 2). In this arrangement, the maximum wind speed rises to $U_r \cong 22.5$ m/s, though the maximum wind speed used in this experiment was $U_r = 15$ m/s. The elevation of the water surface $\eta_{(t)}$ was measured simultaneously at nine stations. In order to check the slope of mean water surface measured with the wave gages, static pressure in the water was measured at three stations (P_1, P_2, P_3). Vertical wind profiles and static pressures over the water surface were measured at fetches X of 6.55, 8.55, 10.55 m. Five mean wind speeds (U_r 5, 7.5, 10, 12.5, 15 m/s) were used for tap water, and three (U_r 5, 7.5, 10 m/s) were used for water containing surfactant.

2.2. Surfactant

The same measurements were repeated in each experiment using clean water (ordinary tap water) and water containing surfactant (sodium lauryl sulfate, $NaC_{12}H_{25}SO_4$). In the first

TABLE I. Data for Water with and without the Surfactant $NaC_{12}H_{25}SO_4$

Expt.	g	$NaC_{12}H_{25}SO_4$ Concentration %	Kinematic viscosity v_w (stakes)	Surface tension γ (dynes/cm)
1	0	tap water	1.8	74
	70	2.3×10^{-3}	1.7	64
	150	5.0×10^{-3}	1.6	49
	220	7.3×10^{-3}	1.6	42
	790	2.6×10^{-2}	1.6	28
2	0	tap water	1.8	75
	790	2.6×10^{-2}	1.8	27
3	0	tap water	1.6	70
	790	2.6×10^{-2}	1.6	27

[a]It took several days to complete experiments 2 and 3. The values of v_w and γ changed slightly each day. Therefore, mean values are shown here.

experiment, the concentration of surfactant was varied to test its effect on wind wave generation. The surfactant concentration and corresponding surface tension and viscosity of the water are shown in Table I. In the second and third experiments the surfactant concentration was fixed at $2.6 \times 10^{-2}\%$; at this concentration, the highest one tested, no wind waves were generated.

2.3. Analysis of Data of Water Surface Elevation

Data of water surface elevation $\eta(t)$ recorded on a magnetic tape were digitized at a sampling frequency of 100 Hz and divided into seven subsamples, each of which contained 2048 data. Mean level of the water surface at each station $\bar{\eta}$ was determined as a mean value of the sampled data relative to that of the still water surface. A frequency spectrum of waves was obtained by the FFT method and the sample mean from seven subspectra. For analysis of the data of regular waves, the sampling frequency was slightly modified such that 2048 sampled data cover an integral number of the regular waves to reduce the leakage in the spectral computation.

3. WIND-GENERATED WAVES

Visual observations showed that wind waves attenuate with increasing surfactant concentration and that they attenuate almost completely at a concentration greater than $10^{-2}\%$.

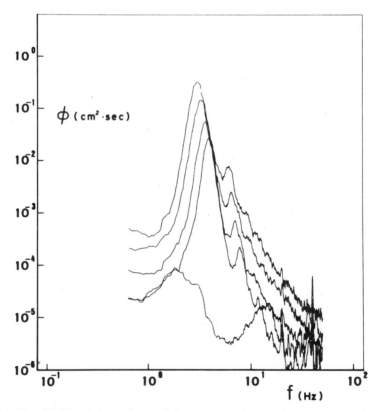

FIGURE 3. The effect of surfactant concentration on the growth of the wave spectrum. Concentration: 0, 2.3×10^{-3}, 5.0×10^{-3}, 7.3×10^{-3}, $2.6 \times 10^{-2}\%$ (from top to bottom); $U_r = 7.5\,\text{m/s}$; $X = 6\,\text{m}$.

However, such a damping effect of the surfactant was limited to wind speeds below $U_r = 12$ m/s. Wind waves were generated intermittently near $U_r = 12$ m/s. Beyond that wind speed, wind waves were generated similarly to those on tap water. These observations are similar to those reported by Keulegan (1951) and Hino et al. (1969). At a lower surfactant concentration, a curious phenomenon was observed: small waves generated by the wind at $U_r \lesssim 10$ m/s attenuate more with an increase of the wind speed. In other words, the wave-damping effect of the surfactant seemed to be greatest at $U_r \approx 10$ m/s.

Figure 3 shows the gradual change of the frequency spectrum of wind waves for the case $U_r = 7.5$ m/s, $X = 6$ m when the surfactant concentration was varied. The uppermost spectrum was measured on tap water (no surfactant), and the lowest spectrum was measured on water containing the maximum surfactant concentration ($2.6 \times 10^{-2}\%$). It is noted that the spectral energy of the wind waves attenuates not only in a high-frequency region $f_m < f$, but also in a low-frequency region $f < f_m$ (f_m: spectral peak frequency). The attenuation pattern of the spectrum in Fig. 3 is much different from our simple expectation that the surface film of the surfactant attenuates a spectral component of wind-generated waves and its damping effect is much larger for high-frequency components than for low-frequency components.

A qualitative explanation for this phenomenon is as follows. The surface film of the surfactant not only attenuates high-frequency components of wind waves, but also decreases the friction velocity of the wind as will be shown in Section 4. In fact, the change of the spectral form

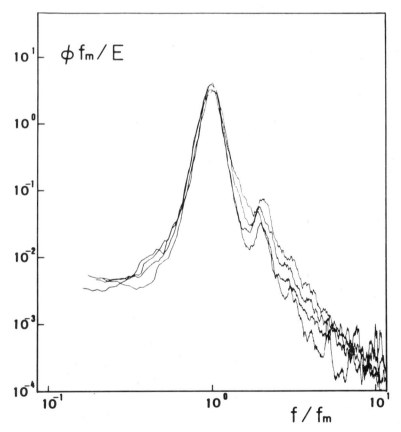

FIGURE 4. Normalized forms of the wave spectra shown in Fig. 3. The data for the concentration $2.6 \times 10^{-2}\%$ are not shown.

in Fig. 3 is quite similar to that observed when the wind speed changes. More detailed analysis of the wind wave spectrum for water containing surfactant will be done in the near future.

Figure 3 also suggests the existence of a similarity form of the wind wave spectrum in a dominant frequency region except for the lowest spectrum measured at the maximum concentration. In fact, the normalized forms of these spectra are quite similar near the dominant frequency (Fig. 4) though the normalized spectral energy at the high-frequency region decreases with increasing surfactant concentration.

Figure 5 shows the growth of the spectra of wind waves generated on water containing the maximum surfactant concentration $(2.6 \times 10^{-2}\%)$, when the wind speed was increased successively from 5 m/s to 12.5 m/s. The wind waves do not develop for $U_r \leqq 10$ m/s, but they develop rapidly into an ordinary wave spectrum at $U_r = 12.5$ m/s.

Figure 6 compares the spectra of wind waves generated by relatively high speed wind $(U_r = 12.5$ m/s) on tap water and water containing various surfactant concentrations. The spectral forms are quite similar at the different concentrations, though the spectral energy for the maximum concentration (the lowest curve) is a little lower than the others.

The relation between the attenuation ratio of the wind waves E/E_T, the surfactant concentration, and the wind speed U_r is summarized in Fig. 7. Here, E/E_T is defined as the ratio of the wave energy attenuated by the surfactant E, to the wave energy in the tap water E_T. At the concentration $5 \times 10^{-3}\%$, E/E_T has a minimum value near $U_r = 10$ m/s, which confirms the visual observations. However, the physical mechanism of this phenomenon is still not clear.

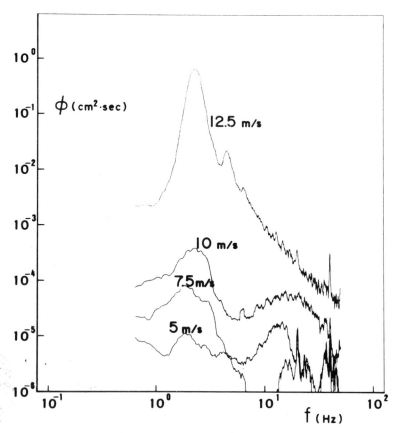

FIGURE 5. The effect of wind speed on the growth of the wave spectrum for water containing the maximum surfactant concentration used $(2.6 \times 10^{-2}\%)$.

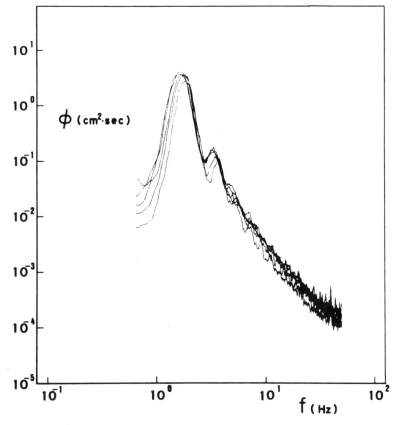

FIGURE 6. Comparison of the wave spectra generated by a high wind speed on tap water and water containing surfactant at various concentrations ($U_r = 12.5\,\text{m/s}$, $X = 10\,\text{m}$).

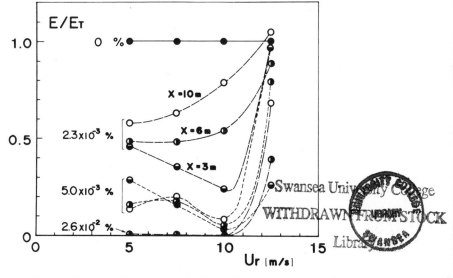

FIGURE 7. Plot of E/E_T versus U_r for various fetches and surfactant concentrations.

4. WIND SHEAR STRESS AND WIND SETUP

4.1. Vertical Wind Profile

Vertical wind profiles over the water surface were measured both for ordinary tap water and for water containing the maximum surfactant concentration ($2.6 \times 10^{-2}\%$). In the latter case, no wind waves were generated for $U_r \lesssim 10\,\text{m/s}$. As shown in Fig. 8, the wind profiles over water containing surfactant differ from those over tap water, although both profiles show a logarithmic distribution. The friction velocity of the wind $u_*[\equiv (\tau_s/\rho_a)^{1/2}$, with τ_s with the wind shear stress and ρ_a the density of the air] and the roughness parameter of the water surface Z_0 were determined from the wind profiles $U(z)$ near the water surface by applying the logarithmic distribution,

$$U(z) = (u_*/\kappa)\ln (Z/Z_0) \tag{1}$$

where κ is the Kármán constant (≈ 0.4). The equivalent wind speed at height $z = 10\,\text{m}$, U_{10}, was extrapolated from (1).

Figure 9 shows the relations between u_* and U_{10} both for tap water and for water containing surfactant. The data for tap water comprise those of our previous study (Mitsuyasu and Honda, 1974) in addition to the present data of the first and third experiments.

The relations determined from the data by a least-squares method are

$$u_* = 4.20 \times 10^{-2} U_{10}^{0.878}, \qquad U_{10} \lesssim 20\,\text{m/s} \tag{2}$$

for water containing surfactant, and

$$u_* = 1.61 \times 10^{-2} U_{10}^{1.327}, \qquad 8 \lesssim U_{10} \lesssim 35\,\text{m/s} \tag{3}$$

for tap water.

For $U_{10} \lesssim 8\,\text{m/s}$, the relations seem to be almost the same for both tap water and water containing surfactant. This means that the water surface for both waters shows aerodynamically similar properties below $U_{10} = 8\,\text{m/s}$, even though small wind waves are generated on the tap water surface. For practical purposes, (2) can be used approximately as the relation for tap water at $U_{10} \lesssim 8\,\text{m/s}$.

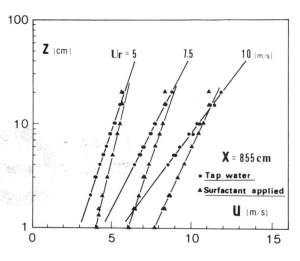

FIGURE 8. Wind profiles over the water surface. ●: data for tap water; ▲: data for water containing surfactant ($2.6 \times 10^{-2}\%$).

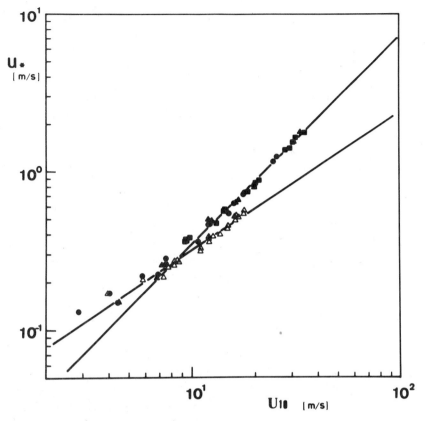

FIGURE 9. Plot of u_* versus U_{10} for tap water ($\bullet$, $\blacksquare$, $\blacktriangle$) and water containing surfactant ($\triangle$). $\bullet$: data of the first experiment; $\blacksquare$: data of the third experiment; $\blacktriangle$: data of Mitsuyasu and Honda (1974).

4.2. Drag Coefficient

By the definition of the drag coefficient,

$$C_D = \tau_s/\rho_a U_{10}^2 = (u_*/U_{10})^2 \tag{4}$$

empirical relations for the drag coefficient of the water surface can be obtained from (2) and (3):

$$C_D = 1.77 \times 10^{-3} U_{10}^{-0.244}, \qquad U_{10} \lesssim 20 \, \text{m/s} \tag{5}$$

for water containing surfactant, and

$$C_D = 2.60 \times 10^{-4} U_{10}^{0.654}, \qquad 8 < U_{10} < 35 \, \text{m/s} \tag{6}$$

for tap water. Again, (5) can be used approximately for tap water at $U_{10} \lesssim 8 \, \text{m/s}$.

In order to determine a linear form of the relation between C_D and U_{10}, the value of $C_D [\equiv (u_*/U_{10})^2]$ was obtained from individual data of u_* and corresponding U_{10} and regressed against U_{10}. The relation is given by

$$C_D = (1.29 - 0.024 U_{10}) \times 10^{-3}, \qquad U_{10} < 20 \, \text{m/s} \tag{7}$$

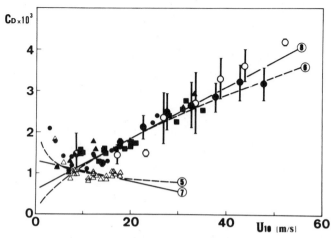

FIGURE 10. Plot of C_D versus U_{10}. ⬥: data of Garratt (1977); other symbols as in Fig. 9.

for water containing surfactant, and

$$C_D = (0.581 + 0.063 U_{10}) \times 10^{-3}, \qquad 8 < U_{10} < 35 \, \text{m/s} \tag{8}$$

for tap water.

The linear forms (7) and (8) and the power law relations (5) and (6) are shown in Fig. 10, where our present data and those of Fig. 4 of Garratt (1977) are also shown, though the latter data were not used for the regression. As shown in Fig. 10, there is little difference between the power law relation and the corresponding linear form within an application range, though the power law relation fits better to the data for water containing surfactant and the linear form fits better to the data of Garratt (1977) at high speed wind.

Many empirical relations have been reported for the drag coefficient of the water surface [e.g., linear form: Deacon and Webb (1962), Garratt (1977), Wu (1980), Large and Pond (1981); power law relation: Wu (1969), Garratt (1977)]. Our present relation for tap water [(6) or (8)] compares favorably with the relation of Large and Pond (1981):

$$C_D = 1.2 \times 10^{-3}, \qquad 4 \leqslant U_{10} < 11 \, \text{m/s} \tag{9}$$

$$C_D = (0.49 + 0.065 U_{10}) \times 10^{-3}, \qquad 11 \leqslant U_{10} \leqslant 25 \, \text{m/s} \tag{10}$$

and the other relations give slightly larger values than our present relation in a range $U_{10} < 30 \, \text{m/s}$.

Van Dorn (1953) studied the effect of a detergent on the wind shear stress in a model-yacht pond, and obtained

$$C_D = 1.04 \times 10^{-3}, \qquad U_{10} \lesssim 14 \, \text{m/s} \tag{11}$$

from his data measured by applying a detergent to the water.

Our present relation (5) or (7) gives a value of C_D not so far from that given by (11) but shows a different trend: C_D given by our relation decreases with increasing wind speed.

4.3. Wind Setup

The water surface elevation $\eta(t)$ under the action of the wind is composed of

$$\eta(t) = \eta_0 + \eta_p + \eta_\tau + \eta_1(t) \tag{12}$$

where η_0 is a still-water surface, η_p and η_τ are the wind setup due respectively to the static pressure and to the wind shear stress, and $\eta_1(t)$ is a surface fluctuation due to the wind waves. Therefore, the wind setup $\zeta \equiv \eta_p + \eta_\tau$ is determined by

$$\zeta \equiv \eta_p + \eta_\tau = \bar{\eta} - \eta_0 \tag{13}$$

where $\bar{\eta}$ is a time average of the water surface elevation. Figure 11 shows the distribution of ζ under the action of the wind $U_r = 10\,\text{m/s}$ both for tap water and for water containing surfactant. It can be seen that the mean surface slope is reduced by the presence of the surfactant and that the mean surface slope is nearly constant in the x direction for both cases except for the initial part where the wind shear stress is not uniform. It should be noted that mean surface slope measured with pressure gages agrees quite well with that measured with wave gages.

When the water depth d is much larger than the surface elevation ζ, the overall momentum

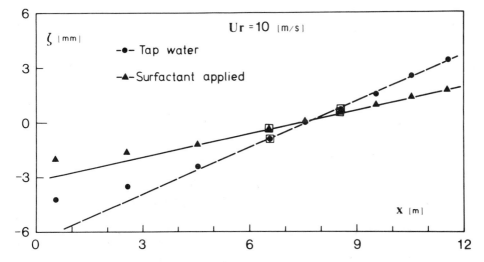

FIGURE 11. Sample data of the wind setup for tap water and water containing surfactant ($2.6 \times 10^{-2}\%$). $\square$: data obtained by pressure gages.

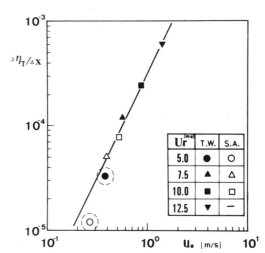

FIGURE 12. Plot of the water surface slope $\Delta\eta_\tau/\Delta x$ versus u_*. The data for $U_r = 5\,\text{m/s}$ are not used for the regression.

balance equation is

$$\frac{\partial \eta_\tau}{\partial x} = n \frac{\tau_s}{\rho_w g d} = n \frac{\rho_a u_*^2}{\rho_w g d} \tag{14}$$

where $n = 1 + \tau_b/\tau_s$, ρ_w is the density of the water, and τ_b is the bottom stress.

Figure 12 shows the relation between the friction velocity of the wind u_* and the water surface slope due to the wind shear stress $\Delta\eta_\tau/\Delta x$, which is obtained by subtracting the surface slope due to the static pressure gradient from the surface slope $\Delta\zeta/\Delta x$. It is interesting that the same relation

$$\frac{\Delta\eta_\tau}{\Delta x} = 1.01 \frac{\rho_a u_*^2}{\rho_w g d} \left(= 1.01 \frac{\tau_s}{\rho_w g d} \right) \tag{15}$$

is satisfied for both cases, i.e., for clean water and for water containing surfactant, if the friction velocity measured for each case is used. The results of this chapter have been described in a recent paper (Mitsuyasu and Kusaba 1984) with additional data.

5. GROWTH OF REGULAR WAVES BY THE WIND

5.1. Viscous Energy Dissipation

In order to determine the net growth rate of the regular waves by the wind, we need to correct the measured growth rate by taking the viscous energy dissipation into account. For this purpose, we first studied the attenuation of the spectral energy E of the fundamental frequency component of the regular waves propagating on tap water without the wind action. The exponential decay of the spectral energy E was observed, and the decay rate Δ was determined from the data by fitting the relation

$$E = E_0 \exp(-\Delta x) \tag{16}$$

to the data.

Similar studies were done for regular waves on water containing the maximum surfactant concentration $(2.6 \times 10^{-2}\%)$. In this case, a larger exponential decay rate was observed due to the effect of surface film of the surfactant. The data of Δ and their analysis have been described in Mitsuyasu and Honda (1982).

5.2. Growth of the Regular Waves on Tap Water by the Wind

Figure 13 shows typical records of regular waves under the action of the wind, which have been measured at fetches $X = 0, 3, 6$, and 9 m. For purposes of comparison, records of wind waves in the absence of the regular waves are shown in the upper part of each record of the regular waves. Figure 14 shows the wave spectra corresponding to the records shown in Fig. 13, though the spectra at all fetches $X = 1$–10 m are also shown. As reported previously (Mitsuyasu, 1966; Phillips and Banner, 1974), wind-generated waves coexist with the regular waves of small steepness (Figs. 13 and 14), though they are much suppressed by the regular waves of large steepness. First, we plotted the spectral energy E for the fundamental frequency component of the regular waves against the distance x in a log-linear scale, and determined an exponential growth rate α' by fitting the equation

$$E = E_0 \exp(\alpha' x) \tag{17}$$

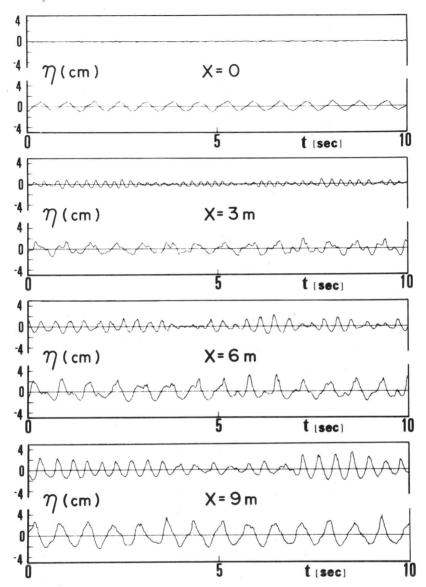

FIGURE 13. Sample records for waves at fetches 0, 3, 6, and 9 m. Upper records: wind waves in the absence
of the regular waves. Lower records: the regular waves under the action of the wind [$T = 0.7$ s, $(H/L)_0 = 0.02$,
$U_r = 10$ m/s].

to the data in an adequate range of fetches. Then we determined a net growth rate α by using the
relation

$$E = E_0 \exp(\alpha' x) = E_0 \exp(\alpha - \Delta)x \qquad (18)$$

where Δ is the decay rate determined in Section 5.1.

In the present analysis, however, we use an alternative method.* Equation (18) is rewritten

---

*The two methods give almost the same result, but the scatter of the data is a little smaller in the latter
  method.

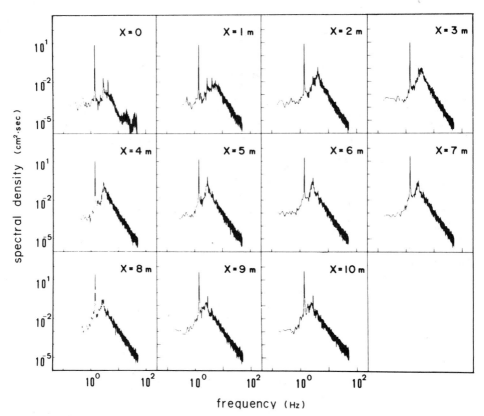

FIGURE 14. Growth of the wave spectrum for tap water [regular waves: $T = 0.7$ s, $(H/L)_0 = 0.02$, $U_r = 10$ m/s].

as

$$E/(E)_0 = \exp(\alpha x) \tag{19}$$

where

$$(E)_0 = E_0 \exp(-\Delta x) \tag{20}$$

is the wave energy affected by the viscous energy dissipation, which is determined by measuring the regular waves in the absence of the wind.[†] The data of $E/(E)_0$ were plotted against x in a log-linear scale (Fig. 15) and the net growth rate α was determined by a least-squares method.[‡]

The growth rate α thus determined was converted to the temporal growth rate β by the relation

$$\beta = \alpha C_g \tag{21}$$

where C_g is the group velocity of the regular waves, which is determined from linear theory. Next, the relation between the dimensionless growth rate β/f and the dimensionless wind speed u_*/c

[†]Although the energy dissipation of the regular waves under the action of the wind may differ slightly from that in the absence of the wind, we have no way to estimate it directly from the wave data.
[‡]Regular waves of large steepness [$(H/L)_0 = 0.05, 0.06$] tend to saturate and break at long fetches under the action of wind. The data at these fetches are eliminated in the regression.

T = 0.7 sec , U = 10 m/s

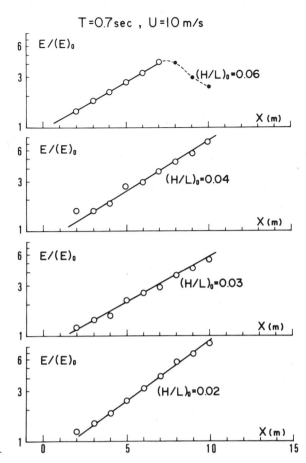

FIGURE 15. Plot of $E/(E)_0$ versus x.

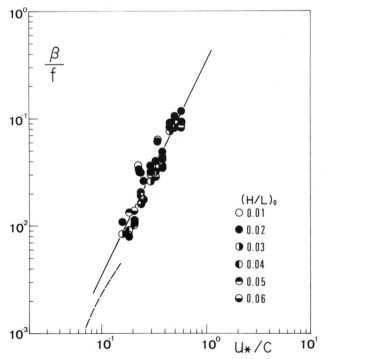

FIGURE 16. Plot of β/f versus u_*/c. The solid straight line is (22); dashed the line is (24).

was examined (Fig. 16), where f is the frequency of the regular waves and c is the corresponding phase velocity determined from linear theory. Although there is a trend that β/f decreases slightly with increasing initial wave steepness $(H/L)_0$, we neglect the effect of $(H/L)_0$ in the present analysis, because its effect is relatively small and the relation between β/f and $(H/L)_0$ is not as clear as when it is determined quantitatively. The best-fit relation for the present data for tap water is given by

$$\beta/f = 0.34(u_*/c)^2 \tag{22}$$

which is shown in Fig. 16 by the solid straight line. The growth rate β/f given by (22) is greater, by a factor of 2, than that given by Miles's (1959) inviscid theory. However, since the assumption in

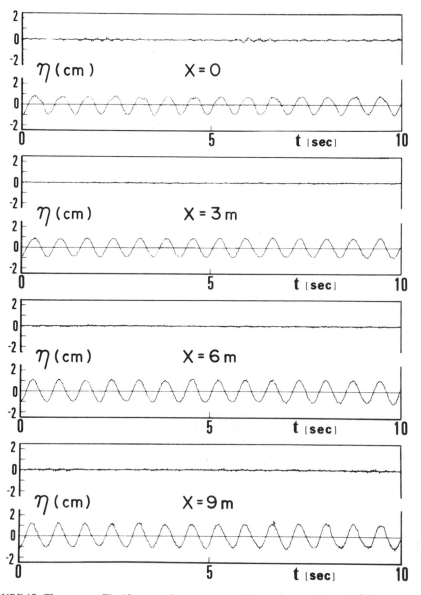

FIGURE 17. The same as Fig. 13 except that water containing surfactant ($2.6 \times 10^{-2}\%$) was studied.

his theory is not satisfied in our experimental conditions,* the disagreement between the theory and the measurement does not mean a failure of the theory.

Recently, Snyder *et al.* (1981) measured pressure fluctuations above water waves in the ocean and obtained the following parameterization for the growth rate parameter Imγ:

$$\text{Im}\gamma = (0.2 \sim 0.3)(U_5/c - 1), \qquad 1 < U_5/c < 3 \tag{23}$$

where U_5 is the wind speed 5 m above the mean water surface. Using the relation $\beta/f = 2\pi(\rho_a/\rho_w)\text{Im}\gamma$, (23) gives

$$\beta/f \sim 0.04(u_*/c - 0.043), \qquad 0.05 < u_*/c < 0.13 \tag{24}$$

where $\rho_a/\rho_w = 0.0012$ and $U_5 = 23u_*$ are assumed. The relation (24) has been included in Fig. 16. The growth rate given by (24) is fairly close to but smaller than that given by our present relation.

5.3. The Effect of Surfactant on Growth of the Regular Waves

The growth of the regular waves by the wind was measured for water containing maximum surfactant concentration (2.6×10^{-2}%). Typical records of waves measured at fetches $X = 0, 3, 6, 9$ m are shown in Fig. 17, which correspond to Fig. 13. It can be seen that wind waves do not

*Critical height does not exist in the logarithmic layer.

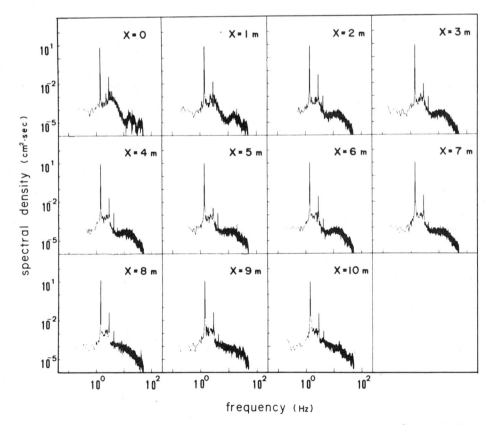

FIGURE 18. The same as Fig. 14 except that water containing surfactant (2.6×10^{-2}%) was studied.

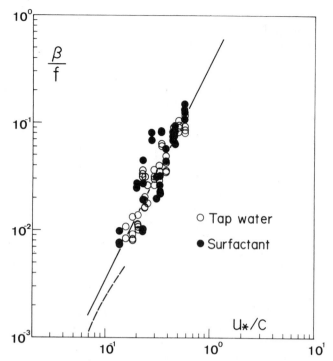

FIGURE 19. Plot of β/f versus u_*/c. The solid straight line is (22). $\bigcirc$: data for tap water; $\bullet$: data for water containing surfactant $(2.6 \times 10^{-2}\%)$. The dashed line is (24).

develop on the surface of the regular waves as well as on the still-water surface. Figure 18 shows the wave spectra corresponding to the wave records shown in Fig. 17, though the spectra at all fetches $X = 1–10$ m are also shown. Higher harmonics of the regular wave can be seen clearly due to the negligibly small energy of wind waves.

The growth rate of the fundamental component of the regular waves was determined in the same way as in the experiment for the waves on tap water. The dimensionless growth rate β/f for the regular waves on water containing surfactant ($\bullet$) was compared in Fig. 19 with that for the regular waves on tap water ($\bigcirc$). In Fig. 19 the data of β/f for both waters are shown as a function of the dimensionless friction velocity u_*/c for each case without showing the difference of the wave steepness in the data. It is interesting that the relation between β/f and u_*/c is little affected by the surfactant, if the friction velocities measured in the respective cases are used in the relation. More detailed results of this chapter have been described in a recent paper (Mitsuyasu and Honda, 1982).

6. CONCLUSIONS

The most interesting finding of this study is summarized as follows. Typical responses of the water to the wind action, such as the wind setup and the growth of regular waves, are related to the friction velocity of the wind over the water surface. The effect of the surfactant on these responses appears through the change of the friction velocity of the wind.

The following conclusions for the individual phenomena may be drawn from the present study:

1. Wind waves are greatly suppressed by the addition of surfactant to the water, but partially suppressed wind waves still hold approximately a similarity form of the spectrum near the dominant frequency.

2. Although the critical wind speed for the generation of wind waves depends on the surfactant concentration, its value for the maximum concentration used $(2.6 \times 10^{-2}\%)$ is in the range $10\,\text{m/s} < U_r < 12.5\,\text{m/s}$ which corresponds approximately to the range $15\,\text{m/s} < U_{10} < 19\,\text{m/s}$.

3. The drag coefficient C_D for water containing surfactant $(2.6 \times 10^{-2}\%)$ decreases with increasing wind speed within the range $U_{10} \lesssim 20\,\text{m/s}$. This trend is different from that in Van Dorn (1953).

4. The drag coefficient C_D for tap water increases with increasing wind speed within the range $8\,\text{m/s} < U_{10} < 35\,\text{m/s}$, and our empirical relation for C_D in this range agrees very well with that of Large and Pond (1981).

5. The surface slope due to the wind shear stress is related to the friction velocity of the wind. The relation holds for water containing surfactant as well as for tap water, if the measured friction velocities of the wind are used for the relations.

6. Growth rate of the fundamental frequency component of regular waves shows a quadratic relation to the friction velocity of the wind, but the value is larger than that predicted by Miles's theory by a factor of 2. The same relation for the growth rate seems to hold for water containing surfactant as well as for tap water, if the measured friction velocity of the wind is used for the relation.

ACKNOWLEDGMENTS. The authors are indebted to Mr. K. Marubayashi and Mr. M. Ishibashi for their assistance in the laboratory experiment, and to Miss M. Hojo and Miss H. Yokobayashi for typing the manuscript. They also express their gratitude to two reviewers for their many invaluable comments and to Dr. A. Masuda for his helpful discussions.

This study was partially supported by a Grant-in-Aid for Scientific Research, Project No. 57460043 and No. 302027, by the Ministry of Education.

REFERENCES

Barger, W. R., W. G. Garrett, E. Mollo-Christensen, and K. W. Ruggles (1970): Effects of an artificial sea slick upon the atmosphere and the ocean. *J. Appl. Meteorol.* **9**, 396–400.

Craik, A. D. D. (1968): Wind-generated waves in contaminated liquid films. *J. Fluid Mech.* **31**, 141–161.

Davies, J. T., and R. W. Vose (1965): On the damping of capillary waves by surface film. *Proc. R. Soc. London Ser. A* **286**, 218–234.

Deacon, E. L., and E. K. Webb (1962): Small-scale interactions. *The Sea*, Vol. 1 (M. N. Hill, ed.), Wiley, New York.

Dorrestein, R. (1951): General linearized theory of the effect of surface films on water ripples. *Ned. Akad. van Wetenschappen Ser. B* **54**, 350.

Fitzgerald, L. M. (1963): Wind-induced stresses on water surfaces: A wind tunnel study. *Aust. J. Phys.* **16**, 475–489.

Garratt, J. R. (1977): Review of drag coefficient over oceans and continents. *Mon Weather Rev.* **105**, 915–929.

Garrett, C., and J. Smith (1976): On the interaction between long and short surface waves. *J. Phys. Oceanogr.* **6**, 926–930.

Goodrich, F. C. (1962): On the damping of water waves by monomolecular films. *J. Phys. Chem.* **66**, 1858–1863.

Gottifredi, J. C. and G. J. Jameson (1968): The suppression of wind-generated waves by a surface film. *J. Fluid Mech.* **32**, 609–617.

Hasselmann, K. (1968): Weak-interaction theory of ocean waves. *Basic Developments in Fluid Dynamics* (M. Holt, ed.), Academic Press, New York, **2**, 117.

Hasselmann, K. (1971): On the mass and momentum transfer between short gravity waves and large-scale motions. *J. Fluid Mech.* **50**, 189–205.

Hino, M., S. Kataoka, and D. Kaneko (1969): Experiment of surface film effect on wind-wave generation. *Coastal Eng. Jpn.* **12**, 1–8.

Hühnerfuss, H. W. Alpers, W. L. Johns, P. A. Lange, and K. Richter (1981a): The damping of ocean surface waves by a monomolecular film measured by wave staffs and microwave radars. *J. Geophys. Res.* **86**, 429–438.

Hühnerfuss, H., W. Alpers, P. A. Lange, and W. Walter (1981b): Attenuation of wind waves by artificial surface films of different chemical structure. *Geophys. Res. Lett.* **8**, 1184–1186.

Keulegan, G. H. (1951): Wind tide in small closed channel. *J. Res. Natl. Bur. Stand.* **46**, 358–381.

Lamb, H. (1932): *Hydrodynamics*, Cambridge University Press, London.

Large, W. G., and S. Pond (1981): Open ocean momentum flux measurements in moderate to strong winds. *J. Phys. Oceanogr.* **11**, 324–336.

Levich, V. G. (1962): *Physico-Chemical Hydrodynamics*, Prentice–Hall, Englewood Cliffs, N. J.

Lombardini, P. P. (1978): Damping effect of monolayers on surface wave motion in liquid. *J. Colloid Interface Sci.* **65**, 387–389.

Longuet-Higgins, M. S. (1969): A nonlinear mechanism for the generation of sea waves. *Proc. R. Soc. London Ser. A* **311**, 371–389.

Miles, J. W. (1959): On the generation of surface waves by shear flows. Part 2. *J. Fluid Mech.* **6**, 568–582.

Miles, J. W. (1967): Surface-wave damping in closed basins. *Proc. R. Soc. London Ser. A* **297**, 459–475.

Mitsuyasu, H. (1966): Interaction between water waves and wind. (1). *Rep. Res. Inst. Appl. Mech. Kyushu Univ.* **14**, 67–88.

Mitsuyasu, H., and T. Honda (1974): The high frequency spectrum of wind-generated waves. *J. Oceanogr. Soc. Jpn.* **30**, 185–198.

Mitsuyasu, H. and T. Honda (1982): Wind-induced growth of water waves. *J. Fluid Mech.* **123**, 425–442.

Mitsuyasu, H. and T. Kusaba (1984): Drag coefficient over water surface under the action of strong wind. *Nat. Disaster Sci.*, **6**, 43–50.

Phillips, O. M. (1963): On the attenuation of long gravity waves by short breaking waves. *J. Fluid Mech.* **16**, 321–332.

Phillips, O. M., and M. L. Banner (1974): Wave breaking in the presence of wind drift and swell. *J. Fluid Mech.* **66**, 625–640.

Scott, J. C. (1972): The influence of surface-active contamination on the interaction of wind waves. *J. Fluid Mech.* **56**, 591–606.

Mitsuyasu, H. and T. Honda (1982): Wind-induced growth of water waves. *J. Fluid Mech.* **123**, 425–442.

Smith, F. I. P., and A. D. D. Craik (1971): Wind-generated waves in thin liquid films with soluble contaminant. *J. Fluid Mech.* **45**, 527–544.

Snyder, R. L., F. W. Dobson, J. A. Elliott, and R. B. Long (1981): Array measurements of atmospheric pressure fluctuations above surface gravity waves. *J. Fluid Mech.* **102**, 1–59.

Valenzuela, G. R., and J. W. Wright (1976): The growth of waves by modulated wind stress. *J. Geophys. Res.* **81**, 5795–5796.

Van Dorn, W. G. (1953): Wind stress on an artificial pond. *J. Mar. Res.* **12**, 249–276.

Wu, J. (1969): Wind stress and surface roughness at air–sea interface. *J. Geophys. Res.* **74**, 444–455.

Wu, J. (1980): Wind-stress coefficients over sea surface near neutral conditions—A revisit. *J. Phys. Oceanogr.* **10**, 727–740.

DISCUSSION

M. WEISSMAN: I would just to add that one must be very careful in experiments of this sort as to what are the local properties of the surface. Even though the detergent is distributed uniformly in the bulk of the fluid, a film can form that changes as a function of fetch and wind speed.

MITSUYASU: I agree with you. We have no way to check the surface conditions under the action of the wind. We sampled the surface water before and after the wave measurements and measured its viscosity and the surface tension. We didn't find the fetch dependence of the properties, but the surface tension showed time dependent values. So the surface phenomenon itself seems to be very complicated. However, our primary concern is to suppress wind waves to measure the growth rate of pure mechanically generated waves.

KERMAN: Presumably, the addition of the surfactant changes the surface tension which in turn alters the minimum phase velocity. According to the theoretical considerations of Banner, Melville, Gent, and Taylor, the occurrence of flow separation and wave breaking when u_* is of the order of c_{min} will be delayed, which agrees with your data of wave spectra and surface roughness and drag. I suggest the key parameter for your analysis is the ratio of u_* and the minimum phase velocity of gravity-capillary waves. I would further expect much of your analysis to follow a general similar form.

MITSUYASU: Thank you for your interesting comments. I would like to check your suggestions after the return to my laboratory.

LONGUET-HIGGINS: (1) This is one of the most interesting sets of experiments that I have seen. To me it appears that the sudden increases in wave and growth at certain wind speeds or surfactant concentrations indicate that the wave generation process is highly *nonlinear*; either the growth rate is very small or, if it gets going, it jumps rapidly to a finite value. This suggests that we look for a *hysteresis* of the rate of growth, as a function of wind speed, i.e., different rates for U increasing or decreasing in time.

(2) With some surfactants one can expect that the wind will blow the surface layer downwind, exposing the upwind surface. Was the surfactant uniformly distributed throughout the fluid, or was it concentrated more highly near the air–water interface?

MITSUYASU: (1) Thank you for your very interesting comment. As you mentioned, the phenomena seem to be highly nonlinear. However, the nonlinearity seems to exist not only in a generation process but also in a damping mechanism.

(2) I am not so sure about the chemical problems. But I think that the maximum concentration I tried was sufficient for generating a stable monolayer on the water surface. For the case of small concentration of the surfactant, the monolayer should be disturbed even by the low speed wind.

TRIZNA: I believe I can answer Professor Longuet-Higgins's question based on results by Garrett of NRL. He has shown that a thin monlayer on the surface is responsible for capillary wave suppression. Once this thin film is broken, the capillary waves grow as usual. Thus, concentration with depth is not an important factor.

MITSUYASU: I agree with you. However, if the concentration of the surfactant is very small, a disturbed (broken) monolayer takes a long time to recover.

LIU: We did a series of experiments to investigate the effects of an oil slick on wind waves. A thin oil film was fed continuously from the upstream end of the wave tank and the wave displacements were measured with a laser displacement gage (which I will describe briefly tomorrow). We found quite similar results in terms of wave damping. For low wind speed ($U_\infty < 10$ m/s), the presence of the oil film is very effective in damping the short gravity-capillary waves. We observed, however, high-frequency capillary waves directly generated by the wind. When intensive wave breaking takes place, the oil film is completely broken into lenses or droplets and no significant wave damping is observed.

MITSUYASU: The phenomena observed by your group seem to be almost the same as ours. Thank you.

PIERSON: For the same wind tunnel wind speed, u_* decreased when the surface-active agent was added compared to pure water. How large was this effect?

MITSUYASU: The ratio of u_* for clean water and with surfactant was approximately 1.5 for a reference wind speed of 10 m/s, though it depends weakly on wind speed.

VALENZUELA: In your results you show the largest damping of wind waves occur for $U_{10} = 15$ m/s. Do you think this is evidence that the indirect input from the atmosphere to long waves is a maximum at these conditions?

MITSUYASU: I don't think so. The largest damping of waves we observed is attributed to the damping effects in a fluid system (wind shear stress, surface film, surface current, etc.)

KATSAROS: That u_* scaling seemed to hold for both clean and contaminated surface indicates to me that it is the small-scale waves which are responsible for the drag on the surface; for constant $\bar{U}$, u_* is proportinal to Z_0—the roughness length. Is it a proper interpretation of your results that they are a commentary on the controversy of whether it is the large- or small-scale waves which are responsible for wind drag?

MITSUYASU: I think that the change of C_D with U_{10} is largely attributed to the high-frequency waves but large waves also affect the value of C_D. The scatter of our data near $U_{10} = 10$ m/s may be due to the effect of dominant waves.

6

Experimental Study of Elementary Processes in Wind-Waves Using Wind over Regular Waves

Yoshiaki Toba, Mitsuhiko Hatori,
Yutaka Imai, and Masayuki Tokuda

ABSTRACT. The evolution of mechanically generated regular waves under the wind has been studied using two wind-wave tanks of different dimensions. There are four distinct stages: (1) the coexistence of local wind waves with unaffected regular waves, (2) the attenuation of wind waves and the simultaneous growth of regular waves, (3) the rapid growth of regular waves after the disappearance of wind wave peak, and (4) the transition of regular waves to wind waves accompanied by a frequency shift to longer periods. In Stage 2 the existence is suggested of some strong nonlinear interactions which effectively transfer energy from the wind to regular waves under a condition where the peak frequency of the local wind waves is close to that of the second harmonics of the regular wave. In Stage 4, the modulation of the regular wave, the amplification of the modulation with its irregularities, and the mutual coalescence of waves, seem essential.

1. INTRODUCTION

The evolution of wave fields has been studied when the wind is applied to a train of regular waves in wind-wave tanks. This chapter synthesizes already published works of Hatori et al. (1981) and Imai et al. (1981) in different tanks, and presents some additional evidence of a new aspect of the processes which effectively transfer energy from the wind to waves. Also, from the results of experiments on the modulation of regular waves by the wind, we discuss how the combination of nonlinear characteristics of water waves with the particular action of the wind should be crucial in the evolution of wind waves.

2. EXPERIMENTS AND OVERALL DESCRIPTION

Two wind-wave tanks of different dimensions were used. Tank I is 20 m long, 0.6 m wide, 1.2 m high containing 0.6 m water, and tank II is 8.5 m long, 0.15 m wide, 0.7 m high containing 0.48 m water. The friction velocity of air u_*, determined from wind profiles, ranged from 0.4 to

YOSHIAKI TOBA, MITSUHIKO HATORI, YUTAKA IMAI, AND MASAYUKI TOKUDA ● Department of Geophysics, Faculty of Science, Tohoku University, Sendai, Japan. *Present address of M. H.:* Maritime Meteorological Division, Japan Meteorological Agency, Tokyo, Japan. *Present address of Y. I.:* Kokusai Kogyo Co., Ltd., Hino Technical Division, Hino, Japan. *Present address of M. T.:* Institute of Coastal Oceanography, National Research Center for Disaster Prevention, Hiratsuka, Japan.

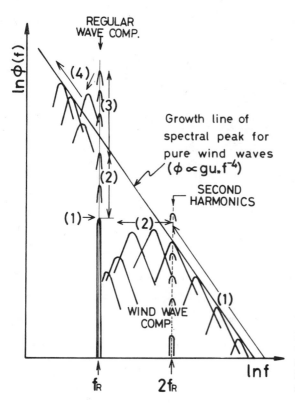

FIGURE 1. Schematic representation of the four stages in the evolution of mechanically generated waves under the influence of wind, in the form of power spectra.

0.7 m/s, the period of regular waves T from 0.40 s to 0.75 s, and the steepness δ from 0.01 to 0.05, where $\delta \equiv H/\lambda$, H being the wave height and λ the wavelength.

From spectral analyses of the wave records, it is seen that there are four distinct stages in wave field evolution (see Fig. 1). In Stage 1, wind waves are generated and grow similarly to the case of no regular waves, along the growth line for the peak point proportional to $gu_* f^{-4}$ as proposed by Toba (1973), where g is acceleration of gravity and f frequency, and there is no significant change in the regular wave component. In Stage 2, wind waves attenuate and simultaneously the mechanically generated wave grows effectively. It is noted here that Stage 2 begins approximately as the peak frequency of the wind wave component approaches twice the frequency of the regular wave f_R. In Stage 3, the regular wave grows rapidly after the disappearance of the enegy peak of the wind wave component. In Stage 4, the regular wave itself transforms to a wind wave after it first experiences overshoot and undershoot, accompanied by the frequency shift to a lower value.

Figure 2a shows an example of the evolution of actual spectra. The dotted lines indicate the case of pure wind waves without regular waves. Three of the four stages described above can be traced in Fig. 2a. At the two largest fetches, a second peak at about 3.6 Hz, representing the second harmonics of the regular wave, is also seen to develop. In Fig. 2b is seen Stage 4.

3. A NEW PHASE OF STRONG INTERACTIONS IN STAGE 2

The attenuation of the wind wave component is now expressed by the ratio

$$R_w \equiv \Phi(f_w)/\Phi(f_{w0}) \tag{1}$$

where $\Phi(f_w)$ is the spectral density of the wind wave component at its peak frequency, f_w, and

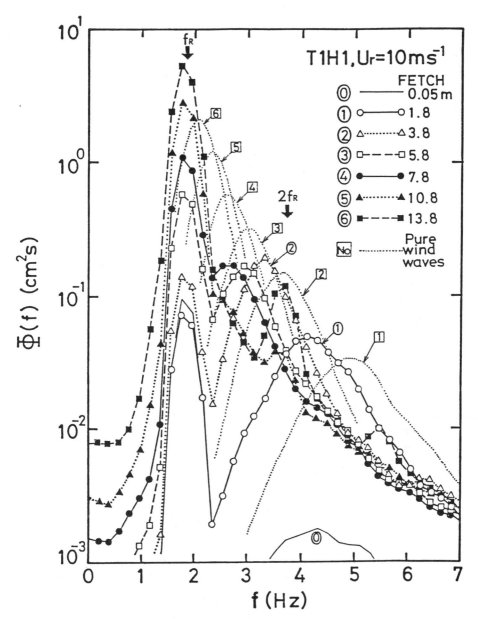

FIGURE 2. (a) An actual example of the evolution of regular waves under the influence of wind. Stages 1 to 3 corresponding to Fig. 1 can be seen. The originally generated regular waves is $T = 0.55$ s, $H = 0.55$ cm, $\delta = 0.012$, and $u_* = 0.55$ m/s.

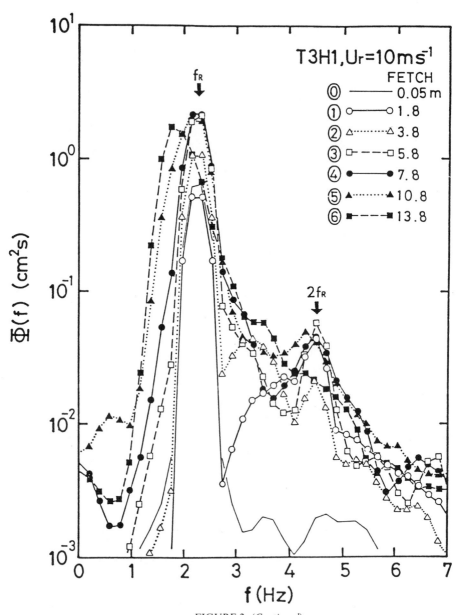

FIGURE 2. (*Continued*)

(b) Stage 4 is also seen. The regular wave is $T = 0.45$ s, $H = 1.45$ cm, $\delta = 0.046$, and $u_* = 0.55$ m/s.

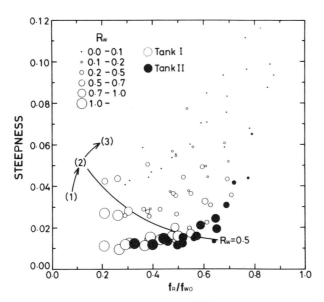

FIGURE 3. The attenuation of the wind wave component relative to pure wind waves. The definition of R_W is given in Eq. (1).

$\Phi(f_{w0})$ that of pure wind waves for the same fetch and the same wind speed (in the case of no regular waves). The value is shown in Fig. 3 according to the size of the circles plotted, as a function of the frequency ratio f_R/f_{w0} between the mechanically generated regular wave and the wind wave (standing for f_R/f_w), and of the steepness of regular waves. Stage 2 corresponds to the region near the curve of $R_w = 0.5$.

As one expression for the growth of the regular waves, we now take the momentum retention rate G_R, namely that part of momentum which is retained as the momentum of the regular wave component M_R, to the total momentum transferred from the wind to the water, defined by

$$G_R \equiv \frac{1}{\tau}\frac{dM_R}{dt}, \qquad \tau = \rho_a u_*^2 \tag{2}$$

where τ is the wind stress, t the time, and ρ_a the density of air. In our case it has been calculated by

$$G_R = (\rho_w g \bar{H}/8\rho_a u_*^2)(\Delta H/\Delta F) \tag{3}$$

where ρ_w is the density of water, ΔH is the wave height increment during the fetch increment ΔF, and $\bar{H}$ is the mean wave height during the increment. This value is shown in Fig. 4. Stage 2 corresponds to the region near $G_R = 6\%$, which is an average value of the momentum retention rate for pure wind waves of short fetches presented by Toba (1978). It is seen from Figs. 3 and 4 that the rapid growth of regular waves corresponds well to the attenuation of the wind wave component, indicating that interactions, including the wind wave component, play a basic role in the growth of regular waves in this stage.

We have further calculated the local growth rate of regular waves β_M defined by

$$\beta_M \equiv \zeta(\rho_a/\rho_w)^{-1}(2.5u_*/c)^{-2}$$
$$k\zeta = \gamma, \qquad \Delta H/\Delta F = \gamma\bar{H} \tag{4}$$

where c is the phase speed and k the wavenumber. The relation between β_M and G_R is expressed by

$$G_R = 0.78 \, (2\pi)^2 \delta^2 \beta_M \tag{5}$$

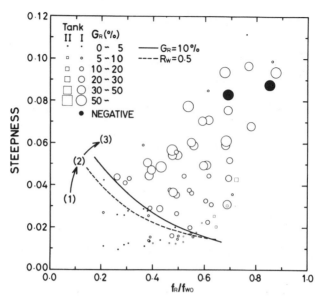

FIGURE 4. The growth of regular waves expressed by momentum retention rate G_R defined by Eq. (2).

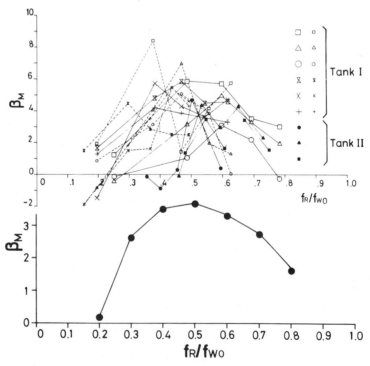

FIGURE 5. The growth rate of a regular wave defined by Eq. (4). The overall average value as a function of the frequency ratio is shown in the lower panel.

Figure 5 shows β_M as a function of the frequency ratio f_R/f_{w0}, for various wind speeds and regular waves. In every case the value has a rather sharp maximum in the vicinity of $f_R/f_{w0} = 0.5$. Though the data are scattered because of the inaccuracy due to differential values, a systematic tendency with the wind and wave conditions is not seen. The maximum values are two or three times larger than the maximum growth rate given by Miles' inviscid model. An overall average of these values as a function of the frequency ratio is shown in the lower portion. The maximum becomes flat but exists at $f_R/f_{w0} = 0.5$.

If there are no particular interactions, the value of β_M should be independent of the frequency ratio, until saturation of the regular wave growth occurs. However, the actual case is that the value is small for small frequency ratios (in Stage 1), and the value becomes large as the frequency ratio approaches 0.5, representing Stage 2. The decrease at larger frequency ratios corresponds to the transition from Stage 2 to Stage 3.

In the case of G_R shown in Fig. 4, the value was a function of steepness δ as well as the frequency ratio. This is understandable since G_R contained a factor of δ^2 from (5), and we consider that β_M is a more basic quantity from the side of waves, whereas G_R is important from the viewpoint of momentum flux.

The fact that attenuation of the wind wave component, R_w has a similar characteristic distribution with G_R, is interpreted as indicating that the existence of the local wind waves plays some significant role in effective energy transfer from the wind to regular waves.

On the other hand, the fact that the centre of Stage 2, which appears as the maximum of β_M, occurs at 0.5 in f_R/f_{w0} suggests that the second harmonic of the regular wave is closely related to the growth of regular waves in Stage 2. This might be a new aspect of strong interactions. Considering that the energy level of the wind wave component is not necessarily comparable with, but sometimes much smaller than, that of regular waves during Stage 2, it is inferred that the wind wave component acts something like a catalyst in the effective energy transfer from the wind to regular waves.

This suggestion is further examined in tank II by detailed observation of the change with fetch in individual waves of the local wind waves in relation to the phase of the regular wave. Figure 6 shows a typical sequence of the surface profiles, traced from a stroboscopic film taken with a flash interval of 0.1 s, and drawn relative to the regular wave component, which is

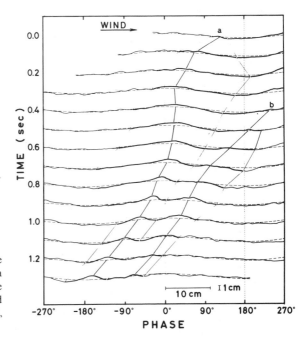

FIGURE 6. A typical sequence of the surface profile relative to the phase of a mechanically generated wave, whose wave profiles are indicated by dashed lines. $T = 0.44$s, $H = 0.42$ cm, $\delta = 0.014$, and $u_* = 0.54$ cm/s.

indicated by dashed lines. The regular wave component in this case does contain its higher harmonics. The local wind wave labeled "a" first proceeds to the crest of the regular wave from the lee side with increasing wave height, wavelength, and phase speed. When it reaches near the crest, it is trapped there for some time, and gains further energy from the wind. When the next wind wave labeled "b" approaches, an interaction seems to occur between "a" and "b": "a" begins to move to the windward side, transferring its energy to "b," and then both "a" and "b" move left, giving their energy to the regular wave. It is speculated that the forced wave of the second harmonic of the regular wave plays the role of a window, through which the energy enters the regular wave. The existence of a strong vortical layer near the wave crest, as studied by Okuda (1982), may be important in interaction of the waves.

This situation is shown more quantitatively in Fig. 7 by the phase–frequency distribution of the local wind wave energy density, reproduced from Imai et al. (1981). The shaded area represents the portion of decreasing energy, and the white area is the reverse. The path of local

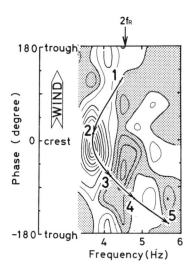

FIGURE 7. The phase-frequency distribution of the local wind wave energy density. The vertical coordinate is the phase of the regular wave, while the horizontal coordinate is the frequency of local wind waves (see the text).

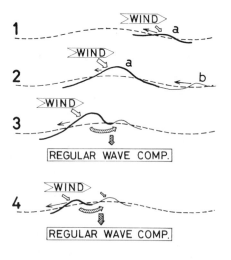

FIGURE 8. A schematic representation of the strong interactions in the presence of local wind waves.

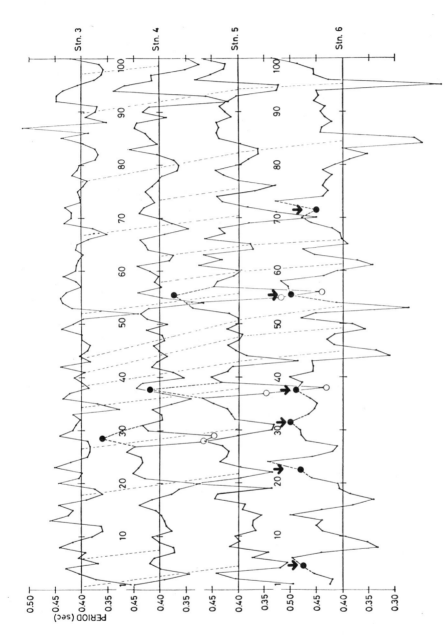

FIGURE 9. The evolution with fetch (6 m to 9 m) of the modulation of the period of regular waves. The original wave is $T = 0.40$ s, $H = 0.28$ cm, $u_* = 0.54$ cm/s. The solid circles indicate mutual coalescence.

wind waves may be traced, in an average sense, along the curve from 1 to 2, 3, 4, and 5. The change in the energy density along the curve corresponds to the above-described process of strong interactions in Stage 2, which is shown schematically in Fig. 8.

It is speculated from the existence of Stage 2 that the coexistence of longer waves and shorter wind waves, with a frequency comparable to that of the second harmonics of the longer wave, is a system which may effectively transfer energy from the wind to the longer wave. The effect of the local wind drift and the second harmonics of the longer wave may be included in these nonlinear interactions.

4. MODULATION OF A REGULAR WAVE AND ITS TRANSITION TO A WIND WAVE IN STAGE 4

The transition of the regular wave to a wind wave is characterized by irregularization and a frequency shift. These characteristics have been investigated in tank I. Figure 9 shows the evolution with fetch of the modulation of the period of individual waves in a train of regular waves of original period 0.4 s. The horizontal coordinate indicates a train of 100 individual regular waves. The dashed lines indicate that the modulated property propagates with the group velocity approximately, at the same time with considerable amplification of the modulation. The modulation is also seen in the wavelength, the wave height, and the phase speed. The period of the modulation is 5 to 6 waves, slightly larger than the value of 4 to 5 waves expected from the theory of sideband instability by Benjamin and Feir (1966) for the case of $\delta = 0.085$ in our experiment. The amplification rate is about 50% larger than the value expected from their theory. However, it should be noted that the modulation is very irregular, presumably due to the effect of the wind. The solid circles indicated by arrows show that mutual coalescence or crest pairing has occurred, after the amplification of the modulation, at the trough region.

Further detailed reports relating to this section is published elsewhere by Hatori and Toba (1983).

5. CONCLUDING REMARKS

There has been much effort to approach wind wave dynamics from the viewpoint of the nonlinear characteristics of "water waves" as reviewed by, e.g., Yuen and Lake (1980). Also, there have been some efforts directed at elucidation of the particular action of the wind. This presumably includes very local, strongly vortical wind drift, as studied by, e.g., Toba *et al.* (1975) and Okuda (1982). However, the results reported above lead us to a conclusion which seems to be very natural: a combination of the nonlinear characteristics of water waves and the particular action of the wind must be responsible for the growth process of wind waves, and further research efforts along this line of approach should be made.

ACKNOWLEDGMENTS. This study was partially supported by a Grant-in-Aid for Scientific Research by the Ministry of Education, Science and Culture, Project Nos. 942004, 254115, and 374020. Valuable discussion by Drs. S. Kawai, K. Okuda, and M. Koga and laboratory assistance by Miss Yoko Inohana are very much appreciated.

REFERENCES

Benjamin, T. B., and J. E. Feir (1966): The disintegration of wave trains on deep water. Part I. Theory. *J. Fluid Mech.* **27**, 417–430.
Hatori, M., M. Tokuda, and Y. Toba (1981): Experimental study on strong interaction between regular waves and wind waves. I. *J. Oceanogr. Soc. Jpn.* **37**, 111–119.

Hatori, M., and Y. Toba (1983): Transition of mechanically generated regular waves into wind waves under the action of wind. *J. Fluid Mech.* **130**, 397–409.

Imai, Y., M. Hatori, M. Tokuda, and Y. Toba (1981): Experimental study on strong interaction between regular waves and wind waves. II. *Tohoku Geophys. J.* **28**, 87–103.

Okuda, K. (1982): Internal flow structure of short wind waves. Part I. On the internal vorticity structure. *J. Oceanogr. Soc. J.* **38**, 28–42.

Toba, Y. (1973): Local balance in the air–sea boundary processes III. On the spectrum of wind waves. *J. Oceanogr. Soc. J.* **29**, 209–220.

Toba, Y. (1978): Stochastic form of the growth of wind waves in a single-parameter representation with physical implications. *J. Phys. Oceanogr.* **8**, 494–507.

Toba, Y., M. Tokuda, K. Okuda and S. Kawai (1975): Forced convection accompanying wind waves. *J. Oceanogr. Soc. Jpn.* **31**, 192–198.

Yuen, H. C., and B. M. Lake (1980): Instabilities of waves on deep water. *Anniu. Rev. Fluid Mech.* **12**, 303–334.

DISCUSSION

Su: I should like to praise your very interesting and important experimental results of random wind wave and regular mechanical wave interactions. From the results you showed and similar experiments in our facility, I may be able to provide a simple physical explanation of the transition from Stage 3 to Stage 4; this transition occurring about $H/\lambda \simeq 0.06$ (or $ak \simeq 0.19$) might be due to the long-time behavior of the Benjamin–Feir instability, as exemplified in wave group/train evolution in my talk last week.

Toba: Thank you for your comments.

Liu: Would you expect similar features of energy transfer to be observed for the higher harmonics, i.e., $f_R/f_w = 0.25, 0.125$, etc?

Toba: It might be expected, but we haven't checked it yet. "The window effect" is in a stage of speculation at present. Further studies are anticipated.

7

An Experimental Study of the Statistical Properties of Wind-Generated Gravity Waves

Norden E. Huang, Steven R. Long, and Larry F. Bliven

ABSTRACT. Statistical properties of wind-generated waves are studied under laboratory conditions. The properties studied include distributions of various zero crossings; crest, trough, and wave amplitudes; local maxima and minima; group length and number of waves per group; and the joint probability distribution of amplitude and period. The results indicate that all the distribution functions are in qualitative agreement with the theoretical expressions derived by Longuet-Higgins in the late 1950s and early 1960s; however, systematic deviations from the results based on a joint Gaussian distribution are apparent. The most likely dynamical reasons for the deviations are: the nonlinear mechanism causing the unsymmetric crest and trough shape, and the breaking of waves. Both of these reasons for deviations are found to be controlled by a single internal variable, the significant slope, defined as the ratio of the rms wave height to the wavelength corresponding to the waves at the spectral peak. The significant slope is also found to be the controlling factor in determining the spectral shape and evolution. Thus, the present set of detailed observational data could be used as the base to link the dynamics and the statistical properties of the wind wave field.

1. INTRODUCTION

Wind-generated waves always form a random surface; consequently, the detailed properties of the resulting sea measures at any fixed position can only be predicted or described in terms of various statistical measures rather than specific deterministic expressions. Using the pioneering works by Rice (1944, 1945), Longuet-Higgins (1952, 1956, 1957, 1958, 1962, 1975) and Cartwright and Longuet-Higgins (1956) published a series of papers that have laid a solid theoretical foundation for the study of the statistical properties of ocean waves. These results govern most of the statistical properties of the wave field consisting primarily of independently propagating small-amplitude, linear waves having a narrowband spectrum and satisfying the Gaussian statistical processes. Restrictive as they are, the comparisons made with the observational data all show good qualitative agreement.

The possible deviations from the Gaussian processes caused by the nonlinearity of the finite-amplitude waves have also been studied theoretically by Longuet-Higgins (1963, 1980a) and Tayfun (1980). However, detailed quantitative comparisons with observational results of sufficient precision were not possible until the data of Huang and Long (1980) became available.

NORDEN E. HUANG • NASA Goddard Space Flight Center, Greenbelt, Maryland 20771.   STEVEN R. LONG • NASA Goddard Space Flight Center, Wallops Flight Center, Wallops Island, Virginia 23337.   LARRY F. BLIVEN • Oceanic Hydrodynamics, Inc., Salisbury, Maryland 21801.

These recent results show that some of the deviations of the statistical properties from the Gaussian forms are caused by nonlinearity and can be successfully parameterized by the significant slope of the wave field defined as

$$\S = (\overline{\zeta^2})^{1/2}/\lambda_0 \tag{1}$$

with ζ as the surface elevation, and λ_0 as the length of the waves with frequency at the peak of the energy spectrum. Unfortunately, previously available data are not detailed enough to allow systematic comparisons with most of the existing theoretical results. This chapter presents a summary of a series of laboratory experiments that were designed to fill this specific gap. It is hoped that the results of this study will provide a clear measure of the effects of nonlinearity on the statistical properties of the wind wave field, and that this evidence will lead to the establishment of a possible parameterization scheme for defining the statistical properties of a wave field based on some simple but crucial measures of the degree of nonlinearity. It is found that the significant slope fills this role nicely.

Before considering the experiments and results, we first introduce a spectral model used in all the calculations. A spectrum is not only a statistical measure of the wave field, it is also the building block of all the statistical expressions derived theoretically.

2. A NEW SPECTRAL MODEL—THE WALLOPS SPECTRUM

According to classical statistical theory, especially in treating Gaussian and joint-Gaussian processes, the parameters in distribution functions are related to the surface elevation correlation functions and their derivatives which, in turn, can be expressed by the various moments of the wave energy spectrum (or combinations of them). Although the nonlinear effects have been shown to cause deviations from the Gaussian assumptions, the classical theory still serves as an excellent reference for comparisons. Deviations from the Gaussian processes, in some cases, can be modeled through dynamical and kinematical calculations as demonstrated by Longuet-Higgins (1963), Tayfun (1980), and Huang et al. (1981a).

Important as the spectrum is in the statistical studies of the random ocean wave field, the existing models such as those by Pierson and Moskowitz (1964) and JONSWAP (Hasselmann et al., 1976) are not applicable for this purpose. The Pierson–Moskowitz spectrum was proposed based on a similarity principle for the fully developed sea; therefore, the bandwidth of the spectrum is fixed. The JONSWAP spectrum, although applicable to a wider range of sea states, is based on curve-fitting of empirical results; therefore, it is not a functional form of direct utility for the parameterization of the statistical properties sought here.

Recently, a new spectral model was proposed by Huang et al. (1981a) as the Wallops spectrum:

$$\varphi(n) = \frac{\beta g^2}{n^m n_0^{5-m}} \exp\left[-\frac{m}{4}\left(\frac{n_0}{n}\right)^4 \right] \tag{2}$$

with $\varphi(n)$ as the frequency spectrum, n as the frequency in radians per second, β and m as functions of the significant slope of the wave field, ie,

$$m = \left| \frac{\log(2^{1/2}\pi\S)^2}{\log 2} \right| \tag{3}$$

$$\beta = \frac{(2\pi\S)^2 m^{(m-1/4)}}{4^{(m-5/4)}} \cdot \frac{1}{\Gamma[(m-1/4)]} \tag{4}$$

To obtain a spectrum, one needs only the rms wave height and the peak frequency. These quantities can be related to such traditional parameters as frictional wind speed, fetch, etc. through empirical formulas (Huang et al., 1981b). Some comparisons of this model with normalized field spectrum data have been reported by Huang et al. (1981a). One of the strong points of the Wallops spectrum is that it satisfies the total energy content of the spectrum automatically. This would guarantee the accuracy of its representation of the energy-containing components in any wave field. To illustrate this point, an example of the spectrum model with field data collected by Houmb et al. (1974) is given in Fig. 1. The agreement between the model and the data is excellent.

Using this spectral model, one can show that the ith moment of the spectrum is a function of only two parameters—the significant slope and the rms elevation of the surface, i.e.,

$$M_i = M_0 n_0^i \left(\frac{m}{4}\right)^{i/4} \frac{\Gamma[(m - 1 - i/4)]}{\Gamma[(m - 1/4)]} \tag{5}$$

with $M_0 = \overline{\zeta^2}$. Then, the various bandwidth measures are given by

$$\nu^2 = \frac{M_0 M_2 - M_1^2}{M_0 M_2} = 1 - \frac{\Gamma^2[(m - 2/4)]}{\Gamma[(m - 3/4)]\Gamma[(m - 1/4)]} \tag{6}$$

$$\varepsilon^2 = \frac{M_0 M_4 - M_2^2}{M_0 M_4} = 1 - \frac{\Gamma^2[(m - 3/4)]}{\Gamma[(m - 5/4)]\Gamma[(m - 1/4)]} \tag{7}$$

Comparisons of the calculated values of ν and ε with observed laboratory data are given in

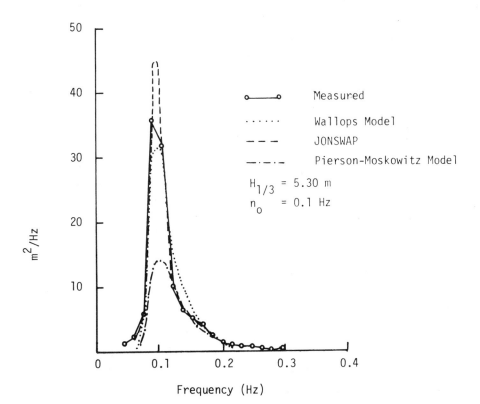

FIGURE 1. Comparison of field-observed spectrum by Houmb et al. (1974) (O—O) with the Wallops spectral mode (·····), JONSWAP spectrum (– – –), and Pierson–Moskowitz (—·—).

Figs. 2 and 3. Furthermore, the ratio v/ε is shown in Fig. 4. The corresponding values of v, ε, and v/ε computed from the Pierson–Moskowitz spectrum and the narrowband spectrum are also given in the figures. Although there are some bias, all the comparisons show excellent agreement between the Wallops spectrum and the observed results. The bias may be induced by the Doppler shift of wave frequency due to the surface drift current. This would cause an overestimate of the significant slope.

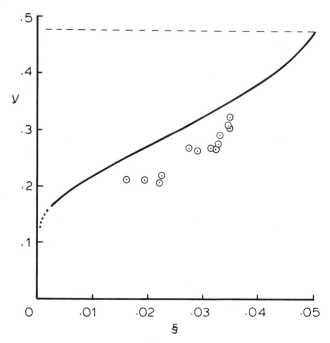

FIGURE 2. Spectral bandwidth parameter v according to the Wallops spectrum vs. observed values from the laboratory wind waves.

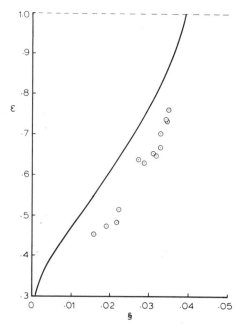

FIGURE 3. Spectral bandwidth parameter ε according to the Wallops spectrum vs. observed values from the laboratory wind waves.

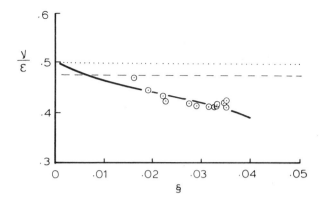

FIGURE 4. The ratio v/ε predicted by Wallops spectrum vs. observed values from the laboratory wind waves.

The fact that the spectral shape and thus the various bandwidth measures are controlled by a single nonlinear parameter, §, strongly suggests the feasibility of modeling the deviations of the statistical properties from the Gaussian processes by §. In the subsequent discussion, the Wallops spectrum will be used to calculate the theoretical distribution whenever the expression exists. Laboratory data will be used to check the theoretical results directly.

3. THE EXPERIMENT

The present series of experiments were carried out in the Wind-Wave-Current Interaction Research Facility at NASA Wallops Flight Center. The basic facility and the wave gage calibrations are described in detail by Huang and Long (1980). For the present experiments, only locally wind-generated waves were used. The wind speed covered a range of u_* from 27 to 72 cm/s. The wave data were collected at a fetch of 8.86 m.

The wave data from the capacitance gages were digitized at an interval of 10 ms and recorded on tape. At this rate, one would have about eight points to define the shape of a 13-Hz wave for the detailed statistical analyses. The presence of a meniscus on contact probes would cause a decrease in response and the introduction of error as the waves measured started to decrease in length toward the width of the meniscus. This begins to occur around the 10- to 13-Hz range, as shown by Sturm and Sorrell (1973). Even though the probe itself is capable of measuring into the hundred hertz range, the meniscus acts as a high-frequency filter. Because of this effect, we chose to stay well under the 10- to 13- Hz range where this begins to occur. We therefore digitally filtered out any response beyond 13 Hz, and confined ourselves to the cases producing a frequency spectrum with a peak frequency under 5 Hz. Filtering out data beyond 13 Hz eliminated waves with a negligible amount of energy, and waves of higher frequency which would be influenced by the meniscus. The resulting data are mostly gravity waves free of the components dominated by surface tension and data distorted by the meniscus. The filtered data were treated as the raw data. All the data processing was carried out on an HP 5451C spectral analyzer which utilized an HP 1000 series minicomputer as the central processor.

The filtered data were processed to produce the energy spectrum and the probability density function of the surface elevation as in Huang and Long (1980). In addition, new statistical quantities derived include various probability density functions of time of zero crossing for the full wave, crest, trough, crest front, crest back, trough front, and trough back; amplitude and time distributions of the crest, trough, the full waves, local maxima, local minima, and local extrema. The definitions of these terms are given in Fig. 5.

The data were also processed for the wave groupiness. The group length is defined here as the interval between the upward zero crossings of the envelope through the mean level of the crest amplitude as shown in Fig. 6. Finally, the data are processed to produce the joint probability distribution of the amplitude and period.

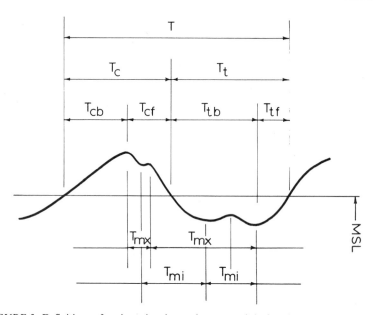

FIGURE 5. Definitions of various time intervals measured during the present experiment.

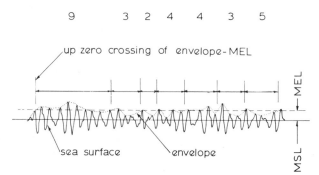

FIGURE 6. Definitions of envelope and group length used in the present experiment.

4. THE RESULTS

The detailed results of this study will be presented later in separate papers. We will present here only a selected summary of the results as follows.

4.1. Wave Amplitude Distribution

A typical amplitude distribution is given in Fig. 7; the theoretical Rayleigh distribution as given by Longuet-Higgins (1980a) is also plotted for comparison. The data, in general, agree quite well with the theoretical curve over most of the range. Once the statistical distribution of the amplitude is known, two tests can be made.

The first is on the relationship between $\overline{a^2}$ and M_0 or $\overline{\zeta^2}$. As discussed by Longuet-Higgins (1980a), the relationship between the mean squared amplitude and M_0 is

$$\overline{a^2}/2M_0 = 1 \tag{8}$$

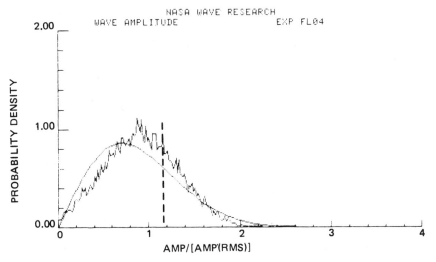

```
PRGM ZERCO REV 810420
PRGM SAVED ON TAPE ZERCO #2 FILE #2
20 APRIL 1981
<FREQ(HZ)>=    2.903
DATA TAPE= PACKED P1  FILE= 10
DELTA X(CM)=     0.019531 #DELTA X=   144
FILTER WIDTH=    1
#OBS=    11339  PEAK LOCATION(CM)=      1.0938
MEAN(CM)=      1.149630  MEDIAN(CM)=       1.124690
1-C-MO(CM)    =   -5.36507E-03  2-C-MO(CM^2)=    .24946
3-C-MO(CM^3)  =    .0130455  4-C-MO(CM^4)=    .163654
ST DV(CM)=      0.4995 SKEWNESS=      0.1047 KURTOSIS=      2.6298
1-MO-WRT-0 (CM)    =    1.14426  2-MO-WRT-0 (CM^2)=   1.55876
3-MO-WRT-0 (CM^3)  =    2.37152  4-MO-WRT-0 (CM^4)=   3.91595
AMP(RMS) [CM]=      1.2485
Breaking Amp=g/(2*(2*PI*(FREQ(HZ)))^2)=AMP(BR) [CM]= 1.47462
AMP(BR)/AMP(RMS)= 1.18111
Chance of AMP>AMP(BR)=  .269402
Curly Pi=Fraction of Energy Lost by Breaking= .179622
CNL=  134 X=    2.0963 Y=   -0.0006
CNL=  137 X=    2.1432 Y=   -0.0006
CNL=  144 X=    2.2527 Y=   -0.0006
Dist. Predicted by Longuet-Higgins,1980. JPO,V85,No C3,1519-1523.
Rayleigh Dist. Y=2X*Exp(-1*X^2)
```

FIGURE 7. A typical observed wave amplitude distribution compared with the theoretical curve given by Longuet-Higgins (1980a).

for a linear wave with narrowband spectrum. The effect of the finite amplitude will cause the ratio to increase as

$$\frac{\overline{a^2}}{2M_0} = [1 - (\bar{a}\bar{k})^2 - \tfrac{19}{2}(\bar{a}\bar{k})^4 - \tfrac{3077}{30}(\bar{a}\bar{k})^6 - \cdots]^{-1} \tag{9}$$

On the other hand, the finite-bandwidth effect will cause the ratio to decrease as

$$\frac{\overline{a^2}}{2M_0} = 1 - \left(\frac{\pi^2}{8} - \frac{1}{2}\right)v^2; \tag{10}$$

To a first order of approximation, the combined effect should be

$$\frac{a^2}{2M_0} = \frac{1 - [(\pi^2/8) - 1/2]v^2(\S)}{1 - 8\pi^2\S^2 - 608\pi^4\S^4 - 52.514\pi^6\S^6 - \cdots} \tag{11}$$

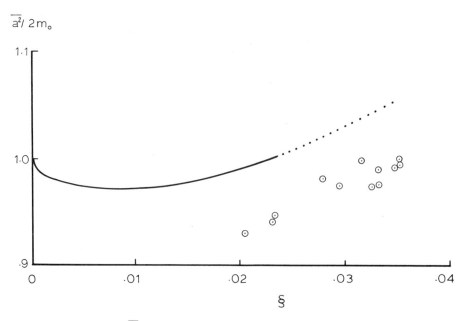

FIGURE 8. $\overline{a^2}/2M_0$ observed vs. predicted by Longuet-Higgins (1980a).

Thus, the ratio of $\overline{a^2}/2M_0$ is reduced to a function of § alone if the Wallops spectrum is used to calculate v according to (6). The functional form of (11) is plotted in Fig. 8 together with the laboratory data. The ratio is not a monotonic function of §, but the value is quite near to unity. The laboratory data follow the trend very well, but there does exist a noticeable bias. One possible explanation for this bias is due to the overestimation of § based on the apparent frequency which included the Doppler shift caused by the surface drift current as discussed by Huang (Chapter 20). But even with this bias, the approximation given in (8) could be used with a high degree of accuracy for any practical purpose. This approximation will be used in the subsequent calculations.

The second test is an indirect check on the wave breaking model proposed by Longuet-Higgins (1969). In that paper, a maximum wave amplitude a_0 for a given wave field was defined as

$$a_0 = \tfrac{1}{2}g(M_0/M_2) \tag{12}$$

Any wave with an amplitude higher than a_0 was assumed to break and reduce its amplitude to a_0. Consequently, the energy loss due to breaking could be calculated by

$$\Delta E = \rho g \int_{a^0}^{\infty} (a^2 - a_0^2)p(a)\,da \tag{13}$$

with $p(a)$ as the probability density function of the amplitude. With these arguments, there seems to be an inconsistency in (13). If the waves with amplitudes larger than a_0 would indeed break, then $p(a|a > a_0)$ should be zero. Thus, the quantity calculated from (13) will be identically zero. In Fig. 7, the value of a_0 is indicated by the dashed line. Clearly, $p(a|a >\cdot a_0)$ is not an empty set. This fact could be used to challenge the validity of the breaking wave model. However, a better explanation for this apparent paradox is the fact that the wave amplitude could be higher than a_0 momentarily just before breaking. Theoretical calculations by Longuet-Higgins and Cokelet (1976) had clearly shown this phenomenon. The accuracy of this model certainly needs an independent check.

4.2. Extrema Distribution

Extrema are either maximum or minimum values. Other than the possibility of inflection points, the extrema distribution can be compared to the specular points' distribution in the case of vertical incidence of light on a wave-covered water surface. A typical local extrema distribution is given in Fig. 9. The data showed a considerable degree of skewness and a tendency of being bimodal. The bimodal tendency could be due to the high probability of the extrema occurring either at the crest or the trough. The skewness of the distribution is not consistent with the theoretical results derived by Cartwright and Longuet-Higgins (1956) in which the density function for either the maximum or the minimum was given by

$$p(\eta) = (2\pi)^{-1/2}\left[\varepsilon e^{-(\eta^2/2)/\varepsilon^2} + (1-\varepsilon^2)^{1/2}\eta e^{-\eta^2/2}\int_{-\infty}^{\eta(1-\varepsilon^2)1/2} e^{-y^2/2}\,dy\right] \tag{14}$$

with $\eta = \zeta/M_0^{1/2}$. If the local maximum and minimum are treated separately, the portion of the maxima (or minima) occurring below (or above) the mean water level is given by

$$r = \frac{1}{2}\left[1 - \frac{M_2}{(M_0 M_4)^{1/2}}\right] = \tfrac{1}{2}[1-(1-\varepsilon^2)^{1/2}] \tag{15}$$

Figure 10 shows the comparison between the data and the theoretical results of (15). The

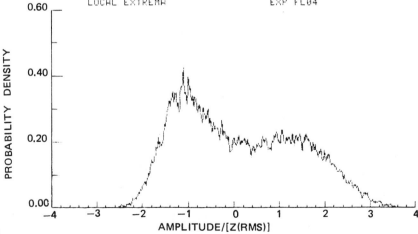

FIGURE 9. A typical observed extrema distribution.

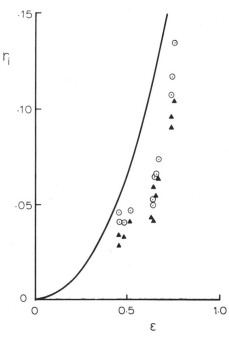

FIGURE 10. The portion of the maxima ($\odot$) [minima ($\blacktriangle$)] below [above] the mean water level that was observed vs. the predicted value by Cartwright and Longuet-Higgins (1956).

agreement in numerical values is quite good. It should be pointed out that there is a noticeable asymmetry between the maxima below zero and the minima above zero.

4.3. Wave Groupiness

The groupiness of the ocean waves is not only a critical parameter in ship operations, but also an important phenomenon serving as an indicator of the nonlinearity of the wave field. However, past attempts to study this problem are limited either to purely theoretical statistical analyses (Cartwright and Longuet-Higgins, 1956; Longuet-Higgins, 1957) or to numerical simulations (Goda, 1970; Ewing, 1973). In all these studies, waves are assumed to be linear. The group formation is produced only by the modulations between totally independent components. Therefore, the group length was shown to be related to the bandwidth of the spectrum only.

A totally different point of view for the group formation is to consider the amplitude modulation as due to the unstable subharmonic perturbations, of which a special case was proposed by Benjamin and Feir (1967). This new interpretation for the groupiness was proposed by Lake and Yuen (1977, 1978) with some support of observational data. Later theoretical analysis by Longuet-Higgins (1980b) extended the range of validity of the original Benjamin–Feir result to a larger wave steepness. Based on these results, Lake and Yuen (1978) proposed a new model for a nonlinear wind wave field in which the effects of nonlinearity dominate the effects of randomness.

To test which of these above-mentioned models is more realistic, a systematic analysis of the laboratory data and parameterization was carried out during the present study. A typical result of the distribution of the number of waves per group is given in Fig. 11. The form of the density function resembles a Rayleigh curve very well. The mean number of waver per group as a function of § is presented in Fig. 12. From this result, it is easily seen that the variation of the mean value is quite small. But the trend of the variation is obvious, i.e., the number of waves per group increases with decreasing § or decreasing bandwidth of the spectrum. This result agrees qualitatively with the classical statistical result which gives

averaged number of waves/group $= (e/2\pi)^{1/2} \, 1/v$ (16)

```
PRGM WAVES PER GROUP REV 810415
PRGM SAVED ON TAPE ZERCO #1 FILE #2
22 APRIL 1981
<FREQ(HZ)>=    2.903   ZRMS(CM)=     0.887
DATA TAPE= PACKED P1   FILE= 14
FILTER WIDTH=    1 .
#WAVES PER GROUP(RMS)=    7.0629
** Moments-WRT-0 Normalized by #Waves per Group(rms) **
1-MO-WRT-0= .859203     2-MO-WRT-0=  1
3-MO-WRT-0= 1.49894     4-MO-WRT-0= 2.79923
CNL=   29 X=    4.1060  Y=    0.0000
CNL=   30 X=    4.2475  Y=    0.0000
CNL=   31 X=    4.3891  Y=    0.0000
CNL=   32 X=    4.5307  Y=    0.0041
**** Rayleigh Dist Y=2X*Exp(-1*X^2) ****
```

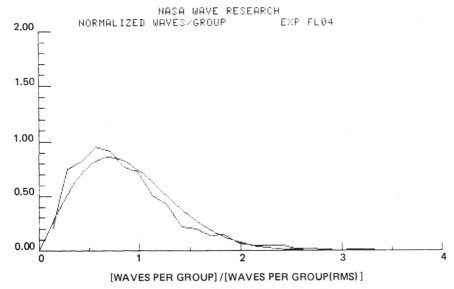

FIGURE 11. A typical observed distribution of the number of waves per group.

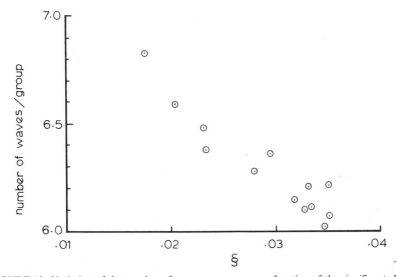

FIGURE 12. Variation of the number of waves per group as a function of the significant slope.

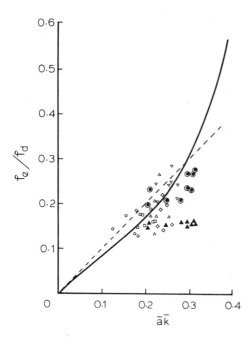

FIGURE 13. Ratio of the envelope frequency f_e, to the dominant wave frequency f_d vs. $\S\uparrow \bigcirc, \square, \triangle, \diamond, \nabla$ from Lake and Yuen (1978); $\odot$ value derived from the model of the present experiment, $\blacktriangle$ mean value. Solid curve according to Longuet-Higgins (1980b), dashed line according to Benjamin and Feir (1967).

But quantitatively, the numerical values are quite different. The laboratory data seem to bear out the old maritime folklore claiming that every seventh wave is the highest one.

To test the effects of nonlinearity, the present results are presented together with those of Lake and Yuen (1978) and Longuet-Higgins (1980b) in Fig. 13. It is clear that the model of the group length does mix well with the results of Lake and Yuen (1978). And they also follow the value predicted based on the nonlinear modulation effects as shown by Longuet-Higgins (1980b). However, there is clearly a wider range of distribution of the f_e/f_0 ratio than indicated by the model. The results could probably be better represented by a contour rather than a set of simple numerical values. The mean values are lower and their trend much flatter. This wide distribution of f_e/f_d, which is proportional to the inverse of the number of waves per group, clearly indicates that the effects of randomness are still a dominating factor in the wind-generated wave field. The true state of the sea is probably somewhere in between the nonlinearity and randomness domination. Consequently, the real ocean is probably best described by a statistical ensemble of waves with the coexistence of a certain number of discrete solitons as suggested by Longuet-Higgins in the discussion session at the NATO Conference on Turbulent Fluxes Through the Sea Surface, Waves Dynamics and Prediction, 1977 (Favre and Hasselmann, 1978, pp. 344–345). More detailed studies are needed here.

4.4. The Joint Probability Distribution of Amplitude and Period

The need for the joint probability distribution of the amplitude and period requires no more stressing here. Past studies of this problem are extremely limited in both theoretical and experimental aspects. The results of Longuet-Higgins (1975) will be used to compare with the laboratory data. A typical result of the joint distribution is given in Fig. 14. Although the form showed some similarity with the theoretical curves, the difference is also very striking. In no case did we see the low-amplitude low-frequency waves distributed symmetrically with respect to the mean values.

To investigate the influence of wave breaking, a smooth line of the breaking limit given by

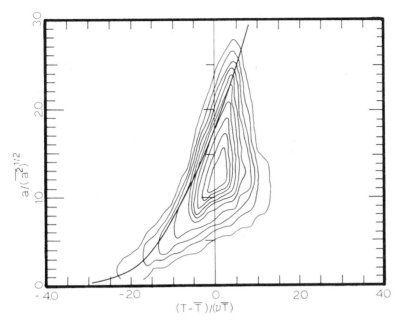

FIGURE 14. A typical joint probability distribution of amplitude and period normalized according to Longuet-Higgins (1975).

the Stokes criteria as

$$a = g/2n^2 \tag{17}$$

is plotted in Fig. 14. The contours of the higher-amplitude short waves are indeed lying parallel to the breaking limit, but with some amplitudes slightly above the limit. This again suggests the importance of the breaking processes in determining the statistical properties. The portion of the area above the breaking limit compares well with the result produced by the simple model proposed by Longuet-Higgins (1969) used in (13). More detailed studies are still under way. The results will be reported in separate papers later.

5. CONCLUSION

In this study, we found that the statistical properties of the laboratory wind waves deviate systematically from the classical results based on linear wave theory under the narrowband and Gaussian assumptions. The deviations increase as the steepness on the waves increases. Thus, the deviations can be modelled by using the significant slope of the wave field as a parameter which gives a good indication of the nonlinearity and finite bandwidth of the spectrum according to the Wallops spectrum model.

In the field, it has been shown by Huang (Chapter 20) that the significant slope cannot be more than 0.02 (or $ak \leq 0.15$) based on an energy consideration. Therefore, the classical statistical results should still work well under these conditions.

ACKNOWLEDGMENTS. We would like to thank Ms. K. Burnett of The Johns Hopkins University who helped greatly in the data collection of the experiment; and Professor C. C. Tung and Mr. Yeli Yuan of North Carolina State University for their suggestions and comments in data analysis and interpretation. LFB is supported by NASA Contract NAS6-2940. This work is part of the NASA support research technology program from the Ocean Processes Office.

REFERENCES

Benjamin, T. B., and J. E. Feir (1967): The disintegration of wave trains on deep water. Part I. Theory. *J. Fluid Mech.* **27**, 417–430.

Cartwright, D. E., and M. S. Longuet-Higgins (1956): The statistical distribution of the maxima of a random function. *Proc. R. Soc. London Ser. A* **237**, 212–232.

Ewing, J. A. (1973): Mean length of runs of high waves. *J. Geophys. Res.* **78**, 1933–1936.

Favre, A., and K. Hasselmann (eds.) (1978): *Turbulent Fluxes Through the Sea Surface, Wave Dynamics, and Prediction*, Plenum Press, New York.

Goda, Y. (1970): Numerical experiments on wave statistics with spectral simulation. Rep. Port Harbor Res. Inst., Min. Transport, Yokosuka **9**.

Hasselmann, K., D. B. Ross, P. Müller, and W. Sell (1976): A parametric wave prediction model. *J. Phys. Oceanogr.* **6**, 200–228.

Houmb, O. G., B. Pedersen, and P. Steinbakke (1974): Norwegian wave climate study. *Sym. Ocean Wave Measurement and Analysis* **1**, 25–39.

Huang, N. E., and S. R. Long (1980): An experimental study of the surface elevation probability distribution and statistics of wind generated waves. *J. Fluid Mech.* **101**, 179–200.

Huang, N. E., S. R. Long, C. C. Tung, Y. Yuen, and L. F. Bliven (1981a): A unified two-parameter wave spectral model for a general sea state. *J. Fluid Mech.* **112**, 203–224.

Huang, N. E., S. R. Long, and L. F. Bliven (1981b): On the importance of the significant slope in empirical wind wave studies. *J. Phys. Oceanogr.* **11**, 509–573.

Lake, B. M., and H. C. Yuen (1977): A note on some nonlinear water-wave experiments and the comparison of data with theory. *J. Fluid Mech.* **83**, 75–81.

Lake, B. M., and H. C. Yuen (1978): A new model for nonlinear wind waves. Part 1. Physical model and experimental evidence. *J. Fluid Mech.* **88**, 33–62.

Longuet-Higgins, M. S. (1952): On the statistical distribution of the height of sea waves. *J. Mar. Res.* **11**, 245–266.

Longuet-Higgins, M. S. (1956): Statistical properties of a moving waveform. *Proc. Cambridge Philos. Soc.* **52**, 234–245.

Longuet-Higgins, M. S. (1957): The statistical analysis of a random moving surface. *Philos. Trans. R. Soc. London Ser. A* **249**, 321–387.

Longuet-Higgins, M. S. (1958): On the intervals between successive zeros of a random function. *Proc. R. Soc. London Ser. A* **246**, 99–118.

Longuet-Higgins, M. S. (1962): The statistical geometry of random surfaces. *Proc. 13th Symp. Appl. Math.* **13**, 105–143.

Longuet-Higgins, M. S. (1963): The effect of non-linearities on statistical distributions in the theory of sea waves. *J. Fluid Mech.* **17**, 459–480.

Longuet-Higgins, M. S. (1969): On wave breaking and the equilibrium spectrum of wind-generating waves. *Proc. R. Soc. London Ser. A* **310**, 151–159.

Longuet-Higgins, M. S. (1975): On the joint distribution of the periods and amplitudes of sea waves. *J. Geophys. Res.* **80**, 2688–2694.

Longuet-Higgins, M. S. (1980a): On the distribution of the heights of sea waves: Some effect of nonlinearity and finite bandwidth. *J. Geophys. Res.* **88**, 1519–1523.

Longuet-Higgins, M. S. (1980b): Modulation of the amplitude of steep wind waves. *J. Fluid Mech.* **99**, 705–713.

Longuet-Higgins, M. S., and E. D. Cokelet (1976): The deformation of steep surface waves on water. I. A numerical method of computation. *Proc. R. Soc. London Ser. A* **350**, 1–26.

Pierson, W. J., and L. Moskowitz (1964): A proposed spectral form for fully developed wind sea based on the similarity theory of S. A. Kitaigorodskii. *J. Geophys. Res.* **69**, 5181–5190.

Rice, S. O. (1944): The mathematical analysis of random noise. *Bell Syst. Tech. J.* **23**, 282–332.

Rice, S. O. (1945): The mathematical analysis of random noise. *Bell Syst. Tech. J* **24**, 46–156.

Sturm, G. V., and F. Y. Sorrell (1973): Optical wave measurement technique and experimental comparison with wave height probes. *Appl. Opt.* **12**, 1228–1233.

Tayfun, M. A. (1980b): Narrow-band nonlinear sea waves. *J. Geophys. Res.* **85**, 1548–1552.

DISCUSSION

Mɪᴛsᴜʏᴀsᴜ: I think this is one of the most comprehensive works done on the statistical properties of wind waves. I would like to make one comment on the spectral width parameter ε. The value of ε depends critically on the high-frequency cutoff, because ε contains the fourth moment of the frequency spectrum. I recommend you to mention f/f_p when you show the value of ε. If the spectra have a similarity form, the value of ε for a fixed value of f/f_p gives a stable value.

HUANG: Thank you for your compliments. The data have been filtered digitally with the high-frequency cutoff at 13 Hz to eliminate any surface tension dominated wave. The value of f/f_p is a variable. As for the similarity form of the spectrum, I personally cannot accept this concept. Similarity in spectral form implies an identical spectrum bandwidth for all wave conditions, and this is just not realistic. We did not see any similarity forms in our spectra either. The proposed Wallops spectral model does not depend on similarity but on the steepness of the waves measured by the significant slope.

HARRIS: (1) Your observations were made at rather short-scaled fetches. This condition is best approached in nature for high waves under the high wind speed of a hurricane. A comparison of your statistics with hurricane statistics may be meaningful. Has it been tried?

(2) Were wave reflections or tank oscillations considered? Measured? What efforts were made to eliminate reflections and tank oscillations or their effects on the data?

HUANG: No. We do not have the field data. But I agree with you that our case will be similar to hurricane cases. The reflection within the tank has been observed visually only. Since the tank is an open-ended looped channel, we do not think reflection would be a serious problem. To minimize any possibility of reflection, beaches made of packing material and parallel sloping plates were installed at the downwind end.

SU: (1) From the significant wave slope $\S = 0.02$, one finds the corresponding $ak = 0.15$ as shown in your estimate for field data. The proper measure of wave slope for strong nonlinear dynamics is different from ak, but rather, estimated based on the mean wave components slightly higher than the peak frequency of the spectrum. This latter slope will be higher than the value you showed. Thus, the effects of strong nonlinear dynamics are really more significant than $ak = 0.15$.

(2) Our experiments on nonlinear evolution of wave trains of mechanically generated waves show that the height probability density, for example, varies considerably at different stages of evolution; from a density for sine waves to a density for near-Gaussian waves. Thus, the interpretation of the experimental results for very steep random waves would be quite complicated.

HUANG: The significant slope defined for the wind wave field is one of many possible ones for measuring the steepness. But we think our definition using the energy-containing waves is a reasonable choice. The conversion of ak to $\S$ is not a trivial problem. The result shown is only an approximation. The probability distribution of paddle waves is of course different from the wind waves. I do not think a comparison of paddle waves with random wind waves is meaningful, especially with the paddle waves measured very close to the generator.

PIERSON: (1) M_4 does not exist in the PM spectrum unless a high-frequency cutoff is assumed. Digitization noise can contaminate spectral elements at high frequency and make moments highly variable.

(2) The PM spectrum especially for high winds is rare over the ocean because such fetch and duration rarely exist. Actual spectra over the ocean rarely have the PM form for many reasons.

(3) How do these effects affect the generalization of your results to field conditions?

HUANG: I agree totally with you on the difficulties one may have measuring wind waves in the field. The PM spectrum is indeed hard to realize. I think the main difficulty of the PM spectrum is the fact that the spectral form is determined by the high-frequency components. Since noise is a problem at the high-frequency range, the results would be unstable. The Wallops spectrum model puts more emphasis on the energy-containing range of the spectrum. Since these waves are large, they are more important in determining the overall statistical properties of the wave field. I do not see any difficulties in the generalization of our results to the field conditions in principle. The elimination of noise is an important practical problem that should be addressed.

LONGUET-HIGGINS: (1) I am glad that the author recognized that his data are done at very high ratios of wave speed to wind speed, so they cannot be regarded as typical of field conditions. He has brought out divergencies from Gaussian statistics as a kind of artistic exaggeration.

(2) In calculating the number of waves in a group by the envelope-crossing definition, one should note that there is just one level ρ for which the number of envelope crossing $N(\rho)$ is a maximum. This is the rms

level, not the mean level. There is some advantage in choosing the rms value, since then N is unaffected by small accidental variations in ρ about this level. Hence we can eliminate one possible source of error or variability.

(3) With regard to the joint distribution of period and amplitude, while I suggested a "cocked-hat" distribution, a slightly different "oyster" distribution was derived by Dr. Cavanie, who is here sitting next to me. In the limiting case of a narrow spectrum, for which both expressions were indeed derived, the two distributions are essentially the same, but in practice the "oyster" generally fits the data better.

HUANG: I agree with you on using the rms value rather than the mean value of amplitude to define the envelope zero crossing for the group. After the discussion with you yesterday, we made a quick check on our data. We found the difference between the mean and the rms values to be very small, typically less than 5%. Therefore, if the rms value of the amplitude will give the most stable definition of waves per group, the mean value will also. In a revision of the paper, however, we will change our definition to use the rms value. I am sorry that we were not aware of the French results by Dr. Cavanie. During my discussion with Professor Mitsuyasu a few days ago, he mentioned the work by Dr. Cavanie but he could not remember the exact reference. I am most glad to meet Dr. Cavanie to discuss our results with him.

CAVANIE: The spectral width parameter, ε, appears both in experimental work and in theoretical models. This may appear disquieting since the fourth moment of the wave spectrum should be infinite. But, in fact, zero upcrossing analysis entails an implicit low-pass filtering of the data, giving finite values of the fourth moment, and values of ε less than one. Since at the present time we do not control this filtering effect, it is logical to adjust the value of ε appearing in theoretical models as a function of observational results.

HUANG: I do not see why (in principle) the fourth moment of the wave spectrum should be infinite, unless, of course, some model is used in which the fourth moment becomes infinite by the analytic form. The existence or the nonexistence of a finite value of the fourth moment has been brought up by Professor Pierson earlier. To be frank with you, this is the first time I have seriously thought about this problem. Firstly, because I have not used the PM spectrum model. The Wallops spectrm always gives a finite value of ε, as do the laboratory data. And they agree with each other quite well. Secondly, I take the physical meaning of the fourth moment as the mean-squared acceleration of the water particles. An infinite fourth moment implies unbounded acceleration that certainly does not make sense. I personally think the problem of noise in the field data as suggested by Professor Pierson is the source of all these difficulties. It may be that for the future applications, a uniform high cutoff frequency should be adopted so that we can compare results more sensibly from the field measurements. Such a cutoff could be defined as, say, 3 Hz. Since all the theoretical results on statistical properties are expressed in terms of higher moments of the spectrum, this uniform definition is a crucial issue that needs to be resolved.

M. WEISSMAN: What probe was used for the wave measurement? Where does it start to fall off in response?

HUANG: Capacitance probes were used. The frequency response starts to roll off when the meniscus width (asymptotic size 6 mm) becomes comparable to the waves to be measured. However, if one were to oscillate the probe wire itself in a rigid fashion in and out the water simulating a wave of infinite length, the frequency response is linear up to 100 Hz. The limiting effect comes from the fluid, not the electronics.

LIU: The statistics of the local maxima and minima depends on the frequency response of the capacitance gage, i.e., the higher the frequency response, the more local maxima and minima will be observed in the wave profiles. What is the frequency cutoff of your capacitance gage?

HUANG: The cutoff of the probe is hard to define; it is a gradual roll-off due to meniscus effects, etc. as shown by Sturm and Sorrell (1973). But the data have been filtered with a cutoff frequency at 13 Hz.

8

ON FINITE-DEPTH WIND-WAVE GENERATION AND DISSIPATION

C. E. KNOWLES

ABSTRACT. The results from the finite-depth transient wave growth and dissipation event discussed in this paper demonstrate the sensitivity of the finite–depth parameter, $k_p h$, in the study of shallow-water wave generation and show that wind-wave spectra and their associated scaled parameters differ most from their deep-water counterparts when $k_p h \leqslant 0.75$. Even a small Doppler shift was shown to cause very rapid changes in the spectrum (and its associated spectral parameters). Energy overshoot was found to be important in the growth of the wave spectrum. There were, for three frequency components, two significant exponential growth periods each followed by energy overshoot which, though not as pronounced as the exponential growth, was consistent with all the transient-event spectral-parameter data and therefore assumed to be real. The second and predominant overshoot was associated with a significant shift in the forward face and peak of the spectrum to lower frequencies and a reduction in the peak spectral density in the manner suggested by Phillips (1977). This overshoot, furthermore, occurred when $k_p h \approx 0.67$ [where the resonant response rate has been calculated by Hasselmann and Hasselmann (1980) to be greater by a factor of more than 3 than the deep-water rate], so was probably the result of resonant interactions. The spectral peak moved back toward higher frequencies during a period when wind speed was nearly constant, $h < 92$ cm and $k_p h \leqslant 0.75$. Initially, this movement (accompanied by a general decrease in wave-height variance) was apparently the result of enhanced resonant interactions; as h dropped below 85 cm, however, the movement and variance decrease was more likely the result of greatly increased dissipation by percolation in the coarse sand near shore and bottom friction at midshelf. Finally, the repeated evidence of enhanced resonant interactions suggests that new criteria more applicable to finite-depth wave generation be established for determining the importance of these interactions.

1. INTRODUCTION

The effect of finite depth on wind-wave spectra recently has received a great deal of attention. Of particular relevance to the present study, which describes a transient wave growth and dissipation event in a shallow estuary, are two theoretical papers (Herterich and Hasselmann, 1980; Hasselmann and Hasselmann, 1980) that relate the nonlinear energy transfer rate between waves to the nondimensional parameter $k_p h$; k_p is the wavenumber associated with the spectral-peak frequency (f_m) by the dispersion relation

$$\omega^2 = gk \tanh(kh) \tag{1}$$

where ω is the angular frequency ($= 2\pi f$), f the frequency in Hz, and h the water depth. Both

C. E. KNOWLES ● Department of Marine, Earth, and Atmospheric Sciences, North Carolina State University, Raleigh, North Carolina 27695.

papers are extensions of previous work for deep-water wave conditions concerning the role of nonlinear energy transfer by resonant wave–wave interactions in the evolution of a wave field (Sell and Hasselmann, 1972, and Masuda, 1980—the SHM model; and Longuet-Higgins, 1976, and Fox, 1976—the LHF model), and both showed that for $k_p h > 0 \sim 1$ there was little difference in the transfer rate from the deep-water conditions, but for $k_p h < 0 \sim 1$ the rate was strongly dependent on $k_p h$. They differ, however, in the applicability of the results for very shallow water when $k_p h < 0.7$.

Herterich and Hasselmann's (1980) extension for finite-depth conditions of the LHF model [which made use of a very narrow-spectrum approximation that Masuda (1980) argued was too narrow for ordinary spectra] was invalid in the range $0.3 < k_p h < 0.7$ for very shallow water because of a discontinuity in the interaction coefficient which contained the depth dependence.

The SHM model provided numerical solutions to the complete Boltzmann integral and had a transfer function that was asymmetric about the spectral-peak frequency f_m (i.e., positive for $f \leqslant f_m$ and negative for $f > f_m$). According to the model, spectral energy is transferred to the steep side of the peak and causes the shift of the peak of a growing wind-wave spectrum toward lower frequencies. In extending the SHM model, Hasselmann and Hasselmann (1980) avoided the complications of the coefficient discontinuity and produced results more applicable to shallow water by providing exact calculations for the nonlinear transfer rate. For $k_p h < 0 \sim 1$ the rate increased rapidly and smoothly from a factor of 1 for $k_p h \approx 1$ to near 10 at $k_p h \approx 0.5$. This author (Knowles, 1982), from finite-depth field data, established trends in spectral parameter plots that obviously were related to depth (particularly for $k_p h < 0 \sim 1$) and that seem to support Hasselmann and Hasselmann's (1980) results; there was a marked departure of the spectral data from the deep-water power-law relations for $k_p h \leqslant 0.7$.

According to linear wave theory, when $k_p h < 3.14$ deep-water wave conditions no longer exist and the generated wave spectrum is a function of depth. As stated above, this author (Knowles, 1982) found trends in finite-depth scaled spectral parameter data to suggest that, while waves generated under conditions where $k_p h < 3.14$ technically are influenced by depth, the dependence on depth does not become critical unit $k_p h \to 0 \sim 1$. Therefore, one objective of the present paper will be to use the transient wave growth event to better establish the critical limit on $k_p h$ that clearly restricts our use of the well known deep-water power-law relations in finite-depth water.

In addition to the sensitivity of $k_p h$ in finite-depth wave generation, other wave-growth criteria established for deep-water conditions also are evaluated in the present study. Phillips (1977) demonstrated that the rapid growth of spectral components during the exponential phase of wave growth is predominantly the result of direct energy input from the wind rather than wave–wave interactions when nondimensional fetch* $\tilde{x} < 10^4$ (where $\tilde{x} = Xg/U^2$, X is the dimensional fetch, g the acceleration of gravity, and U the wind speed at 10 m). The progression of the steep forward face of the spectrum toward lower frequencies is then the consequence of these lower frequencies undergoing transition to an exponential rate of growth. The transition time (T) required to reach exponential growth is given by

$$T = (2\pi f \mu)^{-1} \tag{2}$$

where μ is a dimensionless wind-wave coupling coefficient related to the Doppler parameter c/u_*, and where u_* is the wind frictional velocity and c the wave celerity, i.e.,

$$\mu \sim 0.05(u_*/c)^2 \tag{3}$$

for $c/u_* < 10$.

Phillips (1977) also suggested for deep-water conditions that when $c/u_* \to 20$, resonant

---

*Note that the X used in this study is scaled with wind speed at 10 m rather than by frictional velocity u_*. If u_* is used, $\tilde{x} < 10^7$ [the criterion actually given by Phillips (1977)].

interactions among sets of waves are very important and the oscillatory behavior in these interactions can be attributed to the phenomenon of overshoot, a surprising feature of growing seas under steady winds as fetch or wind duration increases, discussed first for frequency spectra by Barnett and Sutherland (1968). The overshoot effect occurs during wave generation in deep water at the end of the exponential growth phase when the spectral peak loses energy to lower frequencies; and when $c/u_* \to o \sim 20$, the direct input from the wind is unable to overcome the loss. The maximum spectral density of a particular frequency component is attained soon after the spectral peak has moved to lower frequency, but the component continues to lose energy by resonant nonlinear interactions; ultimately, the energy density of the component settles down to its saturation value, which according to Barnett and Sutherland (1968) is about 50% of its peak value.

Therefore, one additional question for finite-depth wave growth will be the degree to which deep-water, direct-wind input, versus resonant-interaction criteria (described for deep water conditions, by Phillips), apply when depth decreases. If, as the studies cited earlier suggest, the finite-depth nonlinear resonant interactions differ from those for deep water when $k_p h < o \sim 1$ then the Phillips criteria will be invalid in that range.

2. RESULTS

The transient wave growth and dissipation event spectral data discussed in this chapter were scaled according to Kitaigorodskii (1962) and Hasselmann et al. (1973) to form the nondimensional parameters of simple fetch, $\tilde{x}_s = gX/U^2$ (where X is the radial distance from the wave gage to the basin boundary), peak frequency, $v = f_m U/g$, and surface displacement variance, $\varepsilon = Eg^2/U^4$ [where $E = \int F(f)df$, $F(f)$ is the spectral density and df the frequency interval in Hz]. Table I summarizes the power-law relations for deep water and finite depth [denoted by astrisks and using these scaling parameters, derived from the finite-depth data—Knowles (1982)]. Atmospheric conditions were strongly unstable for all but the first event time period.

TABLE I. Summary of Nondimensional and Dimensional Parameter Power-Law Relations for Deep and Finite-Depth Conditions

	Relations	Plot symbol	Source
$\varepsilon-\tilde{x}$	$\varepsilon = 1.6 \times 10^{-7}\,\tilde{x}$	JONSWAP	JONSWAP (Hasselmann et al., 1973)
	$\varepsilon_{ru} = 1.2 \times 10^{-7}\,\tilde{x}^{1.1}$	R_u	Liu and Ross (1980—unstable conditions)
	$\varepsilon_{ku} = 4.0 \times 10^{-9}\,\tilde{x}_s^{3/2}$	K_u	*Knowles (1982—unstable conditions)
$v-\tilde{x}$	$v = 3.50\tilde{x}^{-0.33}$	JONSWAP	JONSWAP
	$v_p = 1.78\tilde{x}^{-0.25}$	Phillips	Phillips (1977)
	$v_{ru} = 1.90\tilde{x}^{-0.27}$	R_u	Liu and Ross (1980—unstable conditions)
	$v_{ku} = 2.34\tilde{x}_s^{-0.29}$	K_u	*Knowles (1980—unstable conditions)
$\varepsilon-v$	$\varepsilon_{ru} = 1.64 \times 10^{-6}v_{ru}^{-4.07}$	R_u	Liu and Ross (1980—unstable conditions)
	$\varepsilon_{ku} = 7.70 \times 10^{-7}v_{ku}^{-4.51}$	K_u	*Knowles (1982—unstable conditions)
$E-f_m$	$E = 1.24 f_m^{-4.43}$		*Knowles (1982—unstable conditions)

2.1. Wave Site

It is recognized that the site chosen for this study is not ideal. In the first place, it does not provide an ideal fetch symmetry for all wind directions (especially in the western sector), but for the transient event discussed in this paper (where the wind direction is between 20 and 30 degrees from true north throughout) the deviation from a symmetric fetch is not severe and, most importantly, there is a consistency in the fetch over the event period that should allow for valid comparisons between the spectral densities and their associated scaled parameters.

Secondly, the wind and air temperature data were obtained from Cape Hatteras Weather Facility about 70 km to the south of the site. In an earlier detailed study (Singer and Knowles, 1975), it was demonstrated that there was good agreement between the Hatteras wind data, the more limited USCG Station wind data at Oregon Inlet (20 km southeast of the site) and the continuous-record wind data from a weather station set up by those authors at Stumpy Pt (25 km to the south of the site) for a study of Croatan and Pamlico Sounds; i.e., the study found no consistent bias in speed or direction (when differences were noted) of one station over the other. Some obvious errors were detected between stations, however, during the passage of a slow moving front (where a lag or lead of several hours between stations could exist), but in as much as these fronts did not always approach from the same direction, there also was no bias of one station over the other. While these relatively rare occurrances might have caused some scatter in the composite plots of scaled data (Knowles, 1982), it should not have biased significantly the trend of the scaled data nor the power-law relations derived from them. Furthermore, for the transient event described in this paper, the frontal passage between the site and the Hatteras weather station occurred in less than three hours and, once passed, winds maintained a nearly steady direction over the event period.

Waves were recorded digitally using a subsurface pressure gage near Fort Raleigh (Fig. 1) in SE Albemarle Sound and the pressure-fluctuation time series was depth and frequency

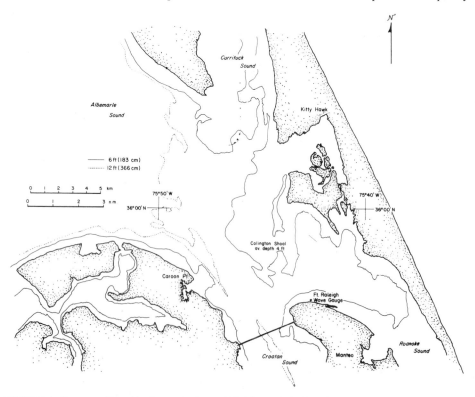

FIGURE 1. Topographic and bathymetric chart of SE Albemarle basin adjacent to Fort Raleigh wave-gage site.

compensated (using linear and empirical correction factors, Knowles, 1983) to obtained estimates of the surface spectra. The power-spectra had 34 degrees of freedom for $f \leqslant f_c$ (where $f_c = f_m + 16df$, and df is the frequency interval between spectral estimates) and 50 degrees of freedom for $f > f_c$; the 90% confidence interval factors are (1.57, 0.70) and (1.44, 0.74), respectively.

The sediment near the site consists of a 3-m-thick sand layer graded from coarse (0.5–1.0 mm) near shore to medium (0.25–0.5 mm) at midshelf and fine (< 0.25 mm) beyond the shelf break. It is expected that wave energy will be dissipated (with a damping coefficient of o ~ 10^{-3}) in the coarse sand by percolation and in the medium and fine sand by bottom friction (Shemdin et al., 1980), especially when the depth becomes very shallow.

2.2. Transient Wave Growth and Dissipation Event

The transient event discussed in this study extends over a 33-h period from 1025 EST February 16 to 1925 EST February 17, 1980, and includes a period of slowly but steadily increasing winds that peak at about 12 m/s at 15 h, decline slightly to a mean of about 10.5 m/s from 18 to 30 h, and then drop to about 8 m/s at 33 h; the wind directions and fetches are nearly the same, as can be seen graphically in Fig. 2 (which includes additional data before and after the event). The changes in $k_p h$ are shown only for the event period (denoted in the text and in this and the remaining figures by the time in hours from the onset), because this parameter could not be determined before the event (wave energies for winds from the lee-side of Roanoke Island were very low) or after the event (water depth had decreased so much—below 78 cm—that the generation of any appreciable wave energy was not possible and k_p could not be estimated). Note from Fig. 2 that $k_p h$ drops rapidly after 9 h, even while h is increasing slightly, and oscillates for the rest of the event period while the depth is in a steady decline. This behavior indicates the sensitivity of k_p (i.e., f_m) to finite-depth wave growth and dissipation mechanisms.

2.2.1. $E–f_m$ and $\varepsilon–v$

Hasselmann et al. (1976) and Liu and Ross (1980) explained that as waves developed under steady winds, the equilibrium state converged to and moved along the $\varepsilon–v$ power-law line in the

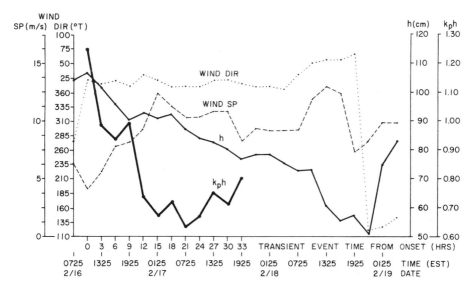

FIGURE 2. Time plot (including 33-h transient event) of wind direction and speed, water depth h, and parameter $k_p h$.

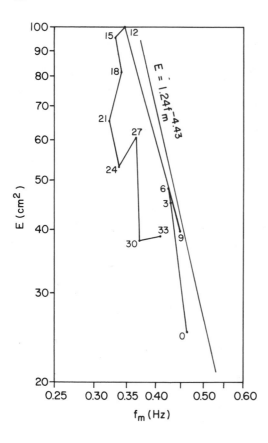

FIGURE 3. Transient-event E–f_m plot including Knowles (1982) equilibrium line.

direction of higher energies and lower frequencies with increasing fetch or duration until the spectrum was fully developed. The same trend should be evident for the dimensional E–f_m relation*; Fig. 3 shows the data from the present study transient event (for conditions of nearly constant fetch and initially for increasing wind speed) and the E–f_m power-law line established earlier (Knowles, 1982). The event data do seem to converge toward the line and then, except for the fluctuation between 6 and 12 h (which occurs even though wind speed continues to increase), move up the line to a maximum E value at 12 h. The energy then declines markedly during the next 12 h as $k_p h$ drops below 0.75 (significantly, the small decline between 12 and 15 h occurs even as the wind increases from 9.3 to 11.3 m/s and the decline continues to the 24th h even though the wind speed remains above the 12th h value). The peak frequency f_m, however, continues its shift toward lower frequencies from 12 to 15 h, shifts back slightly at 18 h, reaches its lowest value at 21 h and shifts dramatically toward higher frequency at 27 h; the large shift in f_m in the last event period (hour 33) is expected because of the relaxation in wind speed to 8 m/s.

The ε–v data are included in Fig. 4 along with the Knowles finite-depth (K_u) and Liu and Ross (R_u) deep-water power-law relations for unstable atmospheric conditions. Note the convergence to and progression along the R_u line (with the same fluctuation as the E–f_m data between 6 and 12 h) for the first 12 h of the event; then the movement is along the K_u line from 15 to 18 h and has a marked departure from it for the 21st, 24th, and 30th h. The trend in the nondimensional scaled parameter data, as expected, is consistent with the E–f_m data discussed earlier.

---

*Obviously, for finite-depth waves, the wave height variance E (and indeed the scaled ε obtained from it) are functions of depth, but as stated above, this dependence may not be critical until $k_p h < o \sim 1$ and the parameters are presented here to show trends.

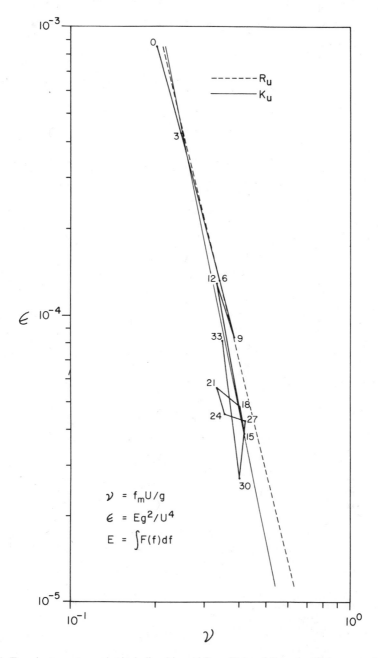

FIGURE 4. Transient-event $\varepsilon-v$ plot including Liu and Ross (R_u) and Knowles (K_u) power-relations for unstable atmospheric conditions.

2.2.2. ε and v as Functions of $\tilde{x}_s$

The same general trend and fluctuations in the ε and v data as a function of fetch $\tilde{x}_s$ are shown in the next two figures. Figure 5 includes the transient event ε and $\tilde{x}_s$ data and the R_u and K_u power-law relations. Note that the simple-fetch $\varepsilon-\tilde{x}_s$ data depart from the R_u line 9 h sooner than the $\varepsilon-v$ data do (i.e., after 3 vs. 12 h) and have the same large fluctuation between 4 and 12 h, but do tend to follow the K_u line.

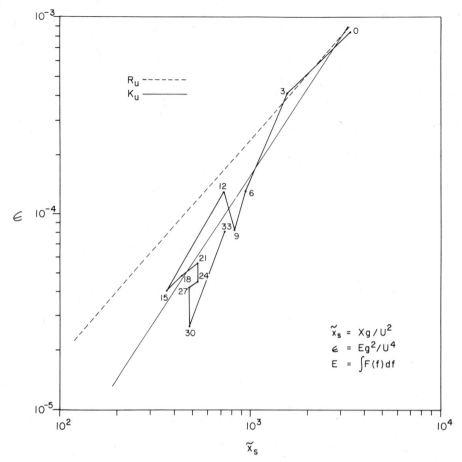

FIGURE 5. Transient-event ε–$\tilde{x}_s$ plot including Liu and Ross (R_u) and Knowles (K_u) power-relations for unstable atmospheric conditions.

Figure 6 includes the v–$\tilde{x}_s$ data and the deep-water R_u, JONSWAP and Phillips, and finite-depth K_u power-law relations. The v–$\tilde{x}_s$ data progress generally along the Ross/Phillips/Knowles lines, starting as in the ε–$\tilde{x}_s$ case near the R_u line, but, moving toward the K_u line, show a much smaller fluctuation during the 6- to 12-h time interval than in the previous cases.

The significant departure (when $k_p h \leqslant 0.75$) of the data from the deep-water power-law relations discussed above confirms trends in the spectral data noted earlier (Knowles, 1982). Figure 7 (which plots ε and v as functions of $k_p h$) provides additional evidence to support this trend; ε and v, in general, do tend to move in the expected way (except for the familiar fluctuation between 6 and 12 h) until $k_p h \to 0.67$, then show oscillations in ε and v that remain (except for the 27th h) for nearly constant wind speed below $k_p h = 0.72$.

2.2.3. Wave Growth and Energy Overshoot

During the event the fluctuation of ε in all plots between 6 and 12 h and the drop from 12 to about 18 h may be the result of energy overshoot. Figure 8 is a plot of $k_p h$ and four frequency components ($f = 0.322$, 0.352, 0.381 and 0.400 Hz) of the normalized spectral density F/F_m (where F_m is the maximum density for each component) as a function of time after the onset of the event; all components except 0.400 are at frequencies less than the frequency of the spectrum's

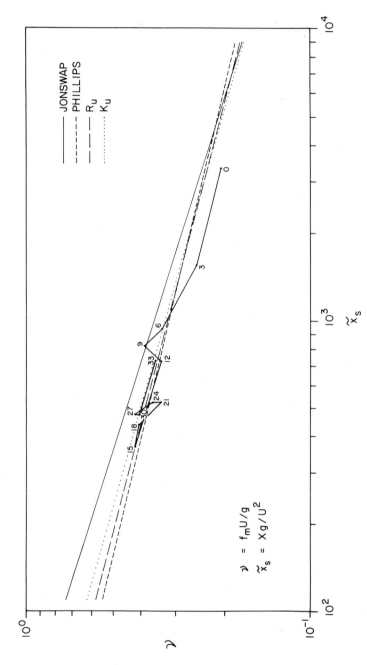

FIGURE 6. Transient-event $\nu - \tilde{x}_s$ plot including the JONSWAP, Phillips, and for unstable atmospheric conditions, Liu and Ross (R_u) and unstable (K_u) power-relations.

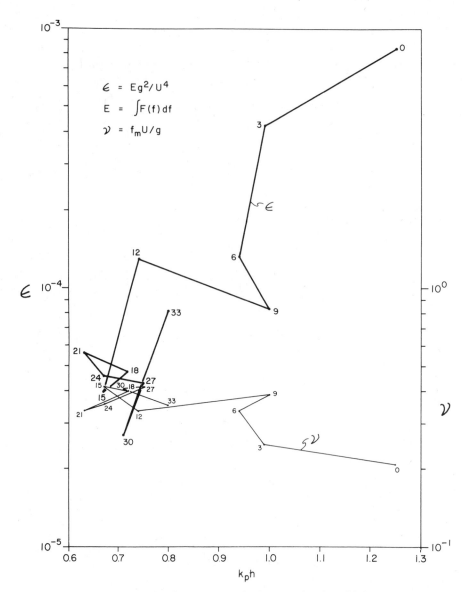

FIGURE 7. Transient-event ε and v data as a function of $k_p h$.

steep forward face (f_b) at the start of the event. Barnett and Sutherland (1968) plotted F/F_m versus normalized fetch X/L (where L was the spectral-peak wavelength), but their data were for constant wind speed and increasing fetch X. The data for the present study are for increasing winds and nearly constant fetch; and while those parameters can be made equivalent for use in plots with other nondimensional parameters by scaling the fetch X with g and wind speed, X/L is not the appropriate parameter for this transient-event plot.

The growth of F/F_m during the first 6 h for all the frequency components and between 9 and 12 h for the smallest three components is nearly exponential and significant ($> 90\%$ C.I.) as can be seen in Fig. 8. The transition time (T) required to reach this exponential-growth phase is inversely proportional to f, so from (2), the lowest-frequency component ($= 0.322\,\mathrm{hz}$) will be associated with the longest transition time ($= 1.2\,\mathrm{h}$); but this T is less than half of the sampling

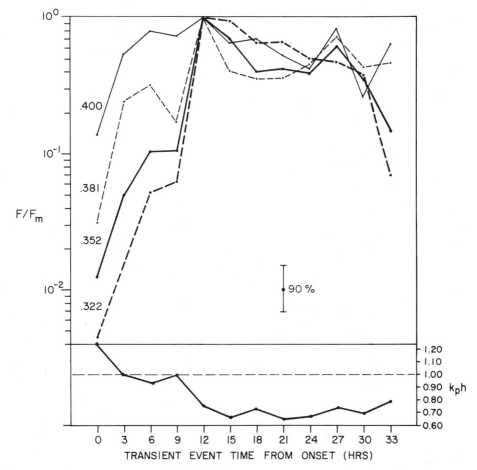

FIGURE 8. Transient-event plot of normalized spectral densities F/F_m (where F_m is the maximal density for each frequency component) for four frequency components and $k_p h$ as a function of time from onset of the event.

interval so the transition to exponential growth at the onset of the event could not be detected for any of the components.

In Fig. 8 there appear to be two distinct periods of energy overshoot (from 6 to 9 h and 12 to 15 h) but only the 0.381–component overshoots are statistically significant. The other components show indications of overshoot and, though not themselves significant, probably are real because they are consistent with the behavior evident between 6 and 15 h in all the earlier transient-event spectral-parameter plots. F/F_m peaks for all components at 12 h and then decreases to a mean for the four components of very roughly 50% of the peak value during the nearly constant-wind period between 18 and 30 h (except for a significant growth in spectal density at 27 h for all but the smallest component). Because the 0.400 component is already on the steep face at the onset of the event, its growth is not so rapid as the other three.

During the second overshoot period (which includes the end of the exponential-growth phase), the three highest-frequency components have progressed through or have reached f_m (at 12 h, $f_m = 0.352$ Hz). Note also from the $k_p h$ plot at the bottom of Fig. 8 that both overshoot periods occur while $k_p h \leqslant 1.00$ ($\leqslant 0.67$ for the second period). According to Hasselmann and Hasselmann (1980), when $k_p h \approx 0.67$ the wave–wave interaction transfer rate increases by a factor of about 3, so while it appears that the exponential-growth phase is obviously the result of direct

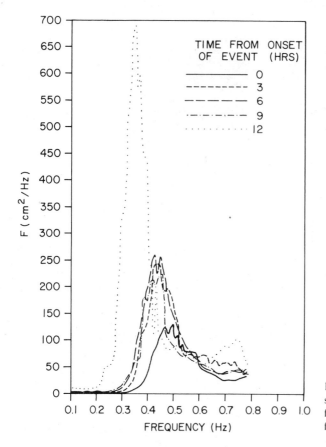

FIGURE 9. Spectral density plots showing spectral progression for the first 12 h of transient event as a function of frequency.

input from a strengthening wind, for finite-depth conditions the data suggest that later in the phase it also may be enhanced by resonant interactions.

2.2.4. Spectral Densities

All the spectra shown in this section are for shallow and intermediate water depths and changes during the transient event are evident in Figs. 9–11.

The 12-h exponential-growth phase of the transient event is shown clearly in Fig. 9; from the onset there is a continual increase in spectral density and, significantly, a steady progression of the steep forward face toward lower frequency in spite of the apparent fluctuations in f_m evident in the parameter plots shown earlier. The statistically insignificant "minipeaks" near the spectral peak make the selection of a precise f_m difficult in nearly half the event spectra. And because the spectra are quite narrow, increasing the degrees of freedom to better resolve f_m from those fluctuations in $F(f)$ also will result in an unacceptable flattening of the spectral peak. This suggests, therefore, that some of the small fluctuations in f_m (e.g., the increase from 6 to 9 h, Figs. 3 and 9) probably are not real. The fluctuations in E or ε and the first overshoot shown in the previous plots at 9 h, however, are real and are apparently the result of a drastic narrowing of the spectral peak and a loss of energy density between 6 and 9 h. Finally, from Fig. 9, note the establishment at 12 h of secondary peaks in the spectra and their general maintenance thereafter (Figs. 10 and 11) as the event progresses. These secondary peaks are at $f \approx 2.1 f_m$, or very nearly the first harmonic of f_m, which according to Thompson (1980) indicates that the waves have a non-Gaussian height distribution.

The second overshoot and maximum and near-equilibrium spectral density phases of the

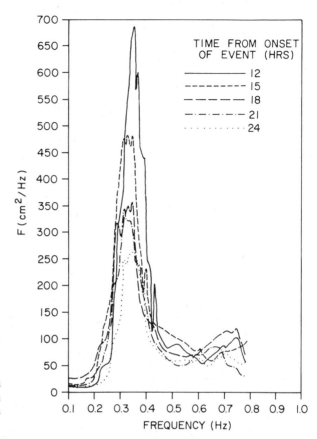

TIME FROM ONSET
OF EVENT (HRS)

——————— 12
----------- 15
— — — — 18
—·—·—·—·— 21
············· 24

FIGURE 10. Spectral density plots showing spectral progression from 12 to 24 h of transient event as a function of frequency.

transient event are shown in Fig. 10. Note in particular the sharp decrease in peak density $F_p(f_m)$ (even while the wind speed continues to increase), and the continued movement of the steep face toward lower frequencies from 12 to 15 h, and the generally continual loss of energy density at $F_p(f_m)$ (and at frequencies both lower and higher than f_m), and movement of f_m and the steep face back toward higher frequencies after 15 h (even while the wind remains constant). Significantly, and as stated earlier for the spectral parameter plots, this distinct movement of the spectrum toward higher frequencies occurs at 18 h when $k_p h < 0.75$. The fluctuations in f_m between 18 and 24 h noted earlier also may be the result of multipeak resolution and probably are not real.

The slight increase in wind speed at 27 h probably accounts for the large increase in $F_p(f_m)$ shown in Fig. 11; with the shallow water encountered during the latter half of this event, $F_p(f_m)$ was very sensitive to wind speed changes. Note that even with an increase in E, the spectral peak at 27 h continues to move toward higher frequency. The spectrum at 30 h has a very flat, narrow peak with low densities at frequencies both higher and lower than f_m, so it is not surprising that E is about 60% of the 27-h value. The significant movement of the peak toward higher frequency at 33 h obviously is the result of the relaxation in wind speed, but note from Fig. 3 that E is nearly the same at 30 and 33 h.

2.2.5. Resonant Interaction Criteria

As stated earlier, the results of the present study also were to be used to evaluate, for finite-depth conditions, Phillips's (1977) deep-water resonant wave–wave interaction criteria; especially for $k_p h < 0 \sim 1$.

All data for this study have $\tilde{x} < 10^4$, which according to Phillips suggests that resonant

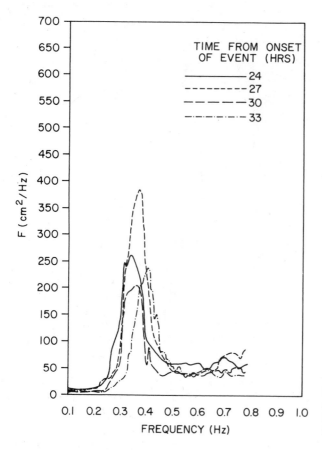

FIGURE 11. Spectral density plots showing spectral progression from 24 to 33 h of transient event as a function of frequency.

wave–wave interactions are not important during the 12-h exponential-growth phase of the transient event. This criterion is at variance, however, with the results of the previous sections and with his Doppler parameter criterion. The Doppler parameters (c/u_*) for the first and second event times (0 and 3 h) for all four components are of o $\sim$ 20 and o $\sim$ 15, respectively. According to Phillips (1977), this would mean that resonant interactions are very important in the growth process at the onset of the event and slightly important through the 3rd h; during the rest of the exponential-growth phase, $c/u_* \leqslant$ o $\sim$ 10, so resonant interactions would be unimportant. The results of the present study, however, suggest that resonant interactions, in direct conflict with the c/u_* criterion, are important later in the growth phase. Phillips's criteria parameters may be different for finite-depth than for deep-water waves either because of the effects of shallow water on c (which would affect the c/u_* criterion) or because for small $k_p h$ the nonlinear wave–wave interaction transfer rates are enhanced (which would affect the $\tilde{x}$ criterion). These discrepancies imply that a different criterion for c/u_* (which includes a depth component) or for $\tilde{x}$ (which considers resonant transfer-rate enhancement) will be needed.

3. SUMMARY AND CONCLUSIONS

The results from the finite–depth transient wave growth and dissipation event discussed in this chapter are consistent with and explain many of the trends shown earlier (Knowles, 1982). As expected, the studies demonstrate clearly that scaled wind-wave spectral parameters are affected significantly by finite depth, but generally differ from their deep-water counterparts only when $k_p h <$ o $\sim$ 1, and that even a small Doppler shift can cause very rapid changes in the spectrum (and its associated spectral parameters).

The scaled nondimensional parameter most affected by finite depth was the wave-height variance ε when plotted versus fetch $\tilde{x}_s$; the power-slope was significantly larger than the deep-water relation, probably because E is an integral quantity that for $f > f_m$ responds rapidly to any process (direct wind input or nonlinear wave breaking, etc.) that will modify that part of the spectrum. The spectral-peak frequency parameter v when plotted versus $\tilde{x}_s$ and ε was the least affected (the slopes were nearly the same as those of Phillips, and Liu and Ross, respectively); and significantly, the finite-depth ε–v data seemed to follow the deep-water line very well until $k_p h \leqslant 0.75$.

The present study also demonstrates that energy overshoot is important in the growth of the finite-depth wave spectrum. There were, for three frequency components, two significant exponential-growth periods, each followed by energy overshoot, which though not as pronounced as the exponential growth was consistent with all the transient-event spectral-parameter data and therefore assumed to be real. The first overshoot was coincident in the spectrum with a reduction in spectral density at $F_p(f_m)$ and at frequencies immediately greater than f_m with a correspondent narrowing of the spectral peak, but no real movement of the steep face of the spectrum toward lower frequency. The second and predominant overshoot was associated with a significant shift in the forward face and peak of the spectrum to lower frequencies and a reduction in $F_p(f_m)$ [i.e., in the manner suggested by Phillips (1977)]. The second overshoot, furthermore, occurred when $k_p h = 0.67$ [where the resonant response rate was calculated by Hasselmann and Hasselmann (1980) to be greater by a factor of about 3 than the deep-water rate]; so this overshoot was probably the result of resonant interactions.

After 18 h (while the wind speed remained at a nearly constant 10.2–10.8 m/s, the water depth decreased steadily from 92 to 77 cm and $k_p h \leqslant 0.75$), f_m and the steep forward face of the spectrum moved back toward higher frequencies. This movement, accompanied also by a general reduction in E from its 9th-h peak value of nearly 100 cm^2, may indicate the increasing presence of stronger interactions not related to the weak wave–wave resonance; i.e., the results of this finite-depth study suggest that while the spectral changes that occur after the energy overshoot may be initially the result of an enhancement of the resonant wave–wave interactions predicted by Hasselmann and Hasselmann (1980), as the depth drops below 84 cm, they are more likely the result of the greatly increased dissipation mechanisms of bottom percolation and bottom friction, which have damping coefficients of greater order $(3 \times 10^{-2}$ and 4×10^{-3}, respectively) than the enhanced interactions.

In conclusion, the most important result of this study is the establishment of the limits on the depth parameter $k_p h$ for use in the study of finite-depth wave generation and dissipation; for $k_p h > 0 \sim 1$, finite-depth spectral parameter data appear to agree with the established deep water power-law relations, but as $k_p h$ decreases below unity, there is a marked departure from these relations. The deviations occur first for the ε–$\tilde{x}_s$ data at $k_p h \approx 1$ but most significantly for all the other parameter data $(v$–$\tilde{x}_s, \varepsilon$–v, ε–$k_p h$, and v–$k_p h)$ at $k_p h \leqslant 0.75$.

As a final conclusion, the repeated evidence of enhanced resonant interactions suggests that new criteria more applicable to finite-depth wave generation need to be established for determining the importance of these interactions.

ACKNOWLEDGMENTS. Financial support for this study was provided by the NOAA Sea Grant Program, UNC Grant NA79AA-D-00048B, and the North Carolina Department of Administration. I wish to thank Duncan Ross for his comments and suggestions, Judy Kapraun for polishing my grammar, Jane Anderson and Brenda Batts for spending so much time typing and retyping the manuscript, and Toni Clay for assistance in the graphics.

REFERENCES

Barnett, T. P., and J. A. Sutherland (1968): A note on an overshoot effect in wind-generated waves. *J. Geophys. Res.* **73**, 6879.

Fox, M. J. H. (1976): On the nonlinear transfer of energy in the peak of a gravity-wave spectrum. II. *Proc. R. Soc. London Ser. A* **348**, 467.

Hasselmann, K., T. P. Barnett, E. Bouws, H. Carlson, N. E. Cartwright, K. Enke, J. A. Ewing, H. Gienapp, D. E. Hasselmann, P. Kruseman, A. Meerburg, P. Müller, D. J. Olbers, K. Richter, W. Sell, and H. Walden (1973): Measurements of wind-wave growth and swell decay during the Joint North Sea Wave Project (JONSWAP). *Dtsch. Hydrogr. Z. Suppl. A* **8** (12).

Hasselmann, K., D. B. Ross, P. Müller and W. Sell (1976): A parametric wave prediction model. *J. Phys. Oceanogr.* **6**, 201.

Hasselmann, S. and K. Hasselmann (1980): A symmetrical method of computing the nonlinear transfer in a gravity wave spectrum. Max-Planck-Institut für Meteorologie, Hamburg.

Herterich, K., and K. Hasselmann (1980): A similarity relation for the nonlinear energy transfer in a finite-depth gravity-wave spectrum. *J. Fluid Mech.* **97**, 215.

Kitaigorodskii, S. A. (1962): Applications of the theory of similarity to the analysis of wind-generated wave motion as a stochastic process. *Izv. Akad. Nauk SSSR Ser. Geofiz.* **1**, 105.

Kitaigorodskii, S. A., V. P. Krasitskaii and M. M. Zaslarskii (1975): O. Phillips' theory of equilibrium range in the spectra of wind-generated gravity waves. *J. Phys. Oceanogr.* **5**, 410.

Knowles, C. E. (1982): On the effects of finite-depth on wind-wave spectra. 1. A comparison with deep-water equilibrium range slope and other spectral parameters. *J. Phys. Oceanogr.* **12**, 56.

Knowles, C. E. (1983): On the estimation of surface gravity waves from subsurface pressure records for estuarine basins. *J. Estuarine, Coasts. and Shelf Sci.*, **17**, 395.

Liu, P. C. and D. B. Ross (1980): Airborne measurements of wave growth for stable and unstable atmospheres in Lake Michigan. *J. Phys. Oceanogr.* **10**, 1842.

Longuet-Higgins, M. S. (1976): On the nonlinear transfer of energy in the peak of a gravity-wave spectrum: A simplified model. *Proc. R. Soc. London Ser. A* **347**, 311.

Masuda, A. (1980): Nonlinear energy transfer between wind waves. *J. Phys. Oceanogr.* **10**, 2082.

Phillips, O. M. (1977): *The Dynamics of the Upper Ocean*, 2nd ed., Cambridge University Press, London.

Sell, W. and K. Hasselmann (1972): Computations of nonlinear energy transfer for JONSWAP and empirical wind-wave spectra. Institute of Geophysics, University of Hamburg.

Shemdin, O. H., S. V. Hsiao, H. E. Carlson, K. Hasselmann, and K. Schulze (1980): Mechanisms of wave transformation in finite-depth water. *J. Geophys. Res.* **85**, 5012.

Singer, J. J. and C. E. Knowles (1975): Hydrology and circulation patterns in the vicinity of Oregon Inlet and Roanoke Island, N.C. *UNC Sea Grant Rep.* SG-75–15, 171 pp.

Thompson, E. F. (1980): Shallow water surface wave elevation distributions. *ASCE, J. Waterw. Port Coastal Ocean Div.* **106**, 285.

9

THE EQUILIBRIUM RANGE FOR WAVES IN WATER OF FINITE DEPTH

Dinorah C. Esteva

ABSTRACT. The validity of Kitaigorodskii and colleagues' finite-depth equilibrium spectrum is investigated by comparing the proposed equation with the best-fit curve, in the least-squares sense, to observed spectra.

Two data sets consisting of simultaneous observations collected off the Corps of Engineers Field Research Facility (FRF) at Duck, North Carolina, are used. One data set was of observations from wave staffs along the FRF pier when the significant height at the end of the pier was either

TABLE I. Results from Power Law Fit to Observed and Proposed Finite-Depth Spectra

	First data	ARSLOE data		Kitaigorodskii and colleagues theoretical data	
Number of spectra	106	161		12	
Water depth range (m)	2.4–9.0	2.4–9.1	17.1–23.5	2.4–9.1	17.1–23.5
Percent of spectra with correlation 0.79 (range)	40–78	43–93	54–92	100	100
Wave steepness[a] range at end of FRF pier ($\times 10^{-2}$)	2.566–14.521	8.976–14.977			
Average power (b) range	3.08–3.67	2.31–3.55	4.10–4.46	3.48–3.81	4.20–4.42
Overall average b ($\times (-1)$)	3.43	3.11	4.25	3.60	4.33
Average coefficient (a) range	1.321–3.730	6.301–17.22	1.944–5.904	1.241–2.304	1.104–1.367
Overall average a $\times 10^{-3}$	2.616	10.67	3.891	2.024	1.118

[a]Defined as ratio of 1/2 the significant wave height to the local wavelength corresponding to the period of maximum energy.

Dinorah C. Esteva ● National Ocean Service, National Oceanic and Atmospheric Administration, Rockville, Maryland 20852.

increasing or remained relatively stationary at around 1 m or above.   The other data set consisted of observations from the staffs and from offshore Waverider buoys at the onset of a gale recorded during the Atlantic Remote Sensing Land Ocean Experiment (ARSLOE).

Measurements were made in water depths ranging from 2.4 to 23.5 m. All the staffs and one Waverider were deployed in less than 10 m of water, while the remaining five Waveriders were deployed in water deeper than 17 m.

A fit to a power law (af^b) is made from f_m, the frequency corresponding to the spectral maximum, to $2f_m$. To eliminate those spectra which are not appropriately described by a power law, only spectra with a correlation above 0.79 in the fit are considered. For those spectra and for each sensor, an average value of the power (b) and of the coefficient (a) are computed.

Table I shows results from the two data sets and stratified according to the two water depth ranges. Results from a power fit to the proposed theoretical curve are also shown.

It can be seen from Table I that: 1) the dependence of the power on water depth is in general agreement with Kitaigorodskii and colleagues' equation; 2) few spectra of observations in less than 10 m of water had a correlation above 0.79, and; 3) a dependence of the coefficient (and thus the Phillips "constant") on wave steepness as well as on water depth is indicated.

These dependences will be discussed in more detail in a paper to be published elsewhere.

II

WAVE PROPAGATION

10

THE 1978 OCEAN WAVE DYNAMICS EXPERIMENT

Optical and in Situ Measurement of the Phase Velocity of Wind Waves

G. B. IRANI, B. L. GOTWOLS, AND A. W. BJERKAAS

ABSTRACT. The Ocean Wave Dynamics Experiment (WAVDYN), an empirical investigation of ocean wind waves in deep, open water, has yielded improved evidence that independently (freely) propagating waves prevail over manifestations of their nonsinusoidal (nonlinear) waveforms. Measurements were made with an array of surface elevation transducers, a two-axis current meter, and a video-based wave-imaging system deployed at an ocean tower during September 1978. Analysis of the array and the video data yielded estimates of directional wavenumber–frequency spectra of signal variance, which were compared with the theoretical wavenumber–frequency relations for free- and bound-wave propagation under the influence of the measured currents. The locations of structure in the spectra indicate the prevalence of free-wave propagation. The video spectra exhibit characteristics indicating advection in excess of that associated with the average measured current. This excess advection is not well resolved in the array spectra, and likely is the orbital flow of the dominant wave near their crests.

1. INTRODUCTION

In their general form, the hydrodynamic equations for water waves are intractably complex. Therefore, various approximations are used that reduce them to tractable forms while preserving, it is hoped, the salient characteristics of the physics. Most formalism describing wind waves employs either linear or weakly nonlinear equations describing wave dynamics. The linear equations exclude harmonics and any interactions among spectral components. Each component travels at a different rate, characterized by its frequency and its wavenumber, and thus represents a dispersive wave. The weakly nonlinear equations, a step away from the simplest theory, permit gradual exchange of energy and momentum between spectral components in order to explain the excitation of waves by wind and the increase, or growth, of the amplitudes of long-wave components. The energy in a system of weakly nonlinear waves is predominantly associated with the freely propagating spectral components.

In recent years, research performed in several countries (Ramamonjiarisoa and Coantic, 1976; Lake and Yuen, 1978; Ramamonjiarisoa and Giovanangeli, 1978; Yuen and Lake, 1978)

G. B. IRANI, B. L. GOTWOLS, AND A. W. BJERKAAS ● Applied Physics Laboratory, The Johns Hopkins University, Laurel, Maryland 20707.

165

has indicated that strongly nonlinear waves—waves with very steep slopes—have substantially more complex dynamics and stronger exchanges of energy between spectral components than nearly linear waves have. Such strong nonlinearity counters the randomness of turbulent generation processes and, with sufficient nonlinearity, can lead to a single, largely coherent wave train that undergoes amplitude and frequency modulations as it propagates. Its temporal spectrum appears much like modeled wind-wave spectra with some strong components, but these components are nondispersive, being phase-locked, or "bound," to the peak (dominant) component and represent its nonlinearity. Measurements show that very steep waves in the laboratory do have these characteristics and suggest that strong nonlinearity may possibly be important in ocean wave systems.

Such wave characteristics sharply contrast with the classical assumption that wind waves are the incoherent superposition of many nearly linear wave trains. The implications of this controversy are both interesting and important: the vast majority of wave research and modeling is based on linear and nearly linear theory and could be misleading.

Stimulated by the impact of the nonlinear model on the validity of ocean wave evolution predictions and by the lack of definitive measurements, APL/JHU undertook an experimental investigation of ocean waves. This effort, called the Ocean Wave Dynamics Experiment (WAVDYN), measured the phase speeds of spectral components in the short to moderate gravity wave regime (wavelengths between 20 cm and 30 m). The objective was to acquire evidence on the prominence of bound- versus free-wave components in ocean wind-wave systems as manifested in estimates of the density of their directional wavenumber–frequency spectra.

Briefly stated, the scheme for meeting this objective involved recording variations of surface elevation and of surface radiance together with current and wind measurements at a deep, open-sea site. The elevation and radiance measurements resolve two spatial dimensions and time. Subsequently, the data were processed at APL so that local maxima in the estimated spectra could be compared with wave propagation theory that included the advective frequency shift associated with the average measured current. This chapter describes the measurements and the results.

2. THE MEASUREMENTS

Two distinct, complementary systems measured the wave dynamics. Both were new designs intended to provide more complete spectral estimates than had yet been reported. One system employed an array of surface elevation transducers, the other a video camera. The array provides the more direct, and thereby the more easily interpreted, information on wave propagation. However, because it measures surface elevation at only 15 locations, its spatial resolution is low, and this limitation is a potential impediment to distinguishing free- from bound-wave spectral density, even when data-adaptive spectral estimation techniques are employed. The video system was designed to fill the requirement for high spatial resolution. The physical process that encodes wave slope in a wave image has been analyzed to first order by Stilwell and Pilon (1974), to second order by Monaldo and Kasevich (1981), and to full order by Chapman and Irani (1981). Gotwols and Irani described the initial results from WAVDYN video data (1980) and also the performance of the video system (1982). The video system is capable of acquiring wave images at a 30-Hz rate with 240×320 resolution elements (*pixels*), a very substantial increase over the spatial resolution of the array. The main drawback of the video technique is the dependence of its response (to surface slope rather than elevation) on sky radiance and reflection. This dependence causes the response of the video camera to be more variable and more nonlinear than that of the array. Nevertheless, the two systems provided complementary data.

Stage I, a Navy research tower standing in 30 m of water 18 km off the coast of Panama City, Florida, was the site for the measurements. Figure 1a depicts the instrumentation layout. The array was mounted 6.6 m from one leg of Stage I by means of the supporting structure shown in Fig. 1b. A three-axis accelerometer package was mounted directly below the array to monitor the stability of the structure. Below the accelerometers at a nominal 4-m depth, a two-axis

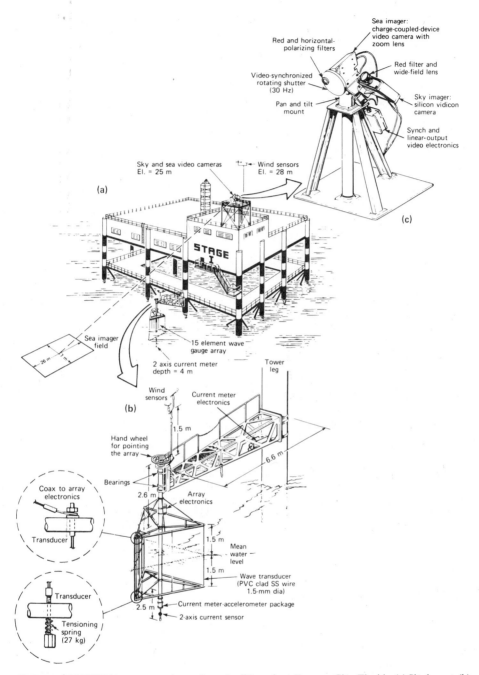

FIGURE 1. WAVDYN instrumentation at Stage I, offshore from Panama City, Florida. (a) Site layout; (b) wave-gage array structure; (c) video cameras for wave imaging and sky imaging.

electromagnetic meter measured water current in a horizontal plane. Above the array, wind speed and direction were monitored. A second, identical array was deployed on the opposite side of Stage I so that, by using the appropriate array, the dominant wind wave could always be measured before it passed under the tower.

The array has 15 vertical capacitance transducers distributed horizontally to form a right

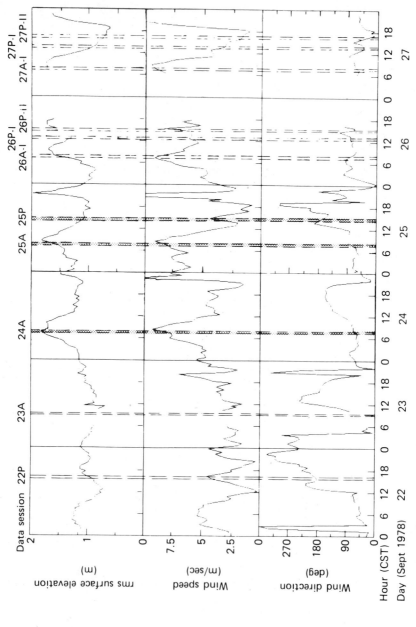

FIGURE 2. Wind and rms surface-elevation histories during the WAVDYN measurements. The traces are interrupted by two intervals lacking data. The data-analysis intervals are noted as pairs of dashed vertical lines. They are called *Sessions* and identified by the code shown at the top of the plot. Spectra representing the shaded intervals appear in this chapter.

angle. Spacings between the elements are 1.5 cm at the apex and increase by a factor of two with each additional transducer (with one exception: 25 cm was used instead of 24 cm) along either leg, so that the maximum spacing is 197.5 cm. Each transducer is a 3-m-long, 1.5-mm-diameter, insulated stainless steel wire that is pulled taut by a spring. A unique electronic circuit converts the capacitance to voltage with better than 1% linearity.

The WAVDYN video system utilized a modified RCA charge coupled device (CCD) camera, a video time-code generator, and a video recorder for data collection at the ocean site. The relation between sea-surface slope statistics and wave-image statistics is linear to a sufficient degree for the nominal conditions of these measurements. The CCD camera stared down at the waves propagating through its 13×25-m field from the highest platform on Stage I. The depression angle was $30°$, and an azimuth was chosen so that the dominant wave propagated nearly toward or away from the camera. This camera was complemented by another video camera, which was aimed at the sky. Figure 1c depicts the cameras. The sky camera had a wide-field lens and provided images of the portion of the sky reflected into the CCD camera by the waves. Both cameras were equipped with red filters and horizontal polarizers. These filters suppress the radiance upwelling from below the sea surface. Additional wind sensors were mounted above the cameras.

Data were recorded during a 6-day period beginning September 23, 1978. Typically, the wind was strongest (8–10 m/s) during the morning and weakest (< 5 m/s) during the afternoon. Figure 2 shows the wind and rms surface elevation histories at Stage I during the 6-day period. The pairs of vertical dashed lines denote intervals of data subjected to wave spectral analysis. Data intervals are identified by a session number, which is listed above the upper scale in Fig. 2. This number indicates when the data were acquired as the date in September 1978, the morning (A) or afternoon (P), and the sequence number (I or II) if more than one session was conducted during the morning or the afternoon.

3. DATA PROCESSING

Separate procedures were implemented for analyzing array and video data. This treatment was dictated by the difference in the sampling densities and in the data encoding (digital vs. analog).

Since the array data were recorded on computer-compatible tape at Stage I, sequences could be directly transferred from tape to APL's general-purpose (IBM 3033) computer system. This step also included demultiplexing the data stream and converting the values from digital to engineering units. In the temporal domain, the array data are not limited either in terms of sampling rate or in terms of sequence length; however, they are limited in both respects in the spatial domain. Therefore, standard fast Fourier transform (FFT) procedures work well for computing the temporal periodicity, whereas the maximum likelihood method (MLM) of spectral estimation was needed to enhance resolution in the spatial domain. For each selected 11-min sequence of data, the result of this computation was an estimate of the directional wavenumber–frequency spectrum of surface elevation variance.

Davis and Regier (1977) described the application of MLM to data from a wave gage array and kindly provided us a copy of their software. Experimentation with WAVDYN data proved to us that MLM provides much better resolution than the traditional beam former technique and that using the nine most widely spaced elements of our array provides the best spatial resolution over the frequency and wavenumbers displayed in subsequent figures. (Of course, we avoided implementing the constraint of linear theory dispersion that Davis and Regier describe.)

The video data were analog-recorded on standard video cassettes together with a time code which identified each image. At APL, a dedicated system built around a PDP 11/34 minicomputer was employed to extract image sequences for analysis, digitize them, and compute directional wavenumber–frequency spectra using a three-dimensional FFT routine. This computer system employs standard components as peripherals, but has special interfaces between the video components and the computer and special software controlling the peripherals.

The video data processing system, with its software, is unique. Given instructions to digitize video data beginning at a prescribed time and at a certain frame rate, the system reads the time code and locates the selected images in the analog data stream as the video cassette is played back on the video recorder. The image sequence is rerecorded on a video disk, where it is held for digitization. This intermediate analog storage slows the data rate (by repeatedly playing one frame for 20 s) to the level required by the video digitizer and the minicomputer. Digitized images are stored on a 260M-byte disk. WAVDYN spectral estimation utilized 256 images spanning 17 s. Each image was represented by a 256×256-pixel array.

The transfer function between sea-surface slope and radiance typically causes the mean and the standard deviation of the radiance to increase from the near to the far field. Thus, a wave image is darker with less contrast in the near field. WAVDYN processing suppressed these trends. For detrending the mean, the signal from each pixel was reduced by its time-averaged level. Detrending the standard deviation required more effort. Standard deviation estimates are statistically less stable than mean estimates, so orthogonal, third-order polynomials were fitted to the set of estimates for an image sequence. Each image was then divided by the resulting polynomial function.

Two further adjustments to the video data were made before computing a spectrum. Perspective distortion arising from the oblique imaging geometry was suppressed by ignoring displacements of the sea surface from the mean sea level and simply linearly interpolating between data to estimate signal levels over a regular grid of points in the sea-level plane. The modulation of the resulting image sequence was tapered to zero at the boundaries. Then, the 3d spectrum was computed from the $256 \times 256 \times 256$-element data set. Stability in this spectrum was improved by smoothing with a $3 \times 3 \times 3$-point moving average.

4. RESULTS

An overview of conditions during the data sessions is provided in Table I, which lists wind, current, and wave parameters by data session for comparison. Typically, seas built during the early and middle morning hours, and then subsided during the afternoon. Consequently, swell from distant wind systems contributed little to the surface elevation variance on the mornings of September 24–27 when the local wind was strong. However, during the other measurement periods the wind at Stage I was light, and swell contributed substantially or dominantly to the elevation variance. This contribution must be removed in order to compare the degree of nonlinearity in the local dominant waves during WAVDYN with theoretical and tank research on the basis of dominant wave slope, measured by the wavenumber–amplitude product $2\pi k_d a_d$. Variance-preserving plots of the elevation frequency spectra were used to identify local wind-wave spectral density and its proportion to the total variance. (Swell was not distinct for Session 25P, although it most probably was the major contributor to the variance.) Assigning the adjusted variance to the local dominant wave and invoking linear wave theory yields the results in the last column of Table I. These values indicate that none of the wave systems during WAVDYN were comparable to the very steep waves studied in tanks or through nonlinear theory, which typically associates $k_d a_d > 0.1$ with significant nonlinearity.

The full spectra estimated from both the array and the video data are difficult to inspect because they are 3d functions. Several procedures were developed to display and to characterize them with reduced dimensionality. An undestanding of their utility can be gained from their relationship to the directional wavenumber–frequency spectral density, $G_0(k, f, \alpha)$, where k is the wavenumber, f is frequency, and α is wave propagation direction. Figure 3 shows two types of reduced spectra calculated from array data for Sessions 24A, 25A, and 25P. In the left column are directional frequency spectra. These are polar plots with frequency assigned to the radial scale and the direction of wave propagation assigned to the angular coordinate. The color-encoded, logarithmically scaled, spectral density is

$$G_1(f, \alpha) = \int_0^\infty G_{0a}(k, f, \alpha)(k/f)\,dk$$

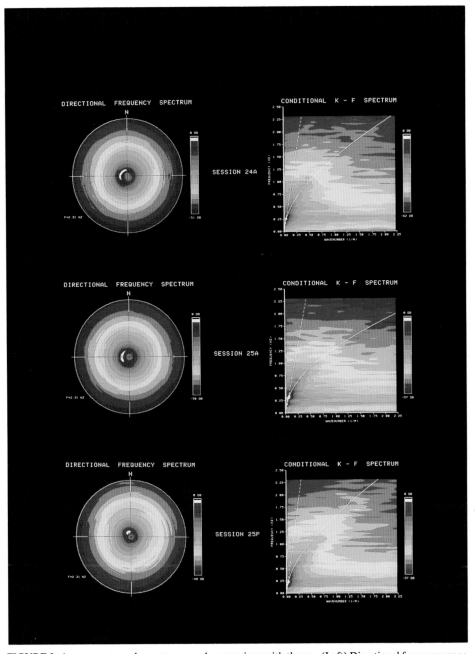

FIGURE 3. Array-spectra color contours and comparison with theory. (Left) Directional frequency spectra $G_1(f,\alpha)$: the radial coordinate is frequency, 0–2.31 Hz; the angular coordinate is direction, 0–360° with north (N) up. (Right) Conditional wavenumber–frequency spectra $G_0(k,f;\alpha_d)$ along the azimuth of the dominant local-wind wave. The solid line is linear theory; the dashed line locates the dominant wave and its harmonics.

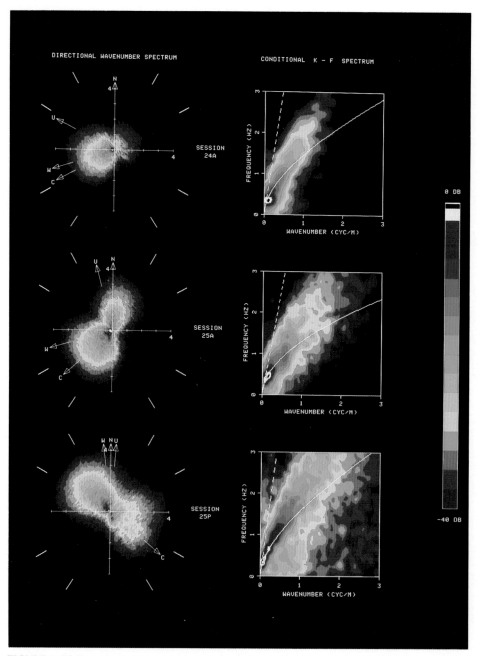

FIGURE 7. Video-spectra color contours and comparison with theory. (Left) Directional wavenumber spectra, $G_2(k, \alpha)$: the radial coordinate is wavenumber, 0–4cyc/m; the angular coordinate is direction 0–360°, with N = north, W = wind vector, U = measured current, C = camera azimuth. (Right) Conditional wavenumber–frequency spectra, $G_0(k, f; \alpha_d)$, along the azimuth of the dominant local wind wave. The solid line is linear theory; the dashed line locates the dominant wave and its harmonics.

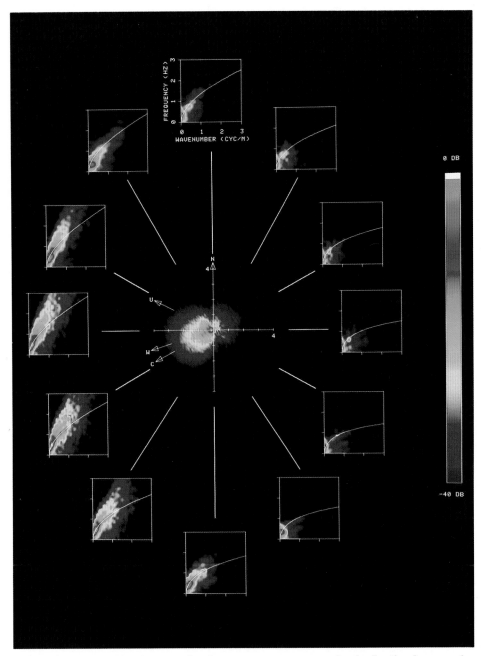

FIGURE 8. Video conditional wavenumber–frequency spectra encompassing the directional wavenumber spectrum: Session 242A.

TABLE I. Wind, Current, and Wave Parameters

Session	Analysis interval (CST)	Wind		Current		rms surface elevation (cm)		Dominant wave		
		Speed (m/s)	Direction (° true)	Speed cm/s	Direction (°true)	Total	Local wave system[a]	Frequency (Hz)[b]	Direction (°true)[b]	$2\pi k_d a_d$[b]
22P	1612:56–1623:56 1620:20–1620:37	4.5	180	14.3	299	7.3	2.4	0.59	0	0.048
23A	0911:40–0911:57	5.8	97	38.3	277	5.6	1.1	0.66	NA	0.027
24A	0743:34–0754:34 0748:13–0748:30	9.5	73	24.6	296	15.5	15.5	0.26	320	0.060
25A	0803:19–0814:19 0807:29–0807:46	8.7	78	23.7	345	15.0	11.3	0.29	270	0.054
25P	1437:31–1448:31 1444:00–1444:17	2.7	170	40.0	5	8.1	8.1	0.22	330	0.022
26A-I	0748:29–0759:29 0752:25–0752:42	9.0	90	34.5	337	14.2	14.2	0.33	270	0.088
26P-I	1232:30–1243:30	6.2	68	31.8	1	11.5	7.7	0.33	310	0.048
26P-II	1456:46–1508:42 1503:19–1503:36	6.2	105	29.6	7	12.2	7.3	0.33	310	0.045
27A-I	0742:30–0753:30	10.0	55	21.2	345	15.7	15.7	0.29	290	0.075
27P-I	1321:00–1332:00	3.7	60	21.8	350	8.0	6.4	0.33	290	0.040
27P-II	1559:31–1610:31 1604:00–1604:17	2.5	78	24.9	349	5.6	2.2	0.45	270	0.026

[a]Estimated from variance-preserving plots of the elevation frequency spectrum.

[b]Values obtained from $G_1(f, \alpha)$ represent the peak of any higher-frequency structure, which presumably measures the local wind-wave system, that is visibly distinct from swell structure. Otherwise, parameters represent the peak of $G_1(f, \alpha)$.

[c]Estimated from linear wave theory: $2\pi k_d a_d = [(2\pi f_d)^2/g][(2\pi f_d)^2/g](\sqrt{2^{1/2}}\sigma_1)$, g = gravitational acceleration, σ_1 = local-wave-system rms elevation.

where the subscript "a" indicates array data and the factor $1/f$ scales $G_{0a}(k,f,\alpha)$ correctly for the polar display. These polar plots make it easy to identify the frequency and the direction of the dominant wave as the black dot in a white region. This direction is the one along which spectral structures are most likely to resolve bound waves.

Conditional wavenumber–frequency spectra, $G_{0a}(k,f;\alpha_d)$, corresponding to the dominant-wave azimuth appear in the right column of Fig. 3. An obvious characteristic of these spectra is that they appear to be horizontally stretched. This is simply a manifestation of greater frequency resolution than wavenumber resolution. The solid line locates linear theory dispersion, while the dashed line locates harmonics of the dominant wave. Advection by the measured current is included in these theoretical relations.

These conditional wavenumber–frequency spectra appear to be largely in agreement with linear theory, although the spectrum for Session 25A has some structure near the bound-wave line at the third and fourth harmonics. Notwithstanding this structure, these spectra show that the waves were predominantly free.

The utility of $G_{0a}(k,f;\alpha_d)$ is that it measures wavelike activity along the direction α_d favoring bound waves. Regions where it has locally high levels are important in this respect. The proximity of these regions to the theoretical k–f relations indicates whether the spectrum better represents bound or free waves. A normalization scheme that enhances structure is useful for inspecting the array spectra. Consider two wavenumber bands, illustrated in Fig. 4, extending 30% on either side of the wavenumbers that the two theories associate with each frequency used for analyzing the data. Normalize the spectral density at each analysis frequency by dividing it by the maximum density at that frequency. This step enhances the lower density levels that occur at higher frequency. For each of the analysis frequencies, plot this normalized density against normalized wavenumbers that deviate from theory by prescribed percentages. Agreement between measurement and theory is indicated by peaking of the normalized spectral density near

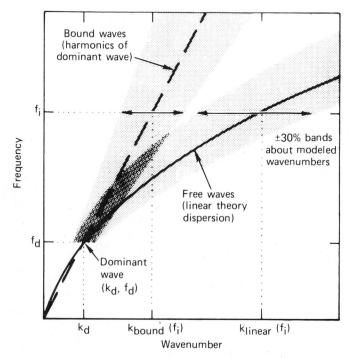

FIGURE 4. Bands for inspecting spectral density structure within $\pm 30\%$ of theoretical wavenumber–frequency relationships.

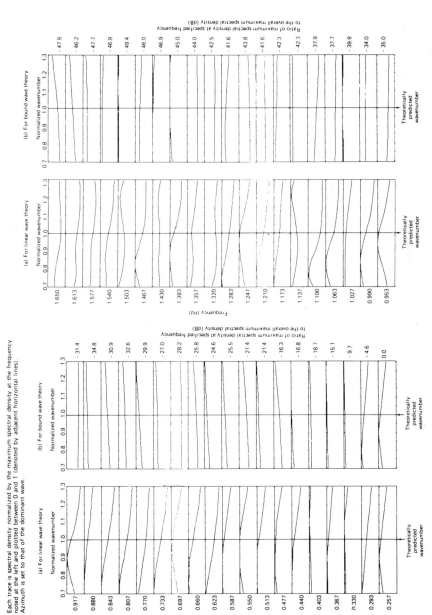

FIGURE 5. Distribution of array spectral density about the wavenumbers theoretically associated with measured frequency: Session 24A.

the normalized wavenumber value of one. Figure 5 shows results obtained for Session 24A. The presence and location of a peak in the density indicates the relative wavenumber at which the greatest wavelike activity occurred. There is substantial scatter in the locations of peaks, but they are predominantly near linear theory.

An overview of all array spectra along the dominant wave direction can be obtained by averaging normalized spectral density like that in Fig. 5 over frequency from the dominant-wave frequency to 1.6 Hz. The 1.6-Hz limit keeps the wavenumbers within the range of measurements. Figure 6 shows the results of such computation applied to 10 different data sequences spanning 6 days. The theoretically predicted wavenumber is 1.0 on these scales. This figure provides information on two topics: The occurrence of distinct wave activity near theory, and the

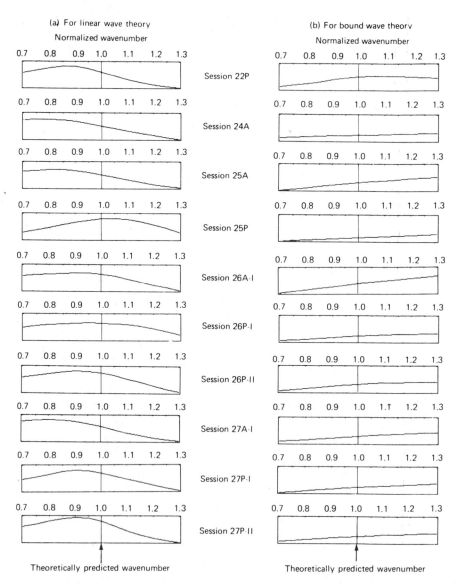

FIGURE 6. Distribution of frequency-averaged array spectral density about theoretical wavenumber–frequency relationships. Shown are results from each interval of array data analyzed (see Fig. 2). The key information is the occurrence of spectral density maxima and their proximity to linear- and bound-wave theories, denoted by a normalized wavenumber value of one in the respective column.

correspondence of distinct wave activity to one of the two theories. The level of the one curve relative to the others is not meaningful. These curves show that the prominent spectral structure for each sequence lies near and, with only one exception, below the normalized wavenumber for linear-wave theory. The weakly nonlinear theory produces similar results, albeit with a smaller offset from the measurements. There are no well-defined peaks in the region of the normalized wavenumber predicted by bound-wave theory.

Spectra from the video data provide information similar to those from the array data, but with much better directional resolution. Figure 7 shows two spectral forms derived from the directional wavenumber–frequency spectra of video data, $G_{0v}(k, f, \alpha)$. Again, spectral density relative to the maximum is logarithmically scaled and color encoded according to the color bar in this figure.

At the left are directional wavenumber spectra. Wavenumber is the radial coordinate; wave direction is the angular coordinate. This spectral form is related to the 3d spectrum through an integral:

$$G_2(k, \alpha) = \int_0^\infty G_{0v}(k, f, \alpha) df$$

As with the directional frequency spectra from the array, these spectra are convenient for visualizing average wave activity, but there are two important differences to keep in mind while comparing array and video spectra. First, slope rather than elevation is the surface property encoded in the data. Thus, the video spectral density rolls off more slowly with both frequency and wavenumber than does the array spectral density. Second, wave-slope visibility varies with the wave azimuth, being generally greatest for waves propagating toward or away from the camera. Clouds can distort this trend, but conditions were such that clouds had minor influence on the directional spectra in Fig. 7. Each of these spectra indicates that waves propagate upwind as well as downwind, in agreement with the array results.

Conditional wavenumber–frequency spectra, $\dot{G}_{0v}(k, f; \alpha_d)$, corresponding to the dominant-wave azimuth resolved by the array, appear in the right column of Fig. 7. Again, the solid line indicates linear wave-theory dispersion, while the dashed line locates harmonics of the dominant wave. Advection by the measured current is included in these theoretical relations. These and similar spectra for other analysis intervals identify no predominant bound-wave activity. The spectra agree generally with free-wave theory, although in every case the trend of the structure is *higher in frequency than predicted by linear theory for azimuths aligned with the dominant wave.* Such behavior is more evident in the video spectra than in the array spectra because the spatial resolution is higher and because the spectral density is more structured over the regions where the theoretical relations are well separated. The breadth of the spectral trends is largely the manifestation of frequency and wavenumber modulations occurring among the short waves as a result of the orbital flow of the dominant wave.

The offset of these spectral trends from linear theory indicates the presence of advection in addition to that produced by the measured current. The possibility of an additional current affecting waves shorter than the dominant wave is strongly suggested by the spectra of Fig. 8. Here, 12 conditional wavenumber–frequency spectra are shown surrounding the directional wavenumber spectrum. These spectra correspond to the 12 directions of wave propagation indicated by the radial lines. North is up. As the wave propagation direction varies, the spectral density trend changes its position relative to the linear-theory curve in a sinusoidal manner. The puzzling aspect of this observation is the identification of a current that could have affected wave propagation, particularly that of the short waves, yet not have been resolved by the current meter. There are at least two possibilities: (1) wind-drift current and (2) the orbital current of the long waves.

The stress of wind blowing over a water surface creates a vertically sheared current very near the surface. Under light to moderate winds, like those during WAVDYN, this wind-drift current is, however, negligible at 4-m depths. Thus, short waves are advected more by this current than are long waves because more of their particle motion is confined to the shallow depths of the

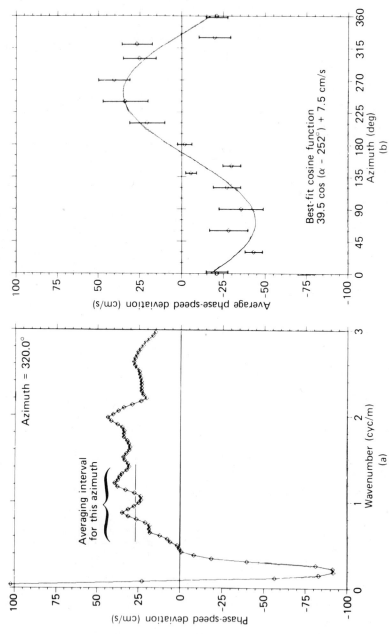

FIGURE 9. Deviation of the video-derived phase speed from linear theory adjusted for measured current: Session 24A. (a) Deviation as a function of wavenumber along the azimuth of the azimuth of the dominant wave. The horizontal line denotes the wavenumber range over which the deviation was averaged to determine the datum at 320° in the adjacent plot. (b) Average deviation as a function of azimuth. The rms error bars and the best-fit cosine curve are shown.

current. Furthermore, at the surface this current is only 3 to 5% of the wind speed, and this level is not enough to account for all of the observed discrepancy.

The second possibility, advection by the orbital flow of the long waves, is a large enough effect provided the short waves are steepened near the crests and flattened near the troughs of the long waves. At the crest, orbital flow is maximum in the direction of the wave propagation. Thus, the steeper short waves near the long-wave crests would dominate the spectrum at high wavenumbers. Their frequencies would appear higher when they are travelling against the long waves. Four-minute averaging was used to process the current data in order to exclude the oscillatory flow of the waves from the measured current. Thus, this second possibility is not taken into account in the dispersion curves shown in the figures.

In spectra from six of the eight wave-image sequences that were processed, the magnitudes and directions of the additional current are resolved. Values for these parameters were calculated by a technique that finds the current vector which best superimposes linear wave theory on the frequency centroids of the conditional wavenumber–frequency spectra taken at 30° increments, such as those in Fig. 8. Figure 9 shows two results from this fitting procedure. At the left is a set of points indicating how the phase speed along the dominant-wave direction and determined from the frequency centroid for each analysis wavenumber deviates from linear wave theory in which advection by the measured current has been included. An experimentally meaningful phase-speed deviation can be calculated by averaging over wavenumber from 0.5 cyc/m up to the wavenumber at which the spectral density is 40 db below its maximum level. Figure 9b displays the average phase-speed deviation as a function of the analysis azimuth. The error bars indicate the standard deviation of the estimate. A function of the form

$$U_v \cos(\alpha + \alpha_v) + B$$

has been fit to the average deviation. Here, U_v represents a current flowing along α_v and not resolved by the current meter. This additional current and its direction were estimated for each of the eight video data sessions: six of these sessions yielded results that are well characterized by a cosinusoidal dependence on azimuth. Table II lists the results of this analysis along with estimates of wind-drift current and of current at the crest of the long wave. The crest current was computed by assuming that measured rms elevation arises solely from a sinusoidal wave having the dominant-wave frequency. This is a gross simplification but can be used as the upper bound of this effect. Table II demonstrates that wind-drift current cannot alone account for the deviation between linear theory and the video spectra, whereas differential steepening and advection of short waves by the dominant wave can explain much of the deviation. Apparently,

TABLE II. Comparison of Additional Current Deduced from Video Spectra with Estimated Current at Dominant-Wave Crest and with Estimated Wind-Drift Current

Session	Additional current		Dominant wave current[a]		Wind-drift currents[b]	
	Speed (cm/s)	Direction (°true)	Speed (cm/s)	Direction (°true)	Speed (cm/s)	Direction (°true)
22P	32.7	58	37.6	0	8.0	0
23A	10.3	284	16.4	NA	10.0	277
24A	39.5	252	34.4	320	17.8	153
25A	28.3	275	38.6	270	15.8	158
25P	20.0	344	15.3	330	5.4	350
26P-II	16.5	285	34.7	310	10.6	285

[a]Maximum horizontal current of sinusoidal wave with rms displacement equal to measured broadband rms elevation and with dominant-wave frequency.
[b]Wind-drift current = 0.55 × wind-friction velocity.

short-wave straining is an important effect and should be taken into account when making high-frequency wave measurements.

5. SUMMARY AND CONCLUSIONS

WAVDYN has substantially extended the empirical basis for developing explanations of how components composing ocean wind-wave spectra propagate and interact. More extensive displays of the spectral estimates appear in a technical report by Irani et al. (1981). Resolution of wave statistics in three dimensions avoids any requirement to assume a relationship between spatial and temporal characteristics of waves. This endeavor required the implementation of modern techniques for resolving wave statistics with greater detail than had previously been achieved and for displaying them in a number of forms. These techniques included two key systems: a wave-gage array system and a wave-imaging system. The array system, with digital data recording and processing, measured surface elevation variations and computed directional wavenumber–frequency elevation spectra using the maximum likelihood method. It also provided measurements of the average currents which advected the waves. The imaging system employed a CCD video camera with a video recorder to acquire sequences of wave images from which directional wavenumber–frequency spectra of surface slope were estimated using specialized processing software, including a three-dimensional FFT. WAVDYN has shown the utility of these two systems.

Results from WAVDYN represent wave characteristics during light to moderate winds ($\leqslant 10\,\text{m/s}$) and for both building and decaying seas. Ten intervals of array data and eight intervals of video data, acquired over a period of 6 days, were processed. In contrast to some laboratory measurements, none of the WAVDYN results shows that nonlinearity is a dominant characteristic. The video data were particularly useful for drawing this conclusion because they have better spatial resolution than array results and greater sensitivity at high wavenumbers and frequencies where the spectral density of bound waves is more separated from that of free waves. At high wavenumbers, the video spectra show an effect similar to advection but not attributable to the measured average current. Rough estimates indicate that this effect can be explained in terms of differential roughening of the short waves near the crests of the dominant wave, where its orbital current is greatest and aligned with its propagation direction.

REFERENCES

Davis, R. E., and L. A. Regier (1977): Methods for estimating directional wave spectra from multi-element arrays. J. Mar. Res. 35, 453.

Gotwols, B. L., and G. B. Irani (1980): Optical determination of the phase velocity of short gravity waves. J. Geophys. Res. 85, 3964.

Gotwols, B. L., and G. B. Irani (1982): A CCD camera system for remotely measuring the dynamics of ocean waves. Appl. Opt. 21, 851.

Irani, G. B., B. L. Gotwols, and A. W. Bjerkaas (1981): Ocean wave dynamics test (WAVDYN): Results and interpretations. APL/JHU Technical Report STD-R-537.

Lake, B. M., and H. C. Yuen (1978): A new model for nonlinear wind waves. Part 1. Physical model and experimental evidence. J. Fluid Mech. 88, 33.

Monaldo, F. M., and R. S. Kasevich (1981): Daylight imagery of ocean surface waves for wave spectra. J. Phys. Oceanogr. 11, 272.

Ramamonjiarisoa, A., and M. Coantic (1976): Loi experimentale de dispersion des vagues produites par le vent sur une faible longueur d'action. C.R. Acad. Sci. Ser. B. 111.

Ramamonjiarisoa, A. and J. P. Giovanangeli (1978): Observations de la vitesse de propagation des vagues engendees par le vent au large. C.R. Acad. Sci. Ser. B. 133.

Stilwell, D., and R. O. Pilon (1974): Directional spectra of surface waves from photographs. J. Geophys. Res. 79, 1277.

Yuen, H. C. and B. M. Lake (1978): Nonlinear wave concepts applied to deep water waves. *Solitons in Action* (K. Lonngren and A. Scott, eds.), Academic Press, New York, 89.

DISCUSSION

EWING: If one measures the two components of horizontal current velocity and compares the total velocity spectrum with the wave height spectrum, one can also check the dispersion relations. Did you try this and find the phase velocity?

GOTWOLS: No, but it sounds like a good idea.

11

TRANSFORMATION OF STATISTICAL PROPERTIES OF SHALLOW-WATER WAVES

B. LE MÉHAUTÉ, C. C. LU, AND E. W. ULMER

ABSTRACT. Directional shallow-water statistical properties of water waves are determined by linear transformation from their deep-water properties. Universal relationships between deep- and shallow-water joint probability distributions of wave height, wave period, and wave direction are established. The formulations are applied to the case of directional narrow-band spectra over a plane bathymetry in intermediate water depth. It is shown that the wave height statistical distributions no longer remain Rayleigh as in deep water, and the wave period distribution is no longer Gaussian but is skewed as a function of water depth. It can be concluded that when waves propagate in intermediate water depth from deeper regions to shallower, the skewness of the marginal probability contours shifts from longer waves to shorter waves.

1. INTRODUCTION

The purpose of this chapter is to establish and present statistical properties of shallow-water waves by transforming their properties in deep water. The transformation of deep-water directional energy spectra has been established for a long time (e.g., Pierson *et al.*, 1953; Longuet-Higgins, 1957; Bretschneider, 1963; Collins, 1972; Krasitskiy, 1974). However, it seems as if less research has been undertaken to establish the transformation of their statistical properties. The joint probability distribution of wave heights and wave periods needs to be characterized in engineering practice as well as, or in addition to, the energy density spectra.

These statistical properties, defined by the joint probability density distribution of wave height and wave period, have been established in deep water (Longuet-Higgins, 1952; Cartwright and Longuet-Higgins, 1956; Bretschneider, 1959; Longuet-Higgins, 1975). In order to obtain their shallow-water equivalents, the directional distribution is also needed.

Ideally, the joint probability distribution of wave height, wavelength, and wave direction should be supported by the statistical compilation of wave heights and zero upcrossing intervals along a number of lines radiating from a given location at a given time. The distribution of wave period will then be obtained from the distribution of wavelength by a simple transformation. Unfortunately, such information is rarely available. Instead, wave recorders typically show the time history of the free surface at a given location and yield little information on the directional function. Therefore, in analogy with directional energy spectrum, it will be assumed that the joint

B. LE MÉHAUTÉ, C. C. LU, AND E. W. ULMER ● Department of Ocean Engineering, University of Miami, Miami, Florida 33149.

probability distribution of wave height H, wave period T, and angle θ is given by the product of two separate independent functions given by the joint probability distribution of wave height and period and the directional function. Subsequently, the directional probability distribution, $p(\theta)$, is identical to the energy spectrum directional function $K(\theta)$.

Finally, nonlinear effects will be assumed to be small so that convective effects and subsequent nonlinear interactions remain negligible (i.e., a case which is valid in intermediate water depth only).

2. THEORETICAL DEVELOPMENTS

The joint probability distributions have been defined by Longuet-Higgins (1975) in terms of relative wave height ξ

$$\xi = a/\mu_0^{1/2} \tag{1}$$

where μ_0 is the zero moment of the deep-water energy spectrum about the mean and a is the amplitude of wave displacement, and relative wave period η

$$\eta = \frac{T - \langle T \rangle}{v \langle T \rangle} \tag{2}$$

where T is the zero crossing wave period and $\langle T \rangle$ is its average value. v characterizes the width of the wave spectrum and is equal to

$$v = \left(\frac{\mu_2}{\mu_0}\right)^{1/2} \left(\frac{\langle T \rangle}{2\pi}\right) \tag{3}$$

μ_2 denotes the second moment of the deep-water energy spectrum about the mean. In addition, defining that subscript "0" denotes deep water and that no subscript indicates intermediate water depth, the transformation of deep-water properties into shallow-water properties is established as follows.

By definition, one must have

$$\int\int\int p(\xi_0, \eta_0, \theta_0)\, d\xi_0 d\eta_0 d\theta_0 = \int\int\int p(\xi, \eta, \theta)\, d\xi\, d\eta\, d\theta = 1$$

That is, by virtue of the rule on the change of variables,

$$p(\xi, \eta, \theta) = p(\xi_0, \eta_0, \theta_0)|J|^{-1} \tag{4}$$

where the Jacobian

$$|J| = \frac{\partial(\xi, \eta, \theta)}{\partial(\xi_0, \eta_0, \theta_0)} = \begin{vmatrix} \partial\xi/\partial\xi_0 & \partial\xi/\partial\eta_0 & \partial\xi/\partial\theta_0 \\ \partial\eta/\partial\xi_0 & \partial\eta/\partial\eta_0 & \partial\eta/\partial\theta_0 \\ \partial\theta/\partial\xi_0 & \partial\theta/\partial\eta_0 & \partial\theta/\partial\theta_0 \end{vmatrix}$$

Therefore, $|J|$ could be determined when the functions

$$\eta, \xi, \theta = F(\xi_0, \eta_0, \theta_0)$$

are known. Note $\partial\theta/\partial\xi_0 = \partial\eta/\partial\theta_0 = 0$. Also in accordance with an assumption of linearity, $\partial\eta/\partial\xi_0 = 0$, since the wave period T is not a function of wave amplitude a. Therefore, the Jacobian $|J|$ is reduced to

$$J = \frac{\partial \xi}{\partial \xi_0} \frac{\partial \eta}{\partial \eta_0} \frac{\partial \theta}{\partial \theta_0} \tag{5}$$

In the case of monochromatic waves,

$$\partial \eta / \partial \eta_0 = 1$$

For irregular waves, the shallow-water spectrum $S(, \theta)$ is given from the deep-water spectrum $S_0(, \theta_0)$ by the application of the Liouville theorem (Longuet-Higgins, 1957; Collins, 1972; Krasitskiy, 1974):

$$S(\omega, \theta) = S_0(\omega, \theta_0) \frac{k}{k_0} \frac{V_0}{V}$$

which yields a shift of energy distribution with frequency. k is the wavenumber and V the group velocity. In deep water the average wave period defined by Longuet-Higgins (1975) is

$$\langle T_0 \rangle = \frac{\displaystyle\iint S_0(\omega, \theta_0)\, d\omega\, d\theta}{\displaystyle\iint \frac{\omega}{2\pi} S_0(\omega, \theta_0)\, d\omega\, d\theta_0}$$

then in shallow water it is

$$\langle T \rangle = \frac{\displaystyle\iint S_0(\omega, \theta_0) \frac{k}{k_0} \frac{V_0}{V} d\omega\, d\theta}{\displaystyle\iint \frac{\omega}{2\pi} S_0(\omega, \theta_0) \frac{k}{k_0} \frac{V_0}{V} d\omega\, d\theta}$$

Consequently, there is a small shift of wave period distribution due to dispersion and refraction. In the case of narrow-band spectrum the error, caused by assuming that $\partial \eta / \partial \eta_0 = 1$, is certainly negligible since the effect of the transform operator $(k/k_0)(V/V_0)$ is small. However, in the case of broadband spectrum, $\partial \eta / \partial \eta_0 \simeq 1$. If it is assumed that $\partial \eta / \partial \eta_0 = 1$, then $\partial \xi / \partial \xi_0 \simeq \xi / \xi_0 = K_S K_R$, where K_R and K_S denote the refraction and shoaling coefficient, respectively; $K_S = (V_0/V)^{1/2}$. Then, referring to Eqs. (4) and (5), one has

$$\frac{p(\xi, \eta, \theta)}{p(\xi_0, \eta_0, \theta_0)} = \frac{1}{K_S K_R} \frac{\partial \theta_0}{\partial \theta} \tag{6}$$

Since $k/k_0 = K_R^2 (\partial \theta_0 / \partial \theta)$ (Dorrestein, 1960; Le Méhauté and Wang, 1982), by eliminating $\partial \theta_0 / \partial \theta$,

$$\frac{p(\xi, \eta, \theta)}{p(\xi_0, \eta_0, \theta_0)} = \frac{k}{k_0} \frac{1}{K_S K_R^3} \tag{7}$$

or

$$\frac{p(\xi, \eta, \theta)}{p(\xi_0, \eta_0, \theta_0)} = \frac{S(\omega, \theta)}{S_0(\omega, \theta_0)} \frac{1}{K_S^3 K_R^3} \tag{8}$$

These relationships are universal since no hypotheses have been made on the form of wave spectrum or probability density distribution.

3. APPLICATION TO A PLANE BATHYMETRY

It is recalled that the joint probability density of wave heights and wave periods has been given in deep water by Longuet-Higgins (1975) for the special case of a narrow energy spectrum and a Gaussian record,

$$p_0(\xi_0, \eta_0) = \frac{\xi_0^2}{(2\pi)^{1/2}} \exp\left[-\xi_0^2(1+\eta_0^2)/2\right] \tag{9}$$

The directional function $K(\alpha_0)$ has been presented in a multiplicity of forms such as that of a normalized cardioid,

$$K_0(\alpha_0) = \cos^s\left(\frac{\alpha_0}{2}\right) \Big/ \int_0^{2\pi} \cos^s\left(\frac{\alpha_0}{2}\right) d\alpha_0$$

One will retain

$$K_0(\alpha_0) = \frac{8}{3\pi}\cos^4\alpha_0 \qquad -\pi/2 \leq \alpha_0 \leq \pi/2 \tag{10}$$

proposed by Barnett (1968) based on JONSWAP experiments. In the case of a plane bathymetry, $\alpha_0 = \theta_0 - \bar{\theta}_0$, where θ_0 is the deep-water wave angle of the wave ray with the bottom contours and $\bar{\theta}_0$ the main wave direction.

In shallow water, $\alpha = \theta - \bar{\theta}$. By application of the Snell law on wave refraction: $k\cos\theta =$ const, one finds $\theta = \cos^{-1}[\tanh kd \cos\theta_0]$ and $\bar{\theta} = \cos^{-1}[\tanh kd \cos\bar{\theta}_0]$. Combining Eqs. (9) and (4) with some simple arithmetic, one obtains

$$p(\xi, \eta, \theta) = \frac{\xi^2}{(2\pi)^{1/2}D^3}\frac{1}{}\exp\left[-\frac{\xi^2(1+\eta^2)}{2D^2}\right] \tag{11}$$

where $D = K_S K_R$ and since

$$K_S = \left(\frac{V_0}{V}\right)^{1/2} = \frac{1}{\left(1+\dfrac{2kd}{\sinh 2kd}\right)^{1/2}(\tanh kd)^{1/2}} \tag{12}$$

$$K_R = \left(\frac{\sin\theta_0}{\sin\theta}\right)^{1/2} \tag{13}$$

then

$$D = \left[\frac{\sin\left(\cos^{-1}\dfrac{\cos\theta}{\tanh kd}\right)}{\sin\theta\left(1+\dfrac{2kd}{\sinh 2kd}\right)(\tanh kd)}\right]^{1/2} \tag{14}$$

and

$$p(\theta) = p_0(\theta)\frac{\partial\theta_0}{\partial\theta} \tag{15}$$

Then introducing Eq. (10), after some manipulation one obtains

$$p(\theta) = \frac{8}{3\pi}\cos^4\left(\cos^{-1}\frac{\cos\theta}{\tanh kd} - \cos^{-1}\frac{\cos\bar{\theta}}{\tanh kd}\right)\frac{\sin\theta}{\tanh kd \sin\left[\cos^{-1}\left(\dfrac{\cos\theta}{\tanh kd}\right)\right]} \tag{16}$$

Finally, the marginal joint probability distribution of wave height and wave period in shallow water is

$$p_\theta(\xi, \eta) = \int_{\theta_1}^{\theta_2} p(\xi, \eta) p(\theta) \, d\theta \tag{17}$$

where $p(\xi, \eta)$ is a function of θ given by Eqs. (11) to (14). $p(\theta)$ is given by Eq. (16). θ_1 and θ_2 are the lower and upper limits for integration. The integration of Eq. (17) has to be done numerically because of the complicated mathematical formulation. In this study, the simple trapezoidal rule is employed for the numerical calculation. θ_1 and θ_2 are determined from the range of the directional function [Eq. (10)] with two criteria: (1) waves have to propagate shorewards in deep water; (2) any arbitrary θ in shallow water has to satisfy $\cos \theta / \tanh kd \leqq 1$ (i.e., the formulation does not account for waves which are reflected seawards).

In the case of a unidirectional spectrum, $K(\theta)$ is a delta function; then $p(\xi)$ is simply

$$p(\xi) = \int_{-\infty}^{\infty} p(\xi, \eta) \, d\eta = \frac{\xi}{K_S^2} \exp\left(\frac{\xi^2}{2K_S^2}\right) \tag{18}$$

which for a constant wave period, the probability of wave height is still Rayleigh in shallow water as well as in deep water. The density of wave period at a specific wave height is

$$p_\xi(\eta) = \frac{\xi}{(2\pi)^{1/2}} \exp\left(\frac{\xi^2 \eta^2}{2K_S^2}\right) \tag{19}$$

which is no longer Gaussian as it is in deep water because K_S is function of the wave period.

4. APPLICATIONS AND RESULTS

The theoretical developments have been applied to the Pierson–Moskowitz spectrum (1964) for fully developed sea. We consider a case in which wind speed is 20.5 m/s (40 knots) and main wave direction in deep water is 45° shorewards. Then, wave height (H) and wave period (T) are determined in terms of relative wave height (ξ) and relative period (η) by the relationship $H = 4.48\,\xi$ and $T = 11.7 + 3.02\,\eta$ obtained from Eqs. (1) to (3).

The results are given in Figs. 1 to 3, which present curves of constant joint probability densities in a Cartesian coordinate system defined by the relative wave height ξ and wave period η. Figure 1 is for deep water and is given for the sake of comparison. It is identical to the curves presented by Longuet-Higgins (1975). Figure 2 is for various water depths, namely, 40, 30, 20, 15, and 10 m, respectively. Miche's breaking limit (1944) is also introduced since it is expected that the probability curves given will fail for significant wave breaking. One should expect that the probability curves $p = 0.05, 0.10, 0.15$ accumulate along the Miche limit yielding a high concentration of breaking waves. It is readily apparent that the effects caused by shoaling and refraction result in asymmetric probability contours. Close examination will show that the contours shift toward larger values of η (i.e., longer wave period) when the water depth is greater than 20 m. Conversely, when water depth is lower than 20 m, the contours skew toward smaller values of η. The longer wave period shift is attributed to the shoaling factor which causes more dispersion on shorter waves, whereas longer waves will refract seawards in shallower water instead of continuously propagating shorewards resulting in the skewness toward shorter waves. Figure 3 shows contours of an identical probability at varying water depths. In the figure, one can see that the probability of high wave height gradually decreases from deep water to $d = 20$ m, and then increases when water depth is 10 m. This is caused by the shoaling effect as waves propagate from deep water to shallow water. Figure 4 show the conditional probability of wave height known to be no longer a Rayleigh distribution as it is in deep water from inspection of Figs. 1 and

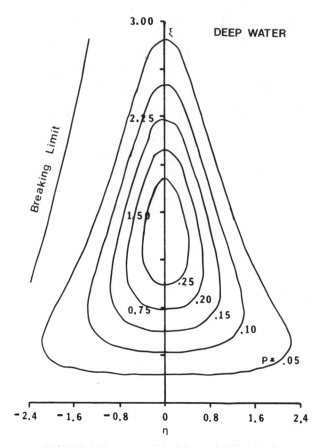

FIGURE 1. Contours of the joint probability density.

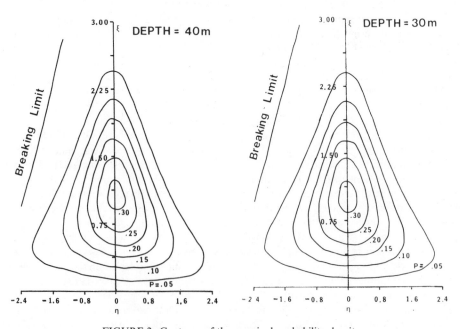

FIGURE 2. Contours of the marginal probability density.

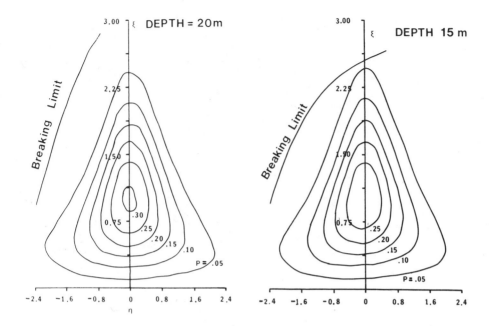

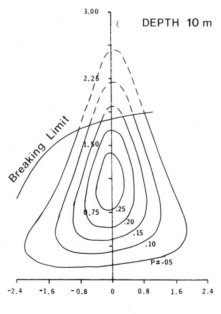

FIGURE 2 (*Continued*).

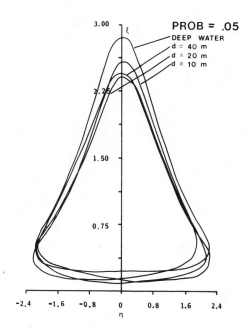

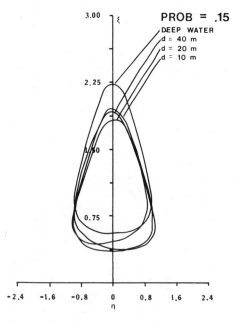

FIGURE 3. Contours of an identical probability for varying depth.

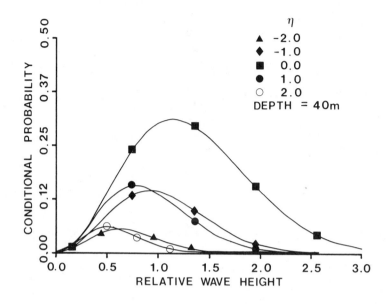

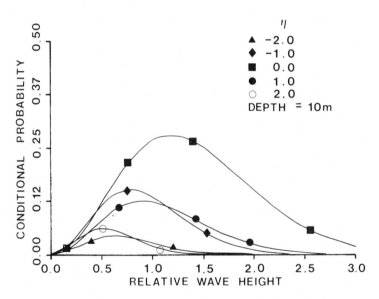

FIGURE 4. Conditional probability of relative wave height.

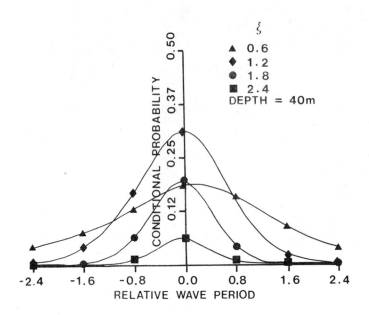

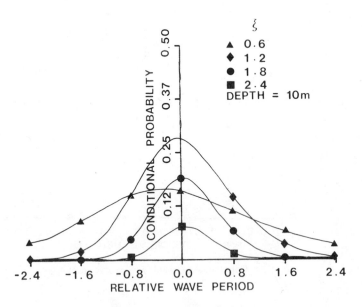

FIGURE 5. Conditional probability of relative wave period.

2. Also, the conditional probability of wave period is no longer a Gaussian distribution as shown by the skewness in Fig. 5.

5. CONCLUSION

A method for the transformation of statistical properties of deep-water waves to their shallow-water equivalents has been proposed, including a directional function and the effects of shoaling and refraction. The effect of shoaling results in the decrease of probability for high wave height in shallow water. Also, it causes the shift of the contours toward longer waves. The directional function included and the effect of refraction theoretically explain the skewness of the contours toward shorter waves. The present theoretical development needs further investigation. A more realistic input for the joint deep-water probability distribution may be that of Cavanie *et al.* (1976) as shown by Huang *et al.* (this volume) and Ochi and Tsai (1981). Nevertheless, it can be concluded that the proposed mathematical formulation indicates a trend of wave statistical properties in intermediate water depth. Under this condition, the probability distribution of wave height is no longer given by a Rayleigh distribution nor can the distribution period be considered Gaussian.

A complete and thorough investigation of this problem is actually very complex as it should include the effect of nonlinearity, weak wave–wave interaction, whitecaps, wave seafloor interaction, and wave generation by wind, in analogy to the radiation transfer theory applied to wave spectrum.

REFERENCES

Barnett, T. P. (1968): On the generation, dissipation and prediction of ocean wind waves. *J. Geophys. Res.* **73**, 513–530.

Bretscheneider, C. L. (1959): Wave variability and wave spectra for wind-generated gravity waves. Technical Memo. 118, U.S. Beach Erosion Board, Washington, D.C.

Bretschneider, C. L. (1963): Modification of wave spectra on the Continental Shelf and in the surf zone. *Proceedings, VIIIth Conference on Coastal Engineering*, 17–33.

Cartwright, D. E., and M. S. Longuet-Higgins (1956): The statistical distribution of the maxima of a random function. *Proc. R. Soc. London Ser. A* **237**, 212–232.

Cavanie, A., M. Arhan, and R. Ezraty (1976): A statistical relationship between individual heights and periods of storm waves. *Proc. BOSS'76* **II**, 13.5, 354–360.

Collins, J. L. (1972): Prediction of shallow water spectra. *J. Geophys. Res.* **77**, 2693–2707.

Dorrestein, R. (1960): Simplified method of determining refraction coefficient for sea waves. *J. Geophys. Res.* **65**, (No. 2) 637–650.

Krasitskiy, V. P. (1974): Toward a theory of transformation of the spectrum on refraction of wind waves. *Isv. Atmos. Ocean Phys.* **10**, 72–82.

Le Méhaute, B., and J. Wang (1982): Review of wave spectrum changes on a sloped beach. *ASCE J. Waterw. Coastal Ocean Div.* **108**, (No. WWL) 33–47.

Longuet-Higgins, M. S. (1952): On the statistical distribution of the heights of sea waves. *J. Mar. Res.* **9**, 245–266.

Longuet-Higgins, M. S. (1957): On the transformation of a continuous spectrum by refraction. *Proc. Cambridge Philos. Soc.* **53**, 226–229.

Longuet-Higgins, M. S. (1975): On the joint distribution of the periods and amplitudes of sea waves. *J. Geophys. Res.* **80**, 2688–2694.

Miche, M. (1944): Movements ondulatoires de la mer en profondeur constante ou decroissant. *Ann. Ponts Chaussees* **7**, 25–28, 131–164.

Ochi, M. K., and C. H. Tsai (1981): Prediction of occurrence of breaking waves in deep water. Unpublished report.

Pierson, W. J., and L. Moskowitz (1964): A proposed spectral form for fully developed wind sea based on the similarity theory of S. A. Kitaigorodskii. *J. Geophys. Res.* **69**, 5181–5190.

Pierson, W. J., J. J. Tuttel, and J. A. Wooley (1953): The theory of the refraction of a short-crested Gaussian sea surface with application to the northern New Jersey coast. *Proceedings IIIrd Conference on Coastal Engineering*, pp. 86–108.

12

Aspects of the Velocity Field and Dispersion Relation in Surface Wind Waves

V. V. Yefimov and B. A. Nelepo

Abstract. The spectra of the wave velocity field and elevations are compared with the relationship of the linear spectral theory. In a first approximation the deviations are correlated with the influence of currents and low frequency components. Experimental tests of the dispersion relation is carried out. The deviations from the linear relation are found. Nonresonance and nonlinear interactions are discussed in relation to these deviations. Tangential wave stresses are measured which differ from zero. Phase lags between wave velocities components are shown.

1. INTRODUCTION

This chapter describes experimental research on wave disturbances in the atmospheric and oceanic boundary layers obtained from simultaneous records of velocity fluctuations at several levels in the air and water and of wave surface elevations at several points with the help of instruments placed upon stable masts. Semisubmerged buoys, such as the one shown in Fig. 1, were used under open sea conditions during research vessel cruises to obtain simultaneous measurements at several levels in the boundary layers of the atmosphere and the ocean. Measurements were made from these buoys to a height of 5 m in the atmosphere and to depths of 15 m in the ocean. Masts, such as the one shown in Fig. 2, placed in the open coastal area of the Black Sea were also used. These were set at a depth of 15 m and at a distance of 300 m from the shore. Quick-response reversible velocity current meters with a propeller sensing element were also used for measurements of the velocity components in water and air. These used electronic circuits for counting the angular velocity of the propeller; three current meters were used for the measurements of three velocity components. The water current meters have time intervals from 0.05 to 0.1 s and their resolution is 2 cm/s. The air current meter has a time decrement of 0.15 s. In this way, the three components of water velocity and the longitudinal component of velocity in the air were measured at several levels. The surface wave elevation at several (five to seven) points was measured by means of a string of capacitance wave gages. Experiments of this kind have been carried out for several years and vast amounts of primary data were treated using standard techniques of spectral analysis.

V. V. Yefimov and B. A. Nelepo ● Marine Hydrophysical Institute, Ukrainian Academy of Sciences, Sevastopol, USSR.

FIGURE 1. Submerged gradient buoy.

FIGURE 2. Gradient mast in the Black Sea.

2. SPECTRA OF ORBITAL VELOCITY COMPONENTS

Figure 3 shows typical estimates of the frequency spectra of surface elevation and the vertical velocity components. These spectra are related to a situation of almost steady waves at a wind velocity U of 11 m/s. Note the correspondence of individual peaks in the elevation spectrum and velocity component spectrum; the influence of instrumental and processing noise is notable at frequencies higher than 1.5 Hz.

The calculated spectrum $S_w^*(f)$ of the vertical velocity is found from the spectrum S_η of the wave surface elevation by using the linear dispersion relation, taking into account the finite depth of the sea at low frequencies. The correspondence between these two spectra in the range of frequencies less than three times that of the spectral peak is in agreement with the linear theory. The differences between experimental and calculated spectra becomes more evident in the range of higher frequencies. This discrepancy is of such a nature that the spectral slope is lower: the measured spectrum of vertical velocity $S_w \sim f^{-3.5}$ and the spectral density are higher than estimated from linear theory. These differences between experimental and calculated spectra are noted at ever-decreasing frequencies as the wave energy and water depth increase. Similar results were obtained in all series of measurements and they support the applicability of linear spectral

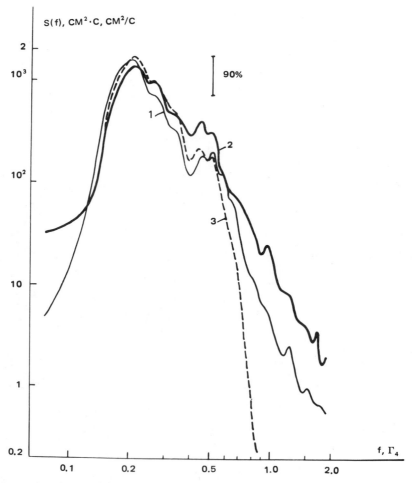

FIGURE 3. Spectra of wave elevation (1), vertical velocity measured (2) and calculated (3) through S_η using simple linear relation.

theory relations to a description of the wave motion in the frequency range near the spectral peak and, at the same time, indicate discrepancies in the high-frequency range.

It is necessary to take into account a number of factors which may influence these interrelationships in any study of the spectral characteristics of the elevation and velocity of wind waves in the ocean. The presence of a surface current and low-frequency spectral components of waves are the principal factors; they are not completely independent. The surface current is influenced not only by the wind and so forth, but is inseparably connected with the period wave motion. Low-frequency spectral components of waves and swell may influence high-frequency short waves. A theoretical calculation of the ratio S_w^*/S_w is shown in Fig. 4 to indicate the influence of a current. The water current velocity which was recorded during the measurements

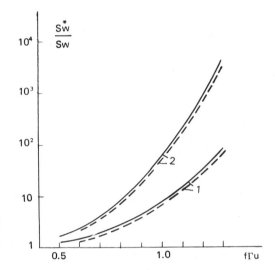

FIGURE 4. Ratio of vertical velocity spectra. S_w^*: calculated by means of S_η considering mean current; S_w: without it. Solid lines: $F(\theta) \sim \delta(\theta)$; dashed lines: $F(\theta) \sim \cos\theta$. (1) Depth 1.2 m; (2) depth 2.25 m.

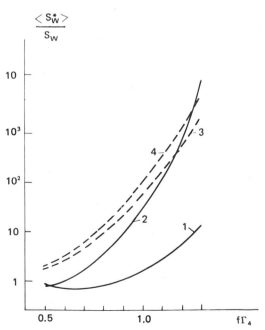

FIGURE 5. Ratio of vertical velocity spectra. S_w^*: calculated by means of S_η considering long wave spectral component; S_w: without it. $F(\theta) = \delta(\theta - \theta_0)$; θ_0: angle between long and short waves. Solid lines: $\theta_0 = 0$; dashed lines: $\theta_0 = \pi/2$. (1, 3) Depth 1.2 m; (2, 4) depth 2.25 m.

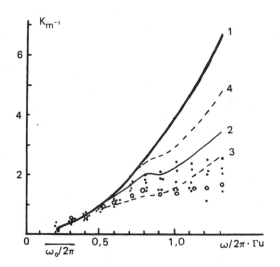

FIGURE 6. Vertical wavenumber calculated from the decrease of vertical velocity spectra with depth. (1) Linear dispersion curve $K = \omega^2/g$. (2) Corrected for mean current dispersion. (3, 4) Corrected for long wave component dispersion. (3: $\theta_0 = 0$; 4: $\theta_0 = \pi/2$). $\bigcirc$, experimental estimates.

was equal to 0.3 m/s. It is clear that corrections for an average current decrease the difference between experimental spectra and those calculated at high frequencies but the difference remains significant, especially in the case of well-developed waves or several overlapping wave systems.

The influence of low-frequency spectral components of the wave field is a second possible reason for the more gradual decrease with depth than would be expected from the usual exponential relationship. The simple kinematic interaction between long and short waves can be invoked to explain the measured spectra of vertical velocity at high frequencies. Figure 5 shows the results of numerical calculation of the correction function at two different measurement depths for a long wave slope $a_0 k_0 = 0.08$ and wavenumber $k_0 = 0.13\,\mathrm{m}^{-1}$. It is clear that the greatest influence of the low-frequency component appears in the case of short wave propagation along the long wave crests, and this effect grows with depth.

Accordingly, for a developed field of wind waves whose parameters are close to those described here, a consideration of the influence of mean current and low-frequency wave components allows us to account for most of the difference between experimental and calculated spectra for $f \gtrsim 1$ Hz. The quantity

$$K = \tfrac{1}{2} z \ln(S_w/S_\eta)$$

contains useful information on the applicability of the linear relationships.

Figure 6 displays experimental values of this quantity for six different cases of wind waves. The frequencies of spectral maxima are in the range 0.18 to 0.35 Hz. It is clear from Fig. 6 that the difference between the experimental data and the linear relation $K = \omega^2/g$ becomes significant in the range of frequencies above 0.7 Hz. A correction with due regard for the average current and long wave components decreases the difference between the experimental points and the linear dispersion curve, but this correction turned out to be insufficient to explain the effects noted.

3. EXPERIMENTAL EXAMINATION OF THE DISPERSION RELATION

The dispersion relation, connecting the frequency ω and the wavenumber k of wind surface waves, is a fundamental characteristic. However, the idea of a dispersion relation becomes conditional for random wave fields with significant nonlinear interactions. The simple connection between the frequency and wavenumber spectrum $F(\mathbf{k},\omega)$ and the wavenumber spectrum $\psi(k)$ in the form (in deep water)

$$F(\mathbf{k}, \omega) = \psi(\mathbf{k})\delta[\omega - (gk)^{1/2}]$$

breaks down since the energy in $(\mathbf{k}, \omega)$ space is distributed not only on the surface $\omega = (gk)^{1/2}$ but over the remainder of the frequency region. Accordingly, it is necessary to know the complete spectrum $F(\mathbf{k}, \omega)$ to describe a random field of nonlinear waves. However, if we accept that the nonlinear interactions in the wind waves are weak, we may conditionally take the dispersion relation, connecting frequencies and wavenumbers, to correspond to the maximum of the section of the frequency spectra $S_\eta(f)$. The theoretical dispersion curve obtained from linear theory $\omega^2 = gk$ is shown by the continuous line. The dotted lines indicate averaged experimental points. The correction for Doppler shift produced by the average current is also introduced here for the frequency spectra. The average experimental curve $\omega(k)$, plotted from the primary data without correction for the average current, is shown as the line $-\cdot-$.

It is necessary to provide sufficiently high-frequency resolution in addition to high wavenumber resolution to trace possible deviations from a linear dispersion relation. If this is not done, the spectral energy will leak from the frequency range to the wavenumber range.

Figure 7 shows the resulting diagram connecting values k_0 and ω_0 which correspond to the maxima in $F(\mathbf{k}, \omega)$. Three series of measurements are shown. Frequencies and wavenumbers along the coordinate axes are normalized on the values f_0 and k_0, which correspond to the peak of the frequency spectra $S_\eta(f)$.

The following conclusions may be drawn from the analysis of Fig. 7. First, the experimental points are in good agreement with the linear theoretical dispersion curve in the frequency range $f_0 < f < 2f_0$. Second, a significant deviation from the theoretical linear relation is observed in the neighborhood of $f = 2f_0$; more precisely, somewhat higher than twice the frequency of the spectral peak. Here, the ratio of the experimental values k_2 to the calculated ones is equal to 0.87 on average and this cannot be explained by possible errors of measurement. Third, the deviation of k_2 from the calculated value is somewhat less at higher frequencies but still exceeds possible

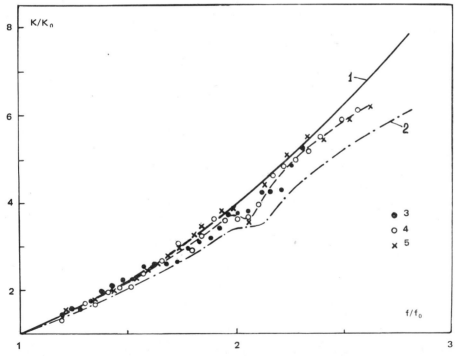

FIGURE 7. Dispersion diagrams. (1) $K = \omega^2/g$; (2) mean experimental curve without correction for mean current; (3–5) experimental estimates with correction for mean current.

measurement errors. Here, the experimental curve is below the theoretical one. The deviations increase at the high-frequency end of the range measured.

The same conclusions follow from an analysis of Fig. 8 which shows frequency–wavenumber spectra in the direction of maximum energy propagation θ_0 by planes parallel to the k/k_0 axis. Relative spectra, normalized for constant sectional areas, are displayed here. They allow us to analyze not only the deviations in the spectral maximum in the plane (k, f) from the linear dispersion curve but also the spreading of energy relative to it. The deviations from linear theory are greatest when $f/f_0 = 2.07$; at higher frequencies one observes the spreading of sections over wavenumbers. Figure 9 exhibits additional important evidence for the fact that the deviation of the maxima from the dispersion curve and the spreading of the energy distribution over wavenumbers for fixed frequencies (the spreading of sections shown in the figure) cannot be attributed to possible errors in the measurements. Figure 10 illustrates the normalized frequency–wavenumber spectra $\tilde{F}(k, \omega, \theta)$ for three frequency values: $f_1 = 0.175$ Hz, $f_2 = 0.362$ Hz, and $f_3 = 0.435$ Hz. These frequencies are close to $f_1 = f_0, f_2 = 2.07 f_0, f_3 = 2.5 f_0$. The wavenumbers corresponding to maxima in $\tilde{F}(k, \omega, \theta)$ are denoted on the horizontal axis: k_1 satisfies the condition $\omega^2 = gk$ and k_e the real values. The deviations are evident.

In addition, spectral sections over $\theta = $ const and $k = $ const are shown. The areas corresponding to spectral spreading due to the effect of a filter in the spectral processing are shown shaded. As is seen, the spectra $F(k, \omega, \theta)$ are even wider over wavenumbers than those over angles θ. For frequency f_0, the difference cannot be considered important and the spreading of

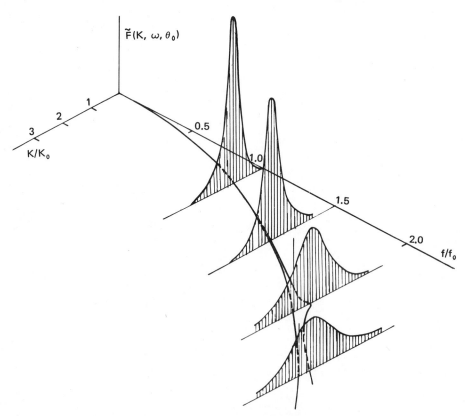

FIGURE 8. Normalized sections of frequency–wavenumber spectrum $\tilde{F}(k, \omega, \theta)$ for $\theta_0 = 0$. Dispersion curve $K = \omega^2/g$ and experimental dispersion diagram are shown.

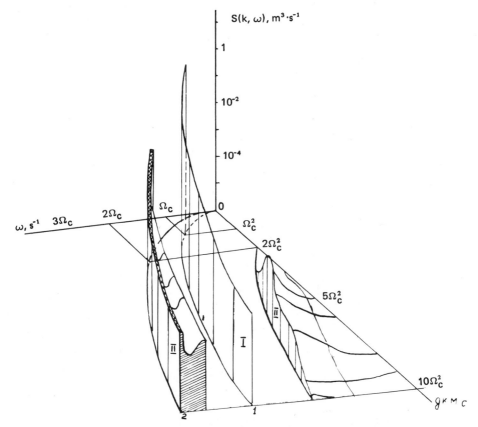

FIGURE 9. Theoretical second-order frequency–wavenumber spectrum using Phillips one-dimensional (collinear) first-order model (following Barrick and Weber theory).

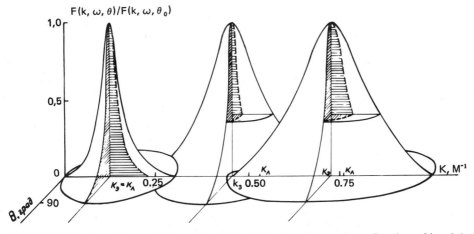

FIGURE 10. Sections of normalized spectrum $F(k, \omega, \theta)$ for three frequencies, as functions of k and ϕ.

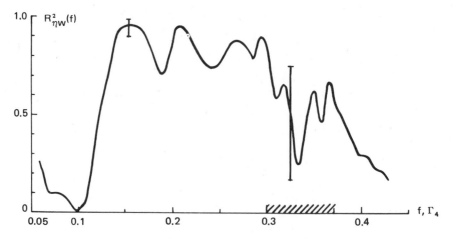

FIGURE 11. Coherence function for vertical velocity and elevation.

the spectral section over $\theta = \theta_0$ as compared with a δ function, is associated with the finiteness of the spectral window over k. However, for frequencies greater than $2f_0$ and especially for frequency $f = 2.07f_0$, considerable spreading of the spectral section over wavenumbers is clear.

The features of the spectra $F(k, \omega, \theta)$ attest to differences between the linear dispersion relation and those actually found in the field; they consist generally of higher propagation velocities of spectral components of the wind wave field over frequencies greater than twice that of the spectral peak. The physical reason for this effect is quite clear: real wind waves are not sinusoidal but contain bound harmonics that contribute to the asymmetry of the wave profiles. For example, in our measurements the skewness of surface elevation records is $+0.29 \pm 0.01$.

It is difficult to estimate from these experimental results a quantitative relationship between the density of harmonics and free wave components. It can, however, be obtained roughly, using the coherence functions between wave surface elevations at different points rather than the spectra F. In Fig. 11, the coherence function $R^2(f)$ between wave elevation and vertical velocity is shown when the distance between gages is 9.6 m and the angle between θ_0 (the dominant wave direction) and the line of wave gages is $12°$. Over the energy-containing frequencies, R^2 is close to unity but the coherence decreases as f increases. The angular distribution of wave energy is, indeed, the main factor contributing to this decrease in coherence, but the magnitude of the decrease at higher frequencies (the range $f \gtrsim 2f_0$) cannot be attributed to this fact alone. The additional decrease in R^2 is a result of a contribution to these bound harmonics to the spectrum. Since estimates of the angular distribution of energy had been obtained, one can derive the ratio of the spectrum of double harmonics S_2 to the free wave spectrum $S_1 = S - S_2$ from this additional decrease in R^2. Conscious that these estimates are rather rough because of the high variance of R^2 values, we present the results without dwelling upon the algebraic procedure of calculating S_2/S_1. Our results indicate that the average value of S_2/S_1 in the frequency range $f \sim 2.1f_0$, determined from 10 measurements, is 0.3 ± 0.2. This is rather higher than a theoretical estimate from potential theory, though the variance of experimental values is high.

These experimental data allow us to formulate the following schematic representation of the spectra of surface elevation and velocity components in the upper ocean. The wave spectrum can be represented as the sum of a linear spectrum (obeying the linear dispersion relationship), a spectrum of nonlinear components, and noise (instrumental and statistical):

$$S_\eta = S_{\eta 1} + S_{\eta n} + S_n$$

and for the spectrum of the vertical component of velocity

$$S_w = S_{w1} + S_{wh} + S_{wt} + S_{wn}$$

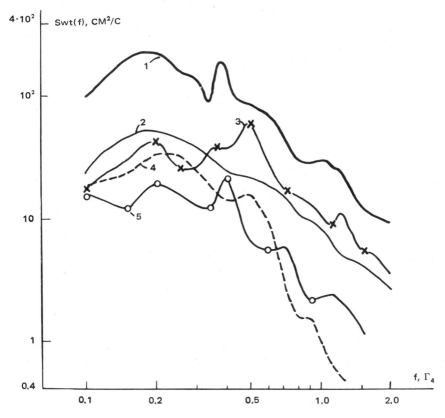

FIGURE 12. Spectra of the noncoherent part of vertical components of velocity.

where S_{wt} is a spectrum of turbulent velocity fluctuations. The nonlinear components, turbulent fluctuations, and noise may be considered uncorrelated. It is evident from the results of these experiments that the wave components are close to linear in the region of the spectral maximum at lower frequencies; in this frequency range, $R_{\eta w}^2$ vary from 0.95 to 0.98, and these deviations from unity cannot be attributed to noise.

Figure 12 shows spectra of the vertical velocity components that are not coherent with surface elevation. The spectral density S_{wt} in the region of the basic energy-containing frequencies varies by more than one order of magnitude and is dependent on wave energy. In these cases, the rms value of the fluctuations in turbulent vertical velocity W_t varies from 2 to 6 cm/s in the frequency range $f < 2f_0$, and the ratio $(w_n^2 + w_z^2)/w^2$ in the range $f < 1.5$ Hz is approximately equal to 3–10% (w_n and w_1 represent the vertical components of the nonlinear and turbulent contributions; w is the total vertical velocity).

The spectra of horizontal velocity components noncoherent with the elevation, are much larger than S_{wt} in the region of the spectral maximum, since they are determined by the directional structure of the wave field and the orientation of the sensors.

4. REYNOLDS STRESSES IN THE UPPER OCEAN

It is well known that the tangential Reynolds stresses in a potential wave field vanish; the normal stresses are connected by the simple relation $u^2 + v^2 = w^2$ and the corresponding phase shifts $\phi_{\eta u}$, $\phi_{\eta v} = 0$ and $\phi_{\eta w} = \pi/2$. Under real conditions, however, wind waves are not free and

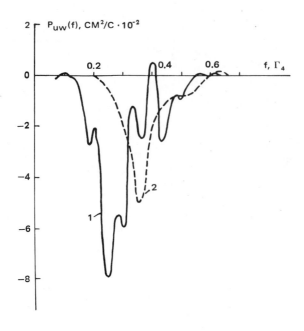

FIGURE 13. Spectra of shear wave stresses for two cases and wind waves.

linear, and the presence of turbulence may cause an increase in the effective fluid viscosity of several orders of magnitude over the molecular value.

Figure 13 shows estimates of the co-spectrum $P_{uw}(f)$ for two wind wave cases. Well-defined peaks at frequencies corresponding to the spectral maxima are indicative of momentum fluxes associated with the waves and the vorticity of the wave motion. The sign of the tangential stresses in the water (note that the z axis is directed upwards from the surface) coincides with that in the air—momentum flux in the water is directed downwards from the free surface. Typical values of the tangential stresses are listed in Table I. The values of Reynolds stress (τ) vary significantly, but in general they fall inside the same limits as those given by Shonting.

TABLE I.   Values of Tangential Stresses in the Ocean Surface Layer[a]

No.	U(m/s)	η^2(cm^2)	f_0(Hz)	$-z$(m)	$\bar{u}^2$(cm^2/s)	$\bar{w}^2$(cm^2/s^2)	K_{uw}	τ (dynes/cm^2)
1	11.0	1000	0.18	1.2	840	1250	−0.10	102
2	10.5	320	0.32	1.2	270	380	−0.23	74
				2.25	130	170	−0.10	15
3*	8.0	350	0.20	1.2	320	370	−0.08	28
			0.34	2.25	130	200	0.03	− 5
4*	6.0	110	0.30	1.2	80	120	−0.11	11
				2.25	30	45	−0.12	4
5*	4.5	230	0.20	1.2	180	250	−0.02	4
			0.47					
6	10.5	150	0.35	1.2	190	240	−0.20	43
				2.25	50	120	−0.11	9
7	10.0	280	0.33	1.2	240	340	−0.24	69
				2.25	110	150	−0.12	15

[a]U is wind speed at a height of 10 m; η^2, $\bar{u}^2$, $\bar{w}^2$ are dispersions of waves and velocity components; f_0 are frequencies of main maxima of wave spectrum; z is depth; K_{uw} is the correlation coefficient of u and w velocity components, τ is Reynolds shear stress. An asterisk denotes several wave systems.

To examine these wave stresses in more detail, let us consider the phase shifts between the surface elevation and the velocity components w, u, and v. Figures 14 and 15 show experimental estimates of the phase relations between the surface elevation and these velocity components. Near the frequency of the spectral maximum, $\phi_{\eta w}$ is close to $\pi/2$, and $\phi_{\eta u}$, $\phi_{\eta v}$ change from 355° to 340° in the case of wind waves and are close to zero for swell. At higher frequencies, the coherence decreases and the phase shifts become unstable. It should be emphasized that the experimental values for phase shifts are close to the theoretical values for potential waves in the case of swell. This is a characteristic feature of our measurements which, indeed, is to be expected since the swell behave most nearly as free waves.

The experimental estimates for $\phi_{\eta w}$ have low variances in the various series of measurements, the coherences ranging from 0.95 to 0.98 in the range of the spectral peak. Figure 16

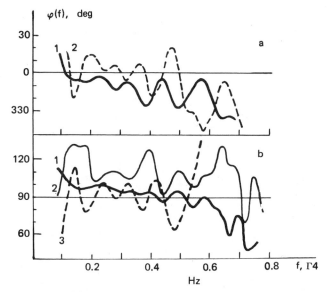

FIGURE 14. Phase shifts between the elevation and the velocity components for wind waves. (a) 1, $\phi_{\eta u}$; 2, $\phi_{\eta v}$. (b) 1, $\phi_{\eta w}$; 2, ϕ_{uw}; 3, ϕ_{vw}.

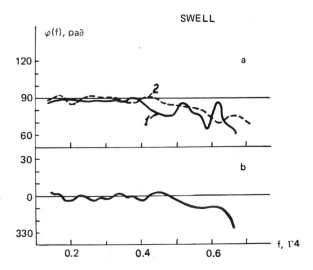

FIGURE 15. Phase shifts between the elevation and the velocity components for swell. (a) 1, $\phi_{\eta w}$; 2, ϕ_{uw}. (b) $\phi_{\eta u}$.

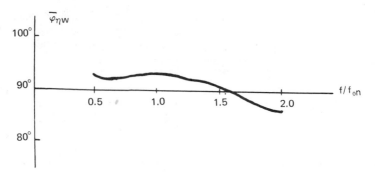

FIGURE 16. Averaged phase shifts $\bar{\phi}_{\eta w}$ for 32 measurements.

displays the spectral phase shift $\bar{\phi}_{\eta w}$, averaged over 32 records in the upper ocean layer at a depth of 3 m, as a function of dimensionless frequency f/f_0. Figure 16 indicates that the deviations of this average phase from $\pi/2$ are not significant (about $\pm 3°$) and are located within the confidence interval.

The scatter of values of corresponding averages $\phi_{\eta u}$ and $\phi_{\eta v}$ is much greater and depends both on the characteristics of the wave, and on the orientation of the current meters relative to the average direction of wave propagation. For this reason, experimental estimates of the normal Reynolds stresses are given in Table I only for the strongest component of horizontal velocity. The maximum values of the coherences $R^2_{\eta w}$ and R^2_{uw} are equal on average to 0.8 to 0.9. In spite of the variance inherent in these data, all deviations of experimental values $\tau = -\overline{uw}$ from zero for the wind waves, have the same sign and indicate that the maximum horizontal velocity component occurs somewhat after the passage of the wave crest.

The fact that the phase difference between u and w is greater than $\pi/2$ indicates that the stresses are nonzero at the dominant wave frequency. It should be noted that the difficulty of measurement of small values of τ in the wave motion is not connected, for example, with errors in measurement and processing, but results from the structure of the wave motion itself. Increased instrumental accuracy will not decrease, in the main, the scatter of estimates since this will be determined as usual by statistical errors owing to the limited values of the coherence R^2_{uw}. This is typical of all experimental estimates of tangential Reynolds stresses in wave motions in which the dominant flow is close to potential. One may overcome this difficulty only by increasing considerably the number of degrees of freedom in the estimate.

It is likely that a simple increase in the length of record will be insufficient because of the inevitable nonstationarity in the wave field. A large number of simultaneous measurements are required, with sensors sufficiently separated and ensemble averaging of realizations. Because of this, the estimates obtained here for the Reynolds stresses of wind waves must be treated as essentially qualitative.

III

WAVE INSTABILITIES AND BREAKING

13

ADVANCES IN BREAKING-WAVE DYNAMICS

M. S. LONGUET-HIGGINS

ABSTRACT. Progress has been made recently not only in numerical methods of calculating steep waves but also in finding analytic solutions for overturning fluid motions. Thus, the tip of the wave can take the form of a slender hyperbola, in which the orientation of the axes and the angle between the asymptotes are both functions of the time. The initial stages of overturning are given approximately by a two-term expression for the potential, with a branch-point located in the "tube" of the wave. Most remarkably, plunging breakers appear experimentally to tend toward certain exact, time-dependent flows (recently discovered) in which the forward face is given by a simple parametric cubic curve. Dynamical aspects breaking waves are discussed, particularly in terms of angular momentum.

1. INTRODUCTION

Breaking waves may play an important part in the transfers of heat, momentum, salt, water, and dissolved gases between ocean and atmosphere, as well as in the generation of turbulence and longshore currents (not to mention the support of recreational facilities in the surf zone). Yet surprisingly, the proper mathematical and physical description of overturning surface waves is still very largely lacking.

The difficulties, as is well known, arise from the highly nonlinear character of a flow in which the particle accelerations are of order g (so that linear or weakly nonlinear theory is inapplicable) and secondly from the time dependence, which prevents the customary reduction to a steady state by choice of moving axes; and in the case of overturning waves, from the multivalued character of the surface elevation, when viewed as a function of the horizontal coordinate (see Figs. 1 and 2).

The purpose of this chapter is to review some recent progress in this field. First, however, it is worth mentioning that the form of the highest steady gravity waves in water of uniform depth has recently been calculated to unsurpassed accuracy by Williams (1981) in such a way as to resolve previous controversies, particularly over the height $2a$ of the steepest solitary wave in water of undisturbed depth h (see Miles, 1980). Thus, Williams finds $2a/h = 0.833197$, which an uncertainty of about one unit in the last decimal place. For waves in deep water, the ratio of height to length is found to be 0.141063.

The form of steep but steady waves of less than the maximum steepness has been calculated to modern accuracy by Cokelet (1977) with independent confirmation by Chen and Saffman

M. S. LONGUET-HIGGINS ● Department of Applied Mathematics and Theoretical Physics, University of Cambridge, Cambridge, England, and Institute of Oceanographic Sciences, Wormley, England GU85UB.

(1980), Olfe and Rottman (1979), and Vanden-Broeck and Schwartz (1979). More interestingly, Chen and Saffman have found that the sequence of gravity waves, considered as a function of their maximum height, bifurcates into a train of waves of steady but *non*uniform amplitudes at the point $2a/L = 0.1289$. This agrees well with the point of neutral stability for subharmonic perturbations, two wavelengths long, as calculated by Longuet-Higgins (1978b).

The process of breaking being essentially *un*steady, it is quite unlikely that a typical breaking wave will pass through any of the steady configurations just mentioned. One approach to the time-dependent problem (see Longuet-Higgins, 1978a) is to consider the breaking waves which arise out of the normal-mode instabilities of steep but steady trains of gravity waves (no extra energy being supplied by surface pressures). Progress in this direction has been described in a previous review paper (Longuet-Higgins, 1978b). In all the cases studied so far, it has been found that normal-mode instabilities, when followed in time by numerical time-stepping (Longuet-Higgins and Cokelet, 1976, 1978), do lead ultimately to an overturning of the free surface and hence to a significant loss of energy, as first observed experimentally by Benjamin (1976).

However, to follow the breaking process as far as overturning (that is to say, beyond the limits of small-perturbation theory), the only tools so far available have been numerical methods. In spite of recent progress in time-stepping techniques for breaking waves (Baker and Israeli, 1981; Vinje and Brevig, 1981), such methods give no clear understanding of the mathematical and physical processes involved. Complementary to such an approach we clearly need an *analytical* description of the later stages of overturning, so as to yield a class of possible asymptotic forms. This must be combined with analytical methods for joining the asymptotic forms to the initial stages of the flow.

Fortunately, we are encouraged by the observation (see Figs. 1 and 2) that a typical plunging breaker does exhibit a rather well-recognizable form. It is the smooth flow associated with this

FIGURE 1. Waves breaking in shallow water, in the Hawaiian Islands.

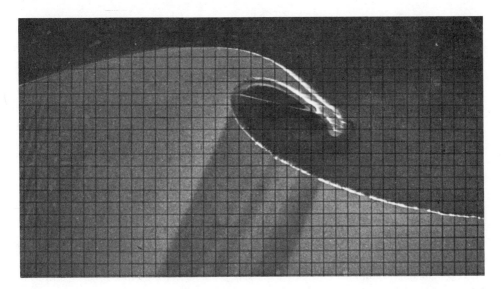

FIGURE 2. Waves breaking in deep water, in a laboratory channel. (From Miller, 1957.)

form that we shall consider, rather than the later stages of breaking associated with strong shear and turbulence, as in a breaker of "spilling" type.

The plan of the chapter is as follows. After stating general equations (Section 2) we describe an approach in which a breaking wave is divided into three parts: (1) the flow near the tip of the jet; here a precise asymptotic solution has been obtained in the form of a slender rotating hyperbola; (2) the time-dependent flow near the crest just prior to overturning; here an approximate solution is obtained which is valid until the surface slope attains an angle of inclination of about 65°; (3) the remaining part of the wave. This line of investigation envisages joining the different parts of the flow together by matching techniques.

The second part of the chapter, Sections 5 to 8, describes a new class of flows which have a free surface closely in agreement with the forward face of typical plunging breakers. The solutions are given in parametric form, which appears particularly suited to flows with a branch-point. It seems possible that a complete analytic solution will ultimately be found in the semi-Lagrangian type of representation that is here used.

Lastly in Section 9 we summarize some dynamical arguments related to wave breaking, which make use of the integral property of angular momentum.

2. GENERAL EQUATIONS

Consider a situation as in Fig. 3. For simplicity, we assume the fluid to be inviscid and incompressible, and the flow to be irrotational and two-dimensional. Surface tension is neglected.

Take rectangular coordinates (x, y) in the plane of motion, with x vertically downwards, and write

$$z = x + iy, \qquad z^* = x - iy \tag{1}$$

so the complex velocity potential

$$\chi = \phi + i\psi \tag{2}$$

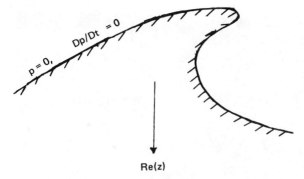

FIGURE 3. Coordinates and boundary conditions.

is a function of z and t only, being independent of z^*, and the velocity is W^*, where

$$W = \chi_z, \qquad W^* = \chi_z^* = \chi_{z^*}^* \tag{3}$$

(we use suffixes to denote partial differentiation).

The boundary conditions are that the pressure p and its rate of change Dp/Dt following a particle must both vanish at the free surface. From Bernoulli's equation we have

$$-2p = (\chi_t + \tfrac{1}{2}WW^* - gz - F) + \text{c.c.} \tag{4}$$

and since

$$\frac{D}{Dt} \equiv \frac{\partial}{\partial t} + W^* \frac{\partial}{\partial z} + W \frac{\partial}{\partial z^*} \tag{5}$$

we find also

$$-2\frac{Dp}{Dt} = [(\chi_{tt} + 2W^*\chi_{zt} + W^{*2}\chi_{zz}) - gW^* - F_t] + \text{c.c.} \tag{6}$$

(see Longuet-Higgins, 1980a).

It is often convenient to express both z and χ as functions of a third complex variable ω, and the time t. The last three equations are then generalized as follows:

$$-2p = (\chi_t - Wz_t + \tfrac{1}{2}WW^* - gz - F) + \text{c.c.} \tag{7}$$

and

$$-2\frac{Dp}{Dt} = [(\chi_{tt} - Wz_{tt}) + 2K(s_{\omega t} - Wz_{\omega t}) + K^2(\chi_{\omega\omega} - Wz_{\omega\omega}) - gW - F_t] + \text{c.c.} \tag{8}$$

where

$$W = \chi_\omega/z_\omega = Dz^*/DT$$
$$K = (W^* - z_t)/z_\omega = D\omega/Dt \tag{9}$$

the general expression for D/Dt being

$$\frac{D}{Dt} \equiv \frac{\partial}{\partial t} + K\frac{\partial}{\partial \omega} + K^*\frac{\partial}{\partial \omega^*} \tag{10}$$

3. TIP OF THE WAVE: THE ROTATING HYPERBOLA

An approximation possibly valid near the tip of the overturning jet (but before it breaks up into spray) has been derived by Longuet-Higgins (1980b). Essentially, this is a generalization of the "Dirichlet hyperbola" discussed earlier (Longuet-Higgins, 1972, 1976).

Take axes accelerating downwards with acceleration g. Then an *exact* solution of the boundary conditions $p = 0$, $Dp/Dt = 0$ is given by

$$\chi = \tfrac{1}{2}Az^2, \qquad A = \alpha e^{i\sigma} \tag{11}$$

where α and σ are real functions of t only, satisfying

$$\sigma_t = \lambda \alpha^2$$
$$\alpha\alpha_{tt} - 4\alpha_t^2 + 2\alpha^4 - \lambda^2\alpha^6 = 0 \tag{12}$$

λ being an arbitrary constant. Equations (12) can be integrated to give

$$\sigma = \int \lambda \alpha^2 dt \tag{13}$$

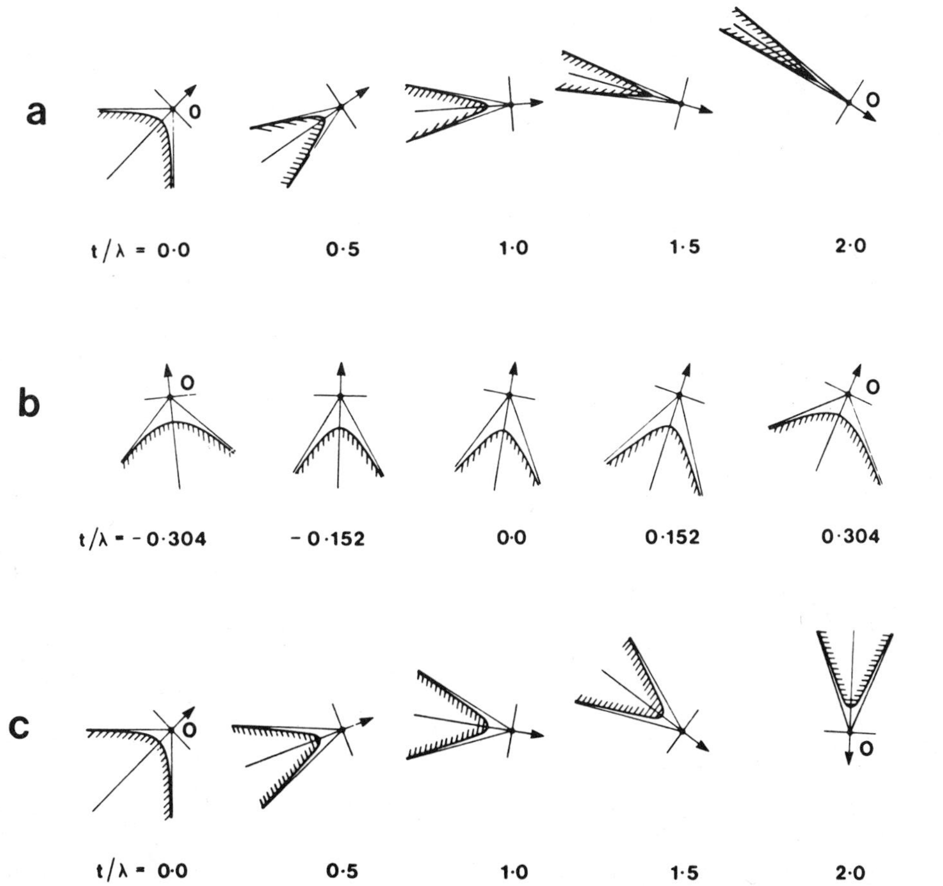

FIGURE 4. The rotating hyperbola, seen in a free-fall reference frame. (a) Class I when $\tilde{\omega} = 0.30$; (b) Class II when $\tilde{\omega} = 0.20$; (c) Class I when $\tilde{\omega} = 0.25$. (From Longuet-Higgins, 1980b.)

and

$$t = \int^c \frac{-\,d\alpha}{\alpha^2(1 - \lambda^2\alpha^2 + P\alpha^4)^{1/2}} \tag{14}$$

P being a second constant of integration. The free surface ($p = 0$, $Dp/Dt = 0$) is then

$$A_t z^2 + 2AA^* zz^* + A_t^* z^{*2} = -2S\alpha^4 \tag{15}$$

S being a third constant. When $S > 0$, this represents a hyperbola. The axes rotate with angular velocity $-\delta_t$ where

$$\delta_t = \lambda\alpha^2/(1 + P\alpha^4) \tag{16}$$

and the angle γ between the asymptotes is given by

$$\cos\gamma = 1/(1 + P\alpha^4) \tag{17}$$

Examples are shown in Fig. 4. The solutions fall into several classes, corresponding to the range of values of $P/\lambda^4 = \tilde{\omega}$. The most interesting case is $\tilde{\omega} > 1/4$, when the asymptotes tend monotonically to zero as $t \to \infty$, so that as time increases the jet becomes sharp-pointed. Meanwhile, the principal axes turn through a finite angle δ (see Fig. 4a).

When $\tilde{\omega} < 1/4$, the angle γ decreases from $90°$ to a minimum value (Greater than $45°$) and then increases again, as in a partly reflected wave (see Fig. 4b).

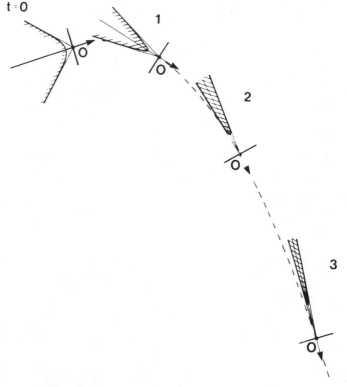

FIGURE 5. The rotating hyperbola, seen in a stationary reference frame $\tilde{\omega} = 0.30$, $U/g\lambda = 1$. (From Longuet-Higgins, 1980b.)

In the intermediate case when $\tilde{\omega} = 1/4$, the angle tends to 45° and the axes rotate indefinitely with angular velocity tending to $1/\lambda$.

In Fig. 4 the origin O was always shown as moving in a horizontal straight line. The reference frame was assumed to be in free-fall. However, relative to a *fixed* reference frame, the origin describes a parabola, with constant horizontal velocity U say, and downward acceleration g. Figure 5 shows the position of the free surface seen in a stationary frame of reference in the typical case $\tilde{\omega} = 0.30$ and when the dimensionless parameter $U/\lambda g$ is taken as unity.

By including in the expression (11) further terms in z^3, z^4, etc., it may be possible to obtain local solutions that have the form of a cusp as $t \to \infty$. For further details, see Longuet-Higgins (1980b).

4. OVERTURNING: INITIAL STAGES

On the other hand, a class of flows representing the initial stages of overturning, not too far from the wave crest, has been derived elsewhere (Longuet-Higgins, 1981). It is argued that because of the evident discontinuity in flow when the jet meets the forward face of the wave, the

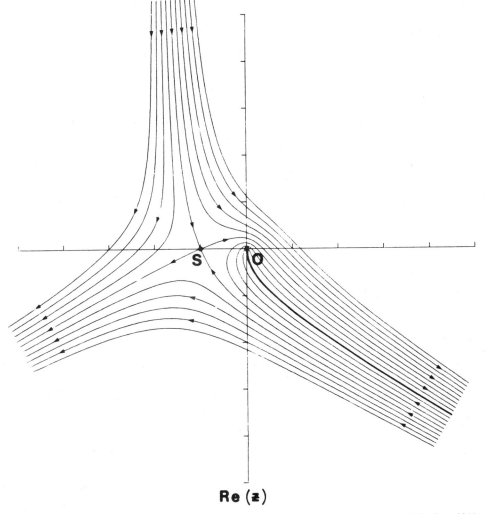

Re (z)

FIGURE 6. Instantaneous streamlines of the flow (19) when arg $A = 180°$. (From Longuet-Higgins, 1981.)

velocity W^* must have a branch-point in the tube of the breaking wave, just outside the fluid. The simplest such branch-point is one of index $\frac{1}{2}$, for example the Stokes corner-flow

$$\chi = \tfrac{2}{3}ig^{1/2}z^{3/2}, \qquad W = i(gz)^{1/2} \tag{18}$$

To obtain a flow which curls over forwards in the observed manner, one can add to χ a term in $z^{1/2}$, which is relatively small at infinity but which dominates near the origin. The streamlines of the resulting flow

$$\chi = \tfrac{2}{3}ig^{1/2}z^{3/2} + 2A(t)z^{1/2} \tag{19}$$

are shown in Fig. 6, for a particular value of $A(t)$. For such a flow it is not generally possible to ensure that the surfaces $p = 0$ and $Dp/Dt = 0$ coincide. But it is shown (in the paper cited) that the surfaces will indeed coincide as $|z| \to \infty$ provided

$$A_{tt} = 0 \quad \text{and} \quad A_t = \tfrac{2}{3}iF_t \tag{20}$$

hence A and F are linear functions of the time t and A_t is pure imaginary. We can then choose F_t and the initial value $F(0)$ of F at time $t = 0$ so as to minimize

$$R \equiv \int \left(\frac{Dp}{Dt}\right)^2 d\sigma \tag{21}$$

(where the integral is taken along the contour $p = 0$). This gives the free surface shown in Fig. 7 (upper curve). The form of the free surface at preceding times $t < 0$ is shown by the curves beneath. (The units are normalized so that $g = 1$, $A_0 = -1$.) However, t cannot be increased beyond about 5 without the value of R becoming unacceptably large.

Further accuracy is obtained by adding to (20) a third term linear in z:

$$\chi = \tfrac{2}{3}ig^{1/2}z^{3/2} + Uz + \tfrac{1}{2}Az^{1/2} \tag{22}$$

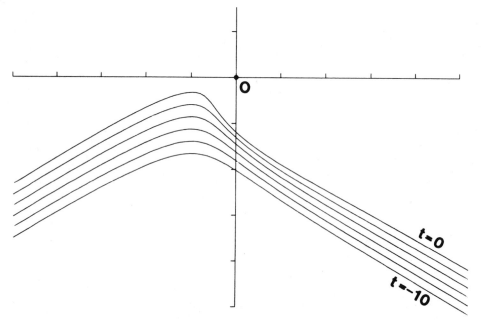

FIGURE 7. Successive positions of the free surface determined by minimizing $R(F_0, F_t)$ at time $t = 0$.

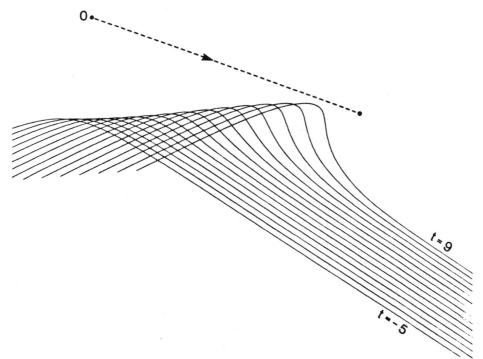

FIGURE 8. Surface profiles corresponding to the three-term expression (23), when R is minimized. (The constant velocity of 0 is chosen arbitrarily.) (From Longuet-Higgins, 1981.)

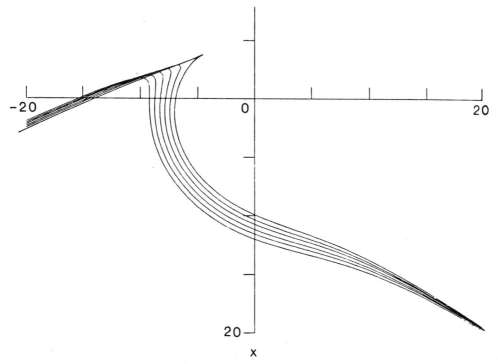

FIGURE 9. A cusped profile: The boundary conditions are satisfied only at the cusp itself and as $|Z| \to \infty$. (From Longuet-Higgins, 1981.)

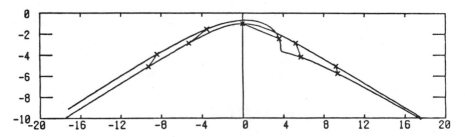

FIGURE 10. The lowest nondegenerate mode of perturbation of the "almost-highest wave." (From Cleaver, 1981.)

U being a complex constant. The conditions for the surface $p = 0$ and $Dp/Dt = 0$ to coincide as $|z| \to \infty$ are then that

$$A_{tt} = 0, \qquad A_t = \tfrac{2}{3}i(F_t + \tfrac{1}{2}U - \tfrac{1}{4}U^*) \tag{23}$$

which generalize (19). Minimizing the integral (21) at time $t = 0$ then gives the sequence of profiles shown in Fig. 8.

Even here, however, comparison with numerical solutions suggests that (22) is not valid much beyond about $t = 7$, when the maximum slope angle of the free surface is about 65°.

Equation (22) is also able to yield approximate solutions in which the free surface takes the form of a sharp cusp (see Fig. 9). Here the minimal condition (21) is replaced by the conditions that both p and Dp/Dt shall vanish at the cusp itself (though not necessarily at other points along the curve, except as $|z| \to \infty$).

In solutions such as shown in Fig. 9, a slight convexity on the forward face appears to be unavoidable. Interestingly, a similar feature is found by Cleaver (1981) in calculating the (linear) instabilities of the almost-highest wave. Figure 10 shows the lowest nondegenerate normal-mode perturbation, exaggerated so as to improve clarity. Cleaver finds that the convexity on the forward lower face tends to disappear when account is taken of the finite-wavelength correction to the inner flow.

5. PARAMETRIC REPRESENTATIONS

A class of flows which seems to represent rather well the forward face of an overturning wave can be derived as follows (see Longuet-Higgins, 1982).

Suppose that χ and z are each expressed in terms of a third complex variable ω, and t, as in Section 2. An immediate consequence is that any simple zero of z_ω corresponds in general to a branch-point of the flow. For if z_ω vanishes when $\omega = \omega_1$, say, then near ω_1 we have

$$(z - z_1) \sim \tfrac{1}{2}z_{\omega\omega}(\omega - \omega_1)^2 \tag{24}$$

and so if $z_{\omega\omega} \neq 0$,

$$(\omega - \omega_1) \propto (z - z_1)^{1/2} \tag{25}$$

we therefore look for solutions in which z_ω vanishes outside, but not too far from, the free surface.

A considerable simplification is introduced if after John (1953) we assume that, for particles in the free surface C,

(1) $\omega = \omega^*$ (ω is real)   (26)

(2) $D\omega/Dt = 0$ (ω is Lagrangian)   (27)

The first condition implies that the tangent to the surface is in the direction of z_ω. The second condition implies that the velocity and acceleration are equal to z_t and z_{tt}, respectively. Thus, the pressure gradient is in the direction $(z_{tt} - g)$. Since this is to be normal to the free surface, the boundary condition then reduces to

$$z_{tt} - g = irz_\omega \qquad \text{on } C \tag{28}$$

where $r(\omega, t)$ is a function which is real when ω is real.

Moreover, the above formulation enables us to construct a velocity potential $\chi(\omega, t)$, for any given $z(\omega, t)$, such that

$$\chi_z = u - iv \tag{29}$$

For let

$$\chi = \int z_t^*(\omega) z_\omega(\omega) d\omega \tag{30}$$

(where, by $z_t^*(\omega)$ we mean $[z_t(\omega^*)]^*$). Then

$$\chi_z = \chi_\omega / z_\omega = z_t^*(\omega) \tag{31}$$

and so

$$u + iv \equiv \chi_z^* = z_t(\omega^*) \tag{32}$$

On the boundary, where $\omega = \omega^*$, we have $u + iv = z_t(\omega)$ as required. In the interior, however, we have only (32) so that the flow is not Lagrangian everywhere. We call such a representation "semi-Lagrangian."

Note that if a particular solution $z = z^{(1)}(\omega, t)$ to (28) can be found with a certain function $r(\omega, t)$, then the general solution is of the from $z = z^{(1)} + z^{(2)}$, where $z^{(2)}$ satisfies the homogeneous first-order equation

$$z_{tt} = irz_\omega \tag{33}$$

with the same function r.

6. OVERTURNING FLOWS

Consider now a particle at a free surface which is inclined at an angle of 30° to the horizontal, as on the forward face of a steep gravity wave. The acceleration of the particle will be equal to $\frac{1}{2}g$, directed down the slope, that is

$$g' = \frac{1}{2}ge^{i\pi/3} \tag{34}$$

If at time $t = 0$ the particle is at rest at $z = 0$, its position at subsequent times will be given by

$$z = \frac{1}{2}g't^2 \tag{35}$$

The simplest nontrivial generalization of (35) is the expression

$$z(\omega, t) = g'(\tfrac{1}{2}t^2 - \omega t) \tag{36}$$

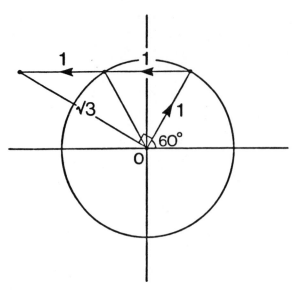

FIGURE 11. Diagram showing the relation $e^{i\pi/3} - 2 = 3^{1/2}ie^{i\pi/3}$.

where ω is real on the surface. Since

$$z_{tt} - g = \tfrac{1}{2}g(e^{i\pi/3} - 2) \tag{37}$$

it is easily seen (Fig. 11) that

$$z_{tt} - g = -3^{1/2}iz_\omega/t \tag{38}$$

in other words the boundary condition (28) is satisfied, with

$$r = -3^{1/2}/t \tag{39}$$

We may add to the particular integral (36) the complementary function, that is to say the general solution of the homogeneous equation (33), which is

$$itz_{tt} = 3^{1/2}z_\omega \tag{40}$$

To find solutions, set

$$z = a_n\omega^n + a_{n-1}\omega^{n-1} + \cdots + a_0 \tag{41}$$

and equate coefficients of $\omega^n, \omega^{n-1}, \cdots 1$. This gives

$$\left.\begin{aligned}
it(a_n)_{tt} &= 0 \\
it(a_{n-1})_{tt} &= 3^{1/2}a_n \\
&\vdots \\
it(a_0)_{tt} &= 2^{1/2}a_1
\end{aligned}\right\} \tag{42}$$

From the first relation, a_n is of the form $(A_n + B_n t)$ where A_n and B_n are constants, and since the equations are linear we may superpose solutions in which $(A_n, B_n) = (1, 0)$ or $(0, 1)$. Calling these P_n and Q_n, respectively, we easily find, up to $n = 3$,

$$P_1 = t\omega - (3^{1/2}/2)it^2$$
$$P_2 = t\omega^2 - 3^{1/2}it^2\omega - (1/2)t^3$$
$$P_3 = t\omega^3 - 3^{3/2}/2\,it^2\omega^2 - (3/2)t^3\omega + (3^{1/2}/8)it^4$$
(43)

and

$$Q_1 = \omega - 3^{1/2}it\ln t$$
$$Q_2 = \omega^2 - 2(3^{1/2})i(t\ln t)\omega - 3t^2(\ln t - 3/2)$$
$$Q_3 = \omega^3 - 3^{3/2}i(t\ln t)\omega^2 - 9t^2(\ln t - 3/2)\omega + 3^{3/2}/2\,it^3(\ln t - 7/3)$$
(44)

(apart from $P_0 = t$ and $Q_0 = 1$).

The complete polynomial solution is of the form

$$z = z_0 + \sum_n (A_n P_n + B_n Q_n)$$
(45)

where the A_n and B_n are arbitrary constants. However, each of the complementary flows P_n and Q_n may be regarded as a solution of the nonhomogeneous equation (38) with $g = 0$, hence as an actual flow seen in a reference frame moving with downwards acceleration g. When the original flow z_0 is viewed in this reference frame, it will be seen to be just a multiple of P_1.

7. INTERPRETATION

Note first that the flows P_n are all self-similar, since they may be expressed in the form

$$P_n(\omega, t) = t^{n+1}f_n(\mu), \qquad \mu = 2\omega/3^{1/2}t$$
(46)

The free surface therefore expands (or contracts) like t^{n+1}.

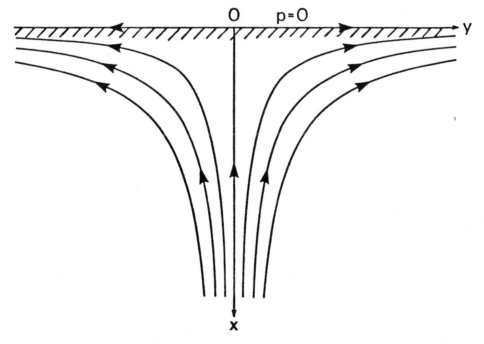

FIGURE 12. Instantaneous streamlines of the linear flow $z = P_1$.

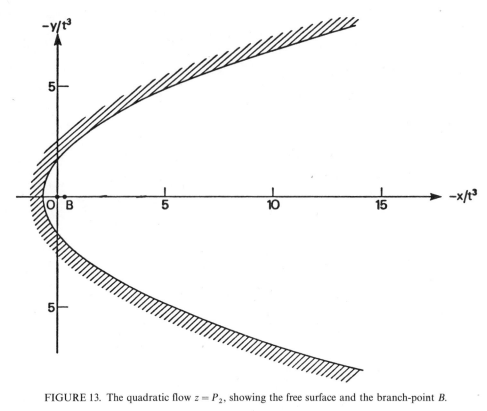

FIGURE 13. The quadratic flow $z = P_2$, showing the free surface and the branch-point B.

The linear flow P_1 is shown in Fig. 12. The free surface is plane and the instantaneous streamlines are hyperbolas. The flow may be described as "a decelerated upwelling," the velocity at any point being dependent on the time t.

The quadratic flow P_2 is shown in Fig. 13. The free surface has the form of a parabola, which expands or contracts about the origin, proportionally to t^3. There is a branch-point at the focus B (not the origin). Hence, the interior of the parabola must be excluded from the flow; the fluid is *outside* the parabola.

Figure 14 shows the cubic flow $z = P_3$. This is the simplest case in which the free surface

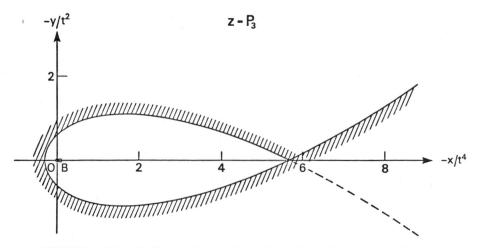

FIGURE 14. The cubic flow $z = P_3$, showing the free surface and one branch-point B.

intersects itself. Writing the equation for z in the form

$$z/t^4 = 3^{3/2}/8(\mu^3 - 3i\mu^2 - 2\mu + \tfrac{1}{3}i) \tag{47}$$

we see that on the free surface, where μ is real we have

$$\begin{aligned} x/t^4 &\propto \mu^3 - 2\mu \\ y/t^4 &\propto 1/3 - 3\mu^2 \end{aligned} \tag{48}$$

Rotating the axes through $90°$ and shifting the origin and scale, we have the equation of the surface in its simplest parametric form, namely

$$\begin{aligned} \xi/t^4 &= 3\mu^2 \\ \eta/t^4 &= \mu^3 - 2\mu \end{aligned} \tag{49}$$

where ξ and η are axes normal and tangential to the curve at its vertex ($\mu = 0$). The curve has a double point where $\mu = \pm 2^{1/2}$, hence $\xi/t^4 = 6$, and its width η/t^4 is a maximum when $\mu = \pm(2/3)^{1/2}$. Hence, the "aspect ratio," i.e., the ratio b/a in Fig. 14, is

$$b/a = (2/3)^{5/2} = 0.363\ldots \tag{50}$$

The flow has branch-point where $z_\mu = 0$, that is

$$\mu^2 - 2i\mu - 2/3 = 0 \tag{51}$$

or $\mu = (1 \pm 2/3^{1/2})i$. One of these is at the point B marked in Fig. 14; the other is outside the loop, but on another sheet of the Riemann surface.

8. COMPARISON WITH OBSERVATION

Figure 15 shows a profile of a plunging breaker from Miller (1957) on which is superposed the curve of Fig. 14. The agreement with the forward face of the wave is remarkably close.

Figure 16 shows some profiles for deep-water waves calculated numerically by Vinje and Brevig (1981). To each of the latter profiles we have added a dashed line to indicate the axis of symmetry of a cubic curve fitted to the lower part of the profile. The lengths of the axes, measured from Fig. 16, are shown in Table I, columns 2 and 3. In column 4 are shown the ratios b/a, which are clearly not far from the theoretical value (50).

According to (47) the linear dimensions of the surface profile vary with time like t^4. In Fig. 17 we have plotted the length of the minor axis b in Fig. 18, as a function of the time t (in arbitrary units) and have fitted this with a curve of the form $C(t - t_0)^4$, where C and t_0 are chosen constants. (The "tube" in this case is contracting, so that t_0 is positive). The agreement appears reasonable.

The orientation of the major axis, shown in Table I, column 5, is not precisely constant, but changes by only $10°$ between $t = 6$ and $t = 13$.

We remark that New (1981) has found numerically calculated profiles to be often quite closely fitted by an ellipse having axes in the ratio $3^{1/2}:1$. In Fig. 18 we show a comparison between the cubic curve of (49) and a "$3^{1/2}$-ellipse" having the same minimum curvature. The two curves agree quite closely over more than half the circumference of the ellipse.

The only part of the free surface described so far is the forward face of the wave. However, if we calculate the complete locus of $p = 0$, we find at least one other branch, labeled II in Fig. 19a. On this branch the second boundary condition $Dp/Dt = 0$ is not generally satisfied (as it is on I). Moreover, the curve is symmetric about the real axis of z. Nevertheless, since the pressure p has a

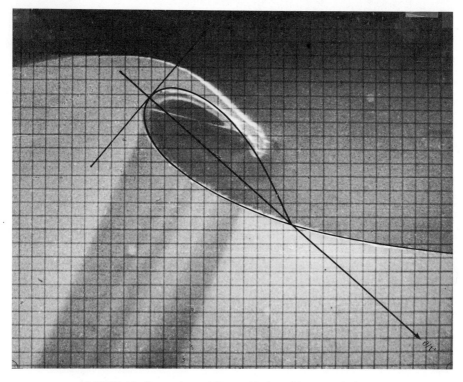

FIGURE 15. Comparison of Fig. 1 with the cubic curve of Fig. 14.

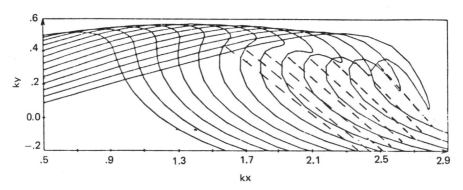

FIGURE 16. Numerically calculated profiles of a breaking wave in deep water, after Vinje and Brevig (1981).

TABLE I. Measured Parameters of Deep-Water Wave Crests, from Fig. 16

t	2a (mm)	2b (mm)	b/a	θ
6	64.0	26.0	0.41	37°
7	55.5	21.0	0.38	38°
8	52.5	19.0	0.37	39°
9	41.5	15.0	0.36	40°
10	40.5	14.0	0.35	41°
11	36.5	12.5	0.34	43°
12	32.5	11.2	0.35	45°
13	29.5	10.5	0.36	47°

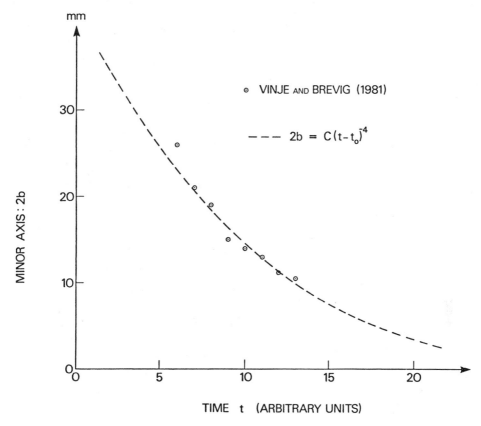

FIGURE 17. The breadth $2b$ of the inner tube in Fig. 16, as a function of the time t.

stationary point (saddle-point) on the real axis between I and II, it appears that only a slight asymmetric perturbation of the flow would be sufficient to alter the free surface to the form III, corresponding very nearly to the surface of a plunging breaker. Near the tip of the jet, the perturbed flow may locally take the form described in Section 3.

Figure 19b shows the same flow $z = P_3$ but at a different time $t = 0.5$. Apart from the perturbation, which is seen to be a small part of the total flow, the surfaces in Figs. 19a and b are precisely similar.

The other polynomials P_n and Q_n may also be useful. For example, by adding to P_3 a small multiple of Q_3, so that the ratio B_3/A_3 in (36) is imaginary, one obtains a flow in which the asymptotes at $|z| = \infty$ rotate in time. In addition, the constants A_n and B_n may be chosen so as to reduce the rms value of Dp/Dt on branch II of the surface $p = 0$.

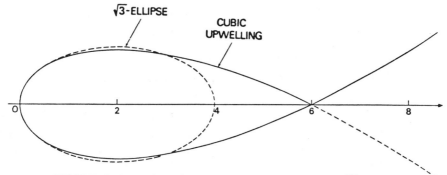

FIGURE 18. Comparison of the parametric cubic of (47) with a "$3^{1/2}$-ellipse."

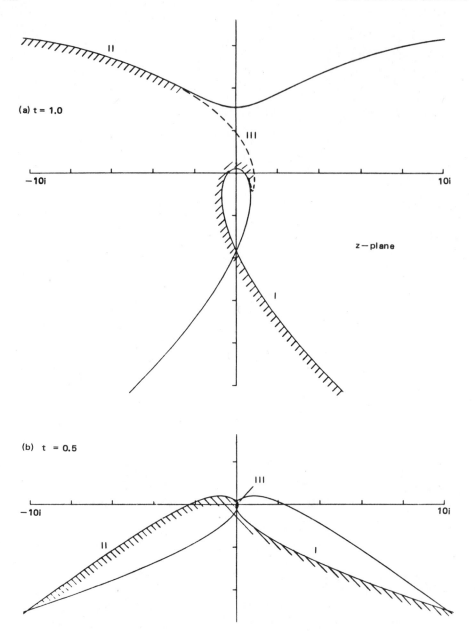

FIGURE 19. Contours of pressure $p = 0$ for the flow $z = P_3$.

Clearly, none of the flows discussed so far represents a complete solution to the problem of overturning. Nevertheless, each may represent a certain asymptotic aspect of the actual flow. It is reasonable to hope that by suitable matching of the different solutions, or by an appropriate generalization of the analytic expressions, a complete description of the flow may soon be found.

9. BREAKING WAVES AND ANGULAR MOMENTUM

So far we have considered the description of a breaking wave from a mathematical point of view. Some discussion in terms of familiar physical quantities also seems desirable. A beginning

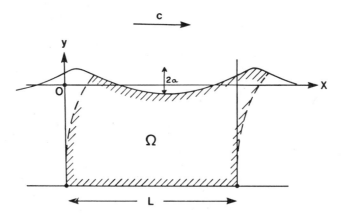

FIGURE 20. Coordinates and notation for considering angular momentum.

has been made (Longuet-Higgins, 1980c) by considering the role of angular momentum in allowing deep-water waves to support whitecaps without drastic change of form.

The argument is as follows. In any space-periodic motion in deep water (see Fig. 20), let Ω denote the fluid bounded by the free surface and two parallel, material surfaces separated by one wavelength L. At infinite depth the motion tends to zero. Hence, the total horizontal momentum of Ω is finite, and will be denoted by LI, so I is the momentum density. Similarly, the angular momentum of Ω about an arbitrary point P in the plane of motion is also finite and may be denoted by LA. It can be shown that A is independent of the horizontal coordinate of P but does depend on the vertical coordinate y, say, measured from the mean level. It is easily shown that

$$A(y) = A(0) - yI \tag{52}$$

Let $\bar{A}$ denote the Lagrangian-mean value of A; then a similar relation holds for $\bar{A}(y)$. Clearly,

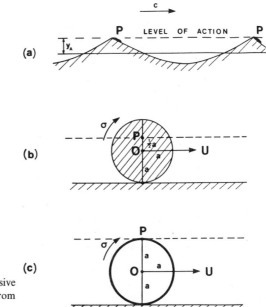

FIGURE 21. Comparison of a progressive wave with a rolling disk or hoop. (From Longuet-Higgins, 1980c.)

there is one particular value of y for which $\bar{A}(y)$ vanishes, and this is given by

$$y_a = \bar{A}(0)/I \tag{53}$$

y_a has been called the "level of action"; it is the level at which a small amount of momentum may be lost without seriously distorting the form of the wave.

An analogy can be drawn with a rolling disk or hoop (Fig. 21). For a uniform disk (Fig. 21b) the level of action is halfway between the center and the top of the disk. A horizontal impulse applied at a point P at this level will cause the disk to gain (or lose) horizontal momentum without slipping or skidding on the plane. For the hoop (Fig. 21c) the corresponding point is at the highest point on the circumference.

Now consider the gravity wave. When the wave breaks, a mass of water is thrown forward at a level near the crest (Fig. 21a) Subsequently, the mass may settle down to form a whitecap in a quasi-steady state. In any case, the loss of mass and momentum to the wave would tend to distort the wave drastically unless the loss took place near the level of action of the wave.

For a wave of speed c and *low* amplitude a, it is easily shown that

$$I = \tfrac{1}{2}ga^2/c, \qquad \bar{A}(0) = \tfrac{1}{4}a^2c \tag{54}$$

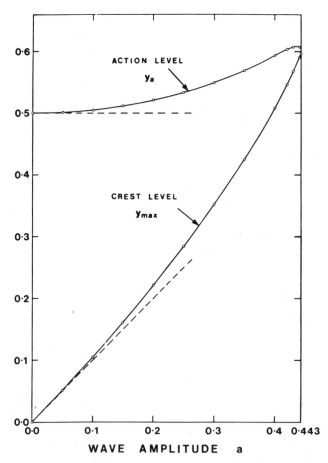

FIGURE 22. The level of action y_a and the crest level y_{max} as functions of the wave amplitude a, for deep-water waves $(L = 2\pi)$.

Hence,

$$y_a = c^2/2g \tag{55}$$

For gravity waves in deep water, where $c^2 = g/k$, this becomes

$$y_a = 1/2k = L/4\pi \tag{56}$$

so that the level of action is at about 1/12th the wavelength above the mean level. Since low waves lie always beneath this level, we have an explanation why low waves cannot support whitecaps.

In gravity waves of finite height $2a$, it is found by calculation (Longuet-Higgins, 1980c; Williams, 1981) that ky_a increases gradually from its value 0.50 for $ak \ll 1$ to the value 0.59 for waves of maximum height (see Fig. 22). Thus, for waves of maximum height, the level of action is very nearly, if not exactly, equal to the crest-height. This helps to explain how steep, irrotational waves can easily support whitecaps.

On the other hand, for Gerstner's rotational waves (see Lamb, 1932, p. 251) in which the particle orbits are perfect circles, the mass-transport is zero. Hence, I vanishes and y_a, by (53), becomes infinite. In other words, there is apparently no finite amplitude at which a Gerstner wave could support a whitecap.

Gerstner waves have of course a strong negative vorticity, and so are quite unlikely to occur in a natural wind-wave field. More to be expected are waves with a positive shear, associated with a positive wind-drift current. Banner and Phillips (1974) have already shown that in such a situation, surface waves will break at a lower wave amplitude than would irrotational waves. An application of the above argument suggests that with positive shear, surface waves can support whitecaps at a lower wave amplitude also.

Although we have discussed angular momentum in detail only for steady waves, it seems likely that the conservation of angular momentum (and of other integral quantities) will play a part in the physical understanding of unsteady waves as well.

REFERENCES

Baker, G. R., and M. Israeli (1981): Numerical techniques for free surface motion with application to drip motion caused by variable surface tension. *Lecture Notes in Physics* **141**, Springer-Verlag, Berlin, 61–67.

Banner, M. L., and O. M. Phillips (1974): On small scale breaking waves. *J. Fluid Mech.* **65**, 647–657.

Benjamin, T. B. (1967): Instability of periodic wavetrains in nonlinear dispersive systems. *Proc. R. Soc. London Ser. A* **299**, 59–67.

Chen, B., and P. G. Saffman (1980): Numerical evidence for the existence of new types of gravity waves of permanent form on deep water. *Stud. Appl. Math.* **62**, 1–21.

Cleaver, R. P. (1981): Instabilities of surface gravity waves. Ph.D. dissertation, University of Cambridge.

Cokelet, E. D. (1977): Steep gravity waves in water of arbitrary uniform depth. *Philos. Trans. R. Soc. London Ser. A* **286**, 183–230.

John, F. (1953): Two-dimensional potential flows with a free boundary. *Comm. Pure Appl. Math.* **6**, 497–503.

Lamb, H. (1932): *Hydrodynamics*, 6th ed. Cambridge University Press, London.

Longuet-Higgins, M. S. (1972): A class of exact, time-dependent free-surface flows. *J. Fluid Mech.* **55**, 529–543.

Longuet-Higgins, M. S. (1976): Self-similar, time-dependent flows with a free surface. *J. Fluid Mech.* **73**, 603–620.

Longuet-Higgins, M. S. (1978a): The instabilities of gravity waves of finite amplitude in deep water. *Proc. R. Soc. London Ser. A* **360**, 471–505.

Longuet-Higgins, M. S. (1978b): On the dynamics of steep gravity waves in water. *Turbulent Fluxes Through the Sea Surface, Wave Dynamics, and Prediction* (A. Favre and K. Hasselmann, eds.), Plenum Press, New York, 199–219.

Longuet-Higgins, M. S. (1980a): A technique for time-dependent, free surface flows. *Proc. R. Soc. London Ser. A* **371**, 441–451.

Longuet-Higgins, M. S. (1980b): On the forming of sharp corners at a free surface. *Proc. R. Soc. London Ser. A* **371**, 453–478.

Longuet-Higgins, M. S. (1980c): Spin and angular momentum in gravity waves. *J. Fluid Mech.* **97**, 1–25.

Longuet-Higgins, M. S. (1981): On the overturning of gravity waves. *Proc. R. Soc. London Ser. A* **376**, 377–400.

Longuet-Higgins, M. S. (1982): Parametric solutions for breaking waves. *J. Fluid Mech.* **121**, 403–424.

Longuet-Higgins, M. S., and E. D. Cokelet (1976): The deformation of steep surface waves on water. I. A numerical method of computation. *Proc. R. Soc. London Ser. A* **350**, 175–189.

Longuet-Higgins, M. S., and E. D. Cokelet (1978): The deformation of steep surface waves on water. II. Growth of normal-mode instabilities. *Proc. R. Soc. London Ser. A* **364**, 1–28.

Miles, J. W. (1980): Solitary waves. *Ann. Rev. Fluid Mech.* **12**, 11–43.

Miller, R. L. (1957): Role of vortices in surf zone prediction, sedimentation and wave forces. *Beach and Nearshore Sedimentation* (R. A. Davis and R. L. Ethington, eds.), *Soc. Econ. Palaeontol. Mineral.*, Tulsa, Okla., 92–114.

New, A. L. (1981): Breaking waves in water of finite depth. *British Theor. Mech. Colloquium*, Bradford, England, pp. 6–9.

Olfe, D. B., and J. W. Rottman (1979): Numerical calculations of steady gravity–capillary waves using an integro-differential formulation. *J. Fluid Mech.* **94**, 777–793.

Vanden-Broeck, J.-M., and L. W. Schwartz (1979): Numerical computation of steep waves in shallow water. *Phys. Fluids* **22**, 1868–1971.

Vinje, T. and Brevig, P. (1981): Breaking waves on finite water depths: a numerical study. *Ship. Res. Inst. Norway*, Report R–111.81.

Williams, J. M. (1981): Limiting gravity waves in water of finite depth. *Philos. Trans. R. Soc. London Ser. A* **302**, 139–188.

DISCUSSION

HOGAN: Did you manage to fit your *bifurcated* solutions to Miller's photographs?

LONGUET-HIGGINS: Not yet, but with 12 free parameters to play with, I am confident that it can be done!

HUANG: In your analysis, the coordinate system is free-falling, therefore noninertial. How will this choice of coordinate system influence the equations of motion and boundary conditions?

LONGUET-HIGGINS: Only in the first solution I described (for the tip of the jet) was a free-fall reference frame chosen. In all other cases the coordinate system was (at most) in uniform motion, and the effects of gravity were therefore included.

WITTING: In casual observations of breaking waves on a beach, I have the impression that the corner (once formed) undergoes horizontal acceleration forward. You have this in free-fall (no horizontal acceleration), which makes sense. Can we definitely depend on this particular conclusion, i.e., there can be no horizontal acceleration of the corner, once formed, in actual waves on a beach?

LONGUET-HIGGINS: Impressions derived from visual observations can sometimes be misleading. I would hesitate to comment, except to say that near the tip of a plunging breaker the pressure gradient must be quite small; hence, the particles must be practically in a free-fall trajectory, with zero horizontal acceleration. Of course, the inertia of the air might produce accelerations in a thin sheet of water.

PIERSON: The word "simple" was used several times during the presentation. The problem of breaking waves has been around for about 100 years. If the problem is really simple, why did it take so long to solve it? (P.S. You really need not answer the question.)

LONGUET-HIGGINS (on reflection): First, the mathematics involved, though not complicated, is perhaps unusual, and it took time to find a fruitful approach. Second, I am fortunate to have been able to devote much of my time over the past 10 years to this seemingly intractable problem. For my appointment as a Research Professor I have to thank the Royal Society of London. If you have any favorable comments, please write to the Executive Secretary, the Royal Society, 6 Carlton House Terrace, London SW1Y 5AG.

14

Experimental Studies of Strong Nonlinear Interactions of Deep-Water Gravity Waves

Ming-Yang Su and Albert W. Green

Abstract. Results of an extensive series of experiments on strong nonlinear interactions of deep and shallow gravity waves are summarized. The experiments are conducted in a large outdoor basin and a long indoor wave tank. The waves are produced by a mechanical wavemaker or by natural wind in some cases in the outdoor basin. The effects of steepness $(0.1 \lesssim ak \lesssim 0.34)$ are analyzed for both wave trains and packets. Waves with moderate to large steepness $(ak \lesssim 0.25)$ are found to be subject to intense subharmonic instabilities. Dynamical processes observed in these experiments include subharmonic instabilities, wave breaking, directional energy spreading, nonlinear energy transfer in narrow spectra, formation/interactions of envelope solitons, and formation of other three-dimensional compact wave groups of permanent forms. Effects of water depth on some of the above nonlinear processes are also considered. The experimental investigations plus analyses of other oceanic measurements and theoretical computations seem to imply that the classical view of weak-in-the-mean interactions of ocean waves can also be explained conceptually by combined effects of intermittent processes of strong three-dimensional nonlinear interactions of various kinds.

1. INTRODUCTION

Recently, several advances have been made in experimental and theoretical analyses of steep, deep-water waves. In this chapter we describe results of several experiments that have demonstrated some new properties of wave group transformations and bifurcations. Many of the features of the observed processes can be described in terms of theories that have not yet been applied to experimental observations, so we give definitions of terminology and brief descriptions of the theoretical results that aid in understanding the observations. We note that during the last two decades the majority of published research on nonlinear, deep-water waves has been theoretical. There have been few controlled experiments that have given results in terms of critical parameters, such as wave steepness. This dearth of experimental results has been partially due to the lack of wave basins and wave generators designed to allow unrestricted evolution of steep waves. The advent of readily available multichannel digital data systems has also enhanced the experimenter's ability to record and analyze the complex evolution of

Ming-Yang Su and Albert W. Green ● Naval Ocean Research and Development Activity, NSTL Station, Mississippi 39529.

nonlinear waves. The results that we report here have directly benefited from improvements of these types. The results of our observations will provide some insights into nonlinear wave processes and will furnish the foundations for new theories.

1.1. Terminology

Some of the terminology used to describe results presented in the following sections may not be familiar to some readers, so we provide definitions.

1. Two- and three-dimensional waves. Unidirectional, long-crested waves are defined to be "two-dimensional," while waves with crestwise variation in waveform are considered "three-dimensional." Waves are generally described in terms of amplitude (a), height ($H = 2a$), frequency (f), period ($T = f^{-1}$), wavelength (λ), wavenumber ($k = 2\pi\lambda^{-1}$), and steepness (ak). A subscript "0" denotes the initial state of the observed waves.

2. Nonlinear interactions: "weak," "strong," "very strong." Nonlinear interactions of wave components can be characterized in terms of their time scales of evolution and steepness. For "weak" interactions the typical e-folding time for wave mode exchanges is $T_1 \sim (a_0 k_0)^{-2} T_0$. The more intense, "strong" interactions progress more rapidly with $T_0(a_0 k_0)^{-2} > T_2 > (a_0 k_0)^{-1} T_0$. Wave evolution may have a mixture of strong and weak processes; consequently, it is not possible to distinguish strong or weak processes solely by the time scales of exchange. When reaction time scales are less than T_2, we rather arbitrarily classify them as "very strong" interactions ($T_3 < T_2$). The time scales are dependent on $a_0 k_0$ in every case, so the steepness is another parameter that characterizes the types of interactions. Table I lists some of the nonlinear phenomena in terms of time scales and characteristic steepness. The categories of classification are arbitrary to some extent, but represent our attempt to organize a complicated set of phenomena that have not been completely described by theory.

3. Bifurcation and stability. Following Iooss and Joseph (1980), we shall introduce the concept of bifurcation. For a given initial-value problem of a nonlinear evolution system governed by a nonlinear differential equation, $[\partial/\partial t + c_g(\partial/\partial x)]U(\mu, x, t) = F(U, \mu, x, t)$, where μ is some parameter and F is a nonlinear operator, there may exist several equilibrium solutions $U(\mu, x, t)$ which are free of the transient effects associated with the initial values. Bifurcating

TABLE I.   Weak, Strong, and Very Strong Interactions

Weak interactions ($T_1 < T_2 = s^{-2} T_0; s = a_0 k_0$)
   1. Resonant wave–wave interactions (3d)
   2. Fermi–Ulam–Paster recurrence (2d)
   3. Benjamin–Feir instability (2d)
   4. Formation of envelope Solitons (2d)
   5. 3d instability (3d)
Strong interactions ($T_2 > T > T_1 = 2s^{-1} T_0$)
   1. Benjamin–Feir instability (2d)
   2. Formation of envelope solitons (2d)
   3. 3d instability (3d)
   4. Frequency downshift phenomenon (2d)
   5. Skew wave bifurcation (3d)
   6. Symmetric wave bifurcation (3d)
Very strong interactions ($T < T_1$)
   1. 3d instability (3d)
   2. Wave breaking (3d)
   3. Formation of whitecapping (3d)
   4. Radiation of oblique wave group (3d)
   5. Long wave–short wave interaction (2d, 3d)
   6. Wave–current interaction (2d, 3d)

solutions are equilibrium solutions which form connected branches in a function space. We say that one equilibrium solution bifurcates from another if there are two distinct equilibrium solutions $U^{(1)}(\mu, x, t)$ and $U^{(2)}(\mu, x, t)$, of the same evolution system, continuous in μ, such that $U^{(1)}(\mu^1, x, t) = U^{(2)}(\mu^1, x, t)$. Not all equilibrium solutions arise from bifurcations; isolated solutions and disjoint separated branches of solutions may exist in these nonlinear systems. Some of the equilibrium solutions are unstable to external disturbances, and may not persist long enough to be observed in nature. So we have to find not only the equilibrium solutions of a given nonlinear system, but the growth rate of the instability must also be determined in order to estimate the probability of observing the process. A common necessary condition for bifurcation is the occurrence of an instability of the equilibrium solution to infinitesimal perturbations.

4. Inverse scattering transform and envelope solitons. A powerful mathematical method, the inverse scattering transformation, has been developed in the past decade. It has been used to solve initial-value problems of nonlinear evolution systems that are of interest in several branches of science (see, e.g., Ablowitz et al., 1974; Scott et al., 1973). One significant result of the inverse scattering theory is the discovery of asymptotic solutions that correspond to envelope solitons; they are localized groups of two-dimensional progressive wave whose envelopes remain unchanged during propagation, and which retain their original envelope shapes after collisions, but with relative phase shifts and positions. The envelope solitons for the surface waves are stable with respect to two-dimensional infinitesimal perturbations, but they are unstable with respect to transverse (three-dimensional perturbations (Zakharov and Shabat, 1972).

2. TWO-DIMENSIONAL NONLINEAR INTERACTIONS

2.1. Evolution of Wave Packets

2.1.1. Zakharov and Shabat's Theory

Zakharov and Shabat (1972) solved the nonlinear Schrodinger equation using the inverse scattering transformation for the initial-value problem with a finite-length wave packet. The nonlinear Schrodinger equation and the Zakharov–Shabat solutions assume explicitly that the mean carrier frequency (f_0) and the wavenumber (k_0) are constant although permitting perturbations of f and k. The Zakharov–Shabat predictions are summarized briefly below for comparison with experiments:

1. An initial wave packet evolves, after sufficiently long time, into a number of envelope solitons and a relatively small oscillatory tail.
2. The envelope solitons assume the shape of a *sech* function.
3. The time scale of formation of the envelope solitons is proportional to $a_0 k_0 N_w$, where N_w is the number of waves in the initial uniform wave packet.
4. The number of envelope solitons formed for a uniform wave packet is $N_s \simeq 2^{1/2} a_0 k_0 N_w$.

2.1.2. Experimental Results and Comparison with Theory

Predictions 1–3 by Zakharov and Shabat agree well with the experimental observations for small $a_0 k_0 < 0.1$ (Yuen and Lake, 1975). Prediction 4 agrees quantitatively with our experiments if $a_0 k_0 N_w < 5.5$; the Zakharov–Shabat theory will overpredict N_s for large values of $a_0 k_0 N_w$.

Figures 1 and 2 show examples of time series of waves passing stations along the propagation path. These cases have similar stages of development in that the initial wave packet separates into a set of two or more distinct envelope solitons by the time the waves pass the 61 m station. Lower-frequency harbinger waves lead the solitons and higher-frequency wave packets trail the envelope solitons, which appear to maintain their shapes from separation onward. The leading soliton has the highest fraction of the total energy of the original packet, and the trailing solitons generally decrease in energy as their position in the sequence. The carrier frequencies for

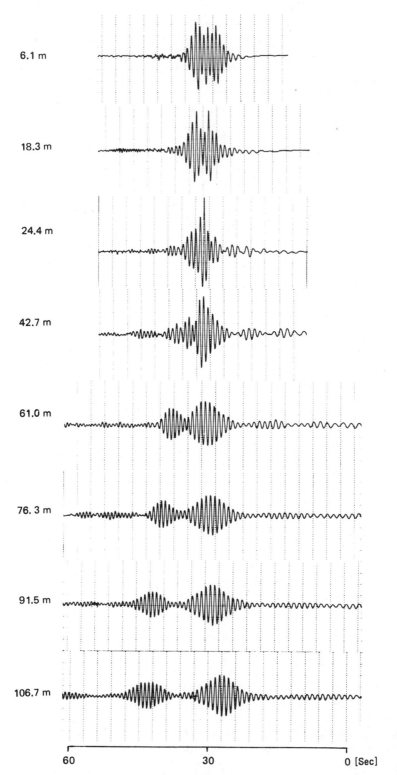

FIGURE 1. Time series of surface displacements at eight stations show the fission of a wave packet ($N_w = 10$, $a_0 K_0 = 0.15$, $\lambda_0 = 1.1\,\mathrm{m}$, $f_0 = 1.15\,\mathrm{Hz}$) and the formation of envelope solitons.

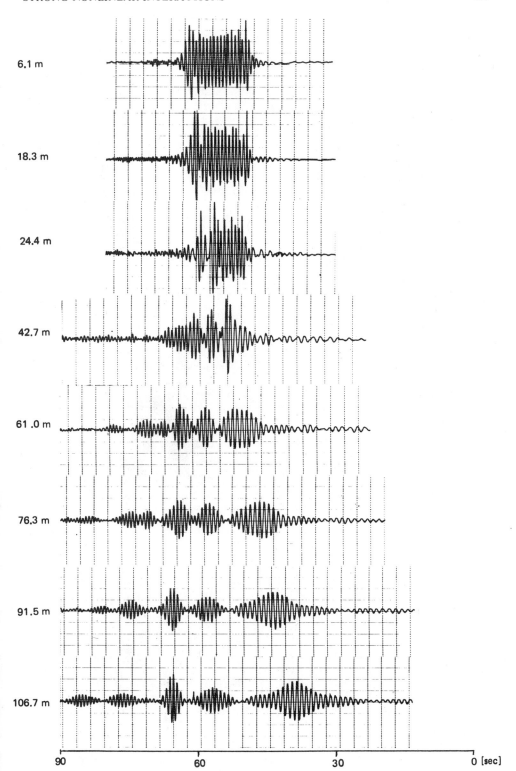

FIGURE 2. Time series of surface displacements at eight stations for a moderate-steepness wave packet ($N_w = 20$, $a_0 k_0 = 0.22$, $\lambda_0 = 0.82$ m, $f_0 = 1.15$ Hz). The packet fissions to form five distinct envelopes (61 m). The leading feature has stabilized to form an envelope soliton, and the trailing packets are still evolving.

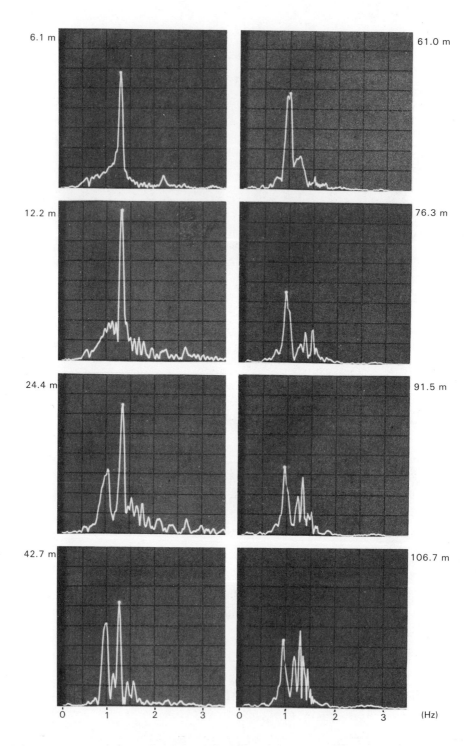

FIGURE 3. The variance spectrum of surface displacements is computed for each station shown in Fig. 2. At the 61.0 m station the wave packets are distinct and the peak variance is at a lower frequency than the initial carrier waves. This peak is associated with the formation of an envelope soliton. Other peaks grow as the trailing packets evolve.

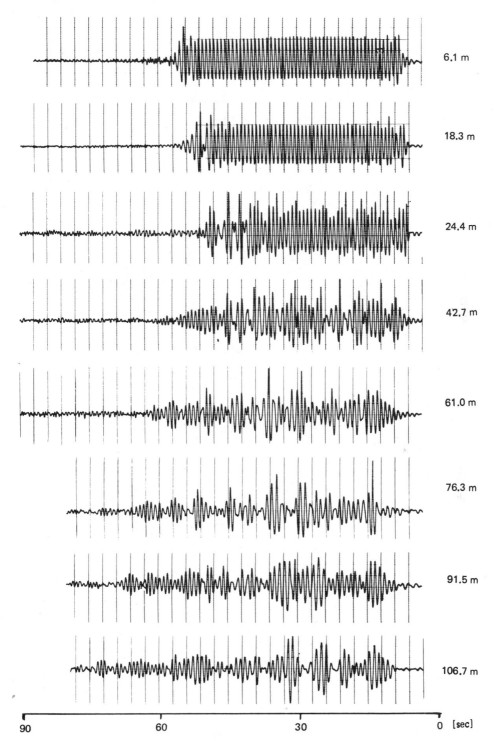

FIGURE 4. The evolution of long wave packets includes packet collisions that occur as the lower-frequency fission products pass through the higher-frequency, slower packets. The sorting of the packets creates stable envelopes that eventually appear at the leading end of the sequence; note the stable envelope forming at the 106.7 m station.

the envelope solitons in Fig. 2 are, respectively, $f_1 = 1.0\,\text{Hz}$, $f_2 = f_3 = 1.2\,\text{Hz}$, $f_4 = f_5 = 1.34\,\text{Hz}$ $= f_0$. f_1, f_2, and f_3 are lower than the original carrier frequency f_0. This shifting of the carrier frequencies to lower values for some of the envelope solitons is a new property which has not been predicted. The evolution of power spectra of surface displacement corresponding to the last example in Fig. 2 is shown in Fig. 3. (See the Appendix for explanation of the computation method of these spectra.) The appearance of rapidly growing spectral peaks coincides with the formation of distinct envelope solitons. In Fig. 3 the growth of the dominant peak ($f_1 = 1.0\,\text{Hz}$) appears to be associated with the leading soliton while the trailing envelope is associated with

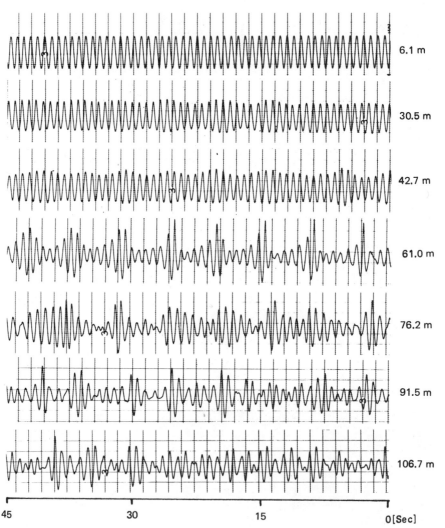

FIGURE 5. The space–time evolution of a continuous wave train with an initial steepness $a_0 k_0 = 0.15$, $f_0 = 1.23\,\text{Hz}$, has three stages: (1) 6.1 m to 42.7 m—the sideband modulations due to Benjamin–Feir instability grow. (2) 61 m to 73.4 m—the Benjamin–Feir modulations grow to about the same amplitude as the primary wave. The wave envelopes are regular and have amplitudes significantly greater than the initial wave train. (3) 91.5 m to 106.8 m—the lower sideband frequency overtakes and surpasses the amplitude of the initial wave component (frequency downshifting). The modulation at this stage is generally less intense; subsequent evolution tends toward a uniform wave train.

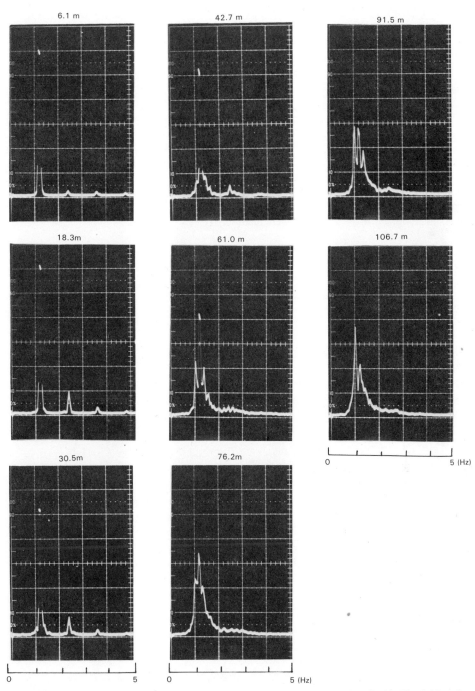

FIGURE 6. The variance spectra of surface fluctuations at the seven stations described in Fig. 5. Note that the bandwidth of the major components of the spectra does not change significantly from the onset of intense modulation (61 m); this indicates that these strong interactions of the sidebands and the initial carrier are contained in relatively narrow ranges of frequency.

waves in a frequency band nearer to the carrier frequency (f_0) of the initial packet. The ratio of the largest frequency shift ($f_0 - f_1$) to the initial carrier frequency is found to be about equal to the initial wave steepness, i.e.,

$$\delta = \frac{f_0 - f_1}{f_0} \cong a_0 k_0$$

The frequency f_1 corresponds to the fastest-growing (most unstable) lower sideband of the Benjamin–Feir instability.

A higher value of N_w increases the probability that some of the wave groups will collide due to differences in group speed. A typical example to illustrate this feature is shown in Fig. 4 for a wave packet with $a_0 k_0 = 0.15$, $f_0 = 1.23$ Hz, and $N_w = 60$. We first notice the fairly symmetric modulation of the wave packet (with regard to its leading and trailing portions) from $x = 6.1$ m to $x = 30.5$ m. We then see the asymmetric modulation from $x = 42.7$ m to $x = 91.5$ m with stronger modulations occurring in the leading portion. This asymmetric modulation is thought to result from collisions of wave groups of different carrier frequencies, as the groups with lower frequencies outrun those with higher frequencies.

2.2. Evolution of Uniform Wave Trains

A representative example of the evolution of a uniform wave train with $a_0 k_0 = 0.15$, $f_0 = 1.23$ Hz is shown in Fig. 5. Temporal records at $x = 61$ m($59\lambda_0$) show that the average group

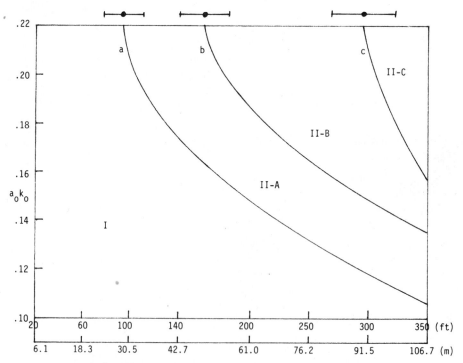

FIGURE 7. The experimentally observed stages of wave train evolution are determined by the initial steepness ($a_0 k_0$). In I, the Benjamin–Feir sideband instabilities grow at equal rates and become comparable to the initial carrier wave. In II-A, the lower sideband energy overtakes the initial wave and the upper sideband diminishes. In II-B and -C, the lower sideband wave dominates, and the upper sideband continues to diminish. Error bars are shown at the top of the figure.

contains about seven wave periods. At this station that largest wave height is 90% greater than the initial uniform wave height at $x = 6.1$ m. The envelope shape of the wave groups is not symmetric; the higher waves occur at the leading edge of a wave group. The intensity of the modulation decreases at $x = 76.3$ m, and returns to a more uniform wave height at $x = 106.7$ m. The power spectra for this case are shown in Fig. 6. The sideband components become visible at $x = 30.5$ m. At the expense of the fundamental peak at $f_0 = 1.23$ Hz, the two sideband components continue to grow. Around $x = 91.5$ m ($90\lambda_0$), the spectral peak of the lower sideband component (f_1) becomes approximately equal to the component at f_0, while the higher sideband component f_2 decreases somewhat. Finally, at $x = 106.7$ m ($104\lambda_0$), the f_1 component has almost twice the energy of the f_0 component; concurrently, the f_2 component diminishes by half. This stage of the largest sideband growth coincides with the strongest modulation in the wave envelope. The majority of wave energy is transferred from a higher-frequency mode f_0 to a lower-frequency mode f_1 due to a long-time instability that occurs after the initial Benjamin–Feir instability. In general, unstable wave trains have four stages of evolution that are characterized by distinct relationships of the carrier frequency (f_0) and the two sideband modulational instabilities (f_1 and f_2) that are due to the Benjamin–Feir instability. The locations of the boundaries of these domains are influenced by the initial waveforms and perturbations. Consequently, the boundaries indicated in Fig. 7 are approximate representations of the observations. These boundaries can be delineated as follows:

1. $\max\left[E(a_0 k_0, f_1)\right] = \max\left[E(a_0 k_0, f_2)\right]$
2. $E(a_0 k_0, f_1) = E(a_0 k_0, f_0)$
3. $E(a_0 k_0, f_1) = 2E(a_0 k_0, f_0)$

The Regime I is characterized by the equal growth of the most unstable sideband modes to their maxima; this could justifiably be called the Benjamin–Feir instability regime. The Regime II including the three stages (II-A,-B, and -C) is the frequency downshift regime.

3. THREE-DIMENSIONAL NONLINEAR INTERACTIONS

3.1. General Characteristics of Three-Dimensional Waves

3.1.1. Short-Crested Waves and Wave Breaking

From casual observations of growing sea and the waves generated in wind-wave tanks, it is readily seen that the mean crest length of dominant waves is of the order of a few wavelengths; i.e., the waves are short-crested. In particular, the crestwise length scale of deep-water breaking waves with whitecapping is of the order of the dominant wavelength, which means that these waves are three-dimensional phenomena. Chappelear (1961), Hui and Hamilton (1979), and Crawford *et al.* (1981) present some theoretical analyses of three-dimensional deep-water waves. Several basic properties of three-dimensional deep-water waves have been identified and explained in the present investigations. We observe that wave trains and packets become more three-dimensional as steepness increases.

3.1.2. Three-Dimensional Instability

McLean *et al.* (1980) and McLean (1981) predicted the existence of two types of instabilities of deep-water wave trains; designated as type I and type II. The type I instability is an extension of the Benjamin–Feir instability for the two-dimensional case with small wave steepness and long wave perturbations. The type I is essentially two-dimensional, since the maximum growth of the instability occurs for a two-dimensional perturbation. The type II instability is predominantly three-dimensional in that the maximum instability always occurs for fully three-dimensional perturbations. For small $a_0 k_0$, the type I instability dominates. At $a_0 k_0$ of about 0.3, the type II overtakes the type I. The type I instability disappears at $a_0 k_0$ of about 0.39, while the type II instability continues to increase in magnitude. At $a_0 k_0$ of about 0.4, the type II instability develops a two-dimensional special case which corresponds to the "explosive," two-dimensional instability near $a_0 k_0 = 0.41$ predicted by Longuet-Higgins (1978). It should be noted that even at this steepness, the maximum growth rate still occurs for fully three-dimensional perturbations.

There is another fundamental difference between type I and II instabilities. The modulational envelope resulting from the type I subharmonic instability travels with the group speed that is approximately half the phase velocity; hence, each wave group displays about twice as many waves in a temporal record as in a spatial record. The type II instability propagates at the carrier wave phase velocity; consequently, the variation in space looks roughly the same as in the time record of surface displacement.

The predictions of three-dimensional instabilities of surface waves are still limited, since the theory is based on linear perturbation analysis; consequently, the predicted instabilities are, strictly speaking, applicable only to observations in the initial stage of instability when the perturbations are still much smaller than the unperturbed waves; long-time, large-amplitude growth of the perturbations is beyond the range of validity of present theories.

3.1.3. Three-Dimensional Wave Bifurcations

Chen and Saffman (1980) and Saffman (1981) have shown that uniform Stokes waves of very large steepness $(a_0 k_0 \geqslant 0.4)$ are subject to subharmonic bifurcations that result in two-dimensional permanent wave trains with unequal wave heights. Experimental evidence of these two-dimensional bifurcated waves has not yet been reported. More recently, Saffman and Yuen (1980) found numerical solutions of the Zakharov (1968) equation that predicted the existence of two new types of three-dimensional permanent waveforms which result from subharmonic bifurcations of uniform two-dimensional Stokes waves. One of the bifurcation types is a steady symmetric wave pattern that propagates with the basic Stokes waves. Meiron, Saffman and Yuen (1982) have made computations based on this theory to predict several configurations of the symmetric wave patterns. The second type is a steady skew wave pattern that propagates obliquely from the direction of the Stokes waves.

3.2. Skew Wave Bifurcation

3.2.1. Experimental Facility

Most of the experiments on three-dimensional nonlinear interactions are conducted in an outdoor basin $1 \times 100 \times 340$ m using a plunger-type wavemaker of 16-m crest length. Some of the experiments on symmetric wave bifurcations and wave breaking are also repeated in the tow tank described in Section 2. More details about experimental facilities, instrumentations, and procedures can be found in Su et al. (1982).

3.2.2. Experimental Observations and Measurements

The bifurcated skew waves show up clearly in the experiments only when $0.16 \leqslant a_0 k_0 \leqslant 0.18$. We shall illustrate common characteristics of the wave patterns observed in the experiments by an example in which $a_0 k_0 = 0.17$. The rather complex three-dimensional wave field, which is schematically represented in Fig. 8, may be divided into five regimes:

1. Stokes waves
2. Skew wave patterns
3. Interactions of skew wave patterns
4. Benjamin–Feir modulations of Stokes waves
5. Skew waves with Benjamin–Feir modulations.

Locations of these regimes are sketched in Fig. 8. Figure 9 is a photograph of four groups of skew

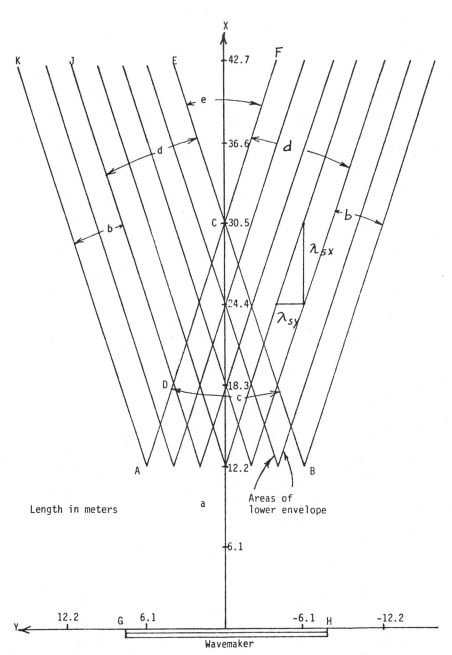

FIGURE 8. Skew wave bifurcations result from bifurcations of wave trains that undergo transition in (a). The skew wave patterns appear in (b) as straight, alternating bands of light and dark (Fig. 9). In region (c), interference bands occur due to superposition of the two skew wave patterns. In zone (d), Benjamin–Feir-type modulations of the skew waves are apparent. In (e), the Benjamin–Feir-type modulations of the primary waves lead to rapid subharmonic transitions.

FIGURE 9. Skew wave patterns are manifested by the bands of lighter (lower amplitude) and darker (higher amplitude) tones.

wave patterns which propagate to the left side of the wavemaker. The areas with lighter tone correspond to the zones where wave amplitudes are smaller than the surrounding (darker) areas. These banded areas are manifestations of skew wave patterns that are propagating with an oblique angle (Ψ) with respect to the primary wave train direction (x axis). The separation distances along the x axis and y axis between two lighter bands are the normal and the crestwise wavelength of the skew wave patterns λ_{xs} and λ_{ys}, respectively. From experimental measurements with the initial wave steepness $0.16 < a_0 k_0 < 0.18$, we find that the propagation angle (Ψ) of the skew wave group has an $18.5°$ mean with small deviations, but the crest wise wavelength is subject to relatively larger variations. The x component of phase velocity of each individual skew wave is almost equal to the phase velocity of the primary wave train; consequently, the skew waves are effectively phase-locked to the primary waves. In the converging cone-shaped area defined by ABC in Figs. 8 and 9c, we can see a diamond-shaped pattern that results from interactions between two sets of the skew wave patterns that propagate in different directions. Note that the diamond-shaped wave pattern is an example of a narrow-band two-dimensional wavenumber spectrum (Longuet-Higgins, 1976, Fig. 1). Furthermore, two sets of the skew wave patterns emerging after the mutual interactions seem to retain their pattern shape without obvious distortion.

More detailed description of experimental measurements of the skew wave bifurcation and a comparison with Saffman and Yuen's (1980) theoretical computations are given in Su (1982).

3.3. Symmetric Wave Bifurcation and Wave Breaking

3.3.1. A Succession of Very Strong Nonlinear Interactions

Figure 10 represents the characteristics of a steep ($a_0 k_0 = 0.32$) uniform wave train as it propagates from the wavemaker. Four distinct stages of evolution are evident (Fig. 10):

1. Symmetric wave bifurcations
2. Wave breaking
3. Radiation of oblique wave groups
4. Frequency downshift.

3.3.1a. Symmetric Wave Bifurcations. From $x = 20\lambda_0$ (Fig. 10), the unstable wave train rapidly bifurcates to a symmetric wave pattern (Fig. 11). These bifurcated symmetric waves are crescent-shaped with the crestwise length

$$\lambda_{BC} \cong (1/1.2)\lambda_0$$

Three configurations of symmetric waves are observed in our experiments. The most frequent form is characterized by the subharmonic scale of $2\lambda_0$, and the shifting by $1/2\lambda_{BC}$ of the crest on successive basic waves. The maximum surface slope on the forward face of the higher crests is found to exceed 1.0, which is much larger than the Stokes limiting slope of $\cos 30° = 0.577$. We surmise that the larger slope is due to the three-dimensionality of the symmetric waves, and that the configuration is caused by the three-dimensional type II instability which has a most unstable mode of subharmonic scale of $2\lambda_0$.

3.3.1b. Breaking Waves. Following the formation of the symmetric bifurcated waves (Fig. 10), we observe a rapid spilling at the middle portion of the highest wave crest (Fig. 12). Capillary waves are generated and propagate away from the breaking wave zones. Strong crestwise interactions between adjacent breaking waves are also evident. Air entrainment and whitecapping are observed, particularly when natural wind is blowing over these breaking waves.

3.3.1c. Radiation of Oblique Wave Groups. Three-dimensional wave breaking is followed

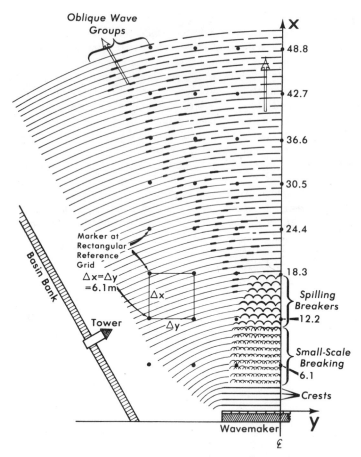

FIGURE 10. The distribution of phenomena observed in the evolution of a steep wave train ($a_0 k_0 = 0.32$, $f_0 = 1.55\,\mathrm{Hz}$, $\lambda_0 = 65\,\mathrm{cm}$).

by a rapid transition to waves that are essentially two-dimensional; breaking radiates wave groups about 30° from the basic wave direction. These oblique wave groups are superimposed on the basic expanding wave trains and give the appearance of distortion on the latter (Fig. 13), where the oblique wave groups propagate along a strip from right to left of Fig. 13. The oblique wave groups retain their shape and propagation direction for a distance of more than 100 wavelengths.

    3.3.1d. Frequency Downshift. As the oblique waves are radiated from the three-dimensional waves, subsequent two-dimensional regime undergoes subharmonic transition accompanied by frequency downshifting by as much as 25%. The two-dimensional modulation associated with this process can be seen in the upper right-hand corner of Fig. 13.

3.3.2. Evolution of Wave Profiles and Power Spectra

    Figures 14 and 15 show the time series of surface fluctuations and the corresponding power spectra at 10 stations. These results were obtained in the tow tank for the case $a_0 k_0 = 0.30$ and $f_0 = 1.15\,\mathrm{Hz}$. Modulations due to the three-dimensional instability are first visible at $x = 12.2\,\mathrm{m}$ in Fig. 15. At stations $x = 30.5\,\mathrm{m}$ and $x = 36.6\,\mathrm{m}$, we see the alternate high and low wave crests which are the indications of symmetric bifurcated waves. In the next two stations ($x = 42.7\,\mathrm{m}$ and $x = 48.8\,\mathrm{m}$), the most dominant modulation is four wave periods; this corresponds to the

FIGURE 11. Three-dimensional symmetric wave patterns ($a_0 k_0 = 0.32$) due to 3d instability.

FIGURE 12. An example of strong interactions and spilling breakers that accompany symmetric wave patterns. In this case, $a_0 k_0 = 0.32$.

FIGURE 13. Oblique groups are radiated from the zone where three-dimensional symmetric waves undergo transition to nearly two-dimensional waves.

transition from three-dimensional to two-dimensional waves. The two-dimensional wave modulations produce the lower-frequency envelope modulations at station $x = 67.1$ m.

In Fig. 15, in addition to the two sidebands, a new peak caused by the three-dimensional instability and breaking waves appears at $f_2 = 1.7$ Hz from the station $x = 12.2$ m to $x = 54.9$ m. For $x = 36.6$ m, the peak at $f_0 = 1.15$ Hz decreases steadily, while a lower-frequency peak at $f_1 = 0.90$ Hz increases rapidly. At $x = 42.7$ m, these two peaks become about the same magnitude. Finally, at $x = 67.1$ m, the f_1 peak is larger than the f_0 peak by about 30%. These observations are consistent with the prediction that the three-dimensional instability grows more rapidly than the two-dimensional instability in this range of steepness.

The average ratio of the final peak frequency (f_1) to the initial peak frequency (f_0) is close to 3/4; i.e., about 25% reduction. The time scale of the phenomenon starting from the generation of the waves is about 50 wave periods for $0.30 \lesssim a_0 k_0 \lesssim 0.34$.

For more detailed information on Section 3.3, readers are referred to Su (1982) and Su et al. (1982).

4. CONCLUSIONS

From the experimental studies of strong nonlinear interactions of deep-water gravity waves in a tow tank and a wide basin, and comparisons with existing theoretical models, we draw the following conclusions:

    1. The frequency downshift phenomenon for wave packets and trains is a long-time

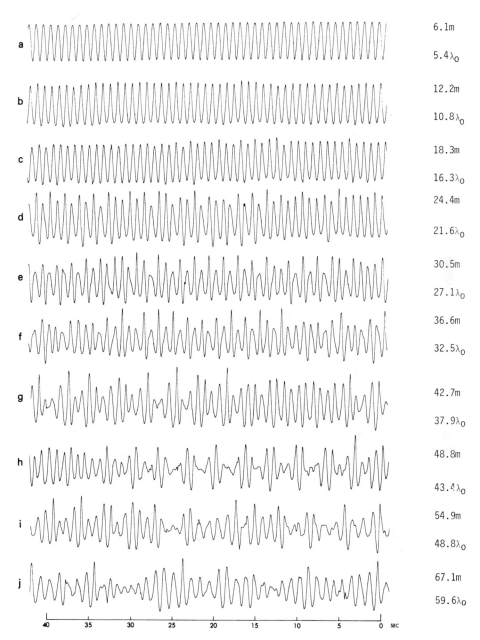

FIGURE 14. Time series of surface displacement of a continuous wave train at stations along a deep tow tank. In this case, $a_0 k_0 = 0.30, f_0 = 1.15\,\mathrm{Hz}, \lambda_0 = 1.12\,\mathrm{m}$.

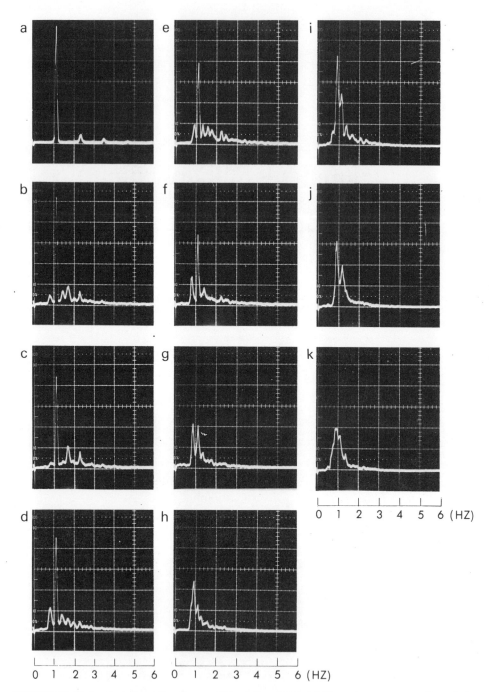

FIGURE 15. Variance spectra at the corresponding stations in Fig. 14. These steep waves pass through the stages of evolution much more rapidly than the less steep examples (Figs. 5 and 6), but the basic features of each stage are the same, except that additional peaks are present due to the three-dimensional crescent-shaped breaking waves.

nonlinear feature of evolution that succeeds the short-time transitions that start with the Benjamin–Feir instability when the initial wave steepness is in the range $0.1 < a_0 k_0 < 0.34$.

2. Two new types of three-dimensional wave bifurcations from uniform finite-amplitude waves are observed: symmetric and skew bifurcations. These are induced by the three-dimensional instabilities that are fundamentally different from the two-dimensional Benjamin–Feir instability.

3. The symmetric bifurcated waves precede a transition to crescent-shaped spilling-type breaking waves that resemble natural wind waves. The spilling waves undergo further transitions to nearly two-dimensional waves while simultaneously radiating two series of oblique wave groups. Finally, the residual two-dimensional waves effect a rapid subharmonic interaction that results in frequency downshifting of the primary waves.

4. The radiation of oblique wave groups and the skew wave bifurcation contribute to the spreading of wave energy away from the dominant wave direction.

5. The Benjamin–Feir-type instability and the newly discovered three-dimensional instability are fundamental in the evolution of wave packets and wave trains.

6. The role of envelope solitons in nonlinear wave dynamics is demonstrated.

All of the results demonstrate that there are close interrelationships among the processes governing wavegrowth, energy redistribution, directional energy spreading, and wave dissipation. It is clear from our work that a comprehensive theoretical description of wave dynamics must include strong nonlinear interactions; weakly nonlinear interactions are not intense enough to produce the rapidly evolving processes reported here. Considerably more experimental work is required to refine the phenomenology of strong interactions so that more accurate theory can be developed and tested.

ACKNOWLEDGMENTS. The authors are grateful to Dr. R. R. Goodman for support of this project since its inception in 1977.

APPENDIX: NOTE ON COMPUTATION OF THE SPECTRA SHOWN IN FIGURE 3

Each of the spectra presented in Fig. 3 is obtained from a corresponding unsteady time series presented in Fig. 2. The total length of each time series is 80 s. The spectrum is computed by an HP-3582A spectrum analyzer with the high-frequency cutoff setting to be 10 Hz. The internal function of the HP-3582A makes use of a 512-point FFT. Comparisons of the spectra from different runs of the experiments under identical initial conditions did show that the significant spectral peaks, both their frequencies and magnitude, are reproducible. As such, we may state that these spectral peaks are indeed significant.

REFERENCES

Ablowitz, M. J., D. J. Kaup, A. C. Newell, and H. Segur (1974): The inverse scattering transform-Fourier analysis of nonlinear problems. *Stud. Appl. Math.* **53**, 249–315.

Benjamin, T. B., and J. E. Feir (1967): The disintegration of wave trains on deep water. Part I. Theory. *J. Fluid Mech.* **27**, 417–430.

Chappelear, J. E. (1961): On the description of short-crested waves. Beach Erosion Board, Office of the Chief of Engineers, T.M. 125. U.S. Army Corps of Engineers, Washington, D.C.

Chen, B., and P. G. Saffman (1980): Numerical evidence for the existence of new types of gravity waves of permanent form on deep water. *Stud. Appl. Math.* **62**, 1–21.

Crawford, D. R., B. M. Lake, P. G. Saffman, and H. C. Yuen (1981): Stability of weakly nonlinear deep-water waves in two and three dimensions. *J. Fluid Mech.* **105**, 177–192.

Hui, W. H., and J. Hamilton (1979): Exact solutions of a three-dimensional nonlinear Schrodinger equation applied to gravity waves. *J. Fluid Mech.* **93**, 117–133.

Iooss, G., and D. D. Joseph (1980): *Elementary Stability and Bifurcation Theory*, Springer-Verlag, Berlin.

Kaup, J., and A. C. Newell (1978): Solitons as particles, oscillators, and in slowly changing media: A singular perturbation theory. *Proc. R. Soc. London Ser. A* **361**, 413–446.

Longuet-Higgins, M. S. (1976): On the nonlinear transfer of energy in the peak of a gravity-wave spectrum: A simplified model. *Proc. R. Soc. London Ser. A* **347**, 331–338.

Longuet-Higgins, M. S. (1978): The instabilities of gravity waves of finite amplitude in deep water. II. Subharmonics. *Proc. R. Soc. London Ser. A* **360**, 489–505.

McLean, J. W. (1981): Instabilities of finite-amplitude water waves on deep water. *J. Fluid Mech.* **114**, 315–330.

McLean, J. W., Y. C. Ma, D. U. Martin, P. G. Saffman, and H. C. Yuen (1980): A new type of three-dimensional instability of finite-amplitude waves. *Phys. Rev. Lett.* **46**, 817–820.

Meiron, D. I., Saffman, P. G., and H. C. Yuen (1982): Computation of steady three-dimensional deep-water waves. *J. Fluid Mech.* **124**, 109–121.

Saffman, P. G. (1981): Long wavelength bifurcation of gravity waves on deep water. *J. Fluid Mech.* **101**, 567–581.

Saffman, P. G., and H. C. Yuen (1980): A new type of three-dimensional deep-water waves of permanent form. *J. Fluid Mech.* **101**, 797–808.

Scott, A. C., F. Y. F. Chu, and D. W. McLaughlin (1973): Solitons. *Proc. IEEE* **61**, 1449–

Su, M. Y. (1982) Three-dimensional deep-water waves. Part 1. Experimental measurement of skew and symmetric wave patterns. *J. Fluid Mech.* **124**, 73–108.

Su, M. Y., M. Bergin, P. Marler, and R. Myrich (1982): Experiments on nonlinear instabilities and evolution of steep gravity wave trains. *J. Fluid Mech.* **124**, 45–72.

Yuen, H. C., and B. M. Lake (1975): Nonlinear deep water waves: Theory and experiment. *Phys. Fluids* **18**, 956–960.

Zakharov, V.,E. (1968): Stability of periodic waves of finite amplitude on the surface of a deep fluid. *J. Appl. Mech. Tech. Phys. (USSR)* **2**, 190–194.

Zakharov, V. E., and A. B. Shabat (1972): Exact theory of two-dimensional self-focusing and one-dimensional self-modulating waves in nonlinear media. *Sov. Phys. JETP* **34**, 62–69.

DISCUSSION

PIERSON: The paper treats two different problems: (1) transients of the form $N(t)$ at $X = 0$ where $N(t) = 0$ outside of some internal $-T/2 < t < T/2$, and (2) spatially varying motions of the form $N(t)$ for $-L/2 < x < L/2$ and $N(t) = 0$ otherwise at $y = 0$. Both of these problems involve Fourier *integral* techniques (see Kinsman and Neumann and Pierson). The first problem is a problem in x, z, t space; the second is a problem in x, y, z, t space. Each yields very complicated Fourier integral spectra that cover a wide range of Fourier frequencies and wavenumbers. I believe that most of the results presented have little to do with nonlinear problems and that most of what was observed can be explained by simple dispersion and diffraction effects. For example, the finite section of a wave generator is more or less the equivalent of the theory for the diffraction of waves through a breakwater gap (except for complicated effects at the corners of the generator) and much of what was observed can be explained by standard coastal engineering techniques (see Wiezel for example).

SU: Most of Professor Pierson's comments on our experimental results presented might have been caused by the lack of time in my talk describing in detail our experimental setup, procedures, and the comparisons made with nonlinear theoretical computations available. The results have been documented in six manuscripts. All the phenomena described involve strongly nonlinear instabilities and bifurcations and further result in substantial transition of wave energy to different modes. As such, the phenomena could not be explained by any linear wave dynamics based on Fourier decomposition.

LONGUET-HIGGINS: (1) Regarding the estimates of energy "loss," I believe you showed calculations of total energy, at different fixed points. But energy conservation applies *not* to the time integral of η^2 at fixed points, but rather to the horizontal integral of η^2 at fixed times. However, if linear theory applies and there is no reflection, there should be conservation of *energy flux* (i.e., integral of spectral density times group velocity).

(2) In nonlinear theory, there will be differences between the potential and kinetic energy densities. Were these negligible?

Su: (1) Your comment on the observation of energy flux is well taken. I have shown here only the total energy density per unit area at fixed points, but I have also computed, but not shown, the total action density at fixed points which, in the linear theory, is equal to the integral of spectra density times group velocity, as you pointed out. In the evolution of the total action density, we have also observed substantial loss, but in lesser amount in comparison with the corresponding loss in the total energy, because of the frequency downshift effect.

(2) If I trust my memory, the difference in potential and kinetic energy densities, for nonlinear Stokes waves, is less than 5% according to Cokelet's (1977) and your exact computations (1975). On the other hand, the loss in energy flux is more than 30% for $a_0 k_0 = 0.22$. Thus, the effect of nonequal partition of potential and kinetic densities could account for, at most, one-sixth of the observed loss in the energy flux. We do not understand yet the physics of the surprising large nonconservation of energy in the strongly nonlinear process, but speculate that it may very well be a new dynamical process, just like the case of the newly found 3d instabilities for large $a_0 k_0$.

15

THE INSTABILITY AND BREAKING OF A DEEP-WATER WAVE TRAIN

W. K. MELVILLE

ABSTRACT. We report measurements of the surface displacement and fluid velocity in a nonlinear deep-water wave train as it evolved to breaking. Two distinct regimes are found. For $ak \leqslant 0.29$ the evolution is sensibly two-dimensional with the Benjamin–Feir instability leading directly to breaking as found by Longuet-Higgins and Cokelet (1978). The measured frequency of the most unstable sideband agrees very well with the predictions of Longuet-Higgins (1978). The surface displacement spectrum is not restricted to a few discrete frequencies but also involves a growing continuous spectrum. Within the accuracy of the measurements, the onset of breaking corresponds to the onset of the asymmetric development of the sidebands about the fundamental frequency. It is suggested that the asymmetric evolution, which ultimately leads to the shift to lower frequency (Lake et al., 1977), may be related to Longuet-Higgins's (1978) breaking instability. For $ak \geqslant 0.31$ a full three-dimensional instability dominates the Benjamin–Feir instability and leads rapidly to breaking. Preliminary measurements of this instability agree very well with the results of McLean et al. (1981). Continuous measurements of the velocity field at the fluctuating surface were made with a laser anemometer and show significant differences between the velocity field in unbroken and breaking waves. In the unbroken waves the measured velocity agrees very well with that inferred by the measured surface displacement. In breaking waves the velocity in the spilling region is comparable to the phase speed of the waves while the perturbation to the surface displacement is small. With the exception of the velocity measurements, this work is reported by Melville (1982). The velocity measuring technique and analyzed data will be reported elsewhere (Melville and Rapp, 1985).

REFERENCES

Benjamin, T. B., and J. E. Feir (1967): The disintegration of wave trains on deep water. Part I. Theory. *J. Fluid Mech.* **27**, 417–430.
Lake, B. M., H. C. Yuen, H. Rungaldier, and W. E. Ferguson (1977): Nonlinear deep-water waves: Theory and experiment. Part 2. Evolution of a continuous wavetrain. *J. Fluid Mech.* **83**, 49–74.
Lo, E. and Mei, C. C. (1985): A numerical study of water-wave modulation based on a higher-order nonlinear Schrödinger equation: *J. Fluid Mech.* **150**, 395–416.
Longuet-Higgins, M. S. (1978): The instabilities of gravity waves of finite amplitude in deep water. II. Subharmonics. *Proc. R. Soc. London Ser. A* **360**, 489–505.

W. K. MELVILLE • Institute of Geophysics and Planetary Physics, University of California, San Diego, California 92037, and Department of Civil Engineering, Massachusetts Institute of Technology, Cambridge, Massachusetts 02139.

Longuet-Higgins, M. S., and E. D. Cokelet (1978): The deformation of steep surface waves on water. II. Growth of normal-mode instabilities. *Proc. R. Soc. London Ser. A* **364**, 1–28.

McLean, J. W., Y. C. Ma, D. U. Martin, P. G. Saffman, and H. C. Yuen (1981): Three-dimensional instability of finite-amplitude water waves. *Phys. Rev. Lett.* **46**, 817–820.

Melville, W. K. (1982): The instability and breaking of deep-water waves. *J. Fluid Mech.* **115**, 165–185.

Melville, W. K. and Rapp, R. J. (1985); The surface velocity field in steep and breaking waves *J. Fluid Mech.* (submitted).

DISCUSSION

YUEN: Your explanation for downshifting and breaking for the situation with $ak < 0.29$ invokes theory of Longuet-Higgins (1978) which applies for $ak \simeq 0.41$, whereas our two-dimensional results (McLean *et al.*, 1981) would predict a $p = 0.5$ (wavelength-doubling) instability for $ak = 0.29$; furthermore, at $ak \simeq 0.29$, the three-dimensional instability has a growth rate as large as the two-dimensional instability; perhaps you can compare your data to the three-dimensional results as well.

MELVILLE: Longuet-Higgins and Cokelet (1978) found that waves that initially underwent a Benjamin–Feir instability finally broke with a localized instability that was similar in every respect to that arising at $ak \simeq 0.41$, i.e., a co-propagating instability. My explanation of the downshift is consistent with these results for $ak \lesssim 0.29$. My observations show a transition from predominantly two-dimensional to three-dimensional instability in the range $0.29 < ak < 0.31$ consistent with the results of McLean *et al.* (1981) that show equal growth rates at $ak \simeq 0.3$. The three-dimensional instability rapidly leads to breaking and no direct measurements of growth rates were made. It is worth emphasizing that my measurements, and the role of breaking in the downshifting process, appear to be consistent with the original work of Lake *et al.* (1977).

ALBER: The Benjamin–Feir instability process is based on the special nonlinear coupling of deterministic wave train systems. In my 1978 paper [Alber, I. E. (1978): The effects of randomness on the stability of two-dimensional surface wavetrains, *Proc. R. Soc. London Ser. A* **363**, 525–546] an instability analysis was developed for weakly nonlinear random surface wave trains (in the absence of breaking). It was found that the B–F instability diminishes and then vanishes as the degree of randomness or spectral bandwidth increases from zero for a Gaussian random wave field. This instability attenuation is due to a detuning of the nonlinear coupling resulting from random phase mixing. Thus, it is not clear at all whether deterministic wave instability features will appear in random wave fields. Do you believe that the special deterministic instabilities you have studied (both at low and high wave slopes) will be significant in an ocean environment when randomness becomes dominant?

MELVILLE: One aspect of these measurements that I want to draw attention to is that the evolution of the wave field involves a growing continuous spectrum as well as the discrete spectrum. Ultimately, I would expect the discrete lines to merge into the continuous spectrum.

I would expect the random phase mixing to have a greater influence on the weaker (lower growth-rate) instabilities. Thus, the Benjamin–Feir instability should be more strongly affected than the much stronger "breaking" instability.

SU: (1) I am very pleased to see that you have reproduced the three-dimensional breaking waves which we first observed two and a half years ago. Your results provided a substantiation of our experiments.

(2) The growth rates of the Benjamin–Feir instability associated with the frequency downshifting and the three-dimensional instabilities associated with the breaking waves are about the same when they grow almost simultaneously. From our experiments, it seems that they develop quite independently. Thus, the three-dimensional breaking might be triggered by the frequency downshifting as your talk seems to imply*.

(3) In answer to the question by the previous commentator, I would like to state that effects of randomness in wind-wave fields in general would suppress the three-dimensional instabilities, but that our experiments in the outdoor basin under windy conditions still show clearly the occurrence of the three-dimensional breaking waves from the initial uniform Stokes waves generated by the plunger.

---

*This discussion of downshifting appears to have been superseded by recent developments (cf., Lo and Mei, 1985)

16

MEASUREMENTS OF BREAKING WAVES

Implications for Wind-Stress and Wave Generation

M. S. LONGUET-HIGGINS AND N. D. SMITH

ABSTRACT. The frequency of whitecapping has been measured with a surface "jump-meter" in wind speeds up to 14 m/s. At this wind speed the proportion of steep or breaking waves was of order 0.05, which is consistent with visual observations of whitecap coverage. Owing to the dispersive property of gravity waves, whitecaps occur intermittently. It is argued that this intermittency, together with the tendency of the air flow to separate preferentially over the steeper waves, causes the horizontal stress exerted by the wind to be both intermittent and patchy. Calculations of the time-dependent flow over steep waves combined with the observed whitecapping frequency suggest that the contribution of intermittent flow separation to the total wind stress is significant.

1. INTRODUCTION

During the MARSEN program of field observations in the North Sea in October–November 1979, the authors deployed a simple instrument which was designed to detect and record the occurrence of breaking or near-breaking waves (see Figs. 1 and 2). A full description of the apparatus and the method of analyzing the observations is given elsewhere (Longuet-Higgins and Smith, 1983) Here we propose to summarize the results, and to discuss their implications for understanding the nature of horizontal wind stresses and the generation of surface waves by wind.

2. METHOD AND RESULTS

The principle is shown in Fig. 1. A spar-buoy of overall length 5.04 m carried a capacitance-wire wave recorder intersecting the free surface nearly vertically. By differentiating the output voltage, any sudden "jumps," or sharp rises in the surface elevation η exceeding a preset critical value R of $\partial\eta/\partial t$ could be detected. By integrating $\partial\eta/\partial t$, it was possible also to measure the height Δ of each "jump." A histogram was calculated automatically, giving the distribution of jump heights in a given length of record, sorted according to size.

For very small values of R, the histogram depended on the particular value of R chosen. But

M. S. LONGUET-HIGGINS ● Department of Applied Mathematics and Theoretical Physics, University of Cambridge, Cambridge, England, and Institute of Oceanographic Sciences, Wormley, England GU8 5UB. N. D. SMITH ● Institute of Oceanographic Sciences, Wormley, England GU8 5UB.

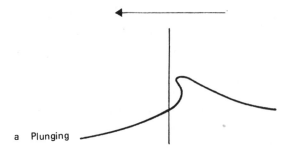

a   Plunging

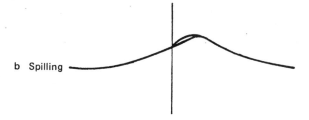

b   Spilling

FIGURE 1. Illustration of (a) a plunging breaker and (b) a spilling breaker.

TABLE I.   Wind and Wave Parameters for Observations at Nordwrijk Tower

Date	Time	U (m/s)	T_s (s)	c (m/s)	U/c	R (m/s)	R/c	N	$\tilde{\omega}$
10/18/79	0922–0932	13.7	4.9	7.6	1.8	6.5	0.86	3	0.02
	0935–0945	13.7	5.9	9.2	1.5	8.0	0.87	1	0.01
	1005–1015	13.7	6.3	9.9	1.4	8.2	0.83	6	0.06
	1035–1045	14.2	6.7	10.4	1.4	8.5	0.82	7	0.08
	1055–1105	14.2	6.8	10.6	1.3	8.5	0.80	5	0.06
10/19/79	1020–1030	14.4	6.1	9.5	1.5	—	—	0	0.00
	1051–1101	14.4	5.1	7.9	1.8	6.5	0.82	5	0.04
11/20/79	1100–1110	6.6	4.3	6.7	1.0	5.0	0.75	2	0.02
	1130–1140	5.2	3.3	5.2	1.0	—		0	0.00
11/23/79	1230–1240[a]	10.8	5.4	8.4	1.3	5.5	0.65	1	0.01
	1255–1305[a]	10.8	5.4	8.4	1.3	5.5	0.65	1	0.01
	1320–1330[a]	10.8	5.4	8.4	1.3	—		0	0.00
11/24/79	1335–1345	6.7	4.2	6.6	1.0	5.0	0.76	2	0.01
	1412–1422	6.9	4.3	6.7	1.0	5.0	0.75	2	0.01

[a]Heavy rain.

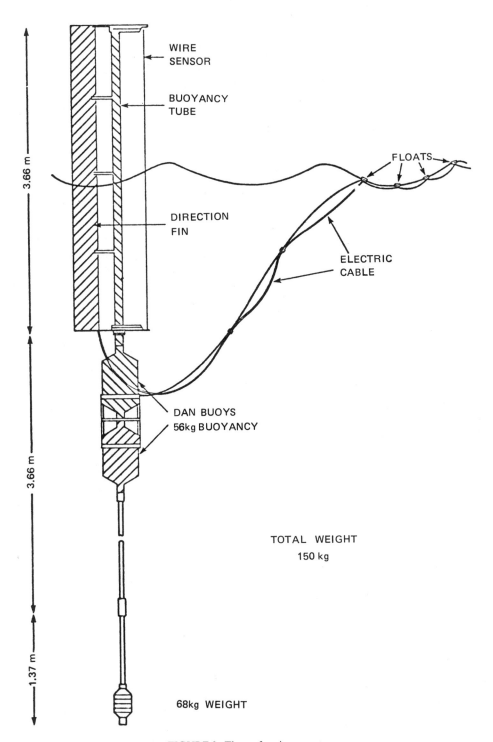

FIGURE 2. The surface jump-meter.

over an intermediate range surrounding some value R_0, say, the histogram was almost independent of R. The events could then be considered as significant and independent of noise level. Table I shows the total number N of "jumps" in representative records each of length 10 min, under a variety of wind conditions.

Now it can be seen that in a progressive wave of speed c, the rate of rise $\partial\eta/\partial t$ of the surface elevation against a fixed vertical line is related to the surface slope $\partial\eta/\partial x$ by

$$\frac{\partial\eta}{\partial t} = c\frac{\partial\eta}{\partial x} = c\tan\alpha$$

where α is the angle of inclination. So we would expect

$$\tan\alpha \geqslant R_0/c$$

approximately. Table I shows the values c of the phase speed corresponding to the period of the "significant" waves in each record, and it appears that all the ratios R_0/c lie somewhat above the value 0.586 for progressive Stokes waves of maximum steepness ($\alpha = 30.37°$; see Longuet-Higgins and Fox, 1977). This confirms our interpretation that all of the events recorded in Table I correspond to events that were on the forward face of steep, unstable waves, hence waves that were either breaking or about to break.

From the next to last column of Table I we see that under the given range of conditions the number N of apparently breaking waves in a 10-min record was typically between 0 and 8. For a wind speed of about 14 m/s, the value of N was of order 5.

3. DISCUSSION

Table I suggests a considerable variability in the value of N, of order $N^{1/2}$ at least. Part of the variability may be associated with rates of change in the wind speed U. The entries under U represent only the mean over each 10-min period. But it was observed, for instance, that over the interval from 0935 to 0945 on October 18, the wind speed actually fell by about 10%, and increased again during the following 10-min period. Thus, whitecapping may be a function of the time history of the wind field, as well as of its instantaneous value. To verify this will require many more observations than are yet available.

The results of Table I may be related to visual observations of whitecap coverage as follows. Let $\tilde{\omega}$ denote the proportion of significant waves observed to be breaking. Since each record had a duration of 600 s, the number of significant waves in the record equals $600/T_s$. Hence, we may take

$$\tilde{\omega} = NT_s/600 \qquad (1)$$

Numerical values given in the last column of Table I suggest that for U of order 14 m/s and $1.3 \leqslant U/c \leqslant 1.8$, the value of $\tilde{\omega}$ is of order 3×10^{-2}.

Now if the foam from a typical "jump" has a persistence time T_p, say, then the proportion of the sea surface covered by foam (the visual whitecap coverage) will be

$$VWC = \tilde{\omega}T_p/T_s = NT_p/600 \qquad (2)$$

For salt water, Monahan and Zietlow (1969) found $T_p = 3.85$ s, whence $VWC = 2 \times 10^{-2}$, which is within the range of observations reported by Monahan (1971, Fig. 2).

4. BREAKING WAVES AND WIND STRESS

We come now to the main point of this chapter: the possible implication of the measurements for horizontal wind stress.

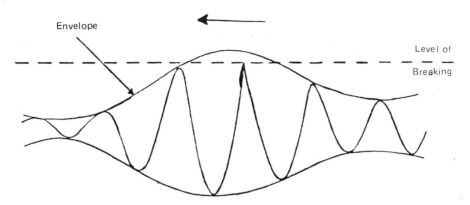

FIGURE 3. Propagation of gravity waves through a wave group, in deep water. Only the waves near the center of the group are breaking.

It was shown experimentally by Banner and Melville (1976) that the occurrence of whitecaps or patches of rough water near the crests of steep gravity waves tended to induce separation of the air flow over the wave. Measurements of the pressure and horizontal velocity in the air stream indicated that the flux of horizontal momentum from the air to the water was increased locally by a factor of order 50. Though separation may also occur when the waves are not breaking, it seems likely that breaking waves induce separation far more strongly and consistently.

It is important to note that in deep water, wave breaking is an essentially intermittent phenomenon (see Fig. 3). This is due to the fact that high waves occur in *groups*, and since the phase velocity c exceeds the group velocity c_g, each individual wave will travel through the group, starting at the rear, growing to its maximum amplitude, and then dying away toward the front of the group. A whitecap is seen only momentarily when the wave is near its point of maximum steepness. Since the speed of the wave relative to its envelope is $(c - c_g)$ or about $\frac{1}{2}c$ in deep water, whitecaps are seen intermittently with a frequency about half that of the dominant wave period, at least in a narrow spectrum (see Donelan *et al.*, 1972).

The separation of the air flow is therefore likely to be intermittent also. No precisely appropriate calculations of flow with a time-dependent boundary are available, but it is reasonable to compare the onset of separation with some recent calculations of the flow started from rest over a steep but regular train of waves (see Fig. 4). In these, a standard vortex-shedding technique was used, in which it was assumed that the vorticity was shed from the boundary layer at a point close to the sharp crest of each wave. In Fig. 4, each small semicircle represents a line-vortex, with circulation proportional to the area of the semicircle. As time increases, a region of vorticity develops in the lee of each crest, simulating a large-scale vortex. Eventually, the vorticity fills a lens-shaped area in the trough of the wave, and further development consists of an occasional slight spilling of vortices from each trough over into the next.

The changes in momentum of the air flow give rise to a mean horizontal component F of the normal pressure on the boundary. Figure 5 shows the resulting drag coefficient $C'_D = F/\rho(U - c)^2$ as a function of time. At first, when the lee eddy is forming, C'_D has the relatively high value of 0.15. As the rate of growth diminishes, C'_D falls, and eventually fluctuates about a mean value which is less by perhaps two orders of magnitude.* The high drag is thus a transient phenomenon, lasting for a time T_D of order $L/(U - c)$ where L is the wavelength and U the wind speed.

---

*The above model does not of course represent accurately all of the features of the flow, including viscous dissipation and mean shear.

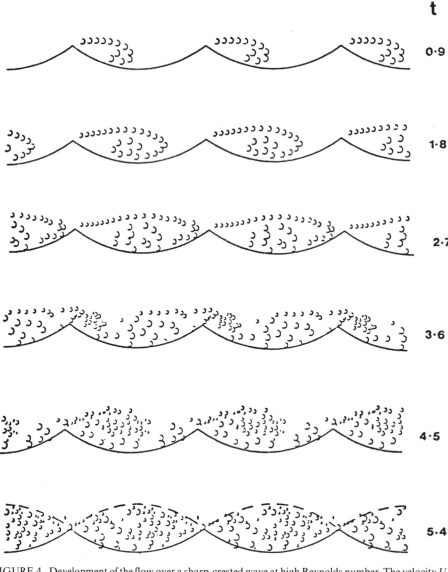

FIGURE 4. Development of the flow over a sharp-crested wave at high Reynolds number. The velocity U at infinity is given by

$$U = \begin{cases} 0 & t < 0 \\ 1 & t > 0 \end{cases}$$

The wavelength L is equal to 2π. (From Longuet-Higgins, 1980.)

Since the wave period T is of order L/c, it follows that

$$T_D/T \simeq c/(U - c) \tag{3}$$

The contribution of the steep waves to the horizontal stress is then

$$\tilde{\omega}(T_D/T)C_0'\rho(U - c)^2 = \tilde{\omega}C_0'\rho(U - c)c \tag{4}$$

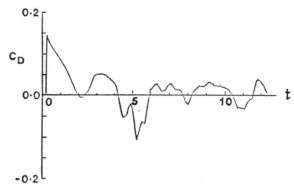

FIGURE 5. The drag coefficient C_D corresponding to the flow in Fig. 4, shown as a function of the time t.

where C_0' denotes the initial drag coefficient, averaged over $0 < t < T_D$. The total drag being $c_D \rho U^2$ by definition, the contribution of the steep waves is equivalent to an effective drag coefficient

$$c_D' = \tilde{\omega} C_0' (U - c)c/U^2 \qquad (5)$$

Taking $\tilde{\omega} = 0.05$ and $U/c = 1.4$ from Table I, and $C_0' = 0.075$ from Fig. 5, we find $c_D' = 0.75 \times 10^{-3}$. This is less than the typical value $c_D = 1.5 \times 10^{-3}$, but is of the same order of magnitude.

We have of course assumed the waves to be predominantly two-dimensional. If the breaking waves are short-crested, the relations (3) to (5) will not necessarily be affected, but the appropriate value of C_0' may be less than given in Fig. 5, though probably not by an order of magnitude.

We conclude that the flow separation induced by steep and breaking waves may indeed make a significant contribution to the mean horizontal wind stress, in agreement with the suggestion of Banner and Melville (1976).

5. WAVE GENERATION BY WIND

The above reasoning suggests that the horizontal stress τ exerted by the wind on the sea surface is very patchy and intermittent, high stresses being confined largely to areas of flow separation induced by steep or breaking waves. Since the energy input to the waves is locally of order τc, we must expect that wave generation is a similarly patchy process, confined predominantly to areas of existing steep waves. (We refer here to the primary input of energy, not the nonlinear transfer.) This would have been true to some extent even with a constant "sheltering coefficient" s.* It is even more pronounced if s itself is a function of the wave steepness. In fact, Fig. 5 implies that for steep waves and separated flow, s is initially of order 1.

The accurate calculation of the time-dependent flow over a propagating nonuniform group of waves presents a challenge to the fluid dynamicist.

Lastly, we note that because of the patchiness of the horizontal momentum transfer, a similar patchiness in the currents generated by wind is to be expected, leading to an input of "turbulence" with horizontal scales comparable to a wavelength, or to a group of breaking waves.

REFERENCES

Banner, M. L., and W. K. Melville (1976): On the separation of airflow over water waves. *J. Fluid Mech.* **77**, 825–842.

*Defined by Jeffreys through the relation $p_s' = s\rho(U - c)^2 \partial\eta/\partial x$ for the effective component of the surface pressure p_s.

Donelan, M., M. S. Longuet-Higgins, and J. S. Turner (1972): Periodicity in whitecaps. *Nature* **239**, 449–451.

Longuet-Higgins, M. S. (1980): Polygon transformation in fluid mechanics. *Lecture Notes in Physics* **141**, 12–30, Springer-Verlag, Berlin.

Longuet-Higgins, M. S., and M. J. H. Fox (1977): Theory of the almost highest wave: The inner solution. *J. Fluid Mech.* **80**, 721–741.

Longuet-Higgins, M. S., and N. D. Smith (1983). Measurement of breaking waves by a surface jump-meter. *Fall Meeting, Am. Geophys. Union*, 10 Dec. (MARSEN Session). *J. Geophys. Res.* **88**, 9823–9831.

Monahan, E. C. (1971): Oceanic whitecaps. *J. Phys. Oceanogr.* **1**, 139–144.

Monahan, E. C., and C. R. Zietlow (1969): Laboratory comparison of fresh-water and salt-water whitecaps. *J. Geophys. Res.* **74**, 6961–6966.

DISCUSSION

GOLDING: If the mechanism you propose for wind energy input to the waves turns out to be of dominant importance, then we shall have to look to whitecapping waves for the primary sources both of energy gain and energy loss in the wave field.

LONGUET-HIGGINS: Yes. I am aware of that possibility.

SU: You explained the periodicity of two wave periods for successive whitecapping in terms of the group velocity being one-half the phase velocity. Could the observed periodicity also be explained in terms of 3d nonlinear breaking waves?

LONGUET-HIGGINS: Possibly, though it is not yet clear to what extent the 3d instability occurs in the ocean. As a matter of fact, visual observations by M. Donelan (reported jointly in our original article in *Nature*) suggested that in the North Atlantic the ratio of whitecapping period to dominant wave period lay more generally between 1 and 2 (it would be 2 only for a narrow-band spectrum).

LIU: The index of the breaking events is a measure of the frequency of occurrence of plunging breakers which are expected to have a shorter duration than that of whitecaps. It appears that the index underestimates the actual frequency of occurrence of the whitecaps, which include both spilling and plunging breakers.

LONGUET-HIGGINS: Perhaps, but most breakers probably are spilling breakers, and the estimated persistence time in any case is very rough.

DONELAN: There may be some observational support for your idea that the momentum transfer is greatly enhanced locally immediately after the appearance of a sharp crest. I refer to the intermittency of momentum transfer over water which is stronger than that over a homogeneous solid surface. This suggests that the momentum transfer is affected by some aspect of the water surface which is intermittent in time or sporadic in space. The appearance of sharp crests or perhaps breaking waves seems a likely candidate.

LONGUET-HIGGINS: Yes, but it may be important to distinguish between intermittency on a short time-scale (comparable to one wave-period) and intermittency on longer time-scales.

DONELAN: Although your instrument yields some useful information on the distribution of sharp slopes, as you yourself have pointed out, one cannot be sure that it can unambiguously detect whitecaps. Perhaps another type of instrument which detects both surface elevation and horizontal velocity would be more revealing. Since the whitecap travels at the phase speed of the wave and sits just ahead of the crest, one would expect that the surface horizontal velocity would lead the surface elevation when and only when a whitecap is swept by the probe.

LONGUET-HIGGINS: We do have some additional information derived from the phase of the "jump" relative to the crest of the wave. From the original record, together with the theoretical phase response of the instrument, we find that the "sharp rise" occurs usually about half-way down the forward slope of the wave, indicating simply a steep wave. Less frequently, the "jump" occurred near the crest, indicating a whitecap, or possibly a plunging breaker.

17

Statistical Characteristics of Breaking Waves

Yeli Yuan, C. C. Tung, and Norden E. Huang

Abstract. The modification of the shape of the wave spectrum in the high-frequency range and the amount of energy loss, due to wave breaking are examined. The original waves are assumed to be Gaussian, stationary, and of finite bandwidth. Breaking is assumed to occur when the vertical acceleration at any point on the surface reaches $g/2$. Based on the wave breaking model, an approximate but accurate spectrum of breaking waves and an exact expression of the amount of energy loss due to wave breaking are derived. It is shown that the spectrum which corresponds to minimum rate of energy loss has an upper limit proportional to ω^{-5} in the high-frequency range.

1. INTRODUCTION

The saturation range spectrum, as proposed by Phillips (1958, 1977), is of the form

$$S(\omega) = \beta g^2 \omega^{-5} \tag{1}$$

in which β is a numerical constant. This spectral form governs the range of frequencies beyond that of the spectral peak but below those of capillary ripples. Assuming that wave breaking is controlled primarily by gravity, the upper limit of the frequency spectrum of the surface elevation of wind waves, found by similarity considerations, was shown to be given by (1).

In 1969, Longuet-Higgins examined, for narrow-band Gaussian stationary random waves, the amount of energy loss by wave breaking in a spectrum of the form (1). He further obtained, under the assumption that energy loss by wave breaking is comparable to that supplied by the wind, an estimate of the value of β. In a saturated state, it was found that the value of β agreed, to at least within an order of magnitude, with those observed in the field.

In this chapter, we shall reexamine the modification of the shape of the wave spectrum in the high-frequency range and the amount of energy loss due to wave breaking. The basic wave breaking mechanism is based essentially on the model used by Longuet-Higgins (1969)

Yeli Yuan and C. C. Tung ● Department of Civil Engineering, North Carolina State University, Raleigh, North Carolina 27650. Norden E. Huang ● NASA Goddard Space Flight Center, Greenbelt, Maryland 20771. *Present address for Y. Y.*: Institute of Oceanology, Academia Sinica, Quingdao, People's Republic of China.

generalized to cover the case in which the wave field may be of finite bandwidth. In this connection, it may be mentioned that Stoker (1957), and more recently, Banner and Phillips (1974) and Phillips and Banner (1974), all have shown that wind stress can affect wave breaking in laboratory conditions. Since in the laboratory the ratio of wind and wave speeds can be as large as 10 and in the field situation this ratio is close to unity, the influence of wind shear on wave breaking must be proportionally reduced (Huang, this volume). Some of the salient points in Longuet-Higgins's paper are given below.

According to Stokes' criterion, for a progressive, sinusoidal wave, breaking only occurs at the crest when the vertical acceleration reaches $g/2$. That is, the critical amplitude is

$$a_b = \tfrac{1}{2}g/\omega^2 \tag{2}$$

in which ω is the wave frequency. Alternatively,

$$a_b = a(\tfrac{1}{2}g/A), \qquad A = a\omega^2 \tag{3}$$

The amplitude of the wave that breaks is reduced according to the ratio of $g/2$ and the vertical acceleration of the original wave at the crest.

When the original waves are random, Gaussian, stationary, and narrow-band, the amplitude is Rayleigh distributed. Under these assumptions, Longuet-Higgins (1969) calculated the expected value of the loss of energy per wave cycle as

$$\Delta E = \int_{a_b}^{\infty} \tfrac{1}{2}\rho g(a^2 - a_b^2)p(a)da \tag{4}$$

in which $p(a)$ is the Rayleigh probability density function of the original wave amplitude a. In (4), a is modified to become

$$a_b = \tfrac{1}{2}g/\bar{\omega}^2 \tag{5}$$

with

$$\bar{\omega} = \left[\int_0^{\infty} \omega^2 S(\omega)d\omega \Big/ \int_0^{\infty} S(\omega)d\omega \right]^{1/2} \tag{6}$$

as the mean frequency. The rate of energy loss per wave cycle, defined as $\tilde{w} = \Delta E/E$ in which E is the expected value of total energy per wave cycle, is computed to be

$$\tilde{w} = \exp(-g^2/8\bar{\omega}^4\mu_0) \tag{7}$$

where $\mu_0 = \int_0^{\infty} S(\omega)d\omega$.

We shall now extend this breaking wave model to the case in which the waves may be of finite spectral width. It is still assumed that the surface elevation $\zeta(t)$ of the original waves is Gaussian and stationary. However, breaking occurs when the vertical acceleration at any point on the surface reaches $g/2$, and the surface elevation is reduced according to the ratio $g/2$ and local acceleration $|\zeta''|$ of the ideal wave. That is, the surface elevation $\zeta_b(t)$ of the breaking waves is given by

$$\zeta_b(t) = \zeta(t)\left\{ \frac{\tfrac{1}{2}g}{|\zeta''(t)|}H\left[|\zeta''(t)| - \frac{g}{2}\right] + H\left[\frac{g}{2} - |\zeta''(t)|\right] \right\} \tag{8}$$

in which $H(\cdot)$ is the Heaviside unit function.

2. MODIFICATION OF SPECTRUM BY WAVE BREAKING

Having established the breaking wave model (8), the spectrum $S_b(\omega)$ of the breaking waves can be obtained in a straightforward manner. To do this, the covariance function of ζ_b is first formed. Using subscripts "1" and "2" to denote quantities evaluated at times $t_1 = t$ and $t_2 = t + \tau$, the covariance function of ζ_b is, from (8),

$$E[\zeta_{b1}\zeta_{b2}] = E\left[\frac{g^2\zeta_1\zeta_2}{4|\zeta_1''||\zeta_2''|}H\left(|\zeta_1''| - \frac{g}{2}\right)H\left(|\zeta_2''| - \frac{g}{2}\right)\right]$$

$$+ E\left[\frac{g\zeta_1\zeta_2}{2|\zeta_1''|}H\left(|\zeta_1''| - \frac{g}{2}\right)H\left(\frac{g}{2} - |\zeta_2''|\right)\right]$$

$$+ E\left[\frac{g\zeta_1\zeta_2}{2|\zeta_2''|}H\left(|\zeta_2''| - \frac{g}{2}\right)H\left(\frac{g}{2} - |\zeta_1''|\right)\right]$$

$$+ E\left[\zeta_1\zeta_2 H\left(\frac{g}{2} - |\zeta_1''|\right)H\left(\frac{g}{2} - |\zeta_2''|\right)\right] \tag{9}$$

in which $E[\cdot]$ denotes the expected value of the quantity in the brackets.

The expected values of the terms on the right-hand side of (9) can be obtained using the joint Gaussian probability function of $\zeta_1, \zeta_2, \zeta_1''$, and ζ_2''. However, noting that ζ_b is a nonlinear function of ζ and ζ'', the covariance function of ζ_b is necessarily nonlinearly related to the auto- and cross-covariance functions of ζ and ζ''. To obtain the spectrum $S_b(\omega)$ of ζ_b, the Fourier transform must be performed numerically. To avoid this, we first represent the joint Gaussian probability density function of $\zeta_1, \zeta_2, \zeta_1''$, and ζ_2'' in terms of Hermite polynomials (details of the calculation may be obtained from the authors). The resulting expression for the convariance function of ζ_b is

$$E[\zeta_{b1}\zeta_{b2}] = \sum_{m=0}^{\infty} \left\{ \frac{g^2}{2}[1 + (-1)^m]\left[2a_1b_1 II_m III_m + (a_1^2 + b_1^2)I_m^2 + \left(\frac{\sigma_{12}}{\sigma''}\right)^2 \rho_{12} III_m^2\right]\right.$$

$$+ 2g\left[a_1b_1\sigma''(II_m I_m' + III_m III_m') + (a_1^2 + b_1^2)\sigma'' I_m II_m' + \left(\frac{\sigma_{12}}{\sigma''}\right)^2 \rho_{12} III_m I_m'\right]$$

$$+ [2a_1b_1(\sigma'')^2 I_m' III_m' + (a_1^2 + b_1^2)(\sigma'')^2(I_m')^2 + \sigma^2\rho_{12}(I_m')^2]\left\}(\rho_{12}')^m \right. \tag{10}$$

In (10), the functions $I_m, II_m, III_m, I_m', II_m'$, and III_m' all have argument $\S = g/2\sigma''$ and are defined in the Appendix. The various other quantities are defined as follows:

$$a_1 = (r''r^{(4)} - r_{12}'' r_{12}^{(4)})/\Delta_{12}, \qquad b_1 = (r_{12}'' r^{(4)} - r'' r_{12}^{(4)})/\Delta_{12}$$

$$r = E[\zeta^2], \qquad r_{12} = E[\zeta_1\zeta_2], \qquad r'' = E[\zeta\zeta''], \qquad r_{12}'' = E[\zeta_1\zeta_2'']$$

$$r^{(4)} = E[(\zeta'')^2], \qquad r_{12}^{(4)} = E[\zeta_1''\zeta_2''], = \rho_{12}, \qquad \Delta_{12} = (r^{(4)})^2 - (r_{12}^{(4)})^2$$

$$\sigma_{12}^2 = \{r\Delta_{12} - r^{(4)}[(r'')^2 + (r_{12}'')^2] + 2r''r_{12}''r_{12}^{(4)}\}/\Delta_{12}$$

$$\rho_{12} = R_{12}/\sigma^2, \qquad R_{12} = \{r_{12}\Delta_{12} + r_{12}^{(4)}[(r'')^2 + (r_{12}'')^2] - 2r''r_{12}''r^{(4)}\}/\Delta_{12}$$

and $(\sigma'')^2 = r^{(4)}$. Here, quantities with subscripts "12" are all functions of τ.

Since (10) is a nonlinear function of r_{12}, r_{12}'', and $r_{12}^{(4)}$, its Fourier transform remains difficult to perform. An approximate, although accurate, solution can be achieved if only terms involving $m = 0$ in the series (10) are retained. After some rearrangement, the linear approximation of the covariance function of ζ_b becomes

$$E[\zeta_{b1}\zeta_{b2}] \doteq r_{12}[2\S III_0(\S) + I'_0(\S)]^2$$
$$+ 2r''_{12}(r''/r^{(4)})[2\S III_0(\S) + I'_0(\S)][2\S II_0(\S) + III'_0(\S) - 2\S III_0(\S) - I'_0(\S)]$$
$$+ r^{(4)}_{12}(r''/r^{(4)})^2[2\S II_0(\S) + III'_0(\S) - 2\S III_0(\S) - I'_0(\S)]^2$$
$$= A_1^2 r_{12} - 2A_1 A_2 r''_{12}(r''/r^{(4)}) + A_2^2 r^{(4)}_{12}(r''/r^{(4)})^2 \tag{11}$$

in which $\S = g/2\sigma''$,

$$A_1 = \frac{1}{(2\pi)^{1/2}} \S E_1\left(\frac{\S^2}{2}\right) + 2\int_0^\S Z(x)dx$$

$$A_2 = \frac{1}{(2\pi)^{1/2}} \S E_1\left(\frac{\S^2}{2}\right)$$

$$Z(x) = \frac{1}{(2\pi)^{1/2}} \exp\left(-\frac{x^2}{2}\right)$$

and

$$E_1(x) = \int_x^\infty \eta^{-1} e^{-\eta} d\eta \tag{12}$$

The linear approximation of $S_b(\omega)$ is the Fourier transform of (11) and is

$$S_b(\omega) \doteq [A_1^2 - 2A_1 A_2 |r''/r^{(4)}|\omega^2 + A_2^2 |r''/r^{(4)}|^2\omega^4]S(\omega)$$
$$= \{[A_2(r''/r^{(4)})]^2(\omega^2 - \omega_1^2)^2\}S(\omega) \tag{13}$$

in which $S(\omega)$ is the spectrum of the original waves and

$$\omega_1^2 = (A_1/A_2)|r^{(4)}/r''| \tag{14}$$

That is, the linear approximation of the spectrum of the breaking waves is the product of the spectrum of the original waves and a factor that accounts for the effect of wave breaking.

To further examine the properties of $S_b(\omega)$, we note that ω_1 is always larger than frequencies of wind waves and that the factor in the braces in (13) is always less than unity in the frequency range $0 \leqslant \omega \leqslant \omega_1$ indicating that the effect of wave breaking is to reduce the spectrum of the original waves.

Now, let us first consider the geometric meaning of the quantity $g/2\sigma''$. Let A be the surface area of waves and A_B that of breaking waves. If it is assumed that the vertical acceleration on the surface of that portion of the waves that break is equal to $g/2$ and is almost equal to zero elsewhere, then

$$g/2\sigma'' = \tfrac{1}{2}g/[(\tfrac{1}{2}g)^2(A_B/A)]^{1/2} = 1/(A_B/A)^{1/2} \tag{15}$$

which is obviously much larger than unity in most cases. Furthermore, consider the asymptotic relation that for large values of x, $E_1(x) \sim 0(1)x^{-1}e^{-x} \ll 0(1)$ and $2\int_0^x Z(\eta)d\eta \sim (2\pi)^{1/2}0(1)$, one obtains $A_1/A_2 \sim 0(1)e^{\S^2/2} \gg 1$ where reference has been made to (12). Noting now that $(r^{(4)}/r'')^{1/2} \geqslant (r''/r)^{1/2} \sim \bar\omega$, we have

$$\omega_1 = (A_1/A_2)^{1/2}(r^{(4)}/r'')^{1/2} \gg \bar\omega \tag{16}$$

In the range $0 \leqslant \omega \leqslant \omega_1$, we note that the factor in the braces in (13) takes on values between A_1^2 and zero and is a monotonically decreasing function of ω. Since $A_1'(\S) = (2\pi)^{-1/2}E_1(\S/2) > 0$ and $A_1(\infty) = 1$, we have $A_1(x) < 1$ for all values of x. That is, the factor is always less than unity in the range $0 < \omega < \omega_1$.

3. RATE OF ENERGY LOSS DUE TO WAVE BREAKING

The present wave breaking model specifies that waves break whenever the vertical acceleration on the surface exceeds $g/2$ and the elevation of the wave is reduced according to (8). Under these circumstances, the amount of energy loss is $\rho g[\zeta^2(t) - \zeta_b^2(t)]$ and the expected value of energy loss per unit time is, from (8),

$$\Delta E = \rho g \int\int_{-\infty}^{\infty} \zeta^2 \left[1 - \frac{\frac{1}{4}g^2}{(\zeta'')^2}\right] H\left(|\zeta''| - \frac{g}{2}\right) P(\zeta, \zeta'') d\zeta d\zeta'' \tag{17}$$

in which $P(\zeta, \zeta'')$ is the joint probability density function of ζ and ζ'', per unit time. The expected value of total energy per unit time is

$$E = \rho g \int\int_{-\infty}^{\infty} \zeta^2 P(\zeta, \zeta'') d\zeta d\zeta'' \tag{18}$$

and the rate of energy loss per unit time is computed (details can be obtained from the authors) to be

$$\tilde{w} = \exp[-D(1 - \varepsilon^2)] - D[\varepsilon^2(1 - \varepsilon^2)/(2 - \varepsilon^2)]E_1[D(1 - \varepsilon^2)] \tag{19}$$

in which $D = g^2\mu_0/8\mu_2^2$, and $0 < \varepsilon^2 = 1 - (\mu_2^2/\mu_0\mu_4) < 1$ is a measure of the width of spectrum $S(\omega)$, μ_i being the ith moment of $S(\omega)$. When the waves are narrow-band, $\varepsilon = 0$ and $\tilde{w} = \exp(-D)$ which agrees with that obtained by Longuet-Higgins (1969).

We wish to show now that $\tilde{w}$ increases with increase of ε. It was mentioned earlier that $g^2/8\mu_4 = D(1 - \varepsilon^2) \gg 1$. Since $E_1(x) \doteq x^{-1}e^{-x}$ when $x \gg 1$, we have

$$\tilde{w} = \exp[-D(1 - \varepsilon^2)] - D\frac{\varepsilon^2(1 - \varepsilon^2)}{2 - \varepsilon^2}\frac{\exp[-D(1 - \varepsilon^2)]}{D(1 - \varepsilon^2)}$$

$$= \frac{2(1 - \varepsilon^2)}{2 - \varepsilon^2}\exp[-D(1 - \varepsilon^2)] \tag{20}$$

If the value of D is held constant,

$$\left.\frac{d\tilde{w}}{d\varepsilon}\right|_{D\text{ fixed}} = \frac{4\varepsilon}{2 - \varepsilon^2}\left[D(1 - \varepsilon^2) - \frac{1}{2 - \varepsilon^2}\right]\exp[-D(1 - \varepsilon^2)] \tag{21}$$

Since $0 < \varepsilon < 1$ and $D(1 - \varepsilon^2) \gg 1$, we have

$$D(1 - \varepsilon^2) - [1/(2 - \varepsilon^2)] \geqslant D(1 - \varepsilon^2) - 1 > 0 \tag{22}$$

That is, $d\tilde{w}/d\varepsilon|_{D\text{ fixed}} > 0$ so that the rate of energy loss indeed increases with bandwidth.

4. ASYMPTOTIC FORM OF THE UPPER LIMIT FOR THE SPECTRUM OF BREAKING WAVES

In the above discussion, especially in (19), it is noted that the random waves must have spectral moments up to the fourth order. For the class of functions $S(\omega) \in \{S(\omega) | \int_0^\infty \omega^4 S(\omega) d\omega < \infty\}$ we offer the following theorem: The spectral function which corresponds to minimum rate of energy loss has an upper limit proportional to ω^{-5} in the high-frequency range. The proof of this theorem may proceed by taking the variation of $\tilde{\omega}$ in (18). That is,

$$\delta\tilde{w} = \frac{1}{\mu_4^2}[B_1 - B_2(1-\varepsilon^2)]\delta\mu_4 + \frac{1}{\mu_2\mu_4}[2B_2(1-\varepsilon^2)]\delta\mu_2$$

$$- \frac{1}{\mu_4}[B_2(1-\varepsilon^2)]\delta\mu_0 \tag{23}$$

in which

$$B_1 = \frac{g^2}{8}\left[\frac{2(1-\varepsilon^2)}{2-\varepsilon^2}\exp\left(-\frac{g^2}{8\mu_4}\right) + \frac{\varepsilon^2}{2-\varepsilon^2}E_1\left(\frac{g^2}{8\mu_4}\right)\right] > 0$$

and

$$B_2 = \frac{g^2}{4(2-\varepsilon^2)^2}E_1\left(\frac{g^2}{8\mu_4}\right) > 0 \tag{24}$$

According to the definition of variation

$$\delta J(s) = \lim_{\varepsilon' \to 0} \frac{J(S+\varepsilon's) - J(S)}{\varepsilon'} \tag{25}$$

so that

$$\delta\mu_i = \int_0^\infty \omega^i s d\omega \tag{26}$$

We thus have

$$\delta\tilde{w} = \int_0^\infty \left\{ \frac{1}{\mu_4^2}[B_1 - B_2(1-\varepsilon^2)]\omega^4 + \frac{1}{\mu_2\mu_4}[2B_2(1-\varepsilon^2)]\omega^2 - \frac{1}{\mu_4}[B_2(1-\varepsilon^2)] \right\} s d\omega \tag{27}$$

in which the arbitrary function $s \in \{S(\omega) | \int_0^\infty \omega^4 S(\omega)d\omega < \infty\}$. Due to the arbitrariness of s, if there exists a minimizing functions S_m, one must have, for $\delta\tilde{w} = 0$,

$$\frac{1}{\mu_4^2}[B_1 - B_2(1-\varepsilon^2)]\omega^4 + \frac{1}{\mu_2\mu_4}[2B_2(1-\varepsilon^2)]\omega^2 - \frac{1}{\mu_4}[B_2(1-\varepsilon^2)] = 0 \tag{28}$$

for any $\omega \in (0, \infty)$. Therefore,

$$\frac{B_1 - B_2(1-\varepsilon^2)}{\mu_4^2} = 0, \quad \frac{1}{\mu_2\mu_4}[2B_2(1-\varepsilon^2)] = 0, \quad \frac{1}{\mu_4}[B_2(1-\varepsilon^2)] = 0 \tag{29}$$

Now, for any large value $M > 0$, we divide $\{S(\omega) | \int_0^\infty \omega^4 s(\omega)d\omega < \infty\}$ into two parts:

$$S_1 = \left\{ S(\omega) \Big| \int_0^\infty \omega^4 S(\omega) < M < \infty \right\} \tag{30}$$

and

$$S_2 = \left\{ S(\omega) | M < \int_0^\infty \omega^4 S(\omega)d\omega < \infty \right\} \tag{31}$$

If $S_m \in S_1$, the quantities μ_0, μ_2, and μ_4 exist and are finite and (29) can not be simultaneously satisfied since $B_1 > 0$ and $B_2 > 0$. That is, S_m does not exist in S_1.

In S_2, according to the theorem of integral convergence, for any small value $\varepsilon_1 > 0$, there is a $\delta > 0$, so that if $S \sim \omega^{-(5+\delta)}$ in the high-frequency range, (31) is satisfied and for $\mu_4 > N/\varepsilon_1$,

$$\frac{B_1 - B_2(1 - \varepsilon^2)}{\mu_4^2} < \varepsilon_1, \qquad \frac{2B_2(1 - \varepsilon^2)}{\mu_2 \mu_4} < \varepsilon_1, \qquad \text{and} \qquad \frac{B_2(1 - \varepsilon^2)}{\mu_4} < \varepsilon_1$$

That is, the minimizing function S_m exists in S_2 and has an upper limit $S_m \sim \omega^{-5}$ in the high-frequency range.

It should be pointed out that due to the finite energy input rate from the wind and limited frequency range of gravity waves, the individual wave spectrum does not automatically approach the asymptotic upper bound given above.

ACKNOWLEDGMENTS. This study is supported by the National Aeronautics and Space Administration Wallops Flight Center.

APPENDIX

The functions I_m, II_m, III_m, I'_m, II'_m, and III'_m are

$$I_m(x) = m^{-1/2} h_{m-1}(x) Z(x), \qquad m \geqslant 1$$

$$I_0(x) = \int_x^\infty Z(\eta) d\eta$$

$$II_m(x) = m^{-1/2} x h_{m-1}(x) Z(x) - m^{-1/2} I_{m-1}(x), \qquad m \geqslant 1$$

$$II_0(x) = Z(x)$$

$$III_m(x) = \int_x^\infty \eta^{-1} h_m(\eta) Z(\eta) d\eta, \qquad m \geqslant 1$$

$$III_0(x) = \int_x^\infty \eta^{-1} Z(\eta) d\eta$$

$$I'_m(x) = m^{-1/2}[1 + (-1)^m] h_{m-1}(x) Z(x), \qquad m \geqslant 1$$

$$I'_0(x) = \int_{-x}^x Z(\eta) d\eta$$

$$II'_m(x) = m^{-1/2}[1 - (-1)^m] x h_{m-1}(x) Z(x) + m^{-1/2} I'_{m-1}(x), \qquad m \geqslant 1$$

$$II'_0(x) = 0$$

$$III'_m(x) = -m^{-1/2}[1 + (-1)^m] x^2 h_{m-1}(x) Z(x) + 2m^{-1/2} II'_{m-1}(x), \qquad m \geqslant 1$$

$$III'_0(x) = -2x Z(x) + I'_0(x)$$

The terms I_0, I'_0 can be expressed as error functions and III_0, in terms of $E_1(\cdot)$ as shown in (11) and (12). Further reduction of III_m to error functions can be done but is quite involved. Since our interest is limited to (11) which does not require the computation of III_m, such derivation is not given here.

REFERENCES

Banner, M. L., and O. M. Phillips (1974): On the incipient breaking of small scale waves. *J. Fluid Mech.* **77**, 825–842.
Longuet-Higgins, M. S. (1969): On wave breaking and the equilibrium spectrum of wind-generated waves. *Proc. R. Soc. London Ser. A* **310**, 151–159.
Phillips, O. M. (1958): The equilibrium range in the spectrum of wind-generated waves. *J. Fluid Mech.* **4**, 426–434.
Phillips, O. M. (1977): *The Dynamics of the Upper Ocean*, 2nd ed., Cambridge University Press, London.
Phillips, O. M., and M. L. Banner (1974): Wave breaking in the presence of wind drift and swell. *J. Fluid Mech.* **66**, 625–640.
Stoker, J. J. (1957): *Water Waves*, Interscience, New York, 372.

DISCUSSION

SU: Phillips's (f^{-5}) power law of the saturation range of the wave spectral density is restricted to the frequencies above the peak frequency. How do you justify the integration of the power law over a range of frequency from zero to infinity in S_1 and S_2 in your last slide?

YUAN: This is an idealized study of the wave spectral form. In the study, we made no assumption on the shape of the spectrum as required by the variational method. We did not start with Phillips's saturation range spectrum. The only assumption made was that gravity-controlled breaking applies throughout the frequency range.

18

ON MICROWAVE SCATTERING BY BREAKING WAVES

LEWIS WETZEL

ABSTRACT. Microwave radars, when viewing breaking waves at low grazing angles with high spatial resolution, observe sharp transient bursts of backscatter having large cross sections (of order $1\,m^2$ or more), and a characteristic polarization dependence, with horizontal polarization giving sharper (briefer) returns down to about $1°$ grazing angle, below which horizontal and vertical returns look much the same. X-band returns have disclosed a modulation with periods of 10–20 ms. The transient nature of breakers, along with the several structurally different forms they may take, makes it difficult to establish a deterministic scattering model having any real credibility from observation alone. However, there is one theoretical model of a breaking wave, the "entraining plume model of a spilling breaker" (Longuet-Higgins and Turner, 1974), that contains structural and dynamic features which can be used as a basis for microwave scattering calculations. This chapter seeks to examine some of the characteristics of the radar return from structures suggested by the plume model. It is shown that several of these plumes, emerging from the wave crest at different times, could produce radar returns with the right cross section (a few square meters) and the right modulation period (10–20 ms). Moreover, the model explains certain observed shifts in the envelope and phase of the modulation produced by small ($\sim 10\%$) changes in radar frequency. In order to bring in polarization, near-field effects of the wave face just ahead of the advancing plume are considered. The results suggest that close to the crest, where wave slopes are steep, vertical and horizontal returns would be roughly the same. As the plume moves down the wave face encountering decreasing wave slopes, the horizontal return drops off sharply, while the vertical return peaks broadly over most of the wave face. This is consistent with the observation of sharp horizontal and broad vertical returns, and with the tendency of the two polarizations to look alike at extreme grazing angles, where only the wave peaks remain visible. While neither rigorous nor definitive, these calculations suggest that accelerating plumes, or something like them, might provide a self-consistent explanation for many of the properties of microwave scattering by breaking waves.

1. INTRODUCTION

When viewing breaking waves at low grazing angles, high-resolution microwave radars observe the sharp, transient bursts of backscatter commonly called "sea spikes." There has been little systematic experimental work done on this topic, but the existing measurements (see, e.g., Long, 1975; Kalmykov et al., 1976; Lewis and Olin, 1980; Keller et al., this volume) suggest that scattering from breakers occurs with large cross sections of $1\,m^2$ or greater, and a characteristic polarization dependence. The returns with horizontal polarization are sharper (of briefer duration) than those with vertical polarization down to grazing angles of about $1°$, below which

LEWIS WETZEL ● Naval Research Laboratory, Radar Division, Washington, D.C. 20375.

the horizontal and vertical returns begin to look much the same. At X-band frequencies (about 10 GHz), Lewis and Olin (1980), using a high-resolution incoherent radar, found envelope modulations with periods of about 10 ms, while the coherent radar measurements of Keller *et al.* (this volume) showed breaking to be accompanied by sharp increases in Doppler frequency and a significant broadening of the Doppler spectrum.

The transient (or as Phillips calls it, "fugitive") nature of breakers, along with the several structurally different forms they may take ("spilling," "plunging," "collapsing," "surging"), make it difficult to establish a deterministic scattering model with any real credibility from observation alone. Observers of sea spikes have used analogies with standard scattering formalisms in attempting to explain what they believe the radar is seeing. Thus, the dominant presence of spray in the air leads to a "rain" model (Kalmykov *et al.*, 1976), a field of small sharp crests suggests a "dielectric wedge" model (Kalmykov and Pustovoytenko, 1976; Kwoh and Lake, this volume), spreading patches of white water look like "very rough surfaces" (Lewis and Olin, 1980). In each case, there is a measure of plausibility to the model and some credibility in the predictions. However, the sum of all of these experiments and their interpretations has failed to provide a convincing self-consistent physical picture of microwave scattering from breaking waves. Ideally, one would find out what breaking waves look like, attempt to construct reasonable scattering models from the actual surface morphology of these wave structures, test the implications of the models against the existing data, and design additional experiments that would move the field forward in an orderly and rational way. Unfortunately, breaking waves are themselves not well understood and the theory is still in a relatively primitive state, although Longuet-Higgins has made significant progress in the modeling of unsteady waves by both numerical (1978) and analytical (this volume) methods. Since our goal here is to understand microwave scattering from breakers, the theory is of interest primarily where it suggests something about the shape and dynamics of breaker features. Just such a feature emerges from the "entraining plume model of a spilling breaker" developed by Longuet-Higgins and Turner (1974), and in what follows we will examine the electromagnetic scattering implications of such plumes in an effort to establish a deterministic relation between the results of radar backscatter experiments and a realistic feature of a breaking wave.

2. SCATTERING FROM DISCRETE PLUMES

In the model proposed by Longuet-Higgins and Turner, a turbulent plume emerges from the unstable wave crest and accelerates down the forward face of the breaking wave, entraining air as it goes. Figure 1 is a sketch based on their Fig. 7, with added parameters that relate to the scattering model: the mean surface of the forward face of the plume is approximated by a cylinder of radius a, the plume thickness is δ, the local slope of the underlying wave is α, and the entry angle of the plume into the wave at the toe is β. The hydrodynamic model is not specific about either a or β so we will conveniently assume that $a = \delta$ and $\beta = 0$; that is, the front face of the plume is modeled as a quarter-cylinder whose radius is equal to the plume thickness. (The scattering implications of a rough plume surface will be discussed later.) The horizontal acceleration of the plume as it moves down the wave face is denoted by the vector **a** attached to the plume's "toe."

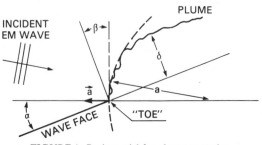

FIGURE 1. Basic model for plume scattering.

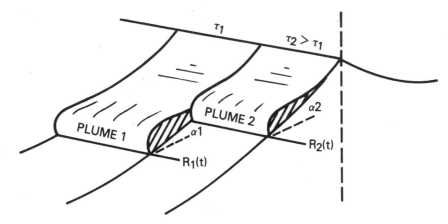

FIGURE 2. Idealization of a two-plume breaker.

In order to develop a breaker model complete enough for scattering calculations, it will be necessary to make some further assumptions. Starting with a single plume as the basic element, we will make an initial assumption that the breaking "event" involves a cascade of discrete plumes emitted along the wave crest at closely spaced times τ_i, which we call their "epochs," and covering the front face of the wave with an ordered sequence of accelerating water masses identified by their individual "ages" and time histories. The entire breaking process spans only a second or two, and the instantaneous radar return will generally include signals scattered from several plumes.

An idealized breaker model including two plumes is sketched in Fig. 2. At time t the toe of a plume emitted at epoch τ_i has advanced down the wave face to a position $R_i(t)$, at which point the inclination of the wave face is $\alpha_i(t)$. If we now assume that the radar wavelength is of the order of the plume thickness (not so long that the plume disappears into the Rayleigh region, nor so short that surface roughness dominates), then we may estimate the broadside scattering cross section by the physical optics approximation for a finite cylinder (Kerr, 1951):

$$\sigma = (ka)L^2 F^2(\alpha) \tag{1}$$

where $k = 2\pi/\lambda$, λ being the radar wavelength, L is the cylinder (plume) length, and $F(\alpha)$ is a wave proximity factor introduced to account for the effects of the wave surface in front of the plume— effects we will find later to be important in the polarization dependence of the scattered field. Although this simple model is unrealistic, it can provide an upper limit to the radar cross section of a plume at common radar frequencies. Assuming a plume about 1 inch thick ($a = 3$ cm) to maintain a cylindrical shape along a toe length of about 1 foot ($L = 30$ cm), we find that at X-band

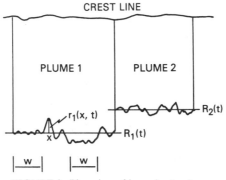

FIGURE 3. Plan view of irregular toe line.

($\lambda = 3$ cm) the factor $(ka) = 2\pi$, and taking the maximum value of $F(\alpha) = 1.5$, the radar cross section according to (1) becomes $\sigma_p = 1.3$ m^2. This is a rather formidable value, especially when compared to the cross section of a 1-square-foot patch of ocean due to Bragg scattering: $\sigma_B = 10^{-4}$ m^2.

A more realistic model of scattering from discrete plumes would account for variations along the toe line of each plume, as shown in plan view in Fig. 3. Retaining the physical optics approximation, we convert the finite cylinder cross section in (1) into a scattering amplitude $\eta = \sigma^{1/2} = (ka)^{1/2}LF$ and use this form to define an incremental scattering amplitude for elements dx along the toe line: $d\eta_{(x)} = [ka(x)]^{1/2}F[\alpha(x)]\,dx$. We then introduce for each such element a phase factor relative to the receiver, $\exp\{i2k[R_p(t) + r_p(x,t)]\}$, where $R_p(t)$ is the mean position of the plume at time t, and assemble all of the scattering elements across each plume into a scattering integral:

$$\eta_p(t) = e^{i2kR_p(t)} \int_{L^p} e^{i2kr_p(x,t)}[ka_p(x,t)]^{1/2}F[\alpha_p(t)]\,dx \qquad (2)$$

The instantaneous radar return derives from the sum of scattering amplitudes of the form of (2) for all of the plumes within the radar footprint. Plume accelerations enter in the time dependence of the mean plume position $R_p(t)$.

$$R_p(t) = v_c t + \tfrac{1}{2}a(t - \tau_p)^2 \qquad (3)$$

where v_c is the crest velocity. The implications of plume acceleration will be covered in a later discussion of the dynamics of the scattered signal. [Note that in the discrete plume model, the plume springs into existence over a length L_p at the instant τ_p. If, instead, it was generated continuously by an instability moving along the wave crest with a certain velocity, then τ_p becomes a function of time, as does x, and the outboard exponential in (2) must be taken inside the integral. We will not pursue this elaboration in this chapter.]

It is tempting to apply statistical arguments to (2) (e.g., assume r_p to be normally distributed along the toe line and find $\langle \eta \rangle$, $\langle |\eta|^2 \rangle$, etc.), but the results are almost self-evident, and since we know nothing of the real statistics of plume behavior they would be of little value. However, photographs of breaking waves in a wave tank (Banner and Phillips, 1974) show a continuous breaking "front" (a "plume"?) to contain several segments along which the toe line is relatively linear. The integral in (2) would select such segments by the stationarity of their phase and read them out as the dominant scattering elements in the plume. Two such segments are indicated between the vertical lines under Plume 1 in Fig. 3. If we assume they have effective lengths w and are located at positions r_1 and r'_1, then the integral is stationary over these segments and the scattering amplitude can be approximated by

$$\eta_1(t) \approx (ka)^{1/2}F(we^{i2kr_1} + we^{i2kr'_1})e^{i2kR_1(t)} \qquad (4)$$

where we have assumed a and F to be fixed. The cross section corresponding to this scattering amplitude is

$$\sigma_1 = |\eta_1|^2 = (ka)(2w)^2F^2\cos^2[k(r_1 - r'_1)] \qquad (5)$$

For effective lengths as little as 10 cm, the X-band cross sections of a plume 3 cm thick could reach 0.5 m^2 (with $F = 1.5$). The frequency sensitivity of the cross section can be illustrated by assuming the two segments to be separated relative to the mean plume position by a few wavelengths [e.g., $(r_1 - r'_1) = 9$ cm]. Then at 10 GHz, σ_1 is a maximum, while at 9.2 GHz, it would vanish. From this we conclude that a plume whose irregular toe line contains a few strong scattering centers could have a microwave scattering cross section that is a sensitive function of radar frequency. Such sensitivities have, in fact, been observed by Lewis and Olin (1980), who found that simultaneous

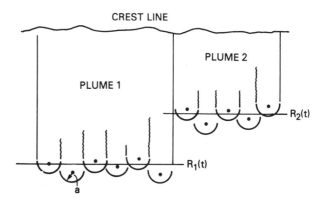

FIGURE 4. Breaker front modeled
by hemisphere-capped fingers.

scattering levels from the same breaker can sometimes be quite different at two X-band frequencies separated by as little as 7%.

A lower limit on X-band scattering from plumes of the size we have been considering may be estimated by imagining the plume to consist of an ensemble of "fingers" capped by hemispheres whose radius is equal to the plume thickness (Fig. 4). In the physical optics approximation, the cross section of each hemisphere will be $\sigma = \pi a^2 F^2$, and if they are located randomly with respect to the mean plume position over a band at least a wavelength or so wide (several centimeters at X-band), the random phase approximation may be used, yielding a plume cross section

$$\sigma_p = N(\pi a^2)F^2 \tag{6}$$

where N is the total number of hemispheres belonging to the particular plume. There would be about 20 such hemispheres across a plume width of 1 m, so with $a = 3$ cm and $F = 1.5$, the plume cross section becomes $\sigma_p \sim 20 \times (0.003) \times 2.3 = 0.14 \, \text{m}^2/\text{m}$. If the scattering amplitudes ($\eta = \sigma^{1/2}$) from several plumes were to add constructively, even this pessimistic estimate could yield peak radar cross sections of the order of $1 \, \text{m}^2$. Moreover, the alignment of several fingers within a single plume would give rise to coherent scattering centers like those responsible for the frequency sensitivity discussed in the previous paragraph.

We have found that one implication of the plume model—the existence of a curved scattering surface rising sharply out of the underlying wave surface—leads to radar scattering cross sections which, with a reasonable assumption of plume thickness, can be as large as those observed experimentally. We note, however, that the scattering results are based on two idealizing assumptions: (1) that the radar wavelength is of the order of the plume thickness; (2) that the front surface of the plume is "smooth." As the radar wavelength increases, the scattering moves into the Rayleigh regime and falls off sharply with the fourth power of frequency. At lower frequencies, therefore, the plumes become invisible and our attention must shift to larger features, such as the wedgelike crest of the underlying wave. On the other hand, as the radar wavelength decreases, the roughness (see Fig. 1) over the scattering face of the plume can no longer be ignored, and the scattering amplitude must be multiplied by a roughness factor $\exp[-2(k\sigma_r)^2]$, where σ_r is the rms roughness of the plume surface. Moreover, at short wavelengths, the likelihood of finding strong-scattering coherent segments along the plume front diminishes. It can be seen, therefore, that plume scattering of the type we have been discussing can occur only within a rather narrow window in the microwave spectrum, the location and width of this window being determined by unknown properties of the plumes, such as their thickness, shape, and roughness. Nevertheless, it is interesting and instructive to sketch the scattering behavior of a (reasonable?) plume having a thickness of 3 cm (about an inch) and an rms roughness of 10% of its thickness. Figure 5 places this behavior in the context of the other scattering mechanisms that might be expected to play a role in microwave scattering from breaking waves. Although this breakdown into scattering domains is speculative, and dependent on as yet poorly understood

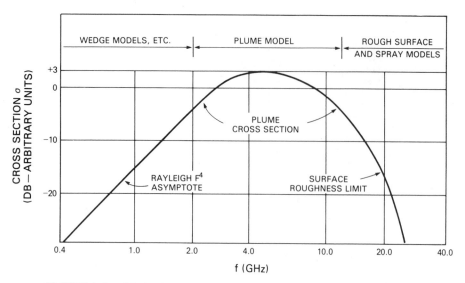

FIGURE 5. Possible breaker scattering mechanisms over the microwave spectrum.

breaker morphology, it has the virtue of differentiating between the kinds of scattering one might expect at different frequencies, and guiding attention to the most appropriate model.

3. WAVE PROXIMITY EFFECTS

It is always a difficult task to include the effects of surrounding objects in a scattering calculation, because the problem becomes one of multiple scattering involving the coupling of fields on obstacles having generally different geometries. In the case of the plumes we have been discussing, the primary interaction must be with that part of the wave surface just ahead of them as they travel down the wave face. Although we know from observation that these regions are likely to be curved, roughened, and of limited extent, it is instructive to consider the local implications of an infinite flat-plane approximation. For a given grazing angle ψ and wave slope angle α, the simple scattering picture will resemble Fig. 6. The reflection coefficient at the wave

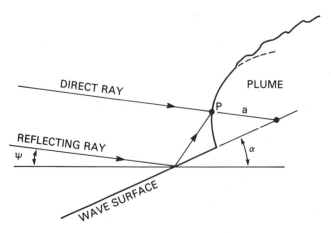

FIGURE 6. Geometry for simple wave-surface/plume interactions.

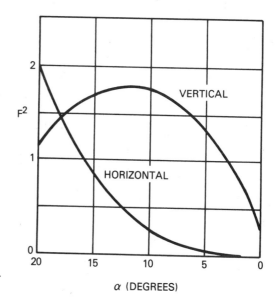

FIGURE 7. Wave-proximity factors for vertical and horizontal polarization.

α (DEGREES)

surface is generally a function of the local grazing angle, $\vartheta = (\psi + \alpha)$, and the polarization of the incident electromagnetic wave. It can be written in the form $R_{H,V}(\vartheta) \exp[i\gamma_{H,V}(\vartheta)]$, where H and V distinguish the horizontal and vertical polarizations. The scattering problem becomes quite complicated even in the simplified geometry of Fig. 6, so we will make a qualitative inference of the effect of the wave surface on plume scattering by simply calculating the total field at the specular point P due to the direct and surface-reflected fields. An elementary geometrical construction yields

$$E_{H,V}(P) \approx E_0 e^{ika}[e^{-ika(1-\cos 2\vartheta)} + R_{H,V}(\vartheta)e^{i\gamma_{H,V}(\vartheta)}], \qquad \vartheta = \psi + \alpha \tag{7}$$

The reflection coefficient for seawater has only a weak dependence on frequency in the microwave spectrum (1–30 GHz), but shows very different behavior for the two polarizations. Figure 7 shows the wave proximity factor $F^2 = |E(P)/E_0|^2$ (for scattering cross sections) plotted against wave slope angles from $20°$ (emission of the plume at an unstable crest) to $0°$ (wave trough), for a grazing angle of $2°$. The values of $R(\vartheta)$ and $\gamma(\vartheta)$ were taken from the classic paper by Saxton and Lane (1952).

The behavior of plume scattering implied by Fig. 7 is clearly quite different for the two polarizations. While both polarizations scatter with about the same intensity when the plume is near the crest, the horizontal return falls off sharply, while the vertical return is spread broadly over the wave surface. It is not unreasonable to view Fig. 7 as representing the temporal behavior of the scattered signal; a plume accelerating away from the crest of a quasi-trochoidal wave is moving slowly where the slope is changing quickly, and rapidly where the slope is changing slowly, which tends to make the wave slope at the toe of the plume a linear function of time. With this interpretation, the scattering behavior represented by Fig. 7 is totally consistent with experimental observations of radar scattering from breaking waves at low grazing angles: horizontal returns are sharp, transient signals, often having the character of scattering from a discrete target, while vertical returns are spread in time, often having an envelope resembling the "vertical" curve in Fig. 7 (Long, 1975; Kalmykov and Pustovoytenko, 1976; Lewis and Olin, 1980; Hansen et al., 1981). Unfortunately, V and H returns have never been measured simultaneously, so the correlation of the instantaneous time behaviors indicated in Fig. 7 cannot be verified with existing data. Such measurements would provide a rather crucial test of the model.

A final implication of Fig. 7 concerns the polarization dependence of breaker backscatter at extremely small grazing angles ($\psi < 1°$). At these extreme angles, most of the surface is in deep shadow, and only the peaks of the breaking waves remain visible to the radar (Wetzel, 1977). For this reason, the plumes will scatter only when they are close to the wave crest (at the left boundary of Fig. 7) where scattering for the two polarizations will be roughly the same. Only a few observers have noted the polarization dependence of breaker (or chop) backscatter at extreme grazing angles (e.g., Long, 1975; Kalmykov and Pustovoytenko, 1976), and their reports tend to support these conclusions.

4. SIGNAL DYNAMICS DUE TO THE PLUME ACCELERATIONS

Generally, the radar footprint over the breaking wave is sufficiently large that several plumes will contribute to the instantaneous radar signal. If we assume for simplicity that the individual plume cross sections are constant and all of their accelerations are the same, then the total scattering amplitude from $N(t)$ plumes coexisting on the front face of a breaking wave at the time t can be written in the form

$$\eta(t) = \sum_{i}^{N(t)} \bar{\eta}_i \exp\{i2k[v_c t + \tfrac{1}{2}a(t - \tau_i)^2]\}, \qquad t > \text{all } \tau_i \tag{8}$$

where v_c, a and τ_i are as in (3), and $\bar{\eta}_i$ is the intrinsic complex scattering amplitude for the ith plume [e.g., the integral term alone in (2)]. A coherent signal-processing radar could deal with a signal of this type and, perhaps, identify the individual plume components. However, most of the radars used thus far in collecting scattering data from breaking waves have been incoherent pulse radars, which measure the instantaneous power returned from the target—i.e., its radar cross section

$$\sigma(t) = |\eta(t)|^2 = \sum_{i,j}^{N} \bar{\eta}_i \bar{\eta}_j^* \exp\{ika[(t - \tau_i)^2 - (t - \tau_j)^2]\} \tag{9}$$

With a few simple manipulations, this expression can be put into the following form:

$$\sigma(t) = \sum_{i=1}^{N} |\bar{\eta}_i|^2 + 2 \sum_{\substack{i<j \\ i \neq j}}^{N} \bar{\eta}_i \bar{\eta}_j^* \cos[\Omega_{ij}(t - \delta_{ij}) + \varphi_{ij}] \tag{10}$$

where

$$\Omega_{ij} = 2ka(\tau_i - \tau_j), \qquad \delta_{ij} = \tfrac{1}{2}(\tau_i + \tau_j) \tag{11}$$

and φ_{ij} is the difference between the phases of complex $\bar{\eta}_i$ and $\bar{\eta}_j$. The frequencies Ω_{ij} are seen to be proportional to the product of the radar frequency (through k), the plume acceleration (taken in the direction of the radar), and the epochal differences for all possible plume pairs.

Although the plume-related parameters are unknown, we can make a few informed guesses based on observation. The wave-tank data analyzed in the paper by Longuet-Higgins and Turner (1974) suggest, when suitably interpreted, that the wave-tank breakers were a few inches high in about 1 foot of water, and the plume accelerations were about 1 m/s². This could be considered a lower limit. An upper limit is provided by the case of a plume emitted from the wave crest and sliding down a 30° wave face under gravitational forces alone. Here the horizontal component of plume acceleration could be no more than $\tfrac{1}{2}g$ (5 m/s²). Consequently, the value $a = 2$ m/s² would seem a relatively safe guess. If a breaking event involves several plumes and extends over a few seconds, we might reasonably take the plumes to be separated in time by about $\tfrac{1}{3}$ s. Taking these estimates together with plume scattering amplitudes of the order of 1 m, we can construct a plausible time history for an X-band radar cross section. Figure 8 shows the result of applying

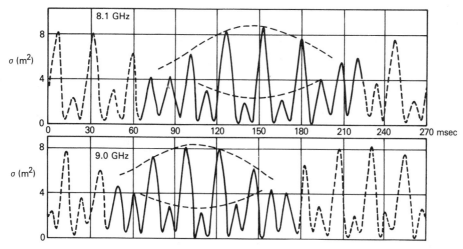

FIGURE 8. Time dependence of radar cross sections for a hypothetical three-plume breaker.

(10) and (11) to a three-plume model at two frequencies 9 and 8.1 GHz, differing by 10%. The plumes were assumed to have equal accelerations of $2 \, \text{m/s}^2$, and real scattering amplitudes ($\phi_{ij} \equiv 0$). The value of η and τ were as follows: $\eta_1 = 1.0 \, \text{m}$, $\tau_1 = 0 \, \text{s}$; $\eta_2 = 0.7 \, \text{m}$, $\tau_2 = 0.310 \, \text{s}$; $\eta_3 = 1.2 \, \text{m}$, $\tau_3 = 0.691 \, \text{s}$ [the epochs were chosen to avoid rational phases in (11)]. The modulation pattern is due, of course, to interference among the different Doppler frequencies for the three plumes. The duration of the pattern and persistence of the envelope shape will depend upon the stability and lifetime of the contributing plumes, so we have shown repetitions of the basic modulation segment in dashed lines. As is characteristic of interference patterns of this type, the envelope shifts its position with a shift in frequency (about 60 ms for a 10% frequency change in the example), and the relative phases of peaks and nulls are periodic (e.g., in phase at 90 and 210 ms, out of phase at 150 ms). Although based entirely on reasonable estimates of the plume parameters, this example is in striking agreement with the only published record of the time behavior of microwave cross sections of a breaking wave recorded simultaneously at two frequencies, obtained by Lewis and Olin (1980) and reproduced in Fig. 9. The dashed lines were

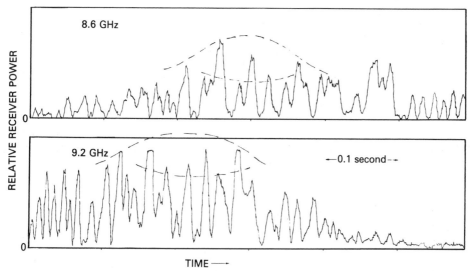

FIGURE 9. Simultaneous time recordings of breaker backscatter at two frequencies. (From Lewis and Olin, 1980.)

added to draw attention to the similarity of the patterns in Figs. 8 and 9. There is even some evidence of the in-phase, out-of-phase periodicities noted above. Figure 9 also shows differences in scattering levels for frequencies differing by only 7%, as mentioned in connection with Eq. (5).

Certainly there are many plume parameters that could affect the appearance of the scattered signal. As we have seen, the accelerations and epoch displacements determine the "frequencies" in the interference pattern, while the total number of plumes contributing, their individual scattering strengths, and their time and frequency variations will largely determine the shape and duration of the pattern envelope. Therefore, if the correspondence between Figs. 8 and 9 is more than accidental—that is, if the multiplume model of breaker scattering has any real validity—we would expect to see certain features appearing rather often in the temporal behavior of the scattering cross section: a regular oscillatory behavior with occasional appearance of multitarget interference patterns; temporal displacement of these patterns with changes in frequency; an oscillating phase relationship between returns at two different frequencies (which should show up as an oscillating component in the cross-correlation function).

The signal spectra recorded by Keller et al. (this volume) with a coherent CW radar are difficult to interpret in terms of the plume model. The spectrum broadens during the breaking process, but its amplitude remains much the same as that produced by the presumed Bragg scattering mechanism before and after breaking. Moreover, the peak velocities inferred from the maximum spectral frequencies are usually only some fraction (1/3 to 1/2) of the expected wave velocity. In view of the large depression angles ($\psi \approx 35°$) used in these experiments, one might conclude that instead of seeing a specular point at the plume front moving with a speed $U = v_c + a(t - \tau_p)$, the radar is responding to orbitally advected Bragg scatterers on the turbulent top surface of the plume as before, but now these scatterers are dragged slowly down the wave face as the plume front accelerates ahead. Whatever the explanation, it is clear that the differences between these results and those obtained at low grazing angles with an incoherent radar must be reconciled before microwave radar can be considered a potentially useful tool for studying the breaking process.

5. DISCUSSION AND CONCLUSIONS

The scattering model developed in this chapter has been called a "plume model" because it derives from the concept of accelerating plumes described by Longuet-Higgins and Turner (1974). It adds certain specific assumptions about how such plumes might enter into the mechanics of wave breaking, and makes some reasonable guesses about their scattering properties and acceleration rates. It should be emphasized that only the *concept* of a plume has been used, not the specific hydrodynamic structure proposed and analyzed by Longuet-Higgins and Turner. Certainly when a wave breaks, *something* moves out from the wave crest and falls down the wave face under gravitational forces. All that we require is that this *something* have a sharp entry into the underlying wave surface, and present a specular point to the incident illumination. The rest depends entirely on questions of geometry, dimensions, acceleration rates, emission sequences, scatterer lifetimes, etc. for which informed guesswork must provide the only clues, since no information is available at the present time. Nevertheless, even though the model is a tissue of assumptions and heuristic devices, it has a certain plausibility because it accounts for the following observed properties of microwave backscatter from breaking waves: peak scattering cross sections of the order of 1 m^2 (or greater) over the most popular part of the microwave spectrum; a possibly sensitive dependence of plume returns on small differences in radar frequency; spikier, more transient, returns with horizontal polarization; broader, more continuous, returns with vertical polarization; similarity in appearance of H and V returns at very small grazing angles; oscillation of the breaker cross section with observed periods; appearance of characteristic modulation envelopes; frequency dependence of position of envelope peaks in time, and varying phase relationship between oscillations at two frequencies.

Lest we be overly impressed by these results, we should recall that the model does not seem to account for the coherent radar observations by Keller et al. (this volume). Moreover, others have found similar polarization dependences and time behavior at various aspect angles away from crest normal, and even when the surface was visually free of breaking waves (although the scattering cross sections were greatly reduced in the latter case) (Long, 1975; Kalmykov and Pustovoytenko, 1976; Lewis and Olin, 1980). In most cases, "ground truth" consisted of visual observation of the sea surface and a narrative or photographic description of what was seen. But we know that the surface of the sea is complex in detail, and cannot be viewed simply as isolated regions of major catastrophic instability (breakers) embedded in a vast statistically homogeneous surface. Observation shows a variety of intermediate structures on the surface—cusped, wedgelike forms that do not break (Long, 1975; Kalmykov and Pustovoytenko, 1976; Kwoh and Lake, this volume), or the small-scale breaking wavelets without air entrainment noted by Banner and Phillips (1974). Such structures might show the enhanced cross sections and/or polarization dependences characteristic of the scattering plume, but without the acceleration effects.

However, there is little point in pushing the plume model for scattering beyond what it is— an interesting and provocative hypothesis. Further careful scattering experiments are necessary to establish (or destroy) its plausibility. Such experiments should certainly result in a better understanding of the hydrodynamic aspects of the problem as well.

REFERENCES

Banner, M. L., and O. M. Phillips (1974): On the incipient breaking of small scale waves. *J. Fluid Mech.* **65**, 647–656.

Hansen, J. P., I. D. Olin, and B. L. Lewis (1981): High resolution radar backscatter from the sea. *Abstracts of the Open Symposium on Remote Sensing, URSI XXth General Assembly*, Washington, D.C., 30.

Kalmykov, A. I., and V. V. Pustovoytenko (1976): On polarization features of radio signals scattered from the sea surface at small grazing angles. *J. Geophys. Res.* **81**, 1960–1964.

Kalmykov, A. I., A. S. Kurekin, Y. A. Lementa, I. Y. Ostrovskiy, and V. V. Pustovoytenko (1976): Scattering of microwave radiation by breaking sea waves. *Gor'kiy Radiofiz.* **19**, 1315–1321 (translation).

Kerr, D. E. (ed.) (1951): *Propagation of Short Radio Waves*, McGraw–Hill, New York, 461.

Lewis, B. L., and I. D. Olin (1980): Experimental study and theoretical model of high-resolution radar backscatter from the sea. *Radio Sci.* **15**, 815–828.

Long, M. W. (1975): *Radar Reflectivity of Land and Sea*, Heath, Boston.

Longuet-Higgins, M. S. (1978): On the dynamics of steep gravity waves in deep water. *Turbulent Fluxes through the Sea Surface, Wave Dynamics, and Prediction* (A. Favre and K. Hasselmann, eds.) Plenum Press, New York, 199–218.

Longuet-Higgins, M. S., and J. S. Turner (1974): An 'entraining plume' model of a spilling breaker. *J. Fluid Mech.* **63**, 1–20.

Saxton, J. A., and J. A. Lane (1952): Electrical properties of sea water. *Wireless Eng.* **Oct.**, 269–275.

Wetzel, L. B. (1977): A model for sea backscatter intermittency at extreme grazing angles. *Radio Sci.* **12**, 749–756.

DISCUSSION

PIERSON: What are the reflection and dielectric properties of the "wedge" as opposed to seawater, both as a reflector of incident EM waves and as a passive emitter of microwaves?

WETZEL: The basic scattering feature in my model is an accelerated "plume," as suggested by Longuet-Higgins and Turner. Such a plume is a mixture of air and water, so its dielectric constant will be smaller than that of the surrounding water—yet sufficiently high to present an effective scattering discontinuity. Since these plumes contain air, I should think they would be radiometrically "warmer" than the surrounding water.

DONELAN: I have made some laboratory observations of "whitecaps" produced by bursts of mechanically generated long waves overtaking bursts of short waves. To the extent that these "whitecaps" approximate natural whitecaps, they provide qualitative support for your chosen model.

WETZEL: As I noted on my last slide, the model parameters were chosen by guesswork and plausible inference. Often what is one man's plausibility is another's total nonsense, so I am grateful for the support provided by your observations. My assumptions were based on several independent sources, but seem sufficiently self-consistent to put the model into reasonable agreement with several different experimental results.

19

OBSERVATION OF BREAKING OCEAN WAVES WITH COHERENT MICROWAVE RADAR

W. C. KELLER, W. J. PLANT, AND G. R. VALENZUELA

ABSTRACT. Coherent microwave radar observations with vertical polarization at 9.375 GHz are performed of breaking ocean waves in deep water (Nordsee Tower in the German Bight) and in shallow water (Jennette's Pier, Nags Head, North Carolina) to illustrate the capability of this instrument to obtain further insight into this nonlinear, unsteady, turbulent, and multiphase process. Coherent microwave radar normally detects the phase (speed) and amplitude (energy) of the short gravity (Bragg resonant) waves (Wright, 1978). In the case of breaking waves, the line-of-sight velocity and amplitude of the backscattered electromagnetic radiation not only contain Bragg scattering contribution, but in addition specular surface scattering and volume scattering if the mixed turbulent fluid during breaking is of low enough density to approach the value in air. In this investigation some very interesting and unusual cases of wave breaking are reported with the radar observations, suggesting that radar may be the ideal instrument to observe breaking waves.

1. INTRODUCTION

Coherent microwave radar (using frequencies of 1.5 to 35 GHz) focused on the water has been used with great success at the U.S. Naval Research Laboratory for at least a decade to study the dynamics (e.g., generation, influence of wind drift, temporal growth, energy balance, and straining) of short gravity-capillary wind waves and their interactions with longer gravity waves in a wave tank (see Wright, 1978). More recently, the instrument has been taken to the field to investigate the properties of ocean waves (see, e.g., Plant et al., 1978; Wright et al., 1980), and presently it is being adapted for aircraft and satellite operation for the measurement of the directional spectrum of ocean waves (see Plant and Keller, 1981).

In this work we use coherent microwave radar at 9.375 GHz and vertical polarization to investigate the process of wave breaking in deep water (during MARSEN from the Nordsee Tower in the German Bight) and in shallow water (from Jennette's Pier, Nags Head, North Carolina) to attempt to gain new insight (e.g., breaker speed distribution, duration, and size) into this elusive transient process. Previous theoretical and experimental studies with conventional oceanographic instruments have not been as definitive as one would like (see Longuet-Higgins, 1973; Longuet-Higgins and Turner, 1974; and others). For the latest developments in analytical modeling of wave breaking, refer to Longuet-Higgins (see Chapter 13).

W. C. KELLER, W. J. PLANT, G. R. VALENZUELA ● Space Systems and Technology Division, Naval Research Laboratory, Washington, D.C. 20375.

Here it will not be necessary to cover all the important dynamical consequences of wave breaking in the ocean; we content ourselves with indicating that wave breaking is a process of the upper ocean which transfers momentum from waves to surface currents and that it plays an important role in the equilibrium of the ocean wave spectrum.

Breaking and near-breaking waves are also of interest in remote sensing applications since they can contribute significant backscatter to microwave sensors (see Alpers *et al.*, 1981; Wetzel, this volume; Kwoh and Lake, this volume). However, this should not be unexpected since earlier controlled wave-tank studies by Keller *et al.* (1974) show that wave breaking yielded unusually broad microwave Doppler spectra.

2. THE COHERENT MICROWAVE RADAR AS A WATER WAVE PROBE

A coherent radar is a contactless (in the mechanical sense) probe which detects simultaneously the amplitude and phase variations of the short gravity waves of the ocean. In contrast, a noncoherent radar only retains the amplitude of the return signal (the phase is lost in the processing). The principal scatterers of electromagnetic (EM) radiation "back" to the microwave radar (i.e., in the backscatter direction) for depression angles θ between 20° and 70° are short gravity waves (also denoted the Bragg resonant waves) of length

$$l = \tfrac{1}{2}\lambda \sec \theta' \tag{1}$$

traveling along the line of sight. λ is the EM wavelength in free space and θ' is the local grazing angle (see Fig. 1).

The short gravity waves are strained (modulated in amplitude) and advected by the orbital velocity of the dominant waves and drift currents if present. For Bragg scattering (see, e.g., Wright, 1978; Valenzuela, 1978), the local backscattered power is proportional to the amplitude-squared (i.e., energy) of the Bragg resonant waves and the resulting phase of the radar return is a measure of the net speed of the Bragg waves including advection. Hence, from the rate of change with time t of the phase of the return signal to a coherent microwave radar, the line-of-sight velocity $v_s(t)$ is obtained:

$$v_s(t) = \frac{f_D(t)\lambda}{2} = \frac{\mathbf{k}_i \cdot \mathbf{u}(t)}{|\mathbf{k}_i|} \tag{2}$$

where $\mathbf{k}_i$ is the propagation vector of the incident EM radiation ($|\mathbf{k}_i| = 2\pi/\lambda$ being its magnitude),

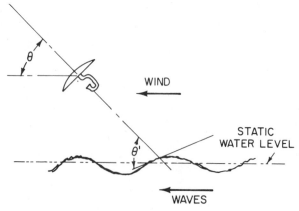

FIGURE 1. Scattering geometry.

$\mathbf{u}(u, v, w)$ is the wave velocity vector in a Cartesian coordinate system $(x, y, z; z$ is positive out of the water), and $f_D(t)$ is the Doppler frequency of the radar return.

To illustrate the use of a coherent microwave radar as a water wave probe, consider the observation of a simple wave system composed of a spectrum of short wind waves superimposed on a purely monochromatic deep gravity wave traveling in the x direction. The pertinent components of velocity on the surface, for constant x, are

$$\left.\begin{array}{l} u = c_s + U_0 \cos \Omega t \\ v = 0 \\ w = U_0 \sin \Omega t \end{array}\right\} \tag{3}$$

where c_s is the phase speed of the Bragg resonant wave, Ω is the radian frequency of the gravity wave, and $U_0 = \Omega A$, A being the amplitude of the gravity wave. The backscattered power $P_B(t)$ to the radar for this wave field may be obtained with the two-scale surface scattering model (see Wright, 1978; Valenzuela, 1980) (assuming $\mathbf{k}_i$ is in the x–z plane):

$$P_B(t) = P_0[1 + (U_0/C)m \cos(\Omega t + \phi)] \tag{4}$$

where m is the modulation transfer function and ϕ is the phase of m, both including hydrodynamic and EM contributions (see Wright et al., 1980). Here C is the phase speed of the gravity wave, U_0/C is the slope, and P_0 is the average backscattered power to the radar.

Clearly, a cross-correlation of (2) and (4), using (3) and proper normalization, yields the modulation transfer function. On the other hand, a Fourier transform of (2), using (3) and proper normalization, yields the wave height spectrum of the dominant wave (see Wright et al., 1980; Plant and Schuler, 1980).

Therefore, the main point to keep in mind from the above example is that coherent microwave radar detects the line-of-sight velocity of the scattering waves (or elements) on the water and the amplitude of the EM return may be predicted from scattering theory, although for breaking waves the scattering is more complex than pure Bragg scattering, since the scattering includes other contributions such as specular surface scattering and volume scattering if there is significant penetration into the air–water mixed turbulent fluid. As is well known, the EM radiation penetration in the mixed turbulent fluid depends on the dielectric properties of the mixture in relation to the air which is a function of the relative concentration of air and water in the fluid.

Also, in practical applications of this technique, one must keep in mind that the illumination area on the water surface has finite dimensions that depend on the antenna beam width of the radar system, so we realize that the measured parameters represent averages over the illuminated area on the water. However, this area size can be minimized by using narrow-beam antennas.

3. DYNAMICS OF BREAKING WATER WAVES

In this section we review briefly some well-known facts on breaking waves. Two main classes of breaking waves exist: "spilling" breakers and "plunging" breakers. Spilling breakers occur mostly in deep water and are caused by instability of the free surface near the wave crest, forming a quasi-steady whitecap on the forward face of the wave. Plunging breakers, on the other hand, are characterized by wave crests toppling forward and falling violently on themselves. This latter type of breaking occurs mostly in coastal regions with sloping bottoms where the speed of the fluid increases with height and the speed of the shoaling waves decreases with decreasing depth. The instability of plunging breakers is a great deal stronger than that of spilling breakers.

To understand the process of wave breaking it is desirable to know about the energetics of solitary waves. Solitary shoaling waves have been investigated analytically and numerically by Longuet-Higgins and Fenton (1974) for a constant depth h. Their maximum phase speed, normalized in terms of $(gh)^{1/2}$ (g is the acceleration of gravity), is $F = 1.286$ for an amplitude-to-water-depth ratio $a/h = 0.827$. The most energetic solitary wave, on the other hand, has a phase speed $F = 1.294$ for $a/h = 0.790$. In preliminary calculations by Witting (1975), an a/h value of 0.8332 was found for the solitary wave of highest amplitude.

Experimental observations in a wave tank with sloping bottoms of 0.023, 0.050, and 0.065 were performed by Ippen and Kulin (1955), who found that spilling breakers are mostly symmetric and occur for $a/h = 0.65$–0.85. In contrast, plunging breakers are asymmetric and have larger a/h ratios of 0.9–3.0.

More recently, Palmer (1976) investigated the velocities of the "overturning" fluid in plunging breakers and found that in the upper one-fifth of the wave, velocities as large as $2(gh)^{1/2}$ can be encountered, while in the lower four-fifths, the velocities are lower than the phase speed of the wave.

4. EXPERIMENTAL RESULTS

Measurements were performed on ocean waves with a continuous wave (cw) microwave radar at 9.375 GHz ($\lambda = 3.2$ cm) using vertical polarization in deep water during the MARSEN Experiment from the Nordsee Tower in the German Bight and in shallow water from Jennette's Pier at the Outer Banks of North Carolina. The data for the shoaling waves were collected during 1975 for the specific purpose of investigating the modulation of short gravity waves. However, one day in particular, on October 2, 1975, large breaking waves 14-s period occurred as a hurricane passed offshore and these data have been used in the present study.

A detailed description of the microwave system has been given in Plant *et al.* (1978), so here we only mention that the radar system used separate adjacent parabolic antennas for the transmitter and receiver, both illuminating the same spot on the water surface. The 3-db beamwidth of each antenna was 4° so the illuminated spot sizes on the water were about 0.5 × 1.5 m for the shallow water measurements. For the measurements from the Nordsee Tower the illuminated spot sizes were 4 times larger for the same depression angle (i.e., 35°). A summary of the environmental conditions and system parameters is given in Table I.

From the microwave return, instantaneous Doppler spectra have been obtained every 0.2 and 0.25 s and the consecutive sequence of these Doppler spectra is shown in Fig. 2 for one set of measurements of September 18, 1979, from MARSEN. A display of this type gives a great deal of physical insight on the scattering mechanisms since the advection of the short waves (the Bragg

TABLE I.   Environmental and Radar System Parameters

	MARSEN 9/18/79	(Nordsee Tower) 9/20/79	Nags Head, N.C. 10/2/75
Wind speed (m/s)	16.5	9–15	<8
Wave period (s)	9	7	14
Water depth (m)	30	30	3.5
Bottom slope	~0	~0	0.014
RMS wave height (m)	1.12	0.7–0.85	—
Wave direction (deg)	256	225	90
Wind direction (deg)	271	239	90 ± 45
Radar look direction (deg)	225	210	90
Antenna height (m)	24	24	6
Depression angle (deg)	35	20	35
Radar frequency (GHz)	9.375	9.375	9.375
Bragg wavelength (cm)	1.95	1.70	1.95
Antenna beamwidth (deg)	4	4	4
Polarization	Vertical	Vertical	Vertical

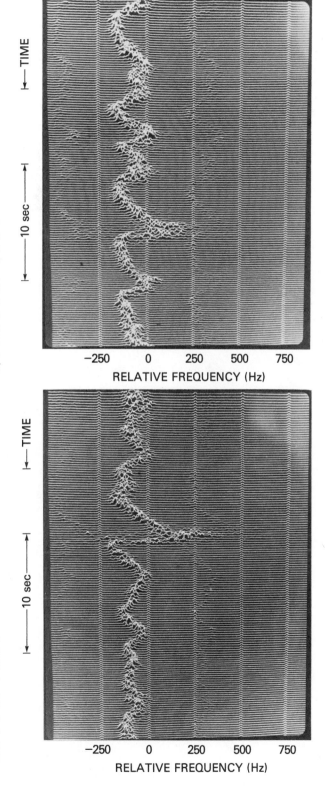

FIGURE 2. Doppler spectra of microwave backscattered power from Nordsee Tower, German Bight (0.20-s averages). Zero Doppler frequency = −116 Hz; September 18, 1979. Case of a well-developed wind-wave system (deep water), the Doppler spectral frequencies for the breaking region approach 250 Hz which corresponds to a horizontal velocity of 7.15 m/s (100 Hz corresponds to a radial velocity of 1.6 m/s).

FIGURE 3. Doppler spectra of microwave backscattered power from Nordsee Tower, German Bight (0.20-s averages). Zero Doppler frequency = −116 Hz; September 18, 1979. Case of strong breaking (deep water), the Doppler spectral frequencies for the breaking region are greater than 300 Hz. For conversion to velocity units see caption of Fig. 2.

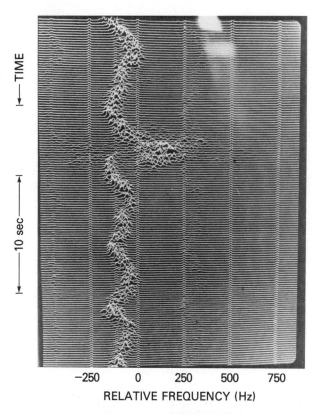

FIGURE 4. Doppler spectra of microwave backscattered power from Nordsee Tower, German Bight (0.20-s averages). Zero Doppler frequency = −116 Hz; September 20, 1979. One wave crest is almost completely obliterated by breaking (deep water), the Doppler spectral frequencies for the breaking region are greater than 250 Hz. For conversion to velocity units see caption of Fig. 2.

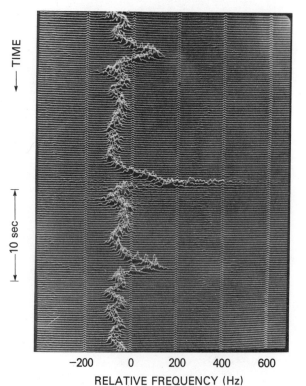

FIGURE 5. Doppler spectra of microwave backscattered power from Jennette's Pier, Nags Head, N.C. (0.25-s averages). Zero Doppler frequency = −16 Hz; October 2, 1975. Case of strong breaking (shallow water), the Doppler spectral frequencies for the breaking region exceed 500 Hz. For conversion to velocity units see caption of Fig. 2.

resonant waves) of 1.95 cm length by the orbital velocity of the dominant waves is clearly visible. Parenthetically, September 18 was a day of fairly constant wind conditions with a well-developed wind wave system. For calibration purposes, frequency markers have been included in Fig. 2 and subsequent figures. The Doppler spectra have been offset in frequency so zero Doppler (i.e., zero velocity) corresponds to -116 Hz in Figs. 2–4 and to -16 Hz in Figs. 5 and 6. Also, in these figures weak "ghost" images of the main display can be observed at offsets of ± 384 Hz. Nonbreaking wave spectra have maximum Doppler shifts of 50 Hz in Fig. 2, while the Doppler spectra for one case of wave breaking have Doppler shifts up to 200 Hz. According to (2), a Doppler frequency of 100 Hz in Fig. 2 corresponds to a line-of-sight velocity of 1.6 m/s.

As mentioned in the Introduction, the bandwidth of the Doppler spectrum is drastically increased by breaking waves. For the measurements reported here, typical bandwidths for nonbreaking waves are on the order of 50 Hz or less, while for breaking waves the bandwidths easily exceed 100 Hz.

In Fig. 3 we display measurements taken later the same day when strong wave breaking was occurring and the Doppler frequencies of the spectral wave approach 400 Hz. Since the Nordsee Tower is in fairly deep water (i.e., 30 m) and the water depth is relatively constant, it is assumed the wave breaking action here is primarily due to "spilling" breakers.

We also performed measurements on September 20, 1979, when the wind was variable and, of course, in this situation the sea waves were not as well developed. On this day there was a very interesting case of wave breaking in which there were two distinct sets of scatterers (Fig. 4). One set is completely removed from the wave (perhaps an indication of flow separation containing windblown spray and foam) while a second set of scatterers is still attached to the dominant wave. In this case it seems the wave breaking process was so intense that the wave crest was almost completely obliterated.

As mentioned previously, we also made radar observations of "plunging" breaking waves at the Outer Banks of North Carolina. As discussed, "plunging" breaking waves may contain

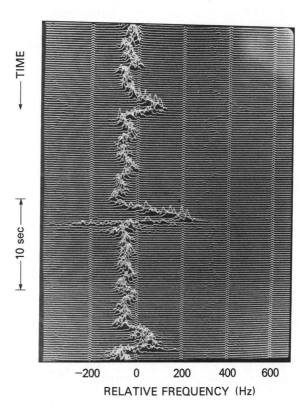

FIGURE 6. Doppler spectra of microwave backscattered power from Jennette's Pier, Nags Head, N.C. (0.25-s averages). Zero Doppler frequency = -16 Hz; October 2, 1975. Unusual case of back-traveling scatterers (shallow water), the Doppler spectral frequencies for the breaking region approach 300 Hz. For conversion to velocity units see caption of Fig. 2.

TIME

10 sec

RELATIVE FREQUENCY (Hz)

velocities as large as $F = 2$ in normalized units. In the observation from Jennette's Pier there was one case in which the maximum Doppler frequencies of the return were greater than 500 Hz (see Fig. 5) approaching the maximum predicted velocites of 11.8 m/s for the 3.5 m water depth at the site. However, the scattering amplitudes beyond 500 Hz are quite small; perhaps this scattering is mostly due to well-aerated spray and foam.

On another occasion of wave breaking an unexpected phenomenon occurred, immediately after breaking took place. Evident in the Doppler spectra were backward-traveling scatterers moving up to 6 m/s (see Fig. 6). Backwash and backrush velocities in wave breaking have been discussed by Kirkgöz (1981).

5. DISCUSSION

In this work we have attempted to present the type of information coherent microwave radar is capable of providing in regard to the process of wave breaking. With a single instrument one is able to detect the line-of-sight velocity and strength of the scattering elements. For nonbreaking waves these scatterers are short gravity-capillary waves of length proportional to the radar wavelength and given by Eq. (1). However, for breaking waves the EM backscattered radiation also contains specular contributions (wave, or mixed turbulent fluid, interfaces normal to the direction of incidence of the EM radiation) and volume scattering if the mixed turbulent fluid is well aerated such as to have a small dielectric constant approaching the air value.

The results presented have combined the amplitude and phase of the backscattered EM radiation in a Doppler spectrum which separates intensity of the return as a function of Doppler frequency. As we have seen, this display can provide a great deal of information on the process of wave breaking. Although our presentation of the results has been of illustrative nature, the measurement itself contains a great deal of quantitative information that can be extracted in a more detailed analysis of the individual Doppler spectra or from the original data, to obtain speed distribution and size of breakers or other parameters related to the wave field.

We realize that, perhaps, one of the limitations in relating radar measurements of this type to the wave breaking process itself rests on our knowledge of the scattering process involved in the backscattering of the EM radiation. However, this handicap can be reduced considerably by performing microwave measurements on wave breaking as a function of various parameters, such as radar frequency, depression angle, and illuminated spot size on the water to identify the scattering process.

6. CONCLUSIONS

Coherent microwave radar may be the ideal instrument to obtain new insight and quantitative information on the strongly nonlinear process of wave breaking, because it is a contactless wave probe and with a single instrument one detects the line-of-sight velocity and the amplitude of scattering elements. Here we have used vertically polarized radar at 9.375 GHz and depression angles of 20° and 35° to avoid shadowing, diffraction by wave crests, and refraction/trapping by the marine boundary layer on the surface. Measurements at other radar frequencies, anywhere in the range 1.5 to 35 GHz, should provide additional very useful information on the process. Therefore, we would like to propose coherent microwave radar as the most appropriate instrument to learn more about the nonlinear, unsteady, turbulent, and multiphase process of wave breaking.

REFERENCES

Alpers, W. R., D. B. Ross, and C. L. Rufenach (1981): On the detectability of ocean surface waves by real and synthetic aperture radar. *J. Geophys. Res.* **86**, 6481–6498.

Ippen, A. T., and G. Kulin (1955): Shoaling and breaking characteristics of the solitary wave. MIT Hydrodynamics Laboratory Report 15.

Keller, W. C., T. R. Larson, and J. W. Wright (1974): Mean speeds of wind waves at short fetch. *Radio Sci.* **12**, 1091–1100.

Kirkgöz, M. S. (1981): A theoretical study of plunging breakers and their run-up. *Coastal Eng.* **5**, 353–370.

Longuet-Higgins, M. S. (1973): A model of flow separation at the free surface. *J. Fluid Mech.* **57**, 129–148.

Longuet-Higgins, M. S., and J. D. Fenton (1974): On the mass, momentum, energy and circulation of a solitary wave. *Proc. R. Soc. London Ser. A* **340**, 471–493.

Longuet-Higgins, M. S., and J. S. Turner (1974): An "entraining plume" model of a spilling breaker. *J. Fluid Mech.* **63**, 1–20.

Palmer, R. Q. (1976): Anatomy of a plunging breaker. *Abstracts of the 15th Coastal Engineering Conference*, Honolulu, 360–363.

Plant, W. J, and W. C. Keller (1981): Measurement of sea surface velocities and directional spectra from aircraft. *IUCRM Symposium on Wave Dynamics and Radio Probing of the Ocean Surface* Miami.

Plant, W. J., and D. L. Schuler (1980): Remote sensing of the sea surface using one- and two-frequency microwave techniques. *Radio Sci.* **15**, 605–615.

Plant, W. J., W. C. Keller, and J. W. Wright (1978): Modulation of coherent microwave backscatter by shoaling waves. *J. Geophys. Res.* **83**, 1347–1352.

Valenzuela, G. R. (1978): Theories for the interaction of electromagnetic and oceanic waves—A review. *Boundary-Layer Meteorol.* **13**, 61–85.

Valenzuela, G. R. (1980): An asymptotic formulation for SAR images of the dynamical ocean surface. *Radio Sci.* **15**, 105–114.

Witting, J. (1975): On the highest and other stationary waves. *SIAM J. Appl. Math.* **28**, 700–719.

Wright, J. W. (1978): Detection of ocean waves by microwave radar: The modulation of short gravity–capillary waves. *Boundary-Layer Meteorol.* **13**, 87–105.

Wright, J. W., W. J. Plant, W. C. Keller, and W. L. Jones (1980): Ocean wave-radar modulation transfer functions from the West Coast Experiment. *J. Geophys. Res.* **85**, 4957–4966.

DISCUSSION

ALPERS: In one slide you showed simultaneous time series of backscattered power and Doppler shift, there seemed to be a good correlation between spikes and breaking waves and in both time series. We did similar experiments at K_a-band (8 mm) from the North Sea platform. We find that high Doppler shifts often do not correspond to a high radar return. Sometimes, large positive Doppler shifts correspond to very low backscattered power. Do you find similar results at X-band or do you find always a good correlation?

VALENZUELA: I have not checked this myself. However, Bill Keller might like to comment.

KELER: At X-band and moderate wind speeds, the radar cross-section modulation transfer function is relatively large. The data in question are for monochromatic swell in shoaling water and under these conditions the coherence is readily visible. At higher wind speeds the modulation transfer function decreases and the coherence would not be obvious for a wide spectrum of large waves.

TRIZNA: The measurement technique when employed from an aircraft will be a convolution of the ocean wave spectrum and the response of the aircraft to air turbulence. Can aircraft motions effectively be removed?

PLANT: The aircraft motion important to the measurement is pitch. Roll measurements are not important and horizontal variations can be removed by employing simultaneous accelerometer measurements. All aircraft motions can be removed, pitch by employing two accelerometers at near opposite ends of the aircraft.

20

An Estimate of the Influence of Breaking Waves on the Dynamics of the Upper Ocean

Norden E. Huang

ABSTRACT. The influence of breaking waves on the dynamics of the upper ocean is estimated using the concept of an energy balance between the wind and waves. The rate of energy release from wave breaking is calculated by the statistical model proposed by Longuet-Higgins (1969). It is found that the result of this calculation depends only on one parameter, the significant slope, defined and used by Huang and Long (1980). Treating the energy released from the breaking waves as the sole source of turbulent energy, various models are constructed to simulate a wide variety of dynamical phenomena in the upper ocean layer. The quantities calculated from these models include surface drift, whitecapping percentage, and mixing efficiency. This treatment using breaking waves as the sole turbulent energy source is only a first-order approximation to reality but the technique does result in the first quantitative estimates of the effect of breaking waves on upper ocean layer dynamics. One of the advantages of using the present approach is that the inputs to the models are all obtainable from remote sensors. Possible future extensions and improvements to the models are also discussed. It is believed that the models presented here offer a new look for the study of the dynamics of the upper ocean. Additional studies and the attention of future investigators are required to bring this technique to full fruition.

1. INTRODUCTION

The random surface waves are the most energetic motions in the upper layer of the ocean. Although these random motions can be described successfully by the irrotational approximation, the effect of direct wind stress and the breaking events are certainly rotational, and thus it is only natural to relate the turbulent motions to breaking waves.

Turbulence generation by waves was first treated by Phillips (1961), who treated the turbulence as an irregular vorticity field generated by the straining associated with the nonbreaking random wave motion and viscous diffusion. According to this reasoning, Phillips found that the random vorticity field was of second order of importance. He further pointed out that the mean vorticity field was also of second order and that the dominant mechanisms of vortex generation by the mean rate of straining associated with the wave field and by turbulence itself were of the same order. The intensity, being second order, is too weak to account for the observed phenomena such as wave attenuation or mixing in the upper ocean.

Norden E. Huang   ●   NASA Goddard Space Flight Center, Greenbelt, Maryland 20771.

A second approach uses a similarity solution. This approach was used by both Navrotskii (1967) and Kitaigorodskii and Miropolskii (1968). The details in the two reports are different. In Navrotskii (1967), the physical model was based on a large correlation between potential wave oscillations and the turbulent oscillation produced mainly by the effect of the horizontal wind shear current on the windward side of the waves. Details of the turbulence generation mechanism were due to the different fluid motions in the two sublayers near the surface. At the surface, a large-impulse velocity is generated by the direct interaction of the wind and the waves at the thickness of the order of wave height, and at the period of the energy-containing waves. This resulting velocity will interact with the orbital velocity and form the turbulence-generating layer. To support this mechanism, a large stress in the main turbulence-generating layer is required. Navrotskii cited the field observations of Shonting (1965) as direct support. But from a dynamical point of view, the field results were highly questionable because the measured stress in the mean turbulence-generating layer was an order of magnitude higher than the wind stress. Balancing these forces is difficult, and this turbulence-generating mechanism would not be effective for smaller values of stress.

The similarity approach used by Kitaigorodskii and Miropolskii was based on the energy flow from the wind to the water surface by potential wave motion through the work by the pressure force. They argued that the wind stress caused the waves to break, and it was the breaking of the wave crests that was regarded as the direct mechanism needed to transfer wave energy into small-scale turbulence. Since they claimed that most of the energy transmitted from wind to water is in the form of wind–wave interaction, they related the initial energy input to wind stress. This energy is then diffused into the mixed layer. Although this model presented a clear and reasonable physical picture, the details are far from being definite. After numerous *ad hoc* assumptions, the final results were expressed in terms of wind stress at the surface and some empirical constants. The major flaw in this approach is that Kitaigorodskii and Miropolskii failed to quantize the energy loss due to wave breaking. Consequently, they could not express the main turbulence-generating mechanism explicitly, but were forced to resort to an implicit relationship between the wind stress and energy loss due to breaking. Therefore, the determination of turbulence intensity was related back to the determination of wind stress. Neither one is a simple task. Thus, this approach only leads us into a hopeless loop of shifting from one difficult task to another.

A method for rigorously quantifying the energy loss due to wave breaking was first proposed by Longuet-Higgins (1969). In that interesting and important paper, Longuet-Higgins outlined a new statistical approach. However, due to the lack of detailed data, no direct calculation was made, but an order of magnitude estimate was given which put the portion of energy loss due to wave breaking in one cycle at about 10^{-4}. Since none of these above methods seems to be able to provide quantitative estimates of energy conversion from wave to turbulence, Monin (1977) was led to state that a quantitative understanding of turbulence energy generation is still the work of the future. Monin's statement was obviously a reflection of the facts, but the method for the solution has already been provided by Longuet-Higgins (1969) without being explored in detail.

2. TURBULENCE ENERGY GENERATION BY BREAKING WAVES

For the sake of completeness, we will briefly summarize the approach given in Longuet-Higgins (1969). First, we have to establish the definition for wave breaking in deep water. A concise review of this subject can be found in Cokelet (1977a). Essentially, the motions of the breaking waves are different from the small-amplitude ones. Free waves tend to break only when the ratio of the amplitude to the wavelength is large. As this ratio increases, the description of the wave motion needs higher and higher orders of solutions. The simplest criteria of the limiting waveforms were established by Stokes (1847, 1880) as: (1) the particle velocity of fluid at the wave crest equals the phase velocity, (2) the crest of the wave attains a sharp point with an included angle of 120°, (3) the ratio of wave height to wavelength is approximately 1/7, and (4) the particle acceleration at the crest of the wave equals $\frac{1}{2}g$. All these conditions are approximately equivalent in steady irrotational flow. They will be used indiscriminately at different parts of this chapter. Although studies by Fenton (1972, 1979), Schwartz (1974), Longuet-Higgins and Cokelet (1976, 1978), Longuet-Higgins and Fox (1977), and Cokelet (1977b) have substantially improved our understanding of the breaking processes of the surface waves, the results are mostly numerical rather than analytical. It should also be pointed out that the surface wind stress could also influence the breaking criteria as shown by Stoker (1957). However, for the sake of clarity, only the simple Stokes criteria will be used here. The analysis, therefore, should be treated as a first-order approximation.

For example, if one uses the condition that wave breaking occurs when the acceleration at the peak reaches $-\frac{1}{2}g$, then the limiting amplitude of a wave form a given frequency is given by

$$an^2 \simeq \tfrac{1}{2}g \tag{1}$$

where a is the wave amplitude, n the frequency of the wave, and g the gravitational acceleration. Hence,

$$a_{\max} = g/2n^2 \tag{2}$$

Now for a nearly Gaussian wave field with a narrow band frequency spectrum, $\phi(n)$, the probability density function of the amplitude, $p(a)$, has been shown by Longuet-Higgins (1962) to be approximately Rayleigh. If a characteristic frequency, $\overline{n^2}$, of the wave field is defined as

$$\overline{n^2} \int_n \phi(n)dn = \int_n n^2 \phi(n)dn \tag{3}$$

one can calculate the characteristic maximum amplitude for a wave field, a_0, from a statistical point of view, as

$$a_0 = g/2\overline{n^2} \tag{4}$$

Any wave having an amplitude higher than a_0 will be unstable and will tend to break. In the process, the wave will lose the portion of energy higher than $\frac{1}{2}\rho g a_0^2$ to reduce the amplitude to a_0 again. Thus, the mean portion of energy loss, $\tilde{\omega}$, per one wave cycle can be calculated as

$$\tilde{\omega} = \frac{1}{\overline{a^2}} \int_{a_0}^{\infty} \frac{a(a^2 - a_0^2)}{\overline{a^2}} \exp(-a^2/\overline{a^2})da = \exp(-E_0/E) \tag{5}$$

with $E_0 = \frac{1}{2}\rho g a_0^2$ and $E = \rho g \overline{\zeta^2}$. This portion of energy will then be converted into turbulence.

Since E_0 is a function of a_0 which is calculated based on the second moment of the spectrum, the values of $\overline{n^2}$, E_0, and $\tilde{\omega}$ will all depend on the spectral function of the wave field. Various spectral models exist, but for this study the Wallops spectrum in its simplified form developed by Huang et al. (1981a) will be used. With that spectral model, the portion of energy loss per one wave cycle is given simply as

$$\tilde{\omega} = \exp\left[-\frac{1}{32\pi^2 \S^2}\left(\frac{m-3}{m-1}\right)^2\right] \tag{6}$$

where

$$m = \left|\frac{\log(2^{1/2}\pi\S)^2}{\log 2}\right|, \quad \S = \frac{(\overline{\zeta^2})^{1/2}}{\lambda_0}$$

This expression indicates that the amount of energy loss in breaking per cycle is determined only by the significant slope of the wave field. The steeper the waves, the higher the probability of breaking will be, and consequently, more energy will be released from wave motion into turbulence for steeper waves.

3. THE EQUILIBRIUM SEA STATE

If the proportion of wave energy lost by breaking per unit cycle is estimated by Eq. (6), the rate of energy loss per unit surface area, D, becomes

$$D = E\tilde{\omega}/T_0 \tag{7}$$

where T_0 is the period of the wave having the frequency at the spectral peak. Now, as the wind blows across the ocean surface, energy is fed into the waves. The rate W at which energy is fed into any wave is the phase velocity times the rate of momentum flux. As discussed by Longuet-Higgins (1969) and Phillips (1977), the rate of momentum flux fed into the waves is only a small portion of the wind stress, τ, with

$$\tau = \rho_a u_*^2 \tag{8}$$

where ρ_a is the density of the air and u_* is the frictional velocity of the wind. On the other hand, the phase velocity of the waves that are breaking in the field is a substantial multiple of u_*. The two numerical factors tend to cancel each other out as discussed by Longuet-Higgins (1969) and Phillips (1977). Consequently, the rate of work done by the wind on the waves can be approximated by

$$W = \rho_a u_*^3 \tag{9}$$

Serious deviations obviously exist especially for the laboratory cases where the phase velocity of the energy-containing waves is of the same magnitude as u_*. A more accurate relationship will have to wait until we gain more understanding of the air–sea interaction processes.

If this approximation is accepted, and with the rate of wave energy dissipation and gain given by (7) and (8), respectively, the integrated energy transport equation of the wave field can be written as

$$\frac{DE}{Dt} = \frac{\partial E}{\partial t} + (\mathbf{C}_g + \mathbf{U}_d) \cdot \nabla E = W - \frac{\tilde{\omega}E}{T_0} \tag{10}$$

where D/Dt is the material derivative, $\mathbf{C}_g$ is the group velocity of the energy-containing waves, and $\mathbf{U}_d$ is the external current velocity. Equation (10) is the integrated form of the energy transport equation proposed by Hasselmann et al. (1976) for a specific frequency range only. Hasselmann's equation also includes a term representing nonlinear energy transfer due to conservative wave–wave interactions. That term is not in (10) because the net effect for that term is to redistribute energy at different frequencies but not a net gain or loss. So the integrated value of the term is zero. It should be pointed out that (10) again represents a first-order approximation; an additional approximation is introduced by considering loss or gain of energy only in the energy-containing wave range. With this approximation, the group velocity could be replaced by $C_0/2$ with C_0 as the phase velocity of the wave with frequency n_0.

The equilibrium state is defined as the state in which

$$DE/Dt = 0 \tag{11}$$

The condition given by (11) can be satisfied only in a stationary and homogeneous wave field. The requirement of stationarity and homogeneity is much stricter here than the commonly adopted criteria in statistical analyses of the wave field. In common application, the correlation length scale is used as a measure of homogeneity—whenever this length scale is much less than the storm scale, the wave field is regarded as homogeneous. But in the present definition of dynamical equilibrium, this correlation length scale is not too meaningful. The only useful length scale becomes the storm scale that measures the energy growth. When dynamical equilibrium is reached, the balance of energy is achieved exactly between the input from the wind and dissipation by wave breaking, i.e.,

$$W = \tilde{\omega} E / T_0 \tag{12}$$

This equation is another form of the definition for an equilibrium sea state. Now, in general, a ratio can be formed by the two terms in (12) as

$$\varepsilon = (\tilde{\omega} E / T_0) / W \tag{13}$$

ε thus measures the relative magnitudes of the rate of work done to the wave field by the wind and the rate of dissipation. Naturally, it follows that when $\varepsilon = 1$, the sea state is in dynamical equilibrium; when $\varepsilon < 1$, the sea state is still developing because the loss of energy is less than the energy input by the wind; and when $\varepsilon > 1$, the sea state is in a decaying stage. Thus, the value ε also measures the growth potential of the sea state. With expressions (6) and (9), ε can be written in physical parameters as

$$\varepsilon = 2\pi \left(\frac{\rho}{\rho_a}\right) \frac{\S^2 \exp\left[-\frac{1}{32\pi^2 \S^2}\left(\frac{m-3}{m-1}\right)^2\right]}{(u_*/C_0)^3} \tag{14}$$

where ρ is the density of water and C_0 is the phase velocity of the waves with frequency at the energy spectrum peak. A contour map of ε is given in Fig. 1.

It should be pointed out that the contour line $\varepsilon = 1$ has some special dynamical significance. For an actively generating wind wave field, $\varepsilon = 1$ actually can be regarded as an upper bound of all the possible ε values. Available data of developing sea from both laboratory measurements (Huang and Long, 1980; Mitsuyasu and Honda, 1975), and field observations from JONSWAP as reported by Müller (1976) were used to test this upper bound. The results are also shown in Fig. 1.

Several observations can be made from the curves and the data in Fig. 1. First, the various upper limits proposed by Stokes, Phillips and Banner (1974), and here as $\varepsilon = 1$ on the steepness of the waves all work, but with a different degree of sharpness. The Stokes limit of

$$\S \leqslant 0.0505 \tag{15}$$

is too high. It can only serve as the absolute upper limit. From a statistical sense, few waves can ever reach such a limit. Consequently, the value of the significant slope is considerably lower. The dynamical equilibrium limit proposed here in Eq. (14) is a much sharper upper bound for the laboratory as well as the field data, when all the waves are strictly locally wind-generated. However, due to the strong influence of surface drift current, the upper values of $\S$ suggested by dynamical equilibrium would never be reached by wind-generated waves because the wind-induced surface skin drift will cause premature breaking. When u_*/C_0 approaches unity, Phillips–Banner's model of drift-induced breaking becomes dominating. Although the data points do not always lie under the Phillips–Banner curve, the trend of the event is unmistakable. Refinement of the drift current correction and the rate of work done by the wind should improve the results.

Based on these considerations, it is suggested that the combination of Phillips–Banner (for $u_*/C_0 \geqslant 1$) and the dynamical equilibrium curve at $\varepsilon = 1$ (for $u_*/C_0 < 1$) offers the sharpest bound of the sea state ranges on steepness for a wind-generated wave field.

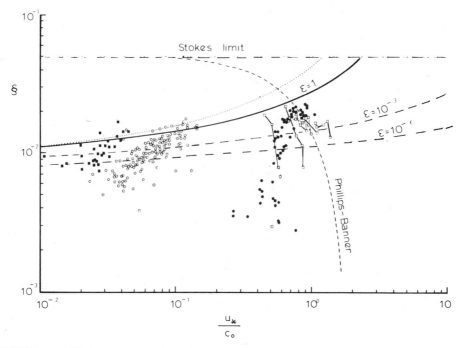

FIGURE 1. Equilibrium sea state as determined by § and u_*/C_0. Data points are: ●, laboratory by Huang and Long (1980); □, laboratory by Mitsuyasu and Honda (1975); ○, ■, field data from JONSWAP series A and E, respectively, by Müller (1976).——, $\varepsilon = 1$;——, contour lines of various ε values;....., $\varepsilon = 1$ in terms of the apparent significant slope calculated from observed frequency without drift current correction;–·–, Stokes limit of § = 0.0505; - - - -, Phillips–Banner limit.

Second, the loci of the points collected at increasing fetches in the laboratory indicate an interesting trend. The trend seems to suggest that under extremely limited fetches, the fresh waves gain energy rapidly in increasing the steepness but with no appreciable increase in wavelength. As the significant slope reaches the upper bound, the waves will grow in length which will cause the value of u_*/C_0 as well as the significant slope to decrease rapidly. Such an initial stage of wave development can only be realized in the laboratory where the initial condition is an absolutely calm surface. In the field, this initial stage of low § will be obscured by the background noise from the existing waves. Consequently, the first visible stage of the field waves will be the fresh choppy short waves with amplitudes above the background. The significant slope of such a wave field would be very near the upper bound already. Further growth of the waves will cause the u_*/C_0 and § values to decrease. Therefore, both high § and u_*/C_0 values are the characteristics of the fresh wave condition in the field. A physical explanation of this trend based on an energy consideration is obvious. As the waves grow bigger, each breaking will release an increasing amount of energy from the wave motion. Therefore, it would take a longer time for the wind to replenish the wave field. To conserve energy, the wave will become gentler to lessen the probability of breaking. This trend can be detected easily from the data collected at different fetches by Mitsuyasu and Honda (1975). Unfortunately, the sequential order with respect to the fetch for the JONSWAP field data is not readily available. Nevertheless, a consistent trend of high significant slope values for short fetches and lower § values for longer fetches in the JONSWAP data compiled by Müller (1976) is quite clear.

Third, the change in sea state is determined primarily by the significant slope of the waves. A 20% increase in § would result in an increase of 10^3 in the ε value for the field cases. Therefore, a subtle change of sea state could result in dramatically different dynamical regimes of upper layer phenomena. Such a highly variable natural state indeed shows up in the subsequent studies. The consequences of these changes will be discussed in detail later.

4. CONTRIBUTION OF BREAKING WAVES TO THE SURFACE DRIFT CURRENTS AND WHITECAPPING

The surface drift current is caused primarily by the direct drag of the wind stress. A classical solution was found by Ekman (1905) by balancing the wind stress and the Coriolis force under a rigid flat surface. Consequently, the contribution from the wave motion was totally neglected. Later calculations by Bye (1967) and Kenyon (1970) found that under high sea states, even the second-order Stokes drift associated with an irrotational wave motion could contribute a substantial part of the total mass transport in the surface layer. Recently, a new solution by considering both the wind stress and the wave motion together was found by Huang (1979). This new solution used only the Stokes drift as the contribution and found that wave motion can influence the dynamics of the surface drift current and produce a variety of flow patterns ranging from Langmuir circulations, to inertial currents, including the classical Ekman flow as one of the possibilities.

In the field, the irrotational wave motion represents only gentle unbroken waves. Under high wind and high wave conditions, the waves tend to break. When a wave breaks, the top of the wave will roll forward at the phase velocity which is two orders of magnitude larger than the Stokes drift. Consequently, for the high sea state conditions considered by Bye (1967) and Kenyon (1970), the contribution of the breaking waves should be included as an important part of the total surface drift. In addition to this direct contribution, there is the indirect influence of breaking waves on the surface drift current. As an indirect contribution, the breaking waves will release turbulence energy in the upper layer and increase the eddy viscosity. According to Huang (1979), the eddy viscosity is an extremely important parameter in determining the dynamical structure of the surface drift current. Since the expression for the eddy viscosity is not known, no attempt will be made here to estimate the indirect contribution.

The direct consequence of wave breaking on the surface drift current can be calculated easily based on the statistical model developed here. This direct contribution comes from the mass and momentum in the part of the waves that breaks from the crests. During breaking, the collapsing crest will move at the phase velocity according to Stokes's criterion. Thus, the breaking crests may contribute greatly to the mass and momentum transport. The calculation of the momentum and mass transport will start with an estimate of the part of the wave amplitude that is lost during breaking.

Following the argument advanced by Longuet-Higgins (1969) and extended here, the portion of the amplitude loss in a unit cycle due to breaking can be calculated by

$$\eta = \int_{a_0}^{\infty} (a - a_0) p(a) da$$

$$= \int_{a_0}^{\infty} \frac{2a(a - a_0)}{\overline{a^2}} \exp\left(-\frac{a^2}{\overline{a^2}} \right) da \tag{16}$$

After some algebra, it can be shown that

$$\eta = (2\overline{\zeta^2})^{1/2} \left[A\tilde{\omega} + \Gamma\left(\frac{3}{2}\right) - \gamma\left(\frac{3}{2}, A^2\right) \right] \tag{17}$$

where substitution of $\overline{a^2}$ by $\overline{\zeta^2}$ was made and

$$A = \frac{a_0}{(\overline{a^2})^{1/2}} = \left[\frac{1}{32\pi^2 \S^2} \left(\frac{m-3}{m-1} \right)^2 \right]^{1/2}$$

from (6), and $\gamma(3/2, A^2)$ is the incomplete gamma function. In this computation, the simplified Wallops spectral form is assumed as in Section 2.

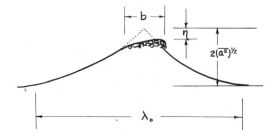

FIGURE 2. A simple model of the breaking wave geometry.

The ratio of η to $(\overline{\zeta^2})^{1/2}$ has special significance. It is proportional to the total area covered by the actively breaking part of the waves. Under a simple geometric argument as shown in Fig. 2, the percentage of the total surface occupied by this active whitecapping area can be estimated. Let the length of the whitecapping area be b. Then

$$\frac{b}{\lambda_0} = \frac{\eta}{2(\overline{a^2})^{1/2}} = \frac{1}{2}\left[A\tilde{\omega} + \frac{\pi^{1/2}}{2} - \gamma\left(\frac{3}{2}, A^2\right) \right] \tag{18}$$

This value is plotted in Fig. 3. Here one can see that the total area of active breaking increases drastically as the significant slope increases. If the breaking waves are the cause of generating the bubbles and foam to form the whitecaps, then the value given in Fig. 3 can be treated as the percentage of foam coverage if the foam lasts exactly one cycle of the wave motion in time. Otherwise, the percentage of foam coverage should be modified by

$$\omega_c\% = \frac{t_1}{T_0} \cdot \frac{b}{\lambda_0} \cdot 100\% \tag{19}$$

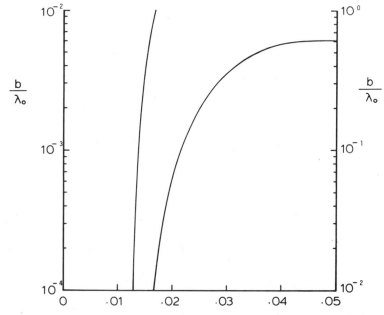

FIGURE 3. Ratio of active wave breaking area to the total sea surface area. The right vertical axis is for the right branch of the curve, the left vertical axis for the left branch of the curve. Horizontal axis = §.

where t_1 is the mean lifetime of the foam generated by the breaking waves. This expression should be treated as an approximation because only the breaking of the energy-containing waves is considered. The actual number in the field could be higher. But qualitatively, the trend makes perfect sense. The sudden increase in the whitecapping area also compares well with the observations by Monahan (1971), Monahan and Muircheartaigh (1980), Ross and Cardone (1974), and Thorpe and Humpries (1979). Unfortunately, in all of these studies the results were presented in such a way that only an empirical formula can be established between the percentage of whitecapping coverage and the surface wind speed. Monahan (1971) found that the highest whitecapping area under wind speeds less than 12 m/s is around 2% of the total surface. This wind value is comparable to data group A of JONSWAP, where the maximum significant slope is around 0.015. This significant slope would give a whitecapping area of a comparable value according to Fig. 3 if t_1 is assumed to be equal to T_0. Ross and Cardone's (1974) data again showed an upper limit of 5% or so. This value would be consistent with the upper limit of § of under 0.02.

It should be pointed out that since the breaking process is controlled by the wind stress and the significant slope of the waves which varies substantially under different stages of wave development, it is unlikely that the whitecaps or the foam coverage should be a function of wind speed alone. Based on this argument, the use in passive microwave remote sensing of the brightness temperature caused by foam coverage as a wind speed indication should be seriously studied before the empirical relationship is accepted.

Next, we can use the result given in (18) to calculate the mean surface drift current due to breaking waves as

$$U = C_0(b/\lambda_0) \tag{20}$$

In order to determine the relative importance of this quantity, U can be normalized with respect

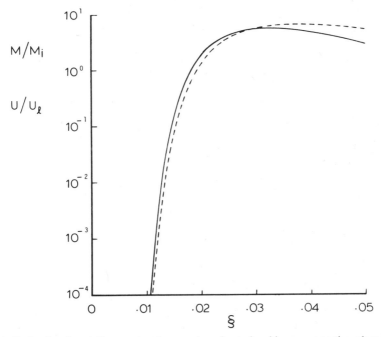

FIGURE 4. Ratio of surface drift current and momentum due to breaking waves to the values based on irrotational wave motion.———, U/U_1 according to (21); -----, M/M_i according to (24).

to the Stokes drift, U_1, which is given by

$$U_1 = an(ak) = 8C_0\pi^2\S^2$$

Thus,

$$\frac{U}{U_1} = \frac{1}{16\pi^2\S^2}\left[A\tilde{\omega} + \frac{\pi^{1/2}}{2} - \gamma\left(\frac{3}{2}, A^2\right)\right] \tag{21}$$

The value of this ratio is plotted in Fig. 4. Over the possible field values of $\S$ (< 0.02), the ratio U/U_1 could reach the order of unity. But the significance of this drift associated with breaking waves is not in the mean value. Rather, the concentrated distribution of the drift current at the breaking crests could induce local convergence of flow in order to maintain the continuity in mass. Such a convergent flow could very well be the cause of cellular circulation patterns in the vertical plane and result in Langmuir circulation. The observed elongated foam streaks in the field under high wind and choppy sea seem to support this argument. Detailed studies are needed to establish a more definite relationship.

The contribution of the breaking waves to the momentum can also be calculated. Since wave breaking occurs when the particle velocity reaches the phase velocity at the crest, the mean value of the momentum is therefore

$$\mathbf{M} = \rho\eta\mathbf{C}_0 \tag{22}$$

The momentum of an irrotational wave motion can be shown to be

$$\mathbf{M}_i = 2\pi\rho(\overline{\zeta^2})^{1/2}\S\mathbf{C}_0 \tag{23}$$

The ratio of M to M_1 is

$$\frac{|\mathbf{M}|}{|\mathbf{M}_i|} = \frac{1}{2^{1/2}\pi\S}\left[A\tilde{\omega} + \frac{\pi^{1/2}}{2} - \gamma\left(\frac{3}{2}, A^2\right)\right] \tag{24}$$

This value is also plotted in Fig. 4. The mean value of momentum from breaking waves only amounts to a small fraction of the momentum from the irrotational wave motion for most practical values of $\S$. The reason for this relatively low percentage of momentum is the shallow depth of the breaking tip in comparison to the wave motion. However, the influence of the breaking wave is by no means confined to the thin layer of the direct drift current. We will discuss this aspect in detail later in this chapter.

Finally, before leaving the subject of momentum, a self-consistent test can be perform to check the rate of energy released from breaking based either on the direct calculation as given in Section 2, or on the rate of momentum flux from the breaking crests as given in this section. To calculate the rate of energy released from momentum considerations, one can start from the mean momentum from the breaking waves as given by (22). The rate of the momentum flux component into the water is then

$$\frac{\mathbf{M}\cdot\mathbf{C}_0}{T_0}\cdot\S = 2^{1/2}\rho C_0^3\S^2\left[A\tilde{\omega} + \frac{\pi^{1/2}}{2} - \gamma\left(\frac{3}{2}, A^2\right)\right] \tag{25}$$

The factor of the quantity in parentheses will make the value of (25) vary over many decades just like the quantities in (21) and (24). A comparison of this quantity with either the rate of work done by the wind as given by (9) or the rate of energy loss by wave breaking as given by (7) can be used as a self-consistent test.

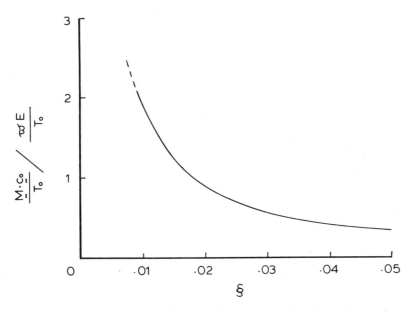

FIGURE 5. Ratio of the rate of energy loss due to breaking waves calculated from the rate of momentum flux to that calculated from energy directly.

If the rate of work done by the wind is used to compare with (25), the ratio is

$$\frac{(\mathbf{M}\cdot C_0/T_0)\cdot\S}{\rho_a u_*^3} = \frac{\rho}{\rho_a}\left(\frac{C_0}{u_*}\right)^3 \S^2\left[A\tilde{\omega} + \frac{\pi^{1/2}}{2} - \gamma\left(\frac{3}{2}, A^2\right)\right] \tag{26}$$

Indeed, by order of magnitude consideration, ρ/ρ_a is 10^3, C_0/u_* is 10, but $\S$ is 10^{-2} and for the realizable $\S$ in the field, the quantity in parentheses is only is 10^{-2}. Thus, the value of (25) is always less than or at most equal to unity. It would equal unity only under equilibrium sea state. This test is only qualitative.

A more rigorous check is to compare the magnitude of (25) with (7). The ratio is

$$\frac{(\mathbf{M}\cdot C_0/T_0)\cdot\S}{\tilde{\omega}E/T_0} = \frac{1}{2^{1/2}\pi\tilde{\omega}}\left[A\tilde{\omega} + \frac{\pi^{1/2}}{2} - \gamma\left(\frac{3}{2}, A^2\right)\right] \tag{27}$$

This value is plotted in Fig. 5. Notice that the value of (27) is derived from the ratio of two highly variable functions of $\S$; each covers many decades but the ratio is never far from unity. This result is evidence of the consistency of the different ways of calculating the energy loss of the breaking waves.

5. THE INFLUENCE OF BREAKING WAVES ON THE FORMATION AND MODIFICATION OF THE MIXED LAYER

The mixed layer is defined as the topmost layer of the ocean (~ 100 m in thickness) where the heat from the sun and momentum from the wind act most directly, causing the water to be rather uniformly mixed most of the time. The state of this layer concerns us for many reasons. Most importantly, this layer is the crucial link between the atmosphere and the ocean. Motions in this upper layer provide the means for the exchange of momentum, mass, heat, and energy between the atmosphere and the large bulk of the deeper ocean. These interactions are not just in one direction. For example, more than 80% of all the solar radiation absorbed by the earth, except for the atmosphere, is stored in the top 20 m of the sea. This heat storage in effect will determine not only the temperature of the low-level wind but also the moisture content of the air and the radiation balance. On the other hand, the wind and the wind-associated processes supply almost all the energy for the general circulation and the turbulence of the whole ocean as discussed in detail by Holland (1977) and Steele (1977).

As crucial as the upper mixed layer is, our understanding of its formation and subsequent modifications due to variable environmental forces is quite incomplete. Past attempts to model this phenomenon have met with some successes for special cases. Such success can almost always be traced to the judicious choice of variable *ad hoc* coefficients which were fixed features of all the models. Reviews of the various models can be found in Niiler and Kraus (1977), Niiler (1977), Garwood (1979) and Kitaigorodskii (1979). Shortcomings of all the models are many. The most critical one is the uncertainty in the adjustable coefficients which appear in all the model equations. For example, a 1d integrated model used by Niiler (1977) is

$$\frac{1}{2}\frac{\partial h}{\partial t}[C_e w_*^2 + \alpha g h(T_s - T_+) - |\mathbf{V}_s - \mathbf{V}_+|^2]$$

$$= m_0 w_*^3 - \varepsilon_0 h - \frac{\alpha g}{2C_p \rho}(Q_0 - Q_+) - \frac{\tau_+}{\rho}(\mathbf{V}_s - \mathbf{V}_+) \tag{28}$$

where h is the depth of the mixed layer; w_* is the frictional velocity at the water surface; T and V are the mean temperature and velocity, respectively, with the subscript s referring to the quantities within the mixed layer of depth h, and the subscript + indicating quantities at the top of the thermocline; C_p is the specific heat of the seawater; α is the heat expansion coefficient of the seawater; Q and τ are the heat and momentum fluxes; and C_e, m_0, and ε_0 are the three adjustable coefficients to be specified.

Physically, these three adjustable coefficients are all related to the turbulence energy intensity in the mixed layer. Therefore, parameterization of the turbulence processes is the central problem in the modeling of the air–sea boundary layer. In this integrated 1d model equation, C_e is defined as the coefficient of the rate of turbulent energy entrainment by

$$\int_{-h}^{0} \frac{1}{2}\frac{\partial e^2}{\partial t} dz = C_e \frac{w_*^2}{2}\frac{\partial h}{\partial t} \tag{29}$$

where $e^2/2$ is the turbulent kinetic energy density within the mixed layer. m_0 is the mixing

TABLE I. Values of Mixing Efficiency m_o

	m_o
Pollard *et al.* (1973)	0
Wu (1973)	0.23
Davis *et al.* (1981)	0.4–0.6
deSzoeke and Rhines (1976)	0.5–1.2
Denman (1973)	0.94
Kraus and Turner (1967)	1
Niiler (1977)	1.2
Kato and Phillips (1969)	1.5
Richman and Garrett (1977)	3–9
Turner (1969)	7.8
Garnich and Kitaigorodskii (1977)	6–40
Miropolskii (1970)	10–50

efficiency and ε_0 is the rate of energy dissipation. The difficulty in accurately determining these coefficients could easily affect the response of any model by a factor of two to three.

The importance of m_0 is its role in parameterizing the wind stress as the source of the turbulent mixing energy. Past effort in the determination of m_0 is summarized in Table I. The wide range of variation for m_0 bears witness to the unsettled state of mixing layer modeling. The difficulty in determining m_0 comes from many sources. In the first place, w_* is extremely hard to determine. Of course, w_* can be computed from the u_* by imposing the continuity condition on stresses across the air–sea interface, but the uncertainty of the u_* value is by no means eliminated. The accuracy in calibrating the wind recorders, and the values of drag coefficient adopted could easily affect the model results by a factor of three as reported by Pollard and Millard (1970). All these uncertainties are thus passed on to the value of m_0 in this empirical approach. Secondly and more fundamentally, the assumption that the energy source of mixing can be parameterized by the frictional velocity of the water and by the frictional velocity alone amounts to neglecting the wave contribution completely. Based on all the discussions presented in this chapter so far, it is obvious that the dynamical characteristics of the surface layer have to be influenced by the wave conditions and especially the breaking ones which are the most energetic at the surface of the ocean. Since mixing events (either the formation or the modification of the mixing layer) are caused primarily by the storms during so-called catastrophic events (Dillon and Caldwell, 1978) when the waves are high and breaking is violent, it would be unreasonable to ignore the wave effect or to bypass it. In fact, the influence of the breaking waves should be the dominant cause of the turbulence energy generation and hence the primary mover of the mixing events. Furthermore, due to the highly nonlinear characteristics of wave development, the relationship between the energy from the breaking waves and the wind stress is by no means a fixed constant of proportionality. Therefore, to use the wind stress to parameterize the air–sea interaction mechanism in mixing is to increase the uncertainty of the model.

The advantage of using the wave properties as input is that the proposed approach could be analytical rather than empirical. The parameterization scheme depends on the sea state which can now be measured readily by remote sensing techniques (see, e.g., Walsh, 1979). Furthermore, the sea state represents an integrated effect of the wind action. Since the water has an inertia 1000-fold greater than that of the air, it is believed that the sluggish response of the sea to the wind is more truthfully represented by the waves rather than by the instantaneous local wind condition.

The influence of breaking waves in mixing has recently been studied by Thompson (1982). He uses the impact momentum of the breaking waves to generate the turbulence field. The approach adopted here will rely on energy considerations. To begin with, not all the energy from breaking waves is available for mixing. A major portion of the energy would be dissipated in the mixed layer in the form of heat before reaching the base of the sharp pycnocline for mixing. The exact portion of this dissipation is still unknown, but various investigators estimate the value at more than 90%. In fact, the adoption of w_* in computing the mixing energy source is a tacit admission of high dissipation. Because of the difference of density between air and water, the friction velocity in the water could only be a few percent of u_*. The value of the dissipation rate has been estimated by Niiler (1977) as $3.4 \times 10^{-3}\,\mathrm{cm}^2/\mathrm{s}^3$, and constant through the mixed layer. From the breaking wave model proposed by Longuet-Higgins (1969), Monin (1977) deduced again a constant value for ε_0:

$$\varepsilon_0 = \frac{ga^2}{2T_0 h} \cdot 10^{-4} \tag{30}$$

where h is the depth of the mixed layer. Sample values of $a = 3\,\mathrm{m}$, $T_0 = 15\,\mathrm{s}$ and $h = 60\,\mathrm{m}$ give an ε_0 of $5 \times 10^{-2}\,\mathrm{cm}^2/\mathrm{s}^3$, which is an order of magnitude higher than Niiler's value. This serves as another reminder of the importance of breaking waves. But should the rate of dissipation be a constant? The available data are indeed scarce, but an important set of data collected by Grant et al. (1968) cited by various authors as the prima facie case of the existence of a uniform layer of energy dissipation could be used to prove the nonconstant value of ε_0 as well.

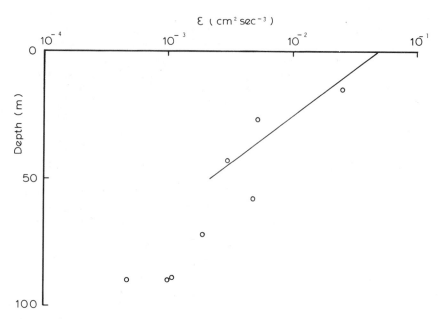

FIGURE 6. Turbulent energy dissipation rate measured by Grant *et al.* (1968) as a function of depth. The solid line represents the $\varepsilon(z)$ calculated according to (35) with a wave 200 m in length and a significant slope of 0.0123. The depth of the mixed layer according to temperature data is 50 m.

The data Grant *et al.* (1968) were collected about 40 miles off the west coast of Vancouver Island where the water depth was about 1000 m. The measurements were made from the bow of a submarine using a hot-film flowmeter and thermistor. The sea state was dominated by a heavy swell with a wavelength of about 200 m. Temperature and salinity data suggested a mixed layer of approximately 50 m in depth. The dissipation rate was calculated from the direct turbulence velocity measurements. The results are presented in Fig. 6. From these data, it is obvious that the dissipation rate is not constant throughout the depth. Of special interest are the points in the mixed layer. Their magnitudes change dramatically. A line representing an exponential decaying with a length scale equal to the surface wavelength is plotted. The slope of the line is quite consistent with the data in the mixed layer. The data of Grant *et al.* (1968) also showed a high dissipation rate that persisted below the mixed layer down to 90 m; a shear current related to tidal motion was suspected as the cause. At any rate, these nonconstant values of ε_0 in the mixed layer strongly suggest the weakness of the past practice of choosing a constant ε_0 for the whole mixed layer. In fact, the trend of the exponential decay makes the choice of wave motion as the mechanism of generating and transporting the turbulence energy an attractive alternative. This type of exponential decay rate has been proposed by most investigators in relating the turbulence generation to waves. Examples can be found in Navrotskii (1967), Kitaigorodskii and Miropolskii (1968), Benilov (1973), and Kitaigorodskii (1973).

Field observations by Thorpe and Stubbs (1979) showed that clouds of bubbles related to breaking waves penetrate to a depth of 5–10 m in lakes for winds of 10 m/s. If so, a deeper penetration of turbulence energy is certainly possible. The exponential decay of the dissipation rate also suggested that wave motion plays an important role in transporting the turbulence energy downward from the surface. Orbital velocity, breaking wave impact, and Langmuir cells are all possible mechanisms to transport the turbulence energy down.

If we assume an exponential decay of the turbulence energy at the same rate as the kinetic energy associated with the wave motion, then the rate of input energy from breaking waves at the surface is given by (7); the actual available energy at the bottom of a mixed layer at depth *h* would be

$$\frac{\tilde{\omega}E}{T_0}e^{-2k_0h} \tag{31}$$

The difference between the input and this value would be the total amount of energy dissipated. Let the specific rate of energy dissipation be $\varepsilon(z)$; then

$$\varepsilon(z) = \tilde{\varepsilon}_0 e^{2k_0 z} \tag{32}$$

By combining (31) and (32), we have

$$\frac{\tilde{\omega}E}{T_0}(1 - e^{-2k_0h}) = \int_{-h}^{0} \varepsilon(z)dz \tag{33}$$

Thus,

$$\tilde{\varepsilon}_0 = \frac{2k_0\tilde{\omega}E}{T_0} = \frac{4\pi g^2 \S^2 \tilde{\omega}}{n_0} \tag{34}$$

Now by (32), we can write

$$\varepsilon(z) = \frac{4\pi g^2 \S^2}{n_0}\tilde{\omega}e^{2k_0 z} \tag{35}$$

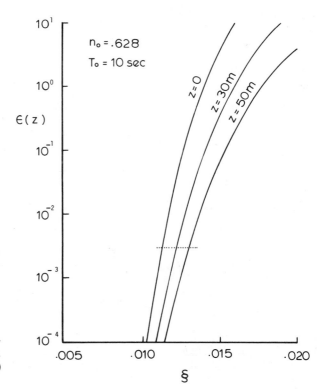

FIGURE 7. Turbulent energy dissipation rate calculated from (35). Dotted line indicates Niiler's (1977) value of 3.4×10^{13} cm^2/s^3.

The integrated value of this dissipation rate should be used to replace $\varepsilon_0 h$ in (28). A typical case of $n_0 = 0.628$ rad/s is chosen to compute $\varepsilon(z)$. The result is plotted in Fig. 7. The $\varepsilon(z)$ value changes very fast with §. However, the value suggested by Niiler (1977) is not unreasonable because of the low § value for most of the field conditions. In fact, Fig. 7 suggests that if the waves were at a peak frequency of 0.628 rad/s, Niiler's value could be real for § around 0.012, which is a typical fresh sea case. Higher values of § are unlikely due to the energy consideration discussed in Section 3. According to this present model, the dissipation rate also varies with depth. The surface $\varepsilon(0)$ would be impossible to obtain. Thus, typical values of ε obtained in the field would be those at a certain depth which would be slightly lower than ε_0. Unfortunately, detailed wave conditions were not recorded by Grant et al. (1968), but according to the present model, the field data obtined by them could be produced by a wave with a significant slope of 0.0123. The computed value is also plotted in Fig. 6 for comparison. Although 0.0123 seems high, it is certainly not unreasonable.

Now, let us consider the problem of the parameterization of the mixing efficiency. After dissipation, the available energy is given by (31). By writing

$$\frac{\tilde{\omega}E}{T_0}e^{-2k_0 h} = \rho m_0 w_*^3 \tag{36}$$

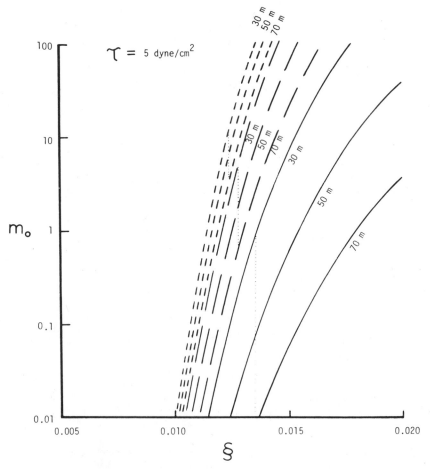

FIGURE 8. The mixing efficiency calculated from (38) for a wind stress of 5 dynes/cm² and different values of mixed layer depth as labeled. - - - - -, $C_0/u_* = 40$;———, $C_0/u_* = 30$;———, $C_0/u_* = 20$.

and also imposing the stress continuity condition,

$$\rho_a u_*^2 = \rho w_*^2 \tag{37}$$

it can be shown that

$$m_0 = \left(\frac{\rho}{\rho_a}\right)^{3/2} 2\pi \S^2 \left(\frac{C_0}{u_*}\right)^3 \tilde{\omega} e^{-2k_0 h} \tag{38}$$

This equation indicates that the mixing efficiency is a function of the significant slope of the waves, the parameter C_0/u_* and the depth of the mixed layer. Sample curves are plotted in Fig. 8 for a wind stress of 5 dynes/cm^2, which corresponds to a U_{10} of ~ 20 m/s. Three values of C_0/u_* and three value of the mixed layer depth are presented. The mixing efficiency values vary over a large range. However, if the equilibrium sea state limit is applied, the cutoff values for $\S$ at each C_0/u_* give the mixing efficiency range that fits the past observed results listed in Table I almost exactly. It should be pointed out that the two largest values in Table I were derived without discounting dissipation. Therefore, they should be treated separately.

Before ending this section, a remark on the present approach should be made. The present parameterization scheme points out a new alternative for mixed layer modeling that reduces the need for adjustable coefficients to a minimum. The results presented here also indicate that the so-called adjustable constants might not be constants after all. The values adopted by various investigators were all correct for their respective cases. Unfortunately, a complete set of mixed layer data with sea state observations is not yet available for detailed comparison.

6. CONCLUSIONS

In this chapter, the contribution of breaking waves to the dynamics of the upper ocean is explored. The approach is based on a statistical method rather than a deterministic method which makes this analysis more relevant in a random wind-generated wave field. It is especially appropriate here to point out that the significant slope $\S$ played a crucial role, just as in the papers on probability distribution functions by Huang and Long (1980), the spectral function by Huang et al. (1981a), and even in the empirical relationships among the nondimensional energy, fetch, and frequency by Huang et al. (1981b). The special significance of this key role played by $\S$ is that this quantity can be measured readily through remote sensing techniques by such instruments as the satellite altimeter and spectrometer.

The expressions given in this chapter, however, should not be regarded as final. They point out the role and relevancy of the wave motion, especially breaking waves, in the dynamics of the upper ocean. The significance lies in that the present approach is based on physical and dynamical principles. Assumptions may be improved as our understanding increases, but the tie between wave motion and all the upper layer phenomena is definite and real. Although further work is urgently needed, this approach could open up a new vista for the application of remote sensing techniques in future oceanographic research.

REFERENCES

Benilov, A. Y. (1973): Generation of ocean turbulence by surface waves. *Izv. Atmos. Ocean Phys.* 9, 293–303.
Bye, J. A. T. (1967): The wave-drift current. *J. Mar. Res.* 25, 95–102.
Cokelet, E. D. (1977a): Breaking waves. *Nature* 267, 769–774.
Cokelet, E. D. (1977b): Steep gravity waves in water of arbitrary uniform depth. *Philos. Trans. R. Soc. London Ser. A* 286, 183–230.
Davis, R. E., R. deSzoeke, and P. Niiler (1981): Variability in the upper ocean during MILE. Part II. Modeling the mixed layer response. *Deep-Sea Res.* 28, 1453–1475.

Denman, K. L. (1973): A time-dependent model of the upper ocean. *J. Phys. Oceanogr.* **3**, 173–184.

deSzoeke, R., and P. B. Rhines (1976): Asymptotic regimes in mixed-layer deepening. *J. Mar. Res.* **34**, 111–116.

Dillon, T. M., and D. R. Caldwell (1978): Catastrophic events in a surface mixed layer. *Nature* **276**, 601–602.

Ekman, V. W. (1905): On the influence of the earth's rotation on ocean-current, *Ark. Math. Astron. Ocean. Phys.* **2**, (11).

Fenton, J. D. (1972): A ninth-order solution for the solitary wave. *J. Fluid Mech.* **53**, 257–271.

Fenton, J. D. (1979): A higher-order cnoidal wave theory. *J. Fluid Mech.* **94**, 129–161.

Garnich, N. G., and S. A. Kitaigorodskii (1977): On the rate of deepening of the oceanic mixed layer. *Izv. Atmos. Ocean Phys.* **13**, 888–893.

Garwood, R. W., Jr. (1979): Air–Sea interaction and dynamics of the surface mixed layer. *Rev. Geophys. Space Phys.* **17**, 1507–1524.

Grant, H. L., A. Moilliet, and W. M. Vogel (1968): Some observations on the occurrence of turbulence in and above the thermocline. *J. Fluid Mech.* **34**, 443–448.

Hasselmann, K., D. B. Ross, P. Müller, and W. Sell (1976): A parametric wave prediction model. *J. Phys. Oceanogr.* **6**, 200–228.

Holland, W. R. (1977): The role of the upper ocean as a boundary layer in models of the oceanic general circulation. *Modelling and Prediction of the Upper Layer of the Ocean* (E. B. Kraus, ed.), Pergamon Press, Elmsford, New York, 7–30.

Huang, N. E. (1979): On surface drift current in the ocean. *J. Fluid Mech.* **91**, 191–208.

Huang, N. E., and S. R. Long (1980): An experimental study of the surface elevation probability distribution and statistics of wind generated waves. *J. Fluid Mech.* **101**, 179–200.

Huang, N. E., S. R. Long, C. C. Tung, T. Yuen, and L. F. Bliven (1981a): A unified two-parameter wave spectral model for a general sea state. *J. Fluid Mech.* **112**, 203–224.

Huang, N. E., S. R. Long, and L. F. Bliven (1981b): On the importance of the significant slope in empirical wind wave studies. *J. Oceanogr.* **11**, 569–573.

Kato, H., and O. Phillips (1969): On the penetration of a turbulent layer into stratified fluid. *J. Fluid Mech.* **37**, 634–655.

Kenyon, K. E. (1970): Stokes transport. *J. Geophys. Res.* **75**, 1133–1135.

Kitaigorodskii, S. A. (1973): *The Physics of Air–Sea Interaction*, Isr. Progr. Sci. Transl., Jerusalem.

Kitaigorodskii, S. A. (1979): Review of the theories of wind-mixed layer deepening. *Marine Forecasting* (J. C. J. Nihoul, ed.), Elsevier, Amsterdam, 1–33.

Kitaigorodskii, S. A., and Y. Z. Miropolskii (1968): Turbulent-energy dissipation in the ocean surface layer. *Izv. Atmos. Ocean Phys.* **4**, 647–659.

Kraus, E. B., and J. S. Turner (1967): A one-dimensional model of the seasonal thermocline. II. The general theory and its consequences. *Tellus* **19**, 98–106.

Longuet-Higgins, M. S. (1962): The statistical geometry of random surface. *Hydrodynamics Stability: Proc. 13th Symp. Appl. Math.* 105–144.

Longuet-Higgins, M. S. (1969): On wave breaking and the equilibrium spectrum of wind-generated waves. *Proc. R. Soc. London Ser. A* **310**, 151–159.

Longuet-Higgins, M. S., and E. D. Cokelet (1976): The deformation of steep surface waves. I. A numerical method of computation. *Proc. R. Soc. London Ser. A* **350**, 1–26.

Longuet-Higgins, M. S., and E. D. Cokelet (1978): The deformation of steep surface waves. II. Growth of normal-mode instabilities. *Proc. R. Soc. London Ser. A* **364**, 1–28.

Longuet-Higgins, M. S., and M. J. H. Fox (1977): Theory of the almost-highest waves: The inner solution. *J. Fluid Mech.* **80**, 721–742.

Miropolskii, Y. Z. (1970): Nonstationary model of the wind-convection mixing layer in the ocean. *Izv. Atmos. Ocean Phys.* **6**, 1284–1294.

Mitsuyasu, H., and T. Honda (1975): The high frequency spectrum of wind-generated waves. *Rep. Res. Inst. Appl. Mech. Krushu Univ.* **22**, 327–355.

Monahan, E. C. (1971): Oceanic whitecaps. *J. Phys. Oceanogr.* **1**, 139–144.

Monahan, E. C., and I. O. Muircheartaigh (1980): Optimal power-law description of oceanic whitecap coverage dependence on wind speed. *J. Phys. Oceanogr.* **10**, 2094–2099.

Monin, A. S. (1977): On the generation of oceanic turbulence. *Izv. Atmos. Ocean Phys.* **13**, 798–803.

Müller, P. (1976): Parameterization of one-dimensional wind wave spectra and their dependence on the state of development. *Hamburger Geophysikalische Einzelschriften Heft* **31**, Wittenborn Sohne, Hamburg.

Navrotskii, V. V. (1967): Waves and turbulence in the ocean surface layer. *Oceanology* **6**, 755–766.

Niiler, P. P. (1977): One-dimensional models of the seasonal thermocline. *The Sea*, Vol. 6 (E. D. Goldberg, I. N. McCave, J. J. O'Brien, and J. H. Steele, eds.), Wiley–Interscience, New York.

Niiler, P. P., and E. B. Kraus (1977): One-dimensional models of the upper ocean. *Modelling and Prediction of the Upper Layers of the Ocean* (E. B. Kraus, ed.), Pergamon Press, Elmsford, New York, 143–172.

Phillips, O. M. (1961): A note on the turbulence generated by gravity waves, *J. Geophys. Res.* **66**, 2889–2893.

Phillips, O. M. (1977): *The Dynamics of the Upper Ocean*, 2nd ed., Cambridge University Press, London.

Phillips, O. M., and M. L. Banner (1974): Wave breaking in the presence of wind drift and swell. *J. Fluid Mech.* **66**, 625–640.

Pollard, R. T., and R. C. Millard, Jr. (1970): Comparison between observed and simulated wind-generated inertial oscillations. *Deep-Sea Res.* **17**, 813–821.

Pollard, R. T., P. B. Rhines, and R. O. R. Y. Thompson (1973): The deepening of the wind-mixed layer. *Geophys. Fluid Dyn.* **4**, 381–404.

Richman, J., and C. Garrett (1977): The transfer of energy and momentum by the wind to the surface mixed layer. *J. Phys. Oceanogr.* **7**, 876–881.

Ross, D. B., and V. Cardone (1974): Observations of oceanic whitecaps and their relation to remote measurements of surface wind speed. *J. Geophys. Res.* **79**, 444–452.

Schwartz, L. W. (1974): Computer extension and analytic continuation of Stokes' expansion for gravity waves. *J. Fluid Mech.* **62**, 553–578.

Shonting, D. H. (1965): A preliminary investigation of momentum flux in ocean waves. *Pure Appl. Geophys.* **57**, 149–152.

Steele, J. (1977): Ecological modeling of the upper layers. *Modelling and Prediction of the Upper Layers of the Ocean* (E. B. Kraus, ed.), Pergamon Press, Elmsford, N.Y., 243–250.

Stoker, J. J. (1957): *Water Waves*, Wiley–Interscience, New York.

Stokes, G. G. (1847): On the theory of oscillatory waves. *Trans. Cambridge Philos. Soc.* **8**, 441–455.

Stokes, G. G. (1880): Considerations relative to the greatest height of oscillatory waves which can be propagated without change of form. *Math and Physics Papers* **1**, 225–228, Cambridge University Press, London.

Thompson, R. O. R. Y. (1982): A potential-flow model of turbulence caused by breaking surface waves. *J. Geophys. Res.* **87**, 1935–1938.

Thorpe, S. A., and P. N. Humpries (1979): Bubbles and breaking waves. *Nature* **283**, 463–465.

Thorpe, S. A., and A. R. Stubbs (1979): Bubbles in a freshwater lake. *Nature* **279**, 403–405.

Turner, J. S. (1969): A note on wind mixing at the seasonal thermocline. *Deep-Sea Res.* **16** (Suppl.), 287–300.

Walsh, E. J. (1979): Extraction of ocean wave height and dominant wavelength from GEOS-3 altimeter data. *J. Geophys. Res.* **84**, 4003–4010.

Wu, J. (1973): Wind-induced turbulent entrainment across a stable density interface. *J. Fluid Mech.* **61**, 275–287.

21

STABILITY OF NONLINEAR CAPILLARY WAVES

S. J. HOGAN

ABSTRACT. Small-scale gravity-capillary waves are present in almost all wind-wave fields. Their importance in limiting the growth of longer waves is well known (Phillips, 1977, p. 171). They can also have very large amplitudes (e.g., Schooley, 1958). In this contribution we present results on the stability of the very shortest capillary waves.

1. INTRODUCTION

We investigate the form and stability of steady, 2d nonlinear capillary waves propagating into inviscid, incompressible, infinitely deep water. The motion is assumed to be irrotational. Previous work by the present author (Hogan, 1979, 1980, 1981) produced profiles of steady gravity-capillary waves for many values of the wavelength, λ.

We present results for the superharmonic instabilities of a train of nonlinear deep water pure capillary waves. These results, apart from their intrinsic interest, act as a guide to instabilities of gravity-capillary waves. The calculation follows very closely the work of Longuet-Higgins (1978a, b) on pure gravity waves.

2. METHOD

If (ϕ, ψ) are the velocity potential and streamfunction of the flow, q the particle speed at the surface, τ the surface tension divided by the density, and R the radius of curvature, then the time-dependent Bernoulli equation for pure capillary waves,

$$\frac{1}{2}q^2 + \phi_t - \frac{\tau}{R} = B(t)$$

is satisfied on the free surface $\psi = F(\phi, t)$, consisting of the basic nonlinear wave ($\psi = 0$) and the perturbation. That is,

$$x(\phi, \psi, t) = X(\phi, \psi) + \xi(\phi, \psi, t)$$

$$y(\phi, \psi, t) = Y(\phi, \psi) + \eta(\phi, \psi, t)$$

S. J. HOGAN ● Department of Applied Mathematics and Theoretical Physics, University of Cambridge, Cambridge, England CB3 9EW.

where ξ, η, and F are small quantities, and (X, Y) is the steady solution for pure capillary waves, due to Crapper (1957):

$$\frac{1}{\lambda}X = \frac{2Ae^{-\psi/c}\sin(2\pi\phi/c\lambda)}{\pi[1 + A^2e^{-2\psi/c} + 2Ae^{-\psi/c}\cos(2\pi\phi/c\lambda)]} - \frac{\phi}{c\lambda}$$

$$\frac{1}{\lambda}Y = \frac{2[1 + Ae^{-\psi/c}\cos(2\pi\phi/c\lambda)]}{\pi[1 + A^2e^{-2\psi/c} + 2Ae^{-\psi/c}\cos(2\pi\phi/c\lambda)]} - \frac{2}{\pi}$$

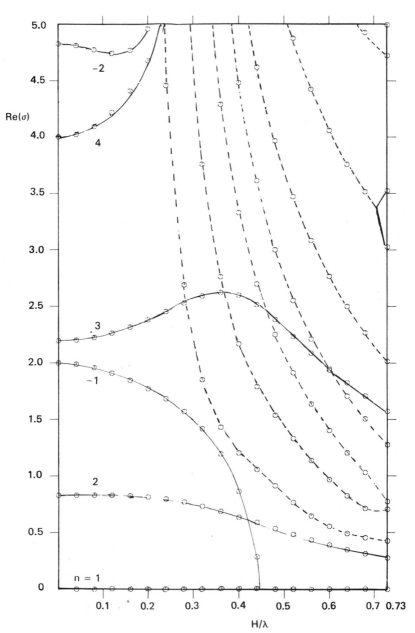

FIGURE 1. Frequency of the perturbation Re (σ) vs. wave steepness a/λ of the unperturbed wave. Dashed lines indicate instability. n is the number of perturbations per wavelength.

where

$$a/\lambda = 4A/[\pi(1 - A^2)]$$

The steepest wave corresponds to $a/\lambda = 0.73$.

For normal mode perturbations, we assume that

$$\zeta = e^{-i\sigma t}\left[a_0 + \sum_{n=1}^{\infty} e^{n\psi/c}\left(a_n\cos\frac{n\phi}{c} + b_n\sin\frac{n\phi}{c}\right)\right]$$

$$\eta = e^{-i\sigma t}\left[b_0 + \sum_{n=1}^{\infty} e^{n\psi/c}\left(b_n\cos\frac{n\phi}{c} - a_n\sin\frac{n\phi}{c}\right)\right]$$

$$F = e^{-i\sigma t}\left[\sum_{n=1}^{\infty} \left(c_n\cos\frac{n\phi}{c} + d_n\sin\frac{n\phi}{c}\right)\right]$$

for a_n, b_n, c_n, d_n real.

We expand the governing equations in Taylor series and substitute the perturbed free surface. We solve the resulting eigenvalue problem in σ by standard techniques, involving truncation after M coefficients and inversion of the resulting matrix.

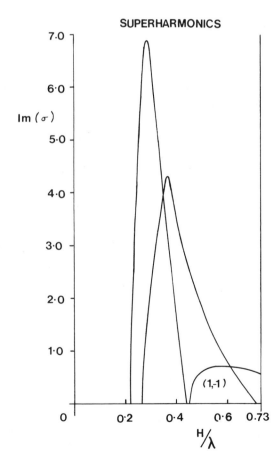

FIGURE 2. Growth rates Im(σ) of the unstable modes vs. wave steepness a/λ of the unperturbed wave.

3. RESULTS

In Fig. 1 we show the results for Re(σ). The designation n indicates the number of perturbation wavelengths in each unperturbed wave. A minus sign indicates that the wave and perturbation are traveling in opposite directions. As the wave steepness $a/\lambda \to 0, \sigma \to \sigma_n$ where

$$\sigma_n = \begin{cases} -n + n^{3/2} & n > 0 \\ n + |n|^{3/2} & n < 0 \end{cases}$$

Figure 2 shows selected growth rates, Im(σ), of the unstable modes in Fig. 1.

The dominant feature in Fig. 1 is the cascade to lower values of Re(σ) of several unstable modes. These originate with coalescences of various modes at larger values of Re(σ). In a manuscript in preparation, the author shows that these instabilities correspond to the appearance of ripples on the surface of the wave. Thus, superharmonic perturbations to a pure capillary wave produce instabilities.

In sharp contrast, pure gravity waves are stable to superharmonic perturbations, almost to the maximum wave height.

REFERENCES

Crapper, G. D. (1957): An exact solution for progressive waves of arbitrary amplitude. *J. Fluid Mech.* **2**, 532–540.

Hogan, S. J. (1979): Some effects of surface tension on steep water waves. *J. Fluid Mech.* **91**, 167–180; Part 2. *J. Fluid Mech.* **96**, 417–445 (1980): Part 3 *J. Fluid Mech.* **110**, 381–410 (1981).

Longuet-Higgins, M. S. (1978a): The instabilities of gravity waves of finite amplitude in deep water. I. Superharmonics. *Proc. R. Soc. London Ser. A* **360**, 471–488.

Longuet-Higgins, M. S. (1978b): II. Subharmonics. *Proc. R. Soc. London Ser. A* **360**, 489–506.

Phillips, O. M. (1977): *The Dynamics of the Upper Ocean*, 2nd ed., Cambridge University Press, London.

Schooley, A. H. (1958): Profiles of wind-created water waves in the capillary–gravity transition region. *J. Mar. Res.* **16**, 100–108.

IV

AIR FLOW OVER WAVES

22

A COMPARISON OF THE WAVE-INDUCED MOMENTUM FLUX TO BREAKING AND NONBREAKING WAVES

M. L. BANNER

ABSTRACT. This chapter describes some initial laboratory measurements which compare the surface pressure distributions over broken and unbroken waves. The resulting normal stress wave-coherent momentum flux levels are found to be enhanced by an order of magnitude due to wave breaking. Estimates are obtained for the strength of this effect as a function of (u_*/c). The findings of Okuda *et al.* (1977) on the distribution of skin friction along wind-wave profiles are reviewed in the present context. The chapter concludes with the possible relevance of these findings to modulated shear stress in the explicably large microwave backscattering modulation levels observed by Wright *et al.* (1980).

1. INTRODUCTION

Wind blowing over the ocean creates a wave spectrum with wave breaking often occurring over a range of scales. For dominant wavelengths greater than of order 1 m, wave breaking is commonly observed as isolated whitecaps. However, patches of short waves often occur locally in a state of saturation without the foam which distinguishes whitecaps. Several factors can contribute to the onset of small-scale wave breaking: these include direct wind input, straining due to the orbital motion of longer wave components and nonlinear transfer from other wave components. The averaged behavior of the short scales is reflected in the high-wavenumber tail of the spectrum, where strong wind-speed-dependent effects on the short scales can be inferred from the field measurements of Cox and Munk (1954a, b) and from the laboratory measurements of Cox (1958) and others.

During the past few years, remote sensing of the sea surface has advanced rapidly, with the development and deployment of various passive and active microwave instruments. In order to interpret properly the return signals they provide, a better understanding is required of the short wave dynamics in relation to the longer wave components. For instance, recent microwave backscatter results using c.w. Doppler radar at a depression angle of 40° reported by Wright *et al.* (1980) indicated that a source other than straining by the horizontal component of the long (10 s) wave scales was responsible for most of the observed local radar return modulation. They suggested that modulated wind stress might be an important additional source to account for their findings. However, at present there are no direct measurements which reveal the local variation of shear stress over the phase of a long wave. Numerical models such as Gent and

M. L. BANNER ● Department of Theoretical and Applied Mechanics, University of New South Wales, Sydney, Australia 2033.

Taylor (1976, Tables 3 and 4) indicate a marked sensitivity of the local shear stress as a function of the assumed roughness (short wave) variation along the long wave. The strength of this effect is strongest for low values of the long wave slope. It is clear that the local short wave behavior really needs to be better understood before numerical models can provide reliable estimates of shear stress variation over a long wave component.

In this contribution, initial results are presented of a laboratory study comparing the wind input levels to breaking and steep nonbreaking waves. Because of the wind-speed-to-phase-speed ratio used, the findings are particularly appropriate to the question of input to the short wave components. We report on pressure measurements in the immediate vicinity of the air–water interface for wind blowing over a breaking wave and a steep, unbroken wave, both embedded in a train of unbroken waves. For a given wind speed, the wave-coherent momentum fluxes due to pressure are calculated for each case. These reveal that the highly nonlinear process of wave breaking, usually associated with augmented dissipation, can concurrently enhance the local wind input level to the wave field.

It is interesting to compare and contrast the present findings with the results of Okuda et al. (1977). Using hydrogen bubble techniques, they reported local skin friction measurements over "representative" wind waves in a wind-wave tank. At a fetch of 2.85 m with a wind speed of 6.2 m/s, the dominant wind waves had characteristic period, wavelength, and wave height of 0.23 s, 8.3 cm, and 0.94 cm, respectively. From their description, these waves were breaking without entraining air. Figure 1 shows a plan view of such wave-tank wind waves (photographed by this author) at a slightly longer fetch (4.27 m) and a comparable wind speed of about 6.5 m/s. The wind is from left to right. Ahead of the breaking dominant wave crests (wavelengths 8–10 cm) are bound capillaries. As the dominant waves lengthen (at longer fetches or higher wind speeds), the capillaries are progressively replaced by more turbulent, aerated breaking zones.

Okuda et al. (1977) report local skin friction values which vary from zero at the windward trough to about 12 dynes/cm^2 at the crest, dropping to effectively zero over the leeward half of the wave. They deduce a mean skin friction of 3.6 dynes/cm^2, slightly in excess of the average wind stress of 3.0 dynes/cm^2 determined from wind profile measurements. They conclude that "the skin friction bears most of the shearing stress of the wind, and that it exerts most intensively around the representative wave crests at their windward faces." They also report that "the pressure drag is only a small fraction of the total wind stress, although its existence is supported

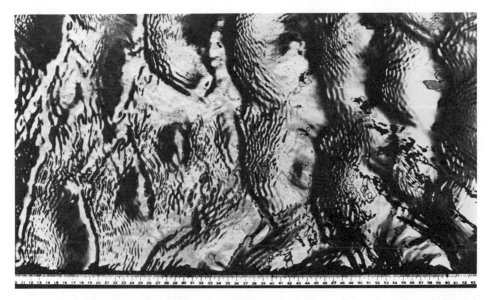

FIGURE 1. Small-scale breaking waves in a conventional wind-wave flume. Centerline wind speed = 6.5 m/s. Fetch = 4.27 m.

from the fact that separation of air flow occurs just past the crest. It may be estimated as at most several percent." It should be noted that their conclusions on the pressure drag were made in the absence of any direct pressure measurements in their study. Their results and conclusions are reviewed critically in Section 4.

Estimates based on the momentum flux levels found for breaking waves in this study suggest that direct wind input effects on the wind-wave spectral peak rapidly become of secondary importance to the spectral evolution process, except possibly under hurricane conditions. However, for the short scales, the same arguments indicate that direct input can continue to be a dominant factor in their dynamics. We conclude with a brief overview of the possible role played by small-scale breaking waves in radar probing of the sea surface.

2. EXPERIMENTAL CONFIGURATION

A schematic of the wind-wave flume used for this study is given in Fig. 2. It is a refined version of the facility used by Banner and Melville (1976). This configuration permits measurements in the vicinity of the air–water interface associated with steady two-dimensional breaking and nonbreaking waves without the need for a wave-follower mechanism. The breaking or steep nonbreaking wave is created within a train of unbroken waves by trial-and-error placement of subsurface slender airfoil sections. Care was taken to maintain successive crests at the same level. The pressure measurements were made in the central region of the wind duct over which the mean wind field was laterally uniform. The roof slope of the wind duct was adjusted to give zero streamwise mean pressure gradient along it. Subsequent checks of the pressure in the free stream along the duct centerline showed negligible pressure variation along the wave profile at that elevation. The surface pressure was sensed by a carefully aligned combination of a contoured disk (10-mm diameter) and a United Sensors kiel tube (shielded total tube) with an outside diameter of 3.175 mm.

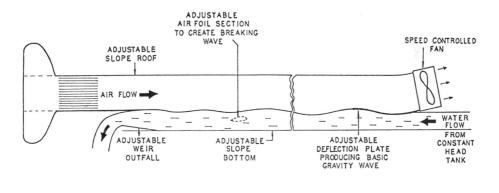

WORKING LENGTH	~	7·3 m
AIR CHANNEL DEPTH	~	36 cm
WATER CHANNEL DEPTH	~	22 cm
CHANNEL WIDTH	~	22·5 cm
MEASUREMENT SITE	~	HALF WAY ALONG CHANNEL

WIND   WAVE   FLUME

FIGURE 2. Experimental configuration.

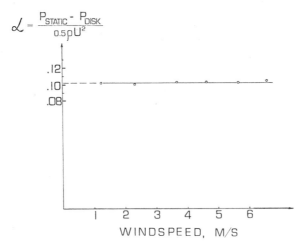

FIGURE 3. Sensor disk calibration.

The free-stream response of these sensors is

$$P_{\text{disk}} = P_{\text{static}} - \alpha \tfrac{1}{2} \rho_a U^2 \tag{1}$$

$$P_{\text{kiel}} = P_{\text{static}} + \tfrac{1}{2} \rho_a U^2 \tag{2}$$

where α is the disk coefficient. For the range of wind speeds used in this study, the variation of α with U was determined by calibration against a standard NPL 2.3-mm-diameter pitot-static tube in the free stream. For all the pressure measurements reported in this study, we used a thermally stabilized Barocel differential pressure transducer on the ± 4 mm H_2O scale. The output voltage was fed to a Thermo Systems 1076 averaging digital voltmeter where a suitable time constant up to 100 s could be selected.

The variation of α was found to be negligible over the wind speed range (0–6 m/s) relevant for this study, as can be seen in Fig. 3. The value of α adopted was 0.102.

For the comparative pressure measurements to be described below, the local measurements using the disk and kiel probes were referenced, in turn, to an NPL standard (pitot)-static probe fixed in the free stream. Typically, averaging times of the order of 5 min, using a time constant of 100 s, were required to obtain reproducible readings.

A preliminary test was conducted which indicated that this combination was capable of estimating the pressure field in the separated quasi-two-dimensional flow associated with breaking waves created in the flume. At a free stream wind speed of 4 m/s, the pressure distribution over a breaking wave was measured by the combination sensor, following the surface profile within 6 mm and by a standard 2.3-mm-diameter NPL elliptic-nosed (pitot)-static tube, traversed at a *fixed* elevation above the mean water level, about 20 mm above the crest level. After correcting the surface-following measurement for the hydrostatic pressure variation, the two pressure distributions are shown in Fig. 4. It is evident that the combination sensor gives pressure variations very consistent with the conventional sensor. Notice that the surface-following measurement gives a larger pressure variation over the wave. This is consistent with the progressive attenuation of the wave-induced pressure with increasing height above the water surface associated with the horizontal traverse of the pitot-static sensor.

For the momentum flux determination, the wave profile was measured using a sighting telescope in conjunction with a sharp pointer. In the breaking zone, an electronic pointer-detector system was used (with a 50% "on" criterion) to define the mean surface height.

The goal of this work was to compare accurate measurements of the pressure distribution over a breaking wave and a steep unbroken wave under similar wind speed conditions. It was

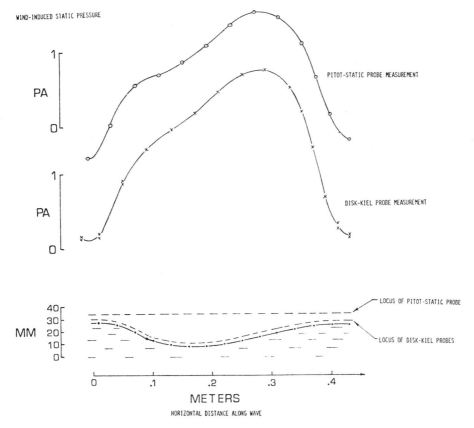

FIGURE 4. Comparison of pressure variation in air flow above a breaking wave using pitot-static tube and disk–kiel tube combination.

also considered desirable to use this probing technique to reexamine the *total* momentum flux level associated with a breaking wave for comparison with the results of Banner and Melville (1976), in view of the potential errors present in their measurements, as described below.

3. RESULTS

A large level of local drag associated with a breaking wave, as reported initially by Banner and Melville (1976), was reproduced in this investigation, mainly as a check on the revised tunnel geometry and measuring system. With the improved sensing system and better-defined momentum integral control volume, a somewhat reduced drag over a breaking wave was obtained here compared to that reported in Banner and Melville (1976). Owing to a mismatch between the two crest levels and to the free-stream pressure gradient, their control volume/differential measurement system may have caused various uncertainties in their total stress estimates for breaking waves. Certain of these effects are also applicable to their unbroken wave stress levels. These are currently being reexamined in an ongoing extension of this work to other wind speeds and wave configurations.

For the free-stream wind speed of 4.0 m/s in this study, the measured momentum flux deficit over a breaking wave was found to be 0.25 Pa. This results from the difference in incoming and outgoing integrated momentum fluxes at the vertical sections of the control surface over the successive crests, as shown in Fig. 5, divided by the wavelength and unit transverse span. In comparison, Banner and Melville (1976) indicate a value of 0.425 Pa for this wind speed. Our

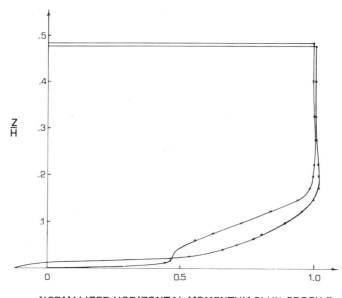

NORMALIZED HORIZONTAL MOMENTUM FLUX PROFILE

FIGURE 5. Momentum flux profiles at the upwind (breaking) crest and subsequent downwind crest. H is the wind tunnel height. The momentum flux is given by $\int p + \rho U^2 dz$ between the appropriate limits, and the upwind centerline local momentum level was used as normalizing factor.

more accurate determination confirms the order of magnitude of the total stress induced by a breaking wave reported by them.

In comparison to friction velocity (u_*) values typical of wind-wave tanks at short fetches, the observed value of about 0.45 m/s for the continuously breaking wave in this study still appears to be about 50% higher than corresponding wind speed estimates of u_* (e.g., Plant and Wright, 1977, Fig. 4). However, for this wind speed the dominant waves are not necessarily breaking continuously in such wind-wave tank realizations, possibly accounting for the lower u_* value observed. In any event, the large drag induced by breaking waves has been reproduced here.

The central results to be presented concern the wave-coherent momentum and energy flux. Traditionally (for small-slope waves), the momentum flux has been taken to be dominated by the pressure–wave slope correlation $\langle p(\partial \eta / \partial x) \rangle$. Here p is the surface pressure field and $z = \eta(x, t)$ is the wave surface, assumed 2d. $\langle \rangle$ denotes an average over the wave profile. In this investigation, the assumption of two-dimensionality was well satisfied, though in the field and in wave tanks, the small-scale wind waves are often observed to be short-crested (e.g., see Fig. 1). The extension to finite amplitude waves is straightforward and it is easily shown that the same result holds. In applying this result to the present data, the breaking wave profile is treated as an unbroken wave with a similar mean profile. For the approximately 2d mean wave profiles of this study, the wave-coherent momentum flux due to normal stress was obtained using the expression

$$\tau_w^{(N)} = \left(\int_0^\lambda p \tan \theta \, dx \right) \Big/ \lambda$$

where the mean wave profile, specified by $z = \eta(x)$, has local slope $\tan \theta = (d\eta / dx)$ and wavelength λ.

Figures 6 and 7 display the measured surface pressures, including the hydrostatic in-phase component, which can contribute up to 20% of the pressure amplitude. Figure 6 shows the pressure distribution over the unbroken wave profile. It is apparent that for this U/c ratio (~ 5), there is a phase shift of the pressure peak forward of the trough of only a few degrees at most. In contrast, for the breaking wave this phase shift approaches 90°, as can be seen in Fig. 7.

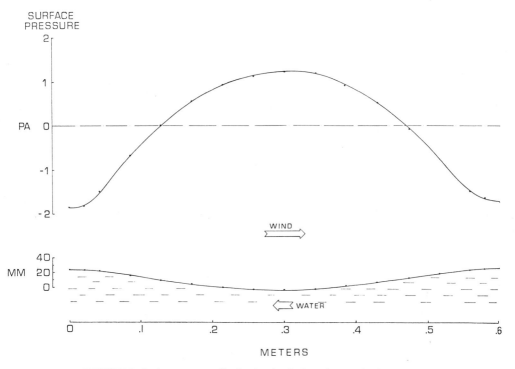

FIGURE 6. Surface pressure distribution in air flow above unbroken wave.

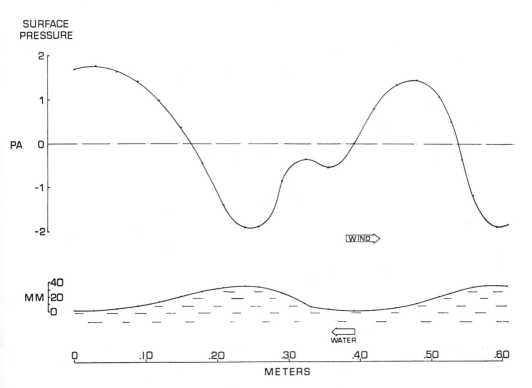

FIGURE 7. Surface pressure distribution in air flow above breaking wave.

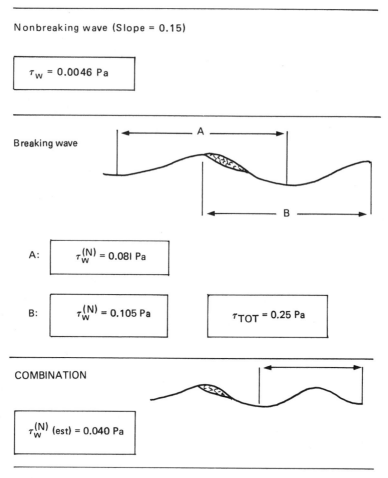

FIGURE 8. Results of wave-coherent momentum flux measurements.

The corresponding momentum fluxes were derived from these results using the measured pressure and wave profile data. The results obtained for the wave-coherent momentum flux τ_w are shown in Fig. 8. Included there is an estimate of τ_w for an unbroken wave immediately ahead of a breaking wave, based on combining the individual breaking and non breaking results.

From Table I the large enhancement of the level of τ_w associated with breaking is apparent. This fundamental result can be interpreted as follows.

Accompanying breaking events, locally large enhancements of the input to the wave field can occur under the influence of wind. Concomitantly, the local total stress is enhanced, as well as the local dissipation rate. The present data are insufficient to assess the dissipation rate so the overall dynamical effect of the breaking remains undetermined. However, the data do suggest a "focusing" mechanism whereby a patch of small-scale breaking waves focusses the available wind input to provide *locally* enhanced levels of input to the short waves, total stress, and dissipation. If the available wind input level is sufficiently high, then the patch of breaking waves tends to be self-sustaining. Equivalently, if the local friction velocity increases due to local breaking of small-scale waves, then the input to the short waves rises by similar proportions. For the limited data set reported here, it was observed that for the wave-coherent momentum fluxes due to normal stress,

$$\frac{\tau_w^{(N)}(\text{unbroken wave})}{\tau_w^{(N)}(\text{breaking wave})} \sim \frac{1}{20}$$

and that the ratio of wave-coherent normal stress to total stress for the breaking wave was

$$\frac{\tau_w^{(N)}(\text{breaking wave})}{\tau_{tot}(\text{breaking wave})} \sim 0.4$$

showing that a significant fraction of the available wind stress appears as wave-coherent normal stress.

4. DISCUSSION

In this section we attempt to assess the significance of the experimental results obtained in this study and relate them to the findings reported by Okuda et al. (1977). We conclude by considering these findings in relation to the inexplicably high radar backscattering modulation levels reported by Wright et al. (1980) in the West Coast Experiment.

4.1. Significance of the Data

From the data, certain comparisons can be made to indicate the significance of the findings reported in Section 3.

a. The wave-coherent momentum flux due to normal stress $\tau_w^{(N)}$ compared with typical wind stress $\tau = \rho_a u_*^2$: The present experiment was conducted at a wind speed of 4 m/s relative to the wave or about 5 m/s relative to still water. Typically, wind-wave tank measurements (e.g., Fig. 4 in Plant and Wright, 1977) yield $u_* \sim 0.25$ to 0.30 m/s for the wind speed considered here. This friction velocity includes the combined effects of breaking and nonbreaking waves of various scales and intermittencies. It would be expected to be higher than the friction velocity appropriate to a wave-free water surface, yet lower than the friction velocity for a continuously breaking wave field (such has been investigated in this study). On this basis, for the data just described,

$$\frac{\tau_w^{(N)}(\text{breaking wave})}{\rho_a u_*^2} \sim \frac{0.1}{1.2(0.25 \text{ to } 0.3)^2}$$

$$\sim 1$$

Thus, *local* to breaking events, the *wave-coherent* momentum flux level can be expected to be close to the *average* total stress level.

b. Ratio of wave-coherent momentum added per period to the approximate momentum density for breaking waves: We assume the momentum per unit surface area of the breaking wave is given by the sum of the momentum density of an unbroken wave of the same slope together with the momentum density of the mass of spilling water transported by the breaking wave, assuming negligible aeration, typical of small-scale breaking waves, i.e.,

$$M_{tot} \sim \tfrac{1}{2}\rho_w \sigma a^2 + \rho_w \bar{h}_b c$$

where $\rho_w =$ water density, $\sigma =$ radian frequency, $c =$ phase speed, $\bar{h}_b =$ vertical scale of the breaking zone, averaged over the wavelength (λ), and $a =$ wave amplitude. An estimate for $\bar{h}_b$

may be obtained by approximating the spilling region as a semielliptic region with overall height h_b and length l_b. Then its area is given by $(\pi/8)h_b l_b$ and so

$$\bar{h}_b \sim \frac{\pi}{8}\frac{h_b l_b}{\lambda}$$

Assuming continuous breaking over a wave period, the ratio of wave-coherent momentum added per wave period (T) to the breaking wave momentum density is given by

$$\frac{\tau_w^{(N)} T}{M_{tot}} = \frac{\rho_a u_*^2 T}{\frac{1}{2}\rho_w \sigma a^2 + \rho_w \bar{h}_b c}$$

from (a) above. Using the $\bar{h}_b$ from above, $\sigma = ck$, $\lambda = 2\pi/k$, and $T = 2\pi/\sigma$,

$$\frac{\tau_w^{(N)} T}{M_{tot}} = \frac{\rho_a}{\rho_w}\left[\frac{2\pi u_*^2}{\frac{1}{2}\sigma^2 a^2 + (\pi/8)\sigma(h_b l_b/\lambda)c}\right]$$

$$= 2\pi\frac{\rho_a}{\rho_w}\left(\frac{u_*}{c}\right)^2\left[\frac{1}{\frac{1}{2}(ak)^2 + (\pi^2/4)(h_b/\lambda)(l_b/\lambda)}\right]$$

Typically, $(\rho_a/\rho_w) \sim 10^{-3}$, (ak)breaking ~ 0.3. Estimates for the length and height of the breaking zone are taken to be $l_b/\lambda \sim 1/4$ and $h_b/\lambda \sim 1/50$. These give

$$\frac{\tau_w^N T}{M_{tot}} \sim 2\pi \times 10^{-3}\left(\frac{1}{0.045 + 0.011}\right)\left(\frac{u_*}{c}\right)^2$$

$$\sim 0.11\left(\frac{u_*}{c}\right)^2$$

This result may be interpreted as the dimensionless rate of addition of momentum (due to normal stress) to a breaking wave field per unit frequency. As such, it provides a measure of the input when the wave component has transitioned through its maximum amplitude. It is of interest to note that this effective "growth" rate falls within the scatter of the relationship proposed by Plant (1981) for wind-wave growth rates:

$$\beta = (0.04 \pm 0.2)\left(\frac{u_*}{c}\right)^2 \omega = (0.25 \pm 0.13)\left(\frac{u_*}{c}\right)^2 f$$

For very short gravity and gravity-capillary waves, under moderate wind stress, u_*/c levels around unity occur frequently. For these conditions, the above estimate suggests about 1/10th of the wave momentum is added per wave period by the enhanced normal stress during wave breaking. This represents a comparable level of wind coupling as existed during the initial growth of these wave components.

It is appropriate to relate these findings to those of Okuda et al. (1977) which have been summarized in Section 3. The following major points arise from an analysis of their findings:

1. The measured average wind stress of 3.0 dynes/cm^2 is about three times higher than the level reported by other wave-tank investigations at comparable wind speeds and fetches (e.g., see Fig. 4 of Plant and Wright, 1977). A value of about 1.0 dyne/cm^2 would be more reasonable under these circumstances.

2. In basing their conclusions on the large average skin friction levels measured in the water, Okuda et al. appear to have failed to distinguish between the average skin

friction and the *wave-coherent* skin friction, i.e., the tangential stress which can do work on the waves. This is given by

$$\tau_w^{(S)} = \frac{1}{c} \langle \tau_s u_s \rangle$$

where τ_s is the local shear stress, u_s is the local tangential fluid velocity component at the surface associated with the wave motion, c is the wave phase speed, and $\langle \rangle$ is an average over the wave profile. The resolution of reported surface velocities given in Table II of their paper into reliable estimates of the wave-associated and mean velocity components is not trivial for such nonlinear (breaking) wave motions. However, an estimate for $\tau_w^{(S)}$ can be obtained by representing their surface velocity data as follow: assume cosine approximations to fit their data for τ_s (dynes/cm^2) and u_s (cm/s) of the form

$$\tau_s = \begin{cases} 6 + 6 \cos \theta & -\pi \leqslant \theta \leqslant 0 \\ 0 & 0 \leqslant \theta \leqslant \pi \end{cases}$$

$$u_s = \begin{cases} 20 + 20 \cos \theta & -\pi \leqslant \theta \leqslant 0 \\ 40 & 0 \leqslant \theta \leqslant \pi/2 \\ 0 & \pi/2 < \theta \leqslant \pi \end{cases}$$

The values for u_s in $0 \leqslant \theta \leqslant \pi$ have been interpolated based on estimations for u_s in the spilling zone and the lower leeward face of the wave ahead of it. The surface velocity distribution u_s includes a wind-driven surface current, which is sporadically disrupted by the breaking zones. The mean (horizontal) surface velocity of about 20 cm/s according to this specification agrees favorably with the usual mean surface drift of 3–4% of the wind speed (6.2 m/s) measured in wind-wave tanks. If the mean current is subtracted from u_s to provide a zero mean variation for the wave-associated surface velocity, then using this revised estimate for u_s, a calculation of

$$\tau_w^{(S)} = \frac{1}{2\pi c} \int_{-\pi}^{\pi} \tau_s u_s \, d\theta$$

yields a mean value of $\tau_w^{(S)} \simeq 0.8$ dyne/cm^2 as the wave-coherent shear stress. If a *wave-induced* drift component is included, then $\tau_w^{(S)}$ would be somewhat higher, but certainly well below the 3.6 dynes/cm^2 implied by Okuda *et al.* (1977). In any event, it seems reasonable to conclude that both normal and tangential stress effects are operating concomitantly and these serve to maintain a significant level of input to short wind waves under saturation conditions. These input levels are consistent with the observation that very short wind waves, under moderate wind stress in wind-wave tanks, are almost continuously breaking (see, e.g., Fig. 1).

In the field, two scales of breaking waves can be observed. At short fetches or durations, the dominant wind waves, with wavelengths in excess of O (0.5 m), break as whitecaps with a reduced frequency of occurrence. Statistics on the level of breaking activity as a function of dimensionless fetch or duration do not appear to have been published. However, with increasing wavelength, the strength of the enhanced input at breaking waves rapidly becomes of decreasing importance, consistent with the $(u_*/c)^2$ dependence deduced above. The direct wave growth efficacy attributable to this effect would decrease appreciably and other mechanisms, such as shear flow instability and nonlinear spectral transfers, might be expected to become increasingly more important in determining the evolution of the wave spectrum. Under more fully developed conditions, with established swell and even light-to-moderate winds, the sea surface is covered with a microstructure of small-scale breaking waves generally in the low-centimeter wavelength range. These structures are periodically modulated by the swell-induced hydrodynamical and aerodynamic effects. This is the environment with which the incident electromagnetic waves interact to produce the backscattered signal in microwave remote sensing of the wave and surface wind fields. We conclude this section with some brief comments on the relevance of this work to remote sensing of the ocean using microwave backscattering techniques.

4.2. Relevance to Remote Sensing

For small-scale waves in open oceanic conditions, the lack of observations limits any immediate conclusions on the occurrence, distribution, and persistence of small-scale breaking waves. Again, casual observation indicates that when a swell is present, small-scale breaking waves are often to be seen, predominantly near the crests of the swell. Figure 2 of Banner and Phillips (1974) is not atypical of the small-scale wave field under moderate swell and wind stress conditions. In the absence of any data or adequate theory, quantitative conclusions as to the role(s) of small-scale breaking waves in this context are speculative. However, this work does establish a reasonable basis for locally enhanced wind stress in the vicinity of patches of small-scale wave breaking activity and suggests a tendency toward increased persistence through the demonstrated strong coupling to the wind. It is suggested that these effects may be linked to the unknown source of strong modulation of the microwave backscatter reported by Wright *et al.* (1980). They point out (p. 4963) that local (Bragg) wave growth rates are proportional to the local wind stress. The effects of localized wave breaking could well increase the local level of wind stress, which modulates the local Bragg wave spectral intensity and hence the local backscattering cross section. These effects clearly warrant further investigation.

5. CONCLUSIONS

The main result of this study is that wave breaking increases not only the local drag of the wind on the water surface but also increases the local level of wave-coherent momentum flux from the wind significantly. For small-scale (centimetric) breaking waves, it is argued that the persistence of breaking is increased, thereby increasing the persistence of the enhanced local wind stress. With wave growth rates directly proportional to the wind stress, the effects reported here seem to provide a physical basis for local wind stress enhancement. This has been suggested in Wright *et al.* (1980) to be a likely source of the unaccountably large microwave backscattering modulation levels they measured in the West Coast Experiment.

ACKNOWLEDGMENTS.   The continued support of the Australian Research Grants Committee is gratefully acknowledged.

REFERENCES

Banner, M. L., and W. K. Melville (1976): On the separation of air flow over water waves. *J. Fluid Mech.* **77**, 825–842.

Banner, M. L., and O. M. Phillips (1974): On the incipient breaking of small scale waves. *J. Fluid Mech.* **65**, 647–656.

Cox, C. S. (1958): Measurements of slopes of high-frequency wind waves. *J. Mar. Res.* **16**, 199–225.

Cox, C. S., and W. H. Munk (1954a): Statistics of the sea surface derived from sun glitter. *J. Mar. Res.* **13**, 198–227.

Cox, C. S., and W. H. Munk (1954b): Measurements of the roughness of the sea surface from photographs of the sun's glitter. *J. Opt. Soc. Am.* **44**, 838–850.

Gent, P. R., and P. A. Taylor (1976): A numerical model of the air flow above water waves. *J. Fluid Mech.* **77**, 105–128.

Okuda, K., S. Kawai, and Y. Toba (1977): Measurements of skin friction distribution along the surface of wind waves. *J. Oceanogr. Soc. J.* **33**, 190–198.

Plant, W. J. (1981): A relationship between wind stress and wave slope. *IUCRM Symposium on Wave Dynamics and Ratio Probing of the Ocean Surface*, Miami, May 1981.

Plant, W. J., and J. W. Wright (1977): Growth and equilibrium of short gravity waves in a wind-wave tank. *J. Fluid Mech.* **82**, 767–793.

Wright, J. W., W. J. Plant, W. C. Keller, and W. L. Jones (1980): Ocean wave-radar modulation transfer functions from the West Coast Experiment. *J. Geophys. Res.* **85**, 4957–4966.

DISCUSSION

PIERSON: How do your results relate to the present understanding of wave generation by wind and in particular that the Miles–Phillips theory *alone*, up to a few years ago at least, does not seem to explain wave generation by the wind?

BANNER: Our results were obtained from steady-state measurements and provide an accurate measure of the momentum flux to certain scales of breaking and nonbreaking waves under a very limited range of conditions.

When attempting to assess the significance of these results, the intermittent character of the wave breaking process needs to be considered carefully. It is not immediately clear how the integrated dynamical effect, taken over a breaking timescale, of the transient separation can be related to the steady-state effects we reported. Our study is intended to demonstrate that breaking waves have the potential to induce strong inputs, complementary to their traditional role as sinks. Transient breaking effects warrant further investigation on the basis of the strong effects reported for the steady case.

If a similarly strong input enhancement was operating in the transient situation, this would constitute an additional mechanism for wind-wave growth. It would be a locally "strong interaction" mechanism, operating when the Miles and Phillips mechanisms were inappropriate. In view of the inability of the latter to account entirely for the observed input levels and wave growth rates, it is possible that additional input due to wave breaking is operating under certain generation situations. It is evident that the effects should be strongest for centimetric waves where small-scale wave breaking occurs with higher probability.

YUEN: Did you observe, in the breaking wave cases, a point of reattachment of the separation streamline? We have calculations of *laminar* flow (thus only a model) over water wave to indicate that the points of separation and reattachment (if any) strongly affects the momentum flux.

BANNER: In these initial experiments, the kinematic structure of the mean air flow field was not investigated in any detail. However, according to fundamental notions as expressed in Fig. 2b of Banner and Melville (1976), we would not expect to find a well-defined mean reattachment point *on* the upwind *unbroken* surface. However, an internal reattachment point is certainly possible.

It is planned to investigate our flow field in much greater detail and vary the wind speed over a range of values.

GREEN: (1) It appeared that there was significant cross-tank variation in the tank flow, so I presume that there was three-dimensional air flow.

(2) The "natural sea" has a three-dimensional character that would create a three-dimensional coupling of the air field; the cross flow variability creates a "patchiness" in the stress that on the average is much smaller than the peak values.

BANNER: (1) The cross tank variation over the measurement plane was minimal. These variations will be published when the data set is complete. Certainly in the separated flow, three-dimensional disturbances were present, but we believe the *mean* field was closely two-dimensional.

(2) This is undoubtedly true. However, I pointed out that the significance of these large, localized strong interaction effects is likely to provide for strong modulation of the short scales and might be a primary factor in the hydrodynamics of radar imaging of long ocean waves.

23

Observations and Measurements of Air Flow over Water Waves

M. A. Weissman

Abstract. Flow separation has been found to occur within the turbulent boundary layer over short gravity waves ($\sim$ 10-cm wavelength) under low-wind conditions ($\sim$ 1 m/s) when the waves are not breaking (wave height/wavelength < 0.06).

1. INTRODUCTION

In order to understand better and model the interaction between wind and waves, we need to know something about how the air flows over the waves. For example, we would like to know whether there is a strong feedback from the waves to the wind, what are the pressure or shear forces that can induce wave growth, and which waves are contributing most to the overall surface shear stress and other transport processes. These things can be better understood if we know whether the air flow "separates" as it passes over the individual wave crests; that is, whether there is a relatively sheltered region in the lee of the crests in which the distribution of pressure and shear forces is quite different from the nonseparated case.

This idea was introduced by Jeffreys in 1925. By analogy to the flow around a sphere, he postulated that in the lee of wave crests there can be a "sheltered" region in which the flow, though turbulent, would be mainly composed of an eddy with a horizontal axis. This would give rise to larger pressures on the windward slopes than on the leeward slopes. This pressure difference, which Jeffreys assumed to be proportional to the wave slope and the square of the velocity difference between the air speed and the phase speed of the crests, contributes to the tangential stress on the surface ("form drag") and can do work on the waves themselves.

Jeffreys did not actually use the term "separation"; however, his concept that "...the air blowing over the waves may be unable to follow the deformed surface of the water.... the main air current, instead of merely flowing steadily down into the troughs and over the crests, merely slides over each crest and impinges on the next wave at some point intermediate between the trough and the crest" is consistent with modern usage. For example, if we take Batchelor (1967) as a guide, an important element of separated flow is "...the departure, from the neighborhood of the body surface, of those streamlines which lay within the boundary layer on the forward portion of the body...."

Of course, the type of "separation" relevant to the air–sea interaction problem is not that discussed by classical texts. Surface waves are "embedded" in the atmospheric boundary layer; that is, their wavelength (which governs the vertical scale of the perturbation) is at most the order

M. A. Weissman ● Microscience, Inc., Federal Way, Washington 98003.

335

of the boundary layer thickness. The "separation" that we are concerned with is that of the shear layer closest to the surface (the viscous sublayer if the waves are smooth), not the separation of the boundary layer as a whole. In addition, since the flow is turbulent, we are referring to the behavior of a time-averaged flow (in a frame of reference moving with a wave), not the actual fluid motions.

However, the concept still appears to be quite relevant for these flows. For example, for turbulent flow over a solid wavy wall, Zilker and Hanratty (1979), and others, have shown the occurrence of "separated flow" in the wave troughs. In other words, there exists, in the mean, a large-scale recirculating region and a high-vorticity layer that has come away from the wall, where it was generated by the no-slip condition. Points of detachment and reattachment can be identified. Their results show that many of the principles that we apply to laminar separation can be applied here: there is separation when the adverse pressure gradient becomes large, the outer flow responds to the new "surface" defined by the boundary of the recirculating region, and so on. Are things very much different when the waves are propagating?

In the steady frame of reference moving with a propagating wave, the surface is moving with speed $-c + u$, where c is the phase speed and u is the orbital velocity, which is generally much less than c. Some authors (Banner and Melville, 1976; Gent and Taylor, 1977) have argued that separated flow can only occur when the dividing streamline (which bounds the recirculating region) intersects the interface, as it does for a solid surface. This would require a stagnation point on the interface and could only occur if $u \geq c$, that is, when the wave is breaking. Thus, these authors conclude that flow separation is uniquely associated with wave breaking.

The requirement for a stagnation point on the surface is clearly not appropriate if the surface is moving. Telionis (1976) gives the example of flow around a rotating cylinder. If the Reynolds number is such that separation occurs when the cylinder is stationary, the addition of a small amount of rotation does not change the flow pattern appreciably. There are still large recirculating eddies in the "sheltered" regions. His flow visualizations confirm this. One can still identify a dividing streamline and a point (fixed in space) where the flow can be said to be separating, but, of course, the dividing streamline cannot intersect the surface. As demonstrated in Telionis's photographs, the dividing streamline now emanates from a stagnation point that lies within the fluid, away from the surface.

The flow over propagating waves is similar. When the surface is not moving ($c = 0$), there is flow separation for certain conditions. If the surface is slowly moving ($|c| > 0$ but small), we do not expect great changes in the flow. Indeed, Kendall's (1970) experiment shows that there is a continuous, smooth change in flow parameters as c changes from positive to negative. Therefore, we expect to find separation even if a stagnation point does not exist on the surface. Banner and Melville and Gent and Taylor may be correct that wave breaking is a sufficient condition for separation, but we shall see that it is not necessary.

Why do we wish to know whether the flow is "separating" or not? In the first place, we expect from our knowledge of other flows that the vertical transports of heat, mass, and momentum will increase substantially. For example, the form drag for a smooth wave will be greatly enhanced if the flow has separated.

Second, how we best model the flow depends upon whether separation is present. "Separation" implies the existence of a large-scale recirculating region, having scales of the order of the wave height and wavelength. If such a flow exists, linear models (e.g., Miles, 1957) will be inadequate. Linear calculations do give recirculating regions, but, as Gent and Taylor (1977) point out, they often lie very near the surface and are necessarily of very small vertical extent. Thus, they do not have a major perturbing effect on the outer flow.

Nonlinear models are necessary, but it is not clear whether the model need be "fully" or "weakly" nonlinear. If the wave slope is not too large, nonlinear critical layer theory may be applied (Benney and Bergeron, 1969; Davis, 1969). The resulting recirculating region has scales the order of the wave height, or larger, but calculations so far do not conform with our concept of separated flow. Davis found that if nonlinear effects dominate over viscous effects at the critical layer, the form drag vanishes.

For turbulence modeling, the concept of separation is also important. Models of turbulence may not be correct if there is large-scale flow recirculation. The model must also be careful how it treats the viscous sublayer. It is the sublayer that is separating over the individual crests, and this should be allowed for in any modeling effort.

The approach taken in the present experiment was to study the air flow as the wave amplitude is increased. For small waves we do not expect separation, for large waves we do, so the concern is that there may be a threshold amplitude beyond which separation occurs. The experiment was not entirely successful at illuminating this point, but measurements were made at the two extremes: over very-small-amplitude waves without separation and over large-amplitude (nonbreaking) waves which had separated flow.

The experiment had two parts. First, a visualization study was performed, using smoke as a flow tracer very near the water surface. This showed the existence and general nature of recirculating eddies in the lee of the crests under certain conditions. These conditions were then repeated with the tunnel fully instrumented to confirm the general picture and to determine the wave-induced velocities and the structure of the boundary layer.

After describing the experimental facility and these two parts of the study and summarizing the main results, we will return to the relationship of this work to other experiments.

2. THE EXPERIMENTAL FACILITY

The experiments were performed in the small wind/wave facility of the National Maritime Institute (see Fig. 1). The air channel, extending the length of the water tank, is 25.4×38.1 cm (10×15 in.) in cross section. The wind is provided by a blower fan and a conditioning section having five screens, two honeycombs, and a 9:1 contraction ratio (see Bradshaw, 1967). The free-stream turbulence level was typically 0.1%. The air flow passes through a connecting section, then over a trip rod and a splitter plate before making contact with the water. The trip rod was used to ensure that the boundary layer flow was turbulent. The wind speeds used were generally low enough (about 1 m/s) that if the rod were not in place, and waves were not generated, the boundary layer would remain laminar for some distance downstream.

The water tank is 4.88 m (16 feet) long and 63.6×38.1 cm (25×15 in.) in cross section. The sides of the tank extend above the water surface to provide the channel for the wind. At the far end of the tank there is a beach to absorb the waves and an overflow tank to aid in keeping the surface clean.

The wavemaker used was a simple flap (length 5 cm—half the typical wavelength) hinged on the edge of the splitter plate. This design was chosen to minimize flow interference as the wind makes contact with the water. The flap was driven by a moving-coil oscillator and sinusoidal motion was guaranteed by a feedback system using a linear position transducer. The design was not entirely successful. At the larger wave amplitudes, variations in wave heights of 10–15% were experienced; this was possibly due to turbulence produced in the water by the motion of the flap. (For accurate flow measurements, this necessitated a conditional sampling technique; see Section 4.)

3. VISUAL OBSERVATIONS

In the first part of the study, the flow very close to the surface was visualized. "Smoke" (condensed "Ondina" oil vapor produced by a Lede smoke generator) was introduced into the boundary layer near to the crests of the waves (see Fig. 2). The wind speeds and wave frequencies were chosen so that the flow pattern could be seen most clearly. For the smoke to remain reasonably coherent in the turbulent flow, wind speeds much in excess of 1 m/s could not be used. The fetch over which the smoke could be clearly seen was also limited to about 50 cm. Thus, the wavelength was chosen to be about 10 cm (4 Hz) in order that a reasonable number of waves were present between the smoke tube and the observation point.

Figure 2 shows the tube used to introduce the smoke. The wind and waves are moving from

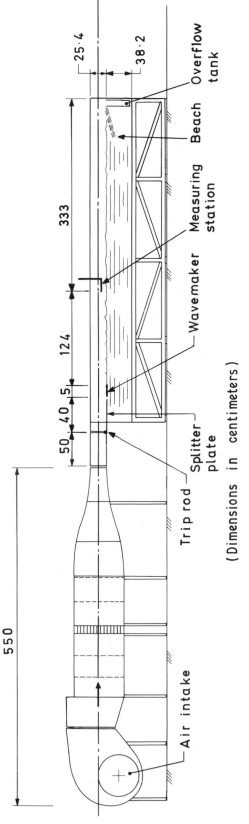

FIGURE 1. Wind tunnel and wave tank.

FIGURE 2. Smoke tube.

right to left and the photograph is taken from the side, looking slightly upstream. The main stem
of the tube (3-mm diameter) was angled at 45° to the vertical in a plane normal to the flow. This
design was chosen in order to minimize any effect the wake of the tubing might have on the
phenomena under investigation. If the tube were vertical, its wake might interfere with any
concentrations of horizontal vorticity produced by the flow over the wave crests; if the tube were
horizontal, it would create horizontal vorticity that may add to or subtract from that inherent in
the original flow. The smoke velocity at the exit of the tube matched the flow velocity at that level
as closely as possible. The smoke was illuminated by a vertical sheet of light, about 2 in. thick,
produced by flood lamps above the air channel.

The presence of separation could be detected by the observation of eddies moving with the
wave crests. For sufficiently large wave amplitude at a given wind speed and wave frequency,
recirculating regions could be seen just to the lee of the wave crests (on the forward face of the
waves). In order to observe these "trapped" eddies, one's eyes naturally followed a particular
wave crest. That is, one automatically put himself in the frame of reference moving with the
wave. A recirculating region is not unexpected, as discussed in the Introduction, but it was
obvious that the flow near the surface, where there must be a thin shear layer (the viscous
sublayer), did not follow the shape of the wave. In other words, the recirculating region was large
enough, of height comparable to the wave height, to deflect significantly the shear layer and to
change the pattern of the flow in the outer region. A "point of separation" could be identified with
a small region just upstream of the wave crests. For the larger amplitude waves, the smoke could
be seen to actually lift vertically off the surface at this point.

Figures 3a and b are photographs of the smoke pattern over waves of steepness* 0.05 and
0.07, respectively, when separation is occurring. The wind is from right to left. Still photographs
of smoke must, of course, be viewed with caution. Smoke marks fluid particles, and the
streakiness seen in a still photograph does not necessarily indicate the flow direction, especially
when the flow is turbulent. However, these photographs were chosen because the pattern
depicted by the smoke does agree with the flow as observed by watching *moving* smoke particles.
They demonstrate the flow separation near the crest and the extent of the recirculating region.
(Note the parallax in these photographs. The right-hand crests are viewed from an angle. The
nearly sinusoidal line in the foreground is the meniscus on the glass wall of the tank. Also, on the
surface there are some reflections of the smoke.)

A sketch of the flow pattern over the larger waves, as determined by "real-time" observation
and later study of slow-motion movies of the flow[†], is given in Fig. 4. This depicts the mean flow

*The steepness is the ratio of wave peak-to-trough height (H) to the wavelength (L).
[†] A 16-mm film was shown at the Miami meeting. It is available for loan from the author or Dr. J. A. B. Wills,
NMI.

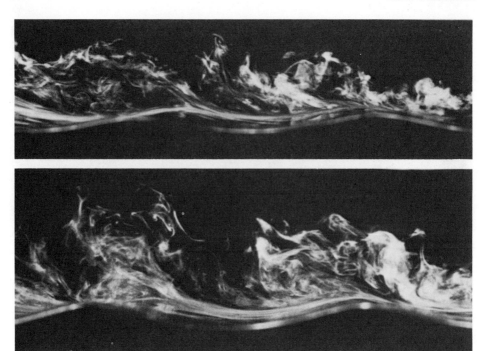

FIGURE 3. Separated flow over water waves.

MEAN AIR FLOW

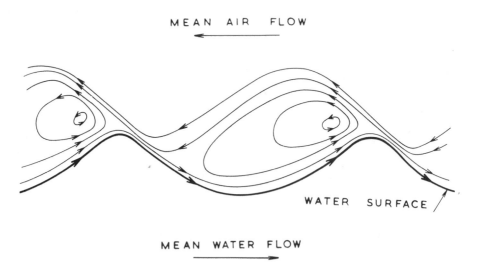

WATER SURFACE

MEAN WATER FLOW

FIGURE 4. Wave-induced flow as determined from smoke visualization. (The vertical scale is exaggerated by a factor of five.)

which is, of course, embedded in the turbulent boundary layer flow, which instantaneously has vorticity in all directions. The separation was unsteady; it did not occur on every crest, even at the higher wave heights. There may even have been shedding of the eddies. But it was striking how well the waves organized the flow. The separated flow and recirculating region readily established itself and became easily observable even in this highly turbulent situation.

4. VELOCITY MEASUREMENTS

4.1. Experimental Apparatus and Method

In the second part of the experiment, in order to confirm the picture of the flow as gathered from the visualization, measurements were made of the flow over the waves. The conditions used were the same as in the first part: free-stream velocity 1 m/s, wave frequency 4.17 Hz, and a range of amplitude. The aim was to determine the boundary layer structure and that part of the flow that was induced by the wave motion. The wave-induced flow was determined by phase-averaging.

At any particular fetch downstream, we assume that the horizontal velocity is composed of three parts: $U(z)$, the overall mean (z is the vertical position from the undisturbed interface); $\tilde{u}(z, \tau)$, a periodic component having the same period as the waves (τ is time as measured from a particular phase of the wave); and $\hat{u}(z,t)$, the turbulence. That is,

$$u(z,t) = U(z) + \tilde{u}(z,t) + \hat{u}(z,t). \tag{1}$$

Note that this really serves as a definition for the turbulence:

$$\hat{u} = u(z,t) - U(z) - \tilde{u}(z,t) \tag{2}$$

where

$$U(z) = \frac{1}{N} \sum_{n=0}^{N-1} \frac{1}{M} \sum_{m=0}^{M-1} u(z, n\Delta t + mT) \tag{3}$$

and

$$\tilde{u}(z, \tau) = \frac{1}{M} \sum_{m=0}^{M-1} u(z, \tau + mT) - U(z) \tag{4}$$

These definitions have been put in the form appropriate for the digital technique that was used; Δt is the sampling interval, T is the wave period, N is the number of samples per period, and M is the number of periods.

The measurements of horizontal velocity were made with a hot-wire anemometer (DISA model 55M). To guard against breakage if immersed in the water, the probe itself was specially made using nickel wire of 12-μm diameter and a current-limiting resistor was added to the anemometer circuitry (see Wills, 1975). The response of the probe was at least 8 KHz as determined by the standard square-wave test.

Since the air velocities were rather low, calibrations of the hot-wire were carefully performed in two ways. Frequent calibrations against a "minivane" anemometer (1-cm-diameter head) were performed, but since this had a limited range (above 0.6 m/s), the hot-wire system was also calibrated in the 18-in. calibration tunnel of the National Maritime Institute. King's law,

$$E^2 = A + BV^n,$$

fit very well over the range 0.1 to 1 m/s with an exponent $n = 0.4$. The minivane was also used to monitor the free-stream velocity, which was held at 1 m/s for all measurements.

The wave probe was of the conductivity type. A bare copper wire was suspended vertically through the surface; the conductivity between it and a ground in the water is linearly proportional to the wave elevation. The wire was frequently renewed and calibrated (statically).

The calibrations of the hot-wire and wave probe were greatly facilitated by the use of a Tektronix 4051 microcomputer, into which data were fed via an analog to digital convertor

(ADC). This computer was also used to monitor the wave amplitude in those cases where the waves were not sufficiently steady. When the change in amplitude was greater than a given amount (typically 10%), the computer would inhibit the collection of data.

The measurements of velocity were made with a single probe, positioned at successive elevations above the surface. Typically, 30 positions were used through the boundary layer, starting as close to the surface as possible without immersion. Thus, measurements were not made in the troughs (the visualization experiments had indicated a large flow disturbance above the crest level). Spacing between positions started at 0.5 mm near the surface and stretched to 1 cm in the outer part of the boundary layer (which was typically 6–8 cm thick).

The data were gathered in "blocks" of length slightly greater than a wave period. The ADC was triggered by a signal from an oscilloscope, which could be adjusted to trigger itself at any desired phase of the wave signal.* The sampling interval was 2 ms, allowing 120 samples of both velocity and wave elevation to be taken every wave period.

The timing, blocking, buffering, and transmission of these data were accomplished using the National Physical Laboratory's Data Communication Network and associated equipment. The data were "captured" by a minicomputer on the Network ("EDIT") and were then passed on to NPL's KDF9 for further processing. Figure 5 shows four typical blocks of raw (uncalibrated) data; the top trace is the hot-wire signal, the bottom the wave signal.

The hope was to study a range of wave amplitudes, from very small to large enough to cause separation. This was not entirely successful; only the very small amplitude case and the larger amplitude cases worked well. The intermediate cases experienced a nonuniformity in wave amplitude across the tank and also slowly with time. It is believed this was due to nonuniform secondary currents. The larger waves, having a strong drift current of their own, were not as sensitive to the underlying currents. The cases studied in detail are indicated in Table I.

*When waves were not present, an oscillator was used to provide a trigger.

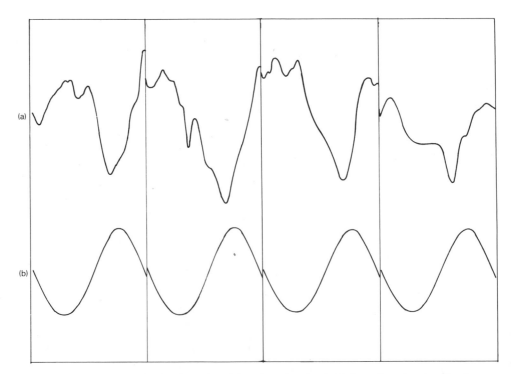

FIGURE 5. Four blocks of raw data: (a) the hot-wire signal; (b) the surface elevation signal.

TABLE I.   Cases Studied and Velocity Profile Results

Run	H^a (mm)	H/L^b	c^b (cm/s)	u_*^c (cm/s)	z_0^c (mm)	δ (mm)	δ_D (mm)	θ (mm)	R_θ
17	0	0	—			60	6.7	5.4	360
13	0	0	—			65	8.1	6.3	420
12	0.17	0.0019	39			64	8.0	6.2	413
15	4.35	0.046	40	7.2 ± 0.2	0.31 ± 0.02	70	13.1	9.5	633
16	5.4	0.057	40	8.25 ± 0.05	0.545 ± 0.005	70	11.9	8.4	560

[a]The peak to trough distance.
[b]L, the wavelength, and c, the phase speed, were determined from the frequency using the dispersion relation for finite-amplitude gravity-capillary waves. This compared well with visual observations.
[c]The variations in these quantities resulted from changing somewhat the range over which the logarithmic fit is applied.

4.2. Phase-Averaged Results

In all cases, the data were gathered in blocks and phase-averaged. If a further average is taken over a wave period, overall mean profiles are obtained, as discussed in the next subsection. Here we consider the phase-averaged results—the periodic part of the flow—for two cases: run 16, for which separation was occurring, and run 12, where it was not.

After linearization of the hot-wire data, the blocks of horizontal velocity signal were averaged together. Figure 6 shows some representative results (not all probe positions are

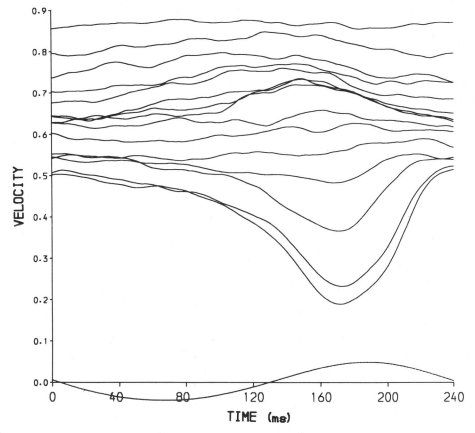

FIGURE 6.   Phase-averaged velocity, run 16.

plotted) for run 16. The vertical scale is the total horizontal velocity; successive probe positions are distinguishable because the mean velocity is increasing. Close to the surface the velocity has a large negative perturbation nearly in phase with the wave crest. This corresponds to the sharp downward peaks seen in Fig. 5, for which the hot-wire position is the same as the lowest trace in Fig. 6 ($z = 3.94$ mm). Note that the averaging over 100 waves has tended to broaden the peak. As we go away from the surface, the velocity perturbation over the peak changes sign. There is a phase change and a null in the horizontal velocity perturbation at an elevation of about 10 mm. A similar structure of the velocity field was found for run 15.

Another way of looking at the velocity field is presented in Fig. 7. Here vertical profiles have been plotted for successive phase points along the wave. The phase speed of the waves (0.40 m/s) has been subtracted from the velocities in order to convert to a frame of reference moving with the waves. Note that there is a region of negative horizontal velocity in this frame of reference. The profiles have also been plotted as if the wind were blowing from right to left. If we assume that the wave and wind fields are moving past the probe without change of form, we may transform from time to space using $x = -ct$. Thus, Fig. 7 gives us a *spatial* picture of the flow if x is positive to the left.

Having made the transformation to spatial coordinates, we may go one step further and calculate streamlines in this (steady) coordinate system. The streamfunction is

$$\psi(x, z) = \int_{z_0}^{z} U_p(x, y)dz + f(x), \tag{5}$$

but if z_0 is chosen to be at the outer edge of the boundary layer where the wave perturbation is

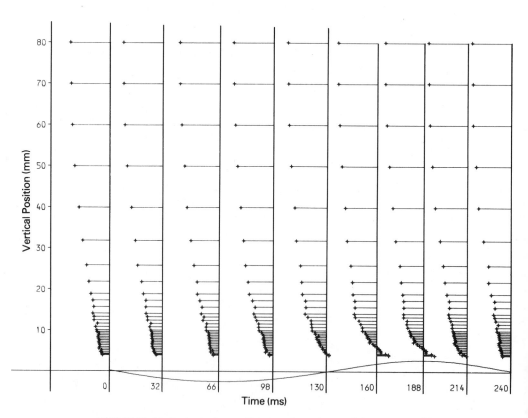

FIGURE 7. Vertical velocity profiles at various phases of the wave, run 16.

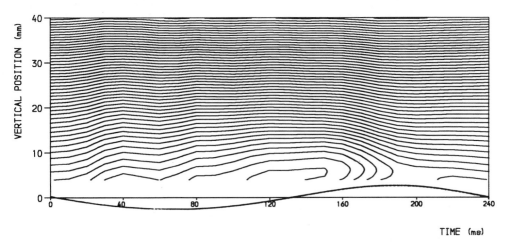

FIGURE 8. Streamlines, run 16.

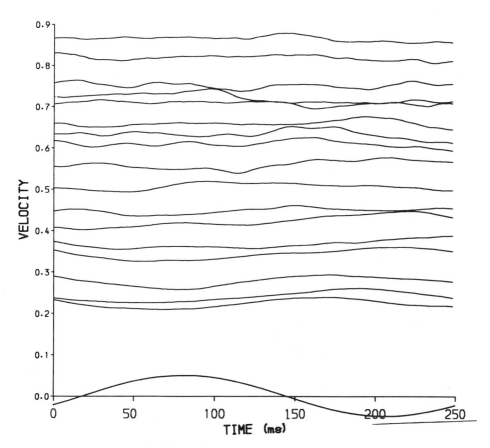

FIGURE 9. Phase-averaged velocity, run 12.

assumed to vanish,

$$f(x) \equiv 0$$

U_p is the phase-averaged velocity—mean plus periodic perturbation—such as plotted in Fig. 7. ψ was calculated by numerical integration using the trapezoidal rule (no smoothing was used) and then a contour plot of ψ was generated to obtain the streamlines. The result for run 16 is shown in Fig. 8 (the contours are at equal intervals of ψ). Run 15 gave a similar pattern.

The wave profile in Fig. 8 has been drawn to scale. We see that the area of recirculation extends to about two wave heights above the undisturbed surface elevation.* A "dividing streamline" could be identified emanating from a stagnation point away from the surface and slightly ahead of the wave crest. Note that this streamline would terminate at the same stagnation point over the next wave, not at a "point of reattachment." (This does not contradict Jeffreys's concept that the flow "impinges" on the windward slope of the next wave. There is probably much turbulent transport in this region and stress may be applied through the recirculation region onto the surface. That is, this part of the separation bubble is not well "sheltered.")

It is interesting to note the position of the "critical layer" in relation to the region of recirculation. From the overall mean profile of this case (i.e., a horizontal average of Fig. 7; see Figs. 10 and 11), the elevation where the mean velocity is equal to the phase speed is very near the bottom edge of the measurements in Fig. 8 (at about $z = 4$ mm). The core of the eddy is close to this level, but perhaps 1 or 2 mm higher (at 5 or 6 mm). Note that the phase change, as seen in Fig. 6, occurs much higher (at about 10 mm), near the outer edge of the recirculating region.

In contrast to this large-amplitude case, flow over a very small wave was also measured. Figure 9 shows the phase-averaged velocity field for run 12. (Note the wave profile is highly exaggerated and the trigger point is at a different phase of the wave as compared to Fig. 6.) The velocity perturbations are much smaller, but a definite wave-induced perturbation can be seen close to the surface. As we go more than about 4 mm away, the signal becomes lost in random perturbations. (This is due to the residual turbulence that would presumably average out if a larger ensemble of waves were used.) There is no evidence for a phase change as we go away from the surface.

Streamlines were also calculated for this case. They also showed a region of recirculation but it was very small and close to the surface. The contour plot did not have sufficient resolution to show it clearly. However, Fig. 9 indicates that the elevation of its core must have been at about 2.7 mm because the velocities below this level are less than the phase speed. (The wave height in this case was about 0.2 mm.) This is very close to the height of the critical layer as determined from the mean velocity profile (Fig. 10 or 11).

4.3. Overall-Averaged Results

To obtain the mean boundary layer structure, the velocity measurements were averaged over all time for each position above the surface. Figure 10 through 13 show the mean profiles, the wave-induced rms, and the turbulence rms for the five cases as given in Table I.

The mean profiles are quite different for the nonseparating cases (17, 13, 12) and the separating cases (15, 16). Figure 10 gives the profiles on a linear–linear plot and Fig. 11 shows them on a linear–log plot in order to test for the "law of the wall." Logarithmic profiles could not be fit to the former cases (because of the low Reynolds number), but in the separating case, logarithmic regions were well established, extending nearly to the outer edge of the boundary layer. Table I gives the standard boundary layer parameters: the friction velocity (u^*), the roughness length (z_0), the boundary layer thickness at the 99% point (δ), the displacement

---

*These are *mean* streamlines. As mentioned above, the formation of eddies was unsteady. When present over an individual wave crest, the recirculation region probably extends to a greater height than indicated in Fig. 8.

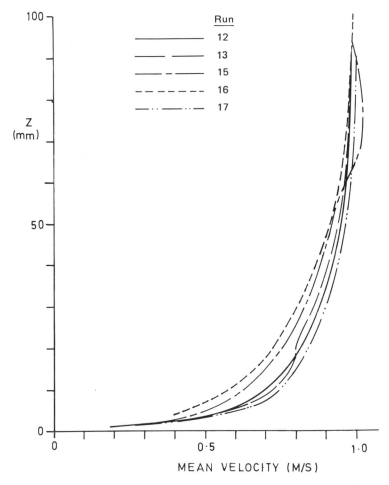

FIGURE 10. Mean velocity profiles; linear–linear plot.

thickness (δ_D), the momentum thickness (θ), and the Reynolds number based on the momentum thickness (R_θ). Assuming the friction velocity to have the usual relationship to the surface shear stress, the drag coefficient turns out to be

$$C_D = \left(\frac{u^*}{U_{FS}}\right)^2 = \begin{cases} 5.2 \times 10^{-3} \text{ for run 15} \\ 6.8 \times 10^{-3} \text{ for run 16} \end{cases}$$

The surface stress could not be measured in the nonseparating cases but the change in momentum thickness is a good indication that the stress is a factor of about 1.5 greater when separation is present.

The rms wave-induced perturbation is defined as

$$\left(\overline{\tilde{u}^2}\right)^{1/2} = \left(\frac{1}{N} \sum_{n=0}^{N-1} [\tilde{u}(n\Delta t)]^2\right)^{1/2}$$

where $\tilde{u}$ is defined by Eq. (4) and the bar is an average over a wave period. Figure 12 shows that the separated flow has maxima near to and away from the surface; the null at about 1 cm agrees with the position of the phase change found in the phase-averaged results. The extent of the wave-induced flow (about 4 cm) also agrees with that found in the phase-averaged case (cf. Fig. 8).

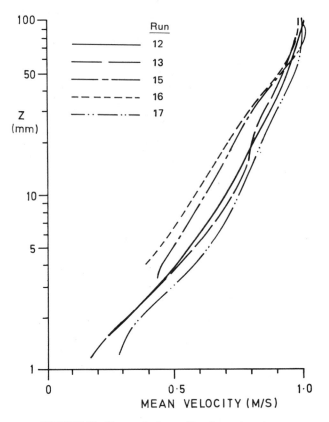

FIGURE 11. Mean velocity profiles, linear–log plot.

The nonseparating cases show a nearly constant background level which is due to the turbulence that has not averaged out. Cases 17 and 13 (no waves) should, of course, average to zero and case 12 (very low waves) should show a small wave-induced motion that falls off roughly exponentially (some evidence of this can be seen very close to the surface in Fig. 12), but, because of the large turbulence levels (cf. Fig. 13), a much longer averaging period is needed to achieve this.

The "turbulence" has been defined as what is left over after the mean and periodic velocities have been subtracted from the instantaneous velocity, i.e.,

$$\hat{u} = u - U - \tilde{u}$$

Therefore, the rms turbulence can be shown to be identically

$$\left(\overline{\hat{u}^2}\right)^{1/2} = \left(\overline{u^2} - U^2 - \overline{\tilde{u}^2}\right)^{1/2}$$

by definition of the other quantities. This quantity has been calculated and plotted in Fig. 13. There are significant differences between the nonseparated and the separated cases. Although the maximum value is roughly the same, the peak has shifted away from the surface and is much broader. In the outer part of the boundary layer, the turbulence level has increased by 50%. Near the surface, this component of the flow, the random part that is not phase-locked to the waves, has actually decreased.

This component *is* the random part of the flow, but it is also closely related to the wave-induced part. In the visual part of the experiment (Section 3), the separation "bubble" could be

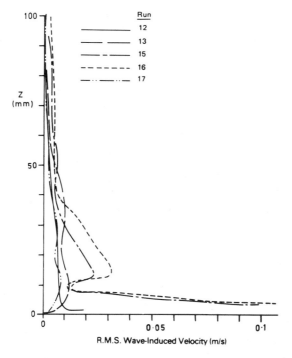

FIGURE 12. Wave-Induced velocity profiles (rms).

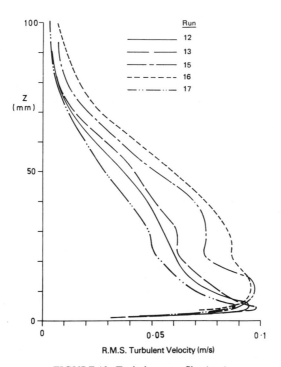

FIGURE 13. Turbulence profiles (rms).

seen to be unsteady. This unsteadiness, even though it is part of the organized structure that is wave-induced, will appear as "turbulence" in this measurement.

5. SUMMARY

The main results of this study are as follows:

- Flow separation—in the Jeffreys–Batchelor sense—can occur over water waves that are not breaking.
- For the conditions studied here—$H/L = 0.05$–0.06, $\delta/L = 0.7$, $R_\theta \cong 600$, $c/U_{FS} = 0.4$ (or $c/u_* = 4.8$–5.6)—the recirculation region, on average, extended two wave heights above the undisturbed water level and stretched from one wave crest to the next.
- From measurements of boundary layer momentum thickness, the increase in total surface stress when separation was present was roughly 50%.
- The level of boundary layer turbulence (i.e., the part of the flow that is not phase-locked to the wave) was also found to increase by about 50%.
- The wave-induced horizontal velocity perturbation extended to about half the boundary layer thickness. It was 180° out of phase with the wave crests near the surface but went through a phase change near the outer edge of the recirculating region.
- Separation did not appear to be present for very-low-amplitude waves; therefore, the occurrence of separated flow could be a function of wave height.
- The presence of the separation bubble affected the entire boundary layer profile.

6. RELATION TO OTHER EXPERIMENTS

One other recent work has reported separated flow over nonbreaking water waves. Kawai (1981) used the smoke-wire technique to visualize the flow over wind-generated waves at moderate wind speed (6 m/s). Observations were made over 41 crests, and separated eddies were present roughly 50% of the time. Separated flow was observed for $H/L = 0.05$–0.1, but nonseparated flow was also found over that range of steepness. Thus, the formation of eddies does appear to be intermittent. This range of steepness is similar to the present experiment, but Kawai's ratio of phase speed to wind speed is much smaller: $c/U_{FS} \cong 0.075$ and $c/u_* \cong 1.2$. The Reynolds number and the thickness of the boundary layer were not reported.

Other experiments have reported evidence of flow separation, but either the waves were induced to break (Banner and Melville, 1976) or the waves were naturally generated by the wind, under which conditions they were probably breaking a fair percentage of the time (Chang et al., 1971; Okuda et al., 1977). Their results are not inconsistent with those presented here.

Banner and Melville report a 40- or 50-fold increase in shear stress when their wave is breaking. This could indicate that the breaking phenomenon is very important for the overall air–sea interaction, but this is a question distinct from whether the flow is separating or not. Note that their experimental conditions for the stress measurements were quite different from those used here. Their ratio of phase speed to free-stream speed (in a fixed frame of reference) was about 0.17 and their ratio of boundary thickness to wavelength was 0.135. (Thus, their waves cannot be said to have been "embedded" in the boundary layer.)

Other measurements of the air flow over nonbreaking waves have also been made. The authors did not draw any conclusions regarding whether the flow was separating but their results are very similar to those reported here.

Kendall's (1970) measurements of flow over a wavy wall (his Fig. 7) show the same characteristics as Fig. 6 here. His flow conditions were close to those used here: $c/U_{FS} = 0.55$ and $H/L = 0.0625$.

Takeuchi et al. (1977) measured flow over water waves under a range of flow conditions. For parameter values similar to those used here ($c/U_{FS} = 0.35$ and 0.63 and $H/L = 0.03$), their velocity traces (their Figs. 9c and 9d) are again similar to Fig. 6. Their measurements were for fixed wave height as the free-stream velocity was increased, and their mean velocity profiles (their Fig. 2) fall into two groups depending on the value of c/U_{FS} (approximately 0.7 appears to be a threshold value). This is similar to the present result that the profiles are very different over small-amplitude and large-amplitude waves.

Finally, Kondo *et al.* (1972) made measurements over ocean waves. They also found that, for any particular frequency component, the phase relation between the horizontal velocity and the sea surface elevation changed from 180° near the surface to 0° away from the surface. The phase change occurred at the elevation where $U = 1.5c$. This agrees almost exactly with the present measurements.

These three experiments plus the present one show similar flow characteristics even though they had widely differing physical scales. This is an indication that the Reynolds number, however it is defined, might not be important once separation of the air flow has occurred. The important parameters appear to be c/U_{FS} (or c/u_*) and H/L. The importance of δ/L (0.7–0.75 here and in Kendall's experiment, 0.2 for Takeuchi *et al.*, and presumably quite large for Kondo *et al.*) is not clear at present.

The Reynolds number can still be expected to play a role in determining the onset of separated flow—assuming such a distinction can be made. Expressed as

$$R = Hu_*/\nu,$$

this compares the wave height to the thickness of the viscous sublayer. If the onset of separation can be thought of as the conditions when the sublayer no longer remains attached, this ratio should be important.

The picture is emerging that there may be three regimes for the flow over water waves:

1. A "small amplitude" stage in which linear or weakly nonlinear theory may be applied
2. A "separation" stage in which separated flow is occurring over nonbreaking waves
3. A "breaking wave" stage in which the air flow is somehow strongly influenced by the breaking process.

The delineation between these regimes is unknown at present. Even the parameter suite needed to define the regimes is unclear [but certainly c/U_{FS} (or c/u_*) and H/L must play important roles]. Much work remains in defining the characteristics of the regimes and the boundaries between them and then in applying the model to environmental conditions.

ACKNOWLEDGMENTS. This work was performed during the period March 1977 to February 1978 as part of the research program of the National Maritime Institute, Feltham, England. The assistance of Mr. Simon Tindall, Ms. Dorothy Goodman, and the rest of my former colleagues at the NMI (especially Drs. Davies, Gaster, Matten, Miller, and Wills) is gratefully acknowledged. I also appreciate the help provided by the Technical Services Department of Flow Industries, Inc., Kent, Washington.

REFERENCES

Banner, M. L., and W. K. Melville (1976): On the separation of air flow over water waves. *J. Fluid Mech.* **77**, 825–842.

Batchelor, G. K. (1967): *An Introduction to Fluid Dynamics*, Cambridge University Press, London.

Benney, D. J., and R. F. Bergeron (1969): A new class of nonlinear waves in parallel flows. *Stud. Appl. Math.* **48**, 181–204.

Bradshaw, P. (1967): A low-turbulence wind tunnel driven by an aerofoil-type centrifugal blower. *J. R. Aeronout. Soc.* **71**, 132–134.

Chang, P. C., E. J. Plate, and G. M. Hidy (1971): Turbulent air flow over the dominant component of wind-generated water waves. *J. Fluid Mech.* **47**, 183–208.

Davis, R. E. (1969): On the high Reynolds number flow over a wavy boundary. *J. Fluid Mech.* **36**, 337–346.

Gent, P. R., and P. A. Taylor (1977): A note on 'separation' over short wind waves. *Boundary-Layer Meteorol.* **11**, 65–87.

Jeffreys, H. (1925): On the formation of water waves by wind. *Proc. R. Soc. London Ser. A* **107**, 189–206. See also *Proc. R. Soc. London Ser. A* **110**, 241–247 (1926).

Kawai, S. (1981): Visualization of airflow separation over wind-wave crests under moderate wind, *Boundary-Layer Meteorol.* **21**, 93–104.

Kendall, J. M. (1970): The turbulent boundary layer over a wall with progressive surface waves. *J. Fluid Mech.* **41**, 259–281.

Kondo, J., Y. Fujinawa, and G. Naito (1972): Wave-induced wind fluctuations over the sea. *J. Fluid Mech.* **51**, 751–771.

Miles, J. W. (1957): On the generation of surface waves by a shear flow. *J. Fluid Mech.* **3**, 185–204.

Okuda, K., S. Kawai, and Y. Toba, (1977): Measurements of skin friction distribution along the surface of wind waves. *J. Oceanogr. Soc. J.* **33**, 190–198.

Takeuchi, K., E. Leavitt, and S. P. Chao (1977): Effect of water waves on the structure of turbulent shear flows. *J. Fluid Mech.* **80**, 535–559.

Telionis, D. P. (1976): Critical points and streamlines in viscous flows. Virginia Polytechnic Institute and State University, Report No. VP1-E-76-28.

Wills, J. A. B. (1975): A submerging hot-wire for flow measurement over waves. National Physical Laboratory, Division of Maritime Science, Technical Memo 107.

Zilker, D. P., and T. J. Hanratty (1979): Influence of the amplitude of a solid wavy wall on a turbulent flow. Part 2. Separated flows. *J. Fluid Mech.* **90**, 257–271.

DISCUSSION

DONELAN: The last two papers have stimulated the controversy over whether or not flow separation occurs over water waves. Although on the face of it the papers appear to contradict each other, it may be that their results are not inconsistent if one allows for the possibility that the separated zone may grow and collapse intermittently as, in fact, was suggested by Weissman's film. This would account for Kawai's observation which captured the flow pattern at discrete intervals over wind waves and revealed separated and nonseparated flows about equally frequently and without any clear dependence on wave slope—all external conditions being constant. The long averaging time of Banner and Willoughby's measurements reveal mean surface pressure patterns which are suggestive of separation occurring much more frequently in the case of breaking waves than otherwise. Since we cannot assume that the phenomenon is steady, it cannot be said that separation occurs in one case and not in the other.

WEISSMAN: I agree with Donelan's remarks. Since it is embedded in a turbulent boundary layer, the separation pattern itself can be expected to be intermittent and it may be present a greater percentage of time when the waves are breaking. However, it is proper and necessary to determine what the *average* flow is so that one can build a model that can be applied to field conditions. There is much to be done in terms of learning what the flow is as a function of the important nondimensional scales of the problem.

BANNER: I basically concur with Dr. Donelan's assessment. I would add the comment that kinematic and dynamical considerations applied to the wind flowing over a breaking wave should guarantee a condition of local (in space and time) air flow separation over that wave as long as it continues to break actively. Our initial measurements indicate the relatively strong "equivalent steady state" dynamical effects which might be expected to accompany these local breaking events.

LONGUET-HIGGINS: It would be interesting to see how the effective drag coefficient varies as a function of the wave steepness. Did you determine the coefficient for any other values of the steepness?

WEISSMAN: I only have two data points at present. For $H/\lambda = 0.046$, $C_D = 5.2 \times 10^{-3}$ and for $H/\lambda = 0.057$, $C_D = 6.8 \times 10^{-3}$. The other important parameters for these cases were $\delta/\lambda = 0.7$ and $c/u_* = 0.4$ (δ = boundary layer thickness).

24

MEASUREMENTS OF WAVE-INDUCED PRESSURE OVER SURFACE GRAVITY WAVES

D. HASSELMANN, J. BÖSENBERG, M. DUNCKEL, K. RICHTER, M. GRÜNEWALD, AND H. CARLSON

ABSTRACT. In the summer of 1977 an experiment was conducted to measure the fluctuating pressure over surface gravity waves. Instruments were mounted on a slim mast, located 27 km offshore, in the North Sea. Instrumentation consisted of two static pressure probes, designed and provided by R. Snyder, two resistance wires and an underwater pressure sensor. Mean wind speed and direction were also measured. The sampling rate was at least 2 Hz for all instruments. The analysis shows that the results obtained by Snyder et al. (1981) can be carried over to conditions encountered in this experiment, which are more representative of open ocean conditions.

1. INTRODUCTION

In this progress report we present preliminary results of an experiment conducted in June 1977 in the North Sea, 27 km west of the island of Sylt. One of the aims of this experiment was to compare our measurements of atmospheric pressure over waves with those obtained by Snyder et al. (1981) in their Bight of Abaco (BoA) experiment, under conditions not nearly as rough as in the North Sea. When properly scaled, ours and the BoA experiment were very similar and one of the few questions left open by Snyder et al. (1981) was whether one may apply such scaling and transform BoA results to obtain North Sea results. One thing was, however, obvious: a simple scaling transformation on the experimental apparatus was not available, and we could not carry out such an extensive and sophisticated experiment as could Snyder et al. (1981). We refer the reader to Snyder et al. (1981) for a discussion of their experiment, an excellent summary of the present state of knowledge, and further references (see also Hsiao and Shemdin, 1983).

Suffice it here to mention the main differences between the two experiments:

1. Our array is smaller, yielding less directional resolution
2. We did not operate a wave-follower
3. Because of the tide range and the higher waves, even the fixed pressure sensors could not be mounted as close to the mean water level as in the BoA experiment.

D. HASSELMANN ● Meteorologisches Institut, Universität Hamburg, Hamburg, West Germany. J. BÖSENBERG AND M. DUNCKEL ● Max-Planck-Institut für Meteorologie, Hamburg, West Germany. K. RICHTER, M. GRÜNEWALD, AND H. CARLSON ● Deutsches Hydrographisches Institut, Hamburg, West Germany.

The standard analysis presented in this chapter rests entirely on spectra and crosspectra. The alternative (or complementary) analysis strategy, to search for typical events (associated with, say, wave breaking), has been postponed and not yet been taken up.

2. SITE AND INSTRUMENTATION

2.1. Instrument Support

The measurements were performed at station 8 of the JONSWAP array (Hasselmann et al., 1973; Günther et al., 1979), where the water depth is about 18 m. Instrument support was provided by a mast, made of cylindrical sections (maximum diameter $d = 2.00$ m at the bottom, minimum diameter $d = 0.62$ m at the top flange, typically 1.50 m above mean water level). A slim mast ("needle") of 9-m length is mounted on the flange of the main mast. The needle has diameters of 0.30 m at the bottom and 0.20 m at the top. The needle then in turn carries spars of 2-m length, which carry the instruments.

2.2. Logistics

RV Gauss served as supporting base for maintenance and repair and lay about 500 m away (crosswind) from the needles. All needle data (i.e., all data except for the pitch and-roll buoy data which were obtained by an independent system) were transmitted by telemetry to Sylt where they were monitored and recorded on tape. An additional monitoring was provided by an extensive first analysis on a larger computer in Hamburg, resulting in a typical delay of 2 days between measurements and control. While not exactly on-line, this control of the experiment was very useful. Finally, RV Regulus maintained connections between RV Gauss and Sylt.

2.3. Instrumentation

Instrumentation consisted of three cup anemometers (2 Hz) at heights of 8.30, 4.10, and 1.10 m above the mast's flange, a wind vane (2 Hz; 8.20 m above flange), two pressure probes (10 Hz), two resistance wires (2 Hz), and an underwater pressure probe (10 Hz) for an additional wave height measurement. Furthermore, a pitch-and-roll buoy (2 Hz) was deployed from nearby RV Gauss. In addition, turbulent velocities were measured, but these shall not concern us here.

2.3.1. Pressure Measurements

The pressure probes were kindly given to us by R. Snyder and are identical in design to those used in Snyder et al. (1981).

The performance characteristics were checked in a wind tunnel (to test for possible manufacturing deviations) and found to agree with those given in Snyder et al. (1974).

The pressure transducer was a Digiquartz (Model 215-A, Paroscientific Sales Inc.). Signal processing with specially developed high-frequency electronics led to absolute (rather than differential) pressure measurements with a resolution of 0.05 Pa (0.5 μbar). Laboratory tests showed the system to be stable with negligible short-term (for times on the order of 10 h) drift. Over longer periods, a long-term drift of about 30 Pa/month at 10^5 Pa, representing a relative variation of 0.0003/month, was observed. Dynamic calibration was performed using a Barocel Type 581 fast-response differential pressure gage as references. The frequency transfer function of the entire system, including pressure probe, shaft, and transducer, is flat and the phase shift is less than 0.3° for frequencies below 5 Hz. As in Snyder and colleagues' arrangement, the pressure probes were mounted on adequately responding wind vanes, thus limiting the angle of attack to small angles (probably less than 9°).

The pressure probes could be mounted at different heights. Normally, they were mounted with about 2-m vertical and less than 10-cm horizontal separation. The height above mwl varied with the tide, but a typical height for the lower instrument is 2 m above mwl. The closest distance to mwl was 0.75 m and it was unsafe to operate at lower levels.

2.3.2. Wave Height Measurements

The resistance wires hung from two diametrically opposed spars and were thus about 4 m apart. They were kept taut by a weight.

Ideally, one of the wires would hang vertically beneath the pressure sensors, but on occasion the wires were observed to slant due to the drag of the current. This causes a mean displacement of the resistance wires from their ideal position which sometimes was observed to be as large as 30 cm. We were unable to measure the mean displacement or even to regularly estimate it by eye. This introduces a considerable element of uncertainty into our analysis of pressure–wave phase shifts. (The apparently obvious solution to the problem, to fasten the wires to the bottom, was not practical, because we had to turn the spars, so that they would be orthogonal to the wind, and also because experience had shown that in such an arrangement the wires would normally snap after a short time.)

The third wave-measuring device, the underwater pressure sensor, was mounted close to the mast (20-cm separation) 3.60 m below the flange.

For all wave measurements we found the system normally to have constant gain (in a few cases the gain factors of the wires did drift by a few percent over periods of 2 h) but a not so constant offset. The offset would sometimes slowly drift, but the usual type of error was a sudden jump to some other level. This was not a frequent error (typically once a day), and thus did not normally interfere with the fluctuation measurements. The rather unexpected offset behavior did, however, cause problems in the determination of the mwl. Our mwl measurements are therefore only accurate to within $\pm$ 6 cm, but this is sufficient for our purposes.

The resistance wires actually produced two output signals: an electronically (Butterworth filter) filtered 2-Hz series, with which we shall be concerned here, and a high-frequency output. The filters were tested before and after the experiment and found to have remained stable to within experimental accuracy. We are thus confident that we have neither aliasing problems nor erroneous phase shifts in our wave data (except for the above-mentioned displacement error).

The pitch-and-roll buoy data were recorded on tape on RV Gauss. While the data were of good quality, severe interfacing problems had to be overcome before they could be entered into the computer. This resulted in a delay of the pitch-and-roll data analysis, so that it was not fully integrated into the analysis scheme. However, the pitch-and-roll data gave a very useful after-the-fact check of the directional analysis obtained from the needle data.

2.4. Disturbing Influences of Mast and Needle

We do not believe the needle to have severely distorted the atmospheric flow. As concerns the waves, however, our mast is a far more effective reflector than the radio masts used by Snyder *et al.* (1981).

The phase shifts in the wave field introduced by scattering are easily calculated. However, the effect of these scattered waves on the atmospheric pressure field is not so readily estimated, even if potential flow is assumed. Presently available estimates indicate that the phase shift will not exceed 5° for waves aligned with the wind and 90° scattering angle and such values will only be found at frequencies close to 0.5 Hz. However, there may be situations such as when in light winds the higher-frequency waves are not aligned with the wind and the mean scattering angle differs appreciably from 90°. It is conceivable that in such situations the corrections due to scattering are more serious, but final conclusions must await the outcome of the calculations.

3. DATA ANALYSIS

3.1. Spectral Analysis

We introduce the notation P_1, P_2 for the two atmospheric pressure signals (10 Hz), ζ_1 for the underwater pressure signal (10 Hz), and ζ_2, ζ_3 for the resistance wire signals (2 Hz). The environmental measurements U_1, U_2, U_3, Θ_u (angle) were simply averaged. The other data were all spectral analyzed according to the same scheme:

1. Short intervals with missing data (transmitting failures) were interpolated.
2. For an analyzing period T the data were multiplied by a cosine-bell function $(2/3)^{1/2}$ $[1 - \cos(2\pi t/T)]$ in order to reduce drift.
3. The whole stretch of data was transformed (FFT) and subsequently spectral estimates were obtained by (top hat) band averages in obtain 32 degrees of freedom for a bandwidth Δf.

The 10-Hz series was analyzed twice: first with $T = 2^{14} \times 10^{-1}$ s ≈ 27.5 min, $\Delta f = 5/512$ Hz; then the 10-Hz data were filtered to a 2-Hz series which was analyzed with $T = 2^{12} \times 1/2$ s ≈ 34 min, $\Delta f = 1/128$ Hz. The high-frequency spectra were not extensively used but were useful for quality control. A typical example is shown in Fig. 1. The spectra from the 2-Hz series

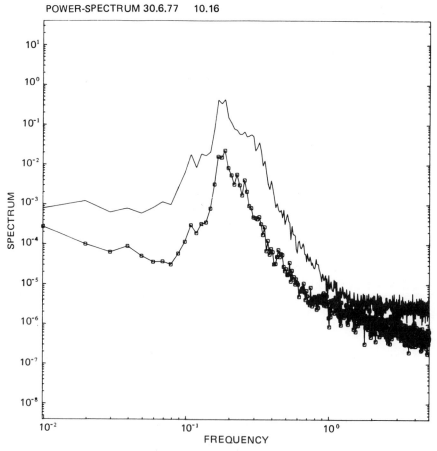

FIGURE 1. Upper and lower curves demonstrate the extent to which the atmospheric pressure spectrum ($\square$) is dominated by wave-induced pressure. The spectrum of the underwater pressure signal (solid) was converted to a length by $p/\rho g$ and the spectrum is given in m²/Hz, the atmospheric pressure in mbar²/Hz (10^2 Pa = 1 mbar), run 45.

formed the material for the further detailed analysis and in fact in this analysis the bandwidth was widened to 5/128 Hz (mainly in order to speed up the analysis).

3.2. Data Rejection

After analyzing our roughly 100 h of data we had about 200 spectral matrices, each size 5×5 (instruments) $\times$ 128 (frequencies). About 30% of the runs were eliminated for one of the following reasons.

1. During the run a sudden change of offset occurred in one of the wave measurements.
2. The wire became entangled.
3. Some of the pressure measurements showed a large broad peak at about 3 Hz. We are certain that this is an instrumental error but have been unable to find its origin (flow distortion due to the mast can be ruled out by comparison with the other pressure sensor). In most cases, however, the low-frequency spectra were not affected (as indicated by comparison with the other pressure measurement and the high coherence with the wave measurements). In some cases the high-frequency peak was extremely large and broad and there was substantial evidence that the low-frequency parts of the spectrum were unreliable. Such cases were rejected.
4. The lower pressure probe had been wetted by a wave. This certainly occurred in one run at $U = 12$ m/s (not shown in Table I). At the end of this run, occasional wetting of the lower probe was observed (protocol entry). The spectra were obviously distorted. The spectra obtained from the higher probe were also very strange and while we are not sure of the reason (perhaps port blockage due to spray), we considered these data too suspect for further analysis.
5. There were cases in which the measurements were sound, but in which some particular geometry for testing purposes had been selected, which was not normally well suited for the standard analysis.

3.3. General Comments on the Data Set

In Table I we have listed a few parameters which are pertinent to the 129 runs selected for the further analysis.

In all runs a considerable swell component was present and in many cases the swell components were the dominating feature of the spectrum. There were hardly any cases in which the spectra looked even similar to the JONSWAP -spectrum (Hasselmann et al., 1973), and in many cases the spectra were very bizarre due to the presence of different swell components. Inspection of Table I shows that while the sea state was definitely higher than in the BoA experiment, we nevertheless had relatively calm seas for North Sea standards, which explains our anomalous spectra.

In the swell region the pressure spectra were clearly dominated by wave-induced pressure, the spectra often rising by factors of 500–1000 over some roughly extrapolated background level. In this range the coherence between waves and pressure was typically above 0.9. For frequencies beyond, say $f \approx 0.3$ Hz, the coherence rapidly dropped to values on the order of 0.3–0.5.

4. DETAILED ANALYSIS

We have adopted the analysis strategy of Snyder et al. (1981) (although details of course differ), both to make a comparison of the results easier, and because we believe the strategy to be sound. We shall use the notation of Snyder et al. (1981). Our analysis consists of three steps:

1. Estimation of the directional wave spectrum

TABLE I. Run Summary^a

Day	Time	N^b	f_p^c	f_{swell}^d	U^e	f_u^f	Θ_u^g	Θ_p^h	Θ_{swell}^i	H_s^j	^{m}	Z_{lower}^k	Δz^l
13.6	10.25–15.10	1–8	0.35–0.28	0.14	3.86–7.21	0.4–0.22	230–260	210–170	160–120	0.32–0.23	NJ^m	2.80–3.60	2.96
13.6	16.14–17.57	9–12	0.43	0.15/0.25	5.68–6.69	0.27–0.23	260	270–10	150/10	0.25–0.28	NJ	1.80–1.90	2.34
14.6	10.54–12.02	13–14	0.34	0.12	4.62–2.91	0.34–0.53	222	237	160	0.36–0.34	NJ	2.39–2.43	0.26
14.6	16.52	15	0.24	0.12	5.28	0.30	230	170	145	0.27	NJ	2.93	0.26
14.6	18.17	16	0.23	0.12	5.70	0.27	235	175	153	0.25	NJ	0.85	2.34
15.6	13.19–14.27	17–18	0.32–0.28	0.10	7.46–7.11	0.21–0.22	155	163–149	71	0.64	NJ	4.82–4.92	0.88
21.6	10.16–18.08	19–24	—	0.10/0.14/0.20	1.09–4.58	1.43–0.34	30–110	—	120 ± 10 / 110–130	0.80–1.12	NJ	2.51–4.00	2.96
21.6	20.20–21.28	25–26	—	0.10/0.14/0.17	5.53–5.09	0.28–0.31	31–123	—	113	0.82–0.79	NJ	1.43–1.61	2.34
22.6	9.14	27	—	0.14/0.17	7.05	0.22	131	—	130	1.17	NJ	2.60	4.26
22.6	9.48–12.36	28–31	—	0.14/0.18	6.20–5.50	0.25–0.28	130–120	—	125–130	1.23–1.12	NJ	4.01–3.60	2.96
22.6	13.44	32	—	0.16/0.23	5.70	0.27	120	—	144	1.05	NJ	1.55	4.26
23.6	13.36	33	—	0.10/0.18	1.62	0.96	48	—	93/144	0.60	NJ	0.86	2.34
23.6	15.08	34	—	0.10/0.19	1.75	0.89	57	—	100/147	0.63	NJ	2.43	0.26
29.6	15.14–19.47	35–42	0.20–0.24	0.14	6.44–9.85	0.24–0.19	126–136	104–70	130	0.56–7.7	NJ	4.10–3.48	2.96
30.6	9.22–20.02	43–59	0.30–0.24	0.18	7.00–10.00	0.22–0.15	30–50	70–40	100–150	0.90–1.20	NJ	2.51–4.14	2.96
3.7	13.09–17.46	60–63	0.35	0.10	4.14–1.37	0.43–1.37	45–90	60	90–130	0.58–0.48	NJ	2.37–2.97	2.96
3.7	18.57–20.39	64–66	—	0.18	1.09–1.69	1.43–0.92	80–110	—	140	0.36	NJ	1.41–1.79	2.34
4.7	11.45–12.27	67–69	0.37	0.17	3.78–4.34	0.41–0.36	181–150	80–110	120	0.28	NJ	2.61–3.10	2.96
5.7	9.19	70	0.26	0.10	5.00	0.26	162	195	120	0.47	NJ	4.16	0
5.7	11.31–13.13	71–72	0.26–0.30	0.12	6.20–5.62	0.25–0.28	150	170	95–120	0.49–0.53	NJ	0.75–1.36	2.34
6.7	9.57–13.21	73–77	0.34–0.26	0.12	5.77–4.88	0.27–0.32	180	160	110–120	0.40	NJ	1.78–1.84	2.34
6.7	19.59	78	—	0.19	3.42	0.46	204	—	110–150	0.76	J	3.27	2.96
7.7	10.37–17.15	79–87	0.25	0.12	5.93–4.18	0.26–0.37	160	150	144	0.41–0.43	NJ	3.92–2.24	0.26
8.7	14.20–15.28	88–89	—	0.24	5.34–4.84	0.29–0.32	169–158	—	150	0.59–0.56	J	1.19–0.90	2.34
8.7	16.36–17.44	90–93	—	0.23–0.18	3.87–4.55	0.40–0.34	189–165	—	167–166	0.56–0.64	NJ	2.59–2.66	2.96
9.7	9.21	94	0.22	—	7.98	0.20	186	158	160–138	0.86	J	3.29	2.96
9.7	10.56–16.03	95–101	0.18–0.22	—	4.26–6.78	0.23–0.37	170–158	160–120	—	0.65–0.76	NJ	0.99–1.74	2.34
10.7	8.49–10.32	102–104	0.17–0.19	0.12	6.01–5.89	0.26–0.27	130	128–132	140–105	0.80–0.72	NJ	2.96–3.12	2.95
10.7	11.27–15.46	105–108	0.18–0.20	0.12	5.59–6.27	0.28–0.25	133–124	141–120	130–110	0.72–0.76	NJ	1.10–1.47	2.34
10.7	16.56-to 11.7.6.43	109–129	—	0.14–0.11	4.75–7.46	0.33–0.21	100–133	—	106–142	0.72–1.17	J	2.65–3.95	2.95

^{a} Ranges given indicate very roughly a development in time, but, in particular for mean wave directions, may also indicate scatter of analysis.
^{b} Run numbers.
^{c} Peak frequency of windsea, if definable (Hz).
^{d} Swell frequencies if present (Hz).
^{e} Wind speed at 8 m (m/s).
^{f} Frequency for which $U = c$ (Hz).
^{g} Mean wave direction at $f = f_p$ (N = 0°, E = 90°).
^{h} Mean swell direction at f_{swell} (N = 0°, E = 90°).
^{i} Significant wave height $4\langle \zeta^2 \rangle^{1/2}$ (m).
^{j} Height of lower pressure sensor above mwl (m).
^{k} Vertical separation of pressure sensors (m).
^{l} Direction toward which wind in blowing (N = 0°, E = 90°).
^{m} NJ, non-JONSWAP-like; J, JONSWAP-like.

2. Estimation of the vertical dependence of the wave-coherent pressure field and its extrapolation to surface values
3. Estimation of the dependence of the surface values on nondimensional parameters (notably $\mu = U/c \cos \Theta$)

The principal range of frequencies we have considered is $9/128 \leqslant f \leqslant 69/128$ Hz, $\Delta f = 5/128$ Hz. The reason for these limits is that (1) the directional estimation fails for lower frequencies and (2) wave pressure coherences are too low at higher frequencies.

Notation

We here introduce the notation as used for a frame of references in which the current **V** vanishes. In fact, the current was not measured and currents obtained from tide tables were used. No attempt was made to estimate a current profile. Calculations were made both for assumed zero current and for tidal currents, which are typically on the order of $V \approx 0.3$ m/s. In view of the rough estimates of the currents, they were only considered in step 3 mentioned above, where they could relatively easily be introduced.

The relevant nondimensional parameters for a linear analysis are $\mu = \mathbf{k} \cdot \mathbf{U}_s/\omega = U/c \cos \Theta$, ω = radian frequency for $\mathbf{V} = 0$; $\lambda_i = kz_i$, z_i = height over mwl of instrument i, with $z_1 < z_2$, Θ = angle between **U** and c, all angles counted with N = 0°, E = 90°.

We realize that our choice of μ is not entirely satisfactory [choices based on the friction velocity u_* or on $U(z = k^{-1})$ are in principle preferable; see also Snyder *et al.* (1981)], but in practical terms the choice of the relevant wind velocity has little influence compared with the uncertainties associated with the wave directional distribution and the currents.

We shall use the same symbol for a variable in the space–time representation and for its Fourier transform and the argument list [e.g.; $p(\mathbf{x}, z, t)$ or $p(\mathbf{k}, z, \omega)$] is given where confusion may arise.

We introduce the nondimensional wave-coherent pressure γ by

$$P_w(\mathbf{k}, z) = \rho_a \cdot g \cdot \zeta(\mathbf{k}, t) \cdot \gamma(\lambda, \mu, \Theta, \ldots) \tag{1}$$

$$\gamma = \alpha + i\beta$$

where $P_w(\mathbf{k}, z, t)$ is the wave-coherent pressure and the most important nondimensional parameters are explicitly shown. Our aim is then to obtain information about $\gamma(\lambda, \mu, \Theta)$ from our cross-spectral data and of particular interest are the values at $\lambda = 0$.

4.1. Directional Wave Spectra

The linear array from which we obtain directional information has the two resistance wires 4 m apart and the pressure sensor nearly in the middle. (In some cases a triangle array was used.) This was the largest array we could install at the needle (at least at moderate cost) and while it was certainly not entirely adequate for our purposes, our choice was simply to try or not to try the experiment with this array.

We thus altogether have three lags, $l_1 = 4$ m, $l_2 \approx l_3 \approx 2$ m, with the difference between l_2 and l_3 depending on the particular setup, but normally $|l_2 - l_3| < 0.50$ m.

In view of the small separation between the instruments, we decided that it was not worthwhile to try to determine more than two parameters at each spectral band and thus we fitted a directional distribution $S(\Theta)$ to the three complex data points, with $S(\Theta)$ given by

$$S(\Theta) = [1/N(s)] \cos 2s[(\Theta - \Theta_m)/2] \tag{2}$$

where $N(s)$ is a normalizing factor.

Thus, if

$$C_{ij} = \langle \zeta_i, \zeta_j^* \rangle = R_{ij} + iQ_{ij} \tag{3}$$

the data were first normalized to

$$d_\alpha = C_{ij} = c_{ij}/(C_{ii} \cdot C_{jj})^{1/2}, \quad \alpha = 1, \ldots 3 \tag{4}$$

$$d_{-\alpha} = d_\alpha^*$$

and then the model values $\hat{c}_{ij} = \hat{d}_\alpha$ obtained from (2) were fitted to the data d_α by minimizing

$$\varepsilon^2 = (d_\alpha - \hat{d}_\alpha)^* W_{\alpha\beta}(d_\beta - \hat{d}_\beta) \tag{5}$$

where $W_{\alpha\beta}$ is the inverse of the known (see, e.g., Müller *et al.*, 1978) covariance matrix $V_{\alpha\beta} = \mathrm{cov}(d_\alpha, d_\beta^*)$, $(\alpha, \beta = \pm 1, \ldots \pm 3)$.

We did not normally achieve a consistent fit to the data at frequencies below $f = 0.25$ Hz, but we do not take this very seriously, because $V_{\alpha\beta}$ is a nearly singular matrix: the data are very highly coherent ($K \gtrsim 0.95$), since $\mathbf{k} \cdot \mathbf{l}_i$ is very small. On the other hand, this high coherence makes phase shifts reliable and offers some hope of determining the mean propagation angle Θ_m.

Since we have a linear array, there still remains an ambiguity for Θ_m. For windsea ($U \geqslant c$) we have resolved this by choosing the propagation direction which is closest to the direction of **U**. For swell we for the moment kept both directions $\Theta_m^{(1)}$ and $\Theta_m^{(2)}$ provided they both corresponded to swell running toward the coast.

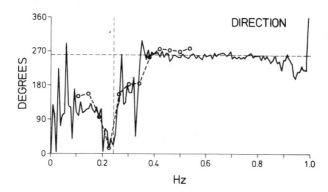

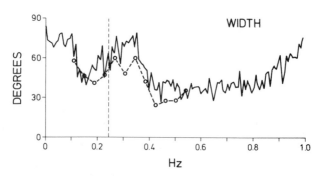

FIGURE 2. Comparison of the directional information obtained from the pitch-and-roll buoy (solid), $\Delta f = 1/128$ Hz, and the needle data (○), $\Delta f = 5/128$ Hz, for run 12. Dashed lines indicate the frequency for which $U = c$ and the direction into which the wind is blowing (N = 0°, E = 90°). The width of the directional distribution is the rms value $\Delta\Theta$ of $\Theta - \Theta_m$ in radians, $\Delta\Theta = [2/(s+1)]^{1/2}$.

At a later stage of the analysis we selected that $\Theta_m^{(i)}$ which gave the better fit to the model of the real part of the pressure transfer function. This will be explained in more detail in Section 4.3. In those swell cases, where both directions $\Theta_m^{(1)}$ and $\Theta_m^{(2)}$ indicated swell coming offshore, the fit was flagged as a failure and for each run such frequency bands were omitted in later analysis steps whenever directional information was required. (Even without an elaborate bookkeeping of the failures, it is safe to say that at most 10% of the swell bands for $f > 0.1$ Hz failed in the directional analysis. For lower frequencies the analysis scheme simply did not work and produced erratic results.)

We have made no attempt to provide error estimates for s and Θ_m (which is not so simple particularly when the fits, as in many cases, are not consistent with the data) and throughout our analysis we have remained skeptical of our directional results.

Thus, we were relieved when, after most of the analysis had been completed, a comparison with the pitch-and-roll buoy data became available and showed good agreement with the array results. One of the better cases, which at the same time shows a fairly complex behavior, is shown in Fig. 2. Most comparisons were of similar quality, but there were a few cases in which the agreement was poor. As yet, these cases have neither been flagged out of further processing nor in any way improved upon, so that some of the scatter shown in Fig. 4 can be attributed to such errors in directional information.

4.2. The Vertical Dependence

The observational material available to determine the vertical dependence are the six cross-spectra

$$C_{1j} = \langle p_1 \zeta_j^* \rangle \quad \text{and} \quad C_{2j} = \langle p_2 \zeta_j^* \rangle, \quad j = 1, \ldots 3$$

which are measured at nondimensional heights $\lambda_i = k z_i$ typically varying between 0.012 and 6.0. Again, some simplification in the analysis was necessary. We wished to eliminate as far as possible the dependence of the surface values $\gamma(\lambda = 0, \mu)$ on $\mu = U/c \cos \Theta$.

Our approach was thus to obtain estimates $\hat{r}$ for the ratio $r = \gamma(\lambda_2, \mu)/\gamma(\lambda_1, \mu)$ by minimizing

$$\varepsilon^2 = d_\alpha^* W_{\alpha\beta} d_\beta \tag{6}$$

where again $W_{\alpha\beta}$ is the inverse of $V_{\alpha\beta}$:

$$V_{\alpha\beta} = \text{var}(d_\alpha, d_\beta^*), \quad \alpha, \beta = \pm 1, \ldots \pm 3 \tag{7}$$

and

$$d_\alpha = (C_{1\alpha} - \hat{r} C_{2\alpha}), \quad \alpha = 1, \ldots 3, \quad d_{-\alpha} = d_\alpha^* \tag{8}$$

This approach is about halfway between the simplified analysis and the complete analysis of Snyder et al. (1981).

To us this appeared to be the easiest way to utilize the information from all three wave measurements with adequate weighting. While it reduces some of the bias problems associated with the simplified analysis of Snyder et al. (1981), it does not completely avoid them. When both pressure measurements have a low coherence with the wave field, the estimated $\hat{r}$ tends to be too high. The reason becomes particularly clear for say a comparison run between the two pressure sensors, which might be mounted at 4.00 m and 4.10 m above mwl. At high wavenumbers the coherence between waves and pressure is very small and may be consistent with zero, but any arbitrary remaining covariance $\langle p_i \zeta_i^* \rangle(\omega)$, due mainly to fluctuations of the sampling error, tends to be the same for both p_1 and p_2 due to the high correlation of p_1 and p_2 through turbulent (as opposed to wave induced) pressure, thus producing a high value of $\hat{r}$. The way to reduce this bias is to model both the wave-coherent and the turbulent pressure, i.e., to make use of the

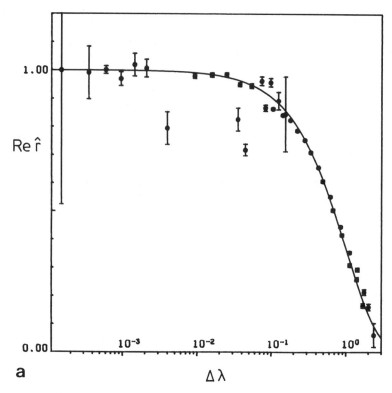

a $\Delta\lambda$

FIGURE 3. Mean values of the real and imaginary parts of $\hat{r}$, the estimator of $r = \gamma(\lambda_2, \bar{u})/\gamma(\lambda_1, \bar{u})$ vs. $\Delta\lambda = \lambda_2 - \lambda_1$, for all data with $\lambda_1 \leqslant 1.6$ and frequencies ranging from 4/128–0.5 Hz. a. Re $\hat{r}$ vs. $\Delta\lambda$, logarithmic scale emphasizes small values of $\Delta\lambda$. Solid line shows Re $\hat{r} = \exp(-\Delta\lambda)$. b. as a, but log-linear scale. c. Im $\hat{r}$ vs. $\Delta\lambda$, solid line shows Im $r = 0$.

information $\langle p_i p_j^* \rangle$ in the least-squares analysis, but this has not yet been done. Such an analysis could not be decoupled from the surface values and would lead to an analysis equivalent in principle to the complete analysis of Snyder *et al.* (1981).

In any case, the procedure is quite satisfactory when the lower pressure sensor is relatively close to the surface ($kz_1 < 1.6$). In the analysis, we first obtained $\hat{r} = \hat{r}(kz_1, kz_2, \bar{\mu}, \Theta, \ldots)$ for each run. For each run we calculated the standard deviations $\Delta\hat{r}$, if the fit was consistent with the data, as determined from a χ^2 test on the residuals. This was not always the case, especially at low frequencies where the high coherences imposed very narrow bands. When the fit was not consistent, we artificially made the errors on the d_α larger to enforce consistency with the deviations $\Delta\hat{r}$ from the mean $\hat{r}$ (at the 95% confidence level). Anticipating a decay close to exponential

$$r(\lambda_1, \lambda_2, \mu) = e^{-\delta(\mu, \lambda_1)\Delta\lambda}, \qquad d\delta/d\lambda_1 \ll 1 \tag{9}$$

we then stratified the data $\hat{r}$ by $\Delta\lambda$ and averaged, weighted by $(\Delta\hat{r})^{-2}$. When the nondimensional λ_1 of the lower pressure sensor p_1 was less than 1.6 m, the resulting data lay so close to $\exp(-\Delta\lambda)$ that we have not even performed a least-squares fit (see Fig. 3). Further stratification by U/c showed no dependence on U/c. A stratification by the average $\bar{\mu}$ at each frequency would in principle have been better and possible, but it was simpler to use U/c. It is unlikely that a marked dependence on μ would not show up in a stratification by U/c. For values of $\lambda_1 > 1.6$ a systematic departure from the exponential decay was observed. This is nearly certainly due to the bias mentioned above, but the analysis is not yet fully completed. In Fig. 3 the points at very small values of $\Delta\lambda$ ($\Delta\lambda < 2 \times 10^{-3}$) arise from intercomparison runs and give practically no profile

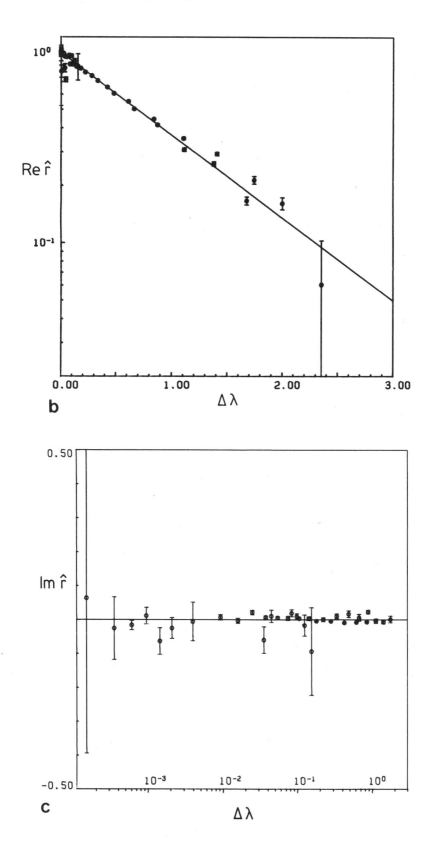

FIGURE 3. (*Continued*)

information but do indicate good agreement between the pressure sensors. The three low-lying outliers are significant, with typically on the order of 40 entries contributing to each point, and all three are associated with the lowest-frequency band $f = 4/128$ Hz at various vertical separations. At present, the interpretation of these points is unclear. We have some doubts whether they are real or some artifact of our experimental arrangement (which appears unlikely) or the analysis.

These points could also be associated with the fact that at $f = 4/128$ Hz, there is very little wave energy. Perhaps the little wave energy present is mainly due to forced waves, which would invalidate the free wave dispersion relation used.

The imaginary part of $\hat{r}$ showed a bit more scatter, but no systematic departure from $\operatorname{Im} r = 0$.

In summary, then, we are confident that we can use (9) to extrapolate our cross-spectra to the surface, with $\delta(\mu) = 1$, but point out that some work remains to be done.

4.3. Dependence of the Surface Values on μ

To obtain information about $\gamma(\lambda = 0, \mu)$ we extrapolated the values of $\langle p_i \zeta_2^* \rangle$, where ζ_2 is the resistance wire under the pressure probes, to the surface by multiplying with $\exp(\lambda_i)$. We did not make use of $\langle p_i \zeta_1^* \rangle$ and $\langle p_i \zeta_3^* \rangle$ in this part of the analysis, since we felt that any estimate of the imaginary part $\beta(\lambda = 0, \mu)$ using these data would be too unreliable, being too sensitively dependent on the directional distribution. We then again at each frequency have two estimates $\hat{\gamma}_1(\lambda = 0, \bar{\mu})$ and $\hat{\gamma}_2(\lambda = 0, \bar{\mu})$. Again using the covariance matrix $\operatorname{cov}(\gamma_i, \gamma_j^*) i, j = 1, 2$, the weighted mean $\bar{\gamma}(\lambda = 0, \bar{\mu})$ was obtained, with standard error. The errors were again artificially enlarged, whenever the mean was not consistent with γ_1 and γ_2. Notice that by this procedure data extrapolated from large nondimensional heights λ_i enter the mean with a low weight.

The problem which now arises can be easily formulated:

Given the roughly 12 (frequencies) × 129 (runs) data of $\bar{\gamma}(\lambda = 0) = \bar{\alpha} + i\beta$ and the associated directional distribution, find the function $\gamma(\lambda = 0, \mu)$ which gives the best fit to these data. This problem we have not yet systematically attacked, but in order to get some insight into the data, we have considered the closely related, yet simpler problem of whether our data are consistent with the formulation suggested by Snyder et al. (1981). For the real part of γ:

$$\alpha_B = \begin{cases} -(\mu - 1)^2 & \text{for } \mu \leqslant 1 \quad\quad\quad (10) \\ -0.65(\mu - 1)^2 & \text{for } \mu \geqslant 1 \quad\quad\quad (11) \end{cases}$$

[These formulas were not explicitly given by Snyder et al. (1981) but are sufficient approximations—well within the leeway allowed by their fits—to their curves.]

For the imaginary part β:

$$\beta_B = \begin{cases} 0 & \text{for } \mu \leqslant 1 \quad\quad\quad\quad\quad\quad\quad\quad\quad\quad\quad (12) \\ b(\mu - 1) & \text{for } \mu \geqslant 1, \quad \text{with } b = 0.2 \text{ to } 0.3 \quad\quad (13) \end{cases}$$

where the index B indicates BoA values.

We first illustrate the procedure for the case of zero current. The data $\bar{\gamma}$ are weighted averages of the extrapolation from the two heights, and are an estimate of $\tilde{\gamma}$, which is the average value of γ integrated over the directional distribution.

One way of comparing the BoA result with ours would simply have been to plot the values of $\bar{\gamma}_B$, obtained by averaging the BoA results over the directional distribution against our $\bar{\gamma}$. We have not followed this track, because from such a plot it is difficult to tell at which μ values the results differ.

Instead, we have calculated for each directional distribution $S(\Theta)$ equivalent values of μ, denoted by $\bar{\mu}_\alpha$ and $\bar{\mu}_\beta$, such that

$$\alpha_B(\bar{\mu}_\alpha) = \int S(\Theta)\alpha_B(\mu)d\Theta \tag{14}$$

and

$$\beta_B(\bar{\mu}_\beta) = \int S(\Theta)\beta_B(\mu)d\Theta \tag{15}$$

and $\bar{\mu}_\alpha$ differs from $\bar{\mu}_\beta$ because $\alpha_B(\mu)$ is not weighted in the same way as $\beta_B(\mu)$, as can be seen from (10) to (13). The formulas for obtaining $\bar{\mu}_\alpha$ and $\bar{\mu}_\beta$ are given in the Appendix. Thus, if the measurements were perfect and in agreement with the BoA results, our values $\bar{\alpha}$ when plotted against the $\bar{\mu}_\alpha$ axis would lie on the curve given by (10)–(11), and similarly, $\bar{\beta}$ plotted against $\bar{\mu}_\beta$ would lie on the curve given by (12)–(13).

Thus, to compare our results with Snyder et al. (1981), we have introduced 200 intervals of equal length on the $\bar{\mu}_\alpha$ and $\bar{\mu}_\beta$ axis from $-1, +3$ and in each interval averaged (again weighted with the inverse variance) the $\bar{\alpha}$ and $\bar{\beta}$ to $\bar{\bar{\alpha}}$ and $\bar{\bar{\beta}}$ and plotted these values against $\bar{\mu}_\alpha$ and $\bar{\mu}_\beta$. We can then again check whether the $\bar{\bar{\alpha}}$ and $\bar{\bar{\beta}}$ are consistent with the individual entries of $\bar{\alpha}$ and $\bar{\beta}$ and whether the $\bar{\bar{\alpha}}$ and $\bar{\bar{\beta}}$ lie on or close to the relations given by Snyder et al. (1981).

Since we accept the uncorrected directional distribution $S(\Theta)$, the introduction of the current corrections is straightforward. We simply have to include the mean wavenumber shift in the vertical extrapolation and the corrections of U and $c(\Theta)$.

We now come back to the choice of the mean propagation direction Θ_m for the swell bands, $U/c < 1$. For any given run and swell frequency, we have taken that value of Θ_m of the at most two allowed values $\Theta_m^{(1)}$ and $\Theta_m^{(2)}$ which gave the best approximation to (10)–(11). By comparing plots of Θ_m for consecutive runs, we could check that this did not result in an arbitrary switching of directions from run to run.

The final results for $\bar{\alpha}$ versus $\bar{\mu}_\alpha$ and $\bar{\beta}$ versus $\bar{\mu}_\beta$ are shown in Fig. 4. (Calculations with

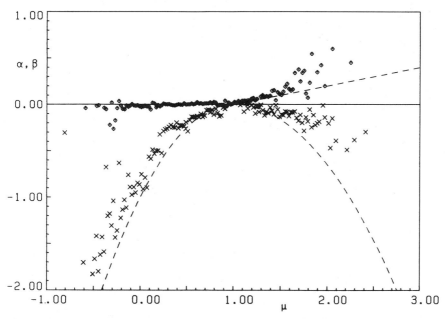

FIGURE 4. Averaged values of α ($\times$) and β ($\blacklozenge$) of the wave-pressure transfer function $\gamma = \alpha + i\beta$, plotted vs. μ. The relations (10)–(13) with $b = 0.2$ obtained by Snyder et al. (1981) for α and β are also shown, and in addition the line $\alpha = \beta = 0$ is shown for orientation. To avoid congestion, the error bars on α and β are not shown, but typical values for the standard deviations are $\Delta\beta = \pm 0.2$ for $\mu < 0$ and $\mu > 1$, $\Delta\beta = \pm 0.05$ for $0 < \mu < 1$, and $\Delta\alpha = \pm 0.2$ for all μ.

assumed zero current produce similar plots, although of course details differ.) Concerning both the real and the imaginary parts, it should first of all be stated that apart from a few exceptions, the $\bar{\bar{\alpha}}$ and $\bar{\bar{\beta}}$ are not consistent with their individual entries $\bar{\alpha}$, $\bar{\beta}$. The errors quoted in Fig. 4 are the observed standard directions of $\bar{\alpha}$ and $\bar{\beta}$, calculated under the assumption that the directional distribution was perfectly known, so that the $\bar{\mu}_\alpha$ and $\bar{\mu}_\beta$ intervals are correctly assigned to $\bar{\alpha}$ and $\bar{\beta}$.

The real part α appears to be in fairly good agreement with the fits of Snyder et al. (1981). Although some systematic deviations are obvious, it is quite possible that these could have been introduced by errors in the directional distributions, especially by too broad directional distributions. Also, any error in the mean direction will bias $\bar{\mu}_\alpha$ to values away from $\mu = 1$, due to the quadratic term.

For the dynamically more interesting imaginary part, we notice that there is no indication of damping over most of the swell range, except for a few points close to $\mu = -0.4$ which indicate very strong damping. Closer inspection shows these points to be associated with waves running against the wind, and with frequencies $f \approx 0.25$ Hz. There are not many data contributing in this range and the possibility that the displacement of the wire has introduced a phase shift must certainly be considered, especially since the damping is so strong. In the windsea domain, our results for β are larger than those of Snyder et al. (1981) and show considerably more scatter. Since in this range relatively high frequencies ($f \approx 0.3$–0.5 Hz) are involved, an influence of a displacement error can again not be ruled out. However, we have more data here, and the chance that displacement errors have averaged out is greater. Although we have found no significant correlation (at least not by eye) of the $\bar{\alpha}$ and $\bar{\beta}$ with $\bar{\mathbf{k}} \cdot \mathbf{V}$, where $\bar{\mathbf{k}}$ is the mean wavenumber for each band and $\mathbf{V}$ the current, this can hardly be taken as sufficient evidence that displacement effects were negligible.

We have also stratified the data by the (dimensional) significant wave height $H_s = 4\langle \zeta^2 \rangle^{1/2}$, by gH_s/U^2, by the slope variance of each band, and by $\cos \Theta$, and while we have not explored the various possibilities in depth, we have found no indication that the data depend systematically on any of these parameters.

We point out that the problem formulated in the beginning of this section has not yet been attacked. Even in the presence of displacement errors it might still be worthwhile to test whether some other parameterization than (12)–(13) gives better agreement with the data, whereas for the real part the agreement seems to be fairly satisfactory considering the uncertainties of the directional distribution.

5. CONCLUSIONS

The detailed analysis of about 70 h of simultaneous wave and pressure measurements in the North Sea has not yet conclusively shown any disagreement with the Bight of Abaco data of Snyder et al. (1981).

The analysis presented here is not yet fully completed and future work will be directed toward obtaining the corrections due to scattering at the mast (which we anticipate to be small), full incorporation of the pitch-and-roll information, some improvement on the study of the vertical dependence of the transfer function, and finally an attempt to better estimate the errors associated with the displacement of the resistance wire. The preliminary nature of our results should therefore be clearly recognized in the following summary:

The analysis indicates that a relatively simple and robust experimental arrangement should yield sufficient information to obtain reliable results, but that unfortunately, in our particular case the displacement error has limited the accuracy of our results concerning the imaginary part of the transfer function at frequencies beyond $f \approx 0.25$ Hz. It would also have been advisable to ensure continuous coverage with pitch-and-roll data and current measurements. Also to quell doubt, whenever possible, a poor reflector, such as a radio mast, should be used as instrument support.

In the low-frequency swell region, we have found no indication of wave damping by atmospheric pressure on the average. We were, however, surprised with the variability of the results, and the indication is that they cannot be consistently modeled by the average pressure transfer function. Doubts with this conclusion can be raised, because the statistical analysis may not be adequate in view of the high coherences in this range. Nevertheless, the point is that we might have arrived at rather different conclusions had we analyzed only 6 or 10 h of swell data. We have been unable to detect any dependence of the results on other nondimensional variables, such as slope variance or gH_s/U^2. Except for the results for $\mu \approx -0.4$ where we have found (in our opinion dubious) damping, we do not anticipate major revisions of our results in the swell range to come from the above-mentioned work yet to be completed.

In the windsea range, our measured growth rates are on average about twice as high and more scattered than the results of Snyder et al. (1981), but we do not yet conclude that there is a systematic disagreement with Snyder et al. (1981) because of the possible influence of displacement errors.

Concerning the vertical dependence, we have so far been unable to detect any departure from a pure exponential decay, $\exp(-kz)$, of the wave-induced pressure amplitude. Again, this is a result which emerged only after averaging a sufficient number of data and the results for individual runs showed considerable scatter.

ACKNOWLEDGMENTS. We express our thanks to R. Snyder for generously providing us with his pressure probes and to F. Dobson who gave us useful advice in many discussions. This work was supported by Deutsche Forschungsgemeinschaft by a grant to SFB 94.

APPENDIX: ALGORITHM FOR OBTAINING $\bar{\mu}_\alpha$ AND $\bar{\mu}_\beta$

The choice for $\bar{\mu}_\beta$ is unique only if $\mu(\Theta) > 1$ for some Θ. If μ_B is in the swell domain, $\mu(\Theta) < 1$ for all angles, the average μ is taken. $\tilde{S}(\Theta) = S(\Theta - \Theta_u)$.

$$\bar{\mu}_\beta - 1 = \begin{cases} \displaystyle\int \tilde{S}(\Theta)\left(\frac{U}{c}\cos\Theta - 1\right)d\Theta & \text{if } U/c < 1 \quad (A1) \\ \displaystyle\int_{\mu>1} \tilde{S}(\Theta)\left(\frac{U}{c}\cos\Theta - 1\right)d\Theta & \text{if } U/c \geqslant 1 \quad (A2) \end{cases}$$

The choice of $\bar{\mu}_\alpha$ is not unique, because α_B is parabolic. We choose $\bar{\mu}_\alpha < 1$ if the weight comes predominantly from the swell region $\mu < 1$, and choose $\bar{\mu}_\alpha > 1$ otherwise; see (10)–(11).

$$\bar{\mu}_\alpha - 1 = \begin{cases} \displaystyle\frac{1}{0.65}(H_1 + 0.65H_2)^{1/2} & \text{for } H_2 > H_1 \quad (A3) \\ -(H_1 + 0.65H_2)^{1/2} & \text{for } H_2 < H_1 \quad (A4) \end{cases}$$

where

$$H_1 = \int_{\mu<1} \tilde{S}(\Theta)\left(\frac{U}{c}\cos\Theta - 1\right)^2 d\Theta \quad (A5)$$

$$H_2 = \int_{\mu>1} \tilde{S}(\Theta)\left(\frac{U}{c}\cos\Theta - 1\right)^2 d\Theta \quad (A6)$$

This representation simply assigns $\bar{\mu}_\alpha$ and $\bar{\mu}_\beta$ to our directional distributions according to the relations given by Snyder et al. (1981) with which we wish to compare our data. No biasing of our data in favor of any particular description is involved—except for the choice of propagation direction in the swell region as explained in the main text.

REFERENCES

Günther, H., W. Rosenthal, and K. Richter (1979): Application of the parametrical wave prediction model of rapidly varying wind fields during JONSWAP 1973. *J. Geophys. Res.* **84**, 4855–4864.

Hasselmann, K., T. P. Barnett, E. Bouws, H. Carlson, D. E. Cartwright, K. Enke, J. A. Ewing, H. Gienapp, D. E. Hasselmann, P. Kruseman, A. Meerburg, P. Müller, D. J., Olbers, K. Richter, W. Sell, and H. Walden (1973): Measurements of wind-wave growth and swell decay during the Joint North Sea Wave Project (JONSWAP). *Dtsch. Hydrogr. Z. Suppl. A* **8**(12).

Hsiao, S. V., and O. H. Shemdin, (1983): Measurements of wind velocity and pressure with a wave follower during MARSEN. *J. Geophys. Res.*, **88**, C14, 9841–9849.

Müller, P., D. J. Olbers, and J. Wilebrand (1978): The IWEX spectrum. *J. Geophys. Res.* **83**, 479–499.

Snyder, R. L., F. W. Dobson, J. A. Elliott, and R. B. Long (1981): Array measurements of atmospheric pressure fluctuations above surface gravity waves. *J. Fluid Mech.* **102**, 1–59.

Snyder, R. L., R. B. Long, J. Irish, D. G. Hunley, and N. C. Pflaum (1974): An instrument to measure atmospheric pressure fluctuations above surface gravity waves. *J. Mar. Res.*, **32**, 485–496.

DISCUSSION

M. WEISSMAN: Please clarify what you used for the reference wind speed.

D. HASSELMANN: U at 5 m height, above mean water level, with $z_0 = 10^{-4}$ m and a reference system in which the mean (not measured) total current vanishes. I have also tried the same analysis with U_{10}; this hardly makes any difference. Of course, a proper nondimensional parameter should be $(\mu_*/c) \cos \Theta$, but for the reduction of our experimental data this only introduces additional scatter.

DONELAN: In your measurements of the waves in an adverse wind, were your measurements made downwave or downwind?

D. HASSELMANN: We always had the spars at right angles to the wind, so that the angle of the spars to the swell was not controlled. (We have looked for signatures typical for reflections but have not seen any. We are a little worried about reflection effects. While the influence of scattering on the waves is easily obtained—at least for a single wave train—the perturbation of the pressure field is difficult to estimate, even for potential flow.)

V

METHODS OF REMOTE SENSING

25

THE SAR IMAGE OF SHORT GRAVITY
WAVES ON A LONG GRAVITY WAVE

ROBERT O. HARGER

ABSTRACT. A SAR imaging model appropriate to oceanographic applications is derived, unifying fundamental models of hydrodynamics, rough surface scattering, and SAR imaging of time-variant scenes.
The sea surface is a sinusoidal long gravity wave upon which short gravity waves propagate and are modified
by the long wave in accordance with a recent theory of Phillips; the electromagnetic scattering is described by
the two-scale approximation appropriate to long wave and short wave ensembles that are, respectively,
smoothly varying and not too rough with respect to the radar wavelength. The resulting model, accurate to
first order in the long wave slope, for the first time fundamentally characterizes the nonlinear hydrodynamic
and scattering interactions of the long and short waves and their effect, along with temporal variation, on the
SAR image. Of particular importance, the long wave enters (among other ways) as a phase-modulated
waveform that, when filtered by the SAR system, can be, for large-amplitude long waves, the principal
determinant of the image nature. The numerical analysis of the model is discussed and an approximation
describing the image of a delimited scene area is derived and exemplified. (1) When the small waves are a
range-directed ensemble and the long wave is azimuth directed, the latter's temporal variation "blurs," in
azimuth, the image due, primarily, to the SAR system's narrowband filtering of the aforementioned phase-
modulated waveform and, secondarily, the nonlinear hydrodynamic interaction; it is shown that this
"blurring" is, at higher long wave amplitudes, due to a quadratic phase error proportional to the phase
velocity of the long wave. That part of the short wave ensemble allowed influential by SAR system
(wavenumber) filtering is approximately nondispersive during the SAR azimuth integration time, its
concerted effect being a rigid azimuth image translation, proportional to the short wave's mean phase
velocity. (2) More briefly treated, when the long wave is ranged directed and the short wave ensemble is
simply a single range-directed sinusoid ("Bragg-matched"), the image is solely range variant and its nature is
primarily determined by the aforementioned narrowband filtering effect, the secondary effects of nonlinear
hydrodynamics and physical optics (i.e., surface slope) being evident. Therefore, the present model, as thus far
elaborated, contradicts predictions of models based on the SAR response to a point scatterer in motion
in accordance with the "orbital motion" of the long wave: e.g., no "azimuth bunching" attributed to such
motion is observed.

1. INTRODUCTION

The frequent occurrences of dramatic images from SAR systems (Beal *et al.*, 1981) demand an
extension of the existing SAR theory (Harger, 1970) to explain and predict image characteristics
which apparently depend on winds, currents, bottom topography, and other phenomena. A

ROBERT O. HARGER ● Department of Electrical Engineering, University of Maryland, College Park,
Maryland 20742.

recent consensus (Zalkan and Wentzel, 1981) stated (1) that an adequate SAR model must account, at the random, sample function level, for the extended, rough, and time-variant nature of the sea surface, (2) that the two-scale approximation to the electromagnetic (EM) scattering from the rough sea surface is suitable, and (3) that a nonlinear hydrodynamic model should be employed to describe the effect of the large-scale sea surface height structure on the small-scale structure, the presence of the latter being generally required for significant backscattering at microwave frequencies and intermediate incidence angles.

A SAR imaging model has been developed accordingly (Harger, 1980a, b): the hydrodynamic approximation employed was due to Longuet-Higgins and Stewart (1961), extended to include surface tension (appropriate to gravity-capillary waves) and general 2d first-order wave height structures. It appears that this hydrodynamic approximation is restricted to the case of a large-scale height less than the wavelength of the fine-scale structure (Phillips, 1981a), and an approximation of wider validity has been presented (Phillips, 1981b). Therefore, the present discussion is based on this latter approximation, here for gravity waves and to first-order in the large-scale slope.

2. A SAR IMAGE OF THE SEA

2.1. The Nonlinear Hydrodynamic Model

The effect of a sinusoidal long wave, characterizing the large-scale structure, on the short gravity waves, comprising the fine-scale structure, can be described to first-order in the slope of the long wave, by following Phillips (1981b). By considering the motion of the short gravity waves in a coordinate system moving with the long wave at its phase velocity C, it is relatively simply seen that the motion of the short gravity waves at a particular location on the long wave is that for a level mean surface with a current u equal to the tangential surface velocity associated with the long wave (at said point) provided the effective gravitational acceleration g' is that component normal to the long wave surface plus the centripetal acceleration due to the long wave (Fig. 1). The "local" intrinsic frequency σ is thereby determined as a function of the "local" wavenumber k by $(g'k)^{1/2}$. Relative to a "local" reference point on the long wave at the mean

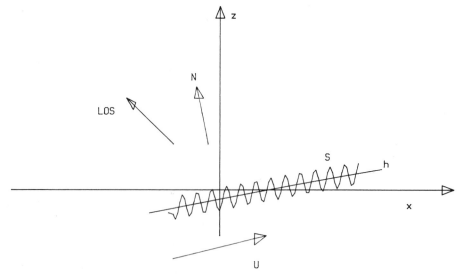

FIGURE 1. The short gravity wave s in a coordinate system (x, y, z) moving at the phase velocity of the long gravity wave h. The line-of-sight to the SAR is LOS, the normal to h is N and the effective current is U.

surface—denoted by the subscript "0"—the apparent frequency n and wavenumber k at other points on the large wave are specified via the "kinematic conservation equation," the relation between n and k that must be satisfied as a consequence of their definition as partial derivatives of the phase of a waveform (Phillips, 1977, Section 2.6). When the long wave is sinusoidal with small slope AK, say, then, to first order in AK, the wavenumber and amplitude of the short gravity waves have components varying sinusoidally over the long wave whereas the intrinsic frequency does not vary.

Noting there is no lateral y dependence of these parameters, we may construct such a short gravity wave as follows. For a long wave of the form

$$h(x, y, t) = A \cos(Kx - \Sigma t), \quad \Sigma \equiv KC \tag{1}$$

we have a short wave wavenumber varying as

$$\mathbf{k}(x, y) = \begin{pmatrix} k_{ox}(1 + AK \cos Kx) \\ k_{oy} \end{pmatrix}$$

and an apparent frequency $n_0 = -\sigma_0 + k_{ox}C$, where σ_0 is the intrinsic frequency (at a point on the long wave at the mean surface). Since by definition $\mathbf{k} \equiv \nabla \chi$ and $n \equiv -\partial\chi/\partial t$ where χ is the phase of the short wave, one integrates to find

$$\chi(x, y, t) = \mathbf{k}_0 \cdot \mathbf{x} + k_{ox} A \sin Kx - (-\sigma_0 + k_{ox}C)t \tag{2a}$$

The amplitude of the short wave varies as

$$a(x, y) = a_0(1 + AK \cos Kx) \tag{2b}$$

in accordance with action conservation (Phillips, 1977, Section 2.7).* Thus, (2) determines a short gravity wave component in the presence of a long, sinusoidal wave—in the aforesaid moving coordinates.

In a fixed reference system, $x \rightarrow x - Ct$, we find the short gravity wave component's structure to be

$$\xi(x, y, t; k_0) = a_0[1 + AK \cos(Kx - \Sigma t)] \cos\{\mathbf{k}_0 \cdot \mathbf{x} - \sigma_0 t + k_{ox} A \sin(Kx - \Sigma t)\} \tag{2c}$$

We observe that the phase contains implicitly the form $k_{ox}[x + A \sin(Kx - \Sigma t)]$: for the deep-water waves under consideration, the x coordinate of a Lagrangian parcel moving on the surface of the long wave is $[-A \sin(Kx - \Sigma t)]$ (Kinsman, 1965, p. 135); therefore, the "x translation to-and-fro" of the small-scale structure by the large-scale structure is present in (2).

It will be convenient to choose, in the remainder, (x, y) coordinates with x parallel to the velocity vector $\mathbf{v}$ of the SAR system. If $\mathbf{K}_{20}$ and $\mathbf{K}_0$ are the wavenumbers, respectively, of the long and short gravity waves—in the latter case on the long wave at the mean water level—then in such a coordinate system

$$\xi(\mathbf{x}, t; K_0) = a_0\{1 + AK_{20} \cos[\mathbf{K}_{20} \cdot \mathbf{x} - \sigma_0(K_{20})t]\}$$
$$\cdot \cos\{\mathbf{K}_0 \cdot \mathbf{x} - \sigma_0(K_0)t + \mathbf{K}_0 \cdot (\mathbf{K}_{20}/K_{20}) A \sin[\mathbf{K}_{20} \cdot \mathbf{x} - \sigma_0(K_{20})t]\} \tag{2d}$$

---

*Dr. S. H. Kim (personal communication) noted that, more exactly following Phillips, the modulation, or nonconstant part, of (2b) is proportional to $(1 - \frac{1}{4}\sin^2 \alpha)$, where α is the angle between k_0 and K's direction x: this factor is approximated by unity here. Similarly ignorable, when the long wave and short wave phase velocities greatly differ, is a variation in the local wavenumber described by Phillips (1981a).

where

$$\mathbf{x} \equiv \begin{pmatrix} x \\ y \end{pmatrix}$$

More generally, the short gravity wave, on the long gravity wave at the mean water level, will be a summation, over K_0, of sinusoids such as (2d), the sum weighted by a spectrum $\tilde{f}_0$ of local wavenumbers. Reasonably assuming that, to first order in AK_{20}, these sinusoids do not nonlinearly interact among themselves—but only each with the long wave—then the general short gravity wave structure is a summation of sinusoids such as (2d), weighted by $\tilde{f}_0$, where $\tilde{f}_0 \equiv \tilde{f}_0(\mathbf{K}_0, 0)$ is the spectrum of $f_0(\mathbf{x}, 0)$, the dispersive short gravity wave ensemble on the long wave at the mean water level at time zero. Thus, one finds a general ensemble of short gravity waves to be, to first order in AK_{20},

$$\xi(\mathbf{x}, t) = a_0 \{ 1 + AK_{20} \cos [\mathbf{K}_{20} \cdot \mathbf{x} - \sigma_0 (K_{20}) t] \}$$
$$f_0 \{ \mathbf{x} + (\mathbf{K}_{20}/K_{20}) A \sin [\mathbf{K}_{20} \cdot \mathbf{x} - \sigma_0 (K_{20}) t], t \} \tag{3a}$$

where

$$f_0(\mathbf{x}, t) \equiv \mathrm{Re} \frac{1}{4\pi^2} \int \int d\mathbf{K} \exp [i\mathbf{K} \cdot \mathbf{x} - i\sigma_0(K) t] \, \tilde{f}_0(\mathbf{K}, 0) \tag{3b}$$

Thus, this "intrinsic" ensemble of short gravity waves, f_0, additive to the long gravity wave, is translated to and fro by the orbital motion of the long gravity wave: this effect is not encompassed by the approximation of Longuet-Higgins and Stewart (1961).

If f_0 generally fluctuates about a nominal wavenumber $\mathbf{K}_{10}$, then it is useful to set

$$\tilde{f}_0(\mathbf{K}, 0) \equiv \tilde{f}(\mathbf{K} - \mathbf{K}_{10})$$

in which case

$$f_0(\mathbf{x}, t) = \mathrm{Re}\{ f_1(\mathbf{x}, t) \exp [i\mathbf{K}_{10} \cdot \mathbf{x} - i\sigma(K_{10}) t] \}$$

where

$$f_1(\mathbf{x}, t) \equiv \frac{1}{4\pi^2} \int \int d\mathbf{K} \exp \{ ix \cdot \mathbf{K} - i[\sigma_0(\mathbf{K} + \mathbf{K}_{10}) - \sigma_0(K_{10})] t \} \tilde{f}(\mathbf{K}) \tag{3c}$$

Thus,

$$\xi(\mathbf{x}, t) = a_0 \{ 1 + AK_{20} \cos [\mathbf{K}_{20} \cdot \mathbf{x} - \sigma_0(K_{20}) t] \}$$
$$\mathrm{Re}[f_1 \{ \mathbf{x} + (\mathbf{K}_{20}/K_{20}) A \sin [\mathbf{K}_{20} \cdot \mathbf{x} - \sigma_0(K_{20}) t], t \}$$
$$\exp (i\{ \mathbf{K}_{10} \cdot \mathbf{x} - \sigma_0(K_{10}) t + \mathbf{K}_{10} \cdot (\mathbf{K}_{20}/K_{20}) A \sin [\mathbf{K}_{20} \cdot \mathbf{x} - \sigma(K_{20}) t] \})] \tag{3d}$$

Note that $\mathbf{K}_{10} \cdot (\mathbf{K}_{20}/K_{20}) = K_{10} \cos \theta$ where θ is the angle between $\mathbf{K}_{10}$ and $\mathbf{K}_{20}$.

2.2. The Two-Scale Electromagnetic Scattering Model

A SAR system model conveniently describes a scene (Fig. 2) by means of a reflectivity density g, equivalently distributed over a plane approximating the mean sea surface. The determination of g is a problem in EM scattering theory: the so-called "two-scale" model

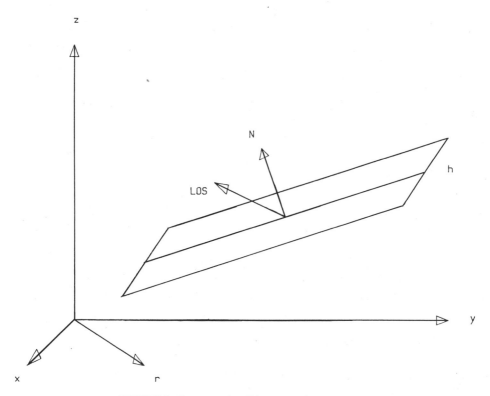

FIGURE 2. Geometry describing sea surface scattering.

(Beckmann and Spizzichino, 1963; Bass $et\ al.$, 1968; Valenzuela, 1968; Wright, 1968) results in an effective reflectivity density

$$g_1 = g_0(1 - i\beta\xi)e^{-i\beta h}(1 + (\partial h/\partial y)\tan\delta_0) \tag{4}$$

where $g_0 = (2\cos\delta_0)\exp(i2k_0R_0)$, $\beta = 2k_0\cos\delta_0$, and h and ξ are the small- and large-scale wave height structures; δ_0 is the incidence angle, $k_0 \equiv 2\pi/\lambda_0$ the mean wavenumber of the microwave radiation, and R_0 the mean range from the SAR vehicle to the scene. The factor $\exp(-i\beta h)$ accounts for the "heaving" of the fine-scale structure, changing the effective round trip path of a ray; the "translation" of the fine-scale structure ξ is included implicitly by the requirement that h and ξ jointly obey nonlinear hydrodynamics, as discussed above. The factor $[1 + (\partial h/\partial y)\tan\delta_0]$ results from the physical optics approximate description of scattering by the large-scale structure: it is zero at grazing incidence when $\partial h/\partial y = -1/\tan\delta_0$ and a maximum at normal incidence when $\partial h/\partial y = \tan\delta_0$. The factor $(1 - i\beta\xi)$ accounts for the perturbation of the surface of the large-scale structure h by an additive, small-scale structure ξ. As ξ and h depend on (x, y) coordinates in the mean plane, and time, t, so does g_1: $g_1 \rightarrow g_1(x, y, t)$. A simple heuristic derivation of g was given in Harger (1980a) along with a recollection of conditions sufficient for the validity of the two-scale model; in particular, horizontal transmit and receive polarizations are assumed.

Therefore, (4) $describes\ the\ sea\ as\ sensed\ by\ the\ SAR$—$indeed,\ any\ microwave\ radar\ system.$ Whereas the fine-scale structure ξ enters linearly, the large-scale structure h enters nonlinearly and multiplicatively with ξ: one expects that generally this "nonlinear scale interaction" can cause a complicated image dependence on their separate natures. Since $\beta h_{\max}$ can be large, $\exp(i\beta h)$ can have a (wavenumber) bandwidth much larger than that of h and, indeed, much larger than the (wavenumber) bandwidth of the SAR.

2.3. The SAR Image of a Time-Variant Scene

The SAR system scans a scene in range (r) by emitting pulses: in effect, the scanning velocity is half the velocity of light so that, during any single such scan, the scene may be here assumed time-invariant. The scanning velocity in the orthogonal, along-track direction (x) is the much slower vehicle velocity v: from scan-to-scan the scene may vary significantly. Omitting consideration of the important details of this sampling scheme (see Harger, 1980), if the SAR vehicle along-track position is $x = vt$, the SAR senses $g_1(x, r/\sin \delta_0, t = x/v)$, where $r = y \sin \delta_0$. The "scanning beam" of the SAR of course has nonzero spatial extent and is given by $F(r)A(x)$, determined, respectively, by the emitted pulse modulation and the far-field azimuth antenna pattern along with the Doppler shift due to relative motion. The actual or virtual (complex) recorded data is then

$$S(x,r) = S_0 \int \int dx' dr' F(r - r') A(x - x') \{ e^{i2k_0 r'} g(x', r', t = x/v) \} \tag{5a}$$

where

$$g(x', r', t = x/v) \equiv g_1(x', r'/\sin \delta_0, t = x/v) \tag{5b}$$

and the constant S_0 depends on k_0 and R_0 among other SAR parameters. The factor $\exp(i2k_0 r')$ accounts for the incremental (to R_0) ray path change to points (x', r') in the SAR's slant range plane [defined by the mapping $(x', y') \to (x', r'/\sin \delta_0)$, where (x', y') is in the mean sea surface]: because it, in effect, "heterodynes"—i.e., translates in wavenumber—g and because $2k_0$ is generally much larger than the SAR bandpass and the bandwidth of $\exp(i\beta h)$, an image will be obtainable only if ξ has wavenumber content near the wavenumber (vector) $(0, -2k_0)$—the so-called Bragg scattering phenomenon.* The presence of such a fine-structure ξ is, at microwave frequencies, most likely to be associated with a propitious local wind (Kitaigorodskii, 1981). We may therefore expect that wind and wave conditions will sometimes exist when SAR imaging is not obtainable.

If the SAR "signal film" (5) is processed with the customary matched filter of impulse response $[F(-r)^* A(-x)^*]$, then the (complex) SAR image is

$$I(x, r) = S_0 \int dx_1 A[-(x - x_1)]^* \int dx_2 A(x_1 - x_2) \int dr_2 q_F(r - r_2)[e^{i2k_0 r_2} g(x_2, r_2, x_1/v)] \tag{6}$$

Here $q_F(r)$, the convolution of $F(r)$ with $F(-r)^\dagger$, is the range-direction impulse response of the SAR *system*. Because of the time dependence of the scene, one cannot similarly carry out the dx_1 integration and identify $q_A(x)$, the convolution of $A(x)$ with $A(-x)^*$, the azimuth-direction impulse response of the SAR *system*: this considerably obscures the analysis of, and intuition toward, (6) (Harger, 1980b).

*There is no implication that solely a single spatial wavenumber is thereby influential: as much of our attendant discussion and Harger (1981) clarifies, a band of spatial wavenumbers is generally influential. For small incidence angles δ_0, the long wave, appearing to the radar as $h(x, r/\sin \delta_0, t = x/v)$, can produce large wavenumbers—particularly via its appearance in $\exp(-i\beta h)$—of the order of $(0, -2k_0)$ so that they themselves, without the assistance of the small waves, are "mixed" into the SAR passband and influence the image—this mechanism includes "specular returns" and is included in our model by the "1" appearing in the factor $(1 - i\beta\xi)$ of (4). We shall here, for simplicity, assume intermediate incidence angles that, in effect, allow us to discard the "1" and, furthermore, require us to consider only the wavenumber content of ξ about $(0, -2k_0)$: ultimately, this assumption is embodied in (7c) being the relevant small wave structure.

*Taking $X = v_0 R_0/D_h$ and $\hat{K} = 4\pi/\lambda_0 R_0$ as given, the SAR system spatial bandwidth, Ω, is defined to be $\hat{K}X$, which is thus seen to be $4\pi/D_h$. The customarily defined SAR system resolution, σ, turns out to be $D_h/2$, as is well known; the actual resolution, however defined, depends on the specifics of the system transfer functions, here taken as truncated Gaussian shapes.

2.4. The SAR Image of the Sea

We may now find the SAR image of short gravity waves on a long gravity wave by combining the SAR image of a time-variant scene (6), with the reflectivity density as approximated by the two-scale model (4), and with the nonlinear hydrodynamic model (3). We find said image to be

$$I(x,r) = I_0 \int dx_1 \, A[-(x-x_1)]^* \int dx_2 \, A(x_1-x_2) \int dr_2 \, q_F(r-r_2)(\mathscr{A}\mathscr{F}\,e^{i\Psi})(x_2,r_2,x_1) \quad (7a)$$

where (i) the amplitude factor

$$\mathscr{A}(x_2,r_2,x_1) \equiv a_0[1 + AK_{20}\cos\phi(x_2,r_2,x_1) - AK_{20y}\tan\delta_0\sin\phi(x_2,r_2,x_1)] \quad (7b)$$

$$\phi(x_2,r_2,x_1) \equiv \mathbf{k}_{20}\cdot\boldsymbol{\rho} - \sigma_v(K_{20})x_1$$

$$\mathbf{k}_{20} \equiv \begin{pmatrix} K_{20x} \\ K_{20y}/\sin\delta_0 \end{pmatrix}, \quad \boldsymbol{\rho} \equiv \begin{pmatrix} x_2 \\ r_2 \end{pmatrix}$$

$$\sigma_v(K_{20}) \equiv \sigma_0(K_{20})/v$$

shows, in its second term, the dependence of the fine scale's amplitude on its location on the large-scale structure and, in its third term, the dependence on the incidence angle as characterized by physical optics; (ii) the factor

$$\mathscr{F}(x_2,r_2,x_1) \equiv f_1\left[\begin{pmatrix} x_2 \\ r_2/\sin\delta_0 \end{pmatrix} + \begin{pmatrix} \mathbf{K}_{20} \\ K_{20} \end{pmatrix} A\sin\phi(x_2,r_2,x_1), \frac{x_1}{v}\right] \quad (7c)$$

exhibits the to-and-fro horizontal translation of the intrinsic fine scale by the long wave, and (iii) the phase

$$\Psi(x_2,r_2,x_1) \equiv 2k_0 r_2 - \mathbf{k}_{10}\cdot\boldsymbol{\rho}_2 + \sigma_v(K_{10})x_1 - (K_{10}\cos\theta)A\sin\phi(x_2,r_2,x_1)$$
$$- \beta A\cos\phi(x_2,r_2,x_1) \quad (7d)$$

$$\mathbf{k}_{10} \equiv \begin{pmatrix} K_{10x} \\ K_{10y}/\sin\delta_0 \end{pmatrix}$$

$$\sigma_v(K_{10}) \equiv \sigma_0(K_{10})/v$$

incorporates, in the last term, the heaving of the fine-scale structure by the long wave and, in the fourth term, the mean wavenumber distortion of the fine scale in accordance with its location on the long wave, in the second and third terms the undistorted mean phase of the fine scale, and, in the first term, the incremental phase due to the ray path to the point r_2 in the slant-range plane of the SAR.

While q_F is a real function, $A(x)$ of course has the quadratic phase $\hat{K}x^2/2$, $\hat{K} \equiv 2k_0/R_0$, inherent to the SAR technique: we note that

$$\text{Arg}\{A[-(x-x_1)]^* A(x_1-x_2)\} = -\frac{\hat{K}}{2}x^2 + \hat{K}(x-x_2)x_1 + \frac{\hat{K}}{2}x_2^2 \quad (7e)$$

The first term does not affect the intensity $|I|^2$ of the SAR image and may be absorbed into $I_1 \equiv I_0\exp(-i\hat{K}x^2/2) = S_0 g_0\exp(-i\hat{K}x^2/2)$.

2.5. Relation to Other Models

As already pointed out, the present SAR image model (7), by incorporating the nonlinear hydrodynamic model of Phillips (1981b) rather than the model of Longuet-Higgins and Stewart (1961), is an improvement over the model of Harger (1980a, b) in that the modification of the short gravity waves by the long gravity wave is more fully evidenced.

The present model appears closely consistent with Wright (1966), where some experimental support is given, Wright (1968), and Plant (1977, 1981). Several quantities appearing in Plant's model can be directly related to—and in fact precisely defined in terms of—the present model.

Valenzuela (1980) has discussed a 1d distributed model in the limit of very large SAR space–bandwidth product, including orbital motion effects and employing an *a priori*-averaged scattering crosssection for the small scale, without phase, related to nonlinear hydrodynamics by a "modulation transfer function." [The search, which has concerned many in this area, for "modulation transfer functions," by usual definition appropriate to linear, coordinate-invariant transformations, seems to us, as the transformation (7) is nonlinear and coordinate-variant, destined to be of limited success.] Jain (1981) has also studied a distributed scene model incorporating, ad hoc seemingly, nonlinear hydrodynamic effects.

A separate class of models (see, e.g., Alpers and Rufenach, 1979, and Swift and Wilson, 1979) are based on the SAR system response to a "point scatterer" moving in accordance with the orbital motion of the long wave but otherwise time-invariant. The applicability of this approach to describing an extended scene given the nonlinear interactions and time variance present, seems problematic. In any case, major predictions of such a model (see, e.g., Alpers *et al.*, 1981) are not borne out in the numerical studies of our model completed thus far and described next: e.g., there is observed no "azimuth bunching" phenomenon.

3. A NUMERICAL ANALYSIS

It appears rather difficult to understand the nature of the SAR image (7) without the assistance of numerical analysis: an efficient computation of (7) is not a simple task for several reasons. First, as noted above, because of the time-variant aspect of the scene, rather than the SAR system (real) impulse response in the azimuth coordinate, q_A, the considerably more complicated (complex) azimuth modulation A must be generated: its complexity is measured by its space–bandwidth product which can be several thousands. Second, the phase βh can make very large excursions, requiring rather dense sampling to adequately represent $\exp(i\beta h)$. Third, the intrinsic short wave ensemble f_0 is most suitably modeled by a sample function of a random field: because of the nonlinear scale interaction, its required bandwidth may be very much broader than the SAR bandwidth. Fourth, as the temporal variation of the sea surface precludes (7) being of convolution form, standard "fast algorithms" cannot be as fully utilized. And, fifth, only a delimited area of the sea can be analyzed: as the "dispersed extent" of A is very large relative to the system resolution and the surface motion might preclude complete "compression" by the processor, the image of a delimited area can affect, and be affected by, the image of other areas; also, global effects such as image skewing may not be as evident.

The first, third, fourth, and fifth cited difficulties are mitigated (first item), avoided (fourth and fifth items), and clarified (third item) if (7) is alternatively expressed in "wavenumber domain form," much as in Harger (1981). While such a form is computationally attractive, it seems considerably more illuminating here to first delimit the area whose mapping will be studied, and then convert to a wavenumber domain description—first in describing the system operation in azimuth and then in the description of the scene, as will now be done.

Here the delimited area will be chosen large relative to the SAR system resolution parameter, defined here as Ω^{-1}, Ω being the SAR system bandwidth in either range or azimuth coordinate—but small, in azimuth (x) dimension, relative to the extent $X \equiv \lambda_0 R_0/D_h$ of the far-field antenna pattern, D_h being the antenna aperture's x dimension.* Then [see (7e)] the

---

*Subsequently, all curves will be normalized to unit maximum value, the actual maximum noted when of interest.

quadratic phase $\hat{K}x^2{}_2/2$ may be neglected: that is, the delimited area is in the far-field region. If the image I is examined only over, roughly, the same delimited area, the quadratic phase $(-\hat{K}x^2/2)$ is also negligible in (7a). Then, with some changes of variables of integration, and letting D_x and D_r be the x extent and r extent, respectively, of the delimited area, (7a) becomes, approximately,

$$I\left(\frac{n}{\Omega},r\right) \approx \frac{1}{2\pi}\int dK e^{i\eta K}|A(\Omega K)|^2 \int_{-\Omega D_x/2}^{\Omega D_x/2} d\eta_2 e^{-iK\eta_2} \int_{-D_r/2}^{D_r/2} dr_2 q_F(r-r_2)$$

$$(\mathscr{A}\mathscr{F}e^{i\psi})\left(\frac{\eta_2}{\Omega},r,T_aK\right), \quad |\eta| \lesssim \Omega D_x/2 \tag{8}$$

where $T_a = \Omega/\hat{K}v$ is the nominal extent (in seconds) of A—i.e., the nominal SAR azimuth integration time. Under the customary SAR large space–bandwidth product condition for A, $|A(\Omega K)|^2 \equiv B(K)$ is just (proportional to) the SAR system transfer function, $q_A(\Omega K)$, in the k_x direction: thus, in the azimuth coordinate, (8) employs what would be in the time-invariant case the standard, alternate, wavenumber-domain description of a linear, coordinate-invariant transformation; however, the time-variant nature of the sea results in, of course, $(\mathscr{A}\mathscr{F}e^{i\psi})$ itself depending upon K. In (8) K is a dimensionless wavenumber, being normalized by the SAR bandwidth Ω; similarly, η and η_2 are dimensionless spatial variables, normalized by Ω^{-1}. [That is, the actual wavenumber variable is ΩK (meter^{-1}) and the actual spatial variable is η/Ω (meter).] As a similar normalization clearly can be done in the range coordinate, (8) yields a "canonical" transformation, with all system and surface parameters appearing in the form $(\mathscr{A}\mathscr{F}e^{i\psi})$.

To exemplify the nature of the SAR image, a study will be discussed of the special cases of an azimuth-traveling long wave (in detail) and range-traveling long wave (more briefly), assuming range-traveling short wave structures.

3.1. Azimuth-Directed Long Wave and Range-Directed Short Waves

Suppose that the long wave is entirely azimuth (x)-directed. Then $K_{20y}=k_{20r}=0$ and $\theta = \pi/2$: therefore, $\mathscr{A}$ is subject only to the nonlinear hydrodynamic perturbation. Suppose also that the short waves are simplified to a purely "range" (y)-directed ensemble: then, for each member of the ensemble of wavenumber k_{1r}, the image (8) is of the form

$$I\left(\frac{\eta}{\Omega},r;k_{1r}\right) \approx \frac{1}{2\pi}\int dK e^{i\eta K}B(K) \int_{-\Omega D_x/2}^{\Omega D_x/2} d\eta_2 e^{-i\eta_2 K} \int_{-D_r/2}^{D_r/2} dr_2 q_F(r-r_2)$$

$$\cdot \mathscr{A}_0(1 + AK_{20}\cos\phi)\exp\{i(2k_0 - k_{1r})r_2 + i[\sigma(K_{1r})T_a]K - \beta A\cos\phi] \tag{9}$$

where

$$\phi \equiv (k_{20x}/\Omega)\eta_2 - [\sigma(K_{20})T_a]K$$

We observe that the dr_2 integration yields

$$\tilde{q}_F(2k_0 - k_{1r})e^{i(2k_0 - k_{1r})r} \tag{10a}$$

$\tilde{q}_F$ being the transfer function of the SAR system associated with the range coordinate. The form (10) indicates clearly those wavenumbers of the short wave ensemble that can affect SAR image: necessarily, k_{1r} is such that

$$(2k_0 - k_{1r}) \in (-\Omega/2, \Omega/2) \tag{10b}$$

Ω the SAR system spatial bandwidth.

Because of the condition (10b), the phase $[\sigma(K_{1r})T_a]K$ appearing in (9) is, for a typical L-band SAR system, within a small fraction of a radian, approximately $[\sigma(K_{10})T_a]K$ where $K_{10} \equiv 2k_0 \sin \delta_0$. For consider

$$\left| \sigma(2k_0 \sin \delta_0) - \sigma\left(2k_0 \sin \delta_0 + \frac{\Omega}{2}\sin \delta_0\right)\right| T_a K$$

$$< (g2k_0 \sin \delta_0)^{1/2} T_a \left|1 - \left(1 + \frac{\Omega/2}{2k_0}\right)^{1/2}\right| \approx \frac{2}{\pi}(\sin \delta_0)^{1/2} \frac{\lambda_0^{3/2} R_0}{v D_h^{\,2}}$$

The ratio R_0/v is roughly 10^2 for satellite or aircraft SAR systems. If $\lambda_0 = 0.3$ m and $D_h = 10$ m—corresponding to a resolution of about 5 m in a stationary scene—then the bound is about 0.1 rad. If $D_h = 5$ m, corresponding to a resolution of about 8 feet in a stationary scene, the bound is about 0.4 rad.

Now, summing over K_{1r}/k_{1r}, the complex SAR image becomes

$$I\left(\frac{\eta}{\Omega}, r\right) = \frac{1}{2\pi}\int dK e^{i[\eta - \sigma(K_{10})T_a]K} B(K)$$

$$\cdot \int_{-\Omega D_x/2}^{\Omega D_x/2} d\eta_2 e^{-iK\eta_2} \cdot a_0 (1 + 2AK_{20}\cos\phi)\exp(-i\beta A \cos\phi)$$

$$\cdot \frac{1}{2\pi}\int \frac{dk_{1r}}{\sin \delta_0} \tilde{\jmath}\left(0, \frac{k_{1r}}{\sin \delta_0}, 0\right) \tilde{q}_F(2k_0 - k_{1r})\, e^{i(2k_0 - k_{1r})r} \qquad (11)$$

a product of range and azimuth dependent factors. The second factor may be written

$$\frac{1}{2\pi}\int \frac{dk}{\sin \delta_0}\tilde{\jmath}\left(0, \frac{k + 2k_0}{\sin \delta_0}, 0\right)\tilde{q}_F(k)e^{ikr} \equiv f_{11}(0, r, 0)$$

the short wave ensemble with wavenumbers centered on $(0, -2k_0)$ and of bandwidth Ω, as filtered by q_F. Since this filtering of the range structure is a familiar operation, we proceed with the examination of the azimuth behavior described by the first factor of (11).

Notice that the phase $[\sigma(K_{10})T_a]K$ simply results in an η translation of $\Delta\eta \equiv [\sigma(K_{10})T_a]$—that is, an x-azimuth translation of

$$\Delta_x \equiv \frac{\sigma(K_{10})T}{\Omega} = \frac{[\sigma(K_{10})/K_{10}]\sin \delta_0}{v}\cdot R_0$$

the familiar (see Harger, 1970) azimuth shift associated with a radial velocity, in this case the mean phase velocity $v_p \sigma(K_{10})/K_{10}$ of the free, short gravity waves, projected onto the SAR's line-of-sight, or range, coordinate. This phenomenon is evident in (8) also where it may be interpreted as a mean x translation.

We may therefore write the first factor of (11) as

$$I\left(\frac{\eta - \Delta_\eta}{\Omega}\right) \approx \frac{1}{2\pi}\int dK e^{i\eta K} B(K)\int_{-\Omega D_x/2}^{\Omega D_x/2} d\eta_2 e^{-iK\eta_2} a_0(1 + 2AK_{20}\cos\phi)e^{-i\beta A\cos\phi} \qquad (12a)$$

where

$$\phi \equiv (K_{20}/\Omega)\eta_2 - [\sigma(K_{20})T_a]K$$

The SAR image (12a) is affected by three phenomena. First is the nonlinear hydrodynamic interaction of the long and short waves evidenced by the contribution $(2AK_{20}\cos\phi)$ to $\mathscr{A}$. Second is the possibly very large deviation of the phase $\psi = \beta A \cos\phi$, resulting in a possibly drastic narrowband filtering by $B(K)$. Third is, of course, the dependence of $\mathscr{A}$ and ψ upon K, reflecting the time variation of the sea surface. In the following numerical study, these effects will be artificially isolated to help understand the image when all three phenomena interact.

Segments of the total wave height and the reflectivity density modulus to be associated with (12) are shown in Fig. 3 for a long wave with a normalized wavelength of $18\pi/\Omega$ and with a height A resulting in a nonlinear hydrodynamic amplitude variation of $2AK_{20} = 1/6.7$. Recall that the total wave height is $[h(x, y, t) + \xi(x, y, t)]$ which here is (at $t = 0$)

$$A \cos K_{20}x + a_0(1 + AK_{20}\cos K_{20}x)\cos K_{10}y$$

therefore, for any t and y, the surface height is proportional to $\cos K_{20}x$: this is shown in Fig. 3, curve a, with an arbitrary normalization. The corresponding reflectivity density has the modulus

$$|a_0(1 + AK_{20}\cos K_{20}x)|$$

this is shown in Fig. 3, curve b.

As it will be seen to be a dominant influence on the image, we observe first, in isolation, the bandlimiting phenomenon mentioned just above: we set $S_2 \equiv \sigma(K_{20})T_a \equiv 0$ and $A_p \equiv 2AK_{20} \equiv 0$—thereby artificially eliminating, respectively, motion effects and nonlinear hydrodynamic effects. Now the image is completely parameterized, in the normalized variables, by $B \equiv \beta A = (2k_0 \cos\delta_0)A$ which will be taken in the set (1, 5, 10, 50, 100): for an incidence angle of $\pi/4$ rad, this corresponds to a long-wave-height-to-radar-wavelength ratio in the set

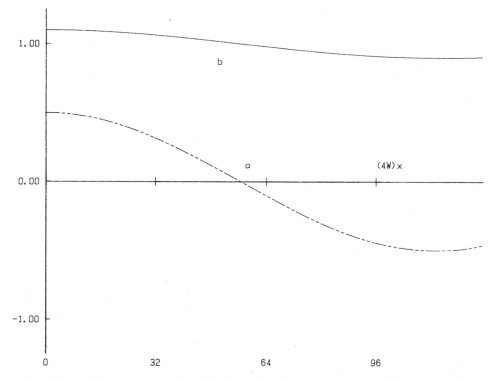

FIGURE 3. (a) The total wave height and (b) the modulus of the reflectivity density for a normalized long wave wavelength of $18\pi/\Omega$ and nonlinear hydrodynamic amplitude variation $2AK_{20} = 1/6.7$.

(0.11, 0.56, 1.12, 5.62, 11.1) and further corresponding to long wave heights—assuming $\lambda_0 = 30\,\mathrm{cm}$—in the set (3.36 cm, 16.9 cm, 33.6 cm, 1.69 m, 3.36 m). [The corresponding long wave slope parameter values AK_{20}—assuming a resolution $\Omega^{-1} = 5\,\mathrm{m}$—are in the set (1/1340, 1/268, 1/134, 1/26.8, 1/13.4) so that the hydrodynamic approximation should be quite accurate.]

It is assumed that $B(K)$ follows a Gaussian shape for $|K| \leqslant 1$ and is zero beyond, as shown in Fig. 8, curve b; the corresponding SAR system impulse response (to a stationary "point scatterer") is shown in Fig. 8, curve a.

The actual numerical computation of (12), which is now, more simply,

$$I\left(\frac{\eta - \Delta_\eta}{\Omega}\right) \approx \frac{1}{2\pi}\int dK e^{i\eta K}B(K)\int_{-\Omega D_x/2}^{\Omega D_x/2} d\eta_2 \exp\left[-iK\eta_2 - iB\cos(K_{20}/\Omega)\eta_2\right] \tag{12b}$$

is efficiently accomplished by using the Jacoby–Anger expansion of $\exp(-i\beta A\cos\phi)$ which is effectively truncated by $B(K)$'s finite normalized wavenumber bandwidth: a "look-up table" for some Bessel functions enables fast computation of the image. The following conditions are reasonable and simplifying: letting $v_{p2} \equiv \sigma(K_{20})/K_{20}$, the phase velocity of the long wave, we assume that

$$\frac{v_{p2}}{v}\frac{D_x}{X} \ll 1 \tag{13a}$$

(the latter already assumed) and (also already assumed)

$$D_x \gg 1/\Omega \tag{13b}$$

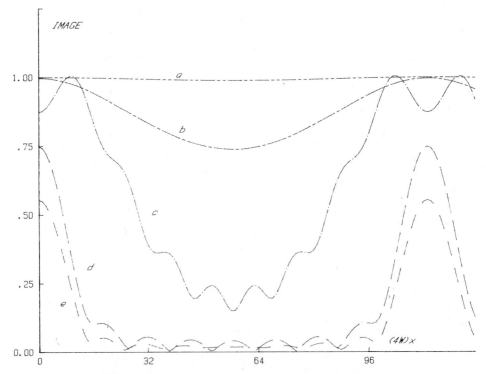

FIGURE 4. The SAR image of (12b) showing, in isolation, the narrowband limiting effect, the long wave relative amplitude parameterized by $B \equiv \beta A$ equal to (a) 1, (b) 5, (c) 10, (d) 50, (e) 100.

Segments of the resulting images are shown in Fig. 4. It is evident that the SAR image modulus bears, generally, *no* relation to the *scene's* reflectivity density modulus—which, in the case under discussion, is constant!—and *little* relation to the height structure, Fig. 3; a spatially variable structure in the image modulus appears *solely* because of the bandlimiting phenomenon. It is also evident that, as the long wave amplitude increases, narrow structures appear centered on the location of the extrema of the long wave. Recalling Fig. 8, curve a, one guesses said structure to be the SAR azimuth impulse response: indeed this is so, as is shown below by a stationary phase evaluation of (12b).

Because of its important role in SAR image formation, we recall that such a "narrowband filtering phenomenon" is simply explained heuristically when A is large. For then the waveform $\exp(-i\beta h)$, near any x_s, has a useful, local, "quasi-stationary" approximation as a sinusoid with a "spatial frequency" $k_s = -\beta h' = \beta A K_{20} \sin k_{20} x$: the SAR image near x_s is then approximately this sinusoid multiplied by $\tilde{q}(k_s)$ (from well-known linear, coordinate-invariant system analysis). For most x_s, $\tilde{q}(K_s)$ is zero, as $\beta A K_{20}$ far exceeds the SAR bandwidth—the support of $\tilde{q}$: it is nonzero only when x_s is near the extrema of h; further, there $\sin k_{20} x_s \approx k_{20} x_s$ and so it can be seen that the image spectrum closely replicates $\tilde{q}(k)$. Below, this explanation is made more precise by the method of stationary phase.

Next, the effect of surface temporal variation is incorporated by taking $S_2 \equiv \sigma(K_{20}) T_a$ nonzero, $\sigma(K_{20}) = (gK_{20})^{1/2}$, g the acceleration of gravity. To specify T_a, some assumptions must be made about radar system parameters: $T_a = \lambda_0 R_0/vD_h$. Typically, we take $T_a = 2$ s, resulting in $S_2 = 0.71$. Thus, as K ranges through $(-1, +1)$, $S_2 K$ ranges through about 2 rad: the apparent variation of $\exp(i\beta \cos \phi)$ is much greater when B is large. The numerical results are shown in Fig. 5. It is apparent that the temporal variation considerably alters the image modulus: in some cases—low B—there is a "period" about that of the long wave and in other cases—large B—there is a rough separation of maxima more of the order of half the long wave wavelength. The

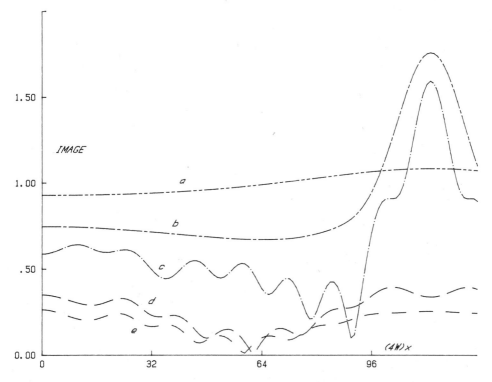

FIGURE 5. The SAR image of (12a), ignoring the nonlinear hydrodynamic interaction, parameterized by the long wave relative amplitude B equal to (a) 1, (b) 5, (c) 10, (d) 50, (e) 100.

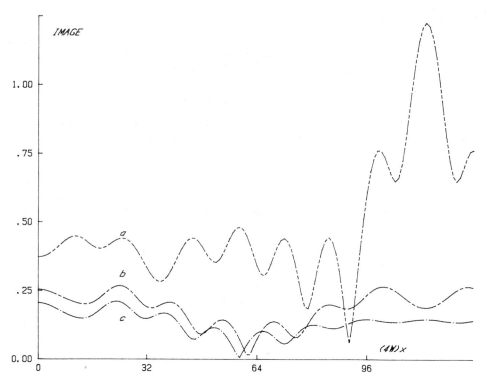

FIGURE 6. The SAR image (12a), parameterized by the long wave's relative amplitude B equal to (a) 10, (b) 50, (c) 100.

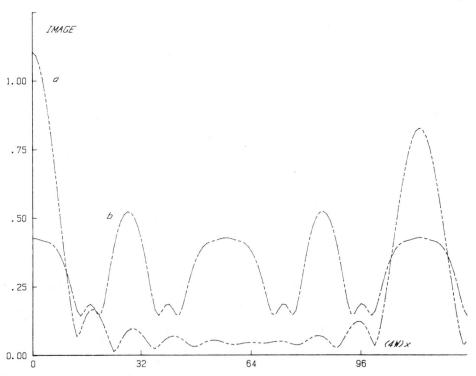

FIGURE 7. (a) The SAR image (12a), incorporating nonlinear hydrodynamics but not motion. ($B = 100$ and $2AK_{20} = 1/6.7$.) (b) The SAR image (12a) but with a long wave wavelength shortened by one-quarter vis-à-vis Fig. 5. ($B = 100$.)

striking effect of bandwidth limiting is now partially obscured. It will be shown below that this "obscuration" is, when B is large, due to a quadratic phase error that is proportional to the phase velocity (v_{p2}) of the long wave: thus, shorter-wavelength long waves yield SAR images whose nature is more dominated by the bandlimiting effect (see Fig. 7, curve b).

Finally, the nonlinear hydrodynamic perturbation of the amplitude $\mathscr{A}$ is incorporated, characterizing the image (12a), whose calculations are shown in Fig. 6. The suppression of those maxima of the image located at the minima of the long wave can be discerned for the larger relative long wave amplitudes B; but for the smaller B, the motion effects dominate.

If, for interest, the motion is removed artificially by setting $S_2 \equiv 0$, the nonlinear hydrodynamic effect is clearly evidenced (see Fig. 7, curve a)

As adumbrated above, the narrowband filtering and nonlinear hydrodynamic effects should dominate the motion effect for shorter long wave wavelengths: this is evidenced in Fig. 7, curve b for $\Lambda_{20} = 18\pi/4\Omega$, one-quarter as long.

The above results indicate that, as $B = \beta A$ becomes larger, the image nature is somewhat more regular and therefore possibly simply predictable: this indeed is so as is now discussed. When $\beta A \gg 1$, the method of stationary phase yields an approximate evaluation of the $d\eta_2$ integral of (12a). The stationary points $\{\eta_n, n = 0, \pm 1, \ldots\}$ are solutions of

$$\beta A(K_{20}/\Omega)\sin\{(K_{20}/\Omega)\eta_n - [\sigma(K_{20})T_a]K\} = K$$

as $K \in (-1, +1)$, if $\beta A(K_{20}/\Omega) \gg 1$, then the stationary points are approximately

$$\frac{\eta_n}{\Omega} = n\frac{\Lambda_{20}}{2} + [\sigma(K_{20})/K_{20}]T_a K, \quad n = 0, \pm 1, \ldots \tag{14a}$$

The stationary phase approximation is accurate provided the integrand does not fluctuate too rapidly: the necessary condition can be seen to be

$$\sqrt{A\beta}\,K_{20}\min(\Lambda_{20}, X) \gg 1 \tag{14b}$$

Then (12a) becomes

$$I\left(\frac{\eta - \Delta y}{\Omega}\right) \approx (2\pi/\beta A(K_{20}/\Omega)^2)^{1/2} \sum_{n=0,\pm 1,\ldots} e^{-i\beta A(-1)^n}[1 + (-1)^n AK_{20}]$$

$$\cdot \frac{1}{2\pi}\int dK e^{i[\eta - n(\Lambda_{20}\Omega/2)]K}\{B(K)e^{-i[\sigma(K_{20})/K_{20}]T_a\Omega K^2}\}$$

$$\cdot \operatorname{rect}\left\{\frac{n(\pi\Omega/K_{20}) + [\sigma(K_{20})/K_{20}]T_a\Omega K}{\Omega D_x}\right\} \tag{14c}$$

The K dependence of the rect function in (14c) may be neglected if $[\sigma(K_{20})/K_{20}]T_a\Omega \ll D_x$, i.e., if $v_{p2}T_a \ll D_x$—or, if

$$\frac{v_{p2}}{v} \ll \frac{D_x}{X}$$

a condition already included in (13). Thus, the role of the rect function is, approximately, simply to truncate the image support to $(-D_x/2, D_x/2)$ (in the unnormalized x coordinate). Therefore, we have

$$I\left(\frac{\eta - \Delta_n}{\Omega}\right) \approx \left[\frac{2\pi}{\beta A(K_{20}/\Omega)}\right]^{1/2} \sum_{n=D_x/\Lambda_{20}}^{+D_x/\Lambda_{20}} e^{-i\beta A(-1)^n}[1 + AK_{20}(-1)^n]a_{A^*}\left(\eta - n\frac{20}{2}\Omega\right) \tag{14d}$$

where

$$q_{A*}(\eta) \equiv \frac{1}{2\pi} \int dK e^{i\eta K} [B(K)e^{-i(v_{p2}T_a\Omega)K^2}] \tag{14e}$$

is *the impulse response of the SAR system subject to a quadratic phase error in the wavenumber domain.*

Reverting to the nonnormalized wavenumber $k \equiv \Omega K$, the quadratic phase error is $(v_{p2}/v)(k^2/\hat{K})$: this corresponds to a spatial domain quadratic phase error $[(v_{p2}/v)\hat{K}]x^2$. At the extremes $k = \pm\Omega/2$, the wavenumber domain maximum quadratic phase error is

$$\left(\frac{v_{p2}}{v}\right)\left(\frac{\Omega X}{4}\right) = \pi\left(\frac{v_{p2}}{v}\right)\frac{\lambda_0 R_0}{D_h^2} \tag{14f}$$

a well-known form (Harger, 1970, p. 36) that states that the long wave's relative phase velocity, v_{p2}/v, must be extremely small—the SAR azimuth space–bandwidth product ΩX is typically thousands—in order that the nominal maximum quadratic phase error be small with respect to, say, 1 rad.

[Alpers *et al.* (1981) discuss (p. 6486) an "azimuth degradiation" due to a quadratic phase error whose parametric dependency differs from (14f) further stating—in their abstract—that "the phase velocity of the long waves does not enter into the imaging process," in direct contradiction to our result.]

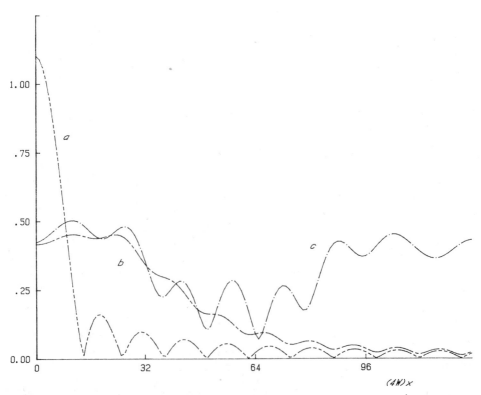

FIGURE 8. (a) SAR azimuth impulse response $q_A(\eta)$ (b) the SAR azimuth impulse response $q_{A*}(\eta)$ and (c) the SAR image constructed using (a) in (14d).

[This asymptotic result also suggests that a processor refocusing might be able to compensate for this quadratic phase error and hence motion effects in this situation: we will not pursue the matter here but merely note that this technique has been employed for some time (Shuchman and Zelenka, 1978).]

The SAR system impulse response, $q_A(\eta)$, in the absence of phase errors, is shown in Fig. 8, curve a. The modulus $|q_{A*}(\eta)|$ of the impulse response, with a phase error consequent to the long wave under discussion, is shown in Fig. 8, curve b: considerable degradation is evident, corresponding to a quite large maximum quadratic phase error of 6.4 rad (in the normalized coordinate K). The SAR image may be approximated by assembling such q_A's according to (14d), the result shown in Fig. 8, curve c, agreeing reasonably well with the more precise image of Fig. 6 curve c.

3.2. Range-Directed Long and Short (Sinusoidal) Waves

We comment briefly on the situation of a y-directed long wave—termed "range-directed"—in the special case of a short wave structure consisting solely of a range-directed sinusoid with $K_{10} = 2k_0 \sin \delta_0$. As now $\cos \theta = 1$ and $K_{20y} = K_{20}$, (8) yields, after performing the dx_2 integration,

$$\overset{*}{I}\left(\frac{\eta - \Delta_\eta}{\Omega}, r\right) = \frac{1}{2\pi} \int dK e^{i\eta K} B_1(K) \int_{-D_r/2}^{D_r/2} dr_2 q(r - r_2) a_0 [1 + AK_{20}5^{1/2} \cos(\phi - \alpha)]$$
$$\cdot \exp[-i2k_0 A \cos(\phi - \delta_0)] \tag{15a}$$

where $\alpha \equiv \arctan(1/2)$, $\phi \equiv k_{20}r - [\sigma(K_{20})T_a]K$, and

$$B_1(K) \equiv B(K)\cdot\Omega D_x \frac{\sin \Omega D_x K/2}{\Omega D_x K/2} \tag{15b}$$

As $\Omega D_x \gg 1$ by assumption all along, $B(K) \to 1$ in (15b). Then, again using the Jacoby–Anger expansion for the exponential of (15a), the integrations may be performed yielding

$$I\left(\frac{\eta - \Delta_\eta}{\Omega}, r\right) = a_0 \sum_{\substack{m = 0,1,\dots \\ v = 0, \pm 1 \\ s = \pm 1}} b_v C_{m,s} \text{rect}\left[\frac{\eta - (m + v)S_2}{\Omega D_x}\right]$$
$$q_F[(sm + v)k_{20}]e^{i(m + v)k_{20}r}, \quad r \in (-D_r/2, D_r/2) \tag{15c}$$

where $S_2 \equiv [\sigma(K_{20})T_a]$, $C_{m,s} \equiv \varepsilon_m(-i)^m J_m(2B)^{1/2} \exp(-ism\delta_0)$, ε_m being 1 for $m = 0$ and 2 otherwise, and

$$b_v \equiv \begin{cases} (AK_{20}5^{1/2}/2)\exp(-iv\alpha) & v = \pm 1 \\ 1 & v = 0 \end{cases}$$

Again because of the finite bandwidth of the SAR, the transfer function in the range coordinate, $\tilde{q}_F$, effectively truncates the summation to indices (m,v,s) such that $|sm + v| < \Omega/2$: then the displacement appearing in the rect functions in (15c) is bounded relative to ΩD_x, the normalized dimension of the delimited scene:

$$|sm + v|S_2/\Omega D_x < (v_{p2}/v)(X/D_x) \tag{16}$$

which is small provided D_x is not too small with respect to X. Assuming (16), the SAR image (15c) is approximately

$$I(r) \approx a_0 \sum_{\substack{m=0,1,\ldots \\ v=0,\pm 1 \\ s=\pm 1 \\ (|sm+v|<\Omega/2)}} b_v C_{m,s} \tilde{q}_F[(sm+v)k_{20}] \quad e^{i(m+v)k_{20}r}, r\varepsilon(-D_r/2, D_r/2), x\in(-D_x/2, D_x/2) \quad (15d)$$

In this approximation there is no azimuth variation and the image appears as if the long wave had no temporal variation at all!

The image (15d) is computed numerically for $B\sqrt{2}$ in the set $(3.36/\sqrt{2}\,\text{cm}, 16.9/\sqrt{2}\,\text{cm}, 3.3.6/\sqrt{2}\,\text{cm}, 1.69/\sqrt{2}\,\text{m}, 3.36/\sqrt{2}\,\text{m})$, assuming, as before, that $\delta_0 = \pi/2$ rad. and $\lambda_0 = 30$ cm. The scene's reflectivity density modulus is, from (15a), at $t=0$,

$$|g(0,y,0)| = |a_0[1 + AK_{20}\sqrt{5}\cos(K_{20}y - \alpha)]|,$$

an example of which is shown in Fig. 9, curve b. The corresponding sea surface height is

$$z(0,y,0) = a_0[1 + AK_{20}\cos K_{20}y]\cos[K_{10}(y + A\sin K_{20}y)] + A\cos K_{20}y,$$

an example of which is shown in Fig. 9, curve a, in segments, to reveal the small wave structure. The underlying long wave is that of Fig. 3a, foreshortened by the projection into the range coordinate.

The set of images is shown in Fig. 10. Because of the effective absence of motion effects, the image's nature, for larger B, is primarily determined by the narrowband filtering effect;

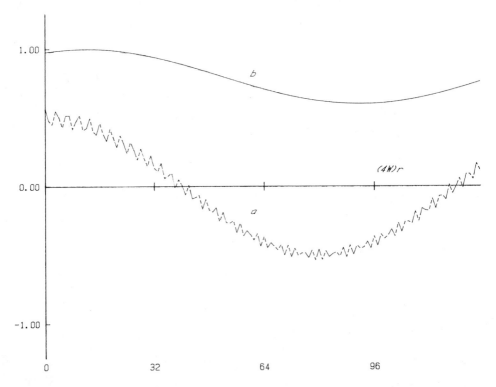

FIGURE 9. (a) A segment of the sea surface height for range-directed, sinusoidal, long and short wave structures. (b) The modulus of the reflectivity density.

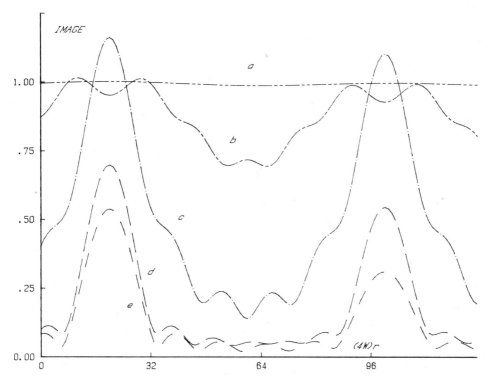

FIGURE 10. The SAR image for range-directed, sinusoidal, long and short wave structures, parameterized by the relative amplitude B equal to (a) $2^{1/2}$, (b) $5(2^{1/2})$, (c) $10(2^{1/2})$, (d) $50(2^{1/2})$, (e) $100(2^{1/2})$. The respective $2AK_{20}$ are (1/848, 1/170, 1/84.8, 1/17, 1/8.48).

secondarily, the combined effect of nonlinear hydrodynamics and the physical optics description of electromagnetic scattering from the long wave—colloquially referred to as, respectively, "straining" and "tilting"—is evident in the displacement of the extrema by the phase δ_0 and the relative suppression of alternate extrema.

Again it is noted that the image behavior is quite regular when B is large and appears to be a suitably modified repetition of the SAR system impulse response in range [recall Eq. (8a)]. And again this behavior is predicted by approximately evaluating (15a) using the method of stationary phase—specifically, to evaluate the dr_2 integral. The stationary points are found as

$$r_n = \left(n + \frac{\delta_0}{\pi} \right) \frac{\lambda_{20}}{2} + (v_{p2} \sin \delta_0) T_a K, \quad n = 0, \pm 1, \ldots$$

that the shift $(\delta_0/2\pi)\lambda_{20}$ is not zero is due to the "to-and-fro" translation of the short wave, as described by nonlinear hydrodynamics, being present along with the "heaving" described, most simply put, by ray optics. The K dependence of r_n—and in fact ϕ—may be neglected if

$$(v_{p2} \sin \delta_0/v)(X/D_x)$$

is small relative to the range resolution $1/\Omega$, a somewhat milder condition than (16). Now the dk integral simply reproduces the "window" in η corresponding to the x-delimited extent $(-D_x/2, D_x/2)$. The image is then proportional to the sum of the translated impulse responses $q_A(r - r_n)$, each weighted by

$$a_0 [1 + AK_{20} 5^{1/2} \cos(\delta_0 - \alpha + n\pi)] e^{-i2k_0(-1)^n}$$

Note, therefore, that while the location of the impulse responses, and hence the extrema of the image, is determined—in this special case—by δ_0, the relative weighting is determined by $(\delta_0 - \alpha)$.

4. SUMMARY

A model of SAR ocean surface imaging has been presented for the first time that unifies fundamental models from nonlinear hydrodynamics, rough surface scattering of electromagnetic waves, and SAR system theory. The resulting SAR image generally does not resemble either the sea height structure or the modulus of the reflectivity density describing the scene as sensed by the SAR. The theory characterizes precisely several "interactions" previously felt to be important: the "nonlinear scale interaction" and, especially, the effect of narrowband filtering of the phase-modulated waveform factor of the reflectivity density seem not to have been sufficiently appreciated before. The nonlinear hydrodynamic and electromagnetic theories used here can now also be applied to other microwave sensors such as "wave probes," "scatterometers," "two-frequency interferometers," "Δk radars," etc., given the development of suitable system models, thereby unifying, to some extent, the important sensors of "radar oceanography." [The theory of "brute-force," side-scanning radar "SLAR" is contained in (7) but not elaborated here.]

The numerical analysis presented here of instances of the fundamental SAR imaging model (7) analyzed a delimited area of the sea: certain reasonable approximations were made that enabled simpler analysis, namely (13) and (16). To verify their innocuousness, additional numerical analysis was performed expressing (7) in the aforementioned, wavenumber-domain representation, the passage of each Fourier component through the SAR system being then simply described, as in Harger (1981) and the corresponding image contributions being then readily summed numerically: the results agree with those presented here. These numerical exemplifications assumed simplified range-directed short gravity wave ensembles: given this caveat we additionally remark upon the results.

It was seen, for an azimuth-directed long wave with a range-directed short wave structure, that the SAR system's finite spatial bandwidth resulted in the (generally restricted) ensemble of influential short waves being approximately, effectively nondispersive during the SAR scanning time—though dispersive over longer time intervals; their thereby concerted rapid fluctuation resulted simply in a (rigid) azimuth translation of the entire delimited image by a distance proportional to the mean phase velocity of the small waves and given by a well-known relation.

The brief discussion of the range-directed long wave situation could be extended to an ensemble of range-directed short gravity waves: it would be seen that the nonlinear interaction, in the two-scale description of the scattering phenomenon, of the long and short waves, means that, as the long wave amplitude B increases, an increasingly broader band of wavenumbers $\{K_{1r}\}$ of the short wave ensemble become influential—because they are "mixed" into the SAR passband by the increasingly greater spectral components of $\exp(i\beta h)$; therefore, the corresponding significant temporal frequencies $\sigma(K_{1r})$ increase. For the specific parameters selected here, at the higher B's, the phase $[\sigma(K_{1r})T_a]$, accumulated during the SAR azimuth scanning time, approaches 1 rad: hence, the influential short waves are only marginally nondispersive while scanned.

The general conjecture, then, is that if the relative long wave amplitude B is not too large, the influential short wave ensemble is effectively nondispersive during the scanning time T_a; as B increases sufficiently, this nature is lost, sooner for those influential short waves directed generally along the long wave direction.

As mentioned earlier, there is no support in these numerical results for the assertion that "orbital motion effects" are an important image-forming mechanism through an "azimuth bunching" phenomenon, especially for higher-amplitude, azimuth-directed long waves: to the contrary, in that case, the temporal variation of the scene "blurs" the image otherwise due, primarily, to narrowband filtering and, secondarily, to nonlinear hydrodynamics; said "blurring" is attributable to a resultant quadratic phase error that is proportional to the phase velocity of the long waves. Consequently, especially for the shorter-wavelength long waves, one may hazard that it is the random, usually rapidly fluctuating (here in range only) nature of the (filtered) short wave ensemble that may tend to obscure the rather regular azimuth behavior.

This "preservation" of the short wave structure, in a direction orthogonal to the long wave direction, if holding suitably generally, could yield useful information about this structure.

To reiterate: the SAR image intensity structure, while clearly informative, generally replicates neither the sea height structure nor the modulus of the reflectivity density describing the sea as sensed by the SAR. Thus, one is led to seek specifically appropriate information extraction procedures. For example, it is observed that the reflectivity density (4) has a phase proportional to the long wave: this leads to the idea that nonlinear processing of the SAR *complex* image could exploit this fact; a preliminary study is reported elsewhere (Harger, 1983, 1984).

REFERENCES

Alpers, W. R., and C. L. Rufenach (1979): The effect of orbital motions on synthetic aperture radar imagery of ocean waves. *IEEE Trans. Antennas Propag.* **AP-27**, 685–690.

Alpers, W. R., D. B. Ross, and C. L. Rufenach (1981): On the detectability of ocean surface waves by real and synthetic aperture radars. *J. Geophys. Res.* **86**, 6481–6498.

Bass, F. G., I. M. Fuks, A. I. Kalmykov, I. E. Ostrovsky, and A. D. Rosenberg (1968): Very high frequency radio wave scattering by a disturbed sea surface. *IEEE Trans. Antennas Propag.* **AP-16**, 554–568.

Beal, R. C., P. S. DeLeonibus, and I. Katz (eds.) (1981): *Spaceborne Synthetic Aperture Radar for Oceanography*, Johns Hopkins Press, Baltimore.

Beckmann, P., and A. Spizzichino (1963): *The Scattering of Electromagnetic Waves from Rough Surfaces*, Macmillan Co., New York.

Harger, R. O. (1970): *Synthetic Aperture Radar Systems*, Academic Press, New York.

Harger, R. O. (1980a): On SAR sea image prediction. *IUCRM Symposium on Oceanography from Space*, Venice.

Harger, R. O. (1980b): The synthetic aperture radar image of time-variant scenes. *Radio Sci.* **15**, 749–756.

Harger, R. O. (1981): SAR ocean imaging mechanisms. *Spaceborne Synthetic Aperture Radar for Oceanography* (R. C. Beal, P. S. Deleonibus, and I. Katz, eds.), Johns Hopkins Press, Baltimore.

Harger, R. O. (1983): A sea surface height estimator using synthetic aperture radar complex imagery. *IEEE Trans. Oceanic Eng.* **OE-8**, 71–78.

Harger, R. O. (1984): A fundamental model and efficient inference for SAR ocean imagery. *IEEE Trans. Oceanic Eng.* **OE-9**, 260–276.

Jain, A. (1981): SAR imaging of ocean waves: Theory. *IEEE Trans. Ocean. Eng.* **OE-6**, 130–139.

Kinsman, B. (1965): *Wind Waves*, Prentice–Hall, Englewood Cliffs, N.J.

Kitaigorodskii, S. A. (1981): Comments. *Spaceborne Synthetic Aperture Radar for Oceanography* (R. C. Beal, P. S. DeLeonibus, and I. Katz, eds.), Johns Hopkins Press, Baltimore, 187.

Longuet-Higgins, M. S., and R. W. Stewart (1961): Changes in the form of short gravity waves on long waves and tidal currents. *J. Fluid Mech.* **11**, 565–583.

Phillips, O. M. (1977): *The Dynamics of the Upper Ocean*, Cambridge University Press, London.

Phillips, O. M. (1981a): The structure of short gravity waves on the ocean surface. *Spaceborne Synthetic Aperture Radar for Oceanography* (R. C. Beal, P. S. DeLeonibus, and I. Katz, eds.), Johns Hopkins Press, Baltimore.

Phillips, O. M. (1981b): The dispersion of short wavelets in the presence of a dominant long wave. *J. Fluid Mech.* **107**, 465–485.

Plant, W. J. (1977): Studies of backscattered sea return with a CW dual-frequency, X-band radar. *IEEE Trans. Antennas Propag.* **AP-25**, 28–36.

Plant, W. J. (1981): The two-scale radar wave probe and SAR imagery of the ocean. U.S. Naval Research Laboratory Report (informally communicated).

Shuchman, R. A., and J. S. Zelenka (1978): Processing of ocean wave data from a synthetic aperture radar. *Boundary-Layer Meteorol.* **13**, 181–192.

Swift, C. T., and L. R. Wilson (1979): Synthetic aperture radar imaging of ocean waves. *IEEE Trans. Antennas Propag.* **AP-27**, 725–729.

Valenzuela, G. R. (1968): Scattering of electromagnetic waves from a tilted, slightly rough surface. *Radio Sci.* **3**, 1057–1066.

Valenzuela, G. R. (1980): An asymptotic formulation for SAR images of the dynamical ocean surface. *Radio Sci.* **15**, 105–114.

Wright, J. W. (1966): Backscattering from capillary waves with application to sea clutter. *IEEE Trans. Antennas Propag.* **AP-14**, 749–754.

Wright, J. W. (1968): A new model for sea clutter. *IEEE Trans. Antennas Propag.* **AP-16**, 217–223.

Zalkan, R., and L. Wentzel (eds.) (1981): *Proceedings of a NORDA Workshop on Describing Ocean Phenomena Using Coherent Radars.* NORDA Technical Note 104, May.

26

THE RESPONSE OF SYNTHETIC APERTURE RADAR TO OCEAN SURFACE WAVES

KLAUS HASSELMANN AND WERNER ALPERS

ABSTRACT. Basic concepts of SAR imaging theory of ocean surface waves are reviewed. The effects of orbital velocity and acceleration on the imaging mechanism are discussed in simple physical terms. The spatially varying orbital velocity contributes to the imaging through "velocity bunching," while the effect is relatively small and still in the linear range, but produces nonlinear image distortions and smearing for larger orbital velocities. The spatially varying orbital acceleration results in a nonuniform azimuthal image smear. Image contrast enhancement by azimuthal focus adjustment of the SAR processor is attributed to this spatial variability of the orbital acceleration effects.

1. INTRODUCTION

Although the response of synthetic aperture radar (SAR) to moving point targets has been studied for more than 10 years (see, e.g., Graf and Guthart, 1969; Kelly and Wishner, 1969; Raney, 1971), the application of these results to ocean wave imaging has created considerable discussion. The difficulties encountered revolve around the proper kinematical description of the sea surface. This in turn depends on the proper multiscale dynamical description of the wave field, extending from long gravity waves down to short backscattering ripples. A number of basic questions have been raised:

1. Can the SAR response to a moving ocean wave field be described to first order by a superposition of the individual SAR responses to independent, moving, single scattering elements (facets) on the ocean surface?
2. Are the coherence times of individual scattering elements sufficiently long that they do not significantly limit the azimuthal SAR resolution?
3. Is the orbital facet acceleration an important factor determining the degradation in azimuthal resolution?
4. Is the wave phase velocity a fundamental wave parameter in the problem of azimuthal focusing?

KLAUS HASSELMANN ● Max-Planck-Institut für Meteorologie, Hamburg, West Germany. WERNER ALPERS ● Institut für Meereskunde, Universität Hamburg, and Max-Planck-Institut für Meteorologie, Hamburg, West Germany; *present address*: Fachbereich Physik, Universität Bremen, Bremen, West Germany.

5. When does the nonlinear distortion of the image through motion effects prevent the detection or interpretation of ocean waves in a SAR image?
6. When can waves be detected as a clear signal above the unavoidable speckle noise of a SAR image?

The answers to these questions have important implications for the interpretation of SAR image spectra in terms of ocean surface wave spectra. Furthermore, the answers must be approximately known to design an optimal SAR for ocean wave imaging. A particular question arising in this context is whether the C-band SAR proposed as payload for the First European Remote Sensing Satellite (ERS-1) is more suitable for ocean wave imaging than the SEASAT L-band SAR.

The problem of azimuthal focusing has been at the centre of much of the discussion on SAR ocean wave imaging.

Controversial views can be found on all of the first four questions in the literature (Larson *et al.*, 1976; Elachi and Brown, 1977; Jain, 1978; Shemdin *et al.*, 1978; Shuchman *et al.*, 1978; Teleki *et al.*, 1978; Alpers and Rufenach, 1979, 1980; Swift and Wilson, 1979; Raney, 1980, 1981; Valenzuela, 1980; Harger, 1980; Rufenach and Alpers, 1981; Alpers *et al.*, 1981; Tucker, 1981; Alpers, 1983).

In the following we summarize a view which follows from a two-scale hydrodynamical and scattering model of the sea surface. A brief outline of the basic elements of SAR theory for ocean wave imaging is presented in a simplified form which attempts to avoid burying the basic physics too deeply in mathematical formalism.

2. BASIC CONCEPTS OF THE SAR OCEAN WAVE IMAGING MECHANISM

Satellite and airborne SARs normally operate at angles of incidence between 20 and 70° for which the microwave return from the sea surface is predominantly due to Bragg backscattering from short surface ripples. Longer waves are seen by the radar because they modulate the backscattering due to the short surface ripples. The modulation can be described in terms of a two-scale model in which the sea surface is represented as a superposition of short Bragg scattering ripples superimposed on longer gravity waves. Bragg scattering theory is applied locally in a reference system lying in the tangent plane ("facet") of the long waves and moving with the local long-wave orbital velocity (see Bass *et al.*, 1968; Wright, 1968; Keller and Wright, 1975; Alpers and Hasselmann, 1978).

Three processes contribute to long-wave imaging:

1. The modulation of the energy of the short Bragg scattering ripples through interactions between the ripple waves and the long gravity waves (hydrodynamic interactions)
2. The changes, through the long-wave slope, in the effective angle of incidence relative to the local facet normal; this modifies the radar return by changing both the Bragg backscattering coefficient and the resonant Bragg wavenumber (electromagnetic interactions)
3. The temporal variations of the facet parameters (facet position, direction of facet normal) and the Bragg backscattering coefficients of the facets during the finite integration time in which the SAR sees a facet (motion effects).

The first two processes are important for both real aperture and SARs, while the third process affects only SARs.

In the following we discuss only the third process, since this part of the imaging mechanism has created the most discussion in the literature.

The effects of facet motions on SAR ocean wave imagery can be evaluated analytically by noting that the SAR integration time (typically 0.1–3 s) is generally short compared with both the period of the long waves (8–16 s), and the intrinsic hydrodynamic interaction time of the backscattering ripples in the reference frame of the moving facets (typically several seconds).

Thus, the temporal variations of the facet parameters and the complex backscattering coefficient can be expanded in a Taylor series with respect to the integration time. The dominant terms in the expansion are found to arise from the radial (slant range) components of the facet (orbital) velocity and acceleration (see Alpers and Rufenach, 1979; Swift and Wilson, 1979; Valenzuela, 1980). We shall neglect here the decorrelation effects arising from the intrinsic hydrodynamic interaction time (see Raney, 1981). These may be expressed in terms of an equivalent Bragg resonance line broadening. Wind wave tank measurements (Keller and Wright, 1975) indicate that the intrinsic Bragg line broadening is generally small compared with the Doppler broadening associated with the facet motion effects.

After correction for motion effects, the SAR image obtained for a moving sea surface can be related to the image of an equivalent time-independent ("frozen") surface whose complex backscattering coefficient $r(x)$ can be represented as the product

$$r(\mathbf{x}) = w(\mathbf{x})m^{1/2}(\mathbf{x})$$

of a random, statistically homogeneous, complex white noise process $w(\mathbf{x})$, which represents the backscattered return due to uniformly distributed, small-scale (Bragg scattering) ripples, and a slowly varying function $m^{1/2}(\mathbf{x})$, which describes the modulation of this return by the long gravity waves. The representation of the small-scale backscattering coefficient $w(\mathbf{x})$ as a zero correlation scale random process is standard. It is valid provided the correlation scale of the backscattering elements is small compared with the SAR resolution, which is normally well satisfied (see Rufenach and Alpers, 1981).

3. THE INFLUENCE OF FACET MOTION ON SAR PERFORMANCE

The phase history of a stationary facet (target) passing through the antenna beam is given by

$$\phi(t) = \tfrac{1}{2}bt^2 \tag{1}$$

where

$$b = 2k_0 V^2/R \tag{2}$$

is a system constant, the "chirp rate," k_0 the radar wavenumber, V the platform velocity, and R the slant range. The phase is recorded over a time interval from $-T/2$ to $+T/2$, where T is the coherent integration time. The SAR processor expects this quadratic phase history, and by matched filtering positions the facet at zero Doppler or zero boresight.

3.1. Azimuthal Image Shift

If the facet has a constant radial velocity component u_r, then the quadratic phase history is not affected. However, the time at which zero Doppler is encountered is changed relative to the stationary facet case. The time difference Δt is determined by the condition

$$\frac{d}{dt}\phi(t)|_{t=\Delta t} = 2k_0 u_r \tag{3}$$

Inserting Eqs. (1) and (2) into (3) yields

$$\Delta t = (2k_0/b)u_r \tag{4}$$

A SAR processor expecting stationary facets therefore erroneously allocates the azimuthal position x of the moving facet to the image position

$$x' = x + \Delta x \tag{5}$$

where

$$\Delta x = V \cdot \Delta t = (R/V)u_r \tag{6}$$

is the azimuthal image shift. Note that Δx is independent of the radar wavenumber and integration time.

Ample evidence of this effect can be found in SAR imagery. For example, a ship traveling in cross-track direction appears displaced in the platform flight direction relative to its wake in the SAR image.

3.2. Azimuthal Image Smear

If the facet is subject to an acceleration in slant-range direction, then an additional quadratic term enters into the phase history:

$$\phi_a(t) = 2k_0(a_r/2)t^2 \tag{7}$$

where $|t| \leq T/2$.

Due to the additional linear chirp generated by the acceleration, a processor which is tuned for stationary facets is mismatched or out-of-focus for this particular facet. The result is a smearing of the azimuthal target position. The magnitude of the azimuthal smearing is given by the variation δx in azimuthal image shift experienced by the facet due to the variation $\delta u_r = a_r T$ of the radial facet velocity during the integration time T:

$$|\delta x| = \left| \frac{R}{V} a_r T \right| \tag{8}$$

The exact formula for the net degraded azimuthal (one-look) resolution ρ'_a can be derived from SAR system theory (Alpers and Rufenach, 1979; Swift and Wilson, 1979; Rufenach and Alpers, 1981) and is given by

$$\rho'_a = (\rho_a^2 + \rho_{acc}^2)^{1/2} \tag{9}$$

where

$$\rho_a = \frac{\pi}{k_0} \frac{R}{V \cdot T} \tag{10}$$

is the full-bandwidth (one-look) azimuthal resolution for stationary targets and

$$\rho_{acc} = \frac{\pi}{2} |\delta x| = \frac{\pi}{2} \left| \frac{R}{V} a_r \cdot T \right| \tag{11}$$

represents the smearing induced by the facet acceleration. We note that the acceleration-induced azimuthal image smear depends linearly on the coherent single-look integration time T.

Equation (9) is obtained formally by computing the width ρ'_a of the convolution of two Gaussian functions of widths ρ_a and ρ_{acc}, respectively. [The factor $\pi/2$ enters in Eq. (11) because we have defined the width of the azimuthal image intensity response function, following the usual definitions for SAR point target imaging, as $\exp\{-(\pi^2/\rho_a^2)x^2\}]$.

Incoherent multilook processing has no effect on the total acceleration smearing. It merely divides it into N azimuthally displaced subsections, where N is the number of looks. The formal expression for the degraded net azimuthal resolution for N looks is given by

$$\rho_{aN}' = [(N\rho_a)^2 + \rho_{acc}^2]^{1/2} \tag{12}$$

3.3. Azimuthal Focusing

Since a target acceleration in range direction generates an additional linear chirp with chirp rate $2k_0a_r$, one can adjust the processor such that it matches the total chirp of the target:

$$b' = b + 2k_0a_r \tag{13}$$

Expressing the adjusted chirp rate b' of the processor in terms of an additional apparent velocity vector ΔV in flight direction, the matching condition reads

$$b' = 2k_0(V + \Delta V)^2/R = b + 2k_0a_r \tag{14}$$

Inserting b from Eq. (2) and assuming $|\Delta V| \ll V$, Eq. (14) yields

$$\Delta V = (R/2V)a_r \tag{15}$$

A focus adjustment of the processor according to Eq. (15) eliminates the azimuthal image smear for this target and minimizes the azimuthal resolution. Thus, the target is in focus. However, for an ensemble of facets experiencing different local radial accelerations, the focusing condition can be satisfied only for a subset of facets which experience the same instantaneous acceleration. Other facets will remain out of focus or will become still more strongly defocused (e.g., stationary facets).

4. THE EFFECT OF ORBITAL MOTIONS ON SAR OCEAN WAVE IMAGERY

4.1. Velocity Bunching

The azimuthal displacement Δx due to radial facet velocities contributes to the imaging of the long waves through the alternating concentration and spreading of the apparent position of the backscatter elements within the long-wave pattern ("velocity bunching," see Alpers and Rufenach, 1979). Velocity bunching is determined by the variation of the number of facet images per unit length in azimuthal direction, which is proportional to

$$c = \frac{R}{V}\frac{du_r}{dx_0} \tag{16}$$

where x_0 is the azimuthal coordinate in the ocean plane. For small $|c|$, the effect is linear and can be characterized in the same way as the hydrodynamic and electromagnetic interactions by a linear transfer function. However, for large $|c|$ (typically $|c| \geq \pi/2$), velocity bunching is a nonlinear process and leads to image distortions. Such situations are often encountered in SAR ocean wave imagery. The effect is largest for azimuth-traveling waves and vanishes for range-traveling waves. The nonlinearity gives rise not only to a distortion of the SAR image spectrum, but also to a shift of the spectral peaks toward lower azimuthal wavenumbers (Alpers, 1983).

4.2. Azimuthal Image Smear

The azimuthal image smear due to radial facet accelerations generally degrades the SAR image by reducing the image contrast and filtering out small-scale features in the image. However, because of the spatially variable nature of the acceleration smearing, it is possible to reduce the smearing for some facets while enhancing it for others by suitable refocusing. Thus, in some situations, the net image quality can be enhanced by suitable empirical focusing adjustments.

In contrast to velocity bunching, the acceleration effects are a basically nonlinear phenomenon, and cannot be expressed by a linear transfer function even for small variations of ρ'_{aN}. The acceleration effects also differ from the orbital velocity effects through their linear dependence on the integration time [see Eq. (11)].

Velocity bunching and azimuthal image smear together constitute, to first order, the motion part of the SAR imaging mechanism. For small integration times, the azimuthal image smear due to acceleration can become negligible compared with velocity bunching (e.g., for ERS-1). However, for the SEASAT SAR it is found generally to be a nonnegligible factor in the imaging process.

4.3. Azimuthal Focusing of Ocean Wave Imagery

As pointed out in Section 3, it is possible to focus on any subset of facets with a common selected slant-range acceleration a_r. It may be expected intuitively that the best image quality will be obtained by focusing on those facets which are associated with the highest image intensity (due to velocity bunching and hydrodynamic and electromagnetic cross-section modulation). In this case, the high image intensity remains concentrated in small azimuthal pixels and a large modulation depth is maintained. (This argument assumes an asymmetry between the spatial distribution of high radar return and low return relative to the mean, which is often observed for steep waves, but may not apply for low-swell fields.)

The parameter range for the focus adjustment is given by [see Eq. (15)]

$$-\left|\frac{R}{2V}a_r^{(0)}\right| \leqslant \Delta V \leqslant \left|\frac{R}{2V}a_r^{(0)}\right| \tag{17}$$

where $a_r^{(0)}$ denotes the maximum of a_r encountered along the long-wave pattern. The factor $(R/2V)a_r^{(0)}$ is often of the same order as the phase velocity of long waves. For example, for the JPL L-band SAR flown at nominal altitude ($H = 10\,\mathrm{km}$) and nominal speed (240 m/s), the ratio $R/2V = 22\,\mathrm{s}$ for an incidence angle of $\theta = 20°$. An ocean wave of 100-m wavelength and 0.9-m wave amplitude would yield $(R/2V)a_r^{(0)} = 12.5\,\mathrm{m}$, which is equal to the phase velocity of this wave in deep water. This may explain why some investigators have associated the focus adjustment ΔV with the wave phase velocity, although the long-wave phase velocity does not enter directly in standard theories of microwave backscattering and SAR processing.

The expression for the degraded azimuthal resolution, including the azimuthal focus adjustment parameter ΔV of the processor, reads [cf. Eqs. (11) and (12)]

$$\rho'_{a,N} = \left\{(N\rho_a)^2 + \left[\frac{\pi R}{2V}\left(a_r - \frac{2V}{R}\Delta V\right)T\right]^2\right\}^{1/2} \tag{18}$$

It may be expected that the larger the variation of $\rho'_{a,N}$ along the wave pattern, the larger will be the sensitivity of ocean wave imagery to azimuthal focusing. The T dependence of the acceleration and focus adjustment term implies that radars with long integration times may be expected to be more sensitive to azimuthal focusing than radars with short integration times, provided other parameters remain equal. For example, an L-band radar is expected (and found) to be more sensitive to focusing than an X- or C-band radar, provided both have the same resolution and same R/V ratio.

Furthermore, we predict that multilook imagery is less sensitive to azimuthal focusing than single-look imagery, since the additional acceleration and focus adjustment term becomes relatively less important than the basic multilook resolution term.

The dependence of the value of the focus shift ΔV on incidence angle is of some interest. For a monochromatic ocean wave

$$\zeta = \zeta_0 \cos(\mathbf{k}\mathbf{x}_0 - \omega t) \tag{19}$$

the slant-range orbital acceleration is given by (see Alpers and Rufenach, 1979)

$$a_r(x_0, t) = -\zeta_0 \omega^2 G(\theta, \phi) \cos(\mathbf{k}\mathbf{x}_0 - \omega t + \delta) \tag{20}$$

where

$$G(\theta, \phi) = (\sin^2 \theta \sin^2 \phi + \cos^2 \theta)^{1/2} \tag{21}$$

and

$$\delta = tg^{-1}(tg\theta \sin \phi) \tag{22}$$

Here θ denotes the incidence angle, and ϕ the azimuth angle.

Expressing the slant range R in terms of the platform height H and the incidence angle θ,

$$R = H(\cos \theta)^{-1} \tag{23}$$

we obtain from Eq. (15)

$$\Delta V = (H/2V)\zeta_0 \omega^2 (\cos \theta)^{-1} G(\theta, \phi) \cos(\mathbf{k}\mathbf{x}_0 - \omega t + \delta) \tag{24}$$

For $\phi = 0$ (azimuthally traveling waves), Eq. (24) reduces to

$$\Delta V = (H/2V)\zeta_0 \omega^2 \cos(\mathbf{k}\mathbf{x}_0 - \omega t + \delta) \tag{25}$$

Thus, for azimuthally traveling waves the azimuthal focus adjustment is independent of incidence angle or slant range. For $\phi = 90°$ (range-traveling waves) the focus shift,

$$\Delta V = (H/2V)\zeta_0 \omega^2 (\cos \theta)^{-1} \cos(\mathbf{k}x - \omega t + \delta) \tag{26}$$

increases with increasing incidence angle.

5. DISCUSSION

On the basis of this simple two-scale scattering and hydrodynamic model, the answers to questions 1–4 posed in the Introduction may now be summarized as follows:

1. The SAR response to a moving ocean wave field can be described to first order by a superposition of SAR responses to independent moving scatter elements on the ocean surface. Thus, the SAR theory for moving point targets (Raney, 1971) is applicable to SAR ocean wave imaging. The dominant influence of the motion of the backscattering surface on SAR imaging is the azimuthal displacement of the apparent positions of individual backscattering elements induced by the component of the facet velocity directed toward the radar.

2. The scene coherence time relevant for SAR ocean wave imaging is the intrinsic hydrodynamic interaction time of the Bragg scattering waves in the reference frame of the moving facet. This may be estimated to be of the order of a few seconds. It follows that the finite scene coherence time of the ocean surface usually contributes significantly less to the total azimuthal image smear than the orbital facet acceleration.

It should be noted that we have considered throughout "infinitesimal" facets of the order of 10–30 Bragg scattering wavelengths, i.e., typically 0.2–1 m. It is also possible to consider significantly larger "macroscopic" backscattering elements of the order of the image resolution cell—typically 10–20 m. In this case the relevant displacements of individual "infinitesimal" facet elements within a pixel, which we attribute to differences in individual facet velocities, would be interpreted as a "smearing" of the single "macroscopic" facet which represents the pixel as a result of the strongly reduced coherence time of the macroscopic facet. Both pictures are clearly equivalent, but the latter description is more cumbersome to carry through quantitatively due to the non-Gaussian nature of the phase decorrelation time of "macroscopic" scattering elements. A more detailed discussion is given in Hasselmann et al. (1985).

3. For SARs with long integration times and large R/V ratios (R = slant range, V = platform velocity), the orbital acceleration is an important factor in SAR ocean wave imaging. In particular, the C-band SAR proposed for ERS-1 may be expected to perform better for ocean wave imaging than the SEASAT L-band SAR because the acceleration effects are reduced by a factor 0.14 corresponding to the ratio of the integration times.

4. The phase velocity of the long ocean waves does not enter directly into the standard theory of SAR ocean wave imaging. However, empirical focusing adjustments chosen to reduce the nonuniform azimuthal smearing induced by orbital slant-range accelerations a_r are governed by the parameter $\Delta V = a_r \cdot R / 2V$, which is often of the same order as the wave phase velocity.

Questions 5 and 6 require a more detailed analysis beyond the scope of this chapter and are discussed in Alpers (1983) and Alpers and Hasselmann (1982).

REFERENCES

Alpers, W. (1983): Monte Carlo simulations for studying the relationship between ocean wave and synthetic aperture radar image spectra, *J. Geophys. Res.* **88**, 1745–1759.

Alpers, W., and K. Hasselmann (1978): The two-frequency microwave technique for measuring ocean wave spectra from an airplane or satellite. *Boundary-Layer Meteorol.* **13**, 215–230.

Alpers, W., and K. Hasselmann (1982): Spectral signal and thermal noise properties of ocean wave imaging synthetic aperture radars. *Int. J. Remote Sensing* **3**, 423–446.

Alpers, W., and C. L. Rufenach (1979): The effect of orbital motions on synthetic aperture radar imagery of ocean waves. *IEEE Trans. Antennas Propag.* **AP-27**, 685–690.

Alpers, W. R., and C. L. Rufenach (1980): Image contrast enhancement by applying focus adjustment in synthetic aperture radar imagery of moving ocean waves. *SEASAT-SAR Processor*, ESA SP 154, ESA Sci. Tech. Publ. Br., ESTEC, Noordwijk, The Netherlands, 25–30.

Alpers, W., D. B. Ross, and C. L. Rufenach (1981): On the detectability of ocean surface waves by real and synthetic aperture radar. *J. Geophys. Res.* **86**, 6481–6498.

Bass, F. G., I. M. Fuks, A. I. Kalinykov, I. E. Ostrowsky, and A. D. Rosenberg (1968): Very high frequency radio wave scattering by a disturbed sea surface. *IEEE Trans. Antennas Propag.* **AP-16**, 554–568.

Elachi, C. E., and W. E. Brown (1977): Models of radar imaging of the ocean surface waves. *IEEE Trans. Antennas Propag.* **AP-25**, 84–95.

Graf, K. A., and H. Guthart (1969): Velocity effects in synthetic apertures. *IEEE Trans. Antennas Propag.* **AP-17**, 541–546.

Harger, R. O. (1980): The synthetic aperture radar image of time-variant scenes. *Radio Sci.* **15**, 749–756.

Hasselmann, K., R. K. Raney, W. J. Plant, W. Alpers, R. A. Shuchman, D. R. Lyzenga, C. L. Rufenach, and M. J. Tucker (1985): Theory of SAR ocean wave imaging: A MARSEN view. *J. Geophys. Res.* **90** (in press).

Jain, A. (1978): Focusing effects in synthetic aperture radar imagine of ocean waves. *Appl. Phys.* **15**, 323–333.

Keller, W. C., and J. W. Wright (1975): Microwave scattering and straining of wind generated waves. *Radio Sci.* **10**, 139–147.

Kelly, E. J., and R. P. Wishner (1969): Matched-filter theory for high velocity accelerating targets. *IEEE Trans. Mil. Electron. Syst.* **MES-5**, 98–105.

Larson, T. R., L. I. Moskowitz, and J. W. Wright (1976): A note on SAR imagery of the ocean. *IEEE Trans. Antennas Propag.* **AP-24**, 393–394.

Raney, R. K. (1971): Synthetic aperture imaging radar and moving targets. *IEEE Trans. Aerosp. Electron. Syst.* **AES-7**, 499–505.

Raney, R. K. (1980): SAR response to partially coherent phenomena. *IEEE Trans. Antennas Propag.* **AP-28**, 777–787.

Raney, R. K. (1981): Wave orbital velocity, fade and SAR response to azimuth waves. *IEEE J. Ocean. Eng.* **OE-6**, 140–146.

Rufenach, C. L., and W. Alpers (1981): Imaging ocean waves by synthetic aperture radars with long integration times. *IEEE Trans. Antennas Propag.* **AP-27**, 725–729.

Shemdin, O. H., W. E. Brown, Jr., F. G. Staudhamer, R. Shuchman, R. Rawson, J. Zelenka, D. B. Ross, W. McLeish, and R. A. Berles (1978): Comparison of in-situ and remotely sensed ocean waves off Marineland, Florida. *Boundary-Layer Meteorol.* **13**, 225–234.

Shuchman, R. A., E. S. Kasischke, and A. Klooster (1978): Synthetic aperature radar ocean wave studies. Final Report No. 131700-3-F, Environmental Research Institute of Michigan, Ann Arbor.

Swift, C. T., and L. R. Wilson (1979): Synthetic aperture radar imaging of ocean waves. *IEEE Trans. Antennas Propag.* **AP-27**, 725–729.

Teleki, P. G., R. A. Shuchman, W. E. Brown, W. McLeish, D. B. Ross, and M. Mattie (1978): Ocean wave detection and direction measurements with microwave radar. *Oceans '78*, Sept. 6–8, 639–648.

Tucker, M. J. (1981): The ability of satellite-borne synthetic aperature radar to measure sea waves: The effects of sea surface motions. Working paper, January.

Valenzuela, G. R. (1980): An asymptotic formulation for SAR images of the dynamical ocean surface. *Radio Sci.* **15**, 105–114.

Wright, J. W. (1968): A new model for sea clutter. *IEEE Trans. Antennas Propag.* **AP-16**, 217–223.

DISCUSSION

ROSENTHAL: Is the acceleration effect you make responsible for the focusing problem dependent on the incidence angle?

ALPERS: Yes, since the relevant acceleration in the component of the facet acceleration is in the range direction, and this depends in general on the angle of incidence θ. The θ dependence is itself a function, however, of the direction of travel of the waves. The strongest dependence ($\sim \cos \theta$) is for waves traveling in the azimuthal direction. Range traveling waves show no incidence angle dependence.

THOMAS: In SAR imaging, Professor Hasselmann is correct in emphasizing image degradations, particularly azimuth image shift. This is illustrated in an image showing the displacement between a ship and its wake in the poster session. The displacement may be used to estimate the ship's velocity. The radar layover problem may also be important. Target S with 23° slope are imaged by SEASAT to a single point. The effects of target velocities and accelerations in radar imagery of the sea need further investigation both from the image distortion point of view and also for an understanding of image speckle statistics.

   Finally, I would ask the author to comment on the choice of a frequency of 5.3 GHz for the proposed European Remote Sensing Satellite ERS-1 (as compared with SEASAT's 1.2 GHz) from the point of view of the physical oceanographer and his observational requirements.

HASSELMANN: I must apologize for talking about the "train off the track" effect. In this audience it would of course have been much more appropriate to refer to the "ship off its wake" effect.

   The 5.3-GHz band proposed for ERS-1 is essentially a compromise between short wavelengths desirable for the wind scatterometers and the longer wavelengths—which SAR designers apparently find it easier to cope with. The essential consideration is to operate both SAR and scatterometer at the same wavelength to obtain mutually supportive data for the sensor algorithms.

PLANT: It occurs to me that the argument over whether phase speed or orbital velocity enters the focusing condition is a bit spurious. The quadratic term which is balanced by focusing results from changing the frame of reference from that of the aircraft to one nearly fixed on earth. If this second frame of reference is, however, moving at the phase speed of the ocean wave, then vertical components of orbital velocity and acceleration are zero. At typical viewing angles, this cancels much (not all) of the effects of orbital velocity. Processing changes necessary to affect any improvement in this focus will be so small as to be negligible.

ALPERS: This will get us into a long discussion which we can't start now. The point is that scatterer motion results from advection by the orbital velocity of long waves and this is the quantity which must be removed by focusing. We cannot remove all motion effects by moving along with the long wave.

PLANT: I fully agree that all motion effects cannot be removed by moving with the long waves.

27

ON THE ABILITY OF SYNTHETIC APERTURE RADAR TO MEASURE OCEAN WAVES

J. F. VESECKY, R. H. STEWART, R. A. SHUCHMAN,
H. M. ASSAL, E. S. KASISCHKE, AND J. D. LYDEN

ABSTRACT. The SEASAT satellite system, using an onboard 23-cm-wavelength synthetic aperture radar (SAR), collected approximately 25- to 40-m-resolution radar images of the ocean in 100-km-wide swaths ranging in length from about 300 to 3000 km. Here we report results from the 18 SEASAT SAR passes during the Joint Air–Sea Interaction (JASIN) experiment conducted off the west coast of Scotland in the summer of 1978. These many SAR images, when coupled with the intensive ship, buoy, and aircraft measurements of the JASIN experiment, provide a unique opportunity to assess the ability of satellite SAR to measure ocean surface phenomena, particularly surface wave fields. In this study we use only optically processed SAR images. Although gravity waves of length approximately 80 to 300 m are often seen in SAR imagery, they are not *always* seen. We find that SAR resolution and wave height are important criteria for determining wave visibility and suggest that wind velocity is also. Comparisons between SAR and buoy estimates of dominant wavelength and direction (including data from several other experiments) agree to within about $\pm 14\%$ and $\pm 10°$, respectively. We use a focus sharpness algorithm to resolve the 180° directional ambiguity of SAR image directional estimates. Correlation of buoy measurements of significant wave height ($H_{1/3}$) with peak signal-to-noise ratio in Fourier transforms of SAR images ($r = 0.7$) suggests that $H_{1/3}$ can be estimated to an accuracy of about ± 1 m. However, a similar, but weaker, correlation exists between signal-to-noise and wave length, i.e., $H_{1/3}$ and wavelength are correlated in the JASIN data set. SAR image power spectra are in rough agreement with buoy measurements of omnidirectional ocean wave height spectra $\Psi(K)$ and correspond less closely with ocean wave slope spectra $\Psi'(K)$. However, the dominant wavenumber (K_{peak}) in SAR spectra typically falls below the dominant wavenumber of corresponding buoy spectra. Further, the slope of SAR spectra for $K < K_{\text{peak}}$ is typically steeper than corresponding buoy measurements of $\Psi(K)$, while the slope of SAR spectra for $K > K_{\text{peak}}$ is typically less steep than corresponding buoy spectra $\Psi(K)$. These differences between SAR spectra and buoy measurements of $\Psi(K)$ form qualitative support for the imaging mechanisms developed by Alpers, Ross, and Rufenach. Analysis algorithms based on this theory, but including empirical modifications, should substantially improve estimates of $\Psi(K, \theta)$ using SAR images.

1. INTRODUCTION

Our primary objective in this chapter is to assemble a number of comparison sets of ocean gravity wave measurements based on both synthetic aperture radar (SAR) and surface buoys. These

J. F. VESECKY AND H. M. ASSAL ● Stanford Center for Radar Astronomy, Stanford University, Stanford, California 94305.   R. H. STEWART ● Scripps Institution of Oceanography, La Jolla, California 92093, and Jet Propulsion Laboratory, Pasadena, California 91103.   R. A. SHUCHMAN, E. S. KASISCHKE, AND J. D. LYDEN ● Environmental Research Institute of Michigan, Ann Arbor, Michigan 48107.

comparison sets are then used to comment upon the ability of SAR to measure ocean gravity waves in the period range from about 7 to 19 s. Comparisons are made with respect to dominant wavelength and direction of the wave field, significant wave height $H_{1/3}$, and the directional wavenumber spectrum of wave height variance $\Psi(\mathbf{K})$. SAR imagery collected by the SEASAT satellite over the Joint Air–Sea Interaction (JASIN) experiment area, west of Scotland, provides an excellent opportunity to make the desired comparisons. During the SEASAT JASIN experiment (August–September 1978), measurements of many air and sea parameters were made by ships, buoys, and aircraft, including the wind and waves which are of most interest here (Pollard, 1979). Swaths of 4 of the 18 SEASAT SAR ·passes over the JASIN area are shown in Fig. 1.

The 23-cm-wavelength SAR carried aboard the SEASAT satellite collected approximately 25- to 40-m-resolution radar images of the ocean in 100-km-wide swaths ranging in length from about 300 to 3000 km along, but displaced to the right of, the subsatellite track. The SEASAT SAR is unique, being the only scientific SAR ever carried aboard a satellite. During the short lifetime of SEASAT (June–September 1978), some 10^8 km^2 (equivalent to about 20% of the earth's surface) was imaged. A description of the instrument is given by Jordan (1980) and preliminary results from all SEASAT instruments were given in the June 29, 1979 issue of *Science*. Beal *et al.* (1981) and Gower (1981) report further SEASAT SAR results, and Vesecky and Stewart (1982) assess and review the ability of SEASAT SAR to sense ocean surface phenomena. Here we report results of SEASAT SAR observations during the JASIN experiment.

SAR images collected by SEASAT are basically high-resolution maps of the radar reflectance (backscatter) of the ocean surface modified by surface motion effects. Because the dielectric properties of the ocean surface are relatively uniform, variations in reflectance are due primarily to variations in surface roughness. In particular, the reflectance is due to resonant backscatter of the $\lambda = 23$-cm radar waves from ocean surface waves of length $\Lambda = (\lambda/2 \sin \theta) \approx 30$ cm, the angle of incidence θ being about 19 to 26°. The resonant mechanism is typical of radiation scattered from a lattice and is often called Bragg scatter (Bragg, 1933). Because SAR uses the phase of the radar echo to achieve high azimuthal resolution in the image (i.e., to correctly locate the origin on a given echo), motion of the ocean surface, which introduces a phase shift in the scattered signal, causes points in the image to be displaced from their true position. Such misplacement is aptly illustrated by moving ships being displaced from their wakes (e.g., see Vesecky *et al.*, 1984 or Vesecky and Stewart, 1982). Spatial modulation of surface roughness and surface motion (giving rise to Doppler shifts) produce the majority of ocean surface features in SAR imagery. Phillips (1981) describes a number of ways in which wind, waves, and current can modulate decimeter-scale surface roughness. Often, the difficulty in interpreting SAR imagery of the ocean is to discover which of the several candidates is really the underlying cause of the surface roughness modulation sensed by the radar. Ocean gravity waves are easily recognized when they are highly coherent and present an easily recognized large-scale pattern as shown in Fig. 2.

The radar echo signal as received on the SEASAT satellite and relayed to a ground station can be processed into an image using any one of a number of different algorithms. Further, these algorithms can be implemented using either optical (analog) or digital processing apparatus. The relative merits of optical and digital processing are discussed by Vesecky and Stewart (1982). In general, optical processing can be done more quickly and at less expense, while digital processing allows a better-quality image to be produced. The SAR images used in this study have all been processed optically either at the Jet Propulsion Laboratory (JPL) or at the Environmental Research Institute of Michigan (ERIM).

2. THE SEASAT JASIN EXPERIMENT

The SEASAT JASIN experiment was conducted some 400 km off the west coast of Scotland during the summer of 1978. As shown in Fig. 1, this area of open ocean is generally deep with numerous banks and sea mounts rising to depths of $\lesssim 500$ m. The oceanic intensive area (OIA), where the majority of buoy measurements were made, has a depth of greater than 1000 m so bottom topography does not modify the deep-water dispersion relation for ocean gravity waves. Moderate seas prevailed during the SEASAT JASIN experiment period, $H_{1/3}$ ranging from about 1 to 5 m. Winds were also moderate, ranging from about 3 to 15 m/s. Locations of

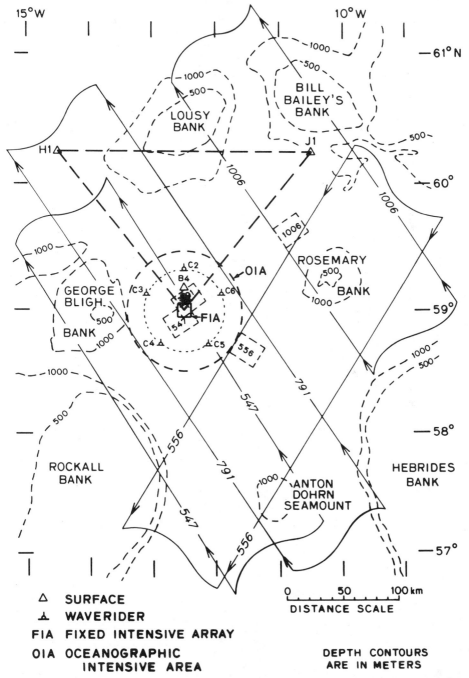

FIGURE 1. Joint Air–Sea Interaction (JASIN) experiment area showing typical SEASAT SAR image swaths and sections of images used for wave analysis. Waverider buoys are denoted by C2, C3, etc. Pitch-and-roll buoys were generally deployed in the FIA area.

TABLE I. SAR and Buoy Estimates of Surface Wave Characteristics during the SEASAT-JASIN Experiment

Date (1978)	Time (UT)	Orbit No.	Surface buoy[a]	Separation (km)	Dominant wavelength (m)		Dominant wave direction (°T)[c]		Wave beamwidth (deg)		Wave visibility (db)		Φ[d] (deg)	$H_{1/3}$ (m)	U (m/s)
					SAR[b]	Surface	SAR[b]	Surface	SAR	Surface	PBR	S/N			
4 Aug.	0615	547	a	0	198	169	266	263	≳32(S)	32	3.6	5.3	116	1.7	4.5
4 Aug.	2136	556	a	63	148	149	274	280	≳40(S)	54	1.8	1.8	248	1.5	3.5
7 Aug.	0622	590	a	0	—	222/89	—	222/329	—	83/77	0		83/183	1.3/1.3	7.2
7 Aug.	2144	599	b(C2)	119	67(S)	71	322(S)	—	≳146(S)	—	0	1.1	289[f]	1.1	6.9
10 Aug.	0629	633	a	56	—	89/58	—	220/220	—	70/40	0	0	74/74	1.1/1.1	9.2
10 Aug.	2151	642	a	204	—	169	—	199	—	29	0	0	167	2.9	12.9
15 Aug.	2235	714	none	—	182(E)	—	276(E)	—	—	—	1.8	0	242[f]	—	—
16 Aug.	0643	719	none	—	160	—	275	—	120(S)	—	1.5	1.6	130[f]	—	7.6
18 Aug.	2241	757	b(C4)	0	275(E)	215	253(E)	—	—	—	2.9	0	221[f]	4.9	15.2
19 Aug.	0649	762	b(C6)	256	265	206	245	—	33(S)	—	5.1	5.1	100[f]	4.4	12.0
19 Aug.	0649	762	b(C2)	266	258	199	239	—	46(S)	—	3.4	8.4	94[f]	4.3	12.0
21 Aug.	0725	791	b(C2)	25	168	132	251	—	75(S)	—	1.8	4.3	104[f]	3.0	13.0
24 Aug.	0730	834	a	53	188	149	266	≈295	76	35	1.8	1.9	≈149	2.8	10.2
1 Sept.	2354	958	b(C4)	83	120(E)	108	296(E)	—	—	68	1.6	0	260[f]	1.3	7.2
5 Sept.	0006	1001	a	76	81(E)	63	125(E)	≈146	—	62	2.5	0	≈110	2.5	12.6
5 Sept.	0815	1006	a	120	—	107	—	≈83	—	61	0	0	≈299	3.6	6.6
8 Sept.	0019[e]	1044	c[e]	22[e]	333	301	256	264	25	19	6.0	6.7	229	3.5	11.5
8 Sept.	0828	1049	c	138	316	301	250	264	21	19	5.2	8.0	117	3.5	6.3
11 Sept.	0031	1087	none	—	312	—	259	—	48(S)	—	4.2	3.8	224[f]	—	13.2

[a] Surface buoys are denoted as follows: a, Atlantis II pitch-and-roll buoy; b, DHI Waverider buoys moored in OIA—see Fig. 1; c, Discovery pitch-and-roll buoy.

[b] (S), Stanford measurement only; (E), ERIM measurement only.

[c] The direction from which waves arrive is used here.

[d] The angle Φ lies along a clockwise direction between the direction of satellite velocity V and the dominant ocean wave vector K.

[e] Buoy measurement made 0742–0755 UT.

[f] Φ estimated from SAR data alone.

Waverider buoys in the OIA are shown in Fig. 1 along with the FIA area, where most of the pitch-and-roll buoy measurements were made. Further details are given in Fig. 1 and Table I.

Both pitch-and-roll and Waverider buoys were used to make wave measurements. Pitch-and-roll buoys were deployed from both the R. V. Atlantis II (Woods Hole Oceanographic Institution) and the M. V. Discovery (Institution of Oceanographic Sciences) on eight and two occasions, respectively. These buoys provide measurements of the omnidirectional wave height variance spectrum $\Psi(\omega)$ as well as estimates of the dominant wave direction and beamwidth. The Atlantis II buoy is described by Stewart (1977). Waverider buoys were moored by the Deutsches Hydrographisches Institut at a number of locations as shown in Fig. 1. These buoys measured only $\Psi(\omega)$ and provided no directional information. The SAR data collected over the JASIN area and used in this study were optically processed into images at both the JPL and ERIM. The JPL images were the standard SEASAT SAR data product having a resolution of 25 to 40 m. The

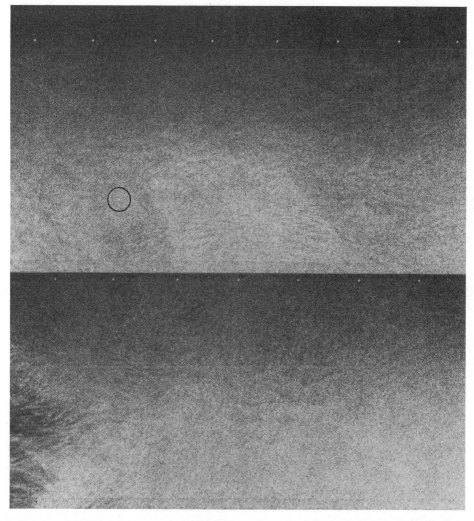

FIGURE 2. SEASAT SAR image including the fixed intensive array (FIA) portion of the JASIN experiment area, near 59°N, 12°W (see Fig. 1). The SAR flight direction is from right to left across the width of the page. Note the ocean waves propagating from lower left toward upper right. The small white crosses are at 1-s intervals and thus are about 7.2 km apart. The small, circled, white dot in the lower left portion of the upper panel is apparently the image of the Research Vessel Atlantis II (Woods Hole Oceanographic Institution). This image was optically processed at JPL.

ERIM images were specially processed to obtain focus information as described below and to achieve consistent 25-m resolution. An example of JPL imagery is shown in Fig. 2 where a field of about 170-m waves is present. Examples of ERIM imagery are given by Kasischke *et al.* (1984).

During the experiment period (August 4 to September 11, 1978), 18 SEASAT SAR passes were made over various portions of the JASIN area. Table I summarizes the data collected during these passes. On 16 of the passes, concurrent or nearly concurrent buoy measurements were made. On three occasions, a buoy was within the SAR swath. The distances between the SAR swaths and the buoy locations in the other cases are noted in Table I. Wind measurements were also made concurrent with most of the SEASAT SAR passes as noted in Table I. The wind speeds are averages of observations by three to five surface instruments within the OIA area of Fig. 1 and hence near the wave buoy locations.

3. VISIBILITY OF OCEAN WAVES IN SAR IMAGES

Although ocean gravity waves of length approximately 80 to 300 m are often seen in SAR images, they are not *always* seen. Wave visibility is not simply a case of insufficient wave height. For example, from Table I, waves with $H_{1/3} \approx 3.6$ m in a 6.6 m/s wind were not seen in SAR images from orbit 1006. Why is this so?

We suggest that three criteria are important in determining wave visibility:

1. SAR resolution (including the degrading effects of wave orbital motion) must be sufficiently good that the Nyquist sampling criterion, i.e., at least two resolution cells per projected ocean wavelength, be satisfied in both the azimuth direction (parallel to the SAR flight path) and the range direction (perpendicular to the SAR flight path).
2. $H_{1/3}$ greater than about 1 m.
3. Wind speed greater than a few meters per second.

In this study we have used only SAR images processed optically at the JPL and ERIM.

Alpers *et al.* (1981) point out that orbital velocities associated with surface gravity waves produce a smearing effect on SAR imagery degrading resolution, especially along the azimuth (y) direction. They also note that the ocean surface coherence time (τ_S) affects resolution (Raney, 1980) as do adjustments (ΔV) in the azimuthal focus relative to the nominal focus for land. Alpers *et al.* (1981) argue that the coherence term (τ_S) can usually be neglected and we do so here. If Δy is the land (unsmeared) azimuth resolution (~ 25 to 40 m for SEASAT SAR), then the degraded azimuth resolution Δy^* for N looks is given by

$$\Delta y^* = N\Delta y\{1 + N^{-2}[(\pi T^2/\lambda)(-A\omega^2 g(\theta,\Phi)\cos(\mathbf{K}\cdot\mathbf{r}+\alpha') + (2V\Delta V/R))]^2 + (T/\tau_S)^2\}^{1/2} \quad (1)$$

where T is the one-look, full-bandwidth integration time (≈ 2.3 s for SEASAT SAR), λ is the radar wavelength, $\mathbf{r}$ is the location of the target in the (x, y) plane, $\alpha' = \tan^{-1}(\tan\theta\sin\Phi)$, and A, ω, and $\mathbf{K}$ are the amplitude, radian frequency, and vector wavenumber of the dominant ocean wave. V is the satellite velocity relative to the earth and ΔV is an azimuthal focus adjustment parameter which we take to be zero here since no focus adjustment was made in imaging the data used here. The function $g(\theta, \Phi) = (\sin^2\theta\sin^2\Phi + \cos^2\theta)^{1/2}$; other variables are defined in Fig. 3. In our calculation of Δy^*, we have taken $1/\pi$ as the typical value of the cosine term. Knowing Δy^* and the dominant ocean wavelength along the y direction $\Lambda_y = \Lambda/\cos\Phi$, the number of SAR resolution elements or samples n_y along the azimuth direction can be estimated. The Nyquist sampling criterion (Bracewell, 1979) demands $n_y \geqslant 2$ for the waves to be adequately sampled. Jain (1978) and Vesecky *et al.* (1981) have developed wave visibility criteria along these lines. While an analogous criterion $n_x \geqslant 2$ must also be met along the range direction, resolution degradation along the x direction is much less severe and we have neglected it here.

Because of the roles played by wave height and/or slope in wave imaging mechanisms (Alpers *et al.*, 1981), one excepts wave visibility to decrease with decreasing wave height $H_{1/3}$ and indeed we find waves are seldom imaged for $H_{1/3} \lesssim 1$ m. This wave height criterion is probably

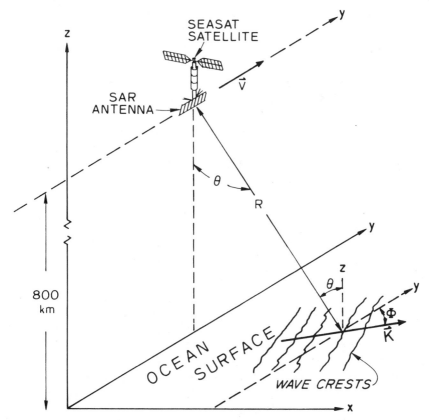

FIGURE 3. Geometry for observation of the ocean surface by SEASAT SAR. This figure defines the geometry for SEASAT SAR images of the ocean surface, (x, y) plane; in particular, observations of long gravity waves at a point (x, y) traveling at an angle Φ with respect to the SAR flight path and having a wave vector $\mathbf{K}$ where $K = 2\pi/\Lambda$ and Λ is the ocean wavelength.

set by the influence of slopes on the wave imaging mechanism, i.e., on the spatial modulation of the normalized radar cross section σ_0. If slopes are the dominant mechanism modulating σ_0, and if slopes are small, then so too is $\Delta\sigma_0$. If $\Delta\sigma_0$ is less than the noise fluctuations of σ_0 in the image, the waves are invisible. Interestingly, theory implies that there may be both a minimum $H_{1/3}$ below which waves are invisible *and a maximum* $H_{1/3}$ above which waves traveling near the azimuth direction are not imaged because azimuthal resolution is severely degraded by large wave orbital velocities. The two criteria $n_y \gtrsim 2$ and $H_{1/3} \gtrsim 1$ m are applied to the SEASAT JASIN data set in Fig. 4. In calculating n_y N was set at 4 and Δy at 6.25 m corresponding to four-look optical processing. Where possible, ocean wave data are drawn from buoy measurements given in Table I. Cases where SAR estimates were used for Φ are noted in Fig. 4.

Since surface winds raise the approximately 30-cm ocean waves to which the radar is sensitive, we expect wave images to disappear when wind speed falls below a few meters per second, and the sea becomes calm. A wind speed of 3.5 m/s is typical of speeds required to maintain a saturated spectrum of gravity waves for $\Lambda \lesssim 30$ cm (Vesecky and Stewart, 1982). A SAR image of a calm region containing no visible waves is given by Beal (1981). During the SEASAT JASIN experiment, waves were detectable over a range of wind speeds of from 3.5 to 15.2 m/s, while at wind speeds of from 6.6 to 12.9 m/s, waves were not detected. Thus, wind speed was apparently not a factor during the SEASAT JASIN experiment mainly because wind speeds were always above 3.5 m/s.

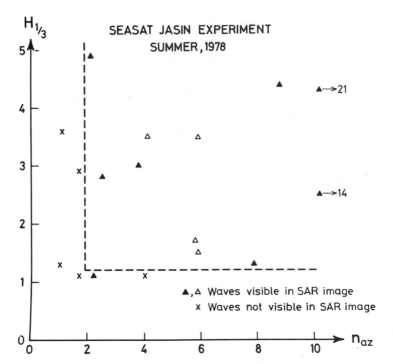

FIGURE 4. Data from the SEASAT JASIN experiment plotted on axes of significant waveheight $H_{1/3}$ and n_y, the number of SAR resolution cells per wavelength along the azimuth. These data fall into two distinct classes according to whether or not waves were visible in SAR images. We have not plotted a point for orbit 590 since visibility was indeterminate. The white triangles use only buoy data, while the black triangles indicate that a SAR estimate for Φ was used in the absence of buoy data.

4. SAR IMAGE ANALYSIS METHODS

SAR image analysis done here is tied to a very simple *assumption*, namely, that SAR image intensity fluctuations δI are proportional to ocean surface height fluctuations δh. This assumption implies that the Fourier power spectrum of the image intensity distribution $|F_I(K, \theta)|^2 = |F\{I(x, y)\}|^2$ is proportional to the directional wave height spectrum $\Psi(K, \theta)$ of the ocean surface corresponding to the image. *As will become evident, the truth is not so simple.* However, this assumption is not far wrong and provides a convenient working hypothesis. Using the assumption, dominant wavelength and direction can be estimated simply by noting the values of K and θ where $|F(K, \theta)|^2$ peaks. We discuss in Section 6 how the constant of proportionality can be estimated and thus $\Psi(K, \theta)$ estimated.

Several complicating factors which qualify the simple assumption can be summarized in the following equation where these factors enter as linear transfer functions:

$$|F_I(K, \theta)|^2 \approx |H|^2 \cdot |B|^2 \cdot |R|^2 \cdot \Psi(K, \theta) \tag{2}$$

These transfer functions H, B, and R are related to the point response of the SAR system, background (non-wave-related) fluctuations in the image, and the radar wave–ocean wave interaction (imaging) machanism, respectively. The magnitude signs are used since the transfer functions are in general complex.

As with all measurements, the response of the measuring instrument is not perfect. Thus, the SAR image of a point target is spread over some region of the image characterized in size by the resolution cell dimensions (Δx and Δy) and in shape by the point spread function. To preserve details, SAR images are digitized at sampling intervals small compared to Δx and Δy. When an image is Fourier transformed and the energy spectrum calculated, one expects the spectral power level to decrease significantly (roll off) as K increases approaching $2\pi/2\Delta x$. Here we take $\Delta x = \Delta y$ since SAR imaging schemes usually strive toward this condition. The transfer function $H(K, \theta)$ can be used to characterize the effects of the point spread function in the wavenumber domain (for reference, see Bracewell, 1979, or Champeney, 1973).

Perusal of SAR images of the ocean quickly reveals that there are numerous features on all size scales resolved. Many of these features are clearly not related to ocean gravity waves and hence constitute a background against which the wave images are viewed. It is well known that the radar backscatter (reflectance) of a uniform rough surface is dependent on angle of incidence. This fact along with the effects of antenna pattern and automatic gain control (AGC) introduce a large-scale trend in image brightness perpendicular to the SAR flight path and hence a low-K background component in the image transform. Although some compensation for this variation can be introduced in the data processing, some effects remain, especially in optically processed images (e.g., see Vesecky and Stewart, 1982). We lump all effects related to background, i.e., non-wave-related, fluctuations in a single transfer function $B(K, \theta)$. Clearly, this is a crude model for these effects, but it proves to be useful.

The physical mechanism(s) which allows ocean gravity waves to be imaged is discussed by a number of authors and summarized by Alpers et al. (1981) and Vesecky and Stewart (1982). The three principal mechanisms considered are as follows:

1. *Tilt mechanism* (R_t) in which the change in large-scale surface slope along the length ($\Lambda \sim 10\,$s to $100\,$s of meters) of a gravity wave changes the angle of incidence at which the short ($\Lambda \sim 30\,$cm) radar-resonant waves are viewed by the radar and hence spatially modulates the radar reflectance as a function of position along the wave
2. *Hydrodynamic mechanism* (R_h) in which the variation in acceleration with position along a long gravity wave, i.e., along the direction of wave travel, induces a change in energy density of the radar-resonant waves and again spatially modulates the radar echo strength as a function of position along the wave
3. *Velocity-bunching mechanism* (R_b) in which the variation in wave orbital velocity with position along a long gravity wave causes the SAR imaging process (which assumes a stationary land surface) to systematically misplace image brightness along the azimuth direction, thus spatially modulating SAR image brightness in a manner corresponding to the variation in wave orbital velocity.

Assuming these imaging mechanisms to be linear, we can write $|R(K, \theta)|^2 = |R_t|^2|R_h|^2|R_b|^2$. As pointed out by Alpers et al. (1981), we can expect the tilt and hydrodynamic mechanisms to be modeled reasonably well by linear transfer functions, but the velocity-bunching mechanism is linear only over a small range of directions nearly perpendicular to the SAR flight path. Alpers et al. have estimated the K dependence of $|R_t|^2$, $|R_h|^2$, and $|R_b|^2$ to be $|R_t|^2 \sim K^2$; $|R_h|^2 \sim K^2$ to $(\omega/\mu)^2K^2$ for μ, the relation time constant, increasing from 0 to $\mu > \omega$, and $|R_b|^2 \sim (K\omega)^2$, where ω is the ocean wave radian frequency. On the basis of these transfer functions alone, we would expect $|F_t|^2 \sim K^\alpha\Psi$, where $\alpha \approx 2$ to 3. This K dependence of R_b applies only to the range of wave directions near $\Phi = 90°$ where the velocity-bunching mechanism is thought to be linear.

The transfer function H can be estimated analytically by knowing the radar system and image processing parameters as done by Beal (1981). In this study we have taken an empirical approach using the variance spectrum of a given SAR image to estimate the product of H and B. Our technique is to first consider the raw image spectrum $|F_I(K, \theta)|^2$ finding the range(s) of θ where wave energy is present. If we further assume that H and B are *isotropic*, then $|F_I(K, \theta)|^2$ averaged over directions away from the dominant wave direction(s) represents background fluctuations B modified by the SAR image response H and we can thus estimate $|H|^2 \cdot |B|^2$ via

$$|H|^2 \cdot |B|^2 \approx \langle |F_I(K, \theta)|^2 \rangle \tag{3}$$

where the average is taken over θ away from dominant wave directions. Having estimated

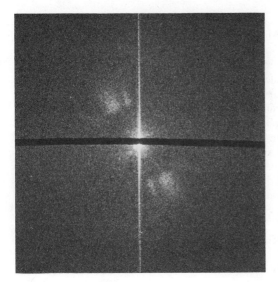

FIGURE 5. Optical Fourier transform (OFT) of an ocean wave field from orbit 547 near the FIA area of Fig. 1. The photographic display is in the form of a polar plot with wavenumber K measured linearly from the bright central spot. The angular coordinate in the display corresponds to wave direction. Note the 180° ambiguity in direction and the 2 directional components. The bright radial lines are due to the rectangular aperture used with the lens system.

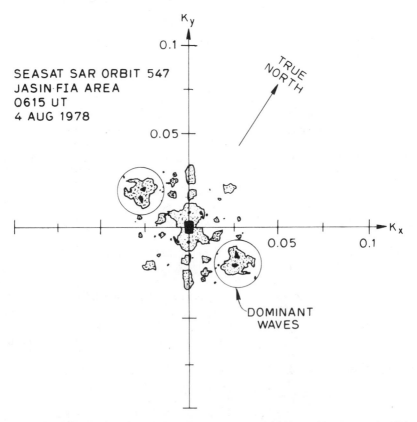

FIGURE 6. Digital Fourier transform (DFT) of an ocean wave field from orbit 547 near the FIA area of Fig. 1. The contour plot is in polar coordinates with the K_x and K_y axes corresponding to the directions shown in Fig. 3. This DFT was computed from a portion of the same image used to form the OFT in Fig. 5. Note the two directional components and the 180° directional ambiguity.

$|H|^2 \cdot |B|^2$, we can remove their effect to obtain the corrected or rectified image K spectrum $|F_R(K, \theta)|^2 = |F_I(K, \theta)|^2/|H|^2 \cdot |B|^2$. Because radar system and processing parameters may change from scene to scene, we estimate $|H|^2 \cdot |B|^2$ separately for each image.

The omnidirectional spectra presented in this chapter (Figs. 7 and 11) were obtained by using the rectified spectrum $|F_R|^2$ averaged over directions near (within about $\pm 15°$) the dominant wave direction(s). Thus, these spectra are not truly omnidirectional. However, to have included background fluctuations from directions well away from the dominant wave directions would only have added noise. We consider $|F_R|^2$ as shown in Fig. 7 to be our best present estimate of $\Psi(K)$ for comparison with buoy estimates of this quantity. Note that we have not made any adjustments for the wave imaging mechanism $|R|^2$ since we feel it is not sufficiently well known at this time. Rather, we hope that comparisons of $|F_R|^2$ with buoy measurements of $\Psi(K)$ will help clarify which if any of the proposed SAR imaging mechanisms correspond to reality.

We consider results from SEASAT orbit 547 (see Fig. 1) to illustrate several techniques of SAR image analysis as well as methods for displaying the results. In Fig. 5 we show an optical Fourier transform (OFT) produced at ERIM. In this method the coherent optical image, formed by a normal SAR processing optical train, is run through a converging lens to obtain the 2d Fourier transform and a piece of film in the focal plane records the intensity spectrum which corresponds to $|F_I|^2$ discussed above. Practice has shown that weakly imaged waves are often more easily detected in a photographic display such as Fig. 5.

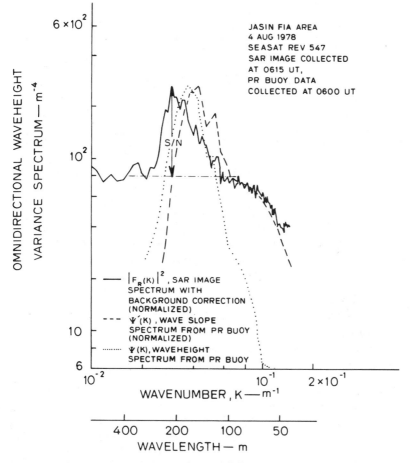

FIGURE 7. Comparison of SAR image spectrum (corrected for background fluctuations and SAR point response) $|F_R(K)|^2$ with omnidirectional wave height and wave slope spectra $\Psi(K)$ and $\Psi'(K)$ as measured by pitch-and-roll buoy. Both $|F_R(K)|^2$ and $\Psi'(K)$ have been normalized along the ordinate to the peak of $\Psi(K)$.

Digital Fourier transforms (DFTs) of SAR images are generated by first digitizing a SAR image and then performing the Fourier transform using the fast Fourier transform algorithm on a digital computer. Generally, only the magnitude squared or variance spectrum $|F_I(K, \theta)|^2$ or $|F_R(K, \theta)|^2$ is examined. In Fig. 6 we see a contour plot of $|F_I(K_x, K_y)|^2$ performed digitally at ERIM using SEASAT data from orbit 547 (see Fig. 1). Figure 7 shows another method of displaying DFT results, which have been corrected as discussed above, i.e., $|F_R|^2$ is displayed. Generally, OFTs have been used for quick-look analysis and to detect weakly imaged waves while DFTs have been used for more quantitative analysis.

Practice thus far indicates that the threshold for detecting waves in SAR images, i.e., estimating dominant wavelength and direction, differs between the OFT and the DFT technique. However, this difference is not fundamental and does not always hold. OFTs, optically displayed as in Fig. 5, are apparently more sensitive because they include data from a relatively large image area. (DFTs usually use a relatively small image area to avoid excessive computation.) Also, the photographically displayed OFT allows the eye and brain to perform the pattern recognition. The $|H|^2 \cdot |B|^2$ correction technique, described above, aids in detecting weakly imaged waves in DFTs, i.e., waves not detected in displays of $|F_I|^2$ are evident in displays of $|F_R|^2$. Waves not detectable in an $|F_I|^2$ plot of orbit 599 data were detected when $|F_R|^2$ was computed and plotted.

An important measurement which is derived from OFT or DFT data is the peak signal-to-noise (S/N) or peak-to-background ratio (PBR) for a given image. We shall use (S/N) to denote the peak signal-to-background-noise level of $|F_R|^2$ as shown in Fig. 7. The PBR ratio refers to OFT results displayed photographically and denotes the ratio of film density at the peak point (corresponding to the dominant wavelength and direction) to the film density at a specific background location in the OFT display, e.g., near the upper left-hand corner of Fig. 5.

5. COMPARISON OF SAR ESTIMATES WITH BUOY MEASUREMENTS FOR THE WAVELENGTH, DIRECTION, AND BEAMWIDTH OF THE DOMINANT WAVES

Comparisons of SAR estimates of ocean wave height spectra with buoy measurements of the same quantity serve two purposes. First, they allow one to judge the usefulness of SAR observations in making wave measurements. Second, the comparisons shed light on the physical mechanisms which allow waves to be imaged by a SAR. Better understanding of the imaging mechanism should lead to more accurate SAR estimates through improvement of the analysis algorithms, i.e., progress beyond the simple assumption discussed above. Table I summarizes the experiment parameters and many of the results of the SEASAT JASIN experiment.

It is clear that buoy and SAR measurements are fundamentally different in that the SAR measurements are largely spatial averages at a point in time while buoy measurements are mainly temporal averages at a point in space. To compare the two, we assume that the ocean surface is statistically homogeneous and stationary over scales of approximately 20 km in space and approximately 0.5 h in time so that spatial and temporal averages are statistically equivalent, i.e., $\Psi(\omega, \theta)d\omega = \Psi(K, \theta)KdK$. The omnidirectional slope spectrum $\Psi'(K)$ is derived from the wave height spectrum simply by multiplication by K^2, i.e. $\Psi'(K) = K^2\Psi(K)$ (Kinsman, 1965).

Estimates of dominant wavelength and direction are obtained simply by noting the (K, θ) location of the wave-related maximum in $|F_I|^2$ or $|F_R|^2$. Two estimates of K_{max} and θ_{max} were made, one using ERIM OFT data corresponding to $|F_I|^2$ and one using Stanford DFT data corresponding to $|F_R|^2$. In making comparisons of dominant wavelength and direction, we have simply averaged these two SAR estimates. As one would expect for random errors, in the two estimates, the average of the OFT and DFT estimates compares more favorably with buoy measurements than either SAR estimate taken alone. This leads one to conclude that the corrections introduced to produce $|F_R|^2$ do not make emphatic improvements in SAR estimates of K_{max} and θ_{max}. However, K spectra, such as those in Figs. 7 and 11, were substantially improved by these corrections especially in terms of distinguishing between the ocean-wave-related component and the background.

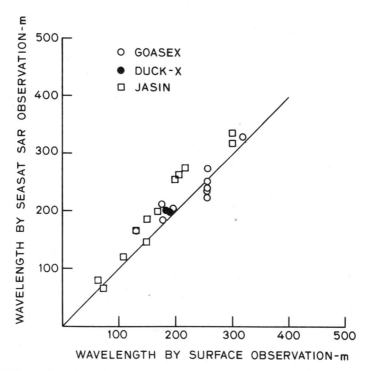

FIGURE 8. Comparison of dominant ocean wavelength as measured by SEASAT SAR and surface buoys. The SAR measurements are biased slightly high with respect to the surface measurements. The average percent difference between the two techniques is $\pm 14\%$.

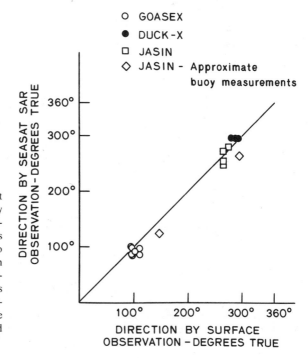

FIGURE 9. Comparison of dominant ocean wave direction as measured by SEASAT SAR and surface pitch-and-roll buoys. The SAR measurements show no significant bias with respect to the surface measurements. The mean difference between the two measurements is $\pm 12°$. If two JASIN cases involving approximate buoy measurements (labeled $\approx$ in Table I) are omitted, the mean difference is reduced to $\pm 10°$.

Figure 8 shows dominant wavelength comparisons for the SEASAT JASIN experiment and includes two similar comparisons from the DUCKEX experiment and 10 from GOASEX (Gonzalez *et al.*, 1981). Using all these data, we find SAR estimates agree with buoy estimates within an average error of about ± 14%. We also note that the SAR estimates of wavelength are biased high by about 10%. The reason for this bias is discussed below. A simple scaling error does not appear to be the cause since the same bias is found in data imaged and analyzed independently by three different research groups—ERIM and Stanford (this chapter) and Gonzalez *et al.* (1981).

In Fig. 9 we compare SAR estimates of wave direction with buoy estimates. Results from GOASEX and DUCKEX are included as in Fig. 8. The SAR measurements show no significant bias and the mean difference between the two estimates is ± 10°. In this error calculation, approximate buoy measurements (marked ≈ in Table I and shown by diamonds in Fig. 9) were excluded and hence only four JASIN cases were used.

SAR estimates of wave direction based on the simple assumption discussed above contain a 180° ambiguity, i.e., one cannot distinguish between waves traveling along Φ and those traveling along $(\Phi + 180°)$. Shuchman and Zeleuka (1978) have shown that this ambiguity can be resolved using SAR data alone by special focusing of a SAR image during the imaging process. The difference in the point of sharp focus for waves relative to the sharp focus point for land reveals the direction of wave travel, either toward or away from the radar. By applying this procedure to the JASIN data set, we are freed from relying on *ad hoc* assumptions (such as waves always traveling toward coasts or downwind) to resolve the 180° ambiguity in dominant wave direction.

As shown in Fig. 6, SAR images yield information on the wave directional distribution. Beside the dominant wave direction discussed above, a "beamwidth" for the distribution can also

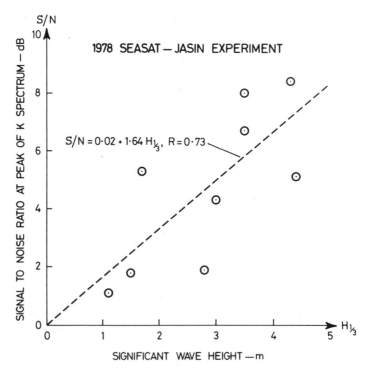

FIGURE 10. Comparison of SEASAT SAR image (S/N) at the peak of the K spectrum with buoy measurements of $H_{1/3}$. The two quantities are correlated with a determination coefficient $R^2 = 0.54$. A linear least-mean-squares fit to the data is shown by the dashed line. Using the linear fit, $H_{1/3}$ could be measured to within ± 1 m using SAR image data. Note, however, that wavelength also correlates with S/N, though less strongly ($R^2 = 0.40$).

be calculated by noting the full width at half-maximum in a SAR image directional distribution such as shown by Vesecky *et al.* (1981, 1983). The SAR directional distribution is calculated by averaging $|F_R(K, \theta)|^2$ over an octave range of K centered on the dominant wavenumber. The SAR estimates of beamwidth are remarkably close to the pitch-and-roll buoy estimates in four cases—average difference between the two is $\sim 4°$. A fifth case (orbit 834) was difficult to interpret since the peak SAR response was only a factor of two above the noise level. We note here that SAR directional distributions often show considerable structure. Vesecky *et al.* (1980) provide evidence that at least some of this structure is physically real. Hence, SAR measurements of wave directional distributions having multiple maxima could be useful in locations where *in situ* wave sensor arrays are impractical.

6. SAR ESTIMATES OF SIGNIFICANT WAVEHEIGHT

As discussed above and illustrated in Fig. 7, the signal-to-noise ratio S/N (at the peak of the wave K spectrum) can be calculated from digital transforms of SAR data. This ratio corresponds to the peak-to-background ratio (PBR) obtained from optically transformed SAR data as shown in Fig. 5. Nine cases during the SEASAT JASIN experiment contain both SAR (S/N) and buoy $H_{1/3}$ estimates. In Fig. 10 we plot these nine cases and note a clear correlation between S/N and $H_{1/3}$—correlation coefficient $r = 0.7$. A linear least-mean-squares fit to these data points yields S/N $= 0.02 + 1.64 H_{1/3}$ with a coefficient of determination $R^2 = 0.53$. Based on this data set, the linear fit estimates the value of $H_{1/3}$ with a mean error of about ± 1 m. The ± 1 m error compares well with the accuracy of wave heights determined from SAR data using observations of speckle diversity (Jain *et al.*, 1982). Further, the correlation of S/N and $H_{1/3}$ could be used to produce SAR estimates of $\Psi(K)$ which yield absolute values along the ordinates as well as the abscissae of Figs. 7 and 11, i.e., no normalization along the ordinates would be required. However, we note that in the JASIN data set, $H_{1/3}$ and Λ are themselves correlated and hence S/N and Λ, though less strongly ($R^2 = 0.40$) than S/N and $H_{1/3}$.

7. COMPARISON OF SAR IMAGE SPECTRA WITH WAVE HEIGHT AND WAVE SLOPE SPECTRA MEASURED BY BUOYS

In Fig. 11 we show six cases in which SAR images from SEASAT and concurrent buoy spectra were collected during the SEASAT JASIN experiment. An additional case is shown in Fig. 7. The SAR spectrum in each case corresponds to $|F_R|^2$, i.e., the correction discussed above is included. In several cases the buoy was located within the SAR image, while in other cases the two are displaced as noted in Table I. The omnidirectional wave height spectrum $M_{00}(K)$, which is equivalent to $\Psi(K)$, is shown by the dotted line and corresponds to the calibration on the ordinate. These values of $M_{00}(K)$ were measured by either Waverider or pitch-and-roll buoys as noted in Table I. To represent the shape of the wave slope spectra $\Psi'(K)$, we have simply multiplied the $M_{00}(K)$ curve by K^2 and normalized the result to the peak level of the M_{00} curve. The SAR spectra are normalized along the ordinate to the peak level of M_{00}. No normalization was done along the abscissa. As discussed above, it should be possible to make a direct comparison of the SAR and buoy estimates of M_{00} without any normalization by using the S/N vs. $H_{1/3}$ correlation.

One of the most salient features of the comparisons is that the peak of the SAR spectrum is often displaced (toward lower K) from the peaks of M_{00} or $K^2 M_{00}$ as measured by buoys. This bias of SAR spectra toward longer wavelengths is also evident in Fig. 8. Although there is only a small range of K above the peak value of K, we can make a rough estimate of the slope of a power law fit (K^β) to $|F_R|^2$ for $K > K_{peak}$. For the cases shown in Figs. 7 and 11, the average slope in this region is $\langle \beta \rangle \sim -2.9$. For the same cases, the average slope of $\Psi(K)$ or M_{00} for $K > K_{peak}$ is $\langle \beta \rangle \sim -3.7$. For wave slope spectra, the slope is of course 2 larger than for wave height spectra, i.e., $\langle \beta \rangle \sim -1.7$. Thus, the SAR spectral slope for $K > K_{peak}$ is approximately midway between

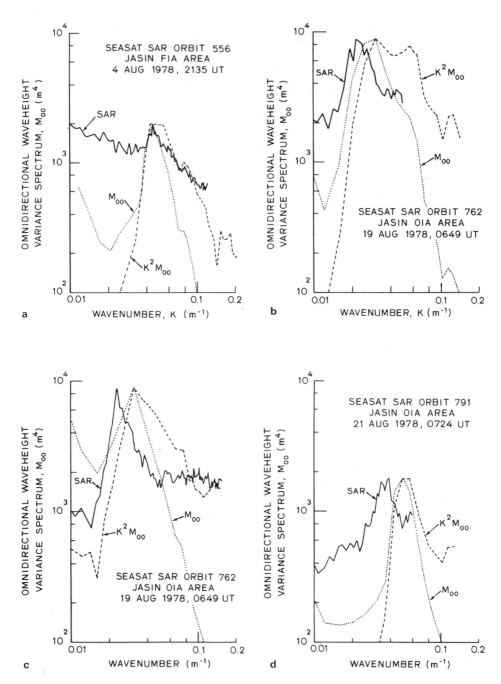

FIGURE 11. Comparisons of SAR image spectra (corrected for background fluctuations and SAR point response) $|F_R(K)|^2$ with omnidirectional wave height and wave slope spectra $\Psi(K)$ (or M_{00}) and $\Psi'(K)$ (or $K^2 M_{00}$) as measured by pitch-and-roll buoys. Both $|F_R(K)|^2$ and $\Psi(K)$ have been normalized along the ordinate at the peak of $\Psi(K)$. Note that the peak of $|F_R(K)|^2$ is typically displaced toward smaller K relative to the peak of $\Psi(K)$. Further notice that the slope of $|F_R(K)|^2$ below the peak (smaller K) is typically more steep than the slope of $\Psi(K)$ and that the slope of $|F_R(K)|^2$ above the peak (larger K) is typically less steep than the slope of $\Psi(K)$. These six comparisons are from orbits 556 (a), 762 (b, c), 791 (d), 1044 (e), 1049 (f). Further details are given in Table I. For orbit 762, (b) refers to the upper entry and (c) to the lower entry in Table I.

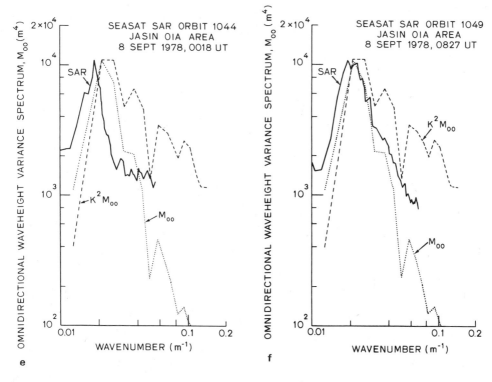

FIGURE 11. (*Continued*)

the average slopes for wave height and wave slope spectra. Overall, the SAR spectra correspond slightly better with the wave height spectra than with the wave slope spectra.

Both of the spectral features described above can be explained, at least qualitatively, by noting that in Eq. (2) there is a transfer function $|R|^2$ related to the radar wave–ocean wave interaction (imaging) mechanism. Work by Alpers *et al.* (1981), discussed in Section 4, indicated that we can expect $|R^2|$ to be proportional to K^2 to K^3. Consider the curves for $\Psi(K)$, or M_{00} in Figs. 7 and 11. If one were to apply the transfer function $|R|^2$ to these curves [as indicated by Eq. (2)], we would expect the slope of the resulting SAR spectrum to be relatively more steep on the low wavenumber side of the peak ($K < K_{peak}$) and relatively less steep on the high wavenumber side of the peak ($K > K_{peak}$) and this is indeed what we generally find in the SEASAT JASIN data of Figs. 7 and 11. Further, the changes in slope from $\Psi(K)$ to $|F_R|^2$ induced by $|R|^2$ could cause the peak wavenumber to be shifted toward relatively lower wavenumber in SAR spectra as is observed. However, the observed change in slope from $\Psi(K)$ to $|F_R(K)|^2$ is not as dramatic as one would expect from $|R|^2 \sim K^2$ or K^3. Thus, we conclude that the theory of Alpers *et al.* is *qualitatively, though not quantitatively*, supported by the JASIN data set. Since the transfer functions derived by Alpers *et al.* are linear and apply rigorously only to monochromatic waves, quantitative agreement is not necessarily expected for a spectrum of real ocean waves in which nonlinear phenomena may play an important role. Further theoretical development along the lines pursued by Alpers et al. together with empirical modifications where necessary should lead to better algorithms for estimating $\Psi(K, \theta)$ from SAR imagery.

8. SUMMARY AND CONCLUSIONS

The principal results and conclusions of the SEASAT JASIN experiment are summarized below. The results are based on optically processed SAR imagery collected by the SEASAT satellite and surface measurements collected by buoys, ships, and aircraft.

1. *Wave Visibility in SAR Images*: Waves observed by surface buoys are often, but not always, imaged by the SEASAT SAR. Three factors appear to be of primary importance for waves to be visible:

    a. SAR system resolution should be good enough to resolve at least two SAR pixels per ocean wavelength along either the along-track (azimuth) or cross-track (range) directions. SAR resolution is adversely affected by ocean wave motion, especially resolution along the azimuth direction.

    b. Significant wave height $H_{1/3}$ should be above some threshold. The data examined here are insufficient to determine this threshold, but suggest it is less than or approximately equal to 1 m.

    c. Wind speed should be above some threshold. In the data examined here, wind speeds do not fall below 3.5 m/s, and in the 3.5 m/s case, waves were visible in the SAR image. Thus, the threshold must be below 3.5 m/s, probably a few meters per second.

2. *SAR-Based Estimates of Dominant Wavelength and Direction*: Unadjusted SAR estimates of dominant wavelength and direction for the SEASAT JASIN, GOASEX and DUCKEX experiments are accurate to about $\pm 14\%$ and $\pm 10°$, respectively. In computing the latter figure, approximate buoy measurements (marked $\approx$ in Table I) were excluded and thus only four JASIN cases were used. The SAR estimates of dominant wavelength are biased high by about 10%. If this bias were to be removed by a linear least-mean-squares fit, the dominant wavelength could be estimated to about $\pm 12\%$ using SAR data. The 180° ambiguity usually present in SAR estimates of wave direction can be resolved by focusing tests during the imaging process. These tests remove the need for *ad hoc* assumptions, such as waves always traveling downwind or toward coasts.

3. *Estimates of $H_{1/3}$ Using SAR Images*: The signal-to-noise ratio (S/N) or peak-to-background ratio (PBR) at the dominant wavenumber of a SAR image K spectrum is correlated with the significant wave height $H_{1/3}$ ($R^2 = 0.53$). On the basis of the data reported here, we suggest that $H_{1/3}$ can be measured to an average accuracy of about ± 1 m using Fourier spectra of SAR images. We note, however, that in the JASIN data set, wavelength is also correlated to S/N, though less strongly ($R^2 = 0.40$).

4. *Estimates of Directional Wave Height Spectra Using SAR Image Spectra*: SAR image spectra are in rough agreement with buoy measuremets of omnidirectional ocean wave height spectra $\Psi(K)$ and correspond less closely with ocean wave slope spectra $\Psi'(K)$. However, the dominant wavenumber in a SAR spectrum typically falls below the dominant wavenumber of a buoy spectrum. Further, the typical slope of the SAR spectra on the high wavenumber side of the peak ($K > K_{peak}$) is $\sim K^{-2.9}$ while the corresponding slopes for buoy spectra are $\sim K^{-3.7}$ and $\sim K^{-1.7}$ for wave height and wave slope, respectively. These differences between SAR and buoy spectra are in qualitative agreement with the theory of Alpers et al. (1981). Analysis algorithms based on this theory, but including empirical modifications, should substantially improve estimates of $\Psi(K, \theta)$ using SAR images.

ACKNOWLEDGMENTS. We gratefully acknowledge Herbert Carlson of the Deutsches Hydrographisches Institut and Trevor Guymer and David Webb of the Institution of Oceanographic Sciences (U.K.) For supplying wind and wave data collected during the JASIN experiment. Kurt Graf, Dennis Douglas, and Dennis Tremain at SRI International provided valuable help in digitizing the JPL imagery. We also thank Harriet Smith, Martha Smith, and Sara Zientek for help in preparing the manuscript. The reviewers' comments contributed substantially to this report and we thank them. The authors gratefully acknowledge financial support from the Office of Naval Research (Physical Oceanography Branch), the National Oceanic and Atmospheric Administration (Ocean Sciences Branch), and the National Aeronautics and Space Administration (Oceanic Processes Branch).

REFERENCES

Alpers, W. R., D. B. Ross, and C. L. Rufenach, (1981): On the detectability of ocean surface waves by real and synthetic aperture radar. *J. Geophys. Res.* **86**, 6481–6498.

Beal, R. C. (1981): Spatial evolution of ocean wave spectra. *Spaceborne Synthetic Aperture Radar for Oceanography* (R. C. Beal, P. S. DeLeonibus, and I. Katz, eds.), Johns Hopkins Press, Baltimore, 110–127.

Beal, R. C., P. S. DeLeonibus, and I. Katz (eds.) (1981): *Spaceborne Synthetic Aperture Radar for Oceanography*, Johns Hopkins Press, Baltimore.

Bracewell, R. N. (1979): *The Fourier Transform and Its Applications*, McGraw–Hill, New York.

Bragg, W. L. (1933): *A General Survey*, Vol. 1 of *The Crystalline State* (Sir L. Bragg, ed.), Bell, London.

Champeney, D. C. (1973): *Fourier Transforms and Their Physical Applications*, Academic Press, New York.

Gonzalez, F. I., R. A. Shuchman, D. B. Ross, C. L. Rufenach, and J. F. R. Gower (1981): Synthetic aperture radar observations during Goasex. *Oceanography from Space* (J. F. R. Gower, ed.), Plenum Press, New York, 459–467.

Gower, J. F. R. (ed.) (1981): *Oceanography from Space*, Plenum Press, New York.

Jain, A. (1978): Focusing effects in synthetic aperture radar imaging of ocean waves. *Appl. Phys.* **15**, 323–333.

Jain, A., G. Medlin, and C. Wu (1982): Ocean wave height measurement with SEASAT SAR using speckle diversity. *IEEE J. Ocean. Eng.* **OE-7**, 103–107.

Jordan, R. L. (1980): The SEASAT A synthetic aperture radar system. *IEEE J. Ocean. Eng.* **OE-1**, 154–163.

Kasischke, E. S., G. A. Meadows and P. J. Jackson (1984): The use of synthetic aperture radar to detect hazards of navigation, Report 169200-2-F, Environmental Research Institute of Michigan, Ann Arbor, Michigan.

Kinsmau, B. (1965): *Wind Waves*, Prentice–Hall, Englewood Cliffs, N. J.

Phillips, O. M. (1981): The structure of short gravity waves on the ocean surface. *Spaceborne Synthetic Aperture Radar for Oceanography* (R. C. Beal, P. S. DeLeonibus, and I. Katz, eds.), Johns Hopkins Press, Baltmore, 24–31.

Pollard, R. T. (coordinator) (1979): Air–sea interaction project: Summary of the 1978 field experiment. The Royal Society.

Raney, R. K. (1980): SAR response to partially coherent phenomena. *IEEE Trans. Antennas Propag.* **AP-28**, 777–787.

Shuchman, R. A., and J. S. Zeleuka (1978): Processing of ocean wave data from a synthetic aperture radar. *Boundary-Layer Meteorol.* **13**, 181–191.

Stewart, R. H. (1977): A discuss-hulled wave measuring buoy. *Ocean Eng.* **4**, 101–107.

Vesecky, J. F., and R. H. Stewart (1982): Observations of ocean surface phenomena by the SEASAT synthetic aperture radar—An assessment. *J. Geophys. Res.* **87**, 3397–3430.

Vesecky, J. F., H. V. Hsiao, C. C. Teague, O. H. Shemdin, and S. S. Pawka (1980): Radar observations of waves in the vicinity of Islands. *J. Geophys. Res.* **85**, 4977–4986.

Vesecky, J. F., H. M. Assal, and R. H. Stewart (1981): Remote sensing of the ocean wave height spectrum using synthetic-aperture-radar images. *Oceanography from Space* (G. F. R. Gower, ed.), Plenum Press, New York, 449–458.

Vesecky, J. F., S. L. Durden, D. J. Napolitano and M. P. Smith (1983): Theory and practice of ocean wave measurements by synthetic aperture radar, in: *Oceans '83* (M. Switzer *et al.*, eds.), IEEE Press, Piscataway, New Jersey, 331–337.

Vesecky, J. F., M. P. Smith, D. J. Napolitano, and S. L. Durden (1984): Models for the imaging of ocean gravity waves by SAR—Comparison of theory and experiment, in: *Oceans '84* (R. Hagen, *et al.*, eds.), IEEE Press, Piscataway, New Jersey, 119–125.

28

LIMITATIONS OF THE SEASAT SAR IN HIGH SEA STATES

F. M. MONALDO AND R. C. BEAL

ABSTRACT. Although the spectral analysis of SEASAT SAR imagery has shown qualitatively good agreement with independent surface spectral measurements, quantitative two-dimensional ocean energy spectra from SAR imagery remain an elusive goal. The exact relationship between the ocean wave energy spectrum and the corresponding SAR image spectrum constitutes the central problem of much recent research. This chapter outlines a systematic approach to estimation of ocean wave spectra from SAR image spectra based on recent analysis of SEASAT SAR digital data. The effect of the various wave-imaging mechanisms, the stationary scatterer resolution response function, and the moving scatterer resolution response function are considered. Corrections for the latter two effects are applied to actual SAR image spectra. Particular attention is paid to the degradation of the azimuth component of SAR spectra in moderate to high sea states by the moving scatterers, which limit the ability of the SAR to image azimuth-traveling waves. This effect is quantitatively modeled and predictions of the model are shown to agree with actual SAR spectra. Corrections for the moving scatterer resolution response function are shown to result in better estimation of the position of remote storm sources.

1. INTRODUCTION

Ocean waves are a manifestation of dynamical processes both above and beneath the surface. The surface wave structure of the oceans is neither entirely random nor homogeneous. Locally, it is dependent on the local wind structure and on storm systems many hundreds of kilometers and tens of hours away whose generated wave systems have propagated to the immediate vicinity. Wave systems generated by storms localized in space and time have generally diverging propagation vectors and a spatial distribution that reveals something about the generating wind field. Although wave propagation in deep water is essentially without loss, waves can lose or gain energy and be refracted by the ocean current structure. Similarly, wave systems entering shallow water will experience refraction in addition to a general energy loss due to bottom friction effects. Consequently, the spatially or temporally evolving wave systems, if they could be accurately monitored on synoptic scales, could reveal not only the storm generation structure but the structure of local currents and bathymetry. If this information could also be provided on a timely basis, it could be used to improve and reinitialize wave forecast models.

Gonzalez *et al.* (1979) were the first to clearly demonstrate that ocean waves could be detected by the SEASAT synthetic aperture radar (SAR). Two-dimensional Fourier analysis of

---

F. M. MONALDO AND R. C. BEAL ● Applied Physics Laboratory, The Johns Hopkins University, Laurel, Maryland 20707.

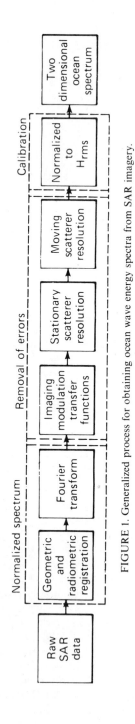

FIGURE 1. Generalized process for obtaining ocean wave energy spectra from SAR imagery.

SAR imagery has shown spectral energy in wavenumber regions closely corresponding to spectral peaks determined from independent surface measurements (e.g., Beal, 1980). Before imagery from spaceborne SARs can be routinely used to provide a synoptic view of long-wave ocean spectra, the circumstances (satellite geometry and surface wind and wave conditions) under which SAR spectra are reliable and accurate must be determined. Moreover, the question of how SAR data should be optimally processed to yield quantitative directional spectra needs to be addressed.

In this chapter we describe a general approach for using SAR data to determine a quantitative ocean spectrum. The process, outlined in Fig. 1, involves the removal of instrument-induced and wave-imaging-mechanism-induced effects to arrive at a filtered version of the actual ocean energy spectrum. The process can be divided into three broad tasks: (1) obtaining a normalized Fourier transform; (2) removing systematic biases introduced by the SAR instrument and imaging process; and (3) calibrating SAR spectra in physical units. These tasks can be further divided into subtasks also shown in Fig. 1. We detail what progress has been made in each subtask and what remains to be done before the entire process is quantified.

For certain situations, the relationship between long-wave slope (or height) and SAR image intensity may be sufficiently nonlinear that there is no unique relationship between the SAR image spectrum and the ocean wave spectrum. In such cases, nonlinear analysis techniques might be useful to obtain estimates of ocean spectra. Even for such nonlinear situations, however, it may be possible to determine at least the dominate wavenumber and direction of the ocean spectrum.

As part of the effort to specify the relationship between the SAR image spectrum and the ocean spectrum, particular attention is paid in this chapter to estimation of resolution losses inherent in the SAR imagery. Because a SAR uses Doppler information to determine the azimuth position of a scatterer, the radial velocity of a scatterer will cause its azimuth displacement in the image. An accelerating scatterer causes a defocusing of the SAR image in the azimuth direction. Therefore, a randomly moving surface like the ocean degrades azimuth resolution in the SAR image spectrum. Using ocean spectra measured with an airborne laser profilometer, the azimuth resolution degradation is calculated and compared to actual SEASAT SAR data. Correction for the azimuth resolution loss is used to improve estimates of storm location from SAR spectra.

In describing SAR imagery, we will assume the applicability of a two-scale ocean model (Wright, 1966). The two-scale model assumes the existence of two widely separated wavelength regimes. The smaller regime has wavelengths in the decimeter and centimeter range and the larger regime has wavelengths of tens or hundreds of meters. It is the large-scale structure that we wish to measure via the structure of the short waves. Specifically, nonvertical incidence microwave radars interact with the ocean surface through the Bragg resonance mechanism involving the short waves. The backscattered microwave power is proportional to the wave-height spectrum evaluated at the Bragg wavenumber. The real or apparent periodic modification of Bragg waves by the longer waves makes the longer waves visible in the SAR image. For the SEASAT SAR geometry and electromagnetic wavelength of 23 cm, interactions occur at a surface wavelength between 30 and 40 cm.

2. NORMALIZED FOURIER TRANSFORM

Raw data from the SAR consist of a signal record giving the two-dimensional phase history of the radar return. By either digital or optical correlation, this signal record is converted to a two-dimensional image. However, this correlation process must be performed carefully so that the images retain both spatial and radiometric fidelity, i.e., so that the image spatial scale is linearly related to scales on the ocean surface, and the image intensity is proportional to the ocean backscatterer power. This radiometric and geometric correction must be performed prior to performing the Fourier transformation. Thus far, MacDonald, Dettwiler and Associates (MDA) and the German Space Operations Center (DFVLR), who use identical digital algorithms, appear to have produced the highest-quality imagery in these respects (Goldfinger, 1980; Beal et al., 1981, 1983).

Once a corrected image is obtained, the square of the Fourier transform yields the normalized image spectrum. Although the transform can be performed optically, digital techniques are much more amenable to removal of systematic error sources. We employ spatial averaging in the wavenumber domain to reduce random noise, but with some cost in local wavenumber resolution.

3. RELATIONSHIP BETWEEN SAR IMAGE AND OCEAN SURFACE SPECTRA—ERROR REMOVAL

The second major task in obtaining a long-wave ocean spectrum involves the correction of the normalized SAR spectrum for systematic imaging effects, most of which are peculiar to the SAR technique. This correction process can be expressed by the equation

$$I^2(\mathbf{k}) = S^2(\mathbf{k})M^2(\mathbf{k})G^2(\mathbf{k})H^2(\mathbf{k}) \tag{1}$$

where $I^2(\mathbf{k})$ is the spatial power spectrum of the image as a function of wavenumber vector, $\mathbf{k}$; $S^2(\mathbf{k})$ is the ocean surface slope spectrum; $M(\mathbf{k})$ is the modulation transfer function due to imaging mechanisms, describing their systematic spectral effects. The spectral resolution falloff for a stationary scatterer is given by $G^2(\mathbf{k})$, whereas the $H^2(\mathbf{k})$ describes the resolution falloff due to moving scatterer effects.

Equation (1) is valid for any system which can be described by linear processes (e.g., Bendat and Piersol, 1971). The justification for invoking linearity is twofold. First, if significant nonlinearities were present in the ocean surface slope to radar cross-section transfer function, harmonics or intermodulation products of the fundamental ocean wavenumber would be present in SAR image spectra. Such distortions have not been found. Second, although the effect of moving scatterers on SAR imagery may be weakly nonlinear, it is an energy-conserving nonlinearity. Raney (1982) points out that the effects of such a nonlinearity can be modeled as a loss of resolution in a linear system.

There may be some circumstances where an assumption of linearity is not appropriate. However, thus far, systematic effects in actual SAR spectra can be explained without resorting to nonlinearity. The ultimate justification of any such assumption, of course, rests in its usefulness in explaining data.

3.1. Image Modulation Transfer Function

The imaging modulation transfer function, $M(\mathbf{k})$, can be thought of as expressing the relationship between image pixel intensity (or radar cross section) and long-wave ocean surface slope. If SAR imaging can be described linearly, then for a simple sine wave of wavenumber $\mathbf{k}_0$, the image intensity can be described by

$$i(\mathbf{x}) = \langle i \rangle [1 + M(\mathbf{k}_0) \, S_0 \cos \mathbf{k}_0 \cdot \mathbf{x}] \tag{2}$$

where $i(\mathbf{x})$ is the pixel intensity at position $\mathbf{x}$ in the image, $\langle i \rangle$ is the spatially averaged image intensity, $M(\mathbf{k}_0)$ is the imaging modulation transfer function at $\mathbf{k}_0$, and S_0 is the wave slope amplitude. For a more realistic ocean with a slope spectrum of $S^2(\mathbf{k})$, then (2) is generalized to

$$i(\mathbf{x}) = \langle i \rangle \left[1 + \int M(\mathbf{k}) \, S(\mathbf{k}) \cos (\mathbf{k} \cdot \mathbf{x}) \, d\mathbf{k} \right] \tag{3}$$

$M(\mathbf{k})$ is the coherent sum of three different imaging mechanisms: tilt modulation, spectral modulation, and orbital velocity modulation (Elachi and Brown, 1977; Alpers *et al.*, 1981).

3.1.1. Tilt Modulation

Because of the variation in the geometric surface area intercepted by a radar beam with surface slope variation, the measured radar cross section and hence image intensity varies in the SAR image. This effect has been described by Elachi and Brown (1977). For the SEASAT SAR geometry (Fig. 2) and HH polarization, the modulation transfer function magnitude, $|M_t| \cong 13$ (slope^{-1}) $\cos\phi$ (Elachi and Brown, 1977), where ϕ is the angle between the long-wave propagation direction and range or cross-track direction. Hence, maximum sensitivity exists for waves traveling in the range direction and minimum sensitivity for azimuth-traveling waves, i.e., for waves parallel to the velocity vector of the spacecraft. The phase of the modulation, $\Delta_t = 90°$. Such a phase means that the maximum slope facing the radar has the highest apparent cross section. Figure 3A is a contour plot of $|M_t|$, the magnitude of the tilt modulation transfer function as a function of vector wavenumber. Note that the function is independent of $|\mathbf{k}|$ and has a minimum in the azimuth direction.

3.1.2. Spectral Modulation

Since, for the SEASAT SAR, the scattering cross section of an ocean patch is proportional to the spectral energy at 30 cm within that patch, the modulation of such spectral energy in the presence of long waves can render long ocean waves detectable by the SAR. Unfortunately, there is no generally accepted complete theoretical treatment of short-wave modulation by long waves. Present theories, as we shall see, are still inadequate to explain the bulk of experimental field measurements.

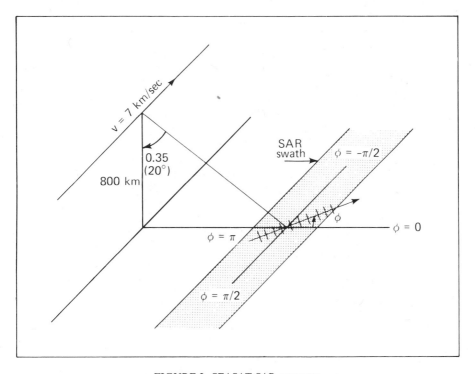

FIGURE 2. SEASAT SAR geometry.

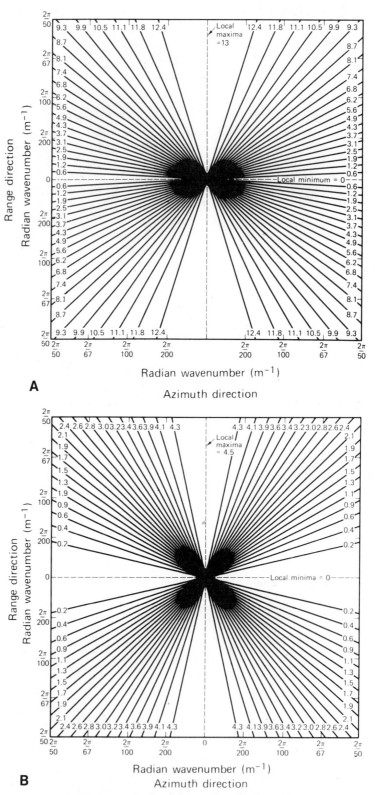

FIGURE 3. Image modulation transfer function magnitudes: (A) tilt modulation, (B) spectral modulation, (C) orbital velocity modulation, and (D) the coherent sum of A, B, and C. Computations assume the SEASAT SAR configuration and the dominant wind being in the range direction.

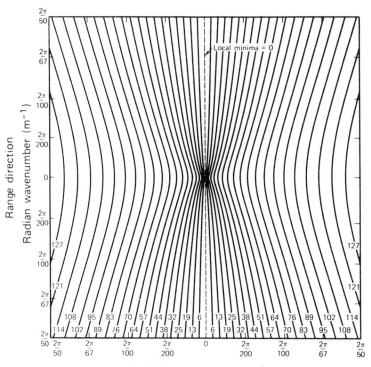

C   Azimuth direction

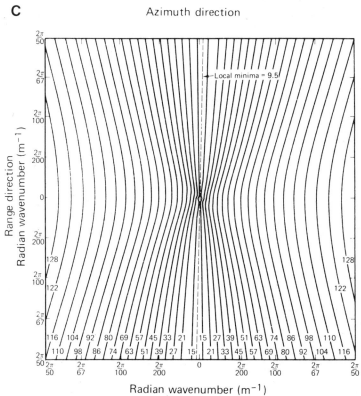

D   Azimuth wavenumber

FIGURE 3. (*Continued*)

The largest body of field data on spectral modulation has been collected by the Naval Research Laboratory (NRL) using stationary, coherent radars (Keller and Wright, 1975; Plant *et al.*, 1978; Wright *et al.*, 1980). The time variation in backscattered power from the ocean surface is presumed to be proportional to the time variation of spectral energy at the Bragg interaction wavelength. This backscattered power is then correlated against long-wave slope to determine the modulation transfer function, M_s, for spectral modulation.

Figures 4 and 5 show the NRL-measured modulation transfer function magnitude and phase as a function of long-wave frequency for 13-cm Bragg waves. Modulation magnitudes vary from 5 to 20(slope^{-1}) and phases vary from 20 to 40° past the long-wave crest.

Field modulation measurements made by a laser slope gage (Evans and Shemdin, 1980) and time series photography (Monaldo and Kasevich, 1981) indicate modulation magnitudes in agreement with radar measurements. Photographically measured phases are somewhat higher at 90°. There seems to be clear experimental consensus on the range of modulation magnitudes. Measured modulation phases can be found at 0 to 90° past the long-wave crest.

Short-wave modulation by long waves has been described theoretically by either an energy (Keller and Wright, 1975; Valenzuela and Wright, 1979) or an action balance (Alpers and Hasselmann, 1978; Phillips, 1981) equation. Wave action is defined as the ratio of wave energy to frequency. Both approaches are equivalent and yield the same results.

Solutions of these equations (Alpers and Hasselmann, 1978; Phillips, 1981) have shown modulation magnitudes of only 4.5(slope^{-1}), with modulation phase for any significant magnitude near 0°. That is, the maximum amount of short-wave energy is found at the crest of the long wave. These theoretically predicted magnitudes and phases, however, are at the lower bound of empirical measurements and do not include any wavenumber or wind dependence. Under most circumstances, the measured magnitudes and phases are significantly larger than predicted by the models (see Figs. 4 and 5). Modified air-flow over the long wave on the short-

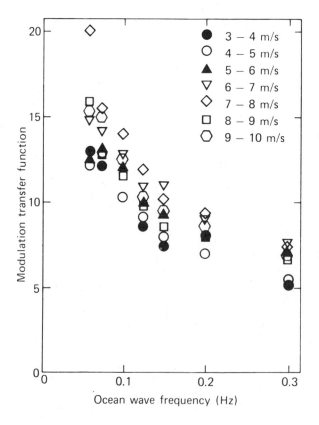

FIGURE 4. Modulus of the spectral modulation transfer function for 13-cm waves as a function of long-wave frequency and wind speed as measured with a coherent microwave radar (Wright *et al.*, 1980).

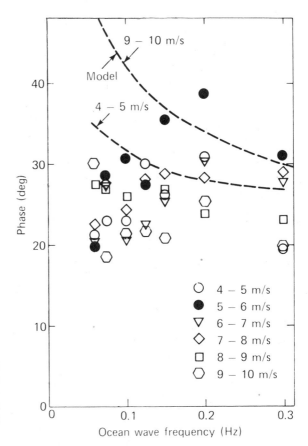

FIGURE 5. Phase of the spectral modulation transfer function for 13-cm waves as a function of long-wave frequency and wind speed as measured with a coherent microwave radar (Wright et al., 1980).

wave structure, including the potentially strong effects of turbulence, could possibly account for this discrepancy.

The directionality of short-wave modulation has not been extensively studied experimentally. By making the simplifying assumption that wave action is conserved, Phillips (1981) estimated the dependence of modulation on the angles between long and short waves and the wind direction. He found that

$$|M_s| = 4.5(\text{slope}^{-1}) \left[\cos(\phi - \Theta) - \frac{2}{9} \frac{1}{f(\Theta)} \frac{\partial f}{\partial \Theta} \sin(\phi - \Theta) \right] \cos \phi \qquad (4)$$

where ϕ is the long-wave propagation direction measured from the SAR range direction, Θ is the dominant wind direction also measured from the range direction, and $f(\Theta)$ is the directional distribution of the short Bragg waves.

For a realistic directionality, $f(\Theta) = \cos^2(\frac{1}{2}\Theta)$ (Tyler et al., 1974), (4) becomes

$$|M_s| = 4.5(\text{slope}^{-1}) \left[\cos(\phi - \Theta) + \frac{2}{9} \tan(\frac{1}{2}\Theta) \sin(\phi - \Theta) \right] \cos \phi \qquad (5)$$

Note that at $\phi = \pm 90°$ (azimuth-traveling waves), the modulation goes to zero. In addition, for some ϕ not in general equal to $\pm 90°$, the bracketed quantity in (5) goes to zero, creating another null in the spectral modulation directional sensitivity.

Despite the inadequacies of the theoretical treatment of spectral modulation, we will assume the magnitude of the spectral modulation transfer function to be given by (5) and the phase, Δ_s, to be equal to $0°$. Although actual measurements are more accurate than the model result, they do not adequately specify the directional dependence of the modulation. Inadequate analytic specification of the spectral modulation transfer function, however, still remains a significant problem in obtaining long-wave ocean spectra from SAR images.

Figure 3B shows a contour of $|M_s|$, the magnitude of the spectral modulation transfer function, as a function of vector wavenumber. In this plot it is assumed that Θ, the dominant wind direction, is in the range ($\phi = 0$) direction. Note that this function is independent of $|\mathbf{k}|$ and the $|M_t|$ has a minimum response for azimuth-traveling waves.

3.1.3. Orbital Velocity Modulation

Long waves impose varying radial velocities on the ocean surface, causing a correspondingly varying Doppler shift in the signal return from the Bragg scatterers distributed across the long wave. The correlation processes used to create a SAR image will cause an apparent azimuth displacement of the position of a scatterer in the image. Small shifts can alter a uniform distribution of scatterers such that their density varies at the same spatial frequency as that on the long wave. This "velocity bunching" is a potential third imaging mechanism. As shall be discussed in Section 3.3, this same mechanism is also apparently the dominant source of azimuth resolution losses.

Larson et al. (1976), Swift and Wilson (1979), Alpers and Rufenach (1979), and Alpers et al. (1981) have studied the orbital velocity modulation mechanism in some detail, and propose that the magnitude of the orbital velocity modulation transfer function is given by [Alpers and Rufenach, 1979, Eqs. (28) and (29)]

$$|M_{ov}| = \frac{R(gk)^{1/2}}{V} a(z)|\sin \phi|[(\sin \theta \cos \phi)^2 + \cos^2 \theta]^{1/2} \tag{6}$$

where R is the range distance between the SAR and the scatterer, V is the velocity of the SAR platform, g is the acceleration due to gravity, z equals $(gk)^{1/2}T/2$, T is the SAR integration time, θ is the SAR nadir look angle and all other variables are the same as previously defined ($R/V \sim 128$ s, $T \sim 0.5$ s for SEASAT). For SAR integration times $\lesssim 1$ s and typical long-wave wavenumbers, the function $a(z) \cong 1$.

Figure 3C is a contour plot of $|M_{ov}|$, the magnitude of the orbital velocity modulation transfer function as a function of vector wavenumber for the SEASAT SAR geometry. Note that $|M_{ov}| \sim |\mathbf{k}|^{1/2}$. Further, unlike tilt and spectral modulation, the orbital velocity modulation is a maximum for azimuth-traveling waves and zero for range-traveling waves.

In addition, the orbital velocity mechanism rests on the assumption that the Bragg waves are uniformly distributed across the long wave. The presence of significant nonuniformity in the distribution of Bragg scatterers could reduce or even eliminate $|M_{ov}|$. Indeed, as we shall see, actual SEASAT SAR spectra do not exhibit the strong azimuth response suggested by this mechanism.

The modulation phase, Δ_{ov}, for the orbital velocity mechanism is given by [Alpers and Rufenach, 1979, Eq. (30)]

$$\Delta_{ov} = \begin{cases} -\tan^{-1}(\tan \theta \cos \phi) & \text{for } 0 \leqslant \phi \leqslant \pi \\ 180° - \tan^{-1}(\tan \theta \cos \phi) & \text{for } -\pi < \phi < 0 \end{cases} \tag{7}$$

where θ is the SAR nadir angle.

For waves whose azimuth component of wavenumber is in the opposite direction to the SAR platform velocity, the maximum effective cross section appears near the crest of the long wave. For waves whose azimuth component of wavenumber is in the same direction as the SAR platform velocity, maximum effective cross section appears near the long-wave trough. This accounts for the $180°$ shifts in (7).

3.1.4. Total Image Modulation Transfer Function

$M(\mathbf{k})$, the *total* image modulation transfer function, is taken as the coherent sum of all individual modulation terms, i.e.,

$$M(\mathbf{k}) = |M_t|e^{i\Delta_t} + |M_s|e^{i\Delta_s} + |M_{ov}|e^{i\Delta_{ov}} \tag{8}$$

Figure 3D shows the coherent sum of all three modulation mechanisms shown in Figs. 3A–C. The salient feature of this plot is that the orbital modulation term as determined by Alpers and Rufenach (1979) clearly dominates both tilt and spectral modulation mechanisms. The tilt and spectral modulation terms serve merely to fill in the range direction nulls left by the orbital velocity term.

Also note that the total modulation transfer function is not symmetric about the azimuth direction. This is because of the asymmetry of Δ_{ov} shown in (7).

Before too many conclusions are drawn from Fig. 3D, we need to consider the terms $G^2(\mathbf{k})$ and $H^2(\mathbf{k})$ in (1).

3.2. Stationary Scatterer Resolution Falloff

Because of finite range and azimuth system resolution for stationary scatterers, there is a Gaussian-type falloff of spectral sensitivity with increasing wavenumber, represented in (1) by the term $G^2(\mathbf{k})$. $G^2(\mathbf{k})$ can be practically determined either by a Fourier transformation of the system impulse response function or by evaluating the falloff in the SAR image spectrum of a white noise image. We have employed the latter strategy in determining $G^2(\mathbf{k})$ for the SEASAT SAR. Beal *et al.* (1981, 1983) detail the determination of $G^2(\mathbf{k})$ and show that after correcting for this falloff, essentially flat response is obtained for white noise images.

3.3. Moving Scatterer Resolution Falloff

3.3.1. Theoretical Development

Due to the Doppler nature of SAR imagery, moving targets affect system resolution. A radial scatterer velocity shifts its apparent azimuth position, whereas a radial acceleration produces defocusing in the azimuth direction.

Consider first the velocity effects on azimuth resolution, already briefly discussed in describing orbital velocity modulation. The azimuth position, x, of a scatterer in a SAR image is defined by the position at which range is a minimum. For a moving scatterer, Raney (1971) expressed the range to the SAR as a function of azimuth position x as

$$r(x) = R + \varepsilon_{\dot{r}}(x - x_0) + [(1 - \varepsilon_{\dot{c}})^2 - \varepsilon_{\ddot{r}}^2]^{1/2}(x - x_0)^2/2R_0 \tag{9}$$

where R is the range to the scatterer at x_0, $\varepsilon_{\dot{r}}$ and $\varepsilon_{\dot{c}}$ are normalized radial and cross-range velocities (velocity/SAR platform velocity), and $\varepsilon_{\ddot{r}}$ is the normalized scatterer radial acceleration (product of R and scatterer acceleration/SAR velocity squared). Azimuth position in the SAR image relative to true position, $x - x_0$, is determined where (9) is a minimum, i.e., when

$$(x - x_0) = \Delta x = \frac{R\varepsilon_r}{[(1 - \varepsilon_c)^2 - \varepsilon_r^2]^{1/2}} \tag{10}$$

For typical sea surface scatterer velocities, $\varepsilon_{\dot{c}}$ and $\varepsilon_{\ddot{r}} \ll 1$, and (10) reduces to

$$\Delta x = (R/V)v_r \tag{11}$$

where v_r is the scatterer's radial velocity.

If an identical velocity could be assigned to every point in a SAR ocean image, (11) would result in a simple displacement of the entire image in the azimuth direction. An ocean surface, however, has a *random* component of velocity in the radial direction at each point on the surface and hence an *uncertain* azimuth position. The distribution of an entire ensemble of scatterer azimuth position displacements therefore should be directly proportional to the distribution of ocean orbital velocity radial to the SAR integrated over a SAR resolution element (= 25 m for the SEASAT SAR). The half-width, for a Gaussian velocity distribution, will be simply defined by its rms velocity, σ_v. This half-width, σ_v, defines the distribution of scatterer azimuth displacements whose half-width, σ_{xv}, is given by

$$\sigma_{xv} = (R/V)\sigma_x \tag{12}$$

Radial acceleration, as can be seen by evaluating (10), does not significantly contribute to the simple displacement of a scatterer, i.e., the acceleration effect on displacement is small. Nonetheless, acceleration can cause azimuth defocusing or blurring of an image thereby decreasing resolution. The defocusing process can be thought of as simply the result of an indefinite velocity. The product of the scatterer radial acceleration and integration time, T, of the SAR represents the total change in velocity of the scatterer during the integration time. Each instantaneous velocity shifts the azimuth position of the scatterer. A continuously changing velocity, associated with an acceleration, continuously displaces the scatterer azimuth position during the integration time, resulting in image blur. The resulting half-width of the resulting azimuth position distribution is given by

$$\sigma_{xa} = \sigma_a T(R/V) \tag{13}$$

where σ_a is the rms of the ocean radial acceleration. For a typical digitally processed four-look SEASAT SAR image, T is approximately 0.5 s.

These moving scatterer effects cause appreciable resolution degradation in SAR image spectra. Figure 6 is the spectrum of a 6.25-km^2 patch of ocean at 59.9°N, 22.2°W collected by the SEASAT SAR on August 24, 1978, and digitally processed to 25-m resolution by the DFVLR. The imagery was collected during a time when the ocean significant waveheight was 4.4 m and surface winds were approximately 14 m/s. Even though the spectrum has been corrected for the stationary scatterer resolution falloff, significant azimuth resolution falloff remains.

Equations (12) and (13) define the width of a system impulse response function due to the movement of scatterers, $h(\mathbf{x})$, which is given by

$$h(\mathbf{x}) = \exp\left[-\frac{1}{2}\frac{x^2}{\sigma_x^2} \right] \tag{14}$$

where $\sigma_x^2 = \sigma_{xv}^2 + \sigma_{xa}^2$. The function $H^2(\mathbf{k})$, the moving scatterer resolution response function, is simply the power spectral density function of (14) and is given by

$$H^2(\mathbf{k}) = \exp\left(-\frac{1}{2}\frac{k_x^2}{2\sigma_k^2} \right)$$

with

$$\sigma_k = (1/2)^{1/2}\sigma_x \tag{15}$$

where k_x is the azimuth component of wavenumber and σ_k is the half-width of the Gaussian falloff in the azimuth direction.

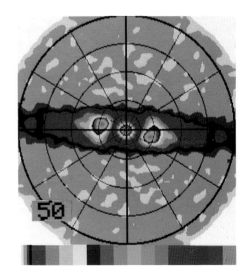

FIGURE 6. Spectrum of SEASAT SAR 6.25-km²
wave image located at 59.9°N, 22.2°W, pass 762,
August 24, 1978. Range is along the horizontal
direction and azimuth along the vertical. Spectrum
has been corrected for stationary scatterer response. Notice the dramatic falloff in the spectrum
for large azimuth wavenumber.

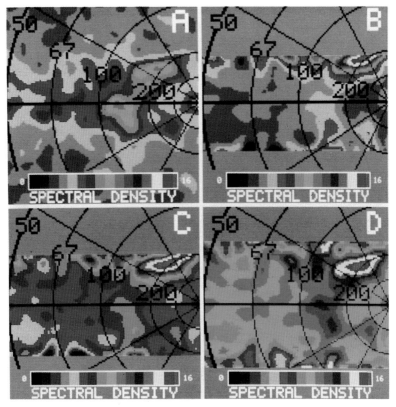

FIGURE 10. Correction of the SAR spectrum for the *moving* scatterer resolution response function from
an image centered at 32.03°N and 73.38°N from pass 1339, September 28, 1978. Range is in the horizontal direction and azimuth is in the vertical direction. (A) no azimuth correction: (B) too severe a correction; (C) best estimate correction; (D) too weak a correction.

3.3.2. Quantitative Comparison with SAR Data

Although the function $H^2(\mathbf{k})$ qualitatively describes a Gaussian azimuth falloff, the real test of any theory is whether quantitative agreement can be found. To test whether the described velocity and acceleration effects quantitatively predict the correct azimuth falloff, we examined digital SAR imagery processed by MDA from pass 1339 off the east coast of the United States. Twelve 6.25×6.25-km images from $29.5°N$ to $34.5°N$ along the satellite path were selected because they were spatially and temporally coincident with an NOAA P-3 aircraft overflight.

A two-dimensional power spectrum transformation was performed on each of the images. Each image was also corrected for the stationary scatterer response function as described earlier. To estimate the azimuth falloff, each of the spectra was averaged along the range dimension to form a single azimuth slice. In averaging along the range dimension, spectral densities more than one standard deviation from the local mean were removed to eliminate wave energy contamination of the falloff estimate, and a revised mean was computed. An example of such an azimuth cut is shown in Fig. 7.

From this series of 12 azimuth slices, least-squares Gaussian curve fits were computed. The half-width of the falloff, σ_k (the azimuth wavenumber where the Gaussian falls to $e^{-1/2}$ of its maximum value), is thus determined from the parameters of the curve fit.

To compare the azimuth falloff estimated from SAR spectra with the model of azimuth falloff due to scatterer velocity and acceleration, we employed laser profilometer wave spectra from the P-3 aircraft overflight provide by D. Ross of NOAA. These wave-height spectra were combined with the gravity wave dispersion relationship to estimate the velocity and acceleration spectra. The integration of these spectra, weighted by the stationary scatterer resolution falloff, yields the SAR-perceived σ_v and σ_a. Figure 8 shows velocity and acceleration spectra for a fully developed model ocean spectrum (Bjerkaas and Riedel, 1979) and the stationary scatterer response.

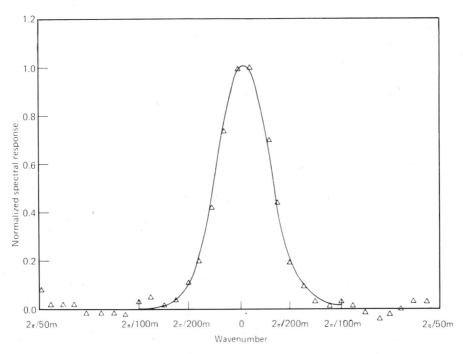

FIGURE 7. A SAR image spectrum from pass 1339, September 28, 1978, collapsed to a single azimuth cut. Each triangle represents the average of eight range rows. The solid line is the least-squares Gaussian curve fit to the points.

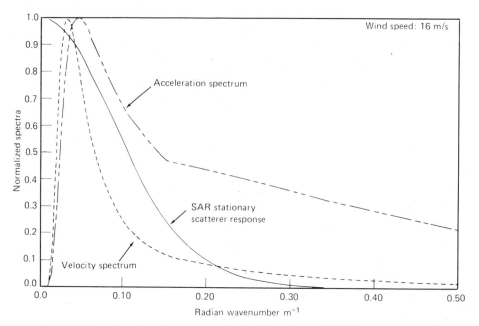

FIGURE 8. Velocity and acceleration spectra calculated from the modified Pierson–Moskowitz wave-height spectrum for a wind speed of 16 m/s. Solid line represents the stationary scatterer resolution response function.

Using (14) and (15), the σ_v and σ_a determined from laser profilometer spectra were used to estimate the σ_k and σ_x [see (15)] associated with the azimuth falloff in SAR spectra. Figure 9 shows a comparison of σ_x estimated from SAR spectra versus σ_x determined from laser profilometer data and the model of the effect of moving scatterers presented earlier. The two estimates of σ_x agree, with a correlation coefficient of 0.75. The agreement is especially good when considering that the laser profilometer remaps waves traveling nearly perpendicular to the

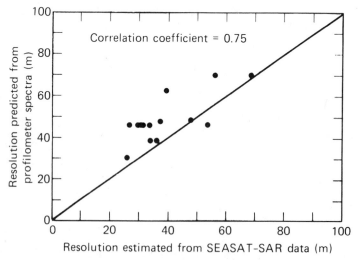

FIGURE 9. Defining σ_x as a measure of system resolution, the resolution obtained from collapsing actual SAR spectra from pass 1339 is plotted versus resolution obtaining from laser profilometer spectra and the model of the effect of moving scatterers presented in the text. For perfect agreement, all points would fall along the solid line.

aircraft flight path. We therefore conclude that the model presented here describing H^2 (k), the moving scatterer resolution response function, is in quantitative agreement with SAR data and independent ocean surface measurements.

3.3.3. Corrections of SAR Spectra for Azimuth Falloff

To obtain useful ocean spectra, the SAR spectra must be corrected for the azimuth falloff due to the moving scatterer resolution. Of course, spectral wave energy which has been attenuated below the background noise cannot be recovered. The azimuth falloff can appreciably distort the wavenumber and direction of a spectral peak. A non-range-traveling wave system with a relatively broad spectral peak, when multiplied by an azimuth falloff, will appear to be rotated toward the azimuth direction and reduced in wavenumber. We shall demonstrate this with SAR spectra from pass 1339.

A digital SAR spectrum on which we perform any subsequent analysis is routinely corrected for the stationary scatterer response function, $G^2(k)$, and smoothed with a two-dimensional Gaussian kernel about 10 pixels in width. In accordance with (1), the remaining spectra should be proportional to the product $H^2(k)M^2(k)S^2(k)$. Using the procedure described in the previous section to estimate $H^2(k)$ from the actual SAR spectra, we applied a correction for the azimuth falloff. The spectra in Fig. 10 (color coded in linear units of image spectral density) provide an illustration of both the effectiveness and the limitations of correcting for the moving scatterer resolution response function in this way.

Figure 10A is the spectrum, $H^2(k)\, M^2(k)\, S^2(k)$, of an actual SAR digital image before any azimuth falloff correction is applied. This spectrum was obtained from a 512×512-pixel image covering a 6.4×6.4-km ocean surface area on September 28, 1978. The image from pass 1339 was off the coast of Hatteras, North Carolina, and is centered at 32.03°N and 73.38°W. Note the existence of two spectral peaks, one in the range direction with an approximate wavelength of 130 m. The second peak represents a 200-m system traveling between range and azimuth directions. The angle between the azimuth direction and the 200-m system peak is 51.8°.

The other three spectra (Figs. 10B–D) represent the effect of applying various corrections for $H^2(k)$, the azimuth falloff. Since $H^2(k)$ approaches zero for large azimuth wavenumbers, the correction for $H^2(k)$ approaches infinity there. Hence, a correction at large azimuth wavenumbers would simply represent the multiplication of a large number with noise. The spectrum is deleted for azimuth wavenumbers where the azimuth falloff equals 10% or less of its maximum value.

Figure 10B is apparently too severe a correction for azimuth falloff. The angle between the azimuth direction and the 200-m system peak is 35.5°. The 200-m system has rotated back toward the azimuth direction. However, the 130-m wave system has been lost in application of the $H^2(k)$ correction. Because this system straddled the range axis, i.e., had the same symmetry as $H^2(k)$, the correction eliminated that peak.

Figure 10C is the correction that is based on the azimuth falloff determined by collapsing the original spectrum along the range direction into a single azimuth slice. Since the 130-m peak, when collapsed onto an azimuth cut, would tend to make the azimuth falloff appear more pronounced, the azimuth correction is probably still too severe. However, unlike Fig. 10B, the correction still retains some of the 130-m system. The angle between the azimuth direction and the 200-m system peak is 37.9°, which is closer to the azimuth direction than the original spectrum.

Figure 10D is a very weak correction for the azimuth falloff. The angle between the azimuth direction and the 200-m system peak is now 48.4°. Note that the peak is not rotated toward azimuth as much in this spectrum as in Figs. 10B and C.

Twelve spectra from 29.4 to 34.2° in latitude from pass 1339 were corrected for azimuth falloff. As in Fig. 10C, the corrections were determined from our best estimate of $H^2(k)$ for each spectrum.

Figure 11 plots the positions, in latitude and longitude, of the 12 spectra that are uncorrected for azimuth falloff. Also plotted are the wave vectors from the 200-m system projected back 4

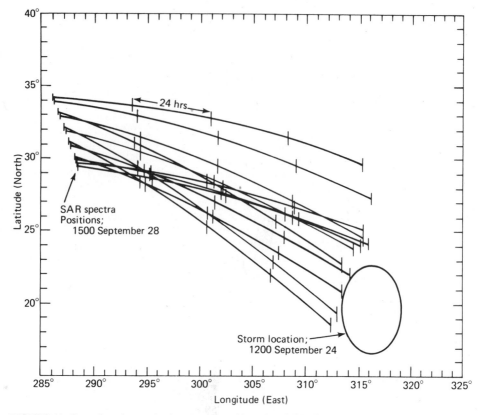

FIGURE 11. Storm location, projection uncorrected for azimuth falloff. The crosses mark the position of 12 SAR images from pass 1339. The plotted rays are the trajectories of a 200-m swell system projected back in time. Each tick mark represents 1 day of travel at the wavelength group speed. The oval represents an estimate of the position and extent of a storm source from GOES IR cloud imagery at 1200 on September 24, 1978.

days along great circle routes. Each tick mark on the return ray represents 1 day of travel at the group speed of the dominant wavelength in each of the 12 spectra. The outlined area in the lower right-hand corner is an estimate of a storm course position taken from GOES IR imagery at 1200 noon 4 days before pass 1339 on September 28, 1978. In general, the projections flow back to the storm source. Of particular interest is the fact that the two spectra at the highest latitude of the 12 were located at the highest sea state (Beal *et al.*, 1983). High sea states produces a more severe azimuth falloff and result in the apparent rotation of the spectral peak toward the range or cross-track direction.

Figure 12 is identical to Fig. 11 except that the spectra have been corrected for azimuth falloff and revised wave projections have been computed. These projections are in much better agreement with the storm source location determined from GOES imagery. Hence, there is a good geophysical basis for concluding that our azimuth falloff corrections substantially improve estimates of the direction of the dominant wave system from SAR data.

3.4. Consideration of the Entire Ocean-to-SAR Spectra Transformation

Earlier in this section we outlined the transfer functions that relate the SAR image spectrum to the ocean surface spectrum via (1). Of these the stationary scatterer resolution response function is clearly the most well understood. We will therefore now presume that this transfer function can always be corrected for and neglect it in further discussion.

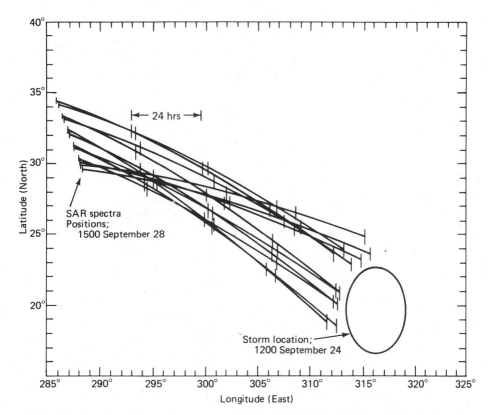

FIGURE 12. Storm location, projection corrected for azimuth falloff. This figure is identical to Fig. 11 except that the wavenumbers (both direction and magnitude) of the 200-m system peak have been adjusted by accounting for the azimuth falloff associated with the moving scatterer resolution response function.

In Fig. 3D, we showed the coherent sum of all three imaging mechanisms that comprise the total image modulation transfer function, as determined from existing literature. Because of the dominance of the orbital velocity modulation transfer function, primary sensitivity is for azimuth-traveling waves. This sensitivity increases with azimuth wavenumber.

Figure 13 is the product of the image modulation transfer function $[M^2(\mathbf{k})]$ with the moving scatterer response function $[H^2(\mathbf{k})]$ for a falloff with $\sigma_k = 2\pi/200$ m. The important result of this computation is that the product of $H^2(\mathbf{k})$, which decreases with azimuth wavenumber, and $M^2(\mathbf{k})$, which increases with azimuth wavenumber, has a local peak in azimuth sensitivity. Hence, if the model is correct, SAR spectra should tend to have an azimuth-traveling component simply due to the local maximum of the transfer function.

The azimuth falloff in the SAR spectrum is a clearly documented and quantified (Beal *et al.*, 1983) phenomenon. Since the azimuth peak produced by the combination of the orbital velocity modulation transfer function and the azimuth falloff does not appear in SAR spectra as predicted, we conclude that the orbital velocity modulation transfer function $[M_{ov}(\mathbf{k})]$ must not be as dominant as predicted by the velocity-bunching theory. This model, as discussed in Section 3.1.3, assumes that the short Bragg waves are uniformly distributed across the long wave. This has not been experimentally verified; consequently, the magnitude of the modulation may be so significantly reduced that the azimuth peak does not appear in SAR spectra. In any case, the absence of the azimuth peak shown in Fig. 13 in actual SAR spectra suggests that the orbital velocity mechanism is not nearly as dominant as theoretical developments suggest.

In spite of the fact that we were able to substantially correct for $H^2(\mathbf{k})$ in regions where the signal exceeded the noise, it still remains as a fundamental limitation for obtaining 2d wave spectra from SAR imagery. Since $H^2(\mathbf{k})$ is directly related to both velocity and acceleration

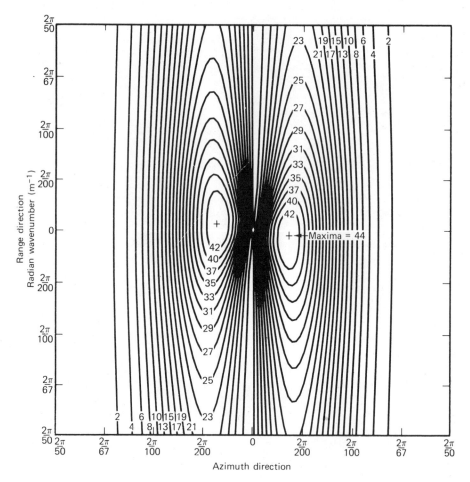

FIGURE 13. The product $H^2(\mathbf{k}) \, M^2(\mathbf{k})$, as calculated from the velocity-bunching model. Note the strong local peak in azimuth wavenumber, in contrast to the actual SEASAT SAR spectra, which show only a monotonically decreasing behavior with increasing azimuth wavenumber.

distributions of the ocean surface, the higher the sea state the more severe the azimuth falloff. The deleted regions in Figs. 10B–D represent areas where the signal-to-noise ratio is insufficient for correction. The higher the sea state, the larger is the region of azimuth wavenumber that will be destroyed by scatterer movement.

Since the extent of azimuth falloff is proportional to the satellite range-to-velocity ratio, R/V, a reduction in satellite altitude will alleviate moving scatterer effects. The European ERS satellite C-band SAR to be launched in the late 1980s will only partially alleviate this problem. The R/V ratio will be approximately the same as that of SEASAT. The use of a C-band SAR will reduce the integration time, and hence reduce the acceleration contribution to the azimuth falloff. However, the improvement may be minimal if the velocity contribution is dominant.

4. CALIBRATION

Through (1), the SAR image spectrum can be transformed to a spectrum proportional to the slope-variance spectrum. The gravity-wave dispersion relationship permits this spectrum to be transformed to one proportional to the height-variance spectrum. However, physical units still need to be attached to the resulting spectral values.

The most practical way of approaching this problem seems to be through an independent measure of total variance. This value can then be equated to the integral of the two-dimensional SAR spectrum to provide absolute variance calibration. For example, a spaceborne radar altimeter can be used to infer rms wave height (Townsend, 1980). However, the spatial resolution of the SAR and altimeter differ. Whereas the SAR can measure waves to one resolution limit, the altimeter may include wave heights on a shorter spatial scale in its estimation of rms height, resulting in a sea-state-dependent bias.

If one could specify the image modulation mechanisms to sufficient quantitative accuracy, the SAR image spectrum would be self-calibrating. Unfortunately, such specification of the image modulation mechanisms is still not sufficient to perform this self-calibration.

5. CONCLUSIONS

We have provided a general approach to the determination of ocean wave spectra from SAR imagery. This process involves the spatial and radiometric correction of SAR imagery, the removal of image-mechanism and stationary and moving scatterer resolution-induced biases in the spectral domain, and the calibration of the SAR spectrum to physical units. Thus far, we have been able to obtain geometrically and radiometrically corrected SAR spectra and have substantially removed the effects of the stationary and moving scatterer resolution response. Although our understanding of the wave imaging mechanisms is not sufficiently complete to totally remove those effects, we conclude that the orbital velocity mechanism must be significantly smaller than predicted by the velocity-bunching model. As improvements in the understanding of each segment of the SAR-spectra-to-ocean-spectra transformation process occur, they will be incorporated within the process without a fundamental change in strategy. Significantly, linear theory has thus far been adequate to explain systematic biases in SAR spectra, offering the hope that one may be able to uniquely go from SAR to ocean wave spectra.

Of particular interest is the overwhelming evidence that a systematic falloff with *azimuth* wavenumber in the SAR spectrum is directly related to the instantaneous radial velocity distribution of scatterers, as well as the radial acceleration distribution, and that this dependence can be quantitatively predicted. Even moderate sea states can result in relatively severe azimuth falloff in the SEASAT geometry, and this should have significant impact on the design parameters of future spaceborne SARs.

ACKNOWLEDGMENTS. This work was supported by the Coastal Processes Branch of the Office of Naval Research and the Oceanic Processes Branch of the National Aeronautics and Space Administration. Figures 6 and 10 were produced by D. G. Tilley of the APL Image Processing Laboratory.

REFERENCES

Alpers, W. R., and K. Hasselmann (1978): The two-frequency microwave technique for measuring ocean-wave spectra from an airplane or satellite. *Boundary-Layer Meteorol.* **13**, 215–230.

Alpers, W. R., and C. L. Rufenach (1979): The effect of orbital velocity motions on synthetic aperture radar imagery of ocean waves. *IEEE Trans. Antennas Propag.* **AP-27**, 685–690.

Alpers, W. R., D. B. Ross, C. L. Rufenach (1981): On the detectability of ocean surface waves by real and synthetic aperture radar. *J. Geophys. Res.* **86**, 6481–6498.

Beal, R. C. (1980): Spaceborne imaging radar: ocean wave monitoring. *Science* **208**, 1373–1375.

Beal, R. C., W. J. Geckle, A. G. Goldfinger, and D. G. Tilley (1981): System calibration strategies for spaceborne synthetic aperture radar. JHU/APL Report CP-084, December.

Beal, R. C., D. G. Tilley, and F. M. Monaldo (1983): Large and small scale spatial evolution of digitally processed ocean wave spectra from the Seasat synthetic aperture radar. *J. Geophys. Res.* **88**, 1761–1778.

Bendat, J. S., and A. G. Piersol (1971): *Random Data: Analysis and Measurement Procedures*, Wiley–Interscience, New York.

Bjerkaas, A. W., and F. W. Riedel (1979): Proposed model of the elevation spectrum of a wind roughened sea surface. JHU/APL TG-1328, Johns Hopkins University/Applied Physics Laboratory, NTIS ADA 08342617.

Elachi, C., and W. E. Brown, Jr. (1977): Models of radar imaging of the ocean surface waves. *IEEE Trans. Antennas Propag.* **AP-27**, 84–95.

Evans, D. D., and O. H. Shemdin (1980): An investigation of the modulation of capillary and short gravity waves in the open ocean. *J. Geophys. Res.* **85**, 5019–5024.

Goldfinger, A. (1980): Seasat SAR processing signatures: Point targets. JHU/APL Technical Report CP-078, Johns Hopkins University/Applied Physics Laboratory.

Gonzalez, F. I., R. C. Beal, W. E. Brown, P. S. DeLeonibus, J. W. Sherman, III, J. F. R. Gower, D. Lichy, D. B. Ross, C. L. Rufenach, and R. A. Shuchman (1979): Seasat synthetic aperture radar: Ocean wave detection capabilities. *Science* **204**, 1418–1421.

Keller, W. C., and J. W. Wright (1975): Microwave scattering and straining of wind generated waves. *Radio Sci.* **10**, 139–147.

Larson, T. R., L. I. Moskowitz, and J. W. Wright (1956): A note on SAR imagery of the ocean. *IEEE Trans. Antennas Propag.* **AP-24**, 393–394.

Monaldo, F. M., and R. S. Kasevich (1981): Optical determination of short-wave modulation by long ocean gravity waves. *IEEE Trans. Geosci. Remote Sensing* **GE-20**, 254–259.

Phillips, O. M. (1981): The structure of short gravity waves on the ocean surface. *Spaceborne Synthetic Aperture Radar for Oceanography* (R. C. Beal, P. S. DeLeonibus, and I. Katz, eds.), Johns Hopkins Press, Baltimore, 24–31.

Plant, W. J., W. C. Keller, and J. W. Wright (1978): Modulation of coherent microwave backscatter. *J. Geophys. Res.* **83**, 1347–1352.

Raney, R. K. (1971): Synthetic aperture imaging radar and moving targets. *IEEE Trans. Aerosp. Electron. Syst.* **AES-7**, 499–505.

Raney, R. K. (1982): Synthetic aperture radar imaging of the sea. *1982 International Geoscience and Remote Sensing Symposium Digest*, WA-2, IEEE Catalog No. 82 CH14723–6.

Swift, C. T., and L. R. Wilson (1979): Synthetic aperture radar imaging of ocean waves. *IEEE Trans. Antennas Propag.* **AP-27**, 725–729.

Townsend, W. F. (1980): An initial assessment of the performance achieved by the Seasat-1 radar altimeter. *IEEE J. Ocean. Eng.* **OE-5**, 80–92.

Tyler, G. L., C. C. Teague, R. H. Stewart, A. M. Peterson, W. H. Munk, and J. W. Joy (1974): Wave directional spectra from synthetic aperture observations of radio scatter. *Deep-Sea Res.* **21**, 989–1016.

Valenzuela, G. R., and J. W. Wright (1979): Modulation of short gravity–capillary waves by longer-scale periodic flows—A higher order theory. *Radio Sci.* **14**, 1099–1110.

Wright, J. W. (1966): Backscattering from capillary waves with application to sea clutter. *IEEE Trans. Antennas Propag.* **AP-14**, 749–754.

Wright, J. W., W. J. Plant, W. C. Keller, and W. L. Jones (1980): Ocean wave-radar modulation transfer functions from the West Coast Experiment. *J. Geophys. Res.* **85**, 4957–4966.

29

Microwave Scattering from Short Gravity Waves

Deterministic, Coherent, Dual-Polarized Study of the Relationship between Backscatter and Water Wave Properties

Daniel S. W. Kwoh and Bruce M. Lake

Abstract. The fundamental mechanisms of microwave backscattering from short gravity waves have been investigated in the laboratory using a CW coherent dualpolarized focused radar and scanning laser slope gauge which provides an almost instantaneous profile of the water surface while scattering is taking place. The surface is also monitored independently for specular reflection by an optical sensor. It was found that microwave backscattering occurs in discrete bursts which are highly correlated with "gentle" breaking of the waves. These backscattering bursts are either completely nonspecular or are partly nonspecular and partly specular. The specular contribution is found to be more important than generally expected, even at moderate to high incidence angle. Its source seems to be the specular facets in the turbulent wake and the capillary waves generated during breaking. Completely nonspecular backscattering bursts have been analyzed by using the method of moments to compute numerically the backscattering complex amplitudes from the measured profiles and then comparing the computed results with the measured results. Using numerical modeling, it can be shown that for a wave in the process of breaking, its small-radius crest is the predominant scattering source in a manner akin to wedge diffraction as described by the geometric theory of diffraction (GTD). The parasitic capillary waves generated during wave breaking also scatter, but their contribution is in general smaller than that of the crest. The relationship and differences between GTD and small perturbation theory (SPT) in the description of wedge diffraction are established. One important implication is that the breakdown of the composite model in the description of scattering from short gravity waves may be traced to the nonlinearity of water waves. Preliminary implications of the result for microwave backscattering from the ocean surface are examined.

X-band microwave backscattering from short gravity waves at moderate incidence angles has been studied in detail in the laboratory with the objective of (1) identifying the different features on the water surface that are responsible for backscattering and (2) understanding the different scattering mechanisms and their relative importance. In the process, the widely accepted composite model was put to a critical test under our particular set of conditions and found to be inadequate.

The work was performed in a laboratory wind-wave tank. Continuous water wave trains with a dominant frequency of 2.5 Hz (25 cm) were generated by a wave paddle at one end of the

Daniel S. W. Kwoh and Bruce M. Lake ● TRW/Space and Technology Group, Redondo Beach, California 90278.

tank. The radar unit is a superheterodyne system with a 9.23-GHz, 100-mW, CW output. The system has two parallel channels, one for vertical polarization, the other for horizontal polarization. Both channels transmit and receive through the same horn, which is a conical corrugated horn with a matched dielectric lens so that the microwave energy is focused in identical circular patterns for both channels with a focal length of 18 inches and a 3-db beamwidth of 8.3 cm. The horn is located inside the wind tunnel, looking "upwind" at incidence angles of 40, 55, and 70°. Both channels have independently, coherently, and linearly detected separate amplitude and phase outputs. Stray reflections from inside the wind tunnel were minimized by covering the interior of the wind tunnel with 40-db microwave-absorbing material to make it effectively an anechoic chamber. Any remnant static reflection is nulled by a static balancing bridge in the radar unit. This ensures that the measured reflected power comes truly from backscattering from the water waves. Measurements of scattering from a series of small-radius metallic spheres were used to test and confirm quantitatively the accuracy of the entire setup.

While microwave backscattering is taking place, the water surface is monitored with a scanning laser slope gage which scans a laser beam at 40 Hz along 13.1 cm of the water surface in a down-tank direction. Thus, almost instantaneous, highly accurate ($\pm 0.6°$), good-resolution (0.5 mm) slope profiles of the surface, colocated with the microwave measurements, are obtained in a noninterfering manner. All measurements were recorded on chart paper and floppy disks via a digital oscilloscope.

For water wave trains with small amplitude, there is hardly any measurable microwave backscattering. As wave amplitude is increased, however, beyond a certain threshold, backscattering quickly appears. This threshold corresponds to the onset of self-modulation in the wave train. At a steepness of $ak = 0.17$, the self-modulation is such that at a fetch of 27 feet, one out of every three or four waves attains a small enough radius of curvature at the crest that it undergoes breaking with capillary waves being radiated down the front face. A turbulent wake may or may not appear behind the crest. We refer to this kind of breaking as "gentle breaking" since it does not involve bubbles or spray. For wave trains under these conditions, we observe that the backscattering occurs as discrete bursts (rather than as a white-noise-like continuous return) and the bursts strongly correlate with the "gentle breaking" events. The discreteness of the bursts implies that the scattering sources are localized on the surface. (This observation is made possible by our highly focused horn.) The proximity of the capillary waves, the high-curvature crests, and the turbulent wakes (all within several centimeters of each other) virtually rule out any timing or sampling measurement techniques to distinguish between the three features as scattering sources. Our observation, that a sharply peaked wave scatters even before it breaks, implies only that the high-curvature crest itself can be a scattering source, but it does not rule out the other two features as possible scatterers. It is conceivable, in fact, that all three surface features contribute during various stages of evolution of the "gentle breaking" process. Aside from the *identity* of the scatterers, the *mechanism* by which scattering occurs is also required in order to have a complete understanding of the scattering. Our measurements with the scanning slope gage suggest that the capillary waves and the turbulent wake can sometimes have large slope, making specular reflections a distinct possibility. We have attempted to resolve these issues by taking the following steps:

1. Among all backscatter events, distinguish specular reflection events from nonspecular events
2. Identify the source of specular reflection
3. Determine the relative strength of specular reflection
4. For nonspecular events, identify source of scattering
5. For nonspecular reflection events, determine the adequacy of the composite model. If the composite model is not adequate, find an alternative explanation.

To accomplish the first step, we deployed a specular reflection sensor which simply consists of placing a projection lamp and a camera, mounted with a photodiode, on both sides of the horn looking at the same backscattering area with the same incidence angle. Any facet on the water surface, normal to the incident microwave radiation, would thus produce a sharp spike in the photodiode output. For a specular reflection burst, the polarization ratio of vv to hh amplitude also approaches unity, which is distinctly different from nonspecular events. This offers an independent, albeit slightly less accurate, way of differentiating specular reflections from nonspecular ones. With the above two methods, we observed that specular reflection is important under our set of wave conditions. At 40° incidence angle, approximately one-third of the events are specular, and even at 70° incidence, about one-sixth are specular. Specular events usually have power two or more times as large as nonspecular events. Subsequent photographs show that specular reflection is associated with the more violent forms of the "gently breaking" waves having highly turbulent wakes. Specular reflection comes either from the wake region where the facets appear as bright dots in the picture and are randomly distributed and/or from the capillary waves which appear as parallel lines in the picture.

Nonspecular reflections, however, remain the main contributor to backscattering, especially at larger incidence angles (55 and 70°). They usually come from those "gently breaking" waves which have little or no turbulent wake activities. In this case, as we have mentioned earlier, it is virtually impossible to determine, by measurement, the relative contributions from the high-curvature crest and from the capillaries. The discrimination can, however, be accomplished by numerical analysis. The slope profiles were measured, recorded, and then transferred directly into a computer. The moments method was then used to compute the complex scattering amplitude in all directions for both polarizations, given the incident wave direction, the measured slope profile, and the measured antenna function on the water surface. The computed amplitude and phase for both polarizations were first compared with the measured values for validation purposes. About 100 events were analyzed for each incidence angle. It was found that the measured power in both polarizations for all three incidence angles were proportional to the computed power over a 20-db range, but were smaller by about 3 db. The measured polarization ratio, however, agrees quite well with the computed ratio over the range of 0 to about 10 db. The measured Doppler shift also agrees remarkably well with the computed values, obtained by computing the backscattering phase of successive scans. The reason for the 3-db discrepancy between computed power and measured power is probably due to the short crestedness of the water wave so that the wavefront is not exactly two-dimensional as assumed. With this one exception, it can be concluded that the quantitative agreement is quite good, validating both our measurement techniques and our numerical procedures. We believe this is the first time that backscattered power from water waves has been accounted for exactly and deterministically.

We then performed numerical modeling to determine the respective scattering contributions due to the small-radius crest and the capillaries. The measured slope profile was first smoothed by removing the capillary waves. This we refer to as the "background waveform" which is macroscopically smooth except for a small-radius crest region. Scattering from the "background waveform" was then computed. The difference between the backscatter from the measured wave and the "background waveform" was then attributed to the capillary waves. A third waveform was then generated by putting the capillary waves on an inclined plane having the same slope as the front face of the "background waveform" on which the capillary waves were originally measured. The scattering from these tilted capillary waves was then computed. It was found that the calculated power from the "background waveform" added to that from the tilted capillaries produced the same calculated power as the original wave in both polarizations. It is thus meaningful to think of scattering from the small-radius crest and scattering from the capillaries separately and the relative contributions can be compared. It was found that scattering from the small-radius crest was approximately between 0 and 3 db higher than that from the capillaries.

Having identified the sources of nonspecular reflections and their relative contributions, the next step was to understand the scattering mechanism. The radius of the crest is typically less than 1 cm, i.e., much less than that of the microwave wavelength. This immediately suggests that wedge scattering as described by the geometric theory of defraction (GTD) should be a good conceptual model for understanding scattering from the small-radius crest. There may still be some suspicion, however, that small perturbation theory (SPT) would describe the scattering from the "background waveform" just as well. To investigate this possibility, we noticed that for a wedge of infinitesimal amplitude, GTD and SPT produce identical results. As the wedge amplitude or angle increased, however, GTD and SPT predictions begin to diverge. Since GTD remains valid for all wedge angles, it is clear that SPT is no longer adequate. Higher-order contributions are obviously required as wedge amplitude or angle increases. Although the small-radius crest is at best a rounded wedge, GTD wedge theory serves to demonstrate graphically the failure of SPT in describing scattering from macroscopic waveforms, i.e., from waveforms which no longer have amplitude small and raidus of curvature large compared to the microwave wavelength. It is no longer valid to assume that scattering from such waveforms is always proportional to the resonant Fourier amplitude of the surface shape. At first sight, it appears that scattering from at least the capillaries should be explained by SPT. Our analysis so far indicates otherwise and we have not yet found a satisfactory explanation. It is possible that the large slope of the capillary waves requires higher-order correction to SPT.

Our detailed study, thus far, has been restricted to very simple wave systems and we do not intend to imply that the results can immediately be applied to complicated ocean wave systems. We feel, however, that a full understanding of a simpler system, such as mechanically generated wave trains, is a necessary first step toward understanding more complicated systems such as wind waves in the laboratory and in the ocean. Our preliminary investigation of wind-wave scattering in the laboratory demonstrates the usefulness of this approach. At low wind (less than 3 m/s), and again a fetch of 27 feet, both the polarization ratio and the Doppler shift indicate that SPT provides the correct description. At moderately high (greater than 7 m/s) wind, however, SPT is no longer adequate. The Doppler spectrum is then doubly peaked. For example, at $55°$ incidence angle, the higher Doppler frequency peak is at approximately 30 Hz and the lower Doppler frequency peak is at approximately 15 Hz. The polarization ratio associated with the high-frequency event is usually about 0 to 4 db, whereas the ratio associated with the lower-frequency event is usually greater than 6 db. An examination of the slope gage output shows that the scattering associated with the higher Doppler shift may be either specular or nonspecular, and is associated with features moving at the phase speed of the gravity wave such as its small-radius crest, parasitic capillaries, and turbulent wake. The scattering associated with the lower Doppler shift is scattering from rough patches moving at much slower Bragg wave speeds which may be created by wind or by previous violent breakdown of waves. The power contained in this lower Doppler shift component is usually 2 to 4 db lower than that in the higher Doppler shift component. Further study is now under way. Previously unexplained scattering phenomena, such as the smaller than expected polarization ratio in the ocean, the relative peakiness of hh scattering relative to vv scattering, and the unexpected broadness and asymmetry of the Doppler spectrum in the laboratory, may all be explainable in view of these findings.

DISCUSSION

M. WEISSMAN: I noticed that the scale on the slope gage output was $\pm 30°$. Was this the actual range of the slope during the breaking event or was this the limits of the gage? What was the size of the laser beam at the surface?

KWOH: The total angular range range of scanning laser slope gage is about $60°$ which can be offset in either direction. We usually use it from about $+20°$ (backslope) to about $-40°$ (frontslope). The actual range of slopes for breaking events is higher than our measuring range. However, the slope of "gently breaking" events, which are associated with the nonspecular part of the return, is usually within the slope gage measurement range.

The size of the laser beam is about 0.2 mm.

HIGHAM: (a) Have you measured polar diagrams of scatter?
(b) What is relative magnitude of specular to nonspecular signals typically?

KWOH: (a) No.
(b) For $\theta_i = 40°$, specular events occur about one-third of the time. For $\theta_i = 70°$, specular events occur

about one-sixth of the time. In both cases, specular events have four times or more power than nonspecular events. These numbers are only intended to provide a rough idea of the importance of specular reflection. It can be seen that at 40°, more than half of the backscattered power comes from specular events. Even at $\theta_i = 70°$, specularly backscattered power is significant.

PLANT: You made the point that the scattering you measured is not Bragg scattering. Can you give us an idea of the magnitude of the backcattered power in this case compared to that observed when you put a mechanically generated Bragg wave on the surface?

KWOH: That depends on the Bragg wave amplitude. Generally, non-Bragg scattering was 10 to 20 db higher than Bragg.

PHILLIPS: This and the previous paper suggest to me that the backscattered returns from the ocean surface at short radar wavelengths may be dominated by these breaking events. If so, X-band radar gives us a useful tool for studying them. Also, the variation in return with wind speed and long wave slope should be interpreted in terms of these events, rather than in terms of trains of gravity-capillary waves. Perhaps Dr. Wetzel could comment?

WETZEL: You raise a good question. I was about to ask the last speaker [Kwoh] whether his results suggest that we might have to abandon the current Bragg theory of microwave scattering by the ocean surface. In regard to the variation of returns with wind speed—it can be shown that at low grazing angles (below a few degrees) the wind dependence of sea backscatter (recorded by Hans Sittrop) is almost identical to estimates of the wind dependence of whitecap coverage (from an early paper by Ross and Cardone). This may mean nothing at all but the close correlation is certainly provocative.

KWOH: I want to point out that the nature of the wave-breaking considered by Dr. Wetzel is different than the gentle breaking we have considered and that he is considering very large incidence angles while our work has been in the range from 40 to 70°.

PIERSON: There will be differences between backscatter from waves and no wind but with parasitic capillaries and waves with wind where large areas of the sea surface are covered by a continuum of capillary waves with a sufficient "area" to provide a backscattered signal.

KWOH: We do not doubt that there will be patches in the ocean that are "slightly rough" and scatter in a Bragg manner. The question is, "how much?" In the laboratory, our first look at wind wave scattering indicates that at wind speed of 3 m/s, scattering is Bragg-like. This is evidenced by a polarization ratio which is roughly correct (i.e., predicted by Bragg) and by the fact that the scattering occurs in a continuous manner. However, at moderate wind speed (4–8 m/s) the scattering again have discrete bursts, similar to what we see from scattering from mechanically generated waves. I think it is still an open question as to how much of the short-wavelength microwave scattering in the ocean comes from Bragg-like patches, how much of it comes from rounded wedgelike crests, how much comes from parasitic capillaries, and how much comes from specular facets.

30

Remote Sensing of Directional Wave Spectra Using the Surface Contour Radar

E. J. Walsh, D. W. Hancock, III, D. E. Hines, and J. E. Kenney

Abstract.   A 36-GHz computer-controlled airborne radar is described which generates a false-color coded elevation map of the sea surface below the aircraft in real time and can routinely produce ocean directional wave spectra with off-line data processing. Data products are shown and the procedures for correcting the encounter spectra for platform effects are indicated. The nondirectional frequency spectrum and Fourier coefficients computed from the radar spectrum are in good agreement with those computed from the NOAA XERB pitch-and-roll buoy.

1. INTRODUCTION

The surface contour radar (SCR) was developed jointly by NASA Wallops Flight Facility (WFF) and the Naval Research Laboratory under the NASA Advanced Applications Flight Experiments program. It is an airborne computer-controlled 36-GHz bistatic radar which produces a real-time topographical map of the surface beneath the aircraft. The SCR is one of the most straightforward remote-sensing instruments in measurement concept. It provides great ease of data interpretation since it involves a direct range measurement.

The system (Kenney et al., 1979) was designed to measure the directional wave spectra of the ocean surface. Figure 1 shows the nominal measurement geometry and the horizontal resolutions in terms of the aircraft altitude, h. An oscillating mirror scans a 0.85° × 1.2° pencil-beam laterally to measure the elevations at 51 evenly spaced points on the surface below the aircraft within a swath which is approximately half the aircraft altitude. At each of the points the SCR measures the slant range to the surface and corrects in real time for the off-nadir angle of the beam to produce the elevation of the point in question with respect to the horizontal reference.

The elevations are false-color coded and displayed on the SCR color TV monitor so that real-time estimates of significant wave height (SWH), dominant wavelength, and direction of propagation can be made. The real-time display allows the aircraft altitude and flight lines to be optimized during the flight even if there was no prior knowledge of the wave conditions.

2. AIRCRAFT VERTICAL MOTION REMOVAL

The aircraft generally has some altitude variation during the time interval in which the data are acquired that contaminates the elevation measurements. An extreme example is shown in the

E. J. Walsh, D. W. Hancock, III, and D. E. Hines ● NASA Goddard Space Flight Center, Wallops Flight Facility, Wallops Island, Virginia 23337.   J. E. Kenney ● Space Systems and Technology Division, Naval Research Laboratory, Washington, D.C. 20375.

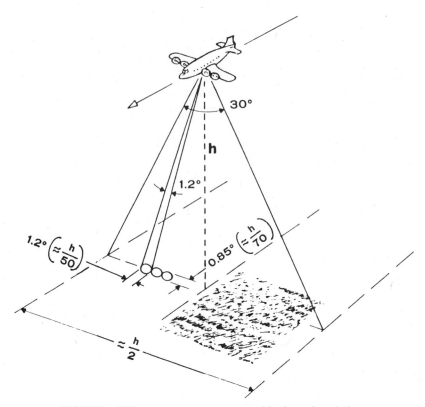

FIGURE 1. SCR measurement geometry and horizontal resolutions.

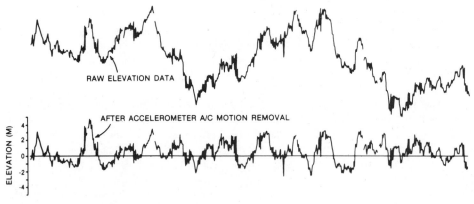

FIGURE 2. Center swath profile of SCR elevation data before (top) and after (bottom) removal of aircraft motion using accelerometer data.

top curve of Fig. 2. It shows the elevation profile from the center of the swath when the aircraft was flying parallel to the crests of a 5.5-m SWH sea whose dominant wavelength was 140 m. In this example, it is apparent that it would be difficult to remove the aircraft motion from the elevation data by high-pass filtering the raw data. The aircraft motion is eliminated by doubly integrating the output of a vertical motion sensing accelerometer to obtain an independent estimate of the aircraft motion. The bottom curve in Fig. 2 shows the result after the accelerometer-determined aircraft motion has been subtracted from the raw elevation data. To

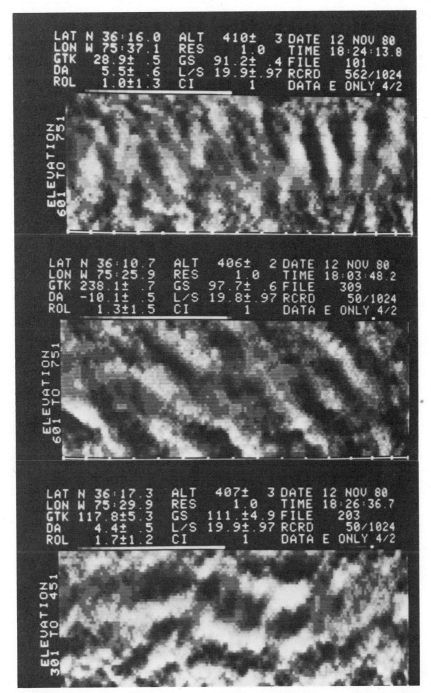

FIGURE 3. Gray-scale coded elevation data taken from 400 m altitude for three different flight directions. The header information indicates latitude (LAT), longitude (LON), aircraft ground track (GTK), drift angle (DA), roll (ROL), altitude (ALT) in m, range resolution (RES) in ns, ground speed (GS) in m/s, and the contour interval (CI) by which the data are divided before being displayed. The angles are in degrees. The data point spacing in the TV monitor display is approximately the same along-track and cross-track. There is some geometric distortion in these pictures due to the disparity between the actual along-track and cross-track data point spacing.

compensate for any uncertainties in the initial values of the aircraft altitude, vertical velocity, and the accelerometer constant, a parabola of the form $a_0 + a_1 y + a_2 y^2$ must also be least-squares fitted over the data span and subtracted from the elevation data. The variable y is the along-track distance. The resulting elevation data may still contain some small residual aircraft motion but it is obvious that it is much lower in amplitude and frequency than the wave data.

3. DATA ANALYSIS

Figure 3 shows close-up views of gray-scale coded SCR elevation data taken during the Atlantic Remote Sensing Land Ocean Experiment (ARSLOE) within minutes of each other. The data were taken at 400 m altitude (ALT) along the three different aircraft ground tracks (GTK) indicated in the headers. The troughs are dark and the crests are light so that the data have the visual appearance of waves illuminated by a low sun angle. Distance along the ground track increases to the right in each case by approximately 5 m for each cross-track sweep of the antenna beam [the ground speed (GS) divided by L/S, the number of radar raster scan lines per second].

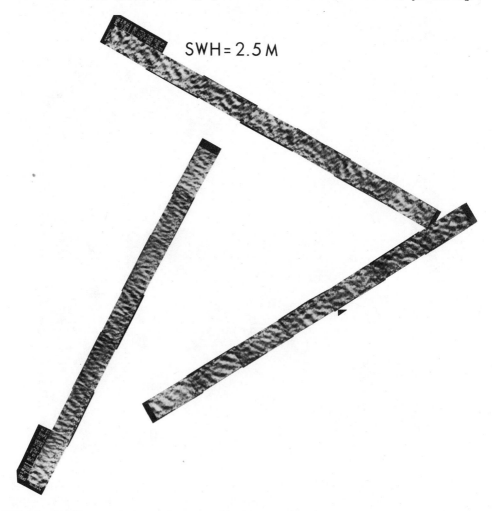

FIGURE 4. Three sets of 1024 scan lines of gray-scale elevation data which were acquired on 11/12/80 at 400 m altitude. North is vertical and the data have been oriented in the respective aircraft ground track directions.

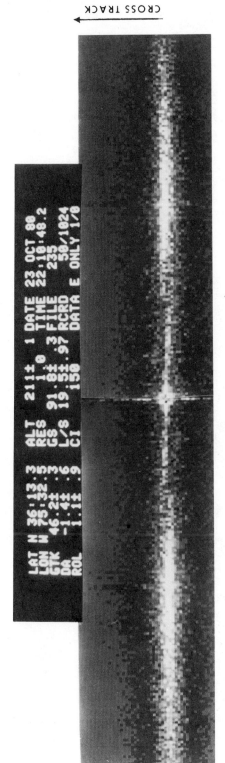

FIGURE 5. Gray-scale coded variance spectrum before corrections for the asymmetric k plane resolution and rotation to orient with respect to true north.

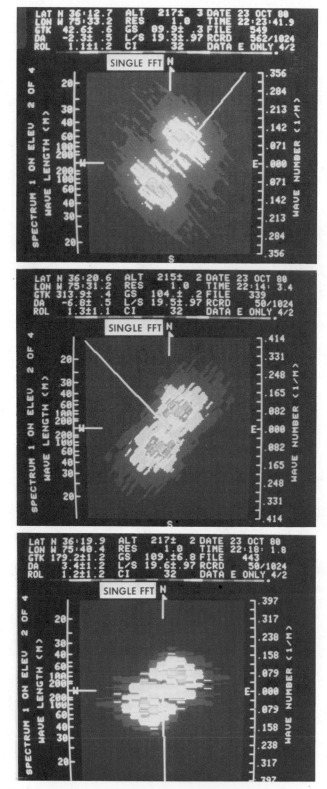

FIGURE 6. Black-and-white presentation of false-color coded variance spectra for three flight directions at 200 m altitude.

The cross-track spacing of the 51 elevation points is the altitude divided by 100, or roughly 4 m. The data point spacing in the TV monitor elevation display is approximately the same both along track and cross track. Therefore, there is a geometric distortion in the display whenever the actual along-track and cross-track data point spacings are not equal. The geometric distortion in Fig. 3 is small and caused principally by the variation of the aircraft ground speed on the various ground tracks.

The directional wave spectra are produced by performing 2d FFTs on sets of 1024 scan lines. Figure 4 shows three sets, taken at approximately the same time and along the same ground tracks as the data of Fig. 3, which have been oriented in the proper ground track directions. The along-track dimension of approximately 5 km produces an along-track spectral resolution of $2\pi/5000\,\mathrm{m}^{-1}$ in wavenumber space, whereas the cross-track resolution is only $2\pi/200\,\mathrm{m}^{-1}$ so the grid of FFT points is badly out of proportion.

Figure 5 shows a raw spectrum which is gray-scale coded. The altitude for the data set associated with the FFT was only 211 m so the swath width was approximately 110 m. Since the along-track distance was 4.8 km, the asymmetric resolution in the k plane is 44:1. Because the spacing of points in the k plane presentation in Fig. 5 is the same both along track and cross track, the spectrum appears very narrow in the cross-track dimension. At the origin of the FFT can be seen a bright region which is due to residual aircraft motion that the accelerometer did not remove.

For the final display on the color TV monitor, the spectrum is oriented with respect to true north and FFT data points are filled-in cross-track so the display is in proper proportion. Figure 6 shows black-and-white presentations of false-color coded spectra for three different aircraft ground tracks (whose directions are indicated by the white radials from the origin) for the same day and altitude as Fig. 5. The reference direction in the k plane is the direction toward which the waves are traveling, not from which they are coming. Since the wind was from the northeast, the actual spectra in Fig. 6 are in the third quadrant. The 180°-ambiguity image in the first quadrant is an artifact of the FFT process since the elevation data could represent waves

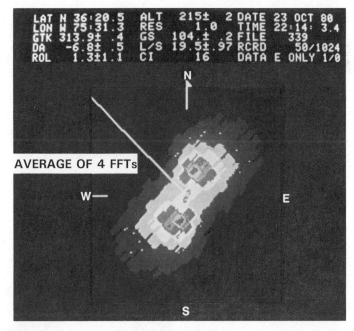

FIGURE 7. Black-and-white presentation of false-color coded variance spectrum for the average of four spectra at 215 m altitude. The color changes at the 1, 4, 9, 16, 25, 36, 49, 64, and 81% levels relative to the spectral peak.

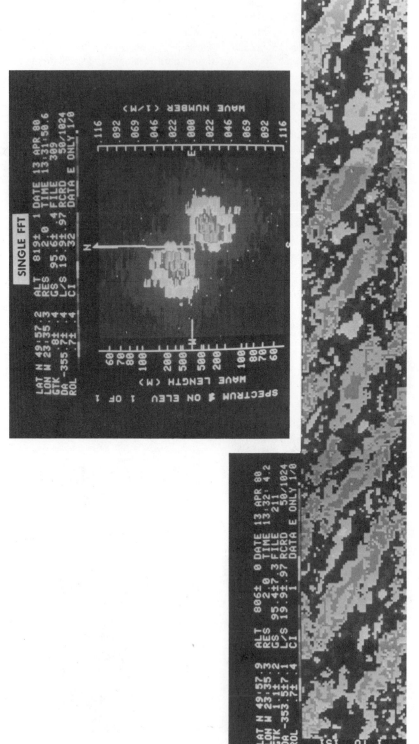

FIGURE 8. Black-and-white presentation of false-color coded 800-m-altitude elevation data and variance spectrum for a day with 5-m SWH and 200-m dominant wavelength.

traveling in either direction. The elimination of that ambiguity will be discussed later when the process for correcting these encounter spectra for the effects of aircraft velocity and drift angle is discussed. Figure 6 shows that essentially the same spectrum was obtained independent of heading. The noise in the spectra can be reduced by increasing the number of degrees of freedom through incoherent averaging of spectra.

Figure 7 shows the results of averaging four spectra taken during the ARSLOE experiment on a 314° ground track at 215 m altitude. It is a black-and-white presentation of the false-color coded variance spectrum whose 10 color levels change at the 1, 4, 9, 16, 25, 36, 49, 64, and 81% levels of the spectral maximum. The banding that is apparent in Fig. 7 parallel to the aircraft ground track is the effect of the poor lateral spectral resolution. Even so, the spectrum is still very well defined.

Figure 8 shows a black-and-white presentation of false-color coded data taken in the North Atlantic from an altitude of 800 m when the SWH was approximately 5 m and the dominant wavelength was 200 m. By varying the aircraft altitude, the spatial resolution and swath width can be adjusted as dictated by the sea state.

4. CORRECTIONS FOR AIRCRAFT GROUND SPEED AND DRIFT ANGLE

Figure 9 shows the average of four variance spectra for each of two different flight directions at 400 m altitude for the same day as the elevation data shown in Fig. 4. The SWH was predominantly swell arriving from the northeast. The color levels change at the 1, 4, 9, 16, 25, 36, 49, 64, and 81% levels of the spectral maximum. The agreement between the two flight directions is excellent. If one examines the spectra carefully, it is apparent that the encounter spectrum (in the third quadrant) and its 180°-ambiguity image (in the first quadrant) are more widely separated for the northeast ground track than for the southeast. Also, the radial joining the spectral peaks is rotated clockwise in the southeast ground track spectrum relative to the northeast ground track. These changes are caused by the aircraft velocity and drift angle and must be corrected to obtain the actual spectrum from the encounter spectrum.

Figure 10 shows the migration of a representative sampling of points in the **k** plane. The base of each vector is the original spectral component position. The arrowhead shows the apparent position to which that component would shift due to an aircraft ground speed of 100 m/s with no drift angle (top) and for a 10° drift angle (bottom). Since it took approximately 50 s to acquire each of the 1024 scan line sets shown in Fig. 4, the data do not represent an instantaneous elevation map of the surface. The waves at the end of the segment would have moved by several wavelengths relative to the positions they were in when the data at the beginning of the segment were recorded. If the waves were traveling in the same direction as the aircraft, their wavelength would appear longer. If they were traveling in the opposite direction as the aircraft, their apparent wavelength would appear shorter. For waves traveling at an angle to the aircraft ground track, there would be changes in both the apparent wavelength and the direction of propagation.

The drift angle is the amount by which the aircraft heading differs from the aircraft ground track. If the aircraft is following its nose, the drift angle is zero. The bottom panel of Fig. 10 represents the situation for a cross-wind from the left. The aircraft heading shown is 10° to the left of the ground track to compensate for the cross-track component of the wind velocity. The aircraft ground track then drifts to the right relative to its heading by 10°. For no drift angle, all the points in the **k** plane migrate antiparallel to the aircraft ground track and the magnitude of the change is proportional to the magnitude of the original **k** vector (Long, 1979). When the drift angle is nonzero, there is also a cross-track component to the displacement in the **k** plane.

Corresponding to the two spectra of Fig. 9, Fig. 11 shows overlays of the plots of the variance spectra before (top) and after (bottom) applying the corrections indicated in Fig. 10. In applying the corrections, no *a priori* knowledge of the direction of propagation was assumed. In effect, all the data were assumed to be real and corrected accordingly. Heavy vertical reference lines have been added to both top and bottom spectra. It can be seen that the corrections have caused the two reference lines to coalesce in the actual spectrum of the lower right, while they are spread further apart in the 180°-ambiguity spectrum in the upper left. The corrections remove the aircraft motion effects in the actual spectrum but are in the wrong direction for the other one and worsen the disparity between the flight directions. Rejection of the ambiguous lobe can be accomplished by examining the absolute magnitude of the differences at the crossing points on the spectral lobes for perpendicular flight lines.

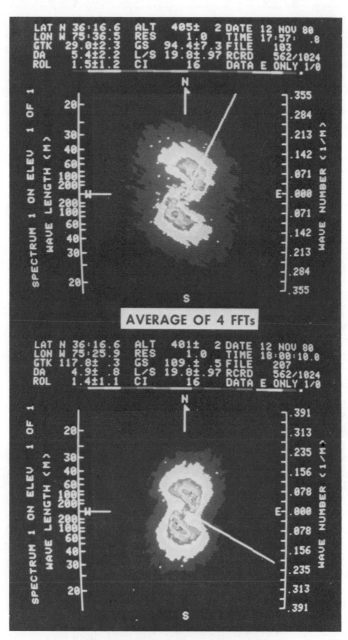

FIGURE 9. Black-and-white presentations of false-color coded averages of four variance spectra for the aircraft ground tracks indicated by the white radial in the displays.

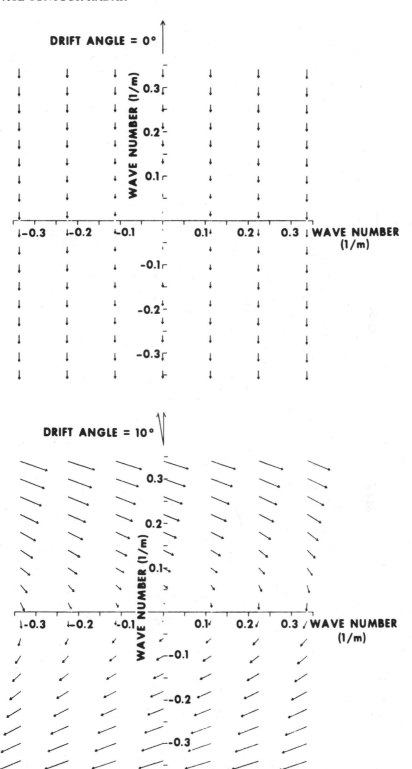

FIGURE 10. The migration of points in **k** space for an aircraft ground speed of 100 m/s and drift angles of 0° (top) and 10° (bottom). The ordinate is the direction of the aircraft ground track.

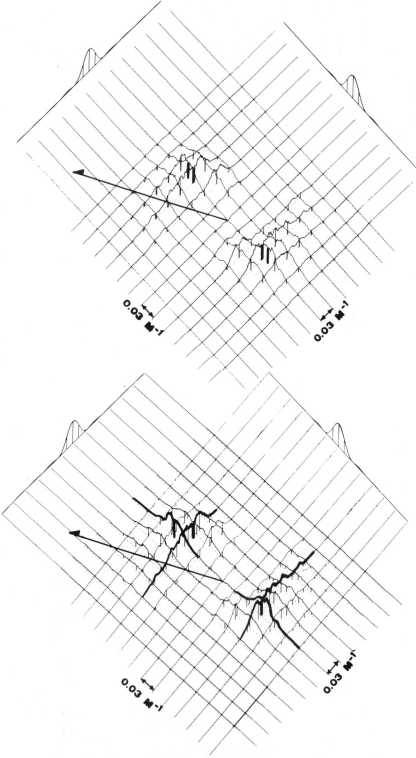

FIGURE 11. Overlay in the **k** plane of the variance spectra of Fig. 9 before (top) and after (bottom) the corrections of Fig. 10. The radials extending from the origin indicate the north direction.

5. SPECTRAL COMPARISON WITH PITCH-AND-ROLL BUOY

Figure 12 shows the approximate ground tracks for the SCR during data acquisition on November 12, 1980 (Figs. 4, 9 and 11). Also shown are the shoreline, the Field Research Facility pier of the Coastal Engineering Research Center, and the XERB pitch-and-roll buoy which was directly offshore at a distance of approximately 35 km. The SCR flight lines occurred half-way between two observations by the XERB which were spaced at 6-h intervals. Figure 12 indicates the wave heights and wind speeds observed by the buoy and the wave height indicated by the SCR. The local wind had been from the north-northwest all day but the dominant waves on the surface were swell coming from the northeast. The wind changed little over the 6-h period of the SCR data.

In an earlier comparison of these data sets (Walsh et al., 1981), the average of the two buoy spectra was compared with the SCR spectrum. The main disparity between the spectra was that the SCR spectrum half-power width at the spectral peak was 40° whereas the pitch-and-roll buoy indicated a width of 160°. This was not surprising since pitch-and-roll buoys have the ability to measure only five Fourier coefficients. Truncating the Fourier series at the second harmonic results in negative side lobes for the directional spectrum and poor directional resolution. Longuet-Higgins et al. (1963) used a nonnegative smoothing function to weight the Fourier series coefficients to remove the negative side lobes and a similar smoothing function had been applied to the XERB data (Kenneth Steele, personal communication). The directional resolution obtained by the buoy would be expected to be poor because the effective width of this smoothed spectrum (width of a rectangle of height equal to peak energy density and area equal to that of the main lobe) is 135°, while the highest resolution attainable with two harmonics without smoothing is 72° (Panicker, 1974). This means that the buoy spectra could have been narrowed if the appearance of negative side lobes had been tolerated. But the spectral width would never approach 40°, which demonstrated the significantly higher resolution of the SCR compared with pitch-and-roll buoys.

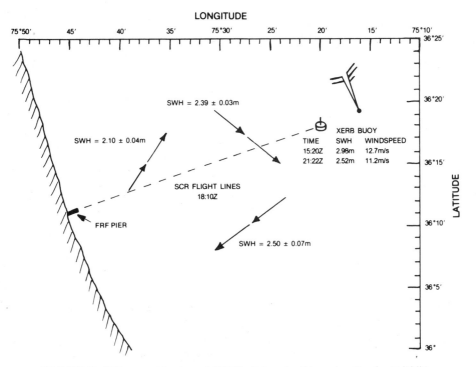

FIGURE 12. SCR ground tracks and XERB pitch-and-roll buoy location for 11/12/80.

It was later discovered (Lau *et al.*, 1982) that C_{33} had been omitted in the a_2 Fourier coefficient computed by the buoy (Longuet-Higgins *et al.*, 1963). Therefore, the buoy spectrum used for the previous comparison was in error by an unknown amount. The buoy problem has been corrected but the a_2 coefficient in the historical data has been irretrievably lost (Kenneth Steele, personal communication).

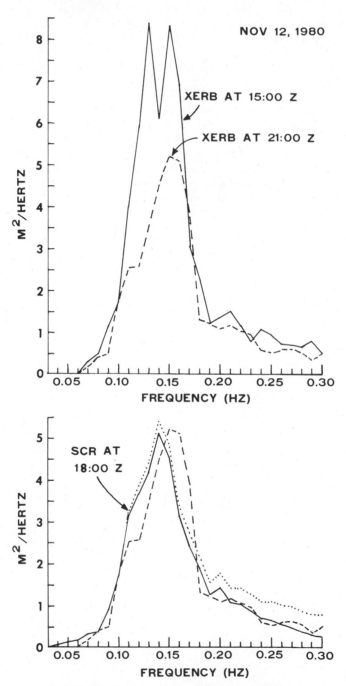

FIGURE 13. Comparison of nondirectional spectra from SCR and XERB buoy. The dotted curve in the bottom plot indicates the SCR data after correction for the spatial filtering effect of the radar footprint.

In this chapter we shall compare the nondirectional spectrum and the three correct Fourier coefficients computed by the buoy with those computed from the SCR. Figures 13 and 14 are based on SCR data from a single 1024 scan line data set taken from 400 m altitude on a 238° ground track at (36.2°, − 74.4°). A linear interpolation was made in the **k** plane to increase the cross-track density of points by a factor of four. Each of the points as then corrected for the effects of aircraft ground speed and drift angle discussed earlier and then transformed into the appropriate interval in the frequency plane.

The top panel of Fig. 13 compares the nondimensional XERB buoy spectra 3h before and after the SCR flight. The wave height reduced over the interval and the disappearance of the peak at 0.13 Hz indicated that the swell coming from the northeast had diminished. The bottom panel

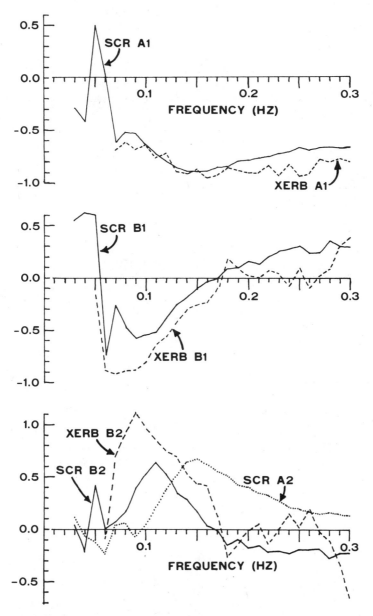

FIGURE 14. Comparison of Fourier coefficients from SCR and XERB buoy.

of Fig. 13 compares the SCR directional spectrum with the buoy spectrum taken 3h after the flight. The agreement is quite good considering the change in the buoy spectra over the period. Also shown is the SCR spectrum after correction for the spatial filtering effect of the radar footprint. The correction becomes significant above 0.20 Hz and pulls the SCR spectrum well above the buoy spectrum in that region. This is probably caused by noise in the radar elevation measurements, and procedures for eliminating it are being developed.

Figure 14 compares the Fourier coefficients in the form used by Hasselmann *et al.* (1980). The magnitudes of the XERB buoy Fourier coefficients were generally larger than unity for 0.05 Hz or below and were not plotted. Kenneth Steele (personal communication) suggested that the error could have been caused by an imperfect vendor calibration on the heave sensor at low frequencies. The agreement between the two systems is once again quite good.

6. CONCLUSIONS

The SCR provides a high-resolution way to easily and directly measure sea surface directional wave spectra. It will be extremely useful in developing oceanographic models as well as validating indirect remote-sensing oceanographic techniques such as side-looking radars and wave spectrometers.

ACKNOWLEDGMENTS.   The XERB buoy data were provided by Kenneth Steele of the NOAA Data Buoy Office and he is thanked for discussions concerning the buoy. Gary L. Donner of Gary Donner Associates developed the real-time software for the radar system as well as the software for the wave spectrum and elevation displays. Robert Swift of EG&G Washington Analytical Services Center, Inc., helped develop the off-line data analysis programs and supervises the routine reduction and archiving of SCR data.

REFERENCES

Hasselmann, D. E., M. Dunckel, and J. A. Ewing (1980): Directional wave spectra observed during JONSWAP 1973. *J. Phys. Oceanogr.* **10**, 1264–1280.

Kenney, J. E., E. A. Uliana, and E. J. Walsh (1979): The surface contour radar, a unique remote sensing instrument. *IEEE Trans. Microwave Theory Tech.* **MTT-27**, 1080–1092.

Lau, J. C., K. E. Steele, and E. L. Burdette (1982): End-to-end testing of NOAA Data Buoy Office directional wave measurement systems. *Oceans '82*, Sept. 20–22.

Long, R. B. (1979): On surface gravity wave spectra observed in a moving frame of reference. NOAA Technical Memorandum ERL AOML-38, June.

Longuet-Higgins, M. S., D. E. Cartwright, and N. D. Smith (1963): Observations of the directional spectrum of sea waves using the motions of a floating buoy. *Ocean Wave Spectra*, Prentice–Hall, Englewood Cliffs, N.J., 111–136.

Panicker, N. N. (1974): Review of techniques for directional wave spectra. *Waves'74*, Sept. 9–10, **1**, 669–688.

Walsh, E. J., D. W. Hancock, D. E. Hines, and J. E. Kenney (1981): Surface contour radar remote sensing of waves. *Proceedings, Conference on Directional Wave Spectra Applications*, Sept. 14–16, 281–297.

31

THE VISIBILITY OF rms SLOPE
VARIATIONS ON THE SEA SURFACE

R. D. CHAPMAN

ABSTRACT. A model relating viewing geometry, sky conditions, and statistical sea surface parameters is presented. The model is used to estimate the relative visibility of surface perturbations manifested by a variation in the rms surface slope. These estimates are presented for a variety of geometries in the solar plane assuming a clear sky and two different wind speeds ($\sim$ 2 and 7 m/s). The results of this analysis, applicable to a unidirectional radiometer with no temporal averaging, show the visibility of surface perturbations to be maximized by geometries with large gradients in the slope-to-luminance transfer function (within the glitter pattern or near the horizon). Detectability of these perturbations, as measured by the luminance SNR sensitivity to rms surface slope variations, is maximized by either large gradients, as for the visibility, or very small gradients in the slope-to-luminance transfer functions (90° away from the glitter pattern). It is shown that improvements in the estimated detectability can be obtained through spatial and temporal averaging. Two methods for quantitatively estimating these improvements are presented.

1. INTRODUCTION

It has been shown that under certain conditions, internal waves, current boundaries, and cyclonic eddies can produce visible patterns on the sea surface (see, e.g., LaFond, 1962; Strong and DeRycke, 1973; Maul *et al.*, 1974). These effects have been observed at various times from ships, aircraft, and satellites (Apel *et al.*, 1975a, b). It has been suggested that the surface patterns induced by these spatially varying currents are manifested by a variation in the ambient rms slope of the surface; a suggestion which has been verified for the internal wave–surface wave case by Hughes (1978) and Hughes and Grant (1978).

This chapter examines the visibility of rms slope variations on the sea surface. A model of the visibility of these surface perturbations assuming an ideal radiometer with no temporal or spatial averaging will be described. The qualitative effects of spatial and temporal averaging will be described, along with two methods for calculating the quantitative effects of averaging.

2. GEOMETRY AND NOMENCLATURE

The observation geometry is shown in Fig. 1. The detector is located on the z' axis at a height h above the mean sea surface. The optical axis of the detector is aimed down at the surface with a

R. D. CHAPMAN • Applied Physics Laboratory, The Johns Hopkins University, Laurel, Maryland 20707; *present address*: Department of Oceanography, Florida State University, Tallahassee, Florida 32306.

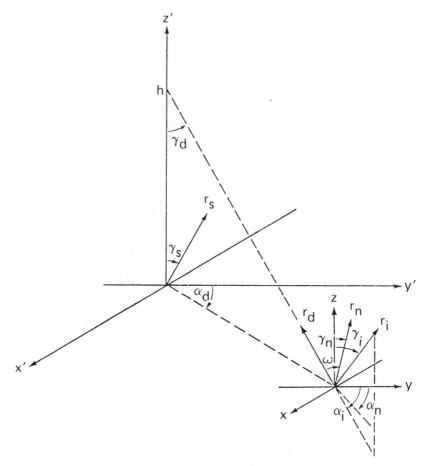

FIGURE 1. Surface observation geometry. Surface normal is designated by $\mathbf{r}_n$. Unit vectors $\mathbf{r}_n$, $\mathbf{r}_i$, and $\mathbf{r}_d$ are coplanar.

nadir angle γ_d and an azimuth α_d with respect to the y' axis. The sun is located at a solar zenith angle γ_s along an azimuth of $0°$. For the calculations to follow, the detector azimuth α_d is $0°$, so that all observations are made in the solar plane, and the solar zenith angle γ_s is fixed at $45°$. The unit vectors $\mathbf{r}_d$, $\mathbf{r}_n$, and $\mathbf{r}_i$ represent the detector direction, the normal of a particular surface facet, and the incident ray direction.

3. VISIBILITY WITHOUT AVERAGING

Three principal sea surface luminance (hereafter, simply referred to as luminance) statistics are required to calculate the visibility of rms slope perturbations: (1) the mean luminance, (2) the standard deviation of the luminance, and (3) the sensitivity of the mean luminance to rms slope variations. The calculation of these statistics, and their implications, are discussed in detail by Chapman (1981). Here, a brief outline of those results will be given.

The mean luminance of the sea surface is calculated by integrating, over all possible surface slopes, the product of the luminance for a given slope and the probability of that slope occurring:

$$\langle L \rangle = S^* \int \int \int^{\infty} RL_s p' Q d\mathbf{s} \tag{1}$$

where

     S^* is a factor statistically describing the shadowing of the surface
     R is the reflectivity of the surface
     L_s is the sky luminance
     p' is the probability density of seen surface slopes
     Q is a shadowing term limiting the integration to slopes which are not tilted so far away
            from the observer as to be self-hiding
     $d\mathbf{s}$ is the differential slope vector

The variance of the luminance, as measured over an infinitesimal field of measurement, is calculated by squaring the reflectivity and sky luminance terms in Eq. (1):

$$\langle L^2 \rangle = S^* \int\!\!\int_{-\infty}^{\infty} R^2 L_s^2 p' Q \, d\mathbf{s} \tag{2}$$

The standard deviation of the luminance is then derived from the mean and variance according to: $\sigma_L = [\langle L^2 \rangle - \langle L \rangle^2]^{1/2}$. The standard deviation thus calculated is an upper limit of any measured standard deviation, since any real measurement must include an integration over a finite area and time.

The luminance sensitivity to rms slope (σ) is given by $\partial \langle L \rangle / \partial \sigma$. The only terms in Eq. (1) that are dependent on σ are the shadowing function, S^*, and the slope probability density function, p'. The luminance sensitivity can thus be calculated from

$$\frac{\partial \langle L \rangle}{\partial \sigma} = \frac{\partial S^*}{\partial \sigma} \int\!\!\int_{-\infty}^{\infty} R L_s p' Q \, d\mathbf{s} + S^* \int\!\!\int_{-\infty}^{\infty} R L_s \frac{\partial p'}{\partial \sigma} Q \, d\mathbf{s} \tag{3}$$

From the above statistics, two additional statistics related to visibility can be derived: the contrast sensitivity, defined as $(\partial \langle L \rangle / \partial \sigma) / \langle L \rangle$, and the lower bound signal-to-noise ratio sensitivity (SNRS), defined as $(\partial \langle L \rangle / \partial \sigma) / \sigma_L$.

For small rms perturbations, $\Delta \sigma$, the actual visibility, is given by: $2 \Delta \sigma (\partial \langle L \rangle / \partial \sigma) / \langle L \rangle$, or alternatively $2 \Delta \sigma$ times the contrast sensitivity. Because of this relationship, the contrast sensitivity is used in place of the visibility in the discussions to follow.

The statistics described above have been calculated for a series of 38 detector nadir angles (γ_d) distributed from horizon to horizon in the solar plane with two different rms surface slopes, $\sigma = 0.1$ and $\sigma = 0.2$. The calculations assumed a clear sky and a solar zenith angle of $45°$. Figures 2–6 are the results of these calculations. In each plot, the dashed curve is for $\sigma = 0.1$ and the solid curve, $\sigma = 0.2$. Each plot includes two pairs of vertical lines which denote the approximate edges of the glitter (specularly reflected sunlight) pattern for the two surface conditions.

In Fig. 2 there are two major trends to note. First, the mean luminance increases markedly toward the horizon. This is principally due to the increased reflectivity at near-grazing angles. Second, the glitter pattern broadens substantially with increasing rms surface slope.

The trends in the standard deviation of the luminance in Fig. 3 are substantially the same as the trends of the mean luminance.

Figure 4 is a rather unconventional plot of the luminance sensitivity. This plot separates the luminance sensitivity into its positive and negative components and displays each along a logarithmic scale. Figure 4 provides a quantitative basis for understanding commonly observed patterns of sea surface luminance. An increase in rms surface slope causes an increase in luminance for geometries with a positive luminance sensitivity and a decrease in luminance for regions of negative luminance sensitivity. At the edges of the glitter pattern, the luminance

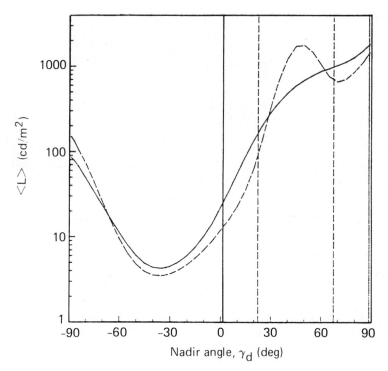

FIGURE 2. Average sea surface luminance for a clear sky and a solar zenith angle of 45°. Dashed curve is for $\sigma = 0.1$; solid for $\sigma = 0.2$. Vertical lines represent the edges of the glitter pattern.

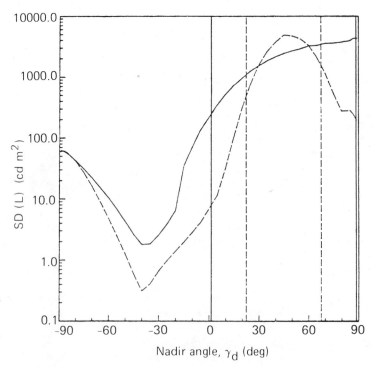

FIGURE 3. Standard deviation of the sea surface luminance for a clear sky.

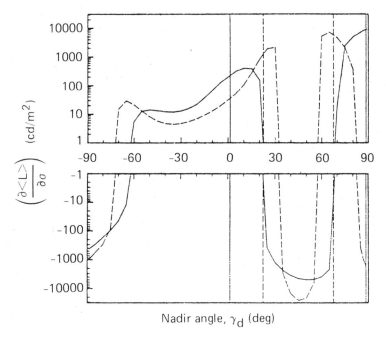

FIGURE 4. Sea surface luminance sensitivity to rms surface slope variations for a clear sky.

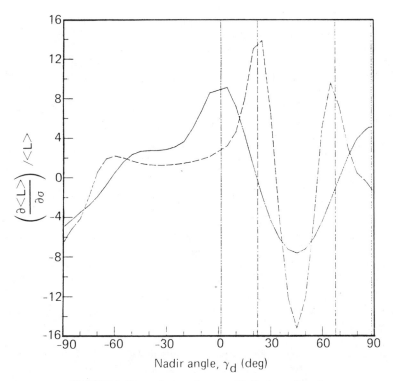

FIGURE 5. Sea surface contrast sensitivity for a clear sky.

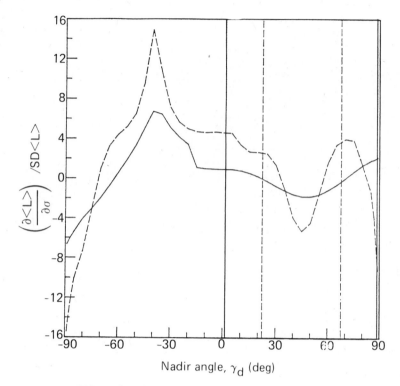

FIGURE 6. Sea surface SNR sensitivity for a clear sky.

sensitivity oscillates from large positive values to large negative values. This effect has been noted and documented by past researchers (e.g., Shand, 1953).

Figure 5 shows this same type of oscillation in the contrast sensitivity. Note the significant contrast sensitivity for a nadir-directed detector with $\sigma = 0.2$. Results of this type, in conjunction with studies of atmospheric attenuation and upwelling, may explain the visibility of internal waves outside of the glitter pattern from satellites.

The SNRS, as previously defined, is displayed in Fig. 6. Note that the SNRS has extrema at both horizons, in the glitter pattern and 90° away from the glitter pattern. These are regions of either relatively large or relatively small gradients in the slope-to-luminance transfer function (the function relating observed luminance to the slope of a surface facet for a given observation geometry).

4. EFFECTS OF AVERAGING

All real measurements necessarily include some form of spatial and temporal averaging. This averaging will tend to reduce both the measured perturbation and the measured noise to an extent dependent upon the spatial and temporal character of the perturbation and the noise as well as the type and extent of averaging used. For extremely small fields of measurement and short integration times, the SNRS with averaging ($SNRS_a$) approaches the SNRS. For fields of measurement and for integration times that are much larger in extent and period than those of surface perturbations, the $SNRS_a$ approaches zero.

Typical surface perturbations due to spatially varying currents are larger in extent and slower moving than many wind waves which contribute significantly to the rms luminance fluctuations. The use of spatial and temporal averaging can therefore increase the $SNRS_a$ of the surface perturbation and hence its visibility. For moderate amounts of spatial and temporal averaging, the reduction in the perturbation amplitude will be small and therefore can be ignored. The SNRS_a can then be calculated according to

$$\text{SNRS}_a = \left(\frac{\sigma_L}{\sigma_{La}}\right)\text{SNRS} \tag{4}$$

where

σ_L is the standard deviation of the luminance without averaging
σ_{La} is the standard deviation of the luminance with averaging

The variance of the luminance with averaging is given by

$$\sigma_{La}^2 = \int\limits_{-\infty}^{\infty}\int\int G_L(\mathbf{k}, w)|F(\mathbf{k}, w)|^2 \, d\mathbf{k}\, dw \tag{5}$$

where

$G_L(\mathbf{k}, w)$ is the luminance variance spectrum
$F(\mathbf{k}, w)$ is the transfer function associated with the spatial and temporal filters

The standard deviation of the luminance with averaging can thus be calculated.

Equations 4 and 5 require that the luminance variance spectrum be known in order to calculate the effects of averaging on the SNRS_a. The next section will describe two methods that can be used to determine the luminance spectrum of the sea surface.

5. METHODS FOR CALCULATING THE LUMINANCE SPECTRUM

The three-dimensional space–time spectrum of the sea surface luminance is difficult to calculate and display. The luminance spectrum is a function of the three-dimensional wind wave spectrum and the generally nonlinear, slope-to-luminance transfer function, both of which are highly variable and difficult to characterize. Two methods for calculating the spatial luminance spectrum will be described in this section. The expansion of these methods to include the temporal domain is straightforward, but the computational expense of such an extension is great.

The first method for this calculation involves the simulation of surface images, from which a luminance spectrum can be estimated. This simulation is described in detail by Chapman and Irani (1981) and thus will only be outlined here. The second method involves a two-dimensional expansion of the luminance autocovariance function in terms of the slope-component autocorrelation functions. This method is currently under development and promises an efficient means of estimating the luminance spectrum.

5.1. Method 1: Simulation of Surface Images

Figure 7 displays the methodology for calculating sea surface radiance spectrum by simulating sea surface images. The analysis displayed in Fig. 7 was developed using a radiometric sky-radiance model, hence the output is in radiometric, and not photometric, units; though with the proper changes of scale, the methodology will apply directly to the photometric problem.

The analysis starts with a two-dimensional model of the wind wave spectrum. At the top left of Fig. 7 is a gray-scale representation of the surface-slope magnitude spectrum. By combining the wave spectral model with an array of uniformly distributed random phases, the Fourier

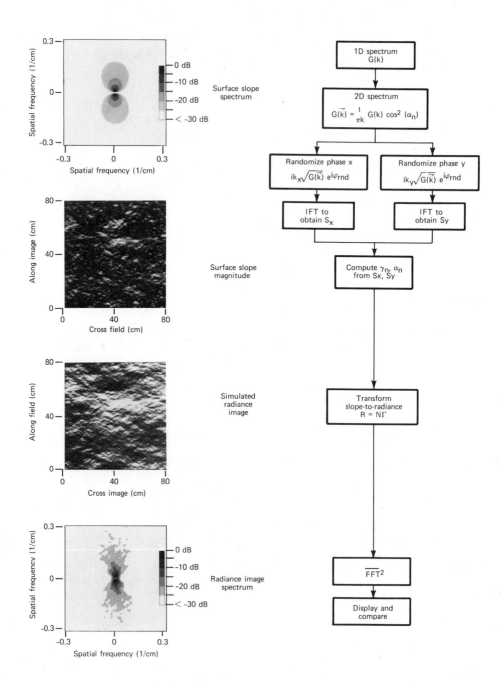

FIGURE 7. Methodology for the simulation of sea surface images. Begin with a spectral representation of the surface (surface slope spectrum). Multiply by a random phase matrix and perform the inverse Fourier transform to obtain a surface realization (surface slope magnitude). Apply the nonlinear slope-to-radiance transform to obtain the simulated radiance image. The radiance image spectrum is then calculated. All of the images displayed are typical results of this simulation.

transform of a single realization of the sea surface is created. In fact, two such transformed arrays are created, one for each of the two surface-slope components. By inverse Fourier transforming these arrays, a model of a section of the sea surface is created. The second panel from the top in Fig. 7 is a representation of a surface-slope magnitude array created by this technique. With a model of the slope-to-radiance transfer function based on viewing geometry and sky conditions, the radiance image corresponding to the simulated sea surface is calculated. From an ensemble of these simulated images, a statistically stable estimate of the surface radiance spectrum can be obtained.

The use of this technique for estimating the reduction in radiance variance achieved through spatial averaging is limited by several factors:

1. The technique is so computationally intensive that the spatial dynamics range of the calculation must be limited. The calculations displayed in Fig. 6 were made using 128×128 arrays, effectively limiting the analysis to a span of 1.8 decades in wavenumber. Under many conditions the spatial dynamic range of surface waves far exceeds this value. Furthermore, since typical glints within the glitter pattern are extremely small, on the order of a millimeter, the radiance field within the glitter pattern cannot be accurately represented by this technique.

2. Shadowing of the surface, a dominant effect at large detector nadir angles, is not taken into account. This restricts the validity of this technique to detector nadir angles of less than approximately $60°$ for moderate to low seas.

3. The expansion of this technique to include the temporal domain would involve the simulation of moving images by advancing the randomized surface phases according to the dispersion relation. This would drastically increase the number of calculations required.

5.2. Method 2: Nonlinear Analysis

A second method for calculating the luminance spectrum that promises to circumvent several of the limitations of the simulation is under development. Unfortunately, the development of this analysis has not proceeded to the point of producing any results, so only the methodology will be described here.

Let (x_1, y_1) be the surface-slope vector at point 1 and (x_2, y_2) be the surface-slope vector at point 2, where point 1 and point 2 are separated by the lag vector $\mathbf{r}$. Furthermore, let the slope-to-luminance transfer function be given by $L(x, y)$. The luminance spectrum can then be expressed as the Fourier transform of the luminance autocovariance function

$$\Phi_{LL}(r) = \langle L(x_1, y_1)L(x_2, y_2) \rangle$$
$$= \int\int\int\limits_{-\infty}^{\infty}\int L(x_1, y_1)L(x_2, y_2)p(x_1, y_1, x_2, y_2)dx_1 dx_2 dy_1 dy_2 \qquad (6)$$

where the slope correlation coefficients, $\rho_{xx}(\mathbf{r})$, $\rho_{yy}(\mathbf{r})$, and $\rho_{xy}(\mathbf{r})$, are embedded in the joint slope probability density function, $p(x_1, x_2, y_1, y_2)$.

The coordinate transformation, defined by

$$\begin{pmatrix} \bar{x}_1 \\ \bar{x}_2 \end{pmatrix} = \frac{1}{\sigma_x}\begin{pmatrix} x_1 \\ x_2 \end{pmatrix}$$

$$\begin{pmatrix} \bar{z}_1 \\ \bar{z}_2 \end{pmatrix} = \frac{1}{\sigma_z}\left[\frac{1}{\sigma_y}\begin{pmatrix} y_1 \\ y_2 \end{pmatrix} + \frac{b_{ij}}{\sigma_x}\begin{pmatrix} x_1 \\ x_2 \end{pmatrix}\right] \qquad (7)$$

where

$$b_{ij} = \left(\frac{1}{1-\rho_{xx}^2}\right)\left[\begin{matrix} \rho_{xx}\rho_{xy} & -\rho_{xy} \\ -\rho_{xy} & \rho_{xx}\rho_{xy} \end{matrix}\right]$$

and

$$\sigma_z = 1 + b_{11}^2 - b_{12}^2$$

yields four, zero-mean, unit-variance, random variables with the property that for all i and j, $\langle x_i y_j \rangle = 0$. This independence of x and y means that the joint slope pdf in Eq. (6) can be separated into a product.

Assuming that L is polynomial of degree n in x and y, then the lagged luminance product can be rewritten in the canonical form

$$L(x_1, y_1)L(x_2, y_2) = \sum_{\alpha=0}^{n}\sum_{\beta=0}^{n}\sum_{\gamma=0}^{3n}\sum_{\delta=0}^{3n} A_{\alpha\beta\gamma\delta}\bar{x}_1^\alpha \bar{x}_2^\beta \bar{z}_1^\gamma \bar{z}_2^\delta \tag{8}$$

With this form for the lagged luminance product and the independence of x and z, we can write

$$\Phi_{LL}(r) = \int\int\int_{-\infty}^{\infty}\int \left(\sum_0^n\sum_0^n\sum_0^{3n}\sum_0^{3n} A_{\alpha\beta\gamma\delta}\bar{x}_1^\alpha \bar{x}_2^\beta \bar{z}_1^\gamma \bar{z}_2^\delta\right) p(\bar{x}_1, \bar{x}_2)p(\bar{z}_1, \bar{z}_2)d\bar{x}_1 d\bar{x}_2 d\bar{z}_1 d\bar{z}_2 \tag{9}$$

Mehler's formula (see Watson, 1933) is now used to express the bivariate Gaussian distributions of Eq. (9) in terms of a double series of Hermite polynomials:

$$p(x_1, x_2) = \frac{1}{2\pi}\exp\left(\frac{x_1^2 + x_2^2}{2}\right)\sum_{k=0}^{\infty} He_k(x_1)He_k(x_2)\frac{\rho_{xx}^k}{k!} \tag{10}$$

Combining (9) and (10) yields the desired form for the luminance autocovariance:

$$\Phi_{LL}(r) = \frac{1}{4\pi^2}\sum_{i=0}^{\infty}\sum_{j=0}^{\infty}\frac{A'_{ij}}{i!j!}\rho_{\bar{x}\bar{x}}^i\rho_{\bar{z}\bar{z}}^j \tag{11}$$

where

$$A'_{ij} = \sum_0^n\sum_0^n\sum_0^{3n}\sum_0^{3n} A_{\alpha\beta\gamma\delta}Q(i,\alpha)Q(i,\beta)Q(j,\gamma)Q(j,\delta) \tag{12}$$

$$Q(i,k) = \int_{-\infty}^{\infty} x^k e^{-x^2/2} He_i(x)dx \tag{13}$$

So, the calculation of the luminance autocovariance function has been reduced from a fourfold integral to an infinite double summation involving $9n^4$ other summations of some strange integral. Believe it or not, this is an improvement. The integral of Eq. (13) has a simple analytic solution. Furthermore, using the orthogonality relation for Hermite polynomials, it is easy to show that if either $i > k$ or $i + k$ is odd, then $Q(i, k) = 0$. This leads to a simplification of the calculation of the coefficients, A_{ij}, as well as the termination of the infinite series of Eq. (11) at $3n$. Combining these simplifications with a reordering of the summations yields

$$\Phi_{LL}(r) = \frac{1}{4\pi^2}\sum_{\alpha=0}^{n}\sum_{\beta=0}^{n}\sum_{\gamma=0}^{3n}\sum_{\delta=0}^{3n} A_{\alpha\beta\gamma\delta}\left(\sum_{i=0}^{\min(\alpha,\beta)} Q(i,\alpha)Q(i,\beta)\frac{\rho_{\bar{x}\bar{x}}^i}{i!}\right)$$

$$\cdot\left(\sum_{j=0}^{\min(\gamma,\delta)} Q(j,\gamma)Q(j,\delta)\frac{\rho_{\bar{z}\bar{z}}^j}{J!}\right) \tag{14}$$

There are only two assumptions required by this analysis: the surface slopes are normally distributed and the slope-to-luminance transfer function can be described by a finite power series. Unfortunately, under certain conditions the error in these assumptions may be severe. For example, a large number of terms would be required in a power series to accurately described the extremely nonlinear slop-to-radiance transfer function in or near the glitter pattern. At the same time, the assumption of a normal distribution of slope components is only valid for reasonably small slopes. The higher-order terms in the transfer-function expansion tend to increase the weighting of the larger slopes. Thus, a high-order expansion would violate the requirement for small slopes.

To avoid these problems in the glitter pattern, alternative forms of slope-to-luminance transfer function are under investigation. One promising possibility is comprised of the sum of a narrow Gaussian shape to model the solar disk and a power series representation of everything else. Initial analysis shows that under certain conditions this transfer function leads to an easily evaluated form for the coefficients, A_{ij}.

Like the simulation, this analysis does not take into account shadowing. This will restrict the validity of analysis to detector nadir angles of less than approximately $60°$ for moderate to low seas. The inclusion of the time domain into the equations, while not difficult, does lead to a sufficient increase in the computational complexity of the analysis to make the utility of such a venture questionable.

6. SUMMARY

A method for calculating the visibility of sea surface perturbations manifested by rms slope variations has been presented. Results from such calculations for a clear sky show the visibility to be maximized for either large gradients in the slope-to-luminance transfer function (within the glitter pattern or near the horizon) or very small gradients in the slope-to-luminance transfer function $90°$ away from the glitter pattern.

The effects of spatial and temporal averaging on the luminance SNR sensitivity to rms slope variations are discussed. Two methods of calculating the reduction of noise due to averaging are presented. The first, based on a simulation of sea surface images, has several limitations including its relatively high cost and its limited spatial dynamic range. The second method, based on an expansion of the luminance autocovariance function in terms of the slope component correlation coefficients, is more efficient than the simulation and promises to be reasonably accurate over a broad range of conditions.

ACKNOWLEDGMENTS. I would like to acknowledge the contributions to this work of G. B. Irani and P. S. Shoenfeld. Dr. Irani was a co-investigator in the analysis of simulated surface images and contributed significantly to those results. Dr. Shoenfeld developed the coordinate transformation [Eq. (7)] that decorrelates the orthogonal slope components.

REFERENCES

Apel, J. R., H. M. Byrne, J. R. Proni, and R. L. Charnell (1975a): Observations of oceanic internal and surface waves from the Earth Resources Technology Satellite. *J. Geophys. Res.* **80**, 865–881.

Apel, J. R., J. R. Proni, H. M. Byrne, and R. L. Sellers (1975b): Near-simultaneous observations of intermittent internal waves on the continental shelf from ship and spacecraft. *Geophys. Res. Lett.* **2**, 128–131.

Chapman, R. D. (1981): Visibility of rms slope variations on the sea surface. *Appl. Optics* **20**, 1959–1966.

Chapman, R. D., and G. B. Irani (1981): Errors in estimating slope spectra from wave images. *Appl. Optics* **20**, pp. 3645–3652.

Hughes, B. A. (1978): The effect of internal waves on surface wind waves. 2. Theoretical analysis. *J. Geophys. Res.* **83**, 455–465.

Hughes, B. A., and H. L. Grant (1978): The effect of internal waves on surface wind waves. 1. Experimental measurements. *J. Geophys. Res.* **83**, 443–454.

LaFond, E. C. (1962): Internal waves, in: *The Sea*, Vol. 1 (M. N. Hill, ed.), Wiley, New York, 745–747.

Maul, G. A., D. R. Norris, and W. R. Johnson (1974): Satellite photography of eddies in the Gulf loop current. *Geophys. Res. Lett.* **1**, 256–258.

Shand, J. A. (1953): Internal waves in Georgia Strait. *Trans. Am. Geophys. Union* **34**, 849–856.

Strong, A. E., and R. J. DeRycke (1973): Ocean current monitoring employing a new satellite sensing technique. *Science*, **182**, 482–484.

Watson, G. N. (1933): Notes on generating functions of polynomials. 2. Hermite polynomials. *J. London Math. Soc.* **8**, 194–199.

VI

SEA SURFACE MEASUREMENTS

32

SOUTHERN OCEAN WAVES AND WINDS DERIVED FROM SEASAT ALTIMETER MEASUREMENTS

NELLY M. MOGNARD, WILLIAM J. CAMPBELL, ROBERT E. CHENEY, JAMES G. MARSH, AND DUNCAN ROSS

ABSTRACT. SEASAT altimeter measurements of significant wave height in the Southern Ocean during August and September 1978 have been examined. Significant wave heights of greater than 10 m were observed. Background levels of 2–4 m indicate generally rougher seas than occur in the North Atlantic during a comparable season. The altimeter data set is examined statistically and the spectrum of the worst case situations is determined parametrically based upon a JONSWAP form of the spectrum. Results of a statistical analysis of the altimeter data are presented together with an evaluation of the probable spectral characteristics based upon a parametric derivation of the spectrum.

1. INTRODUCTION

The Southern Ocean is thought to have one of the most hostile environments in the world. Snodgrass *et al.* (1966) and Munk *et al.* (1963) measured swell generated by Southern Ocean storms, some of which propagated across the entire Pacific until they impinged on the coast of Alaska. Cartwright (1971) and Cartwright *et al.* (1978) observed Southern Ocean-generated swell at St. Helena Island in the Atlantic at 16°S which, on certain occasions, caused extreme surf ("roller") that damped ships and structures within the harbor. Until SEASAT, however, there have been no reported measurements of waves within the Southern Ocean generation areas. Furthermore, due to a paucity of ship traffic, few observations of the surface wind or pressure fields have been obtained, and the intensity of Southern Ocean storms has been poorly known. Recently, Guymer and Le Marshall (1981) have reported on pressure fields observed by FGGE drifters in the Southern Ocean. From these data, central pressures of storm systems have frequently been found to be in the vicinity of 950 mbars and lower. Furthermore, high-pressure systems at lower latitudes have often been found to be in excess of 1030 mbars. Such extreme coexisting systems ought to be capable of generation of severe wind and wave conditions.

NELLY M. MOGNARD ● Centre National d'Etudes Spatiales, Groupe de Recherche de Geodesie Spatiale, 31055 Toulouse, Cedex, France. WILLIAM J. CAMPBELL ● Cryosphere Interactions Project, USGS, University of Puget Sound, Tacoma, Washington 98416. ROBERT E. CHENEY AND JAMES G. MARSH ● NASA Goddard Space Flight Center, Greenbelt, Maryland 20771. DUNCAN ROSS ● Atlantic Oceanographic and Meteorological Laboratories, National Oceanic and Atmospheric Administration, Miami, Florida 33149.

In 1978, the SEASAT satellite obtained extensive measurements of surface wind and wave fields in the Southern Ocean by means of a radar altimeter. In this chapter we present an analysis of the altimeter wave height and wind speed measurements from July to October 1978, which is during the austral winter.

Using the SEASAT algorithms we plotted the significant wave height ($H_{1/3}$) and surface wind speed for all longitudes in the area south of 35°S latitude to the ice edge and compared these fields with surface weather maps provided to us by the Australian weather service. The verification of $H_{1/3}$ and surface wind speed determinations using altimeter data and algorithms is given in Tapley *et al.* (1979) and in Fedor and Brown (1982) where comparisons of SEASAT radar altimeter-inferred estimates of $H_{1/3}$ and wind speed with buoy measurements yielded a mean difference of 0.07 m with a standard deviation of 0.29 m over a range of 0.5 to 5.0 m $H_{1/3}$, and a mean difference of -0.25 m/s and a standard deviation of 1.6 m/s over the range of 1 to 10 m/s wind speed. Those numbers are very similar to GEOS-3 data comparisons (Mognard and Lago, 1979; Brown *et al.*, 1981), and provide a corroboration of the wave height and wind speed measurement capability of a short-pulse spaceborne radar altimeter.

Throughout the Southern Oceans we find good qualitative agreement between the surface winds derived from the SEASAT altimeter data and those derived from the synoptic surface pressure fields of the southern hemisphere given in the weather charts. However, this agreement is not as good as that obtained in analyses of SEASAT wind and wave data in areas when more accurate weather data are available (Guymer and Le Marshall, 1981).

From September 21 through 23 we took every available ship surface wind speed report in the Southern Ocean and compared them to the altimeter wind speed measurement within 100 km and 12 h of the ship reports. The number of ship reports is few, but as is shown in Fig. 1,

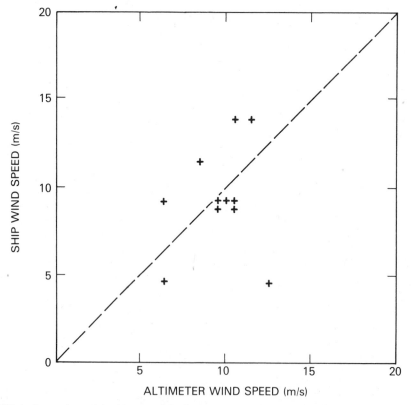

FIGURE 1. Comparison of the ship-reported wind speed versus the nearest SEASAT altimeter wind speed measurement.

the 11 ship observations agree reasonably well with the altimeter-deduced wind speeds. Where they disagree, generally the ship wind speed values exceed the satellite values.

An interesting aspect of this study is that in the $H_{1/3}$ fields that we have derived throughout the Southern Oceans, the values of $H_{1/3}$ are much higher than those derived from GEOS-3 altimeter data for the North Atlantic in winter reported by Mognard and Lago (1979). Indeed, we have found $H_{1/3}$ values as high as 10–12 m occurring in some part of the Southern Ocean every few days. A reasonable explanation for this would be that in the Southern Oceans, with their great unbounded expanse and more intense meteorological regime, the wave field is generated by higher winds with greater fetch. Moreover, the wave conditions in general seem to consist of large-amplitude swell often mixed with strong windseas. This thesis can be evaluated by comparing the local wind wave fields as deduced from the altimeter-derived wind speed fields and assumption of a fully developed sea, with the $H_{1/3}$ fields deduced from the altimeter return pulse shape.

2. THE SEASAT ALTIMETER

The SEASAT altimeter is a short-pulse, nadir-looking radar. Operating at 13.5 GHz (X-band), the radar transmits a 3-ns square-wave pulse of microwave energy. Owing to the curvature of the earth, the reflected microwave energy received at the satellite increases linearly with time until a plateau is reached, and then decreases approximately linearly. For a rough surface, the slope of the initial increase ramp is decreased since the beam is first reflected from the elevated roughness elements of the wave crests, and last reflected by the lowest roughness elements within the wave troughs. Thus, by means of an empirically based algorithm, the slope of the return pulse is used to extract the significant wave height. The total amount of reflected radar energy is related to the surface wind speed which roughens the surface by increasing the value of mean square slope and thereby decreases the backscatter cross section σ_0. Again, an empirical algorithm is imposed to extract a surface wind speed. Webb (1981), in a comparison of SEASAT-derived $H_{1/3}$ against JASIN wave measurements, reports a 4% low bias of the altimeter estimate for waves which varied between 0.7 m and 2.3 m. For further information, the reader is referred to Brown et al. (1981).

3. THE ANALYSIS PROCEDURES

Neumann and Pierson (1966), by examining an extensive set of simultaneous measurements of surface winds and waves observed at Atlantic weather stations "I" and "J," established a wind speed to wave height relationship for fully developed wave conditions. This relationship, $H_{1/3} = 0.022U^2_{19.5}$, provides a specification of the average of the one-third highest waves ($H_{1/3}$) referenced to the wind speed as observed by a cup anemometer mounted on the ship's mast 19.5 m above mean sea level. As described above, the altimeter radar backscatter cross section$_\sigma$ has been empirically related to a surface wind speed as measured at a height of 10 m. Using a logarithmic model of the boundary layer, the 10-m wind can be related to the 19.5-wind. Although many studies have shown (Garrett, 1977) that this relationship is not a constant, we have adjusted the altimeter wind upward by 8% to approximate the 19.5-m wind and calculated maximum expected significant wave heights (Hww) according to the equation

$$\text{Hww} = 0.022 \, (1.08U_{10})^2 \qquad\qquad (2)$$

where U_{10} is the wind speed derived from the altimeter.

Where Hww is found to be below the observed $H_{1/3}$, i.e., when the waves are too large for the observed windspeed, we assume swell is present. We then calculate the total energy (E) of the particular wave field from

$$E = \frac{H_{1/3}^2}{16} \qquad (3)$$

Subtracting the energy of the wind waves Hww from the total energy established from the observed $H_{1/3}$ then leaves us with an estimate of swell energy which can be used to calculate the significant height of the swell.

In the absence of swell energy, we can perceive a given $H_{1/3}$ is representative of either a fully developed condition or a duration-limited situation. By imposition of a model of the spectrum, we can estimate the characteristics of the spectrum which might be associated with the most severe cyclones. Nearly all spectra in the absence of swell can be fit by a JONSWAP (Hasselmann *et al.*, 1973) form of the spectrum

$$E(f) = \alpha g^2 (2\pi)^{-4} f^{-5} \exp\left\{ -\frac{5}{4}\left(\frac{f_m}{f}\right)^4 + \ln\gamma \cdot \exp\left[-\frac{(f-f_m)^2}{2\sigma^2 f_m^2} \right] \right\} \qquad (4)$$

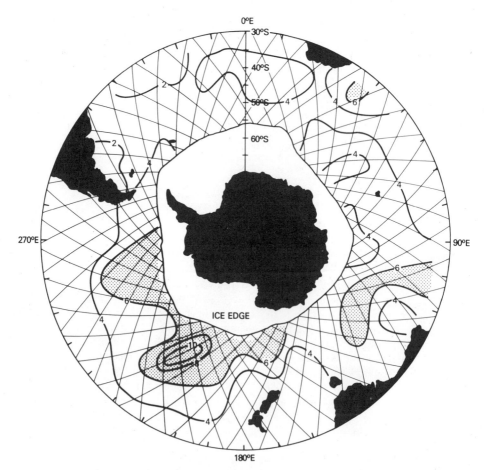

FIGURE 2. Southern Ocean map of $H_{1/3}$ (meters) for September 21 to 23 (1978) derived from the SEASAT altimeter measurements and showing the altimeter ground tracks.

This function contains two scale parameters (Phillips' α and the peak frequency f_m) and two shape parameters (γ, a peak enhancement factor, and σ which defines the width of the peak region of the spectrum). For fully developed wave conditions, $\gamma = 1$. For underdeveloped conditions, Hasselmann $et\ al.$ (1973) found γ to average 3.3 and the transition to fully developed conditions was not observed. Using Eq. (4) and an assumption of the stage of development, a spectrum can be estimated. The peak frequency of the spectrum can be determined by means of the measured wind speed and wave height by use of (3) and the dimensionless energy frequency (ε–v) relationship of Hasselmann $et\ al.$ (1976):

$$\varepsilon = 5.3 \times 10^{-6} v^{-10/3} \tag{5}$$

where $\varepsilon = Eg^2/U^4$ and $v = f_m U_{10}/g$. For the case of September 23–26, the observed wind of 25.5 m/s and wave height of 8.7 m leads to $v = 0.18$ and $f_m = 0.07$. Such a low peak frequency is consistent with the measurements of Southern Ocean-generated swell by Munk $et\ al.$ (1963) along the southern California coast (see also Cardone $et\ al.$, 1981).

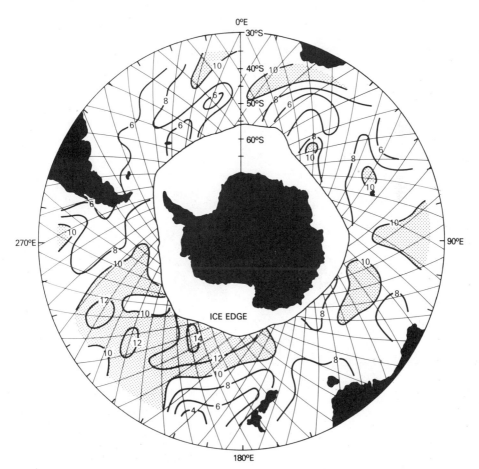

FIGURE 3. Southern Ocean map of surface wind speed (m/sec) derived from the SEASAT altimeter for September 21 to 23 (1978) with altimeter ground tracks superimposed.

4. RESULTS

A comparison of the wave field derived from (2) with that calculated from the altimeter radar return waveform shows that most of the time throughout the Southern Ocean, the $H_{1/3}$ values are greater than the wind wave values indicating the usual presence of significant swell. Comparisons of the Hww fields with the surface pressure charts show that they generally move to the northeast, namely, in the general direction of the motion of the atmospheric cyclones. The calculated wind waves, as one would expect, respond rapidly to variations in the surface wind speed.

During the month of September the satellite pattern repeated every 3 days. For each day, the distribution of four ocean parameters was mapped throughout the Southern Ocean: surface wind speed, $H_{1/3}$, Hww, and swell. From September 21 to 23, between the latitude of 35°S and the ice edge, a composite map of each of the four ocean parameters was plotted (Figs. 2–5). On the each maps the values of the corresponding ocean parameters are contoured.

In Fig. 2, very few $H_{1/3}$ contours are under 2 m. These low values are mainly located in the

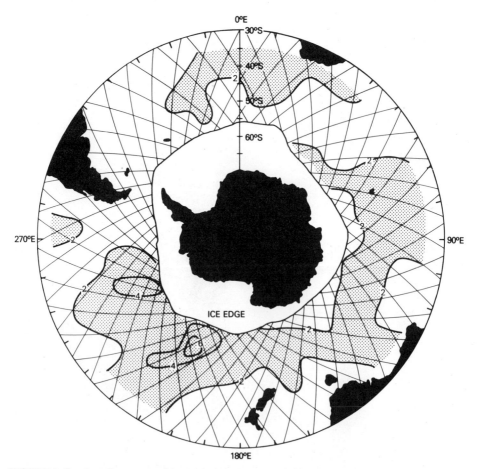

FIGURE 4. Southern Ocean map of the maximum wind wave heights (in meters) deduced from the $H_{1/3}$ and surface wind speed SEASAT altimeter measurements for September 21 to 23 (1978).

Atlantic Ocean whereas in the Pacific Ocean values as high as 10 m were recorded. The Pacific Ocean exhibits the highest $H_{1/3}$ values in two different regions respectively located at longitude 200°E and 260°E. In the Indian Ocean, also, $H_{1/3}$ values as high as 6 m were found along two satellite tracks on September 21. The Atlantic Ocean has the lowest sea state with $H_{1/3}$ generally varying between 2 and 4 m.

Figure 3 shows the corresponding surface wind speed map. Generally, the surface wind speed is lower than 10 m/s. As in the $H_{1/3}$ map, the Atlantic Ocean is relatively calm with wind speed less than 10 m/s. In the Indian Ocean, the wind speed is mainly between 8 and 12 m/s with only limited regions lower than 8 m/s. In the Pacific Ocean, by far the roughest, at a longitude of 200°E, wind speeds up to 25 m/s are measured along individual satellite tracks; this corresponds with the region of 14 m/s wind in our smooth map. Wind speeds higher than 12.5 and 15 m/s are also observed along several tracks in the Pacific Ocean.

The map of the maximum wind waves (Fig. 4) reproduces the characteristic features of the wind speed map with the highest wind waves occurring in the Pacific Ocean in the location associated with the maximum observed wind speed. In the sequential daily maps (not shown), the

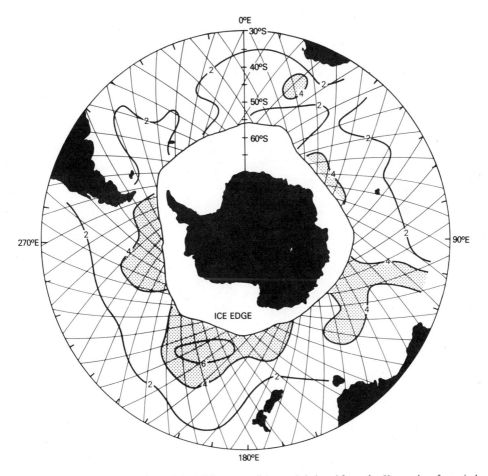

FIGURE 5. Southern Ocean map of the minimum swell (meters) deduced from the $H_{1/3}$ and surface wind speed SEASAT altimeter measurements for September 21 to 23 (1978).

motion of the wind wave families corresponds directly to the movement of the atmospheric forcing field.

For swell the situation is more complex. However, the swell minimum map (Fig. 5) exhibits many of the same characteristics as the former map: low levels in the Atlantic Ocean, higher levels in the Indian Ocean with one region higher than 6 m, whereas in the Pacific Ocean the swell level is consistently above 6 m. A background level of significant height of about 2 to 4 m is found to be characteristic of the Southern Ocean.

The swell of the Southern Ocean is the result of the action of the atmospheric forcing field. High, localized seas are generated every few days in regions where surface winds with speed in the order of 15 m/s and higher have occurred. These regions are associated with developing, rapidly moving cyclones, or with a steep pressure gradient associated with the conjugate action of a subtropical high and of an intense Antarctic low (Guymer and Le Marshall, 1981). A time delay of 1 day is observed between the measurement of high wind speed and the formation of a high-amplitude swell. During this day, the significant height reaches an amplitude of 8 to 10 m for cases with a generating surface wind speed higher than 20 m/s. Once such seas have escaped the generation region as swell, we observe that they usually move northward and not in the direction of the migrating cyclone that formed them. One case we analyzed in detail concerned an intense storm that formed on September 22, 1978, in the region of 55°S latitude and 200°E longitude and was characterized by surface wind speeds as great as 25 m/s. The following day, as the cyclone moved toward the northeast, the local seas had grown from 3–4 m to 8 m significant height. A swell family was generated along the cyclone trajectory and persisted on the sequential swell maps derived from the altimeter data until September 25 with a significant height on the order of 7 to 8 m. After this date it could still be followed on the SEASAT altimeter swell maps propagating north of latitude 35°S with a slowly decreasing amplitude.

For each of the three days considered, we show (Fig. 6) the histograms of the surface wind speed and $H_{1/3}$. In the southern oceanic regions during the Antarctic winter, surface wind speeds

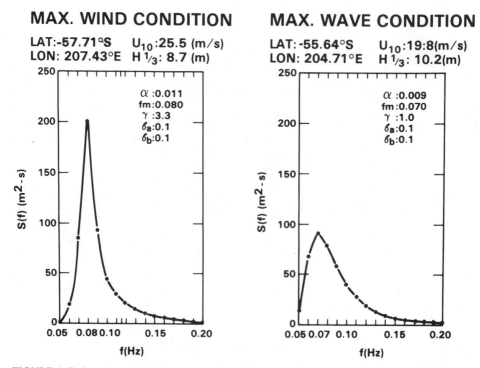

FIGURE 6. Estimated wave spectrum associated with (left) observed maximum wind conditions and (right) observed maximum sea conditions.

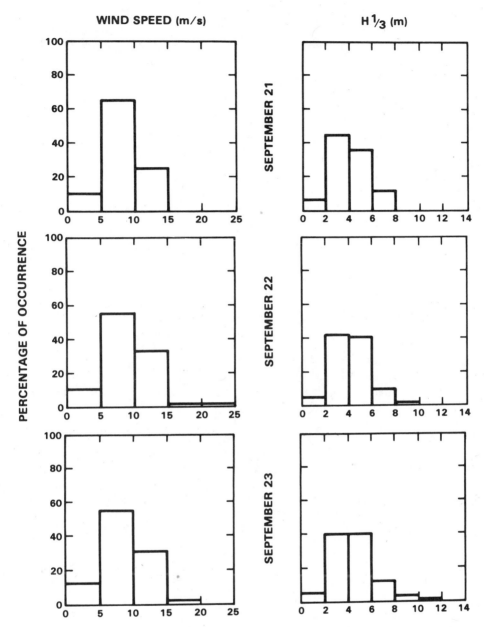

FIGURE 7. Histograms of the surface wind speed and $H_{1/3}$ SEASAT altimeter measurements for September 21 to 23 (1978).

in the order of 15 m/s and higher occur every few days. Consequently, at the same time scale, medium to high-amplitude swell families (6 to 10 m) are generated. These swells rapidly interfere with one another and are responsible for the 2-m background swell level generally measured.

5. SPECTRA

We used the parametric wave model described by Hasselmann *et al.* (1973) to study the spatial and temporal behavior of a characteristic spectral parameter, the nondimensional peak frequency: $v = f_m U_{10}/g$, where U_{10} is the local 10-m wind speed and g is the gravity constant. This parameter, a function of both the surface wind speed and the wave peak frequency, is a sensitive indicator of swell. When there is no, or minimal, swell, v is equal to or greater than the fully developed condition ($v \geqslant 0.13$). In swell-dominated situations, $v \leqslant 0.13$. Figure 4 shows a composite map of the wind wave conditions of Antarctic regions for the 3-day period derived using the parameter v to separate the swell-dominated regions from the regions with fully- and under-developed sea conditions.

In stormy regions, where the swells were generated, the use of the parametric model allows us to derive characteristic wave spectra. On September 22, in the region 55°S latitude and 200°E longitude, where a severe storm was situated, we computed two characteristic wave spectra corresponding to (1) a maximum wind speed condition with underdeveloped sea, and (2) a maximum $H_{1/3}$ condition with fully developed sea, respectively characterized by a peak enhancement factor of 3.3 and 1 (Fig. 7). These two wave spectra characterize a storm region where the background swell of 3 to 4 m reaches a significant height of 8–10 m a day later.

6. CONCLUSIONS

The SEASAT altimeter has been used to investigate surface wind and wave conditions in the Southern Ocean. From the wind speed data, we find high winds consistent with very intense cyclones with central pressures of 950 mbars as reported by Guymer and Le Marshall (1981). The waves generated by these storms are found to be correspondingly high and more severe than observed by Mognard and Lago (1979) in the North Atlantic for a similar situation. Finally, imposition of a model of the wave spectrum allows calculation of the one-dimensional wave spectrum for situations which are underdeveloped, and estimation of swell significant wave height for overdeveloped conditions. The SEASAT altimeter thus can provide unique and valuable climatological observations of surface wind and wave conditions in hostile and generally inaccessible regions.

REFERENCES

Brown, G. S., H. R. Stanley, and N. A. Roy (1981): The windspeed measurement capability of spaceborne radar altimeters. *IEEE J. Ocean. Eng.* **OE-6**, 59–63.

Cardone, V. J., H. Carlson, J. A. Ewing, K. Hasselmann, S. Lazanoff, D. Ross, and W. McLeish (1981): The surface environment in the Gate B/C scale-phase III. *J. Phys. Oceanogr.* **11**, 1280–1293.

Cartwright, D. E. (1971): Tides and waves in the vicinity of St. Helena. *Philos. Trans. R. Soc. London Ser. A* **270**, 603–649.

Cartwright, D. E., J. S. Drives, and J. E. Tranter (1978): Swell waves at St. Helena related to distant storms. *Q. J. R. Meteorol. Soc.* **103**, 655–683.

Fedor, L. A., and G. S. Brown (1981): Wave height and wind speed measurements from the Seasat radar altimeter. *J. Geophys. Res.* **87**, C5, 3254–3260.

Garrett, J. R. (1977): Review of drag coefficients over oceans and continents. *Mon. Weather Rev.* **105**, 915–929.

Guymer, L. B., and J. F. Le Marshall (1981): Impact of FGGE buoy data on southern hemisphere analyses. *Bull. Am. Meteorol. Soc.* **62**, 38–47.

Hasselmann, K., T. P. Barnett, E. Bouws, H. Carlson, D. E. Cartwright, K. Enke, J. A. Ewing, H. Gienapp, D. E. Hasselmann, P. Kruseman, A. Meerburg, P. Müller, D. J. Olbers, K. Richter, W. Sell, and H. Walden (1973): Measurements of wind-wave growth and swell during the Joint North Sea Wave Project (JONSWAP). *Dtsch. Hydrogr. Z. Suppl. A* **8** (12).

Hasselmann, K., D. B. Ross, P. Müller, and W. Sell (1976): A parametric wave prediction model. *J. Phys. Oceanogr.* **6**, 200–228.

Mognard, N., and B. Lago (1979): The computation of wind speed and wave heights from GEOS-3 data. *J. Geophys. Res.* **84**, B8, 3979–3986.

Munk, W. H., G. R. Miller, F. E. Snodgrass, and N. J. Barber (1963): Directional recording of swell from distant storms. *Philos. Trans. R. Soc. London Ser. A* **255**, 505–584.

Neumann, G., and W. J. Pierson, Jr. (1966): *Principles of Physical Oceanography*, Prentice–Hall, Englewood Cliffs, N.J.

Snodgrass, F. E., G. W. Groves, K. Hasselmann, G. R. Miller, W. H. Munk, and W. H. Powers (1966): Propagation of ocean swell across the Pacific. *Philos. Trans. R. Soc. London Ser. A* **259**, 431–497.

Tapley, B. D., G. H. Born, H. H. Hagar, J. Lorell, M. E. Parke, J. M. Diamante, B. C. Douglas, C. C. Goad, R. Kolenkiewicz, J. G. Marsh, C. F. Martin, S. L. Smith, III, W. F. Townsend, J. A. Whitehead, H. M. Byrne, L. S. Fedor, D. C. Hammond, and N. M. Mognard (1979): Seasat altimeter calibration, initial results. *Science* **204**, 1410–1412.

Webb, D. J. (1981): A comparison of Seasat I altimeter measurement made by a pitch-roll buoy. *J. Geophys. Res.* **86**, 6394–6398.

33

Marineland Aircraft Observations of L-Band Radar Backscatter Dependence upon Wind Direction

T. W. Thompson, D. E. Weissman, and W. T. Liu

Abstract. Radar observations of the ocean during the Marineland Experiment have been examined to establish whether L-band backscatter varies with surface wind direction. This complements recent aircraft and SEASAT observations which indicate that L-band backscatter from the ocean at angles of incidence near 20° varies as the square root of the wind speed. The Marineland data are particularly well suited for this since flights over instrumented surface sites were conducted in a number of directions for several different days.

The Marineland data show two distinct types of behavior. In all cases except one, the data suggest that there is little or no dependence of L-band backscatter upon wind direction. In the exceptional case where there was a moderate wind and unstable atmosphere, there was a strong directional dependence of L-band backscatter with wind. In particular, these data suggest a 40% change in backscatter between the weakest echoes (when the radar is looking cross-wind) and the strongest echoes (when the radar is looking up- or downwind).

Unusual synthetic aperture radar (SAR) images are also associated with the one case with the strong dependence upon wind direction. These SAR images show bright kilometer-sized splotches which are likely surface expressions of turbulent eddies imbedded in the unstable atmosphere. These splotches appear only when the radar is looking near the up- or downwind directions. These splotches do not appear in SAR images when the radar is looking cross-wind. Also, these splotches are most intense on the warmer Gulf Stream waters and disappear near shore where the fetch is small. The features in the SAR image correlated with cumulus clouds observed with simultaneous aircraft photography.

1. INTRODUCTION

Several recent radar observations of the ocean near the Gulf Stream (Weissman *et al.*, 1980), near hurricane Gloria (Weissman *et al.*, 1979), and during the SEASAT GOASEX experiment (Thompson *et al.*, 1983) indicate that L-band radar backscatter from the ocean near angles of incidence of 20° is modulated by surface winds. These data indicate a strong dependence upon surface wind speed since L-band backscatter power is approximately proportional to the square root of the wind speed.

The fact that L-band backscatter from the ocean varies with surface wind is not surprising. K_u-band backscatter from the ocean has a strong dependence upon surface winds. This has been studied over the last 15 years and provided the basis for the very successful operation of

T. W. Thompson ● Jet Propulsion Laboratory, Pasadena, California 91101. D. E. Weissman ● Department of Engineering Science, Hofstra University, Hempstead, New York 11550. W. T. Liu ● Earth and Space Sciences Division, Jet Propulsion Laboratory, Pasadena, California 91109.

SEASAT-1 Windspeed Scatterometer (Johnson et al.,1980). It is well established that this K_u-band backscatter from the ocean has a dependence upon wind speed and wind direction (Moore and Fung, 1979; Jones and Schroeder, 1978) Bracalente et al. (1980). A question remains whether L-band backscatter at SEASAT SAR angles of incidence has a wind dependence similar to that at K_u-band.

The primary goal of this chapter is to address questions about the azimuthal dependence of L-band backscatter by a further study of data from the Marineland Experiment. In particular, aircraft overflights of the Gulf Stream and adjacent coastal waters of Marineland, Florida, were conducted on 5 days in December 1975 (see Shemdin, 1980; Thompson, 1976, Weissman and Thompson, 1977). These aircraft flights were conducted such that the JPL L-band SAR observed instrumented surface sites from a number of different directions.

A number of radar backscatter measurements at both L-band and K_u-band suggest that L-band scatter from the ocean is related to surface wind by the following:

$$P(\lambda = 25\,\text{cm}) \propto |U|^a [1 + b\cos(2\phi)] \tag{1}$$

where

$\quad\quad P$ = radar backscatter power

$\quad\quad \lambda$ = radar wavelength = 25 cm

$\quad |U|$ = wind-speed magnitude

$\quad\quad \phi$ = wind–radar angle

$\quad\quad a$ = wind-speed coefficient

$\quad\quad b$ = wind-direction coefficient

The wind–radar angle is defined such that $\phi = 0°$ or $180°$ if the radar is looking up- or downwind; $\phi = 90°$ or $270°$ if the radar is looking cross-wind.

A number of aircraft and SEASAT-1 observations indicate that Eq. (1) is a good empirical relation for the measurements that have been reported to date. The wind-speed exponent appears to be approximately 0.5 (see Weissman et al., 1979, 1980; Thompson et al., 1981, 1983). The wind-direction coefficient has been reported to be near zero for SEASAT-1 observations during GOASEX, the Gulf of Alaska Surface Experiment (Thompson et al., 1981). A wind-direction coefficient of 0.5 has been reported for aircraft observations of hurricane Gloria (Weissman et al., 1979). A primary goal here is to review aircraft observations during the Marineland Experiment to see whether they provide further insights into the value of b, the wind-direction coefficient.

2. INSTRUMENTATION

The data for this study were obtained by the JPL L-band SAR of the NASA CV-990 aircraft. Flights were conducted in December 1975 while three ocean sites were instrumented with a number of surface instruments (Shemdin, 1980). The collection of scatterometer data with this radar is described in detail by Weissman et al. (1980); only a brief outline of that equipment is included here. There are two main points concerning the data collection. The first point concerns the radar equipment while the second point concerns the aircraft tracks. Each of these points will be discussed separately below.

The data for our study were supplied by radar equipment added to the L-band SAR as shown in Fig. 1. In particular, the video signal was enveloped detected at a fixed delay and recorded on a strip chart. This provides a "Scatterometer mode" in addition to the normal SAR operation which produces images of the surface. This scatterometer mode was implemented during the Marineland Experiment by sampling the radar's video signal, followed by a square-

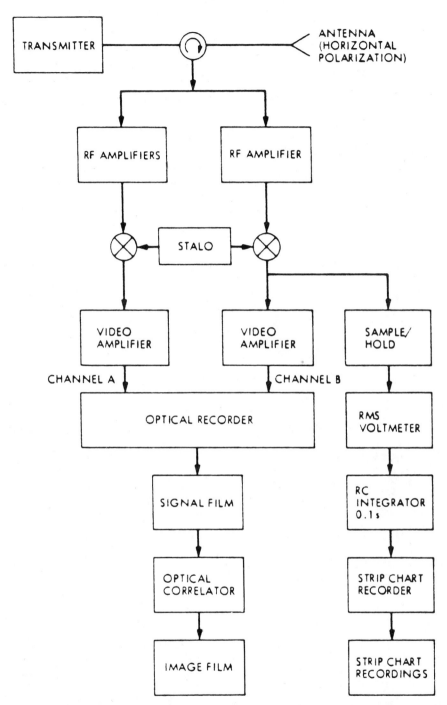

FIGURE 1. Block diagram for JPL L-band synthetic aperture radar for the Marineland observations. In the normal imaging mode, video radar signals are recorded on a film, which is later optically correlated providing images like those shown in Figs. 8 and 9. In an auxiliary scatterometer mode, video signals are sampled at a fixed delay. The sampled signals are square law detected, integrated with a 0.1-s time-constant, and recorded on a strip chart. These strip charts are then a continuous recording of ocean radar backscatter at a given angle of incidence. In this study, we used the strip chart recording to produce the data shown in Figs. 4–7; SAR images are shown in Figs. 8 and 9.

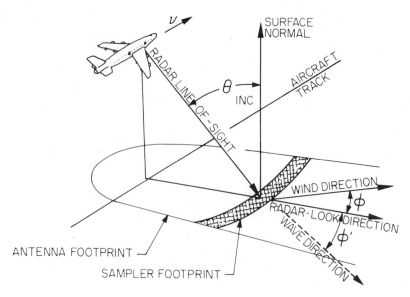

FIGURE 2. Geometry for aircraft measurements of L-band backscatter from the ocean. All echoes for a given delay are square law detected and recorded on a strip chart. Typical recording parameters are angle of incidence, Θ, equal to about 20°. Typical aircraft velocities are 250 m/s. Several different directions were flown, providing samples of random power at several radar look-angles (ϕ's).

TABLE IA. Operating Parameters for JPL L-Band SAR Radar

Parameter	Values
Center frequency	1220 MHz
Wavelength	24.6 cm
Pulse length	1.25 μs
Bandwidth	10 MHz
Time–bandwidth product	12.5
Peak power	4 kW
Antenna azimuth beamwidth	18°
Antenna range beamwidth	90°
Antenna beam center gain	12 db
Nominal altitude	3 to 12 km
Nominal ground speed	400–500 knots
Nominal pulse frequency	1000 pps at 500 knots

TABLE IB. Typical Values for the Strip Chart Scatterometer

Altitude	12 km (40,000 feet)
Angle of incidence	20°
Range	12.8 km
Azimuth beamwidth	18°
Transmitter pulse length	1.25 μs
Azimuth (along-track) footprint length	4.0 km
Range (cross-track) length	0.55 km
Antenna gain at beam center ($\theta = 45°$)	12.0 db
Antenna gain at $\theta = 20°$	10.5 db

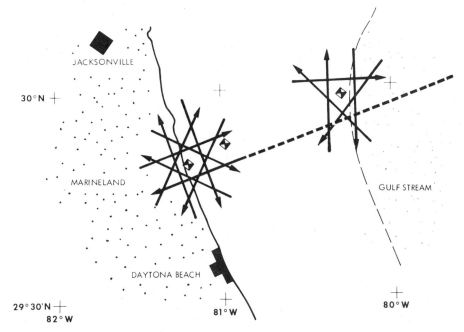

FIGURE 3. Flight paths for the eight-sided and five-sided patterns during the Marineland Experiment (see Thompson, 1976; Shemdin, 1980). Instrumented surface sites are indicated. The long-dash flight line indicates the track of the Gulf Stream Transit, which had SAR images shown in Figs. 8 and 9.

law detection, an integration (with 0.1-s time-constant), and a recording on a strip chart.

This strip chart recording provides a continuous recording of radar backscatter for a fixed angle of incidence. The geometry of these aircraft observations is shown in Fig. 2. The radar footprint for the strip chart recording is approximately rectangular and parallel to the aircraft's ground track. The footprint length is controlled by the antenna width; the footprint width is controlled by the radar's pulse length. Typical operating parameters for the Marineland observations are given in table I.

The primary importance of the Marineland data is that the aircraft was flown over the same instrumented site in various directions as shown in Fig. 3. There are two major flight patterns— the eight-sided pattern at the near-shore site and a five-sided pattern at the site on the edge of the Gulf Stream. Both patterns have ground tracks in four different directions, each separated by 45°. All four of these directions are repeated in the eight-sided pattern, whereas only one direction is repeated in the five-sided pattern. Each leg of these patterns was 44.4 km in length, which, on average, was flown in 3 min. Maneuvering between turns consumed about 7 min. The eight-sided patterns were flown in 80 ± 5 min and the five-sided patterns in 50 ± 5 min. Thus, these two patterns permitted observations of four different wind–radar angles in about 60 min.

3. DATA

The primary output from all of these Marineland strip chart recordings are measures of backscatter power versus wind–radar angle. For convenience, we plot a relative power since we are interested in small changes for the legs at different directions. To investigate whether Eq. (1) is valid, these relative powers are plotted versus $\cos(2\phi)$, where ϕ is the angle between the radar-look and wind directions.

All Marineland strip chart recordings, reduced to plots of relative powers versus $\cos(2\phi)$, are shown in Figs. 4–7. The vertical bars indicate the measured variation in the strip chart

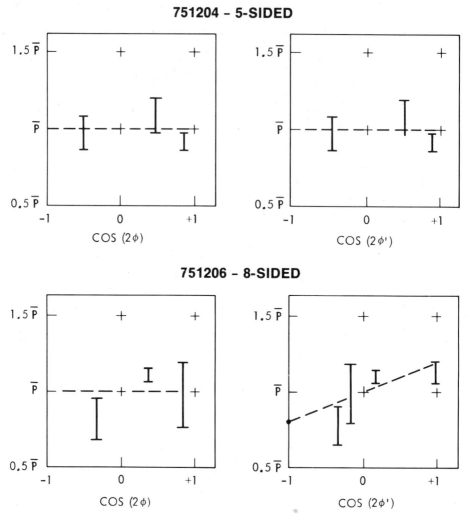

FIGURE 4. Plots of observed powers versus $\cos(2\phi)$ and $\cos(2\phi')$ where ϕ = radar look–wind direction angle and ϕ' = radar look–wave direction (see Fig. 2). Vertical bars indicate maximum and minimum values of observed powers; the dashed line is a least-squares fit to the midpoints of these bars.

signals. The left-hand plots show measured powers versus $\cos(2\phi)$, where ϕ is the radar–wind direction angle described above. The right-hand plots show measured powers versus $\cos(2\phi')$, where ϕ' is a radar–wave direction angle defined similarly to the radar–wind direction angle. The dashed lines indicate least-squares fit to the midpoints between maximum and minimum powers. For the five-sided patterns the stronger radar echoes from the Gulf Stream are plotted separately from the weaker echoes from continental shelf waters.

The data for the Marineland Experiment are also summarized in Table II, which gives the least-squares fit in terms of b and b' and their rms errors. (Both of these were computed using the midpoint values of backscatter.) For 2 days, Dec. 6 and 14, 1975, we have data for both eight-sided and five-sided patterns. An eight-sided pattern was flown on Dec. 4, but no data were

751214 – 8-SIDED

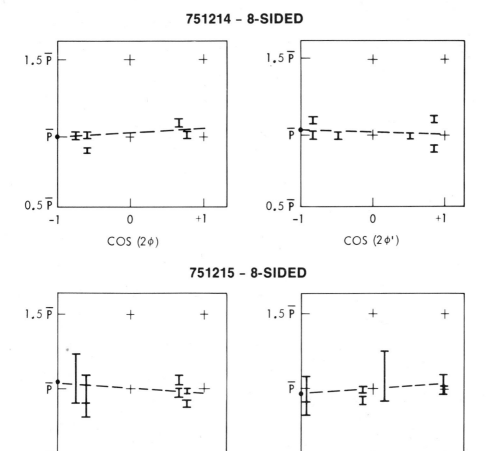

751215 – 8-SIDED

FIGURE 5. Plots of observed powers versus $\cos(2\phi)$ and $\cos(2\phi')$ where ϕ = radar look–wind direction angle and ϕ' = radar look–wave direction (see Fig. 2). Vertical bars indicate maximum and minimum values of observed powers; the dashed line is a least-squares fit to the midpoints of these bars.

recorded. Also no eight-sided pattern was flown on Dec. 10 and no five-sided pattern was flown on Dec. 15. There is no plot for the five-sided pattern on Dec. 6 since the surface was calm during the observation; the strip chart recording here shows no change in power for the five legs of this pattern.

The data given in Table II show two distinct types of behavior. All of the data except for Dec. 10 indicate that b is near zero, while the data for Dec. 10 indicate that b is near 0.2 and significant. The large value of b for December 10, 1975 appears to be associated with an unstable atmosphere that existed on that date, as will be discussed latter.

In addition to the dependence upon wind direction, we also searched for a possible dependence upon dominant wave direction. The results are also shown in Figs. 4–7 and Table II where ϕ' is the radar–wave direction angle and b' is the direction coefficient similar to the

751214 – 5-SIDED – GULF STREAM

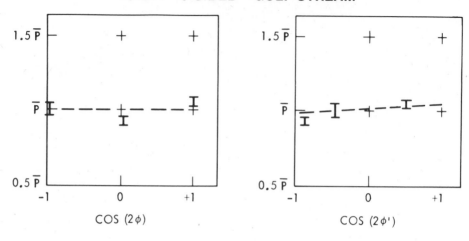

751214 – 5-SIDED – SHELF

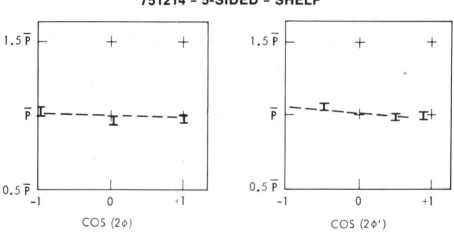

FIGURE 6. Plots of observed powers versus $\cos(2\phi)$ and $\cos(2\phi')$ where ϕ = radar look–wind direction angle and ϕ' = radar look–wave direction (see Fig. 2). Vertical bars indicate maximum and minimum values of observed powers; the dashed line is a least-squares fit to the midpoints of these bars.

coefficient b described above. There is a strong wave direction dependence for the observations of Dec. 10 when the winds and waves had the same direction. Also, there is a strong dependence upon wave direction for the observations of the eight-sided pattern on Dec. 6 where there is a large scatter in backscatter values for the different legs of the pattern.

We feel that these data suggest that wave direction has little or no effect on average L-band backscatter. The data for Dec. 6 may be an anomaly caused by intermittent land breezes at the near-shore site. The data for Dec. 10 should be ignored since wind and waves were in the same direction. The remainder of the data suggest little or no dependence upon wave direction (although our data set may be somewhat incomplete for searching for a wave direction dependence).

751210 - 5-SIDED - GULF STREAM

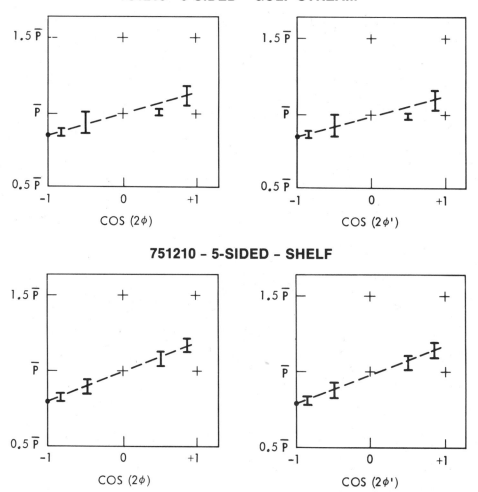

751210 - 5-SIDED - SHELF

FIGURE 7. Plots of observed powers versus $\cos(2\phi)$ and $\cos(2\phi')$ where ϕ = radar look–wind direction angle and ϕ' = radar look–wave direction (see Fig. 2). Vertical bars indicate maximum and minimum values of observed powers; the dashed line is a least-squares fit to the midpoints of these bars.

TABLE II. Summary of Marineland Five-Sided and Eight-Sided Pattern Results

| Date/pattern | Wind | | | | Waves | | | | |
	b	rms error	θ_{wind}	$|U|$	b'	rms error	θ_{wave}	$H_{1/3}$	T
751204/5-sided	−0.008	0.074	60°	4.5 m/s	−0.008	0.074	60°	0.7 m	7.0 s
751206/8-sided	−0.007	0.146	80°	2.0 m/s	0.189	0.060	120°	1.8 m	7.0 s
751206/5-sided	—	—	—	0	—	—	no data		
751210/5-sided[a]	0.144/0.203	0.038/0.010	300°	7.0 m/s	0.144/0.203	0.038/0.010	300°	1.8 m	5.5 s
751214/8-sided	0.033	0.059	90°	4.0 m/s	−0.021	0.061	100°	1.2 m	9.0 s
751214/5-sided[a]	0.013/−0.031	0.043/0.014	90°	3.0 m/s	−0.040/0.048	0.038/0.007	120°	1.8 m	7.5 s
751215/8-sided	−0.034	0.050	140°	5.0 m/s	0.024	0.052	110°	0.9 m	8.5 s

[a]Higher Gulf Stream backscatter analyzed separately from lower shelf backscatter.

The Data for Dec. 10

Since the data for Dec. 10 are distinctly different from those of the other days, we will consider them in more detail. The meteorological conditions for Dec. 10 differed from those of the other days. A cold front had passed through the Marineland area on the previous day and there were cold northwest winds blowing over the warm offshore and Gulf Stream waters. This led to a 10°C air–sea temperature difference and an unstable atmosphere at the ocean surface.

The SAR images for the five-sided pattern are also distinctly different from those of the other days. The SAR images for Dec. 10 have large kilometer-sized splotches of radar-bright areas which appear or disappear depending upon the mean wind direction as shown in Fig. 8. These

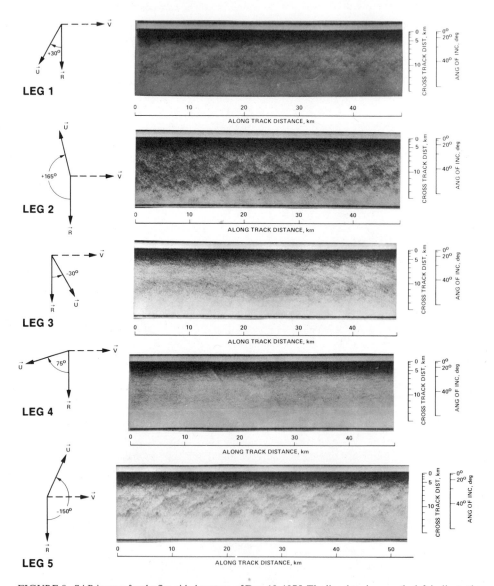

FIGURE 8. SAR images for the five sided pattern of Dec. 10, 1975. The line drawings on the left indicate the wind direction ($\vec{U} = 10$ m/s at 330°), radar look direction ($\vec{R}$), and aircraft velocity ($\vec{V}$). Negative SAR images are used, which show where radar-bright areas are darker. Note the most intense splotches (the darkest areas) occur on Leg 2 when the radar is looking nearly upwind, while no splotches occur on Leg 4 when the radar is looking nearly cross-wind. Radar tracks for this five-sided pattern are shown in Fig. 3.

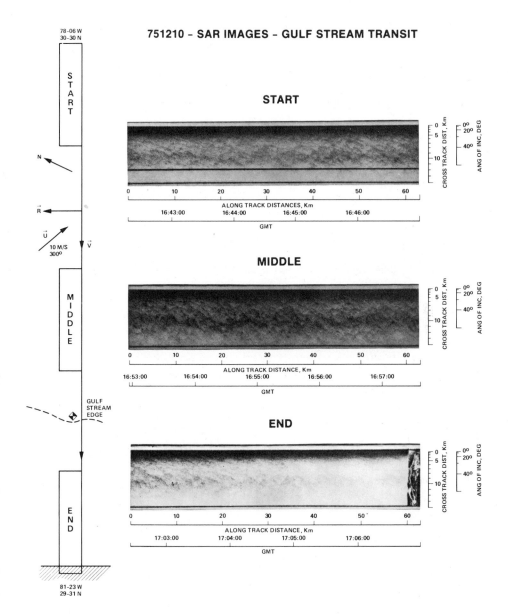

FIGURE 9. SAR images for the start, middle, and end of the Gulf stream Transit run of Dec. 10, 1975. Image positions with respect to the Gulf Stream and the coast of Florida are shown in the left-hand drawing (the last third of this track is also shown in Fig. 3). Negative SAR images are shown where the radar-bright areas are photographically darker. The horizontal line across the bottom of the "start" image is an artifact and should be ignored. Note that the splotches are most intense on the Gulf Stream waters and disappear at the shore.

radar-bright splotches appear vividly when the radar is looking up- or downwind. These splotches disappear when the radar is looking cross-wind.

Another SAR image for the Gulf Stream Transit run on Dec. 10 shown in Fig. 9 also has interesting features. This observation was conducted along a 334-km flight track which started well east of the Gulf Stream, crossed the Gulf Stream at midrun, and ended on land. SAR images for the beginning, middle, and end portions of this run are shown in Fig. 9. Note that the kilometer-sized splotches are most intense near the Gulf Stream, where the warmer Gulf Stream waters have a greater air–sea temperature difference than the adjacent ocean waters (to the east) and shelf waters (to the west). Also, these data indicate that the weakest splotches occur near shore, suggesting that the development of the features in the SAR image are related to the fetch of the wind.

In both SAR observations, radar backscatter is controlled primarily by short ocean wavelength which satisfy the Bragg condition where ocean wavelengths = [radar wavelength/2 sin (angle of incidence)]. For the SAR images shown in Figs. 8 and 9, the angle of incidence varies from 20° to 50°; the Bragg ocean wavelength varies from 36 cm to 16 cm, respectively, for the L-band radar wavelength of 24.6 cm.

4. DISCUSSION

The important observation to be made concerning the images obtained on Dec. 10 is that most of these SAR images displayed random "splotches" of cross-section variation whose structure and dimensions vary widely, but most often are in the range from about 1 to 10 km. With the sea temperature as much as 10°C above that of the atmosphere, the air closest to the interface becomes warm enough to rise, causing sizable vertical air currents that generate instabilities upon interaction with the strong horizontal winds and shear stresses in the boundary layer. These instabilities produce turbulent fluctuations in the vertical and horizontal wind vectors within the boundary layer. Their effect on the sea surface is generally referred to as "turbulent shear stresses," and they affect the large-scale momentum transfer between the atmosphere and the ocean. For this reason, there have been numerous studies of their statistical and aerodynamic parameters (Roll, 1965). Their detection with SAR images offers a new way to study some of their spatial properties. They are visible in these images because of the dependence of the local surface roughness (wind waves) on the friction velocity, through the surface stress. Of course, this friction velocity depends on the resultant of the mean and random wind vectors, and if a relatively significant ocean current exists, it should also be included.

It is probable that these splotches are "cat's paws," patches of short gravity waves moving over otherwise smooth water due to intermittent turbulence in the air, as described by Dorman and Mollo-Christensen (1973) and Liu (1978). Also, Khalsa and Businger (1977) found that these types of intermittent turbulences are caused by stability-dependent plumes. These types of plumes over the strong temperature gradients associated with the Gulf Stream are expected to line up with secondary flows (longitudinal rolls) of the atmosphere boundary layer as described by LeMone (1976).

Surface photographs obtained simultaneously with the SAR imagery also suggest that the SAR features are related to thermally induced plumes in the boundary layer. Fluffy, cumulus clouds are observed in the Gulf Stream Transit run (Fig. 9) where the patches in the SAR image occur, while near-shore waters are essentially cloud-free. The cumulus clouds observed over the Gulf Stream are kilometer-sized and approximately equal in area to the SAR features. Also, patches of these cumulus clouds are aligned with the mean wind. Thus, both the clouds and the SAR features appear to be the result of the same phenomenon—thermally induced plumes in an unstable boundary layer. This association of differences in radar backscatter from the ocean with these atmospheric features, is described in greater detail by Thompson et al. (1983).

5. SUMMARY

The data from the Marineland Experiment were used here to investigate the dependence of L-band backscatter upon wind direction. This was accomplished by reviewing backscatter recorded in a scatterometer mode for five-sided and eight-sided patterns over instrumented sites. This is a good data set for this particular investigation since four wind–radar directions could be examined in about 1 h.

All the data except for one case suggest that there is very little or no dependence of L-band backscatter on wind direction. The unusual data set for Dec. 10, in contrast, indicated a strong dependence of L-band backscatter on wind. The SAR images for Dec. 10 are also unusual since they show kilometer-sized radar-bright splotches which appear or disappear depending on whether the radar was looking up; down; or cross-wind. These radar-bright splotches may indicate local areas with short gravity waves aligned with the mean wind. These may result from local increases of surface stress induced by thermal plumes in the atmosphere.

These data suggest that: (1) L-band backscatter from the ocean may depend upon boundary layer conditions. The data show no dependence on direction for stable atmospheric conditions, but a strong dependence on wind direction during unstable conditions. (2) The SAR may be an important tool for studying the spatial scales, intensities, and alignment of boundary-layer turbulence features at the air–sea interface. (3) The radar backscatter features associated with turbulence features have kilometer scales and depend upon both fetch and air–sea temperature differences.

ACKNOWLEDGMENTS. All of us were supported by the Jet Propulsion Laboratory, California Institute of Technology, where research is carried out under NASA Contract NAS7-100, sponsored by the National Aeronautics and Space Administration. The basic measurements by the JPL L-band radar were due to the efforts of Elmer McMillan, Tom Anderson, and Don Harrison of the Jet Propulsion Laboratory. Operations of the NASA CV-900 aircraft were conducted by the very capable Airborne Science Office of the NASA Ames Research Center.

REFERENCES

Bracalente, E. M., D. H. Boggs, W. L. Grantham, and J. L. Sweet (1980): The SASS scattering coefficient algorithm. *IEEE J. Ocean Eng.* **OE5**, 145–154.

Dorman, C. E., and E. Mollo-Christensen (1973): Observation of the structure on moving gust patterns over a water surface ("cat's paws"). *J. Phys. Oceanogr.* **3**, 120–132.

Johnson, J. W., L. Williams, E. M. Bracalente, W. L. Grantham, and F. B. Beck (1980): Seasat-A satellite scatterometer instrument evaluation. *IEEE J. Ocean. Eng.* **OE5**, 138–144.

Jones, W. L., and L. C. Schroeder (1978): Radar backscatter from the ocean: Dependence on surface friction velocity. *Boundary-Layer Meteorol.* **13**, 133–149.

Jones, W. L., V. E. Delmore, and E. M. Bracalente (1981): The study of mesoscale ocean winds. *Spaceborne Synthetic Aperture Radar for Oceanography* (R. Beal, P. S. Deleonibus, I. Katz, eds.), Johns Hopkins Press, Baltimore, 87–97.

Khalsa, S. J. S., and J. A. Businger (1977): The drag coefficient determined by dissipation method and its relation to intermittent convection in the surface layer. *Boundary-Layer Meteorol.* **12**, 273–297.

LeMone, M. A. (1976): Modulation of turbulence energy by longitudinal rolls in an unstable planetary boundary layer. *J. Atmos. Sci.* **33**, 1308–1320.

Liu, W. T. (1978): The Molecular Effects of Air–Sea Exchanges. Ph.D. dissertation, University of Washington, Seattle.

Moore, R. K., and A. K. Fung (1979): Radar determination of winds at sea. *Proc. IEEE* **67**, 1504–1521.

Roll, H. U. (1965): *Physics of the Marine Atmosphere*, Academic Press, New York.

Shemdin, O. H. (1980): The Marineland Experiment: An overview, *Eos* **61**, 625–626

Thompson, T. W. (1976): JPL Radar Operations: Marineland-GASP Test Series, 25 November to 16 December 1975. *JPL Internal Report*, April 15.

Thompson, T. W., D. E. Weissman, and F. I. Gonzalez (1981): SEASAT SAR cross-section modulation by surface winds: GOASEX observations. *Geophys. Res. Lett.* **8**, 159–162.

Thompson, T. W., D. E. Weissman, and F. I. Gonzalez (1982): L-band radar backscatter dependence upon surface wind: A summary of SEASAT-1 and aircraft observations. *J. Geophys. Res.* **88**(C3), 1727–1735.

Thompson, T. W., W. T. Lin, and D. E. Weissman (1983): Synthetic aperture radar observations of ocean roughness from rolls in an unstable marine boundary layer. *Geophysical Research Letters* **10**(12), 1172–1175.

Weissman, D. E., and T. W. Thompson (1977): Detection and interpretation of ocean roughness variations across the Gulf Stream inferred from radar cross section observations. *Oceans '77 Conference Record* **1**, 14B1–14B10.

Weissman, D. E., D. B. King, and T. W. Thompson (1979): Relationship between hurricane surface winds and L-band radar backscatter from the sea surface. *J. Appl. Meteorol.* **18**, 1023–1034.

Weissman, D. E., T. W. Thompson, and R. Legeckis (1980): Modulation of sea surface radar cross-section by surface stress: Wind speed and temperature effects across the Gulf Stream. *J. Geophys. Res.* **85**, 5032–5042.

34

MICROWAVE MEASUREMENTS OVER THE NORTH SEA

G. P. De Loor, P. Hoogeboom, R. Spanhoff, and J. Bruinsma

ABSTRACT. As reported in earlier IUCRM meetings, we investigate radar remote sensing as a tool for the control and study of the Dutch part of the North Sea. This is done in a program which comprises ground-based microwave measurements, and flights with a real-aperture digital SLAR. This program is called Project Noordwijk. We investigate the average (noncoherent) backscatter of the sea as a function of incidence angle, wind speed, and look direction, but also the modulation of the backscatter by sea waves and other phenomena such as: bottom topography, currents, eddies, and oil films. Some results of this program are reported.

1. INTRODUCTION

Within Rijkswaterstaat (Department of Water Control) the potential of microwave remote sensing is investigated as a tool for the observation, study, and control of the Dutch part of the Continental Shelf (North Sea). To provide a sound foundation for this work, a program of basic research (object–sensor interaction studies) is carried out. This program comprises ground-based as well as airborne experiments. In the following, some results for these experiments are reported; where appropriate, references are given to more extensive reports.

For the *ground-based* experiments a research platform is available in the North Sea (Noordwijk tower; see Fig. 1). It offers the opportunity to perform experiments with ground-based reflectometers in open sea. This tower is permanently instrumented for the measurement of pertinent meteorological and oceanographic data.

For the experiments with *airborne* equipment, an EMI real-aperture X-band side-looking airborne radar (SLAR) was used originally, providing images on film. As the program proceeded, a more accurate system was necessary and therefore a SLAR system with digital recording was developed.

To obtain experience with the *operational use* of SLAR, the purchase of a commercial SLAR is considered for the monitoring of oil spills.

G. P. De Loor and P. Hoogeboom ● Physics Laboratory TNO, The Hague, The Netherlands. R. Spanhoff and J. Bruinsma ● Ministry of Transport and Public Works, Rijkswaterstaat, Directorate for Water Management and Hydraulic Research, The Hague, The Netherlands.

FIGURE 1. Tower Noordwijk. The radars are mounted on the small platform under the helicopter deck on the left.

2. GROUND-BASED EXPERIMENTS

2.1. Equipment

The ground-based program is called Project Noordwijk (De Loor and Brunsveld van Hulten, 1978). Within this program, radar backscatter measurements have been made in the autumn of the years 1977, 1978, and 1979. In 1979, these experiments were associated with the MARSEN Experiment in the North Sea and the German Bight. Two other institutes have participated in these radar backscatter measurements on the Noordwijk tower by placing their own reflectometers there. They were: the Institut Francais du Pétrole in 1978 and 1979 with a four-frequency FM/CW scatterometer (1.5, 3, 4.5, 9 GHz) and the Remote Sensing Laboratory of Kansas University in 1979 with an FM/CW system operating in the band from 8 to 18 GHz. This cooperation offered the unique opportunity of intercomparison of the three systems. Our own radar is a short-range X-band FM/CW system. Its properties are given in Table I and a full description of this scatterometer is given by Smith (1978).

All radars were mounted at a height of 16 m above mean sea level on a small platform (under the helicopter deck in Fig. 1) which protruded from the tower to the NW, the prevailing wind direction.

2.2. Radar Backscatter Measurements

The radar data were recorded on magnetic tape (in analog form) together with simultaneous wave height and wind speed measurements over periods from 5 to 6 min. Occasionally, the

TABLE I.   The FM/CW X-Band Scatterometer

Frequency	X-band, used at 9.5 GHz
Frequency sweep	Variable, used at 200 MHz (1977), 75 MHz (1978), 100 MHz (1979)
Modulation waveform	Triangular
Modulation frequency	25–200 Hz; used at 50 Hz (1977), 150 Hz (1978), 200 Hz (1979)
Power transmitted	1 W
Polarization	VV, HH, VH, HV
Range	5–100 m
Antenna beamwidth	4.4°
Recording	Analog

recording was extended to 30 min. in order to achieve a better spectral resolution at the lower frequencies (< 1 Hz). With the aid of a spectrum analyzer and a microprocessor, the values for $\gamma\,(\sigma^\circ = \gamma \sin\theta)$ were available immediately after each measurement. Measurements were made for grazing angles θ between 20 and 90° at upwind, downwind, and cross-wind for a large variety of sea conditions and wind speeds (De Loor and Hoogeboom, 1982). Figure 2 gives an example of the results obtained. In this example, some of the French data ("+") are given for comparison (Fontanel $et\ al.$, 1979). Looking at the data we find good agreement with the French and the US data.

A considerable number of measurements of γ as a function of wind speed are now available. A linear relation of the form

$$\log \gamma \sim c \log v$$

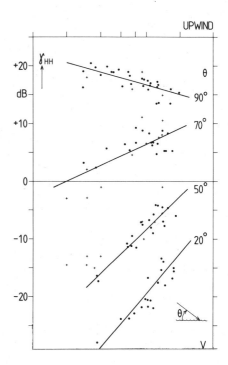

FIGURE 2. γ vs. wind speed v.   ● : our measurements; + : French data (Fontanel $et\ al.$, 1978).

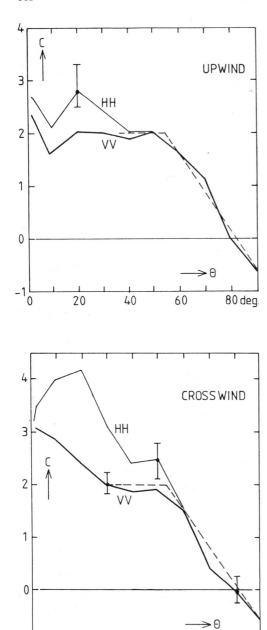

FIGURE 3. Exponent c of wind dependency of γ as a function of grazing angle.

provides a good fit to the data. Figure 3 shows c as a function of grazing angle θ for upwind and cross-wind conditions. Apart from our own measurements, use is made of data given in the literature to produce these curves (Moore and Fung, 1979; Kaupp and Holtzman, 1979; Fung *et al.*, 1978; Jones *et al.*, 1977; Sittrop, 1976). The data suggest that the exponent c increases linearly from -0.6 to 2 for grazing angles going from 90° (normal incidence) to 55° (the dashed line in Fig. 3), and is equal to 2 or larger for angles smaller than 55°, dependent on polarization and look angle.

Due to the coherent illumination and the movement of the sea surface, the reflection of the

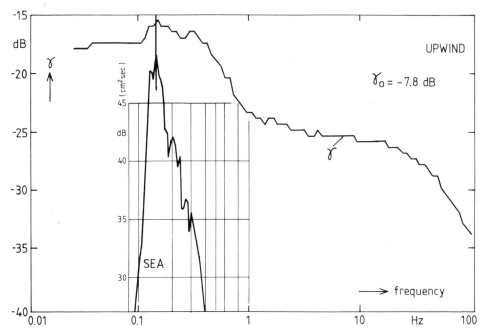

FIGURE 4. Spectrum of $\gamma(t)$ (example); inset: power spectrum of the sea as measured simultaneously with a wave staff. Measurement of Oct. 18, 1979, 1140–1210 MET; horizontal polarization, $\theta = 50°$; upwind, wind speed: 14.3 m/s, wind direction: 300°; $H_{1/3} = 2.5$ m.

radar waves comes from combinations of scatterers in the illuminated patch which vary continuously. Therefore, the instantaneous value of the backscattered, signal, γ, varies with time. By taking the FFT of a time series of $\gamma(t)$, the spectral distribution of $\gamma(t)$ can be determined. Figure 4 gives a typical example; also plotted is the power spectrum of the sea as determined from the simultaneously measured wave height.

To correlate the backscattered signal with the sea wave spectrum, we introduced the modulation index A (De Loor and Hoogeboom, 1982). Assuming only one (sinusoidal) sea wave with frequency f_0, we may write

$$\gamma(t) = \gamma_0[1 + A(f_0)\sin(2\pi f_0 t + \phi_0)]$$

where A is the modulation index. Figure 5 gives $A(f_0)$ as a function of grazing angle. $A(f_0)$ is now the ratio of the spectral density at the frequency f_0 of the main sea wave to that at frequency 0 Hz ($= \gamma_0 =$ the average value of γ). For grazing angles between 60 and 90°, $A(f_0)$ is practically independent of polarization whereas for grazing angles less than 60°, $A(f_0)$ is smaller for VV polarization than for HH polarization.

We used this (limited) approach for the rapid analysis of all of our data series, since it gives a quick insight into the relation between the backscattered signal and the sea wave spectrum. Work is now under way to determine the modulation transfer function as used by Alpers and Hasselmann (1978) and Wright et al. (1980).

To gain an insight into the times associated with the rearrangement of the scatterers in the illuminated patch, we determined the autocorrelation function. Figure 6 gives this function for the same example as in Fig. 4. Three regions can be distinguished. Since the time scales of these three regions are so different, it seems reasonable to assume that the mechanisms associated with them are uncorrelated; thus, we can treat them separately and determine the associated decorrelation times (De Loor and Hoogeboom, 1982). Region I with a decorrelation time of a little less than 10 ms seems to be associated with the capillary and small gravity waves in

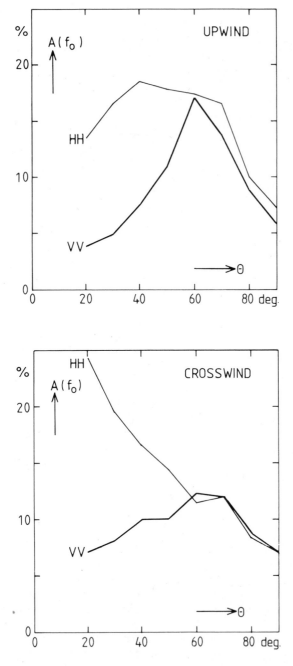

FIGURE 5. Modulation index $A(f_0)$ in % as a function of grazing angle.

resonance with the radar wave; region II with a decorrelation time between 0.3 and 1 s with the facets (Sittrop, 1976; Long, 1975); and region III with the main sea wave.

2.3. Oil Spills

Project Noordwijk also included measurements on oil spills. To monitor all measurements, an 8-mm ship radar was used. Its properties are given in Table II. It was mounted on the

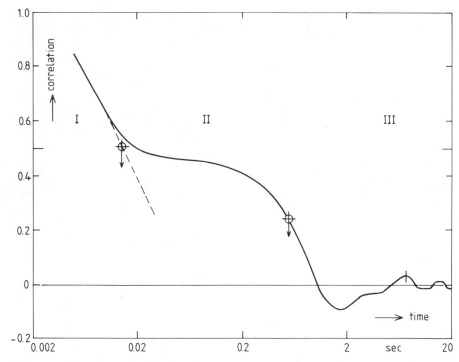

FIGURE 6. Autocorrelation function for the same measurement as in Fig. 4. (From De Loor and Hoogeboom, 1982.)

helicopter deck (Fig. 1) just above the scatterometers. Sea wave spectra and oil spills were monitored on the PPI (image) screen of this radar and recorded by photographing this screen. Figure 7 gives a typical example of such recordings. Both images were taken shortly after each other during an oil spill experiment. By photographing a single sweep of the radar (one antenna rotation, 1.5 s), the wave pattern is seen (Fig. 7a), but the oil spill does not stand out clearly. In the other image (Fig. 7b), the camera remained open for 100 sweeps (150 s); the sea waves are now blurred but the oil spill becomes distinct.

The data obtained in Project Noordwijk on oil spills are still in the process of evaluation. The results obtained so far confirm older work in this field in The Netherlands (De Loor and Brunsveld van Hulten, 1978; Van Kuilenburg, 1975).

TABLE II. Properties of the K_a-Band
Navigation Radar Phillips 8GR260/00

Frequency	36 GHz
PRF	2500 Hz
Antenna	Double "cheese":
beamwidth	hor. 36′, vert. 17°
Polarization	HH
Rotation rate	40 rev./min
Output	20 kW
Receiver MF	60 MHz
Pulse length	40 ns

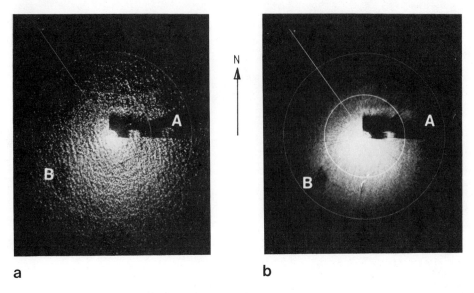

a   b

FIGURE 7. Examples of images obtained by photographing the PPI screen of the 8-mm ship's navigation radar. Dec. 12, 1978; wind speed 12.5 m/s, wind direction 194°; range setting 1500 m, range rings each 500 m. A: radar shadow due to screening by the platform; B: oil spill. (a) 1 sweep (rotation of the antenna): spill is vaguely visible; (b) 100 sweeps over each other (150 s): sea waves are blurred, but the oil spill stands out.

3. AIRBORNE MEASUREMENTS

SLAR imagery of sea "clutter" usually contains a lot of detail (De Loor, 1981). The capillary and short gravity waves—which cause the echo—can be modulated by the underlying longer sea waves, by swell, and other phenomena which thus become visible in the radar image. In Fig. 8

TABLE III.   The Dutch Digital SLAR

Frequency	9.4 GHz
Transceiver	
Peak power	25 kW
Pulse length	50 ns/250 ns
PRF	1 kHz
Antenna	
Length	2 m
Beamwidth	hor. 54', vert. 23°
Polar diagram	$cosec^2$ (approximately)
Dynamic range	70 db (special amplifier)
Digitalization	
Video bandwidth (after filter)	dc—5 MHz
Sampling frequency	20 MHz (50 ns)
Samples per line	2000
Digits	8
Lines/s	128
Pixel size	$15 \times 15\,m^2/30 \times 30\,m^2$
Range	15 km/30 km
Resolution	
Range (across track)	30 m
Azimuth (along track)	16 mrad

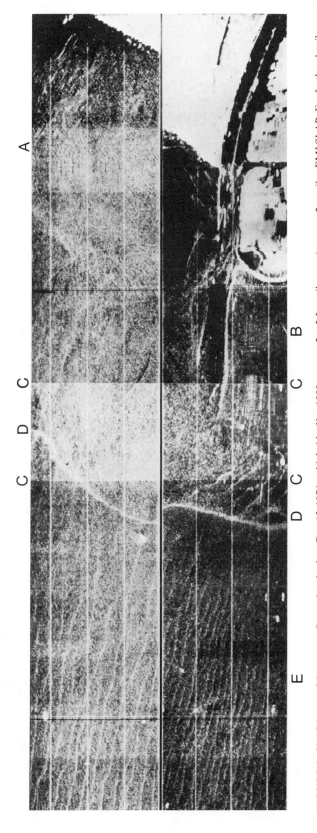

FIGURE 8. SLAR image of the sea near Rotterdam harbor; Dec. 13, 1974, at high tide. $H = 1000$ m; range 2×3.5 n. miles, gap in center: 2 n. miles. EMI SLAR. For further details see text.

a whole group of such phenomena are observed:

  A. Wave pattern, waves traveling to the coast
  B. Edge of current: the Rhine flowing into the North Sea. This edge moves with the tide
  C. At C the gain setting of the equipment was varied. By increasing the gain, details in the water become visible: sea waves, ship trails, etc.
  D. A boundary between the fresher water of the Rhine and the salt water of the North Sea
  E. Sand dunes on the sea bottom. They have a height on the order of 4 m and are at a depth of about 30 m. They become visible by their effect on the surface through the tidal current.

For more details the reader is referred to De Loor (1981) and De Loor and Brunsveld van Hulten (1978).

Phenomena as shown in Fig. 8 are in fact a modulation of the average backscatter of the sea.

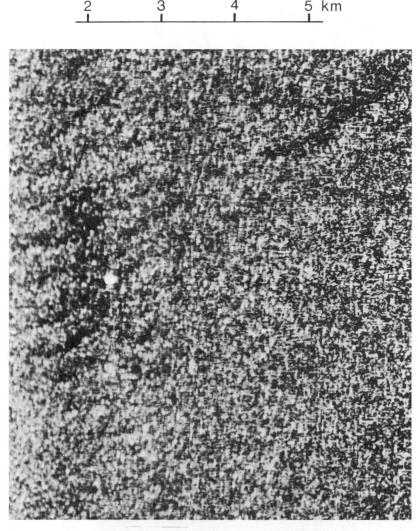

FIGURE 9. SLAR image of an oil spill with the new digital SLAR; June 17, 1980. The spill enhances the pattern made by the underlying bottom structure.

This modulation is usually small in intensity and thus a radar with a high radiometric resolution is necessary for the observation of these phenomena. In our new digital SLAR, particular attention has been given to radiometric resolution. A high radiometric resolution can be obtained by averaging over sufficient independent observations (looks) in one pixel. Our new SLAR (Table III) thus uses 30 independent observations in one pixel (Hoogeboom, 1981). This results in an important reduction in speckle as compared with most SAR systems which use one to four looks only.

Figure 9 gives an example of an image made with the new SLAR system during an oil spill experiment over the North Sea. The underlying bottom is visible also (compare feature E in Fig. 8). A remarkable feature is seen in this image. The lowest echoes are found parallel with the dune pattern: the oil spill enhances this pattern. Whether this also means that the oil has spread out in streaks along that pattern has not been checked.

4. TOWARD OPERATIONAL USE

The final goal of all this work is the operational use of microwave remote sensing for the control of the Dutch part of the North Sea. The experience obtained in the experiments summarized in the foregoing sections will be used in an experiment in routine operation which is now being planned. Within Rijkswaterstaat there exists an operational branch: the Oil Watch. This branch now uses a small aircraft and monitors the sea by visual inspection. To extend the capabilities of this branch, working at longer ranges, at night and under conditions of poor visibility, the purchase of a commercial SLAR is being considered as a very useful first step toward the routine use of microwave remote sensing for the observation, study, and control of the North Sea.

REFERENCES

Alpers, W., and K. Hasselmann (1978): The two-frequency microwave technique for measuring ocean-wave spectra from an airplane or satellite. *Boundary-Layer Meteorol.* **13**, 215–230.

De Loor, G. P. (1981): The observation of tidal patterns, currents, and bathymetry with SLAR imagery of the sea. *IEEE J. Ocean. Eng.* **OE-6**, 124–129.

De Loor, G. P., and H. W. Brunsveld van Hulten (1978): Microwave measurements over the North Sea. *Boundary-Layer Meteorol.* **13**, 119–131.

De Loor, G. P., and P. Hoogeboom (1982): Radar backscatter measurements from platform Noordwijk in the North Sea. *IEEE J. Ocean. Eng.* **OE-7**, 15–20.

Fontanel, A., N. Lannelongue, and D. de Staerke (1979): *Etude de l'état de la mer par utilization des hyperfréquences—Expériences RANO—Mer du Nord.* Premier rapport, Institut Francais du Pétrole.

Fung, R. K., A. K. Fung, G. J. Dome, and I. J. Birrer (1978): *Estimates of oceanic wind speed and direction using orthogonal beam scatterometer measurements and comparison of recent scattering theories.* NASA Contractor Report 158908, Remote Sensing Laboratory, Kansas University.

Hoogeboom, P. (1981): Preprocessing of airborne remote sensing data. *Proceedings, Ninth Annual Conference of the Remote Sensing Society—Matching remote sensing technologies and their applications,* London, Dec. 16–18, 410–417.

Jones, W. L., L. C. Schroeder, and J. L. Mitchell (1977): Aircraft measurements of the microwave scattering signature of the ocean. *IEEE Trans. Antennas Propag.* **AP-25**, 52–61.

Kaupp, V. H., and J. C. Holtzman (1979): Skylab scatterometer measurements of hurricane AVA: Anomalous data correction. *IEEE Trans. Geosci. Electron.* **GE-17**, 6–13.

Long, M. W. (1975): *Radar Reflectivity of Land and Sea,* Lexington Books, Lexington, Mass.

Moore, R. K., and A. K. Fung (1979): Radar determination of winds at sea. *Proc. IEEE* **67**, 1504–1521.

Sittrop, H. (1976): Characteristics of clutter and targets at X- and K_u-band. *Proceedings, AGARD Symposium—New devices, techniques and systems in radar,* The Hague, June 14–17.

Smith, M. K. (1978): Radar reflectometry in The Netherlands: Measurement system, data handling and some results. *Proceedings, Conference on Earth Observation from Space and Management of Planetary Resources,* 377–387, ESA Publication ESA-SP 134.

Van Kuilenburg, J. (1975): Radar observation of controlled oil spills. *Proceedings, Xth International Symposium on Remote Sensing of Environment*, Ann Arbor, Oct. 6–10, 243–250.

Wright, J. W., W. J. Plant, W. C. Keller, and W. L. Jones (1980): Ocean wave-radar modulation transfer functions from the West Coast Experiment. *J. Geophys. Res.* **85**, 4957–4966.

DISCUSSION

D. WEISSMAN: A regular pattern appeared in the slick area (Fig. 9) that may be due to swell. Please comment on why it appears there and not in the surrounding area.

DE LOOR: The regular pattern is not due to swell, but again you see the underlying dune pattern of the sea bottom. Such patterns appear under specific combinations of wind and current. Here the oil spill enhances the pattern. Whether this also means that the oil has finally spread out in streaks along that pattern has not yet been checked.

TRIZNA: The autocorrelation function (Fig. 6) shows three different regions. Can you give an explanation?

DE LOOR: Region I is associated with the capillary and short gravity waves in resonance with the radar wave; region II with the facets; and region III with the main sea wave.

KOMEN: I like to look at the spectrum (Fig. 4). There you see the occurrence of a flat shoulder to the right of the peak. I think this shoulder calls for an explanation. W. Rosenthal and I have looked into that problem and we found a possible explanation in terms of the Doppler shift due to the orbital motion in a random wave field. It leads to an additional broadly peaked contribution to the spectrum, the width of which is related to the width of the Gaussian probability distribution of the orbital motions. Its position is at such a frequency that it could perhaps explain the observed shoulder.

ALPERS: (1) Why did you calculate the modulation index only at the peak frequency? Was the coherence too low for all other frequencies?

(2) How large was the footprint in your measurements? The size of this area strongly influences the statistical behavior of the backscattered signal, in particular the shape of the autocorrelation function.

BRUINSMA: (1) We have only taken the peak frequency because we have not had enough time. We intend to calculate the modulation transfer function also for other frequencies. As you saw, we use a slightly different definition for the modulation index than you do. Work is under way to determine the modulation transfer function as used by Alpers and Plant. The coherence between the orbital velocity and the backscatter cross section is unknown since we have an incoherent radar.

(2) The size of the footprint is 2×2 m. The sensitivity of the size of the footprint to the correlation function has been studied extensively. It turned out that it hardly influences the slope of the correlation function.

35

SOME SKYWAVE RADAR MEASUREMENTS OF WIND VECTORS AND WAVE SPECTRA

Comparison with Conventional Data for JASIN 1978

P. E. DEXTER AND S. THEODORIDIS

ABSTRACT. Observations of ocean backscatter spectra were made with a skywave radar during the Joint Air Sea Interaction (JASIN) Experiment in July and August 1978. This chapter is concerned with a detailed analysis of the results from four days of the experiment, July 20–21 and August 7–8, and with a comparison of radar-derived ocean parameters (sea surface wind vectors and nondimensional wave spectra) with equivalent parameters determined by conventional ship and buoy instruments. Because of the ionospheric propagation mode employed, contamination of the radar return signal due to ionospheric motions causes some degradation of radar Doppler spectra and radar-derived ocean parameters. In general, however, mean agreement to within 10% for all parameters except wave spectra was found over the four days, while the radar-derived surface wind fields fitted well to the mesoscale flows determined by conventional meteorological analysis.

1. INTRODUCTION

Observations of ocean backscatter spectra were made with an HF skywave (or over-the-horizon, OTH) radar during the Joint Air Sea Interaction (JASIN) experiment, from July to September 1978, in 18 resolution cells (pixels) covering the Rockall Bank area in the northeast Atlantic. Details of the radar facility and an overall view of the experiment may be found in Shearman *et al.* (1977), Shearman (1980), and Shearman and Wyatt (1982). This chapter concentrates on a detailed analysis of the results from four days of the experiment, July 20–21 and August 7–8. It contains a comparison between radar-derived sea surface parameters (wind vectors, significant wave heights, spectral model frequencies, and wave spectra) and the same parameters obtained from *in situ* ship and buoy observations.

During JASIN, radar data were recorded on three bearings spaced 7° apart in azimuth, and from six range cells per bearing with a slant-range increment of 75 km. Propagation to the

P. E. DEXTER AND S. THEODORIDIS ● Department of Electronic and Electrical Engineering, University of Birmingham, Birmingham, England. *Present address for P.E.D.:* Bureau of Meteorology, Melbourne, Australia. *Present address for S.T.:* Chair of Telecommunications, School of Electrical Engineering, University of Thessaloniki, Thessaloniki, Greece.

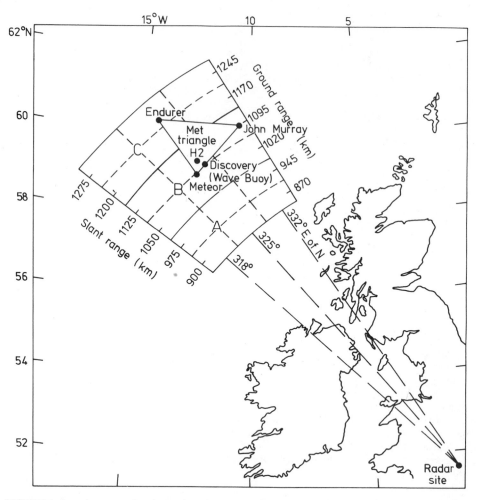

FIGURE 1. Location map showing radar observation cells in relation to the JASIN meteorological triangle. Radar bearings, slant and ground ranges are indicated, as well as the positions of the principal observing platforms during JASIN.

scattering area was generally via the ionospheric E_s layer. There was thus a total spatial coverage of around $450 \times 450 \, km^2$ which included the main JASIN meteorological triangle (Fig. 1). Data were recorded in blocks of approximately 10 min per azimuth on a total of 17 days during JASIN, at 1000, 1300, and 1600 GMT each day. Each 10 min of analog data was digitized into 15 overlapping blocks (50% overlap) of 1024 data points, spectral analysis performed, and the resulting 15 spectra averaged incoherently for each scattering cell, with a final spectral resolution of 0.03 Hz. Examples of such spectra, for 1012 GMT on July 21 (radar bearing 325°) and 1001 GMT on August 8 (radar bearing 318°), are shown in Figs. 2a–d. The spectra in each figure represent scattering cells at various slant ranges as indicated. Such spectra formed the basic radar data set for further analysis.

2. WIND AND WAVE PARAMETER EXTRACTION

While two of the Doppler spectra in Fig. 2 have particularly well-defined second-order structure in the region of the first-order (Bragg) peaks and thus are likely to provide good

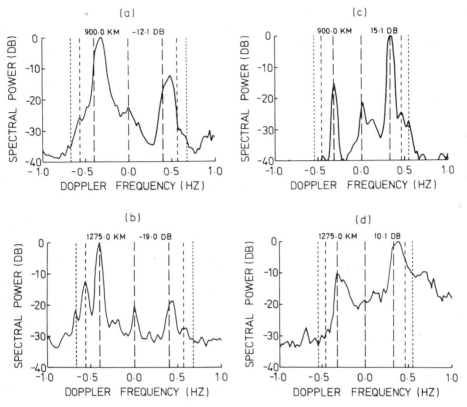

FIGURE 2. Doppler spectra for two range cells obtained at 1012 GMT on July 21, 1978, for radar bearing 325° [(a) and (b)], and for two range cells obtained at 1001 GMT on August 8, 1978, radar bearing 318° [(c) and (d)]. Spectra represent 15 time averages (with 50% overlap). Slant ranges are indicated.

estimates of the required sea surface parameters, the others show evidence of (often considerable) ionospheric contamination. This typifies the principal problem in the utilization of HF skywave radars: how often may we obtain estimates of particular ocean parameters with a skywave radar, and how good are these estimates, in the (inevitable) presence of some ionospheric contamination? This question has been discussed in detail by Georges (1980), who provides a hierarchy of sea surface parameters which may be extracted from radar Doppler spectra. In order of increasing difficulty of extraction, these are:

1. Surface wind and high-frequency wave propagation direction
2. Significant (or rms) wave height and spectral mode frequency
3. Surface wind speeds
4. Nondirectional ocean wave spectra
5. Directional wave spectra.

Such a hierarchy has been borne out from our results so far analyzed from JASIN. A total of 174 spectral measurements were made during the four days under consideration here. Of these, some 91% yielded wind direction estimates, 31% significant wave height and wind speed estimates, and 15% nondirectional wave spectral estimates. In each case, spectra were rejected, for the purposes of further analysis, on the basis of the degree of apparent ionospheric contamination. Thus, the spectra from which wind direction estimates could not be obtained (9%) were without identifiable first-order peaks; those from which significant wave height and wind speed estimates could not be obtained (69%) had first-order peak widths (at − 10 db) greater than 0.15 Hz and peak-to-noise ratios less than 35 db; and those from which nondirectional wave

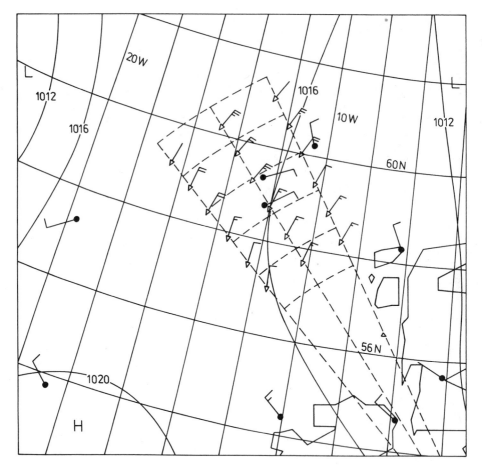

FIGURE 3. Comparison of radar-derived surface wind vectors obtained for 1000–1030 GMT on July 20, 1978, with surface meteorological analysis for 1200 GMT the same day. Ship winds in the JASIN area are for 1000 GMT. Feathers on the arrows follow the usual meteorological notation, with each half-feather indicating 2.5 m/s wind speed, each full-feather 5 m/s wind speed.

spectral estimates could not be obtained (85%) were without identifiable second-order features at $2^{1/2}f_B$ and $2^{3/4}f_B$ (where f_B is the frequency of the first-order or Bragg peak).

Wind directions have been extracted from the radar Doppler spectra by employing the technique of Stewart and Barnum (1975). This technique utilizes the ratio of the two first-order (Bragg) lines (at positive and negative Doppler frequencies; Fig. 2) only, and is therefore much less subject to ionospheric contamination than procedures for other parameters which must depend on measurements of the second-order spectral structure (Barrick, 1972). In this case, Stewart and Barnum's (1975) "s" parameter has been set at 2 or 4 depending on whether the mean wind speed over the whole area was less than or greater than 8 m/s.

Wave heights (rms), spectral mode periods, and nondirectional wave spectra have been extracted using Barrick's (1977a, b) inversion procedure. Barrick has derived closed-form inversion equations relating these parameters to the ratio of weighted second- to first-order powers in the Doppler spectrum. These equations allow the rapid computation of rms surface displacement H (and hence significant wave height H_s), mean wave period T (and hence spectral mode period T_m), and the wave spectrum $S(f)$.

Surface wind speeds have been estimated through a wind-wave model which uses the derived H_s, T_m from Barrick's inversion formulas to determine both mean surface (10 m) wind

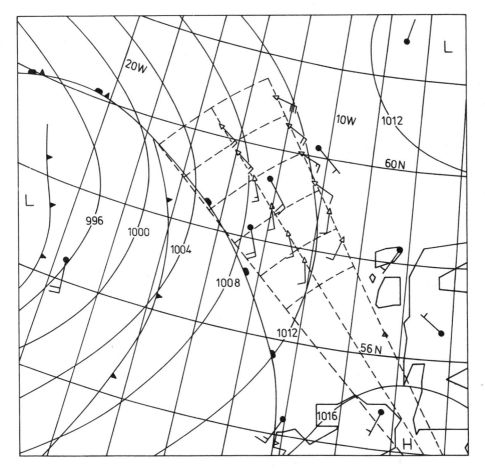

FIGURE 4. As for Fig. 3, July 21, 1978.

speed, and space and time scales for generation of the observed sea state (Dexter and Theodoridis, 1982). The model does not depend on the type of inversion procedure used to obtain H_s and T_m, and assumes only that the measured values apply to a simple wind-driven sea.

3. COMPARISONS WITH JASIN SURFACE DATA

Comparison data from the JASIN surface data set were kindly supplied by Dr. T. Guymer (surface winds), Dr. D. Webb (directional wave spectra from "Discovery" pitch-and-roll buoy), and Dr. R. Stewart (averaged significant wave heights and spectral mode periods from wave buoys). All the comparison data were available only from indicated positions within the meteorological triangle (Fig. 1). The surface winds were all corrected to the 10-m level and standardized to a single platform. Other available data were the normal six hourly meteorological surface analyses, with scattered wind observations in the general vicinity.

General wind-field comparisons for each of the four days are shown in Figs. 3 to 6 (July 20, 21 and August 7, 8, respectively). Radar observations were obtained between 1000 and 1030 GMT on each day, with the surface meteorological pressure analyses being for 1200 GMT in each case. JASIN ship wind observations for 1000 GMT are also shown, where available. The radar-derived wind vectors correspond well with the general surface flow in all cases except in the near-calm conditions occurring at the outermost ranges on August 8 (Fig. 6). Indeed, it is

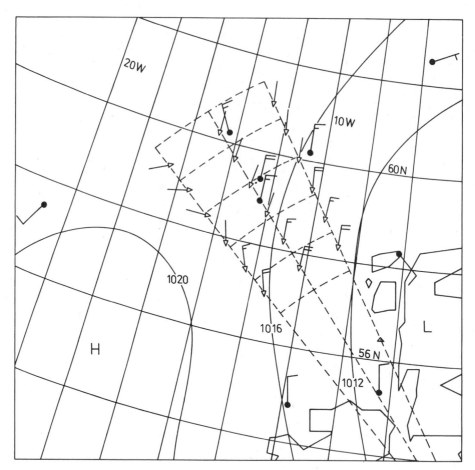

FIGURE 5. As for Fig. 3, August 7, 1978.

apparent that considerably more detail on mesoscale wind-field structure is contained in the radar data than may be obtained from conventional sources. Principal errors occur for the radar-derived wind speeds where ionospheric contamination results in some overestimation.

Summaries of all radar wind observations made during the four days are shown in Figs. 7 and 8. These show, respectively, scatter diagrams for radar-measured wind speeds and wind directions versus surface observations. Radar winds are, of course, averaged over the scattering cell of approximately 75 × 150 km. Ship observations for comparison have also been averaged— over corresponding scattering cells, where possible, and over a 100-km fetch upstream of the scattering cell. [The length and time scales appropriate to such averaging have been discussed in detail by Dexter and Theodoridis (1982).] Solid lines in each figure represent the lines of perfect fit, with the dashed lines being, respectively, ± 2 m/s and ± 15° from perfect fit. These latter numbers have been selected as those claimed for SEASAT surface wind vector measurements.

The directional fit is obviously good (Fig. 8), the one very poor comparison occurring for the light wind situation discussed previously (Fig. 6). The wind speed fit is poorer, as expected—the generally low sea state conditions prevailing make the effects of ionospheric contamination on wind speed estimates relatively greater. In addition, if a spectral quality indicator such as that of Maresca and Georges (Georges, 1980) is employed, or some similar procedure adopted to reject additional contaminated spectra, the general agreement may conform more closely to the ± 2 m/s shown.

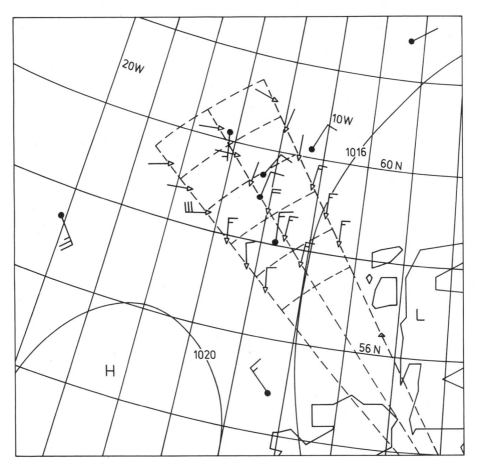

FIGURE 6. As for Fig. 3, August 8, 1978.

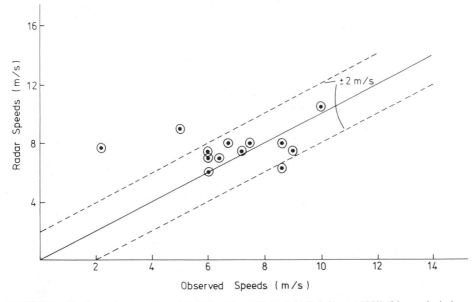

FIGURE 7. Wind speed comparisons for the four days (20–21 July, 7–8 August 1978). Observed winds are averaged over the coincident scattering cell, where possible, and over fetches of 100 km upwind from the cell. The dashed lines are at ± 2 m/s from perfect fit (solid line).

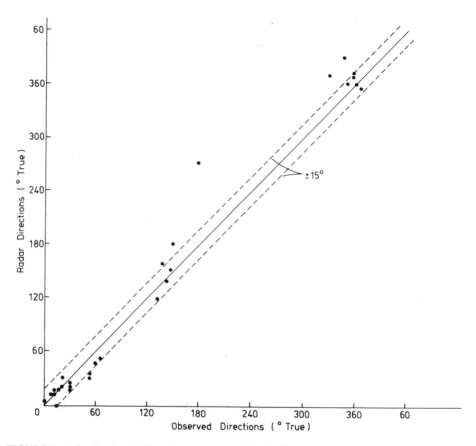

FIGURE 8. As for Fig. 7, wind direction comparisons. Dashed lines are at $\pm\,15°$ from perfect fit (solid line). Directions are given in the meteorological sense, i.e., from which the wind is blowing.

Wave spectral comparisons are available only for 1000 GMT on July 20 and August 8, and these are shown in Figs. 9 and 10, respectively. Conventional wave data were obtained from a pitch-and-roll buoy operated by D. Webb of the Institute of Oceanographic Sciences from "Discovery" (Fig. 1). In the first comparison (Fig. 9), the buoy was located within the radar scattering cell. Spectral amplitudes and total spectral energies are almost identical, but there is a discrepancy in spectral mode frequencies of around 0.05 Hz which is difficult to explain physically. However, spectral energies are very low (Significant wave heights less than 1.5 m), and the effects of ionospheric distortion likely to appear disproportionately large—in this case resulting in an error in mode frequency.

In the second comparison (Fig. 10), the radar scattering cells were located some 100–200 km downwind of the buoy. Again, winds were light, and spectral energies very low (significant wave heights less than 1.0 m). The increased energy levels in the radar-derived spectra resulted from both the increased fetch lengths and the probable effects of contamination (increasing with radial range from the radar). However, spectral mode frequencies are nearly coincident and a second, low-frequency, spectral peak is apparent in both sets of spectra.

4. CONCLUSIONS

A summary of all comparisons made between conventional and radar-derived significant wave heights (H_s), spectral mode frequencies (f_m), and surface wind speeds (V) for August 7 and 8

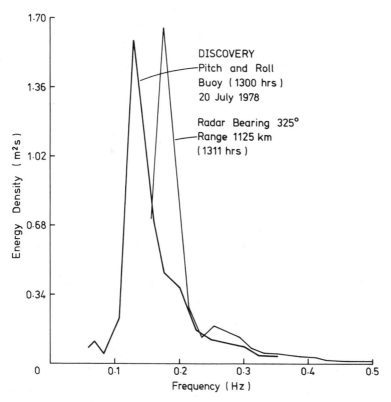

FIGURE 9. Comparison of radar-derived and buoy-measured nondirectional wave spectra for 1300 GMT, July 20, 1978. Wave buoy was located within the radar scattering cell.

is given in Table I. Radar observations were made between 1000 and 1030 GMT each day, and are averaged over all scattering cells from which data could be extracted (or over the two innermost cells on each bearing in the case of V, since the data were considered most reliable there). Wave buoy data (supplied by R. Stewart of Scripps) were averaged over all buoys in the JASIN area on each day. H_s is available as a whole-day average or for the time 1000–1030 GMT,

TABLE I. Summary of Comparisons—August 7 and 8, 1978[a]

	H_s(m)			f_m(Hz)		V(m/s)	
	Av. radar 1000–1030	Av. buoys 1000–1030	Av. buoys Whole day	Av. radar 1000–1030	Av. buoys Whole day	Av. radar 1000–1030	Av. JASIN 0600–1000
August 7	1.69	1.60	1.50	0.153	0.155	8.7	7.5
August 8	1.29	1.25	1.35	0.149	0.125	7.0	6.35

[a]In this configuration the comparisons are obviously good, being within 10% in most cases. The exceptions are wind speed on August 7, when the inclusion of poorer-quality (contaminated) Doppler spectra has obviously degraded the radar average, and f_m on August 8 when winds were lighter and the effects of contamination more severe. In all cases the trends in the observations, from August 7 to 8, agree well.

In general, results from the limited portion of the JASIN data set so far analyzed show good agreement between radar and conventional observations. While it is perhaps fortuitous that it was possible to make the radar observations using stable sporadic E-layer propagation, these results do demonstrate clearly the possibilities for using an HF skywave sea state radar in a general observational role, particularly when proper radar management techniques are employed (e.g., Georges, 1980).

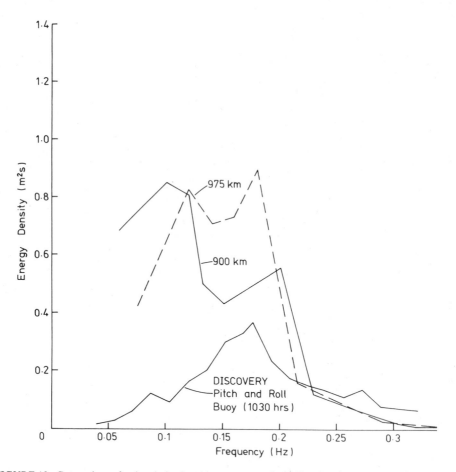

FIGURE 10. Comparison of radar-derived and buoy-measured nondirectional wave spectra for August 8, 1978. Radar scattering cells were at slant ranges of 900 and 975 km from the array, on bearing 318°, these being some 100–200 km downwind of the buoy location. Observations were made at 1000 GMT (radar) and 1030 GMT (buoy).

while f_m is given as a whole-day average only. Following Dexter and Theodoridis (1982), JASIN surface wind speeds for comparison were averaged over the time period 0600–1000 GMT each day, and over fetches of 100 km upwind of the radar scattering cell in each case.

ACKNOWLEDGMENTS. Skywave radar measurements during JASIN were made as part of the joint project of the University of Birmingham and the Appleton Laboratory, with the support of the U.K. Science Research Council. We thank colleagues in both institutions for their contributions to this work, in particular Professor E. D. R. Shearman, Dr. W. A. Sandham, and Mr. D. J. Bagwell (B.U.), and Dr. E. N. Bramley and Mr. P. A. Bradley (A.L.). One of us (P.E.D.) acknowledges the support of the Australian Bureau of Meteorology and the Royal Society, the other (S.T.) that of the Science Research Council.

REFERENCES

Barrick, D. E. (1972): Remote sensing of sea state by radar. *Remote Sensing of the Troposphere* (V. Derr, ed.), U.S. Government Printing Office, Washington, D.C.

Barrick, D. E. (1977a): The ocean wave height nondirectional spectrum from inversion of the HF sea-echo Doppler spectrum. *Remote Sensing Environ.* **6**, 201–227.

Barrick, D. E. (1977b): Extraction of wave parameters from measured HF radar sea-echo Doppler spectra. *Radio Sci.* **12**, 415–424.

Dexter, P. E., and S. Theodoridis (1982): Surface wind speed extraction from HF skywave radar Doppler spectra. *Radio Sci.* **17**(3), 643–652.

Georges, T. M. (1980): Towards a successful skywave sea-state radar. *IEEE Trans. Antennas Propag.* **AP28**, 751–761.

Shearman, E. D. R. (1980): Remote sensing of the sea-surface by dekametric radar. *Radio Electron. Eng.* **50**, 611–623.

Shearman, E. D. R., and L. R. Wyatt (1982): Shore-based dekametric radar for surveillance of sea-state and oceanic winds. *J. Inst. Navig.* **35**(3).

Shearman, E. D. R., D. J. Bagwell, and W. A. Sandham (1977): Progress in remote sensing of sea-state and oceanic winds by HF radar. *Proc. Conf. "Radar 77,"* London, IEE Conf. Publ. No. 155, 41.

Stewart, R. H., and J. R. Barnum (1975): Radio measurements of oceanic winds at long ranges: An evaluation. *Radio Sci.* **10**, 853–857.

36

STUDY OF THE MODULATION BY CORRELATION IN THE TIME AND FREQUENCY DOMAINS OF WAVE HEIGHT AND MICROWAVE SIGNAL
Preliminary Results

DANIELLE DE STAERKE AND ANDRÉ FONTANEL

ABSTRACT. This chapter presents some preliminary results on the modulation of backscattered microwave signals over a range of wave heights and wind speeds, as a function of several radar parameters (angle of incidence, polarization and frequency). The spectrum of the backscattered signal generally has a peak at the dominant ocean wave frequency, though significant peaks are observed at other frequencies as well. These, presumably, result from the non-linearities in the relationship between wave spectrum and radar return spectrum.

Backscattering of microwaves at the ocean surface at oblique incidence angles is primarily caused by Bragg scattering from short ripple waves, which have wavelengths comparable to the wavelengths of the microwaves. Therefore, to first order, the backscattered microwave signal only contains information about the small-scale structure of the ocean surface. However, these small-scale structures are modified by the long ocean waves and thus cause a modulation of the backscattered microwave power. This modulation permits the detection of long ocean waves by microwave techniques.

Measurements obtained from techniques such as imaging radars may be converted into ocean surface data (wave spectra) if the modulation transfer function is known.

1. THE THEORY

The theoretical model presented by Wright *et al.* (1980) and Alpers *et al.* (1978) describes the modulation as linear, and is essentially based on the two-scale ocean wave model of Wright (1966) in which Bragg scattering by short waves is the dominant mechanism for radar return from the sea surface. The radar echo is determined by the number of ocean waves (Bragg waves) with component of wavelength along the radar line of sight equal to one-half the radar wavelength. The modulation of these Bragg waves by the long waves is attributed to two effects:

1. The *tilt modulation* which is due to the purely geometric effect that Bragg scattering waves are seen by the radar at different local incidence angles depending on their

DANIELLE DE STAERKE AND ANDRÉ FONTANEL ● Institut Francais du Petrole, PP 311 92506 Rueil Malmaison, France.

location on the long wave. This tilt modulation is practically the same for L and X band, but is found to be larger for HH than for VV polarization

2. The *hydrodynamic contribution* is characterized by a nonuniform distribution of the short waves with respect to the long ocean wave field, which is attributed to interactions between short and long waves. The result is a modulation of the amplitude and the frequency of the small-scale waves (for example, the straining of the short waves by the horizontal component of the orbital speed of the larger waves).

This model is valid for low to moderate sea states. But it is evident, from data gathered during the experiments at sea—JONSWAP Experiment, West Coast Experiment, and from the preliminary results of MARSEN—that the modulation for higher sea states is nonlinear and might be largely due to some other complementary nonlinear sources such as breaking waves.

2. THE EXPERIMENT

One of the objectives of our experiments was to study the modulation (and to calculate the modulation transfer function) for a large range of wave heights and wind speeds and as a function of several radar parameters (e.g., angle of incidence, polarization, frequency). The principal set of data processed was acquired during MARSEN experiments (fall 1979); an additional set of data was gathered during fall 1980. In both cases, measurements were made from the Noordwijk tower. This tower, which is permanently instrumented for the measurement of meteorological and oceanographic data, is located (41°18′E, 52°16′N) off the coast of Holland in water depth of about 20 m (Fig. 1).

FIGURE 1. Noordwijk tower.

3. THE SENSOR: RAMSES SCATTEROMETER

The scatterometer RAMSES (Radar Multifrequences Sol pour l'Etude des Signatures Spectrales) was operated from a small platform, under the helicopter deck, located at an altitude of 16 m above the mean sea level. As the wind and wave conditions at the site are from the westerly direction, this platform has been built outside the tower to the NW, so the radars were operating away from the legs of the tower by at least 8 m when looking vertically downward. The scatterometer RAMSES is a ground-based radar, frequency modulated, continuous wave type. It can be operated at the following frequencies: 1.5, 3, 4.5, and 9 GHz, for the four combinations of polarization and at incidence angle varying from 0 to 80° (Fig. 2).

The wave height was measured simultaneously with the microwave backscattered power by two wave staffs, a vertical and an oblique one (Fig. 3), but the oblique one ceased to function after a while so we had to use the vertical one. Each record was of about 20 min. They were recorded on an analog magnetic tape recorder.

4. THE RESULTS*

4.1. Signal Processing

The time series analyzed have a length of 15 to 20 min. The analog wave staff and radar cross section series were sampled at a frequency of 2.5 Hz; 256 points were used for each FFT. In this way, between 8 and 10 spectra were obtained for each case. These spectra were then averaged.

*Preliminary results are presented in this section.

FIGURE 2. RAMSES scatterometer.

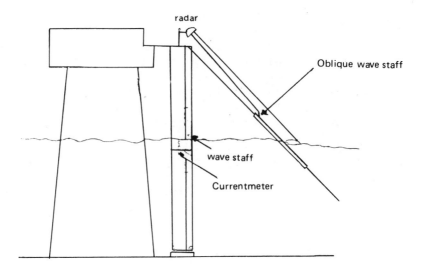

FIGURE 3. Experimental configuration of radar and wave staffs on Noordwijk.

4.2. Time Domain Results

4.2.1. Time Series

Examples of time series for the oblique wave staff and the backscattered radar signal are shown in Fig. 4. In these two cases, the wave staff is in the radar beam, at an angle of 50° to the vertical.

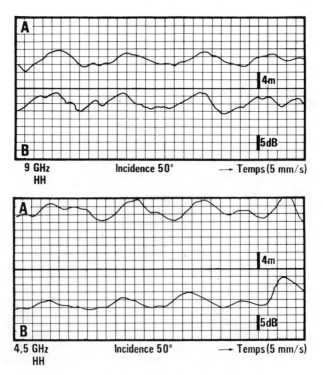

FIGURE 4. Time series. (A) Waves; (B) backscattered signal.

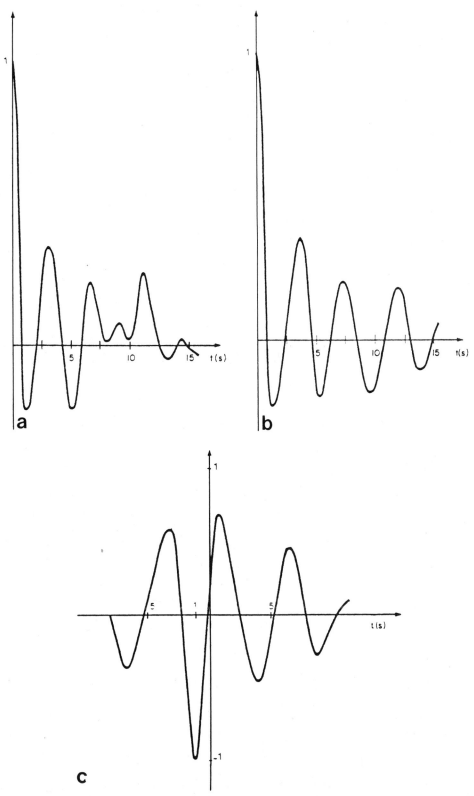

FIGURE 5. (a) Normalized backscattered signal autocorrelation ($\theta = 20°$, $F = 9$ GHz, upwind, VV).
(b) Normalized wave autocorrelation. (c) Normalized crosscorrelation of the two time series.

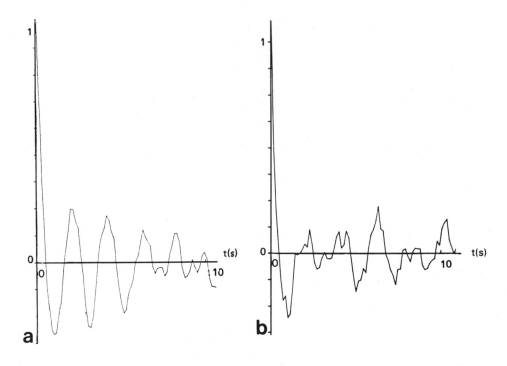

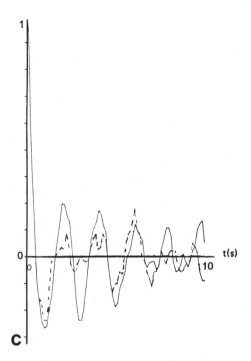

FIGURE 6. (a) Normalized autocorrelation of backscattered signal for upwind measurements ($\theta = 30°$, $F = 9\,\text{GHz}$, VV). (b) Normalized autocorrelation of backscattered signal for cross-wind measurements ($\theta = 30°$, $F = 9\,\text{GHz}$, VV). (c) Comparison of the two autocorrelations.

For these cases, the two traces are clearly of the same form, although the radar signal (B) leads the wave staff signal by *less than 1 s*. For waves with period of 6 s, this is equivalent to a phase difference of *less than 60°*.

4.2.2. Autocorrelation

Autocorrelation is another way of presenting the time records, in which the periodicity of the data is emphasized. We can see in Figs. 5a and b that the periodicity of the wave and radar records is of the same order, about 4 s, and that the periodicity of the radar record is destroyed faster than that of the waves. The time delay between the two records can be obtained by cross correlation (Fig. 5c). This delay is found to be of the order of 1 s. But as the wave staff was vertical, it is necessary to take into account the time delay resulting from the 9-m separation of the two measurements. This is about 2/3 s* so that after this correction, the time delay between the two records at the same point is about 1/3 s, equivalent to a phase difference of *about 30°* (the wave period being 4 s).

*In shallow water the wave velocity (v) is given by $v = (gH)^{1/2}$. H, the depth, is equal to 20 m, so $v = 14$ m/s. The time necessary for the wave to propagate a 9-m distance is equal to $\frac{2}{3}$ s, as the wave propagation was parallel to the line of sight of the radar, from the radar to the wave staff.

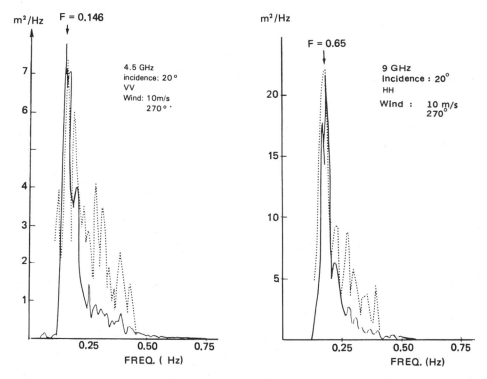

FIGURE 7. Comparison of wave height spectra and backscattered signal spectra for upwind measurements. (—) Wavestaff spectrum, (····) Backscattered signal spectrum.

Figure 6 compares the autocorrelation of the backscattered signal of an upwind measurement (a) and a cross-wind one (b). We can see that the periodicity of the upwind case cannot be found in the crosswind one, though the zero crossing (c) are the same for the two cases.

4.3. Frequency Domain Qualitative Results

4.3.1. Upwind Cases

The examples of upwind cases shown in Fig. 7* were recorded on the same day. We can see that (1) the wave height and radar signal spectra have a main, nearly monochromatic (narrow) peak, which occurs at the same frequency, (2) the backscattered power spectrum is broader than the wave height spectrum, and (3) secondary peaks occurring at frequencies higher than the main peak seem to be more pronounced in the backscattered power spectrum. This is particularly obvious in Fig. 7a where although there is a peak in the radar spectrum corresponding to the main one in the wave height spectrum, another radar peak occurs at a higher frequency.

However, for the cross-wind and downwind cases, the main peaks do not occur at the same frequency as for wave height.

---

*The vertical scale shown for the figures is that for the wave height spectra. The radar spectrum has been shown at the same size for comparison.

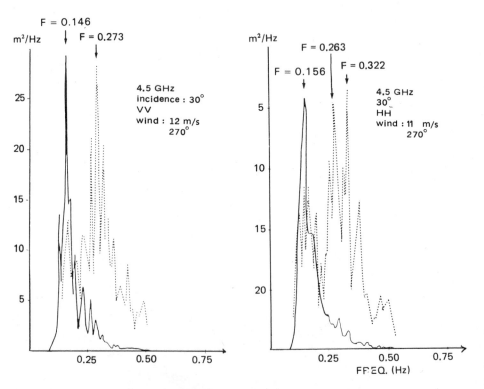

FIGURE 8. Comparison of wave height spectra and backscattered signal spectra for cross-wind measurements. (—) Wavestaff spectrum, (····) Backscattered signal spectrum.

4.3.2. Cross-Wind

The main peak in the radar spectrum occurs at a frequency almost twice that of the wave frequency (Fig. 8) and seems to be a harmonic of the wave frequency.

4.3.3. Downwind

The picture here is more complicated.* The main peak in the wave height spectrum occurs at a frequency of 0.127 and is most likely a swell peak. The two main peaks (of equal intensity) in the radar spectrum occur at higher frequencies: one at 0.175 which corresponds to one of the secondary peaks in the wave height spectrum and the second at 0.283 (Fig. 9).

*Downwind measurements were made in the shadow of the platform, so the waves might be disturbed in these cases.

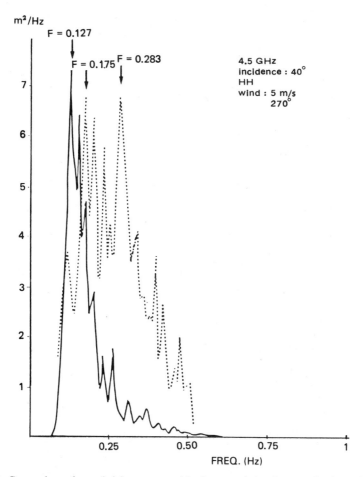

FIGURE 9. Comparison of wave height spectra and backscattered signal spectra for downwind measurements. (—) Wavestaff spectrum, (····) Backscattered signal spectrum.

Finally, although for upwind cases the most powerful frequency peaks in the backscattering spectra are at the dominant wave frequency, this is not true for cross-wind cases where powerful peaks might be found at higher frequencies (mostly at twice the dominant wave frequency). Thus, for cross-wind cases it does not seem possible to infer the wave spectrum from the backscattering spectrum, using a *linear relationship*, so that in the following quantitative results we only take into account upwind cases.

4.4. Frequency Domain Quantitative Results

Coherence: The coherence of the measurements is particularly high for the main frequency peaks, often higher than 0.9.

Phase: The phase difference varies, when corrected for the distance between the radar beam axis and the wave staff, between 40 and 70°.

Modulation transfer function: The modulation transfer function we calculated is the dimensionless form, usually denoted by m. From Figs. 10 and 11 we can see that m decreases with

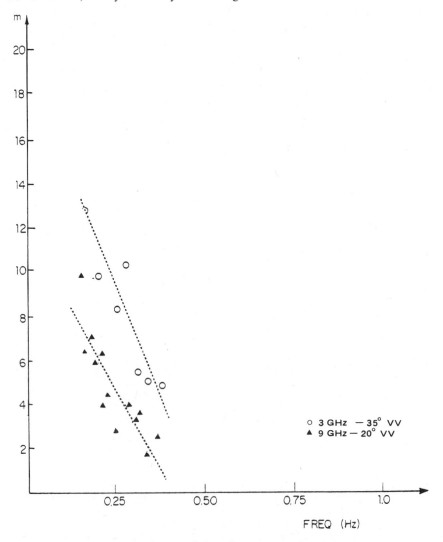

FIGURE 10. Variation of $|m|$ for upwind measurements at 3 and 9 GHz.

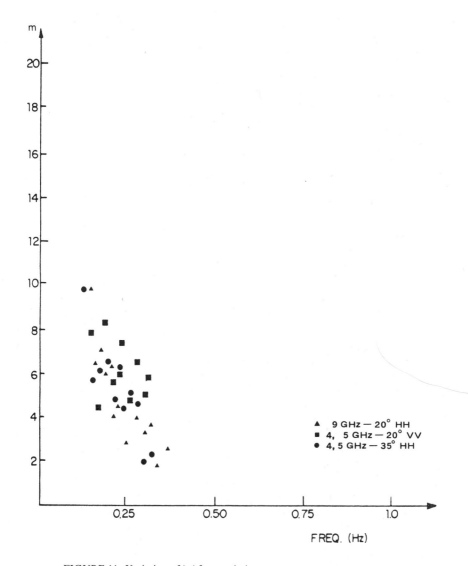

FIGURE 11. Variation of |m| for upwind measurements at 4.5 and 9 GHz.

wave frequency, decreases with radar frequency (m is higher at 3 GHz than at 9 GHz, as we can see in Fig. 10), but does not seem to be very different for 4, 5, and 9 GHz (as shown in Fig. 11). However, the values we get for m are of the same order as the values obtained by Wright et al. (1980) at X band and W. R. Alpers (personal communication) at C band.

5. CONCLUSION

Peaks generally occur in the spectrum of the backscattered signal at the dominant ocean wave frequency. However, the existence of peaks at other frequencies (harmonics) appear to depend on radar look direction (as evidenced by cross-wind and downwind measurements), and may perhaps be explained in terms of nonlinear modulation. It is clear from these preliminary results that more measurements are needed to improve our understanding of the modulation mechanism.

REFERENCES

Alpers, W. R., D. B. Ross, and C. L. Rufenach (1981): On the detectability of ocean surface waves by real and synthetic aperture radar. *J. Geophys. Res.* **86**, 6481–6497.

Wright, J. W. (1966): Backscattering from capillary waves with application to sea clutter. *IEEE Trans. Antennas Propag.* **AP14**, 749–754.

Wright, J. W., W. J. Plant, W. C. Keller, and W. L. Jones (1980): Ocean-wave-radar modulation transfer functions from the West Coast Experiment. *J. Geophys. Res.* **85**, 4957–4966.

DISCUSSION

ROSENTHAL: (1) What was the distance between radar footprint and wave staff position?

(2) Do you have phase spectra between radar and wave staff?

DE STAERKE: (1) About 6 m for 20° and 11 m for 35°.

(2) We plan to do them but have not done so yet (we have only preliminary results).

37

HF Radar Measurements of Wave Spectral Development

Dennis B. Trizna

Abstract. Measurements are reported of ocean spectral wave growth rates parametric in wave frequency and azimuth using measurements of first-order Bragg lines of HF radar Doppler spectra. Preliminary analysis indicates that: (1) growth rates versus angle show the same narrow shape that the directional spectra did, between cosine 16th to 32nd about the wave direction; (2) magnitudes were slightly larger than those by Stewart and Teague (1980) and were in the range predicted by the wave–wave interaction contribution as reported in the JONSWAP results (Hasselmann et al., 1973); (3) the wave growth rate did not approach zero as the ratio of wind speed to phase velocity, U/c, approached 1, as did the atmospheric contribution measured by Snyder et al. (1981), implying a wave–wave interaction contribution; (4) a pronounced overshoot in time was observed for values of U/c greater than 1, with weak overshoot for this ratio less than 1. For wave frequencies in the region of U/c nearly equal to 1, the exponential growth period had finished and a steady-state level was reached before the data collection had begun.

1. INTRODUCTION

Measurements of directional wave spectra are difficult and expensive to make with virtually all instrumentation available to the oceanographer, and the problem of getting such a measurement over an extended period of time to determine directional spectral development is even more difficult. Remote sensing of ocean waves by means of radar techniques offers a new opportunity to make such measurements. Land-based remote sensing techniques offer a less expensive technique than those requiring use of a platform, such as an aircraft or satellite, because of simplicity in station keeping, fuel costs, and maintenance. The HF radar offers in addition the capability to remotely measure waves at a distance from the shore under open ocean conditions, using either surface wave or skywave propagation via the ionosphere. Three types of HF radar systems are available for such measurements: (1) the synthetic aperture technique, which uses an available omnidirectional transmitter and a receiver system carried aboard a vehicle, which is used to synthesize a receive antenna (e.g., Stewart and Teague, 1980); (2) the NOAA CODAR, being developed by Don Barrick at the Wave Propagation Laboratories in Boulder, Colorado, which makes use of a relatively broad antenna bream (Barrick and Lipa, 1981)—allowing a compact portable system, and which employs analysis algorithms used for pitch-and-roll buoy data; and (3) the application of a large-aperture HF radar set up at the seashore, its beam steered over a series of azimuthal bearings, illuminating a very small patch size, but requiring no assumptions regarding zero currents in the area of interest such as is required for the other two

Dennis B. Trizna ● Radar Division, Naval Research Laboratory, Washington, D.C. 20375.

HF systems. The data we shall present here are of this third type, collected with a radar on San Clemente Island, off the coast of San Diego, California. Note that the HF first-order measurements are the only ones which can produce unambiguous ocean wave spectral amplitude information; microwave SAR and other similar techniques require a transfer function between radar spectral terms and ocean spectral terms, which constitutes an area of current investigation.

The details of the data collection are described in an NRL report (Trizna *et al.*, 1980a) and in a more condensed form (Trizna *et al.*, 1980b). In those papers, results were presented regarding the measurement of very narrow directional ocean wave spectra for three successive data collection periods, which also showed the presence of dual peaks on either side of the direction of the wind. These were interpreted to be manifestations of the Phillips resonance mechanism of wave generation. The variation with angle to wind as a function of wave frequency agreed with the predicted values for the wind speed which prevailed. In this work, we wish to present the results of additional periods of data of that set which have been analyzed, with radar system parameter changes from one data collection period to the next accounted for, thus allowing amplitude measurements versus time to be considered. The technique which we employ is measurement of the strongest of the pair of first-order Bragg line peaks in the Doppler spectrum. This spectrum is simply the average of several power spectra of time series of range samples of the beat frequency due to the backscattered signal mixed with the local oscillator signal. The Bragg lines are the major contribution to the Doppler spectrum and represent scatter from ocean wave spectral components of the directional wave spectrum with wavelength one-half the radar wavelength, and directions approaching and receding from the radar pointing bearing. This interpretation was first made by Crombie (1955) and calculated by Wright (1966), and by Barrick more rigorously (1972). For this work the antenna gain versus azimuthal bearing was calculated absolutely, and measured in a relative sense versus bearing. Absolute antenna gain can be variable at HF, with tidal variations, soil moisture, etc., but gain versus angle is a geometric function of the array structure and should not vary. Hence, relative amplitudes of wave spectra versus radar bearing and time can be retrieved. Wave spectra were collected continuously with a Waverider buoy in the radar cell at radar antenna boresight throughout the entire experiment, simultaneously with the radar data, and provided a real time indication of wave development. In this work, we shall present a synopsis of the results of the measurements, with further details to be published elsewhere. Comparisons with other open ocean results which have recently been published are also made.

2. EXPERIMENTAL PROCEDURE AND RESULTS

In brief, the experiment consisted of operating the radar using 10 different radar frequencies sequentially pulse-to-pulse, at each of five different azimuthal bearings, also in sequence. Two different sets of five bearings were used so as to scan two sections of radar coverage in two successive time periods of roughly 30 min each. The central section of the radar coverage, 300 to 340° bearing, was covered with one set of steer angles for the first half hour; then the pair of sections, 280 to 300° and 340 to 360°, were covered during the second period. Other types of data were then collected before the scan sequence was repeated. The Santa Barbara Channel Islands some 90 miles north of San Clemente Island provide a fetch-limited situation for winds from the reported direction for this experiment. This has been seen to be the case by observing radar data in patches along this bearing on either side of the island chain. The data on the side of the islands indicated very high seas, while the waters behind the island appeared very calm, for several range cells into the fetch. This fetch limitation was confirmed in Trizna *et al.* (1980b) by Kitaigorodskii scaling of peak frequency with fetch, and agreed well with the final peak frequency which was observed. This peak frequency did not change throughout this night even though the winds blew continuously. The data considered for this experiment end at local midnight.

The Doppler spectra used 512 input samples, so that the data are essentially simultaneous for a given azimuthal section, since a full set of frequencies and azimuths are sequentially sampled for each of the 512 input FFT points. Spectra collected in each half-hour period were incoherently averaged to form the Doppler spectrum from which the strongest Bragg lines were measured for analysis. Typically, the weaker of the two Bragg lines did not respond to the winds within the data periods as did the stronger line. This was due either to the lack of reflection of wave energy from the narrow end of San Clemente Island, or to the lack of wave–wave interactions because of insufficient fetch, which has been postulated as a possible source of wave energy in the half-plane opposite the wind commonly observed with HF radar (Crombie *et al.*, 1978). The radar frequencies used ranged from 2.01 to 11.65 MHz, with respective Bragg resonant ocean wave frequencies between 0.145 and 0.348 Hz. Actual bearings used extend from 285 to 357°, in steps of 8°.

The Waverider buoy signal was received and spectrum analyzed in real time on a Ubiquitous spectrum analyzer for 20-min periods, with these averaged for 1 h with 90% time overlap from one spectrum to the next for smoothing. The 1-h averages were then plotted in real time automatically so that wave development could be monitored. Wave spectral development is shown in Fig. 1 for the case under considration for four 1-h periods covering about 12 h. The spectral peak near 0.05 Hz is probably long-range swell from the southern hemisphere propagated into the region near San Clemente Island, similar to that measured by Munk *et al.* (1963). This type of feature is common in spectra we have collected in this area. The broad region from 0.06 to 0.10 Hz in the spectrum is due to a storm some 200 miles west of San Clemente Island, reported by the Fleet Numerical Weather Service, and measured independently by a pressure sensor array operated by the Shore Processes Laboratory of Scripps Institution of Oceanography. The wave energy observed to change over the period covered by the four spectra is due to local winds blowing from 310 to 320° during the period, and are the focus of interest for

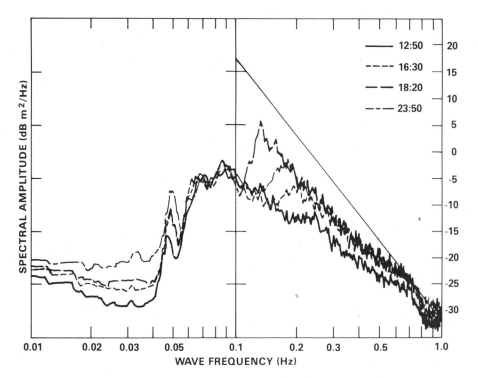

FIGURE 1. Waverider spectra consisting of 1-h averages are shown for several time periods during which radar measurements were made. The developing portion of the spectrum is due to local winds, while both a long-period monochromatic swell component (possibly from a southern hemisphere storm) and a broad-frequency spread of swell from a storm a few hundred miles to the west contribute at lower frequencies.

this work. Note that the wave spectra do not follow a frequency to the fifth power, which again is typical for moderate winds in this area. On a day when the winds blew very strongly, the resulting spectrum was found to follow the fifth power law. [Similar results have been reported by others in this regard (Forristall, 1981).] Winds were estimated to be of the order of 6 m/s based upon the angles of the Phillips resonance peaks in the directional spectra, which agreed with values measured at several points in the area, including San Nicholas Island and several shore points. Also, the peak frequency generated in the spectrum did not change throughout the night as wave spectra were continuously collected. The phase speed of this component is of the order of 11 m/s.

Local winds began to blow at San Clemente Island about 0700 on the test day, from a direction estimated to be very nearly from the boresight direction of the radar antenna, 320°. Data obtained after the fact from the Naval Weather Service Detachment at the National Climatic Center, Asheville, North Carolina, collected at the weather station on San Nicholas Island, were in close agreement, with the reported wind speeds ranging from 10 to 14 knots. At the radar site, winds were observed to decay about midnight, so that a 17-h period of sustained winds was experienced at the minimum. These winds generated an ocean wave spectrum (as measured by the Waverider buoy) which began to become noticeable in the spectrum collected at 1250 local time, and which is seen rising in successive spectra in Fig. 1. The spectrum for the period in which the winds ceased to blow at the radar site shows the maximum development, with the lowest frequencies excited of the order of 0.14 Hz, probed by the lowest radar frequency used, 2 MHz. Spectra collected during the night showed the same general character, with no lower frequencies excited. The midnight spectrum can then be considered a bimodal spectrum in the sense that two different major ocean wave spectra with two primary directions are superimposed to produce the observed spectrum. The wind-driven portion can be considered a transient one, in light of the results which follow, rather than a fully developed steady state typically reported by other workers.

Examples of variation with time of wave spectral amplitudes for a given frequency are shown in Figs. 2–4, which illustrate a series of time samples measured in local time for the 0.231, 0.284, and 0.334-Hz wave components for all 10 different antenna pointing directions. Each time sample represents roughly a half-hour average, and the amplitude character is drawn at the center of the time period. Each of the time series sets shows a different overall character, based upon the region of U/c of which each is representative, as well as generally similar interesting characteristics versus azimuth. Here, U is the wind speed determined from the Phillips resonance angles in Trizna et al. (1980b), 6 m/s, and c is the wave component phase velocity. Lines are drawn through the first two points of the component time samples to determine a slope, or absolute growth rate, $\beta(f, \theta)$. The values of the growth rates are then normalized by dividing by the wave frequency to give a dimensionless value. These are written for each azimuth above the antenna bearing in the lower right-hand corner of each plot.

The data of Fig. 2 are representative of data for which $U/c < 1$, relative wave amplitude versus local time. All amplitude values are scaled correctly relative to one another in time and azimuth, since antenna gain versus angle has been removed. The data show the effect of swell propagating into the region from the west, as discussed earlier, as the leftmost bearing generally have a higher amplitude in the first time period than the rightmost. (This was observed to be the case for the lower wave frequencies in the directional spectrum measurements.) The magnitude of the growth rates versus angle peaks along the direction 317°, that nearest the prevailing wind in the first time period. The growth rates calculated for 341° and greater exhibit large values again, which is probably spurious because of the lower level of wave amplitude in the first time period for these northernmost bearings. It is noted that the final amplitudes reached at these bearings are lower than those around 317°. The growth of wave amplitude along the direction 317° represents a 40-db increase from the first time period to the maximum value, which then decreases to the final steady-state value. A weak overshoot in time can be inferred for the 301° bearing. It may also appear at some of the other bearings as well, but there are not a sufficient number of time samples to verify such behavior from the data points available. The lack of data points for some of the azimuths is due to too weak a signal relative to a large noise value for that particular bearing.

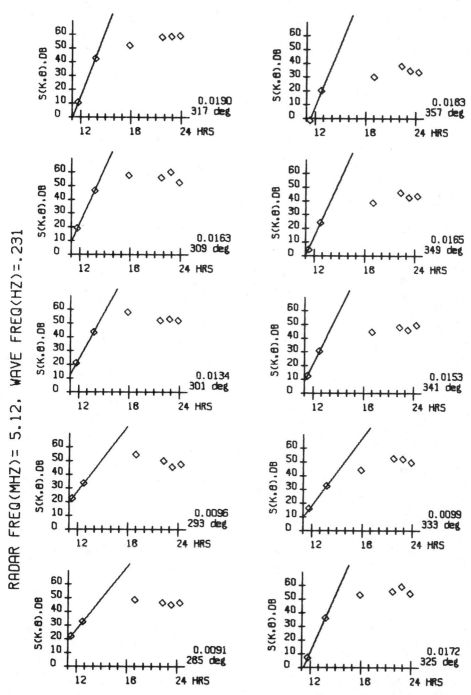

FIGURE 2. A series of wave spectral amplitude time samples are shown for each of 10 different radar bearings, for one of 10 ocean wave components satisfying $U/c < 1$. The abscissa is local time in hours, while the ordinate is in decibels, on a relative scale but consistent between azimuths. The slope and wave direction are written to the right of each plot.

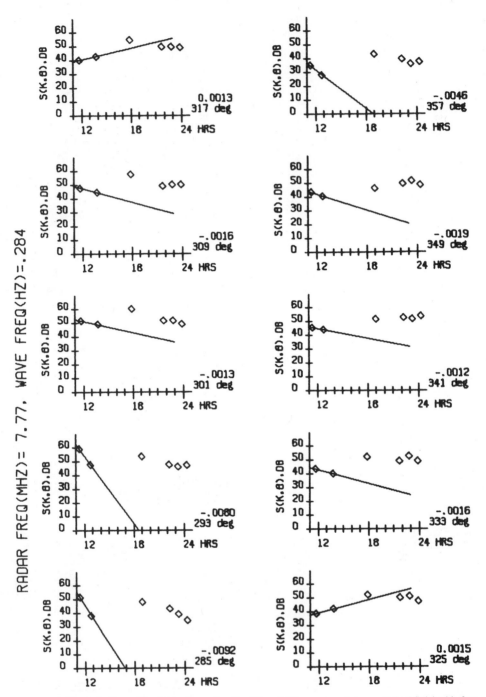

FIGURE 3. Data similar to Fig. 2 are shown, but for $U/c \sim 1$. The exponential growth had finished before radar data collection began, and the time period appears to be just after the ov•·shoot phenomenon has occurred.

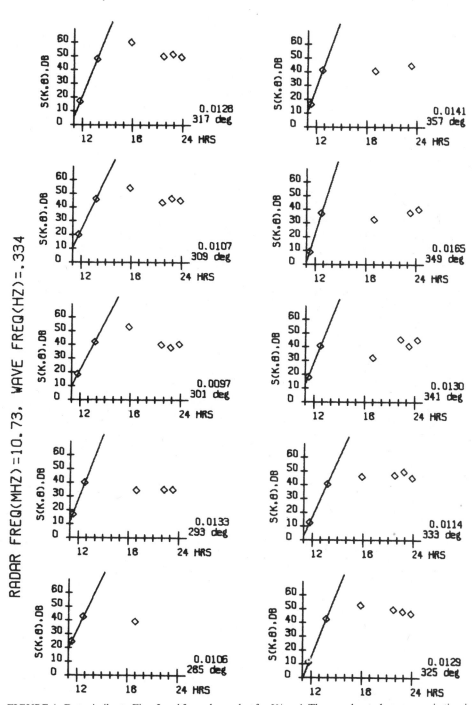

FIGURE 4. Data similar to Figs. 2 and 3 are shown, but for $U/c > 1$. The overshoot phenomenon in time is quite apparent for this case over a wide selection of angles.

The second set of amplitude histories shown are for a value of U/c just greater than 1, and indicate that the wave growth to equilibrium has already occurred. The negative slopes for several of the azimuths are interpreted as due to the downside of the overshoot effect in time, since the exponential growth had occurred prior to collection of radar data, and the waves are settling asymptotically to their final steady state. As frequencies with successively greater values of U/c are considered, just the peak of the overshoot is observed, then the transition from exponential growth to the overshoot period. For these cases the slopes are not representative of an exponential growth phase, and hence are omitted here since this phase had finished before data collection began.

The final range of U/c to be considered is for values much greater than 1, such that exponential growth is again observed. Figure 4 is an example of some of the available data which satisfy this criterion. The overshoot of the wave amplitude for angles near the wind direction is very pronounced. It may also be present at angles further away from the wind direction near 320° as well, but again insufficient data samples exist in the appropriate time period to be sure. Suffice it to say that the overshoot is apparently more strong an effect for $U/c > 1$ than it was for $U/c < 1$. This was generally found to be the case for other frequencies which were considered in these regions as well. A general picture of the wave growth was for wave components near $U/c = 1$ to grow very rapidly to equilibrium, and those with values of U/c much greater or lower than 1 to begin exponential growth phase at a much later time. Because wave energy already existed in this frequency range traveling from another direction, it is not known whether this delay in onset of exponential growth is in some way connected with the initial wave field characteristics, and whether or not the delay pattern would be different versus frequency for a different set of initial conditions.

Another interesting feature of this last series of amplitude histories is the slope or growth rate as a function of azimuth. Note that a maximum growth rate exists near the direction of the wind, at 317° bearing. The growth rate falls off on either side of this bearing, but then reaches local maximum at the angles 293 and 349°. These do not quite correspond to the Phillips resonance angles for this frequency which were observed and reported in Trizna et al. (1980b), but do indicate a growth pattern with angle similar in behavior. In that work, very narrow spreading prevailed around the wind direction and azimuthal maximum of the angular spreading, and minor peaks occurred near the Phillips resonance angles for cases of $U/c > 1$. Similar behavior is observed here for growth rate versus angle. It was conjectured that, because of this fine structure in angle for a given wave frequency, the wave development observed was in a

TABLE I.   Radar Frequencies (F) and Ocean Wave Frequencies (f) along with the Cosine Exponents and Primary Wave Directions, N and θ, as Determined from Wave Growth Measurements (Subscript β) and Directional Wave Spectrum Measurements (Subscript S), respectively, as Described in the Text Regarding Angular Dependences[a]

F (MHz)	f (Hz)	N_β/θ_β	N_S/θ_S
2.01	0.145	32/316	16–32/315
2.42	0.159	16/318	32/315
4.04	0.205	16/319	16–32/315
5.12	0.231	16/317	16–32/315
6.76	0.265	(4)/312	16–32/315
7.77	0.284	—	8/321
8.15	0.291	—	32/321
9.25	0.310	(32)/317	16–32/325
10.73	0.348	8/322	32/325
11.65	0.348	16–32/325	32/325

[a]Values in parentheses are best estimates from scattered data points.

state preliminary to the onset of wave–wave interactions which would mix energy at the Phillips resonance angles with that along the wind direction, thus resulting in the more traditional cosine-squared angular spreading observed in steady-state open ocean conditions. [In fact, one referee has pointed out that a plot of the Phillips resonance angle versus U/c provides a rather nice envelope to the angular spread of the data of Regier and Davis (1977).] This picture fits rather well the observations of Regier and Davis (1977) that for wave frequencies beyond $U/c < 1$, wave spreading was found to be much narrower than for $U/c > 1$.

As a summary of the angular dependence of the growth rates measured for those cases where the data permitted, Table I shows an estimate of the fit of cosine-N functions to $\beta(f, \theta)$ comparing N for wave growth rates and wave energy spread for the same data reported in Trizna *et al.* (1980b). It is apparent that they are very similar. This result then offers an independent verification of the narrow spreading of wave energy measured by azimuthal variation in Bragg line amplitudes in Trizna *et al.* (1980b), and does not require that the relative antenna gain be known from bearing to bearing, but only that the gain for a given bearing be constant over the total measurement period of all data samples. This is made clearer in the discussion of the next section.

3. DISCUSSION OF RESULTS

One can write the radiative transport equation for a given Fourier component of the ocean wave directional spectrum as

$$\frac{\partial S(f, \theta)}{\partial t} + c \cdot \frac{\partial S(f, \theta)}{\partial x} = A(f, \theta) + N(f, \theta) + D(f, \theta) \tag{1}$$

where A, N, and D are the atmospheric source function, the nonlinear wave–wave interaction transfer term, and dissipative sinks, respectively. As written by Snyder *et al.* (1981), the atmospheric source function has two terms: one represents effects of the turbulence spectrum of the atmosphere at the surface, $F(k, \omega)$, and the second the exponential growth term:

$$A(f, \theta) = 2\pi\rho^{-2}g^{-4}(2\pi f)^5 F(\bar{k}, \omega) + 2\pi f \beta(f, \theta)S(f, \theta) \tag{2}$$

They have measured β using a technique and analysis which assumes the form of Eq. (2) for A, and which involves simultaneous measurements of wave amplitude and overhead air pressure fluctuations and calculation of their cross-spectra. The growth rate defined in (2), due to atmospheric effects only, has one significant difference in predicted behavior (and measured by Snyder *et al.*): the growth rate goes to zero at $U/c = 1$. Our measurements, as well as those of Stewart and Teague, indicate nonzero values near $U/c = 1$. The implication is that there are wave–wave interactions which can cause waves to grow with exponential dependence near $U/c = 1$, which must be accounted for in $N(f, \theta)$. That is, there must be a term proportional to $S(f, \theta)$ in $N(f, \theta)$ like the second term of Eq. (2). Since Snyder *et al.* did not report on wave amplitude growth with time explicitly, although the data apparently exist to allow such a calculation, the total wave growth was not given, but only the atmospheric interaction contribution calculated for the cross-spectra. If one calculates the angular spread for the curves in their Fig. 16, the results show spreading of the order of cosine 8th to cosine 16th, just slightly broader than the results of Trizna *et al.* (1980).

We shall limit our discussion to a final comparison with the work of Stewart and Teague (1980) and Snyder *et al.* (1981), since both of these review most of the measurements made previously. There are two main results of Stewart and Teague (1980) which are relevant to our results: (1) the wave growth parameter variation with angle to wind was cosine to the first power, as determined from total wave amplitude growth rate, much broader than our results for all frequencies; (2) the magnitude of the growth rate did not approach zero at $U/c = 1$, as did the results of Snyder *et al.*, who measured only the atmospheric source contribution to the growth rate. This is similar to our observations.

Addressing the second point first, the implication is that a wave–wave interaction contribution exists which also has an exponential growth form. Other data summarized by Stewart and Teague (1980) (their Fig. 9) show a similar behavior near $U/c = 1$. The source of the other data was not indicated, so it is impossible to know the conditions under which those measurements were made. Hasselmann *et al.* (1973, p. 52) give the following form for the wave–wave interaction contribution to an exponential growth rate:

$$\beta \approx 6(U_{10}/c)^n f \times 10^{-3} \tag{3}$$

where n is 1.03 or 1.33, depending on the choice of scaling laws used in their nondimensional scaling relationships. These equations are plotted in Fig. 5, along with our data, and the radar results of Stewart and Teague (1980), which were collected for one radar (and wave) frequency, but for different wind speeds. Our results are somewhat higher than those of Stewart and Teague

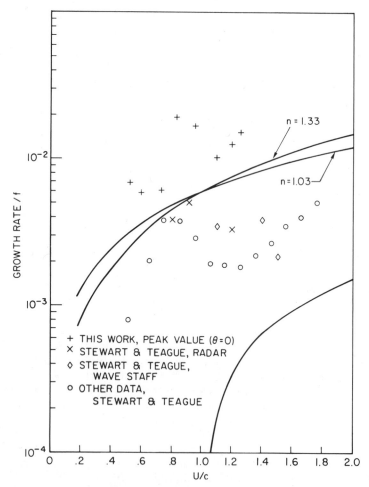

FIGURE 5. Wave growth rates along the wind direction ($\theta = 0$) are plotted for this work ($+$), as well as the radar results of Stewart and Teague ($\bigcirc$). The two upper, continuous curves are the predicted contributions of an exponential growth due to wave–wave interactions, and the lower curve is the air–sea interaction term of the Miles theory.

(1980), but the directional energy spectrum spreading dependence for the two cases were quite different as well. Hence, the differences may be real, perhaps reflecting the different wind conditions which prevailed at the two locations. The data are not inconsistent with Eq. (3), which is in itself only an approximation. The plot of the function, $\alpha(U/c - 1)$, is also shown in Fig. 5, with α equal to 1.51×10^{-3}, as used by Snyder *et al.* (1981). It is seen to be at most 10% of the wave–wave interaction growth rate at the highest frequency shown.

Regarding the first point of the Stewart and Teague (1980) paper, their wave growth rates measured as a function of angle showed a cosine to the first power spread. For waves starting from a uniform small amplitude, this would imply a final angular dependence of the directional energy spectrum which reflects the angular dependence of the wave growth rate as well. Wave energy spread was not reported in their paper, although such data were available because of the nature of the measurement technique. One may assume that the wave energy spread was also cosine-first, which is quite broad compared to most reports in the literature. One implication from such a broad directional spread is that the winds prevailing for their measurements made off Galveston were very turbulent, since they were wintertime offshore winds. These might be expected to have a much more turbulent character than winds blowing over a long fetch over water, since the much larger scale sizes of land features (cities, trees, etc.) would leave their imprint as much larger scale sizes of turbulence than would be expected from onshore or open ocean winds, via the Miles mechanism.

Let us consider some concepts about wind coherence and wave spreading which can be expressed as an analog to general antenna theory. The half-power beamwidth of a filled array is inversely proportional to the aperture of the array of omnidirectional elements (e.g., pressure sensors in the case of oceanographic instrumentation, or monopoles if considering radio systems). This aperture is the largest area over which one can measure a coherent phase front. If the phase front is coherent over a much larger number of wavelengths than the aperture size, the coherence of the front is not determinable to a greater accuracy than the beamwidth of the antenna. If it is coherent over a smaller front, the measured angular width will be a convolution of the antenna pattern and angular spread of the coherence (i.e., the inverse of the coherence length of the phase front measured in number of wavelengths).

One can consider the generation of wave spectra in a similar sense, regarding the Miles linear instability theory and very large scale turbulence features, much larger than the longest ocean wavelengths, beyond those turbulence components responsible for the linear growth term first proposed by Phillips. The Miles theory is a one-dimensional one, assuming perfect phase coherence along the wind front. If the wind field is very coherent with little turbulence, then the ocean surface will experience the same forcing function at points perpendicular to the direction of the wind, and very narrow wave spectra will be generated. If the coherence of the wind field is not perfect, but has some associated coherence length along the phase front, then the ocean surface will experience a similar forcing function only across this coherence length. Hence, spectral wave packets will be produced with a similar coherence width across the wind field, resulting in an effective angular spread which is the reciprocal of this coherence length in number of ocean wavelengths. One obvious source of finite coherence lengths in the wind field is large scale turbulence Fourier components associated with offshore winds. Assuming homogeneous and isotropic turbulence, with the frozen hypothesis on a time scale long relative to ocean surface features, the coherence lengths are similar in the cross-wind direction as they are in the along-wind, with the most effective coherence length that of the largest scale size of turbulence present. Such an intuitive picture could describe the rather broad angular dependence in growth rates (and implied directional spectrum spread) measured in Stewart and Teague (1980).

4. SUMMARY

We have measured total growth rates of ocean waves parametric in frequency and angle to wind. Growth along the wind direction was somewhat larger than predicted by exponential growth wave–wave interaction mechanisms discussed in the JONSWAP work of Hasselmann *et al.*

However, the expression they derived assumed cosine-squared spreading in wave energy, whereas we found between cosine 16th and 32nd to prevail for both wave energy spread and wave growth rates, measured using independent assumptions. These measured growth rates are about four times those measured by Stewart and Teague (1980), but the angular spread in growth rates which they reported were only cosine to the first power in angle to wind, so that the differences may be real. The fall off the omnidirectional power spectrum of wave energy was only f^{-4}, which may be related to our very narrow directional spectra and growth rates. A speculation is that perhaps as long as wave–wave interactions provide transfer of energy to the forward face of the frequency spectrum only along the wind direction, wave spectra exhibit f^{-4} shapes. Only after energy is distributed in angle as well, among "unfilled energy states," does true equilibrium begin to exist, allowing f^{-5} slope spectra. In addition, under conditions when waves growth exists preferentially along the wind direction, growth rates would be expected to be higher since off-angle states are not available to draw energy from other lower frequency components. Energy at the peak of the spectrum is able to be transferred only upwind and downwind to other frequency components and is not available for distribution to angles off the wind. Such an intuitive picture might explain the differences between the growth rates of Stewart and Teague, the JONSWAP growth rate, and our measured values, which narrow with angle in respective order, but increase in magnitude.

One final comment shall be made along this line regarding the existence of wave energy opposite the wind direction. Wave energy and components of turbulence fluctuations opposite the wind were reported in Snyder (1974), of the order of a few percent of the upwind value as an estimate. Such wave energy is always observed with HF radar measurements of ocean waves, due to the large dynamic range capability inherent in an HF radar. For open ocean conditions, in which 10-m waves are measured, the upwind component ocean wave is typically of the order of -24 db of the downwind component, or about 0.4% (Long and Trizna, 1973). In that work, an empirical relationship was determined from a large data set which indicated the product of spectral energy in a given direction times that opposite in direction was a constant as angle to wind is varied. Hence, a maximum–minimum pair in wave energy was found along the wind direction, with the energy rising in the rear half-plane in a like manner as the forward half-plane energy fell as angle to wind was varied. For the data considered in this work, as discussed in Trizna et al. (1980b), the energy in the rear half-plane was roughly a constant value, perhaps indicating the lack of wave–wave interactions at angles off the wind direction. One may speculate that the transition from f^{-4} to f^{-5} occurs as energy is distributed in angle to wind and into the back half-plane, allowing more energy states to be filled per unit frequency.

REFERENCES

Barrick, D. E. (1972): First order theory and analysis of MF/HF/VHF scatter from the sea. *IEEE Trans. Antennas Propag.* **AP20**, 2–9.

Barrick, D. E., and B. J. Lipa (1981): Codar observations of onshore directional ocean wave height spectra in shallow water. *IUCRM Symposium on Wave Dynamics and Radio Probing of the Ocean Surface*, Miami.

Crombie, D. D. (1955): Doppler spectrum of sea echo at 13.65 Mcs. *Nature* **176**, 681–682.

Crombie, D. D., K. Hasselmann, and W. Sell (1978): High frequency radar observations of sea waves traveling in opposition to the wind. *Boundary Layer Meteorol.* **13**, 45–54.

Forristall, G. Z. (1981): Measurements of a saturated range in ocean wave spectra. *J. Geophys. Res.* **86**, 8075–8084.

Hasselmann, K., T. P. Barnett, E. Bouws, H. Carlson, D. E. Cartwright, K. Enke, J. A. Ewing, H. Gienapp, D. E. Hasselmann, P. Kruseman, A. Meerburg, P. Müller, D. J. Olbers, K. Richter, W. Sell, and H. Walden (1973): Measurements of wind-wave growth and swell decay during the Joint North Sea Wave Project (JONSWAP). *Dtsch. Hydrogr. Z. Suppl. A* **8**(12).

Long, A. E., and D. B. Trizna (1973): Mapping of North Atlantic winds by HF radar sea backscatter interpretaion. *IEEE Trans. Antennas Propag.* **AP21**, 680–685.

Munk, W. H., G. R. Miller, F. E. Snodgrass, and N. F. Barker (1963): Directional recording of swell from distant storms. *Philos. Trans. R. Soc. London Ser. A* **255**, 505–584.

Regier, L. A., and R. E. Davis (1977): Observations of the power and directional spectrum of ocean surface waves. *J. Mar. Res.* **35**, 433–451.

Snyder, R. L. (1974): A field study of wave induced pressure fluctuations above surface gravity waves. *J. Mar. Res.* **32**, 497–531.

Snyder, R. L., E. W. Dobson, J. A. Elliott, and R. B. Long (1981): Array measurements of atmospheric pressure fluctuations above surface gravity waves. *J. Fluid Mech.* **102**, 1–59.

Stewart, R. W., and C. Teague (1980): Dekameter radar observations of ocean wave swell growth and decay. *J. Phys. Oceanogr.* **10**, 128–143.

Trizna, D. B., R. W. Bogle, J. C. Moore, and C. M. Howe (1980a): HRL Report 8386, U.S. Naval Research Laboratory.

Trizna, D. B., R. W. Bogle, J. C. Moore, and C. M. Howe (1980b): Observation by HF radar of the Phillips resonance mechanism for the generation of wind waves. *J. Geophys. Res.* **85**, 4946–4956.

Wright, J. W. (1966): Backscattering from capillary waves with application to sea clutter. *IEEE Trans. Antennas Propag.* **AP14**, 749–754.

38

PASSIVE MICROWAVE PROBING OF ROUGHENED SEA

A. M. SHUTKO

ABSTRACT. Typical data obtained from experimental observations made in the laboratory and under natural conditions on microwave radiation characteristics of a roughened sea, are presented in this chapter. The influence of different constituents of a wind-driven sea—from capillary to gravity waves, from ripples to stormy rollers including patches of foam—on the characteristics of microwave radiation is analyzed. Original data of brightness temperature variations for different sea state conditions are given as an example. The relationship between characteristics of microwave radiation and wind speed is discussed.

1. INTRODUCTION

It has been shown during experimental investigations in both the USSR and the USA that the characteristics of sea surface microwave radiation are dependent on sea state—on the presence of small- and large-scale waves and foam formations. The degree of influence of these constituents is dependent on electromagnetic wavelength, angle of observation, and polarization. In this chapter the relationships between the characteristics of microwave radiation and the above-mentioned parameters are analyzed on the basis of experimental data obtained by the author and his colleagues under laboratory and natural conditions. The effectiveness of passive microwave sensing of roughened sea is discussed.

2. THE INFLUENCE OF DIFFERENT CONSTITUENTS OF ROUGHENED SEA ON THE CHARACTERISTICS OF MICROWAVE RADIATION

The influence of small-scale roughness—ripples and short gravity waves—on the characteristics of microwave radiation has been studied under laboratory conditions and from aircraft at wavelengths of 2.25, 3.4, 8.5, 18, and 30 cm. It has been determined that the maximum brightness temperature increase due to the occurrence of small-scale roughness takes place in the short-centimeter wavelength range (Shutko, 1978b; Shutko et al., 1978). Its value is about 4–6°K for steep ripples of turbulent character generated by a fan or by sudden fluxes of wind (Fig. 1a). Similar phenomena were observed by Au et al. (1974). The longer the wavelength the smaller is this effect and at wavelengths longer than 20–30 cm it is practically not observed. Classic ripples with a spatial size smaller than 2 cm cause a brightness temperature increase of about 0.5–1°K in the wavelength range 2–3 cm.

A. M. SHUTKO ● Institute of Radioengineering and Electronics, Academy of Sciences of the USSR, Moscow, USSR.

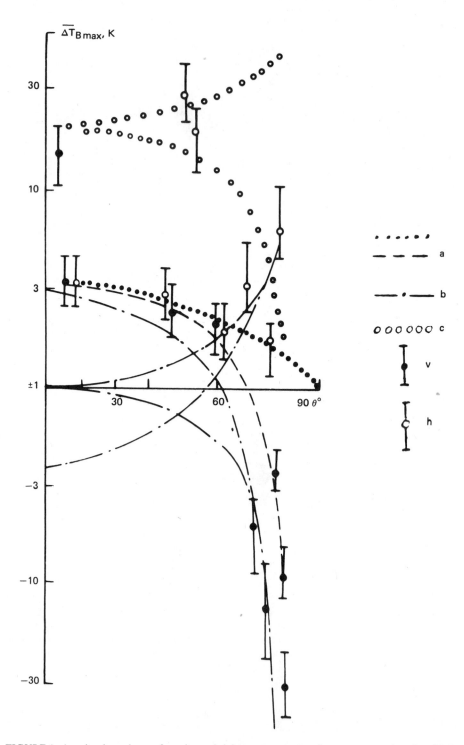

FIGURE 1. Angular dependence of maximum brightness temperature increase at wavelengths of 2–3 cm due to steep ripples (a), big gravity waves (b), and foamy breaker (c) at vertical (v) and horizontal (h) polarization.

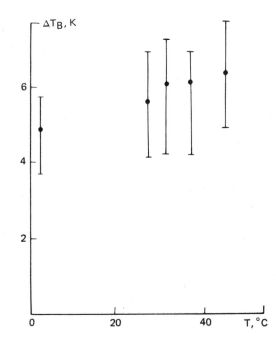

FIGURE 2. Maximum brightness temperature increase at 2.25-cm wavelength due to steep ripples for different values of water surface temperature.

It was shown by Eklund *et al.* (1974) that the scattering coefficient is dependent on sea surface temperature over the wind speed range of 0–8 m/s. In our experiments conducted under laboratory conditions, the data have shown a certain dependence of the brightness temperature increase on physical temperature, employing artificial generation of steep short waves by a fan (Fig. 2), but the sensitivity of emissivity changes caused by ripples to brightness temperature variations is much smaller than that for changes of the scattering coefficient.

The influence of large sea waves—either covered by ripples or not—on microwave radiation was investigated under natural conditions by making measurements from seashore

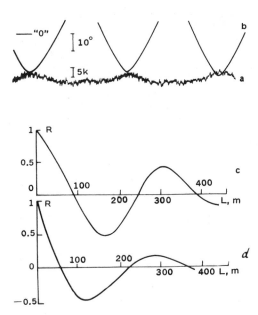

FIGURE 3. Examples of horizontally polarized brightness temperature data at 3-cm wavelength (a) for variations of viewing angle (b) and corresponding correlation functions (c) and (d).

and from aircraft at wavelengths of 0.8, 1.35, 3.4, 8.5 cm (Shutko, 1968, 1978a, b; Shutko et al., 1978; Basharinov et al., 1980; Armand et al., 1980) (Figs. 1b, 5, 8–10). Experiments were made as well on modeling large waves by temporally varying the angle of observation (Fig. 3). It is seen that large sea waves affect the radiation mostly at grazing angles of observation. With growth of their steepness, the brightness temperature decreases by about 10–30°K at vertical polarization and increases by about 5–10°K at horizontal polarization (Figs. 1b, 11). The coefficient of polarization decreases (Fig. 12). The influence of large-scale waves under vertical observation does not exceed a few degrees Kelvin and is observed mainly in the case of 2d sea roughness. There is a strong correlation between the local angle of observation and variations of emissivity (Fig. 3). The spectral peculiarities of these phenomena are not notable.

Ripples on large-scale waves cause an increase in brightness temperature with respect to the level of radiation for smooth waves, an increase that is practically additive.

The emissivity of whitecaps of foam (breakers) is about 0.7–1 in all centimeter wavelength ranges (Fig. 1c). Strips of thin foam on slopes of waves cause a brightness temperature increase of about 10°K. The influence of foam formation on radiation is negligible at wavelengths of 20–30 cm.

An increase of brightness temperature of about 10–20°K is observed upon interaction of water droplets (due to rainfall or spray under stormy conditions) with the sea surface. At wavelengths shorter than 3 cm, the influence of atmospheric components, mainly clouds, becomes of the same degree or even stronger than that of sea surface roughness.

3. EXAMPLES OF DATA OBTAINED UNDER NATURAL CONDITIONS

Under calm sea conditions, the brightness temperature variations of about 1–3°K in the centimeter wavelength range are caused by spatial (or temporal) alternation of areas with ripples and slicks (Fig. 4a). Patches of thin foam that form due to the presence of chemical pollutants or a high concentration of plankton (La Fond and Bhavanarayana, 1959) cause a brightness temperature increase of about a few degrees Kelvin (Fig. 4b).

Large-scale swells cause periodic brightness temperature variations if the spatial resolution of the observed area is smaller than the length of swell (Fig. 5).

Observations of stormy areas with a large area of resolution (from high altitude) show a brightness temperature increase (Fig. 6) the mean value of which is dependent on wind speed in a fully developed sea practically as a linear function.

Inside this area, spatial–temporal brightness temperature variations have a quasiperiodic character similar to that of local sea surface elevations and slopes (Figs. 3, 7–10). The mean period of the brightness temperature oscillations corresponds with an accuracy of 10–20% to the mean period (spatial length) of sea waves. Increase of the sea surface roughness causes an increase

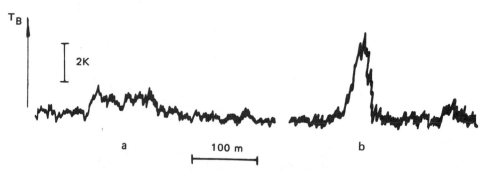

FIGURE 4. Example of brightness temperature data at 3-cm wavelength for calm surface with interchange of areas with ripples (a) and smoothed zones and small patches of thin foamy layer (b).

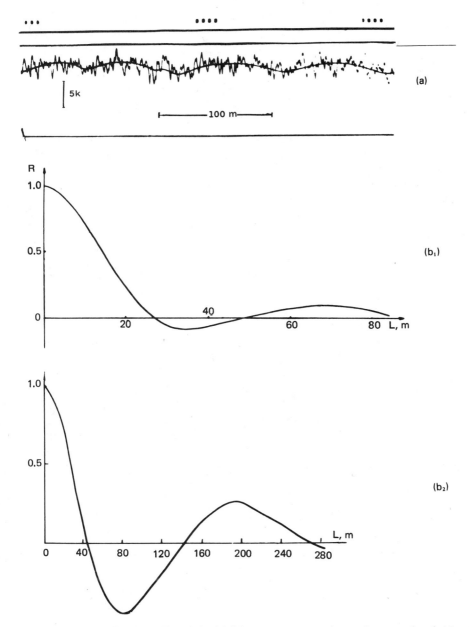

FIGURE 5. Example of horizontally polarized brightness temperature data at 3-cm wavelength (a) and corresponding correlation function (b). Observations at 25° nadir viewing angle.

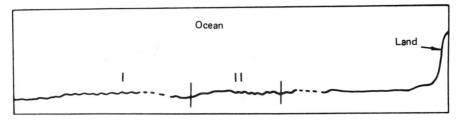

FIGURE 6. Example of brightness temperature data at 3.4-cm wavelength in stormy areas (I) and (II).

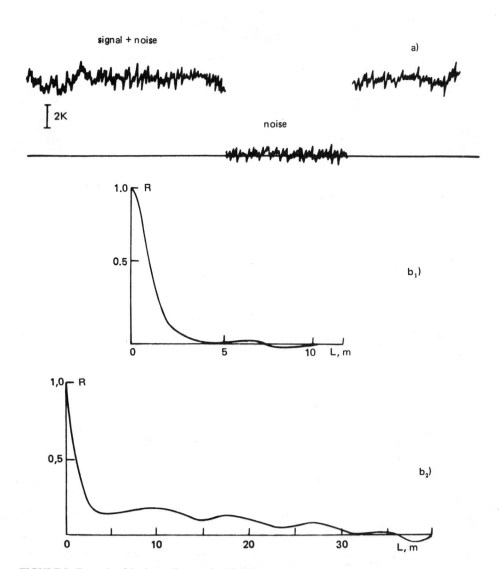

FIGURE 7. Example of horizontally polarized brightness temperature data at 3-cm wavelength (a) and corresponding correlation functions (b) for wind force 0.5–1 (b_1) and 2–3 (b_2). Observations at 25° nadir viewing angle.

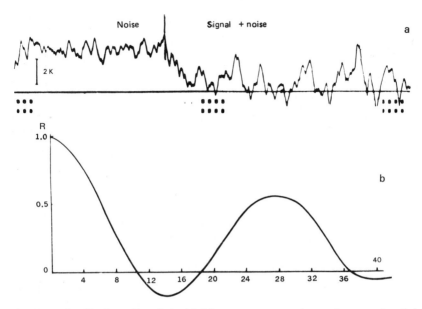

FIGURE 8. Example of horizontally polarized brightness temperature data at 3-cm wavelength (a) and corresponding correlation function (b) for wind force 3–4. Observations at 25° nadir viewing angle.

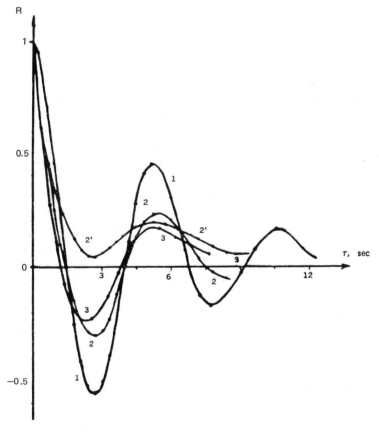

FIGURE 9. Correlation function of brightness temperature variations for wind force 4–5 at wavelengths of 3.4 cm (1), 3.2 cm (2, 2′); 1.6 cm (3) at vertical (1–3) and horizontal (2′) polarization.

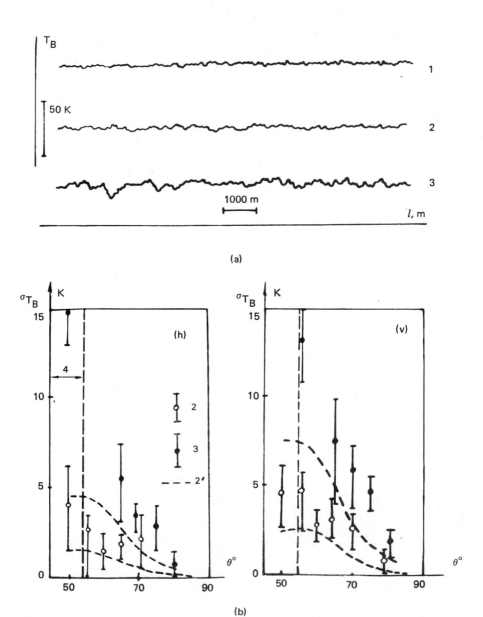

FIGURE 10. Example of horizontally polarized brightness temperature data (a) at 3.4-cm wavelength for viewing angle of 30° and wind force 0–1 (1), 2–3 (2), and 3–4(3), and angular dependence of effective values of brightness temperature variations (b) at horizontal (h) and vertical (v) polarization. Beamwidth is about 3°, altitude of observation is about 200 m at (a) and 10 m at (b). 2′ = calculations; data between curves 2′ = wind force 2–3.

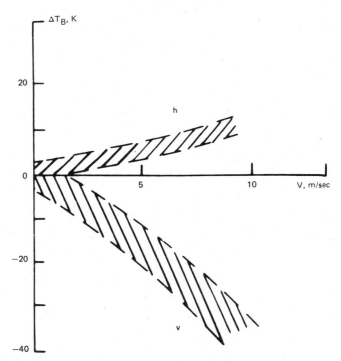

FIGURE 11. Brightness temperature changes at 3.2-cm wavelength vs. wind speed. Observations at 80° nadir viewing angle. Polarization: h, horizontal; v, vertical.

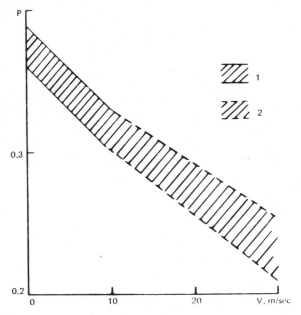

FIGURE 12. Coefficient of polarization at 3.2-cm wavelength for nadir viewing angle of 55° vs. wind speed. (1) Experimental data; (2) calculations for given angle with consideration of experimental data of Shutko (1978b) on brightness temperature increase for nadir observation.

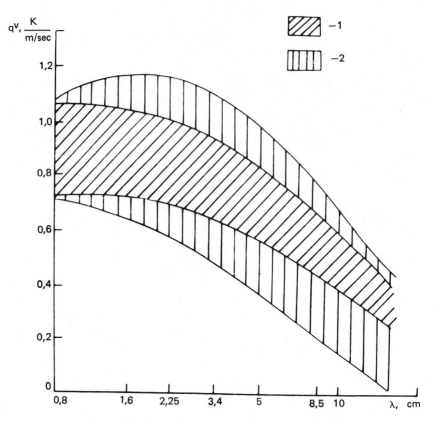

FIGURE 13. Spectral dependence of microwave radiation sensitivity at nadir observation to wind speed variations for $V \geqslant 10$ m/s. (1) Generalized with consideration of the experimental data of Shutko (1978b) and the "Bering" Experiment (1975); (2) generalized model calculations.

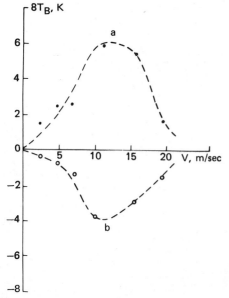

FIGURE 14. Maximum (a) and minimum (b) deflection of brightness temperature from averaged dependence $T_B^{av} \cdot (V)$ vs. wind speed.

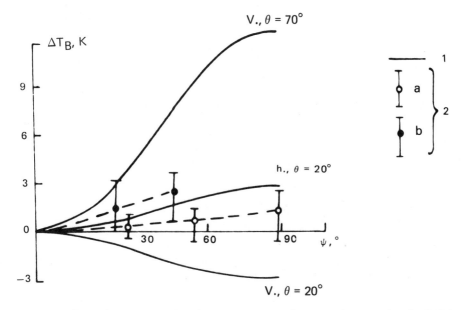

FIGURE 15. Azimuthal dependence of brightness temperature changes at 3-cm wavelength. (1) Calculations for model of 2d roughness. (2) Experimental data; angles of observations: 25° (a), 70° (b). Polarization: h, horizontal; v, vertical.

of the spatial–temporal interval of correlation as well as the intensity of brightness temperature variations (Figs. 10, 16, 17).

4. GENERALIZED WIND FORCE DEPENDENCES

The generalized dependences of mean values, degree of polarization, and effective variations of microwave radiation on wind speed (or sea state) are given in Figs. 11–19.

At grazing angles of observation, a monotonic change of brightness temperature and polarization occurs with an increase of wind speed (Figs. 11, 12). The longer the electromagnetic wavelength, the smaller is the sensitivity of radiation under vertical observation to sea state variations (Fig. 13). Less scattering is observed in data obtained experimentally than in data from model calculations (Armand et al., 1980). The characteristic feature here is the dependence of

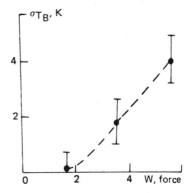

FIGURE 16. Effective values of horizontally polarized brightness temperature variation at 3.4-cm wavelength vs. wind speed. Spatial resolution is about 10 m. Observations at 30° nadir viewing angle.

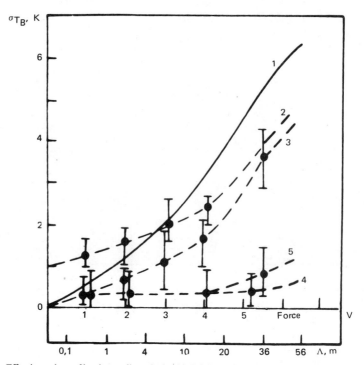

FIGURE 17. Effective values of horizontally polarized brightness temperature variation at 3-cm wavelength vs. wind force for different sea state situations and spatial resolution. Observations at 25° nadir viewing angle. (1) Calculations. (2–5) Experimental data; spatial resolution: 5 m (2, 3); 100 m (4, 5); situation: wind waves (3, 4); wind waves and swell (2, 5).

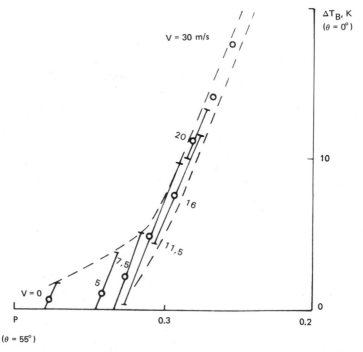

FIGURE 18. Two-dimensional characteristics of radiation in "increase of mean values—polarization" coordinates at 3-cm wavelength for wind speed changes.

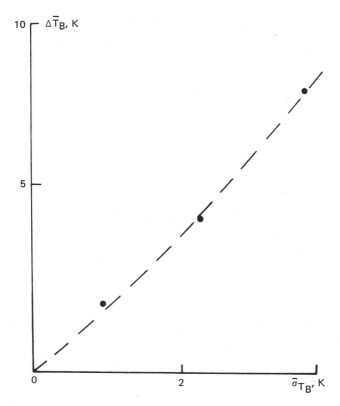

FIGURE 19. Two-dimensional characteristics of radiation in "increase of mean values—effective values of variation" coordinates at 3-cm wavelength for wind speed changes.

maximum (and minimum) difference between observed and averaged brightness temperature data—$\delta T_B(V) = \Delta T_B^{obs.}(V) - \Delta T_B^{av.}(V)$ $[\Delta T_B^{av.}(V) \simeq q(V - V_0); V_0 = 7-8\,m/s]$—on wind speed (Fig. 14). Maximum differences are found for $V \simeq 10-15\,m/s$. This can be explained by the dependence of the conditions that produce foam at the sea surface on some physical and chemical parameters of the water in a certain region—on temperature, salinity, and presence of plankton (La Fond and Bhavanarayana, 1959).

An azimuthal brightness temperature dependence is observed for near-nadir and grazing observation (Fig. 15) (Shutko, 1978a). The effect of large-scale waves occurs here mainly at grazing angles of observation. At near-nadir angles of observation, 0–30°, at centimeter wavelengths the main reason for this effect is the influence of small-scale waves. A certain dependence of emissivity at these wavelengths on wind speed for $V \leqslant 5-7\,m/s$ can be explained as well by the influence of small-scale waves.

Data in Figs. 10b, 16, and 17 illustrate the effective values of brightness temperature fluctuations for different wind speed and linear size of the observed area (resolution).

Two-dimensional diagrams in Figs. 18 and 19 represent the relative sensitivity of mean values, variations, and polarization at centimeter wavelengths to wind speed.

5. CONCLUSIONS

The main parameters which determine the characteristics of microwave radiation from a roughened sea are:

1. Effective steepness of small- and large-scale waves
2. Relative coverage of sea surface by foam formations (or even relative volume of foam, considering the dependence of emissivity on the thickness of foam patches)
3. Spatial anisotropy (inhomogeneity: the 2d or 3d character) of sea surface roughness
4. Relative concentration of droplets due to rainfall or storm spray which interact simultaneously with the sea surface.

At wavelengths shorter than 3 cm, the characteristics of microwave radiation are dependent on geometrical sea surface parameters for $V \leqslant 7–10$ m/s and on foam formations for $V > 10$ m/s.

At wavelengths longer than 20 cm, the characteristics of radiation are dependent mainly on the geometry of the surface.

A field of microwave radiation over a roughened sea is characterized by spatial–temporal variations, quasi-periodic oscillations of which are caused by variations of wave slope and by the different character of ripples on the windy and sheltered sides of large-scale waves.

Thus, objectively, the spectral data of microwave radiation give information about the effective steepness of waves, in some manner on the anisotropy (inhomogeneity) of roughness, on relative coverage by foam (relative volume of foam), and about mean length (mean period) of waves. The accuracy of wind speed (sea state) estimates using these data is about 3–5 m/s in the range from 0 to 20–30 m/s (or 1–2 of wind force). In the absence of a complex wave structure, there is the possibility of wind direction determination for $V \leqslant 5–7$ m/s and the main direction of waves for stronger wind speeds with an accuracy of about $\pm 20–30°$. The accuracy of determination of mean length (period) of sea waves is about 20% for high resolution.

ACKNOWLEDGMENTS. The author thanks his colleagues Dr. S. V. Pereslegin, Dr. G. I. Chuhray, postgraduates M. A. Antipychev and A. G. Grankov for their help in experiments and, posthumously, Professor A. E. Basharinov, for fruitful discussions and advice which stimulated the author in these investigations.

REFERENCES

Armand, N. A., A. E. Basharinov, L. F. Borodin, E. N. Zotova, and A. M. Shutko (1980): Radiophysical methods in remote sensing of environment. *Problems of Modern Radioengineering and Electronics*, Nauka, Moscow, 95–138 (in Russian).

Au, B., J. Kenney, L. U. Martin, and D. Ross (1974): Multi-frequency radiometric measurements of foam and a mono-molecular slick. *Proc. 9th Int. Symp. Remote Sensing of the Environment*, Ann Arbor, 1974, **3**, 1763.

Basharinov, A. E., and A. M. Shutko (1980): Research into the measurements of sea state, sea temperature and salinity by means of microwave radiometry. *Boundary-Layer Meteorol.* **18**, 55–64.

Eklund, F., J. Nilson, and A. Blomquist (1974): False alarm risks at radar detection of oil spill. *Proceedings, URSI Commiss. II Specialist Meeting on Microwave Scattering and Emission from the Earth*, Berne, 1974, 38–45.

La Fond, E. C., and P. V. Bhavanarayana (1959): Foam on the sea. *J. Mar. Biol. Assoc. India* **1**, 228–232.

Shutko, A. M. (1968): Experimental investigations of the characteristics of sea surface microwave radiation. *Trans. Gener. Geophys. Observ.* **222**, 19–21 (in Russian).

Shutko, A. M. (1978a): Characteristics of background microwave radiation of sea surface. *Theses Rep. at 12th All Union Conference on Radio Waves Prop.*, Nauka, Moscow, **2**, 206–210 (in Russian).

Shutko, A. M. (1978b): Investigations of water areas by means of microwave radiometry. *Radiotekh. Electron*, **23**, 2107–2119.

Shutko, A. M., B. G. Kutuza, O. I. Yakovlev, A. I. Efimov, and Paveliev (1978): *Radiophysical Investigations of Planets. Itogi Nauki Tekh. Radiotekh.* **16**, Moscow, VINITI, 1978 (in Russian).

Soviet–American Experiment "Bering" (1975): *Proc. Final Symp.* (K. Y. Kondratiev, Y. I. Rabinovich, and W. Nordberg, eds.), Hydrometeoizdat, Leningrad, 1975.

VII

Wave Modeling

39

INVERSE MODELING IN OCEAN WAVE STUDIES

Robert Bryan Long

ABSTRACT. A fundamental description of sea state is provided by the surface wave directional spectrum. Efforts to measure the directional spectrum with *in situ* instrumentation generally yield data which represent estimates of integral properties of the spectrum. Extracting an estimate of the full, two-dimensional spectrum from such data presents a typical example of the so-called "inverse problem." The problem is solved by finding a model spectrum which is statistically consistent with the observations and, at the same time, satisfies a set of externally imposed constraints required to make the problem determinate. We review here the theory of the linear inverse problem and its application to the estimation of surface wave directional spectra. The principles developed are then used to investigate the optimal analysis procedure proposed by Long and Hasselmann (1979).

1. INTRODUCTION

A fundamental and most useful description of sea state is provided by the surface wave directional spectrum, $F(\theta, f)$, which defines the distribution of surface displacement variance density as a function of wave propagation direction θ and cyclic wave frequency f. In the linear approximation, all other statistical properties of the wave-induced water motion can be calculated from knowledge of the directional spectrum.

Unfortunately, $F(\theta, f)$ cannot be measured directly but must be deduced from observations of the evolution in space and time of other directly measurable properties of the water motion, such as surface elevation, or pressure or velocity in the water column. With a few notable exceptions (e.g., stereophotographs and measurements made from moving platforms), these observations take the form of time series of wave properties at a few discrete points in space. As we demonstrate in the next section, auto- and cross spectra of these time series are weighted integral properties of the directional spectrum. The problem of estimating the directional spectrum from these integral relationships is a typical example of the so-called "inverse problem."

The inverse problem arises whenever the physical observables of a system under study are given by known functionals of a set of unknown parameters of the system. The known functionals must be defined by a valid mathematical model of the system, so that values of the physical observables (the "exact data") can be calculated once the parameters of the system are known. Calculating the data from the system parameters is the forward problem; fixing the parameters from observations of the data is the inverse problem.

ROBERT BRYAN LONG ● Atlantic Oceanographic and Meteorological Laboratories. National Oceanic and Atmospheric Administration, Miami, Florida 33149.

Clearly, solving the forward problem is trivial once the physics of the system are properly understood. The inverse problem, on the other hand, is inevitably fraught with difficulties: Physical limitations restrict the data to a discrete and finite set of observations which, moreover, always contain errors. Generally, the unknown parameter set is infinite-dimensional, so that the data, besides being inaccurate, are also generally insufficient to fix a solution. In those cases where the physics of the system restrict the unknown parameters to a finite set, the indeterminacy of the problem may be overcome by taking just enough observations, but then one often finds the results to be unstable. Small errors in the data may cause large and unrealistic variations in the solution, indicating that the data set is, to some extent, insensitive to changes in the system parameters. Finally, if the number of observations exceeds the number of unknown parameters, the data set may be inconsistent due to errors in the observations and, possibly, to approximations in the mathematical models. The rational interpretation of such inaccurate, insensitive, insufficient, and/or inconsistent data is the province of inverse modeling theory.

The theory of the inverse problem is now well developed, at least for the linear case, due largely to the efforts of geophysicists and mathematicians attempting to deduce the internal structure of the earth from seismic data. In recent years, the rapid development of remote sensing has provided additional impetus, and explicit applications now abound in many different fields, including physics, chemistry, astronomy, engineering, medicine, meteorology, and oceanography.

It is not the intent of this chapter to present a comprehensive review of the vast and rapidly expanding literature on this subject, although a brief list of references which I have found useful is included (Franklin, 1970; Jackson, 1972; Munk and Wunsch, 1979; Parker, 1972, 1977; Twomey, 1963; Wiggins, 1972; Wunsch, 1978). Instead, I shall review in some detail the estimation of the surface gravity wave directional spectrum from multicomponent wave measurements, a problem of immediate interest in the present context which draws upon many aspects of inverse modeling theory for its solution and for evaluation of the results obtained. In the process, we shall derive the general form of the directional spectrum estimation problem for observations consisting of time series of wave properties at fixed points (Section 2). The derived data are shown to be integral properties of the unknown directional spectrum, and solving the resulting set of integral equations presents a typical linear inverse problem. In Section 3, we review the fundamentals of linear inverse modeling theory to identify the principles to be applied to finding a solution and evaluating its significance. In Section 4, the variational approach of Long and Hasselmann (1979), which introduces a nonlinear model spectrum, is reviewed, interpreted, and evaluated in the light of these principles. Pitch/roll buoy data are used for illustration, and some additional comments and conclusions are presented in Section 5.

2. ESTIMATING THE DIRECTIONAL WAVE SPECTRUM—DEFINING THE PROBLEM

Let $\zeta_i(\mathbf{x}, t)$ be some directly measurable property of the wave-induced water motion at horizontal vector position $\mathbf{x}$ and time t. Obvious examples are displacement of the surface, pressure and velocity in the water column, and temporal and spatial derivatives of any of these. In the linear approximation, these properties may be represented by Fourier–Stieltjes integrals,

$$\zeta_i(\mathbf{x}, t) = \int dA(\mathbf{k}, \omega) L_i(\mathbf{k}, \omega) e^{i(\mathbf{k} \cdot \mathbf{x} - \omega t)} \tag{1}$$

where $2|dA(\mathbf{k}, \omega)|$ is the differential amplitude of the surface wave with vector wavenumber $\mathbf{k}$ and radian frequency ω, and $L_i(\mathbf{k}, \omega)$ is the transfer function relating ζ_i to surface displacement. At this level of approximation, the statistics of the wave properties are jointly Gaussian with means

$$\langle \zeta_i \rangle = 0 \tag{2}$$

and covariances

$$\langle \zeta_i(\mathbf{x},t)\zeta_j(\mathbf{x}+\mathbf{r},t+\tau)\rangle = \int \langle dA(\mathbf{k},\omega)dA^*(\mathbf{k},\omega)\rangle L_i(\mathbf{k},\omega)L_j^*(\mathbf{k},\omega)e^{-i(\mathbf{k}\cdot\mathbf{r}-\omega\tau)} \tag{3}$$

The angle brackets $\langle\ \rangle$ denote ensemble averages, and in (3) we have invoked the reality, homogeneity, and stationarity of the wave properties. The existence of a dispersion relation allows associating each incremental contribution to the sum in (3) with a linear surface wave having propagation direction θ and cyclic frequency $f(=\omega/2\pi)$; this is done by introducing the *directional spectral density function,*

$$F(\theta,f)=\frac{\langle dA(\mathbf{k},\omega)dA^*(\mathbf{k},\omega)\rangle}{d\theta df} \tag{4}$$

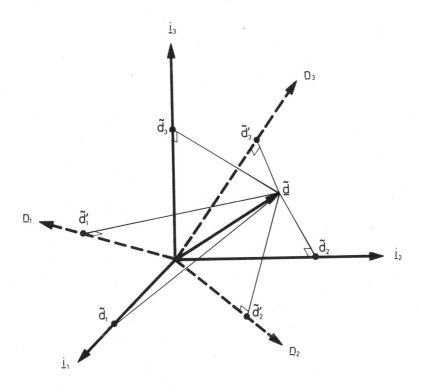

Standard Basis (Cols. of $\underline{\mathbf{I}}_n$) $\longrightarrow$ $\underline{\mathbf{i}}_j$

Alternative Basis (Cols. of $\underline{\mathbf{N}}$) $\dashrightarrow$ $\underline{\mathbf{n}}_j$

FIGURE 1. Definition of a vector in alternative coordinate systems. Each element of an $n \times 1$ matrix may be interpreted as the projection of an n-dimensional vector onto the corresponding axis of a "standard" coordinate system. Thus, the columns $\mathbf{i}_j$ of $\mathbf{I}_n$ are orthogonal unit vectors defining the standard basis. The columns $\mathbf{n}_j$ of $\mathbf{N}$ also represent orthogonal unit vectors projected onto the standard coordinates; these form an alternative reference system, and each element of any vector in the new coordinates is given by the scalar product of the corresponding column vector (in the standard basis) with the corresponding column of $\mathbf{N}$. A three-dimensional example is illustrated.

Substituting this into (3) and Fourier transforming with respect to τ then gives

$$E_{ij}(f,\mathbf{r}) = \int_0^{2\pi} d\theta [L_i(\theta,f)L_j^*(\theta,f)e^{-i\mathbf{k}\cdot\mathbf{r}}]F(\theta,f) \qquad (5)$$

where $E_{ij}(f,\mathbf{r})$ is the *frequency cross spectrum* between the ith and jth wave property at frequency f and horizontal displacement $\mathbf{r}$. The transform with respect to τ is feasible because realizable instrumentation generally provides densely sampled time series of wave properties, but only at a few discrete points in space. For that reason, the additional transform with respect to $\mathbf{r}$, which would yield an explicit expression for $F(\theta,f)$, is generally not possible.

For example, the pitch/roll buoy, widely used in recent years to make directional wave observations, records time series of surface vertical acceleration $(L_1 = -4\pi^2 f^2)$ and two components of surface slope $(L_2 = ik\cos\theta$ and $L_3 = ik\sin\theta)$ at a single location $(\mathbf{r} = 0)$ (Longuet-Higgins *et al.*, 1963). The cloverleaf buoy (Cartwright and Smith, 1964), which resembles three pitch/roll buoys tied together by a rigid triangular framework, measures, in addition to the above, three components of surface curvature $(L_4 = -k^2\cos^2\theta$, $L_5 = -k^2\sin^2\theta$, and $L_6 = -k^2\sin\theta\cos\theta)$ at the same point $(\mathbf{r} = 0)$. An array of N wave gages (Barber, 1963) measures surface elevation $(L_1 = 1)$ at N discrete locations, yielding cross spectra for $N(N-1)+1$ displacements $\mathbf{r}_j$. Arrays containing wave gages and current meters have been used (Forristall *et al.*, 1978), as have arrays of bottom-mounted pressure gages (Snyder, 1974). Arrays containing any mix of such instruments are conceivable, yielding data which can always be placed in the form of Eq. (5).

The cross spectra E_{ij}, then, are the physical observables of the system (the wave field); the unknown parameters are the values of the directional spectrum F for each value of θ. Since θ is continuous, the parameter set is infinite-dimensional, while any realizable instrumentation can provide only a finite number of cross spectra; moreover, the available time series are inevitably short, so that estimates of E_{ij} always contain statistical sampling error. The data are, therefore, both inadequate and inaccurate. They may also be insensitive if the instrumentation is not optimally designed. Solving Eq. (5) for F is clearly an archetypical inverse problem.

In order to obtain a meaningful solution, we must (1) add sufficient external constraints to make the problem determinate and (2) account properly for the sampling errors in the observations. Having constructed a solution, we measure our success by (1) how well the solution reproduces the observations (data fit), (2) how well the solution resolves details in the directional properties of the spectrum (parameter resolution), and (3) how sensitive the solution is to small variations in the data (parameter stability). In the next section, we will develop the tools needed to carry out this analysis.

3. THEORY OF THE GENERALIZED INVERSE

In the past, constructing solutions to sets of integral equations such as (5) generally began with a more-or-less arbitrary, finite-dimensional parameterization of the system unknown. This casts the integral equations into a matrix form which subsequently determines the steps to be taken to find an acceptable solution. Recently, a new approach, based on the work of Backus and Gilbert (1967), has been developed which applies a variational principle to the problem *a priori* and yields an optimal parameterization, a statistically acceptable fit to the data, the maximum possible solution stability, and the best parameter resolution consistent with these properties (Long and Hasselmann, 1979). We will examine the former approach first in order to introduce relevant aspects of inverse modeling theory; we will then use those ideas to examine the variational method in Section 4.

Equation (5) is a specific example of the inverse problem having general form

$$\underset{n\times 1}{\mathbf{d}} = \int_R d\theta\, S(\theta)\, \underset{n\times 1}{\mathbf{b}}\,(\theta) \qquad (6)$$

where $\mathbf{d}$ is an $n \times 1$ column vector composed of the n physical observables, $\mathbf{b}$ is an $n \times 1$ column vector of kernel functions, R is the range of the independent variable θ, and S is the system unknown which is to be estimated. In this case, the functional is linear; more complicated inverse problems involving nonlinear functionals are also possible, but, with few exceptions, these are solved by iterative procedures involving linearization about some approximate solution, which recasts the problem at each iteration into the above form. Equation (6) is, therefore, of even more general interest than is initially apparent.

The problem is converted to matrix form by replacing S with a model function containing a finite number (say, m) of free parameters, i.e., we set

$$S(\theta) = \underset{1 \times m}{\boldsymbol{\phi}^{\mathrm{T}}(\theta)} \underset{m \times 1}{\mathbf{p}} \tag{7}$$

where $\boldsymbol{\theta}$ is a column vector of basis functions, T indicates transpose, and $\mathbf{p}$ is a column vector of coefficients. This is a linear parameterization; nonlinear models are also sometimes used (Olbers et al., 1976), but again, solutions are generally obtained by linearizing about a zero-order state, so that the linear model is still the appropriate form at each iteration. Substituting (7) into (6) then yields

$$\underset{n \times m}{\mathbf{A}} \underset{m \times 1}{\mathbf{p}} = \underset{n \times 1}{\mathbf{d}} \tag{8}$$

where

$$\underset{n \times m}{\mathbf{A}} = \int_R d\theta \, \underset{n \times 1}{\mathbf{b}} \, \underset{1 \times m}{\boldsymbol{\phi}^{\mathrm{T}}} \tag{9}$$

To the extent that (7) holds, (8) is an exact statement of the forward problem. For the inverse problem, the exact data $\mathbf{d}$ are replaced by the available estimates $\hat{\mathbf{d}}$, and the problem to be solved is

$$\underset{n \times 1}{\hat{\mathbf{d}}} = \underset{n \times m}{\mathbf{A}} \underset{m \times 1}{\mathbf{p}} \tag{10}$$

Given a reasonable set of basis functions, the error in (7) may be made arbitrarily small by increasing the number of elements in $\boldsymbol{\phi}$. As we shall see, this does not imply that the error in recovering S by solving (10) can be made arbitrarily small in the same way.

Depending on the number of equations n, the number of unknown parameters m, and the rank q of the matrix $\mathbf{A}$, the problem (10) may or may not have an exact solution, and if one exists, it may or may not unique. But exact solutions are not necessary, since $\hat{\mathbf{d}}$ is known to contain errors, and we can always invoke additional external constraints to fix a nonunique solution. Therefore, we can always construct a *generalized inverse* matrix $\mathbf{H}$ which, operating on the observed data $\hat{\mathbf{d}}$, will generate a model solution $\hat{\mathbf{p}}$ to (10),

$$\underset{m \times 1}{\hat{\mathbf{p}}} = \underset{m \times n}{\mathbf{H}} \underset{n \times 1}{\hat{\mathbf{d}}} = \underset{m \times m}{(\mathbf{HA})} \underset{m \times 1}{\mathbf{p}} \tag{11}$$

and a corresponding set of model data $\hat{\mathbf{d}}$,

$$\underset{n \times 1}{\hat{\mathbf{d}}} = \underset{n \times m}{\mathbf{A}} \underset{m \times 1}{\hat{\mathbf{p}}} = \underset{n \times n}{(\mathbf{AH})} \underset{n \times 1}{\hat{\mathbf{d}}} \tag{12}$$

The matrix $\mathbf{HA}$ is called the *resolution matrix*; each element of $\hat{\mathbf{p}}$ is the result of "viewing" the unknown vector $\mathbf{p}$ through the "window" defined by the corresponding row of $\mathbf{HA}$. If $\mathbf{HA} = \mathbf{i}_m$

(the $m \times m$ unit matrix), parameter resolution is complete and the model solution is *unique*. The matrix $\mathbf{AH}$ is called the *information density matrix* and plays a similar role for the data. If $\mathbf{AH} = \mathbf{I}_n$ (the $n \times n$ unit matrix), the observed data are completely resolved and the model solution is *exact*.

To construct a generalized inverse requires specifying what we consider to be desirable properties. This is done by identifying some property η of the solution which is to be minimized. Obvious candidates are:

1. Best possible data fit, e.g.,

$$\eta = |\hat{\mathbf{d}} - \mathbf{d}|^2 = |(\mathbf{AH} - \mathbf{I}_n)\mathbf{d}|^2 = \min \tag{13}$$

2. Best possible parameter resolution, e.g.,

$$\eta = |\hat{\mathbf{p}} - \mathbf{p}|^2 = |(\mathbf{HA} - \mathbf{I}_m)\mathbf{p}|^2 = \min \tag{14}$$

3. Maximum possible solution stability, e.g.,

$$\eta = \operatorname{var}\{|\hat{\mathbf{p}}|\} = \min \tag{15}$$

where η is the variance induced by variability in $\mathbf{d}$

4. Maximum possible solution simplicity, e.g.,

$$\eta = |\hat{\mathbf{p}}|^2 = \min \tag{16}$$

The first three are related to the "measures of success" mentioned in the previous section; the fourth is a reasonable external constraint which may be used when needed to resolve the nonunique case. These properties are not mutually compatible, and some trade-off is always necessary.

If, for example, we demand an exact fit to the data at all costs, then (13) becomes

$$\eta = |(\mathbf{AH} - \mathbf{I}_n)\mathbf{d}|^2 = 0$$

from which

$$\mathbf{H} = \mathbf{A}^{-1} \tag{17}$$

if $\mathbf{A}^{-1}$ exists. It does, as we shall shortly show, if $q = m = n$, which defines the class of problems labeled *determinate*. Then,

$$\mathbf{AH} = \mathbf{HA} = \mathbf{I}_n$$

so that the solution is both exact (perfect fit) and unique (complete parameter resolution).

If $\mathbf{A}^{-1}$ does not exist, we may have to make do with a least-squares fit to the data; then, from (13), setting

$$\frac{\partial \eta}{\partial \hat{p}_k} = \frac{\partial}{\partial \hat{p}_k}[(\mathbf{A}\hat{\mathbf{p}} - \mathbf{d})^{\mathrm{T}}(\mathbf{A}\hat{\mathbf{p}} - \mathbf{d})] = 0$$

yields

$$\mathbf{H} = (\mathbf{A}^{\mathrm{T}}\mathbf{A})^{-1}\mathbf{A}^{\mathrm{T}} \tag{18}$$

if $(\mathbf{A}^{\mathrm{T}}\mathbf{A})^{-1}$ exists. It does if $q = m < n$, which defines the class of problems labeled *overconstrained*.

The matrices

$$\mathbf{AH} = \mathbf{A}(\mathbf{A}^T\mathbf{A})^{-1}\mathbf{A}^T \neq \mathbf{I}_n$$

and

$$\mathbf{HA} = \mathbf{I}_m$$

indicating an inexact but unique result.

A third class of problems, called *underdetermined*, exists for which $q = n < m$. In this case, an infinity of exact solutions exist, and we select one by applying (16) subject to the constraint $\mathbf{A}\hat{\mathbf{p}} - \hat{\mathbf{d}} = 0$, i.e., by minimizing

$$\eta = \hat{\mathbf{p}}^T\hat{\mathbf{p}} + \boldsymbol{\alpha}^T(\mathbf{A}\hat{\mathbf{p}} - \hat{\mathbf{d}})$$

where $\boldsymbol{\alpha}$ is a vector Lagrange multiplier. This yields

$$\mathbf{H} = \mathbf{A}^T(\mathbf{A}\mathbf{A}^T)^{-1} \tag{19}$$

if $(\mathbf{A}\mathbf{A}^T)^{-1}$ exists, which it does in this case. Then

$$\mathbf{AH} = \mathbf{I}_n$$

and

$$\mathbf{HA} = \mathbf{A}^T(\mathbf{A}\mathbf{A}^T)^{-1}\mathbf{A} \neq \mathbf{I}_m$$

indicating an exact but nonunique result.

A fourth and last class of problems exists for which $q < m$ and n (*overconstrained and underdetermined*); no obvious generalized inverse immediately suggests itself in this case.

For each of the first three classes, we have asserted without proof that the required inverse matrix exists. Further, we have not constructed a generalized inverse for the fourth class of problems. And finally, we have not examined the stability of the resulting solutions. These gaps are neatly filled and a useful geometric interpretation provided by introducing the *natural inverse* (Lanczos, 1961) as follows.

The matrix $\mathbf{A}$ can a always be factored into the following product,

$$\underset{n \times m}{\mathbf{A}} = \underset{n \times q}{\mathbf{N}} \; \underset{q \times q}{\boldsymbol{\Lambda}} \; \underset{q \times m}{\mathbf{M}^T} \tag{20}$$

where

$$\mathbf{A}\mathbf{A}^T\mathbf{N} = \mathbf{N}\boldsymbol{\Lambda}^2$$

and

$$\mathbf{A}^T\mathbf{A}\mathbf{M} = \mathbf{M}\boldsymbol{\Lambda}^2$$

The $q \times q$ matrix $\boldsymbol{\Lambda}^2$ is diagonal and contains the q nonzero eigenvalues common to the matrices $\mathbf{A}\mathbf{A}^T$ and $\mathbf{A}^T\mathbf{A}$, arranged for convenience in the order of decreasing eigenvalue. The q columns of $\mathbf{N}$ and $\mathbf{M}$ are the corresponding eigenvectors. Each set of eigenvectors is orthogonal and normalized so that

$$\mathbf{N}^T\mathbf{N} = \mathbf{I}_q = \mathbf{M}^T\mathbf{M} \tag{21}$$

N is not square and has no inverse unless $q = n$; similarly, M has no inverse unless $q = m$. Hence, from (21),

N^{-1} exists and equals N^T if $q = n$

M^{-1} exists and equals M^T if $q = m$

Then,

$A^{-1} = (N\Lambda M^T)^{-1}$ exists if $q = m = n$

$(A^T A)^{-1} = (M\Lambda^2 M^T)^{-1}$ exists if $q = m$

$(AA^T)^{-1} = (N\Lambda^2 N^T)^{-1}$ exists if $q = n$

which are just the conditions defining the determinate, overconstrained, and underdetermined problem classes, respectively.

Geometrically, the elements $\tilde{d}_i$ of the column vector $\tilde{d}$ may be interpreted as the components of the vector $\tilde{d}$ in an n-dimensional, *standard coordinate system* specified by the set of orthogonal unit vectors forming the columns of I_n. Any other set of orthonormal column vectors in this standard basis may be used to define an alternative coordinate system. The components $\tilde{d}'_i$ of $\tilde{d}$ with respect to the new coordinates are given by the scalar products of the original column vector with the corresponding unit vectors of the new set. For example,

$$N^T \tilde{d} = d'$$
$$\scriptstyle q \times n \quad n \times 1 \qquad q \times 1$$

gives the projection of the vector $\tilde{d}$ onto the coordinates established by the columns of N (see Fig. 1). The expansion

$$N\tilde{d}' = \tilde{d}'_1 \begin{bmatrix} N_{11} \\ \vdots \\ N_{N1} \end{bmatrix} + \cdots + \tilde{d}'_q \begin{bmatrix} N_{1q} \\ \vdots \\ N_{Nq} \end{bmatrix} = NN^T \tilde{d} \tag{22}$$

is, then, the projected vector expressed in the original, standard coordinates. Note that the original vector is recovered exactly if N contains all n eigenvectors (i.e., if $q = n$, in which case $N^T = N^{-1}$). Otherwise, $\tilde{d}$ may have components orthogonal to the q-dimensional space spanned by the columns of N which cannot be recovered. Similarly,

$$M^T \quad p \quad = p'$$
$$\scriptstyle q \times m \quad m \times 1 \qquad q \times 1$$

gives the projection of the parameter vector p, defined in the standard coordinates specified by the columns of I_m, onto the alternative coordinates defined by the columns of M. The projected vector expressed in the standard coordinates,

$$Mp' = MM^T p \tag{23}$$

is the same as p only if M contains all m eigenvectors ($q = m$) or if p has no components orthogonal to the subspace spanned by the columns of M.

Note, however, that in both cases, the projected vector is a least-squares approximation to the original in that no other vector constrained to lie in the q-dimensional subspace exhibits a smaller displacement from the original vector (see Fig. 2).

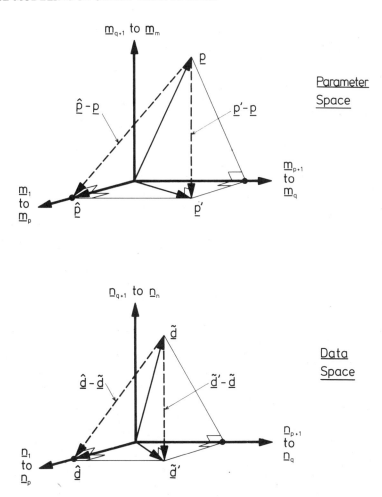

FIGURE 2. Projection of a finite-dimensional vector onto a smaller subspace. Greater-than-three-dimensionality is indicated schematically by telescoping subsets of orthogonal dimensions onto the three axes shown. Thus, the horizontal plane has dimensionality q; the remaining dimensions, represented by the vertical axis, correspond to eigenvectors of $\mathbf{A}^T\mathbf{A}$ or $\mathbf{A}\mathbf{A}^T$ having zero eigenvalues. Projections $\mathbf{p}'$ and $\tilde{\mathbf{d}}'$ are least-squares approximations to $\mathbf{p}$ and $\tilde{\mathbf{d}}$, respectively, since no other vector lying in the q-dimensional hyperplane can yield a smaller difference vector. In particular, projections onto the p-dimensional subspace (hatted vectors), where $p < q$, are poorer approximations to the original vectors than are the primed vectors.

Returning to the inverse problem, we now substitute the factorization (20) into (10) to get

$$\mathbf{Ap} = \mathbf{N}\Lambda(\mathbf{M}^T\mathbf{p}) = \tilde{\mathbf{d}} \tag{24}$$

By inspection, the statement $\mathbf{Ap} = \tilde{\mathbf{d}}$ is seen to contain information only on the projection of $\mathbf{p}$ onto the subspace spanned by the eigenvectors in $\mathbf{M}$; components of $\mathbf{p}$ orthogonal to this subspace may have any value, since they are annihilated by the matrix $\mathbf{A}$ and contribute nothing to the product $\mathbf{Ap}$. The simplest (minimum modulus) model solution would, therefore, have those undetermined components set to zero, equivalent to taking the projection of $\mathbf{p}$ expressed in the standard coordinates as the model solution [see (23)],

$$\hat{\mathbf{p}} = \mathbf{M}\mathbf{M}^\mathsf{T}\mathbf{p} \tag{25}$$

Further, the left-hand side of (24) is seen to be an expansion in the eigenvectors forming $\mathbf{N}$; thus, the product $\mathbf{Ap}$ cannot generate any components of $\hat{\mathbf{d}}$ orthogonal to the subspace spanned by these eigenvectors. The best job (in the least-squares error sense) that any solution can do is to reproduce as model data the projection of $\hat{\mathbf{d}}$ onto that subspace, expressed in the standard coordinates [see (22)],

$$\hat{\mathbf{d}} = \mathbf{N}\mathbf{N}^\mathsf{T}\hat{\mathbf{d}} \tag{26}$$

Comparing (25) and (26) with (11) and (12), we see that the optimal resolution matrix (in the minimum-modulus sense) is

$$\mathbf{HA} = \mathbf{M}\mathbf{M}^\mathsf{T} \tag{27}$$

and the optimal information density matrix (in the least-squares-data-fit sense) is

$$\mathbf{AH} = \mathbf{N}\mathbf{N}^\mathsf{T} \tag{28}$$

The generalized inverse

$$\underset{m \times n}{\mathbf{H}} = \underset{m \times q}{\mathbf{M}} \underset{q \times q}{\mathbf{\Lambda}^{-1}} \underset{q \times n}{\mathbf{N}^\mathsf{T}} \tag{29}$$

has exactly these properties. This is Lanczos's so-called "natural inverse."

Clearly, the natural inverse has most of the desirable properties proposed above. Specifically:

1. It always exists (hence, provides an optimal generalized inverse for the fourth class of problems; moreover, it reduces to $\mathbf{A}^{-1}$, $(\mathbf{A}^\mathsf{T}\mathbf{A})^{-1}\mathbf{A}^\mathsf{T}$, and $\mathbf{A}^\mathsf{T}(\mathbf{A}\mathbf{A}^\mathsf{T})^{-1}$ whenever these forms exist)
2. It maximizes data fit (as argued geometrically above, or one can show analytically that $\sum_j [(\mathbf{AH})_{ij} - \delta_{ij}]^2 = \min$ for each i when $\mathbf{AH} = \mathbf{N}\mathbf{N}^\mathsf{T}$)
3. It maximizes parameter resolution (as argued geometrically above, or one can show analytically that $\sum_j [(\mathbf{AH})_{ij} - \delta_{ij}]^2 = \min$ for each i when $\mathbf{HA} = \mathbf{M}\mathbf{M}^\mathsf{T}$)
4. It maximizes solution simplicity (all undetermined components of $\hat{\mathbf{p}}$ are set to zero)
5. Finally, it enables the study and management of solution stability.

The stability of a given model solution is characterized by the covariances of the errors in the elements of $\mathbf{p}$. These are related to errors in the data by

$$\delta\hat{\mathbf{p}} = \mathbf{H}\delta\hat{\mathbf{d}}$$

whence

$$\mathrm{cov}\{\hat{\mathbf{p}}\} = \langle \delta\hat{\mathbf{p}}\delta\hat{\mathbf{p}}^\mathsf{T} \rangle = \mathbf{H}\mathbf{V}\mathbf{H}^\mathsf{T} \equiv \mathbf{U} \tag{30}$$

where

$$\mathbf{V} = \langle \delta\hat{\mathbf{d}}\delta\hat{\mathbf{d}}^\mathsf{T} \rangle$$

is the covariance matrix describing the statistical properties of the data errors. For the natural inverse, the parameter variances are

$$U_{ii} = \sum_{j,k} M_{ij} \Lambda_{jj}^{-1} V'_{jk} \Lambda_{kk}^{-1} M_{ik} \tag{31}$$

where $\mathbf{V}' = \mathbf{N}^T \mathbf{V} \mathbf{N}$. If some of the eigenvalues Λ_{ii}^2 are very much smaller than the others, terms in the sum (31) containing these as factors dominate, and variability becomes intolerably large whenever any eigenvalue becomes vanishingly small. Clearly, the source of solution instability is the existence of small (but nonzero) eigenvalues of the matrices $\mathbf{AA}^T$ and $\mathbf{A}^T\mathbf{A}$. These, in turn, are the consequence of the nature of the system and the quantities being measured (which fix the kernel functions $\mathbf{b}$) and the choice of basis functions $\boldsymbol{\phi}$, which together, through (9) determine $\mathbf{A}$.

The fact that it is difficult to differentiate numerically between identically zero and very small eigenvalues suggests an appropriate way to improve solution stability, namely, treat very small eigenvalues as if they were zero. This reduces the dimensions of Λ to $p \times p$, where $p < q$, and drops the corresponding eigenvectors from the $\mathbf{N}$ and $\mathbf{M}$ matrices. The sums in (31) then extend over $1 \leqslant (j,k) \leqslant p < q$, so that the largest contributors have been "filtered out." As is usually the case, the price paid for improved stability is diminished parameter resolution and data fit; in effect, additional components of $\mathbf{p}$ and $\bar{\mathbf{d}}$ have been lost (become undetermined or unrepresentable by using the projections onto the p-dimensional subspaces for $\hat{\mathbf{p}}$ and $\bar{\mathbf{d}}$ instead of their projections onto the larger q-dimensional subspaces (see Fig. 2). The choice of cutoff index p depends on the trade-off one is willing to make between solution variability and data fit. A suitable rule is to seek maximum parameter stability consistent with a model data vector which is not statistically inconsistent with the observations. We need to make these statements precise and quantitative, but this is best done in the context of the variational approach, so we postpone further discussion of this point for the present.

Note that solution variability and resolution, expressed so far in terms of parameter characteristics, may also be displayed in the space of the variable θ. Since

$$\delta \hat{S}(\theta) = \boldsymbol{\phi}(\theta)^T \delta \hat{\mathbf{p}}$$

from (7), then

$$\langle \delta S(\theta)^2 \rangle = \boldsymbol{\phi}(\theta)^T \langle \delta \hat{\mathbf{p}} \delta \hat{\mathbf{p}}^T \rangle \boldsymbol{\phi}(\theta)$$

$$= \boldsymbol{\phi}(\theta)^T \mathbf{HVH} \boldsymbol{\phi}(\theta) \tag{32}$$

gives the variance of the solution $\hat{S}(\theta)$ produced by $\mathbf{H}$ as a function of θ and the known data covariances. Further, if we let $\tilde{S}$ be the sample value of S corresponding to $\bar{\mathbf{d}}$,

$$\bar{\mathbf{d}} = \int_R d\theta' \, \tilde{S}(\theta') \mathbf{b}(\theta')$$

then

$$\hat{S}(\theta) = \boldsymbol{\phi}(\theta)^T \hat{\mathbf{p}} = \boldsymbol{\phi}(\theta)^T \mathbf{H} \bar{\mathbf{d}}$$

$$= \int_R d\theta' \, \tilde{S}(\theta') W(\theta, \theta') \tag{33}$$

where

$$W(\theta, \theta') = \boldsymbol{\phi}(\theta)^T \mathbf{H} \mathbf{b}(\theta') \tag{34}$$

is the window through which any $\tilde{S}(\theta')$ will be viewed in constructing $\hat{S}(\theta)$ at each θ. These forms are more instructive in some cases than equivalent statements of parameter variability and resolution.

Summarizing: (1) The traditional approach to the inverse problem (6) starts with an arbitrary linear finite-dimensional parameterization. (2) The "geometry" of the resulting matrix problem places absolute limits on the ability to recover a given parameter vector or to reproduce a given data vector. (3) The Lanczos natural inverse provides solutions which achieve this limiting performance. (4) These solutions may be unstable but can be stabilized by filtering out components associated with small eigenvalues at the expense of parameter resolution and data fit. (5) The resulting trade-off is constrained by the requirement that the residual difference between model and observed data must not be "improbably large," a term which will be defined quantitatively in the next section.

The discussion here is, of course, far from complete. There are any number of alternative sets of desirable properties which might be used instead of (13) to (16) to construct generalized inverses. For example, given some knowledge of the statistics of $\mathbf{\bar{d}}$ and the anticipated statistics of $\mathbf{p}$, better choices for (13) and (14) would be

$$\eta = (\mathbf{\hat{d}} - \mathbf{\bar{d}})^T \mathbf{V}^{-1}(\mathbf{\hat{d}} - \mathbf{\bar{d}}) = \min, \qquad \mathbf{V} = \langle \delta\mathbf{d}\,\delta\mathbf{d}^T \rangle$$

and

$$\eta = (\mathbf{\hat{p}} - \mathbf{p})^T \mathbf{U}^{-1}(\mathbf{\hat{p}} - \mathbf{p}) = \min, \qquad \mathbf{U} = \langle \delta\mathbf{p}\,\delta\mathbf{p}^T \rangle$$

(if the necessary inverses exist), so that elements with large expected variability contribute proportionately less to the sums than elements known to be more accurately determined. While the algebra is a little messier, the results remain conceptually the same; we may, in fact, transform the problem to new sets of coordinates for the data and parameter spaces in which $\mathbf{V}$ and $\mathbf{U}$ are unit matrices, after which everything that we have already done follows exactly. Further, we might choose to define the simplest solution to be the smoothest (Twomey, 1977) rather than the solution with minimum modulus, or we might ask for the minimum-modulus solution consistent with the constraint that $|\mathbf{\hat{d}} - \mathbf{\bar{d}}|^2$ equals some constant to be fixed by the statistics of the data errors. Each alternative leads to a different generalized inverse, and to cover all the interesting possibilities would require far more space than is available here. But, in any case, the principles presented in this section are sufficient to analyze and evaluate the performance of any generalized inverse arising from applying a strictly linear model solution such as (7) to a strictly linear problem such as (5). In the next section, we examine the variational technique of Long and Hasselmann (1979) and encounter a nonlinear model parameterization. Although the linear theory no longer strictly applies, the concepts developed here will still prove useful in understanding the performance of this procedure.

4. THE VARIATIONAL TECHNIQUE

Returning to the integral form, the forward problem under consideration is defined by

$$\mathbf{d} = \int_R d\theta\, S\mathbf{b}$$

where S is the "true" function characterizing the system and $\mathbf{d}$ is the resulting "true" data. Realizable observations of $\mathbf{d}$ inevitably contain errors ε, so that

$$\mathbf{\bar{d}} = \mathbf{d} + \varepsilon \tag{35}$$

Then,

$$\mathbf{d} = \varepsilon + \int_R d\theta \, S\mathbf{b} \tag{36}$$

is an exact statement of the inverse problem [whereas the equality implied by (10) was, in general, only approximate].

As we have seen, finding a solution to (36) requires the parameterization of S and a set of additional constraints sufficient to make the problem determinate. In the previous section, the constraints were applied to the matrix form of the problem arising from a more-or-less arbitrary parameterization. As pointed out by Backus and Gilbert (1967), the necessary constraints may also be applied directly to the integral form of the problem. This procedure has two outstanding advantages: (1) a complete set of constraints defining desirable properties of the solution may be incorporated into the problem in one step; and (2) the resulting variational principle specifies the optimal parameterization. If the optimal model solution is strictly linear in the unknown parameters, the theory of Section 3 may then be applied to completely characterize the resulting analysis. However, in the case of directional spectrum estimation, the true solution is known to be nonnegative; including this constraint yields a nonlinear optimal parameterization which, as we shall see, has far-reaching consequences.

In the following, we accept a restriction on the generality of the problem (36) that was not imposed in the previous section, namely, that the elements of the data vector (or equivalently, of the kernel function vector) must be linearly independent; this is to permit the existence of the inverse data covariance matrix $\mathbf{V}^{-1}$, needed for the statistical evaluation of the results. For directional spectrum estimation, this is not an intolerable restriction, since it is always possible to design the instrumentation or formulate the data in a way which avoids exact linear interdependences.

Following Long and Hasselmann (1979) (with some important changes in notation), we define the optimal model solution to (36) as that choice $S = \hat{S}$ for which

$$\eta = \varepsilon^T \mathbf{Q} \varepsilon + \alpha \int_0^{2\pi} d\theta (\hat{S} - S_0)^2 = \min \tag{37}$$

where, from (36),

$$\varepsilon = \mathbf{d} - \int_0^{2\pi} d\theta \, \hat{S}\mathbf{b} \tag{38}$$

Here, $\mathbf{Q}$ is an $n \times n$ symmetric positive-definite matrix, and S_0 is some favored distribution expressing our preconceived notions about what the solution should look like. The two components of (37) are analogous to the best-data-fit and maximum-simplicity criteria proposed for the matrix form of the problem in the previous section. Both cannot be simultaneously minimized unless, by chance, $\hat{S} = S_0$ yields $\varepsilon = 0$. The constant weights α and Q_{ij} determine the balance between the two components at the minimum of η. We will return to how the weights are chosen later.

Any known data-independent constraints on the solution may be incorporated at this stage into (37) using Lagrange multipliers. For example, if S is the directional wave spectrum, then $\hat{S}$ must be nonnegative, whence

$$\int_0^{2\pi} d\theta (\hat{S} - |\hat{S}|)^2 = 0 \tag{39}$$

The quantity to be minimized then becomes

$$\eta = \varepsilon^T \mathbf{Q} \varepsilon + \alpha \int_0^{2\pi} d\theta (\hat{S} - S_0)^2 + \beta \int_0^{2\pi} d\theta (\hat{S} - |\hat{S}|)^2 \tag{40}$$

Taking variations with respect to $\hat{S}$ and requiring $\delta\eta$ to vanish for arbitrary $\delta\hat{S}$ yields

$$\hat{S} = \begin{cases} S_0 + \alpha^{-1}\boldsymbol{\varepsilon}^{\mathrm{T}}\mathbf{Q}\mathbf{b} & \text{if } \hat{S} \geqslant 0 \\ \dfrac{\alpha}{\alpha + 4\beta}[S_0 + \alpha^{-1}\boldsymbol{\varepsilon}^{\mathrm{T}}\mathbf{Q}\mathbf{b}] & \text{if } \hat{S} < 0 \end{cases} \tag{41}$$

The constraint (39) requires that the Lagrange multiplier $\beta \to \infty$, whence the model solution is

$$\hat{S} = [S_0 + \mathbf{b}^{\mathrm{T}}\hat{\mathbf{p}}]G \tag{42}$$

where the unknown parameter vector

$$\hat{\mathbf{p}} = \alpha^{-1}\mathbf{Q}\boldsymbol{\varepsilon} \tag{43}$$

and the masking function

$$G(\theta) = \begin{cases} 1 \text{ where } [S_0 + \mathbf{b}^{\mathrm{T}}\mathbf{p}] \geqslant 0 \\ 0 \text{ otherwise} \end{cases} \tag{44}$$

Thus, the optimal model solution is the sum of the favored model and an expansion in the kernel functions of the data equations, with negative excursions set to zero by the factor G. The n unknown coefficients are fixed by the n data equations (38), which become

$$\mathbf{\hat{d}} = \alpha\mathbf{Q}^{-1}\hat{\mathbf{p}} + \int_0^{2\pi} d\theta[S_0 + \mathbf{b}^{\mathrm{T}}\hat{\mathbf{p}}]G\mathbf{b}$$

or equivalently,

$$(\mathbf{\hat{d}} - \mathbf{d}_0) = [\alpha\mathbf{Q}^{-1} + \mathbf{A}]\hat{\mathbf{p}} \tag{45}$$

where

$$\mathbf{A} = \int_0^{2\pi} d\theta \, \mathbf{b}\mathbf{b}^{\mathrm{T}}G \tag{46}$$

$$\mathbf{d}_0 = \int_0^{2\pi} d\theta \, S_0\mathbf{b}G \tag{47}$$

and the corresponding model data are

$$(\mathbf{\hat{d}} - \mathbf{d}_0) = \mathbf{A}\hat{\mathbf{p}} \tag{48}$$

Solving (45) is a nonlinear problem because G depends implicitly on $\hat{\mathbf{p}}$. The solution is found by an iterative procedure which yields at the ith step

$$\hat{S}^{(i)} = [S_0 + \mathbf{b}^{\mathrm{T}}\hat{\mathbf{p}}^{(i)}]G^{(i)}$$

$$\mathbf{d}_0^{(i)} = \int_0^{2\pi} d\theta \, S_0\mathbf{b}G^{(i)}$$

and

$$\mathbf{A}^{(i)} = \int_0^{2\pi} d\theta \, \mathbf{b}\mathbf{b}^{\mathrm{T}}G^{(i)}$$

Since both $\mathbf{Q}$ and $\mathbf{A}^{(i)}$ are positive definite ($\mathbf{Q}$ by definition, $\mathbf{A}^{(i)}$ because the elements of $\mathbf{b}$ are linearly independent),

$$\hat{\mathbf{p}}^{(i+1)} = [\alpha \mathbf{Q}^{-1} + \mathbf{A}^{(i)}]^{-1}(\check{\mathbf{d}} - \mathbf{d}_0^{(i)})$$
(49)

gives the $(i+1)$st approximation to the solution. On convergence,

$$\hat{\mathbf{p}} = \mathbf{H}(\check{\mathbf{d}} - \mathbf{d}_0)$$

where the "equivalent generalized inverse"

$$\mathbf{H} = (\alpha \mathbf{Q}^{-1} + \mathbf{A})^{-1}$$
(50)

and the absence of a superscript now denotes converged value.

We would like now to examine the performance of this procedure in terms of the same measures of success, i.e., data fit, resolution, and stability, which were applied to the strictly linear problem in the previous section. Unfortunately, the principles developed then do not extend without reservation to the present, nonlinear case; the factor G in (46) and (47) has made the problem itself dependent on the solution in a complicated, implicit way which severely impedes analysis. The best that we can do, short of a full-scale Monte Carlo study of a particular case, is to examine the behavior of the equivalent linear problem posed by the final iteration of the procedure outlined above. With G and, therefore, $\mathbf{A}$ and $\mathbf{d}_0$ fixed at their converged values, the principles of the previous section may then be used to gain at least qualitative insight into how the procedure performs.

Data Fit: With the remaining degrees of freedom provided by the so far undetermined weights α and Q_{ij}, the "equivalent generalized inverse" (50) can be made to reproduce the data as accurately as we wish; if we set $\alpha = 0$, for example, the converged information density matrix,

$$\mathbf{A}\mathbf{H} = \mathbf{A}(\alpha \mathbf{Q}^{-1} + \mathbf{A})^{-1} = (\alpha \mathbf{Q}^{-1}\mathbf{A}^{-1} + \mathbf{I}_n)^{-1}$$
(51)

reduces to $\mathbf{I}_n$, whence, from (48),

$$\check{\mathbf{d}} - \mathbf{d}_0 = \mathbf{A}\hat{\mathbf{p}} = \mathbf{A}\mathbf{H}(\check{\mathbf{d}} - \mathbf{d}_0) = \check{\mathbf{d}} - \mathbf{d}_0$$

Nonzero α, on the other hand, causes the appearance of nonzero off-diagonal elements in the rows of $\mathbf{A}\mathbf{H}$, broadening the windows through which the observed data are viewed in producing elements of the model data. Recognizing the inevitability of data sampling errors and the likelihood that $\mathbf{H} = \mathbf{A}^{-1}$ will yield unstable results, we ask only that the error in data fit not be "improbably large." This is quantified by formulating the hypothesis that *the derived model solution $\hat{S}$ is the true value of S*. If this is true, the residual error in the data fit, ε, is due entirely to instrumentation and statistical sampling error in the observations. By suitable calibration, the former can be made negligibly small compared to the latter. Then, given sufficient degrees of freedom in the cross spectra (5), the errors ε_i are approximately jointly Gaussian, whence,

$$\text{pdf}(\varepsilon; \mathbf{d}) = (2\pi)^{-n/2}|\mathbf{V}|^{-1/2}\exp[-\tfrac{1}{2}\rho^2]$$

where

$$\rho^2 = \varepsilon^{\mathrm{T}}\mathbf{V}^{-1}\varepsilon$$
(52)

and

$$\mathbf{V} = \langle \varepsilon\varepsilon^{\mathrm{T}}\rangle = \mathbf{V}(\mathbf{d})$$

The random variable ρ^2 is an appropriately scaled measure of the size of ε and is distributed as chi-square with n degrees of freedom. If the available data realization $\mathbf{\hat{d}}$ and the hypothetical true data $\mathbf{\hat{d}}$ imply a realization of ρ^2 greater than some threshold value, ρ_γ^2, then ε is "improbably large"; we must then reject the hypothesis at the γ-level of confidence, where

$$\int_0^{\rho_\gamma^2} d\rho^2 \chi_n^2(\rho^2) = \gamma \tag{53}$$

Otherwise, the model solution is accepted as statistically consistent with the data.

Clearly, we facilitate this statistical test of the solution and maximize our chances of passing it if we take the constant weights $\mathbf{Q} = \mathbf{V}^{-1}$ in the definition of the variational principle (37). The first component of η then converges to ρ^2 at the minimum and attains a value which varies with the weight α assigned to the favored model component. If the solution obtained implies $\rho^2 > \rho_\gamma^2$, the α used was too large; conversely, if $\rho^2 < \rho_\gamma^2$, the α used was satisfactory but could have been larger without generating a statistically unacceptable answer. In practice, an outer cycle of iterations is employed to fix that α for which $\rho^2 = \rho_\gamma^2$ for the one available data realization. The resulting solution has the smallest mean-square deviation from the favored model (and the largest α) possible without becoming statistically inconsistent with the available observation at the γ-level of rejection confidence.

The value chosen for γ measures our unwillingness to risk erroneously rejecting a model solution. Long and Hasselmann (1979) chose $\gamma = 0.8$ as a satisfactory compromise between discrimination and statistical confidence. As we shall see shortly, this choice eventually establishes the data-fit/parameter-resolution/parameter-stability trade-off.

The matrix $\mathbf{V}$ is determined by the true data which, by the hypothesis being tested, is the same as the converged model data corresponding to the available observation. The fact that $\mathbf{V}$ is not known before the problem is solved is accommodated in practice by rewriting the recursion formula (49) to read

$$\mathbf{\hat{p}}^{(i+1)} = [\alpha \mathbf{V}^{(i)} + \mathbf{A}^{(i)}]^{-1}(\mathbf{\hat{d}} - \mathbf{d}_0^{(i)})$$

where $\mathbf{V}^{(i)}$ is evaluated at $\mathbf{d} = \mathbf{\hat{d}}^{(i)}$. Long (1980) describes how formulas for $\mathbf{V}$ are derived and gives explicit results for pitch/roll buoy observations.

Once α and $\mathbf{V}$ are fixed (by the one available observation), they are subsequently treated as constants independent of the solution in evaluating the performance of the analysis scheme. This is necessary in order to be conceptually consistent with their treatment as constants in doing the variational calculus.

The effect of having chosen the largest α permitted by the statistical consistency test is best demonstrated by introducing the factorization (20), which, since $\mathbf{A}$ is symmetric in the present case, reduces to

$$\mathbf{A} = \mathbf{N}\mathbf{\Lambda}\mathbf{N}^{\mathrm{T}} \tag{54}$$

here, the columns of $\mathbf{N}$ are eigenvectors and the elements Λ_{ii} of the diagonal matrix $\mathbf{\Lambda}$ are the related eigenvalues of $\mathbf{A}$. As $\mathbf{A}$ is positive definite, $\mathbf{\Lambda}$ is $n \times n$ and $\mathbf{N}$ contains a complete set of eigenvectors (whence $\mathbf{N}^{\mathrm{T}} = \mathbf{N}^{-1}$). Then,

$$(\mathbf{\hat{d}} - \mathbf{d}_0) = \mathbf{A}\mathbf{H}(\mathbf{\hat{d}} - \mathbf{d}_0)$$

$$= \mathbf{N}\mathbf{\Lambda}(\alpha\mathbf{V}' + \mathbf{\Lambda})^{-1}\mathbf{N}^{\mathrm{T}}(\mathbf{\hat{d}} - \mathbf{d}_0) \tag{55}$$

where $\mathbf{V}' = \mathbf{N}^{\mathrm{T}}\mathbf{V}\mathbf{N}$. This simplifies greatly if we imagine having formulated the problem *a priori* in coordinates in which $\mathbf{V} = \mathbf{I}_n$. Then $\mathbf{V}' = \mathbf{I}_n$ also and (55) may be written

$$(\mathbf{\hat{d}}' - \mathbf{d}_0')_i = \frac{\Lambda_{ii}}{\alpha + \Lambda_{ii}}(\mathbf{\hat{d}}' - \mathbf{d}_0')_i \tag{56}$$

where the primes denote projections onto the eigenvectors in $\mathbf{N}$ (e.g., $\hat{\mathbf{d}}' = \mathbf{N}^T \hat{\mathbf{d}}$). Thus, in coordinates defined by the columns of $\mathbf{N}$, components of the observed data corresponding to small eigenvectors ($\Lambda_{ii} \ll \alpha$) are attenuated in the model data by the factor $\sim \Lambda_{ii}/\alpha$. As α is as large as possible, consistent with the available data realization, we have apparently allowed the maximum acceptable data filtering.

Solution Resolution: Similarly, the converged parameter resolution matrix is

$$\mathbf{HA} = (\alpha \mathbf{A}^{-1} \mathbf{V} + \mathbf{I}_n)^{-1}$$

Again, nonzero α leads to nonzero off-diagonal elements in the rows of $\mathbf{HA}$, broadening the windows through which $\mathbf{p}$ is viewed in forming the elements of $\hat{\mathbf{p}}$,

$$\hat{\mathbf{p}} = \mathbf{HAp} \tag{57}$$

Introducing the factorization (54) and the special initial coordinates in which $\mathbf{V} = \mathbf{I}_n$, (57) becomes

$$\hat{p}_i' = \frac{\Lambda_{ii}}{\alpha + \Lambda_{ii}} p_i'$$

where, as before, primes denote projection onto the eigenvectors in $\mathbf{N}$ (e.g., $\hat{\mathbf{p}}' = \mathbf{N}^T \hat{\mathbf{p}}$). Thus, components of the parameter vector $\mathbf{p}'$ corresponding to small eigenvalues ($\Lambda_{ii} \ll \alpha$) are filtered out of the model solution. As α is as large as possible, consistent with the available data realization, we have effectively allowed the maximum acceptable parameter filtering.

In the space of the variable θ, the directional window equivalent to (34) is

$$W(\theta, \theta') = G(\theta) \mathbf{b}(\theta)^T \mathbf{Hb}(\theta') \tag{58}$$

where

$$\hat{S}(\theta) - G(\theta) S_0(\theta) = \int_0^{2\pi} d\theta' [\tilde{S}(\theta') - G(\theta') S_0(\theta')] \times W(\theta, \theta')$$

Thus, $W(\theta, \theta')$ defines the ability of the model to resolve details in the difference between any exact sample spectrum $\tilde{S}$ and the favored model S_0 in those θ-regions where $\hat{S}$ is nonzero. Because the nonlinearity of the problem makes $\mathbf{H}$ data-dependent, the window itself is data-adaptive. This enables better resolution than a strictly linear model would permit, as we show by numerical example in the next section.

Solution Stability: In keeping with the notion of examining the converged value of $\mathbf{H}$ as if it were an "equivalent generalized inverse," we ask what the statistics of the model parameters would be if the derived $\mathbf{H}$ were applied, without further change, to the ensemble of data realizations corresponding to the hypothetical true spectrum. Then (30) applies, and

$$\mathbf{U} = [(\alpha \mathbf{V} + \mathbf{A})^{-1}] \mathbf{V} [(\alpha \mathbf{V} + \mathbf{A})^{-1}]^T$$

Introducing the factorization (54) and the special initial coordinates in which $\mathbf{V} = \mathbf{I}_n$, this becomes, e.g.,

$$U_{ii} = \sum_k (N_{ik})^2 (\alpha + \Lambda_{kk})^{-2}$$

Thus, components of the sum for which Λ_{kk} is very small ($\ll \alpha$), and which otherwise would have contributed very large amounts to parameter variance, are suppressed. As α is as large as possible, consistent with the one available data realization, this suppression has apparently been maximized.

The effect of nonzero α is clearly analogous to the solution stabilization procedure proposed in Section 3 except that the present filter is tapered rather than sharply cut off at the pth eigenvalue. By choosing α as large as possible, we have effectively chosen the most stable, statistically acceptable model solution (at least in this specific sense) and accepted the consequences elsewhere. The size of α is ultimately fixed by the rejection confidence level at which the statistical test operates; e.g., lowering the confidence level γ results in a smaller ρ_γ^2, which requires a smaller weight α on the favored model to achieve $\rho^2 = \rho_\gamma^2$ at the minimum of η for the available data realization, which results in better data fit and parameter resolution at the cost of increased solution instability.

These conclusions, needless to say, are not unequivocal, since our quasi-linear analysis has ignored the inevitable perturbations in G (and hence in A) which arise in response to varying α or perturbing $\mathbf{d}$. Nevertheless, extensive experience with applications to pitch/roll buoy and wave gage array data strongly supports the interpretation this analysis has provided.

Summarizing: (1) The variational technique of Long and Hasselmann (1979) incorporates model properties and known external constraints into a functional η (which they call "model nastiness"). (2) The minimization of this functional defines the optimal model parameterization. (3) The parameters of the model and the weights assigned to the various components of η are then determined by the data and the external constraints. (4) Among the latter is the requirement that the solution must be nonnegative (which leads to a nonlinear optimal parameterization) and the requirement that the observed-data/model-data mismatch must not be improbably large; these together result in a data-adaptive "equivalent generalized inverse." (5) The parameters and weights which result apparently yield the maximum model stability consistent with the above statistical constraint (although the implicit nonlinear character of the model makes it difficult to establish this unequivocally). (6) The level of confidence at which the statistical test defining "improbably large" operates establishes the final trade-off between data fit, solution resolution, and solution stability.

TABLE I. Parameter Resolution Matrices for the Problem of Fig. 4

1. *Linear problem* (negativity allowed):

$$HA = \begin{bmatrix} 0.94 & -0.04 & 0.12 & -0.05 & 0.00 \\ -0.04 & 0.94 & 0.12 & 0.06 & 0.00 \\ 0.12 & 0.12 & 0.71 & 0.00 & 0.00 \\ -0.05 & 0.06 & 0.00 & 0.73 & 0.00 \\ 0.00 & 0.00 & 0.00 & 0.00 & 1.00 \end{bmatrix}$$

2. *Nonlinear problem* (negativity prohibited):

$$HA = \begin{bmatrix} 1.90 & -0.41 & 0.98 & 3.77 & 0.00 \\ 0.41 & 1.84 & -0.52 & -3.62 & 0.00 \\ 0.45 & 0.42 & -0.24 & 0.07 & 0.00 \\ -0.41 & 0.44 & 0.07 & -1.31 & 0.00 \\ 0.60 & -0.56 & -0.24 & 3.42 & 1.00 \end{bmatrix}$$

● True spectrum: $S = K \cos^{20}((\theta - \theta_0)/2)$

$$\theta_0 = 137°; \qquad K = \left(\int_0^{2\pi} d\theta\, S \right)^{-1}$$

● 64 degrees of freedom in spectra and cross spectra
● 80% rejection confidence ($\rho_{0.8}^2 = 6.2$)

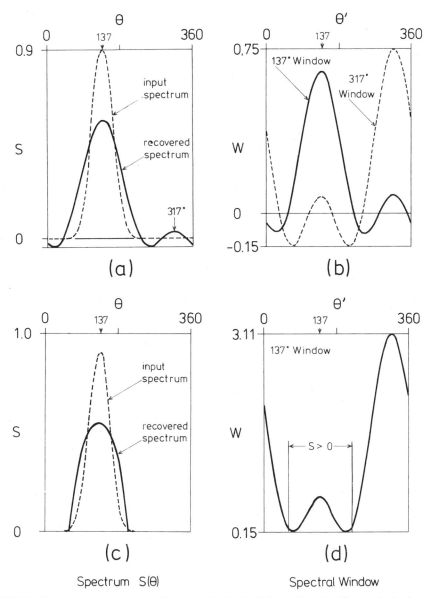

Spectrum S(θ)   Spectral Window

FIGURE 3. Numerical experiments with synthetic pitch/roll buoy data. (a) The synthetic data were computed by numerically integrating (6) for the input spectrum indicated. The recovered spectrum was obtained using the optimal model (42) with $S_0 = (2\pi)^{-1}$ and $G \equiv 1$ (nonnegativity constraint omitted). The data covariance matrix $\mathbf{V}$ assumed 64 degrees of freedom in the observed cross spectra. (b) Spectral windows are shown for look directions $137°$ and $317°$ (the locations of the positive lobes in the recovered spectrum). The windows always have a large positive lobe in the look direction but sizable positive and negative side lobes also appear. (c) Results for the same input data when the nonnegativity constraint is imposed are significantly better. The recovered distribution is narrower and the anomalous side lobes have vanished. (d) Including the nonnegativity constraint results in a spectral window which, for the look direction matching the peak of the input spectrum, has a positive lobe significantly narrower than that of Fig. 3b. This is accomplished at the cost of large positive side lobes; but these occur at angles for which the input spectrum is essentially zero and, hence, do not affect the result.

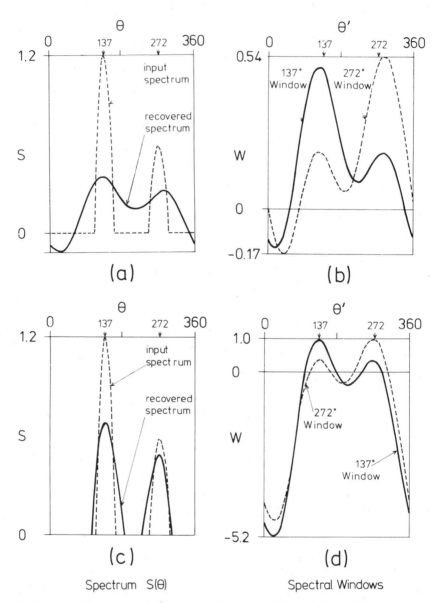

FIGURE 4. Numerical experiments with synthetic pitch/roll buoy data. (a) Same as Fig. 3a except that the input test spectrum is the indicated bimodal distribution. Without the nonnegativity constraint, the two lobes smear together in the recovered spectrum, and negative spectral densities occur. (b) The reasons for this are apparent in the windows for 137° and 272° look directions shown here. (c) The improvement in the recovered spectrum is dramatic when the nonnegativity constraint is imposed. (d) The corresponding windows for 137° and 272° look directions show narrow positive lobes in the principal direction and, to a lesser extent, at the location of the other peak in each case. Large negative side lobes occur in θ' ranges where no spectral density exists in the input spectrum, hence do not disturb the result. The cross-coupling between the two wave trains permitted by the window accounts for the reduced difference between the lobes in the recovered spectrum relative to that in the input spectrum.

5. CONCLUSIONS AND ILLUSTRATIONS

The variational technique of Long and Hasselmann represents a rational approach to designing optimal solutions to sets of linear integral equations of the form (36). Whenever the model parameterization obtained by this means (or any other means, for that matter) is also linear, the ideas of Section 3 are sufficient to clearly understand and evaluate the consequences. However, if the parameterization is nonlinear, we are on shakier ground. The best that we can do, short of Monte Carlo studies of specific cases, is to linearize the problem about some zero-order state, then apply the linear theory to the resulting approximation. The procedure suggested in Section 4 is equivalent to having linearized the problem about the solution obtained at the final iteration and, further, having neglected subsequent perturbations in $\mathbf{A}$ and $\mathbf{d}_0$ as being of higher order than those in $\hat{\mathbf{d}}$ and $\hat{\mathbf{p}}$. If the nonlinearity were explicit, the adequacy of this approximation could be evaluated. In the case of directional spectrum estimation as formulated here, the nonlinearity is implicit, making it hard to establish beyond doubt the validity of conclusions drawn from the linearized analysis (though experience with the procedure strongly supports these results). The difficulty arises from the imposed constraint that the solution not be negative; in view of the violence this does to our theoretical understanding, we might reasonably ask if it is worth the cost.

To demonstrate that it is, the procedure of Long and Hasselmann has been implemented (both with and without the nonnegativity constraint) for pitch/roll buoy observations and applied to synthetic data generated by numerically integrating (6) for two different hypothetical true spectra. The data were formated and normalized as in Long (1980) and provide four linearly independent data equations and one additional absolute constraint, namely, that the solution must have unit area under the curve. Formally, this additional constraint is treated as a fifth data equation which has zero variability associated with it. The statistical constraint was based on 80% rejection confidence and data covariance matrices which assumed 64 degrees of freedom in the auto- and cross spectra forming the basic data. The optimal model (42) was then fitted to the data using an isotropic favored model $[S_0 = (2\pi)^{-1}]$.

The results for the single-lobed true spectrum are shown in Table I and Fig. 3. For the linear problem, the parameter resolution matrix (which, in this case, is the transpose of the information density matrix) is diagonally dominant, but the nonzero $\alpha(= 114)$ produces sizable off-diagonal terms, as expected. The corresponding model solution in Fig. 3a is a smeared version of the true spectrum and exhibits anomalous positive and negative side lobes. The reason for this is apparent in the shape of the directional window $W(\theta, \theta')$, which is shown in Fig. 3b for the two look directions at which peaks occur in the recovered spectrum. When the nonnegativity constraint is imposed, the anomalous spectral side lobes are inhibited, and the directional resolution is improved (Fig. 3c). Inspection of the corresponding window (Fig. 3d) shows a narrower opening in the vicinity of the true peak direction but at the cost of very large side lobes; these, however, are located in θ' ranges where the solution indicates zero spectral density. Similarly, the parameter resolution matrix (Table I) develops large off-diagonal terms which, though more difficult to interpret, reflect this window shape in the space of the five model parameters.

The ability of the nonlinear procedure to adapt to the data is demonstrated even more clearly by the results for the bimodal true spectrum displayed in Fig. 4. This characteristic allows the pitch/roll buoy to perform better in resolution than its commonly accepted limit, provided the true directional distribution is very small over some significant range of θ. If this is not so, the nonnegativity constraint contributes no additional useful information, the model solution reverts to the linear form, and the commonly accepted limitations apply.

REFERENCES

Backus, G. E., and J. F. Gilbert (1967): Numerical applications of a formalism for geophysical inverse problems. *Geophys. J. R. Astron. Soc.* **13**, 247–276.

Barber, N. F. (1963): The directional resolving power of an array of wave detectors. *Ocean Wave Spectra*, Prentice–Hall, Englewood Cliffs, N.J., 137–150.

Cartwright, D. E., and N. D. Smith (1964): Buoy techniques for obtaining directional wave spectra. *Buoy Technology*, Mar. Tech. Soc., Washington, D.C., 112–121.

Forristall, G. Z., E. G. Ward, L. E. Borgmann, and V. J. Cardone (1978): Storm wave kinematics. *Proceedings of Offshore Technology Conference*, May 8–11, 1978, Houston, 1503–1509.

Franklin, J. N. (1970): Well-posed stochastic extensions of ill-posed linear problems. *J. Math. Anal. Appl.* **31**, 682–716.

Jackson, D. D. (1972): Interpretation of inaccurate, insufficient, and inconsistent data. *Geophys. J. R. Astron. Soc.* **28**, 97–109.

Lanczos, C. (1961): *Linear Differential Operators*. Van Nostrand, Reinhold, New York, 564 pp.

Long, R. B. (1980): The statistical evaluation of directional spectrum estimates derived from pitch/roll buoy data. *J. Phys. Oceanogr.* **10**, 944–952.

Long, R. B., and K. Hasselmann (1979): A variational technique for extracting directional spectra from multicomponent wave data. *J. Phys. Oceanogr.* **9**, 373–381.

Longuet-Higgins, M. S., D. E. Cartwright, and N. D. Smith (1963): Observations of the directional spectrum of sea waves using the motions of a floating buoy. *Ocean Wave Spectra*, Prentice–Hall, Englewood Cliffs, N.J., 111–136.

Munk, W., and C. Wunsch (1979): Ocean acoustic tomography: A scheme for large-scale monitoring. *Deep-Sea Res.* **26A**, 123–161.

Olbers, K. J., P. Muller, and J. Willebrand (1976): Inverse technique analysis of a large data set. *Phys. Earth Planet. Inter.* **12**, 248–252.

Parker, R. L. (1972): Inverse theory with grossly inadequate data. *Geophys. J. R. Astron Soc.* **29**, 123–138.

Parker, R. L. (1977): Understanding inverse theory. *Annu. Rev. Earth Planet. Sci.* **5**, 35–64.

Snyder, R. L. (1964): A field study of wave-induced pressure fluctuations above surface gravity waves. *J. Mar. Res.* **32**, 497–531.

Twomey, S. (1963): On the numerical solution of Fredholm integral equations of the first kind by the inversion of a linear system produced by quadrature. *J. Assoc. Comput. Mach.* **10**, 97–101.

Twomey, S. (1977): *Introduction to the Mathematics of Inversion in Remote Sensing and Indirect Measurements*, Elsevier, Amsterdan.

Wiggins, R. A. (1972): The general linear inverse problem: Implication of surface waves and free oscillations for earth structure. *Rev. Geophys. Space Phys.* **10**, 251–285.

Wunsch, C. (1978): The North Atlantic general circulation west of 50°W determined by inverse methods. *Rev. Geophys. Space Phys.* **16**, 583–620.

DISCUSSION

HOLTHUIJSEN: I understand that in your technique, data points which have been observed with high confidence have a relatively large influence on the result of the fitting procedure. These points are not necessarily the physically most important ones. How do you increase the influence of the physically important data points (e.g., those related to energetic parts of a spectrum)?

LONG: I think you must have misunderstood my presentation in this respect. Analysis schemes such as I have suggested here sort out the frequency dependence *a priori* at the stage where the cross spectra are computed. Thereafter, the directional properties of each frequency band are derived, independently from all other frequency bands (we have ignored background currents or, perhaps, instrument motion which would prohibit this decoupling, of course) using the elements of the data vector corresponding to that frequency band. The influence of each element on the eventual result decreases as its variability increases as a consequence of the weighting in the inverse covariance matrix, but except in this respect, none is more important physically than any other.

SEGUR: How can you assure that the errors in the data are linearly independent? It would seem that with real data, one would have no control over whether or not the errors are related.

LONG: We have considered, in those aspects of the analysis relating to directional wave spectra, only statistical variability in the basic set of cross-spectral estimates as the source of data error. In this case, each realization of the data is, in effect, a set of integral properties of a particular realization of the spectrum, just as the "true" (expectation) values of the data are integral properties of the true spectrum. If the data are linearly independent (equivalently, if the kernels in the integrals defining the data are linearly independent), then the errors will be linearly independent (though not necessarily *statistically* independent). In this case, the covariance matrix V will not be singular, and V^{-1} will exist. If there are other sources of error, the analysis technique can still be applied as long as the appropriate (and nonsingular) covariance matrix can be constructed.

THOMAS: The ordering of the series involving the eigenvalues ($\lambda^{1/2}$) of the Ω matrix and its truncation is the basis of the method of Singular Value Decomposition of matrices for achieving image data compression. It is capable of giving high data compression factors and is clearly related to the theory you describe. Am I right in thinking that the reconstituted image from the truncated series effectively shows an improved signal-to-noise ratio?

LONG: The analogy is close but not exact. The data compression technique (also called factor analysis, the method of empirical orthogonal functions, and other names in other fields) involves assembling a series of data realizations (column vectors) into a matrix, say A. The eigenvectors of AA^T and A^TA "decompose" A (i.e., factor it), and each eigenvector in the subsequent reconstruction contributes to the variance of the reconstructed data an amount proportional to the corresponding eigenvalue. Filtering out those corresponding to small eigenvalues (usually the most oscillatory) hopefully filters out those components representing noise, hence improving the signal-to-noise ratio. In the present case, the objective of the filtering is not to smooth A but to stabilize the solution of the inverse problem, effectively filtering out the oscillatory response to noise in the data. But, in this sense, the answer to your question is yes.

40

COMPARISONS OF HURRICANE FICO WINDS AND WAVES FROM NUMERICAL MODELS WITH OBSERVATIONS FROM SEASAT-A

DUNCAN ROSS, LINDA M. LAWSON, AND WILLIAM McLEISH

ABSTRACT.   Several types of satellite data collected during SEASAT-A overpasses of hurricane Fico are used to adjust input parameters in a hurricane wind prediction model. Derived winds are then used in a complex discrete spectral wave prediction model to calculate wave heights and directional spectrum distributions. Three different areas of the storm with winds ranging from 12 m/s to 23 m/s and hindcast wave heights varying from 4 m to 8 m were used to compare model normalized directional wave spectra with those from SEASAT synthetic aperture radar (SAR) imagery. The two-dimensional normalized spectral energy results from the SAR and wave model showed an average difference in peak directions of 0.9° with a standard deviation of 6.8°. Wind and wave model results are also compared with altimeter data along a subsatellite track. This study shows the value of satellite observations in model validation and, conversely, the value of the model results in verification of the efficacy of the satellite data sets. Especially noteworthy is the snapshot view of the characteristics of the storm, a demonstration of the potential power of satellite remote sensing.

1. INTRODUCTION

The state of the art in wave prediction modeling has changed considerably since World War II when Sverdrup and Munk (1947) first applied a singular technique with success. The early approach of Sverdrup and Munk, although still in use in certain applications, has been superceded by so-called spectral models which treat the process of energy exchange between the wind field and the wave field as a function of the existing wave energy spectrum, as opposed to integrated properties of the spectrum such as the significant wave height and period. The "discrete" spectral approach allows growth, propagation, and decay of each energy component independent of those adjacent to it. In principle, the added complexity of the models suggests their suitability for handling more complex situations such as the sharply curving wind fields associated with hurricanes and extratropical fronts. This spectral approach was first described by Pierson and Marks (1952) and resulted in several operational models (Pierson *et al.*, 1955, 1966; Gelci *et al.*, 1957).

Ross and Cardone (1978), for example, found significant discrepancies between models when they were tested for extreme situations such as stationary or fast-moving hurricanes. A more in-depth review of the history of wave modeling and the suitability of various approaches to

DUNCAN ROSS, LINDA M. LAWSON, AND WILLIAM McLEISH ● Atlantic Oceanographic and Meteorological Laboratories, National Oceanic and Atmospheric Administration, Miami, Florida 33149.

595

the problem is given by Cardone (1974) and Cardone and Ross (1977). A principal observation of the latter study was that none of the contemporary models have been tested extensively over a wide range of real and hypothetical conditions.

One of the most critical problems now facing wave model development is that of properly describing the actual sea surface conditions during extreme events that have traditionally been difficult to study. The major problem lies in getting sufficient observations under adverse weather conditions to properly describe wind and wave fields in order to determine the validity of model hindcasts of wave fields.

This study is based on the 1978 SEASAT-A satellite passes over Northeastern Pacific tropical cyclone Fico, one of several tropical storms observed by the satellite. These overpasses provided useful data from an array of instruments including a radar scatterometer (SASS), a scanning multichannel microwave radiometer (SMMR), an altimeter (ALT), and a synthetic aperture radar (SAR). Data collected by these instruments were used to derive several geophysical parameters characteristic of the sea surface conditions. In particular, these parameters were wind speeds from the SASS, SMMR, and ALT, wind directions from the SASS, significant wave heights from the ALT, and wavelength and directional properties from the SAR.

Several recent studies (Gonzalez *et al.*, 1979) indicate that SAR data can in some cases be used to determine the mean length and direction of the dominant surface waves. The principal focus in the present study is on the directional wave spectra obtained from the SAR in comparison with the directional properties obtained from wave model hindcasts. The characteristics of the directional redistribution of waves in various regions of cyclone-type wind fields and the correct directional relaxation parameterization are two of the most pressing problems of wind-driven wave models (Forristall *et al.*, 1978). The importance of this aspect underscores the lack of a complete understanding of the mechanisms involved in the transfer of energy from the wind field to the wave field and between different components of the wave field. As suggested above, this is partially due to the difficulty in obtaining reliable data to use for testing theories. Although data similar to the satellite data can be obtained by aircraft (Elachi, 1978; Schwab *et al.*, 1981), little has been accomplished to date. In addition to revealing, to some extent, the adequacy of the physics incorporated in the wave model, the comparisons herein reveal some properties of the satellite measurements and lead to a better understanding of the ability of satellites to return useful measurements of severe conditions.

In order to apply the surface wave model, it is necessary first to generate a satisfactory wind field history. To this end, all available information including satellite photographs was used to estimate storm characteristics of location, radius to maximum winds, and central pressure. These parameters were then used as input into a numerical hurricane wind field prediction model. SEASAT wind data obtained from revolutions 222, 229, and 251 were used to specify the general characteristics of the surface winds. Using the known characteristics of the storm and the satellite wind measurements, an hourly wind field was derived and used to drive a discrete-parametric hybrid wave model for deep water for a 61-h period terminating at 1300 GMT on July 14, coincident with rev 251. The model-generated directional wave spectra were compared with SAR spectra in three different areas of the storm. Significant wave heights were compared with altimeter measurements.

2. INSTRUMENTATION

The SEASAT sensor complement consisted of a radar altimeter (ALT), a SEASAT-A scatterometer system (SASS), a synthetic aperture radar (SAR), a visible and infrared radiometer (VIRR), and a scanning multichannel microwave radiometer (SMMR).

The short-pulse (3 ns) ALT viewed the earth at nadir and operated at 13.56 GHz. It provides an estimate of significant wave height, $H_{1/3}$, through evaluation of the shape of the reflected radar pulse. A rough reflecting surface extends the time period of the return pulse resulting in a decreasing slope to the leading edge of the return pulse ramp as wave height increases. Furthermore, as the rms slope of the roughness elements increases, total radar return decreases approximately as a logarithmic function of the wind speed (Brown *et al.*, 1982). Fedor and Brown (1982) have verified the sensitivity of the ALT to wind speed and significant wave height; they found mean wind speed errors of about 0.25 m/s with a standard deviation of 1.6 m/s and mean significant wave height errors of 0.07 ± 0.29 m.

The SASS is also an active radar system and is capable of observing surface wind speed and direction. It illuminates the ocean surface with four fan-shaped beams; two look forward at an azimuth of $\pm 45°$ to the track of the spacecraft at incidence angles from 24° to 34° off nadir. The two aft beams view the surface at azimuth angles of $\pm 135°$ to the ground track at incidence angles which vary from 25° to 50° from nadir. Thus, a given earth location is first viewed by one of the forward beams and then, a short time later, by the corresponding aft radar beam with the azimuth viewing angle shifted by 90°. The two views of the earth location allow estimation of speed and direction of the wind through exploitation of the dependence of the radar return σ_0 on both wind speed and direction. Unfortunately, the solution is not unique and several choices of wind speed and directions result, which necessitates the need for additional information to make the proper selection. Using high-quality surface measurements as the additional information, Jones *et al.* (1982) found wind speed accurate to ± 2 m/s and direction accurate to better than $\pm 10°$.

The VIRR scanned in the visible and near-infrared region (0.49 to 0.94 μm) and in the far-infrared region (10.5 to 12.5 μm) producing images with a resolution of about 4 km. The instrument was used principally to aid in feature identification, but produces useful sea surface temperature maps in cloud-free areas.

The SAR, a coherent L-band imaging radar, produced a 100-km-wide image of the surface at a location centered about 250 km to the right of the satellite ground track. The SAR data were transmitted in real time to a ground receiving station and recorded digitally. Subsequently, the data were optically processed at an azimuth resolution of 25 m to yield images. Selected time periods could be reprocessed digitally at a resolution of 6 m in azimuth. The SAR was found by Gonzalez *et al.* (1979) to observe ocean swell length and direction with an accuracy of 15 and 25°, respectively. The ability to image locally generated sea, however, is limited by resolution which is unfortunately degraded with increasing wave height (see Alpers *et al.*, 1981). The limitations of SAR in this regard remain an area of intense interest.

The SMMR, a passive microwave device, operated at frequencies of 6.6, 10.7, 18, 21, and 37 GHz with both horizontal and vertical polarization. The SMMR viewed aft and imaged a swath of 600 km to the starboard side of the satellite with a constant incidence angle of 48°. The SMMR observes the naturally emitted and reflected microwave energy of the earth and atmosphere. Because this emission depends on both frequency and various geophysical parameters, the multifrequency capability of SMMR allows estimation of several geophysical variables including sea surface temperature, surface wind speed, rainfall rate, and liquid water content of the atmosphere.

3. STORM DESCRIPTION AND WIND FIELD CONSTRUCTION

Hurricane Fico was a moderate Pacific tropical cyclone lasting 19 days. The track positions and intensities of the storm were based on information from the *Monthly Weather Review* (Gunther, 1979), SEASAT observations, and NOAA satellite observations. Figure 1 shows the Fico eye positions; Fig. 2 is an NOAA satellite photographs of the storm on July 14 at 1644Z. It was necessary to estimate the storm intensity and size from photographs of this type and SEASAT data since there was no aircraft reconnaissance prior to and during rev 251, our principal source of verification wave data. The period of the storm that was modeled was between 0000Z on July 12 and 1300Z on July 14. During this time the movement of Fico was essentially westward at a latitude of 14.8°N between longitude 114°W and 122°W. Figure 3 shows the 49×36 point grid used in the model calculations; the points are spaced at approximately 40 km. Figure 3 also shows the eye positions of the storm every 6 h during the 61-h hindcast period and the eye position at the last time step of 1300Z on July 14. It can be seen that the forward velocity is

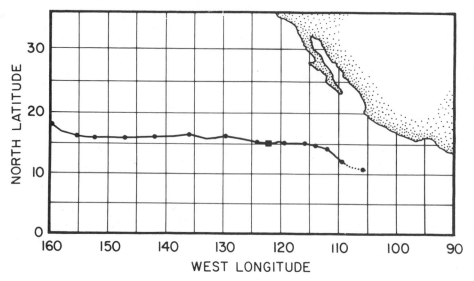

FIGURE 1. Track of hurricane Fico from *Mon. Weather Rev.* **107**, 914.

SFD 195:16:44:36 8850 V3F4474 14JUL78 N5 71.3E 1-2

FIGURE 2. NOAA satellite photograph of Fico, July 14, 1978, 16:44:36 GMT.

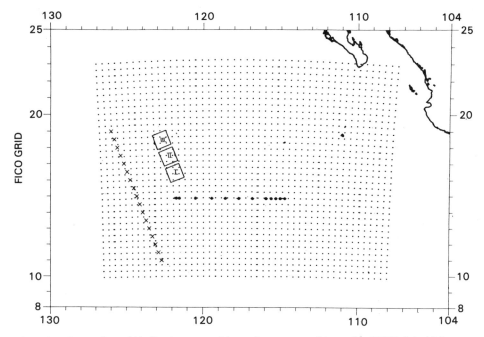

FIGURE 3. Computing grid indicating eye positions of storm every 6 h starting 0000Z, July 12, last eye position 1300 GMT, July 14 (●). SAR-imaged areas indicated as I, II, III. Track lower left (x) is location of ALT data.

quite low during the initial 24-h period, being around 2–m/s, and that the velocity increases to about 5 m/s during the last 12 h. Two other features of interest in Fig. 3 are the locations of the three 100-km-square SAR-imaged regions selected for analysis. These areas are referred to as I, II, III in order of increasing distance from the eye. It should be noticed that area I is about 160 km (90 n miles) from the storm center, but still in a very high wind speed region (23 m/s), while area III is considerably farther away in a region of much lower wind speeds (~ 12 m/s). The track indicated in the lower left-hand corner of the gridded area shows the positions of the ALT data and represents the nadir track of rev 251.

The wind model used to generate hourly wind speeds and directions is a planetary boundary layer wind model that has been applied extensively in studies of wind distributions in Gulf of Mexico and North Atlantic hurricanes by Cardone and Ross (1977) and Cardone et al. (1977). These studies have shown the model to be normally accurate to within $\pm$ 2 m/s in reproducing wind fields under reasonably steady-state conditions such as those that occur in slow-moving, stable storms. The model is based on the theoretical work of Chow (1971) dealing with the horizontal air flow in the boundary layer of a moving vortex and involves the numerical integration of the vertically averaged equations of motion that describe a boundary layer experiencing vertical and horizontal stresses on the rotating earth. It is possible to apply this wind model with knowledge of only a minimal amount of meteorological data typical of that available through analysis of satellite visible region imagery, available ship measurements, or aircraft reconnaissance. These measurements include central pressure, far-field pressure, radius to maximum winds, magnitude and direction of the ambient large-scale pressure field, and the position of the storm. Hourly model wind fields are generated from a prescribed pressure field and then transformed onto the wave model computing grid in real earth coordinates incorporating the storm forward velocity. In general, the model is able to reproduce satisfactorily wind fields for a variety of hurricane types. It should be pointed out, however, that this model cannot reproduce small-scale variability in storms such as rain band-type disturbances found often in the eye–wall region of hurricanes, nor can anomalous situations such as "eye wobble"

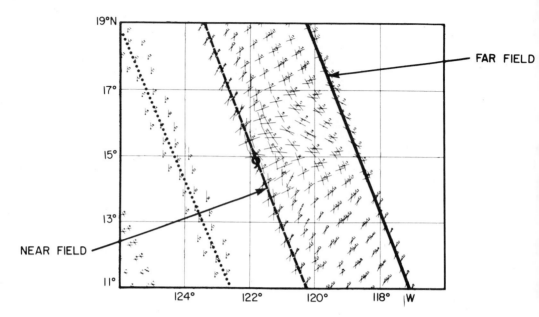

FIGURE 4. Plot of SASS wind field from rev 251. The dotted line indicates the location of the satellite track; the data lying along the dashed line are designated as the near-field, while the data along the solid line are far-field data.

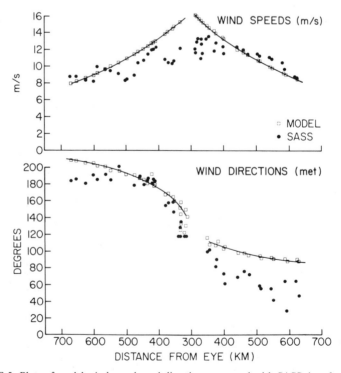

FIGURE 5. Plots of model wind speeds and direction compared with SASS data for far field.

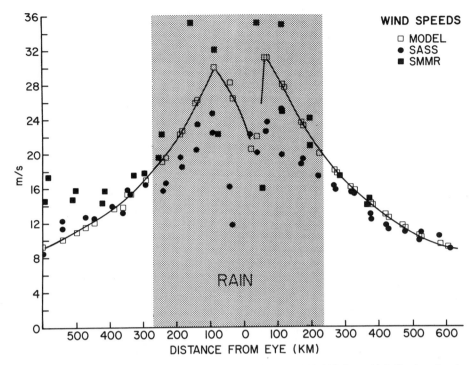

FIGURE 6. Near-field model wind speeds compared with SASS and SMMR data with indication of region containing areas of heavy rainfall.

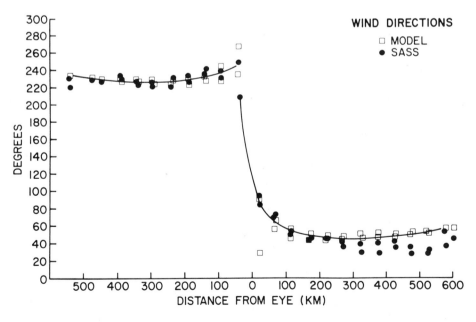

FIGURE 7. Near-field model wind directions compared with SASS data.

and double eye formations be modeled. It is obvious that these irregularities in storms will affect wind patterns and hence the local sea states.

There are some general circulation characteristics found to be typical of most tropical cyclone wind fields. These include asymmetries in the winds due to a combination of storm motion, latitude, and asymmetries in the gradient pressure field with the result that maximum wind conditions usually occur in the right quadrant (relative to the storm track) of the storm. In addition, tropical cyclones exhibit varying amounts of wind inflow angle depending primarily on the storm intensity which is related to the pressure gradients. These characteristics are reproduced in the wind model by quadrant-dependent pressure gradient parameters, the prescribed radius to maximum wind, and the forward velocity of the storm.

Since there were no surface measurements of the pressure field in Fico, the central and-field pressures were estimated from the SASS data obtained during rev 222 on July 12 around 1200Z, rev 229 on July 13 at 0130Z, and rev 251 on July 14 at 1330Z. An example of the available SASS data is shown in Fig. 4 along with an indication of two areas within this data set that were used extensively to compare with model wind fields; these two areas are termed near-field and far-field regions. The nadir position of the satellite, along which ALT data were obtained, is also shown in this plot. Plots of this type and the satellite VIRR imagery were used to estimate the radius to maximum winds to be approximately 63 km. These data revealed an unusual degree of wind speed symmetry, a feature not typical in hurricane wind fields and difficult to obtain with the wind model while maintaining other storm features, such as inflow angle of the wind streamlines.

The lack of asymmetry found in the Fico wind field was reproduced in the wind model by treating the storm as being stationary (no forward velocity) and embedded in a calm ambient flow. The wind model parameters were taken to be 975 mbar central pressure and maximum 10-m-height wind speeds of about 30–32 m/s, with a far-field pressure of 1013 mbars, and radius to maximum winds of 63 km. A 10% variation in the overall pressure field gradient results in wind field uncertainties of less than 5%, based upon past experience. This implies an uncertainty of about 10%, or less, in model-predicted wave parameters. The model outputs were compared with SMMR and SASS data from two different areas of rev 251 as shown in Figs. 5–7. Several things are apparent from these figures, but the most noticeable characteristic is the wind speed symmetry indicated by the SASS winds, especially in the near-field data. Also, the model wind speeds are in quite good agreement with the SASS and SMMR winds except for regions close to the eye where SMMR measurements of rain rates, atmospheric water vapor content, and two-way transmittance clearly indicated the presence of rain bands. The model and SASS wind speeds for the near field had a mean difference of 2.03 ± 3.9 m/s and for the far field differed by 1.39 ± 1.88 m/s, including the points where there was contamination of the satellite data by rain. It is apparent from Figs. 5 and 6 that the comparisons are considerably better outside the heavy rain band area. It is also noted in these plots that the directional characteristics of the modeled winds are in good agreement with SASS data with the exception of the northern area of the far field where the SASS data indicate considerably more inflow than was obtained in the model. This discrepancy could be due to a local disturbance but certainly represents an effect that was not reproduced in the wind model. The final wind field was the result of many iterations of the wind model and represents the best subjective fit to the data in terms of general characteristics.

4. WAVE MODEL

The wave model applied in this study is a descendent of Pierson and colleagues' (1966) discrete spectral approach and was primarily developed by Oceanweather, Inc. (Greenwood et al., 1985). This model uses directional wave energy spectra to describe wave growth and propagation while incorporating implicit parameterization of dissipation and wave–wave interactions. The structure of the model is basically discrete, due to the fact that the entire energy spectrum is distributed over 15 frequency bands and 24 direction bands. Blended into this discrete scheme is a parametric relationship between nondimensional energy and fetch used to incorporate nonlinear wave interactions. The growth is a combination of B-term growth consistent with the Bight of Abaco results (Snyder *et al.*, 1981), and a parametric growth based on a nondimensional energy (ε) vs. nondimensional fetch (ξ) relationship, where the energy is a weighted sum of the difference between downwind and upwind traveling energy. This parameterization results in a growth somewhat faster than given by Hasselmann *et al.* (1976) and slower than from Phillips (1957). At any growth step the larger of the two mechanisms is applied. This growth from the ε–ξ law and an α–ε relationship, where ε is nondimensional total wave energy and α is the Phillips constant, results in a weak coupling of the frequency bands. A frequency-dependent angular redistribution is incorporated in frequency bands higher than the P–M peak when variances differ from the reference spectrum, while frequencies lower than the P–M peak are treated as swell and allowed to propagate independently. A more extensive description of the model can be found in Greenwood *et al.* (1985).

The rectangular grid is the result of a transverse mercator projection which gives essentially uniform spacing. The propagation scheme uses downstream interpolation, and the convergence of meridians results in the propagation formula coefficients being latitude dependent. The frequency bands range from 0.04 Hz to 0.22 Hz with ratios of adjacent bands being $3^{1/9}$.

5. RADAR PROCESSING

Three 100×100-km regions of the ocean surface north of the Fico storm center were recorded with the SEASAT-A SAR at 1330Z on July 14. A description of the instrument performance during this period is given by Jordan (1980). The SAR imagery was prepared by the Jet Propulsion Laboratory through the implementation of a digital correlation technique (Wu *et al.*, 1981). The radar data, spaced at approximately 17-m intervals, represent a resolution of 25 m. Twelve 9×9-km locations were selected for analysis in each of the three regions. Sixteen 128×128 arrays of radar intensity values then were used to calculate the average two-dimensional wavenumber spectra at each of the 12 locations. These 12 spectra were averaged to obtain a single spectrum from each region. At 384 degrees of freedom, the 80% confidence limits of an estimate are approximately $\pm 9\%$ of the value. The radar spectral energy was interpolated at the wavenumber corresponding to each of the wave model frequencies in a series of directions. Plots were made of the directional energy distributions for each frequency and normalized by each peak spectral value for comparison with model results. In addition, wavenumber spectra were calculated by integrating the spectral energy over a $25°$ directional band centered at the peak with the background energy level removed.

6. RESULTS

Figure 3 shows the area imaged by the ALT and the location of its measurements relative to the SAR regions and storm position. The 10-m wind speeds and significant wave heights derived from ALT data are compared with the model results in Fig. 8. These two plots show the ALT to indicate much lower wind speeds in one section, differing by as much as 7 m/s, and to furnish correspondingly lower wave heights. Although the significant wave heights from the ALT showed considerable variability, many of the values were lower than the model results. It is noted that there is very little change in ALT wind speed or wave heights along the subtrack although the distance from the storm center varies from 250 km to 650 km. This feature would not commonly occur in slow-moving storms and the discrepancies between the model and the ALT suggest a local anomaly in the wind field structure that was not present in the model. Since there was no other source of reliable wind data is this region (SASS data rely upon sidelobes of the antenna pattern to extract wind speed), it is difficult to specify otherwise the general wind-field structure in the area of the ALT track. It is noted that in the northwest region where the model wind speeds agree with the ALT data there is also agreement in significant wave heights.

In Fig. 9 comparison of the energy levels of the SAR and model spectra is made by plotting the integrated radar wavenumber spectra (normalized by the maximum model energy density)

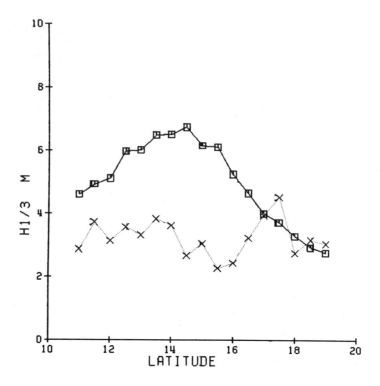

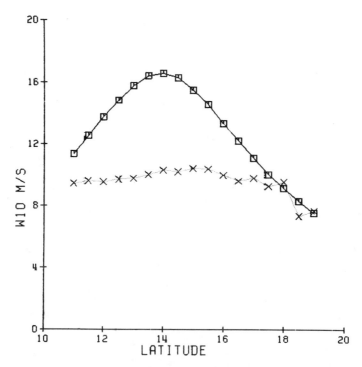

FIGURE 8. Comparison of significant wave heights and wind speeds from wave model (——) and ALT (.....).

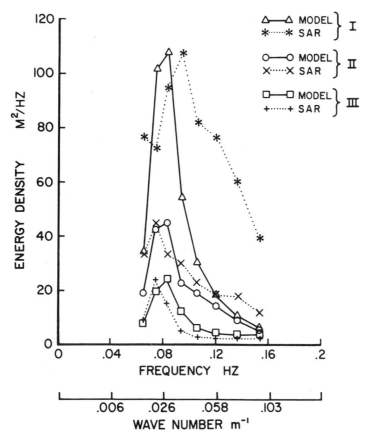

FIGURE 9. Radar wavenumber spectra and model frequency spectra for SAR I, II, and III.

along with the corresponding model frequency spectra at each of the SAR-imaged areas. The peaks of the model and radar spectra differ by less than one model frequency band in all three regions. The radar peak energy is at a lower frequency than the model in regions II and III and higher in region I. These results are in general agreement with previous findings (Gonzalez *et al.*, 1979; Shuchman and Kasischke, 1981) that the approximate wavelengths of dominant ocean waves can be determined from radar spectra.

A noteworthy characteristic of the one-dimensional spectra is the relative energy level of the peak in area III as compared to areas I and II vis-à-vis the "imageability" of the waves. Area III was strongly dominated by local swell conditions and the energy level of the spectral peak was 8000 arbitrary units squared whereas peak levels of about 3000 were obtained in regions I and II. Images constructed from the raw data showed poor visual registration of the waves in region III, and virtually indiscernible registration of the waves in regions I and II. This reduced image quality suggests that significantly less modulation of short wave structure occurs in areas I and II, despite higher significant wave heights, and motion effects.

Alpers *et al.* (1981) describe a simple modulation transfer function

$$R = R^{\text{Tilt}} + R^{\text{hydr}}$$

where R^{Tilt} is associated with the actual local slope, and R^{hydr} is due to nonuniform distribution of the scattering short waves caused by wave–wave interaction with the long wave field. Alpers and

Hasselmann (1978) predict the increased relative importance of R^{hydr} in relatively smooth seas as is approximately the case for region III. In the presence of strong energy transfer by the local wind, highly nonlinear effects become important, reducing the effect of R^{hydr} and, therefore,

$$R^{hydr} = f(U_{10})$$

Alpers *et al.* (1981) predict a degradation in imageability of the waves due to increasing nonlinear effects and smearing of azimuth-traveling wave components due to increased motions in high sea states. Certainly, a degradation of the imagery was observed in the case of Fico. When properly examined, however, substantial directional wave characteristics were observed as is discussed below.

It is important at this point to reflect on the analog optical processing of SAR imagery. One of the most critical factors in the quality of optically processed (correlated) images is the ability of the operators of the optical bench to focus on the waves. In the case of strong nonlinear effects, virtually no *visual* registration of the waves will be possible. The likelihood of improper focus and loss of the directional properties of the wave information therefore increases considerably. This may be one reason for the lack of wind wave information present in much of the optically correlated SEASAT imagery and strongly argues for the objectivity of digital processing when ocean waves are to be studied.

The directional properties of the radar and model are compared in Figs. 10–12. These show the two-dimensional energy spectra of the model averaged over the grid points within each particular SAR-imaged region along with the two-dimensional spectra of the radar averaged as described above. These energy densities are normalized so that the maximum values for each frequency have the same amplitude, allowing a more detailed study of the directional properties. McLeish and Ross (1983) examined the relationship between the normalized radar directional spectra and ocean waves. Under appropriate restrictions, the shape of a radar peak in a directional energy distribution was found to be similar to that for surface wave spectra. For a detailed comparison, a directional peak was described by its mean direction and half-power width at each frequency, where the directions were weighted averages calculated between the half-power points. The calculated values for the model and radar results are shown in Table I. In some cases the half-power points of the radar spectra were clearly not identifiable with the dominant peak due to high levels of extraneous variance. Contamination at low frequencies may be attributed to swell from distant storms, and high-frequency interference may result from the radar performance. In cases of excessive contamination the derived values of Table I were excluded from further calculations. The model and radar peak energy directions differed by 0.9° over all three areas with a standard deviation of 6.8°. The average width of the peaks from the model is 45.6° and from the radar is 36.1°, differing by 9.5° with a standard deviation of 20.6°. In general, the directional characteristics of the model and SAR spectra agree well in all three imaged areas, despite the fact that the sea states were considerably different.

One of the most noticeable features in Figs. 10–12 is the shift in the peak direction toward a more southerly direction with an increase in frequency in both the radar and the model. This migration of wave direction toward the local wind direction is seen quite clearly in Fig. 13, where the mean direction vs. frequency of the model is compared with the radar in each of the three areas. The plots agree with previous studies that indicate that high-frequency waves adjust to the local wind direction at a faster rate than low-frequency waves (Hasselmann *et al.*, 1980; Gunther *et al.*, 1981). The peak directions observed by the radar are consistent with the redistribution of wave energy now currently incorporated in the wave model. The development of the directional wave spectrum in strongly curving hurricane wind fields has not been studied extensively due to limited available data; these results, therefore, are heartening.

Figures 10–12 indicate and Table I demonstrates that the width of the model directional spectrum (as measured to the half-power points) is greater in the region of highly variable wind directions of region I, near the storm center, than at region II or III. The average of the model widths decreases from 58° in region I to 32° in region III. Such a difference is in accord with the expectation that a wind changing direction either in time or in space will produce a broader

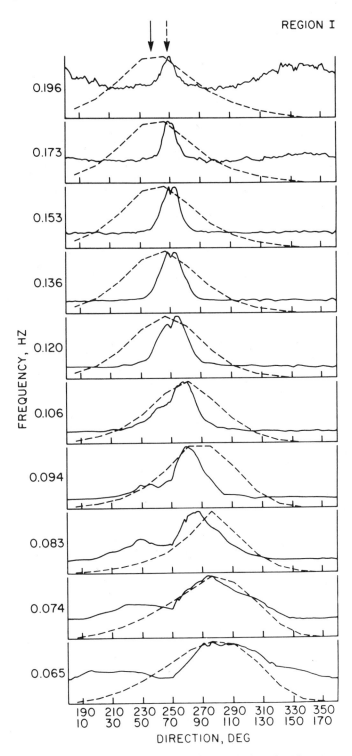

FIGURE 10. Directional distribution of waves of several frequencies in region I from wave model and from SAR. Solid arrow at the top indicates the model-determined wind direction; dashed arrow the range look direction of the SAR.

REGION II

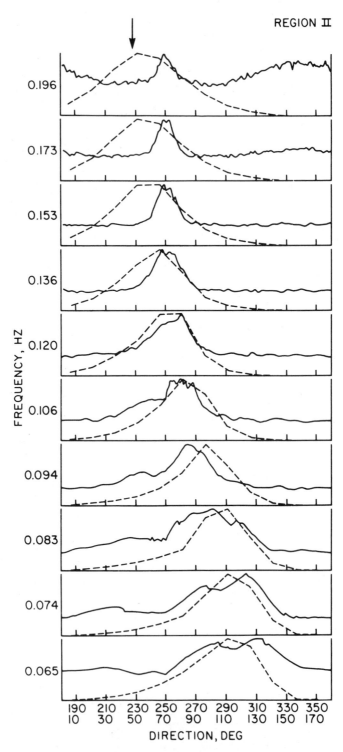

FIGURE 11. Directional distribution of waves of several frequencies in region II from wave model and from SAR.

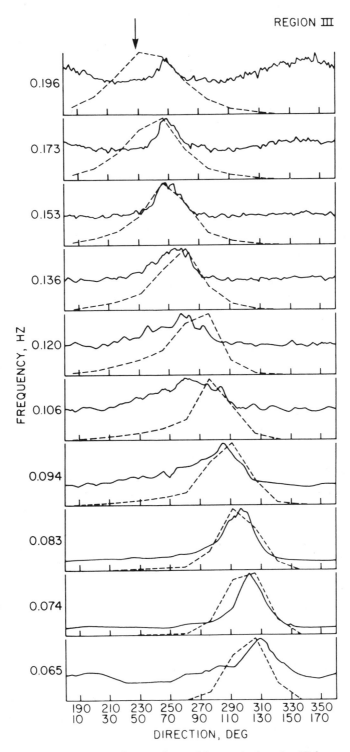

FIGURE 12. Directional distribution of waves of several frequencies in region III from wave model and from SAR.

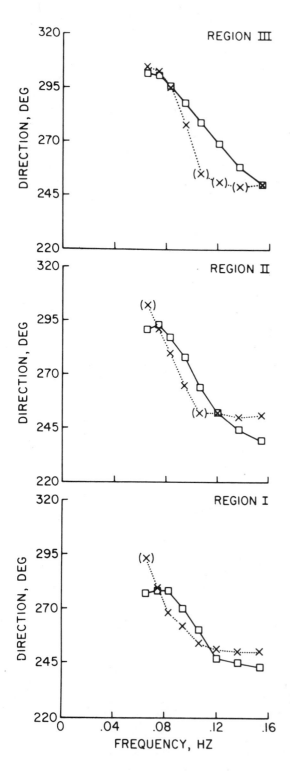

FIGURE 13. Mean direction vs. frequency for SAR (.....) and wave model (———) in regions I, II, and III. The radar data indicated by (X) contained large background levels that interfered with the direction calculation.

TABLE I.   Mean Wave Directions and Directional Peak Widths
from Model and from Radar

		Model		Radar	
	Freq	Direction	Width	Direction	Width
Region I	0.065	277	72	293	84[a]
	0.074	278	65	280	61
	0.083	278	42	268	36
	0.094	270	58	262	24
	0.106	260	53	254	30
	0.120	247	57	251	25
	0.136	245	57	250	22
	0.153	243	58	250	16
Region II	0.065	291	52	302	101[a]
	0.074	293	44	291	64
	0.083	287	42	280	60
	0.094	278	37	265	39
	0.106	264	41	252	54[a]
	0.120	252	45	252	38
	0.136	244	46	250	29
	0.153	239	54	251	21
Region III	0.065	302	31	305	52
	0.074	301	34	303	21
	0.083	296	30	295	26
	0.094	288	33	278	50
	0.106	279	28	255	92[a]
	0.120	269	36	251	79[a]
	0.136	258	33	249	66[a]
	0.153	250	34	250	36

[a]Large background level interfered with calculation.

directional distribution of waves than will a wind constant in direction. However, the averages of radar width values in Table I are somewhat near 35° in all three regions and do not show a broadening near the storm center.

7. CONCLUSIONS

This SEASAT data set provides a unique opportunity to test in part the ability of contemporary wave models to predict accurately the wave properties of a severe hurricane. The availability of the satellite data was especially important in view of the unusual wind field associated with this storm and the scarcity of direct observations.

The two-dimensional wave field generated by the model proved to be in good agreement with the directional properties of the wave field observed by the SAR. The SAR provided important data supporting a very narrow distribution of wave energy about the local peak direction often narrower than the model predictions, a departure from the previously accepted concept of a $\cos^{2P}[(\theta - \bar{\theta})/2]$ distribution of wave energy, where the maximum value of P is on the order of 10. Most significantly, the SAR results confirm that the model algorithm for turning the waves is highly accurate with respect to mean direction.

One of the most interesting aspects of the SAR data was the validity of the wave direction information up to a frequency of 0.2 Hz. In no other data set is there such clear evidence of the ability of the satellite SAR to image wind waves. This suggests the need for digital processing and a reevaluation of SAR data for other experiments such as JASIN and GOASEX.

In summary, this study has clearly demonstrated the utility of satellite techniques to observe the surface vector wind field synoptically and the normalized two-dimensional wave field for all significant wavelengths. Clearly, an operational satellite system would collect a massive climatological data set in just a short period. Furthermore, if satellite-observed data sets were examined using iterative procedures together with numerical prediction models involving surface wind and wave models, the accuracy and application of the data could be significantly extended.

ACKNOWLEDGMENTS.   Financial support was furnished by the SEASAT project office of NESS. The authors are indebted to V. J. Cardone for advice in determining the wind field, and along with J. A. Greenwood for developing and programming the numerical model calculations. C. Wu of the Jet Propulsion Laboratory provided the digitally correlated SAR data.

REFERENCES

Alpers, W. R., and K. Hasselmann (1978): The two-frequency microwave technique for measuring ocean wave spectra from an airplane or satellite. *Boundary-Layer Meteorol.* **13**, 215–230.

Alpers, W. R., D. B. Ross, and C. L. Rufenach (1981): On the detectability of ocean surface waves by real and synthetic aperture radar. *J. Geophys. Res.* **86**, 6481–6498.

Brown, G. S., H. R. Stanley, and N. A. Roy, (1981): The wind speed measurement capability of spaceborne radar altimetry. *IEEE Oceanic Eng.*, OE–6, 59–63.

Cardone, V. J. (1974): Ocean wave prediction: Two decades of progress and future prospects. *Seakeeping 1953–1973. Sponsored by panel H-7 (seakeeping characteristics) at Webb Institute of Naval Architecture, Glen Cove, N.Y., October 28–29, 1973.* Society of Naval Architects and Marine Engineers, New York, 5–18.

Cardone, V. J., and D. B. Ross (1977): State of the art wave predictions and data requirements. *Ocean Wave Climate* (M. D. Earle and A. Malahoff, eds.), Plenum Press, New York, 61–91.

Cardone, V. J., D. B. Ross, and M. R. Ahrens, (1977): An experiment in forecasting hurricane generated sea states. *Preprints 11th Technical Conference on Hurricanes and Tropical Meteorology*, Miami, Florida, Amer. Meteor. Soc., 688–695.

Chow, S. (1971): A study of the wind field in the planetary boundary of a moving cyclone. M. S. thesis, Department of Meteorology and Oceanography, New York University.

Elachi, C. (1978): Radar imaging of the ocean surface. *Boundary Layer Meteorol.* **13**, 165–179.

Fedor, L. S., and G. S. Brown, (1982): Waveheight and wind speed measurements from the SEASAT radar altimeter. *J. Geophys. Res.*, **87**, 3254–3260.

Forristall, G. Z., E. G. Ward, V. J. Cardone, and L. E. Borgmann (1978): The directional spectra and kinematics of surface gravity waves in tropical storm Delia. *J. Phys. Oceanogr.* **8**, 888–909.

Gelci, R. H. Cazale, and J. Vassal (1957): Prevision de la houle: La methode des densites spectroangularies. *Bulletin d'information du comite central d'oceanographie et d'etudes des cotes* **9**, 416.

Gonzalez, F. I., R. C. Beal, W. E. Brown, P. S. DeLeonibus, J. W. Sherman, III, J. F. R. Gower, D. Lichy, D. B. Ross, C. L. Rufenach, and R. A. Shuchman (1979): SEASAT synthetic aperture radar: Ocean wave detection capabilities. *Science* **204**, 1418–1421.

Greenwood, J. A., V. J. Cardone, and L. M. Lawson (1985): Intercomparison test version of the SAIL model. *Ocean Wave Modeling* (The SWAMP Group, eds.), Plenum Press, New York, 221–233.

Gunther, E. B. (1979): Eastern North Pacific tropical cyclones of 1978. *Mon. Weather Rev.* **107**, 911–927.

Gunther, H., W. Rosenthal, and M. Dunckel (1981): The response of surface gravity waves to changing wind direction. *J. Phys. Oceanogr.* **11**, 718–728.

Hasselmann, D. E., M. Dunckel, and J. A. Ewing (1980): Directional wave spectra observed during JONSWAP 1973. *J. Phys. Oceanogr.* **10**, 1264–1280.

Hasselmann, K., D. B. Ross, P. Muller, and W. Sell (1976): A parametric wave prediction model. *J. Phys. Oceanogr.* **6**, 200–228.

Jones, W. L., L. C. Schroeder, D. H. Boggs, E. M. Bracalente, R. A. Brown, G. J. Dome, W. J. Pierson, and F. J. Wentz, (1982): The SEASAT–A satellite scatterometer: The geophysical evaluation of remotely sensed wind vectors over the ocean. *J. Geophys. Res.*, **87**, 3297–3317.

Jordan, R. L. (1980): The SEASAT-A synthetic aperture radar system. *IEEE J. Ocean. Eng.* OE-5, 154–164.

McLeish, W., and D. B. Ross, (1983): Imaging radar observations of directional properties of ocean waves. *J. Geophys. Res.*, **88**, 4407–4419.

Mitsuyasu, H., F. Tasai, T. Sahara, S. Mizuno, M. Ohkusu, T. Honda, and K. Rikuski (1975): Observations of the directional spectrum of ocean waves using a cloverleaf buoy. *J. Phys. Oceanogr.* **5**, 750–760.

Phillips, O. M. (1957): On the generation of waves by turbulent wind. *J. Fluid Mech.* **2**, 417–445.

Pierson, W. J., and W. Marks (1952): The power spectrum analysis of ocean wave records. *Trans. Am. Geophys. Union* **33**, 834–844.

Pierson, W. J., G. Neumann, and R. W. James (1955): Practical methods for observing and forecasting ocean ·waves by means of wave spectra and statistics. *H.O. Publication 603*, U.S. Naval Oceanograpic Office, Washington, D.C.

Pierson, W. J., L. J. Tick, and L. Baer (1966): Computer-based procedures for preparing global wave forecasts and wind field analyses capable of using wave data obtained by a spacecraft. *Sixth Naval Hydrodynamics Symposium*, Office of Naval Research, Washington, D.C., 499–532.

Ross, D. B., and V. J. Cardone, (1978): A comparison of parametric and spectral hurricane wave prediction products. *Turbulent Fluxes through the Sea Surface, Wave Dynamics, and Prediction*, (A. Favre and K. Hasselmann, eds.), Plenum Press, New York, 647–664.

Schwab, D. J., R. A. Shuchman, and P. C. Liu (1981): Wind wave directions determined from synthetic aperture radar imagery and from a tower in Lake Michigan. *J. Geophys. Res.* **86**, 2059–2064.

Shuchman, R. A., and E. F. Kasischke (1981): Refraction of coastal ocean waves. *Spaceborne Synthetic Aperture Radar for Oceanography* (R. C. Beal, P. S. DeLeonibus, and I. Katz, eds.), Johns Hopkins Press, Baltimore, 128–135.

Snyder, R. L., F. W. Dobson, J. A. Elliott, and R. B. Long (1981): Array measurements of atmospheric pressure fluctuations above surface gravity waves. *J. Fluid Mech.* **102**, 1–59.

Sverdrup, H. U., and W. H. Munk (1947): *Wind sea and swell: Theory of relation for forecasting.* H.O. Publication 601, U.S. Naval Oceanographic Office, Washington, D.C.

Wu, C., B. Barkan, B. Huneycutt, C. Leang, and S. Pang (1981): An introduction to the interim digital SAR processor and the characteristics of the associated SEASAT SAR imagery. Publication 81–26, Jet Propulsion Laboratory, California Institute of Technology, Pasadena, Calif.

41

MODELING WIND-DRIVEN SEA IN SHALLOW WATER

J. W. SANDERS AND J. BRUINSMA

ABSTRACT. A wind-driven sea in shallow water can be described by a spectral shape given in terms of a small number of parameters. These parameters are a function of the stage of development, defined in terms of the total energy, local wind speed, and depth. The total energy is derived from the energy balance equation in which the dissipation term is nonlinear. The results of the model, in which these concepts are incorporated, are compared with measurements for shallow water in the North Sea in an extreme situation.

1. INTRODUCTION

In 1976, Sanders introduced a concept for modeling the wind-driven sea in deep water in terms of the stage of development of the windsea. This concept, together with a swell-scanning procedure, is the basis of the KNMI (Royal Netherlands Meteorological Institute) wave prediction model for the North Sea. The southern part of the North Sea has depths of 50 m or less, and wind fields in that area frequently generate waves with wavelengths of 150 m and greater. Wave forecasting in the southern North Sea therefore requires modeling of shallow-water effects.

In this chapter an extension of the original concept of Sanders with respect to depth-dependent wave growth and spectral distribution is discussed. To this end, a review of modeling windsea in deep water is given in Section 2. This is followed by the derivation from empirical arguments of a model for depth-limited wave growth and the modeling of the shape of the wave spectrum in shallow water through a redefinition of the stage of development in Section 3. The above developments were incorporated into the KNMI operational wave model GONO (de Voogt et al., 1985). Hindcast results from this wave model have been tested against wave measurements made in shallow water in the North Sea for an extreme storm situation. The results are presented in this Chapter. A verification study for operational purposes has been reported by Bouws et al. (this volume).

2. WINDSEA IN DEEP WATER

From measurements it has been found that the spectral shape of a wind-driven sea in deep water can be described by a small number of parameters. For the parameterization used in this Chapter, it is assumed that the directional spreading is independent of frequency so that only the

J. W. SANDERS AND J. BRUINSMA ● Ministry of Transport and Public Works, Rijkswaterstaat, Directorate for Water Management and Hydraulic Research, The Hague, The Netherlands.

1d spectrum will be discussed. Throughout the text, the word "energy" instead of "variance" is used to omit the factor ρg (ρ = water density). The high-frequency part of the wave spectrum, up to the peak, is proportional to f^{-5}, while the forward face is approximated by a straight line (see Fig. 1).

From this spectral shape, it follows that the total energy E is given by

$$E = \frac{g^2}{4(2\pi)^4} \cdot \frac{\hat{\alpha}}{f_p^4} \cdot (3 - 2\mu) \tag{1}$$

The nondimensional parameters $\hat{\alpha}$ and $\hat{\nu}$, where $\hat{\nu} = u f_p / g$ (u = wind speed at 10-m height), depend on the stage of development of a wind-driven sea (Sanders, 1976). This stage of development ξ is defined as

$$\xi = (H_s / H_s^{max})^{1/2} \tag{2}$$

where $H_s = 4(E)^{1/2}$ (significant wave height) and H_s^{max} is the maximum value of H_s for infinite fetch and duration and an uniform wind field with wind velocity u.

From the asymptotic value β of the nondimensional growth curve (Fig. 2), one has

$$H_s^{max} = \beta u^2 / g$$

The value of β depends on the behavior of the atmospheric boundary layer. Fluctuating wind conditions with strong gusts may lead to greater values of β than stable conditions.

At present, theoretical information about the value of β is lacking although it is anticipated that β can be calculated from the lowest layers of an atmospheric model. Model tests have shown that for the North Sea, the best average value for β is 0.22.

In nonuniform wind fields or in fetch-limited situations, one can still define the stage of development according to Eq. (2). The dependence of $\hat{\alpha}$ and $\hat{\nu}$ on the stage of development ξ (Figs. 3 and 4) were examined from KNMI Waverider measurements of windsea recorded at 55° 06'N, 3° 54'E (water depth 40 m) and from the JONSWAP measurements (Hasselmann et al., 1973) made in the German Bight.

From these data the following relations were obtained:

$$\hat{\alpha} = 4.93 \times 10^{-3} \xi^{-1.944} \tag{3a}$$

$$\hat{\nu} = 6.89 \times 10^{-2} \xi^{-1.376} / \beta^{1/2} \tag{3b}$$

The relations are valid in the range $0.5 \leqslant \xi \leqslant 1$. Situations where $\xi < 0.5$ are not taken into account, because they arise from winds with very short fetches or unrealistic sudden and large increases in wind velocity.

The two data sets fit the same relations (3), assuming that the ratio of β for the KNMI data and β for the JONSWAP data is 0.22/0.14. The smaller value of β for the JONSWAP data can be expected, because these data were recorded during very stable meteorological conditions and were selected with regard to high wind field uniformity. It is possible to fit the JONSWAP data to a growth curve which is not in contradiction with $\beta = 0.14$.

For a given significant wave height (total energy), the position of the peak will depend on the value of β. For a low atmospheric input, when the value of β is small, nonlinear wave–wave interactions will have had more time to transfer the peak to the left before the same significant wave height is reached when compared with a large value of β and the same mean wind velocity. A comparison of these two data sets clearly shows this effect (see Fig. 4).

Equations (1)–(3) give the spectral shape once the total energy E is known. If the local wind speed is given, then ξ can be calculated from (2). Using Eqs. (3), the spectral parameters $\hat{\alpha}$ and $\hat{\nu}$ are determined, whereupon μ is derived from (1). The total energy is found from the energy

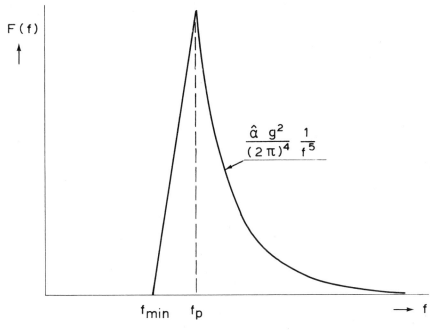

FIGURE 1. Model spectrum of a wind-driven sea. g = gravitational acceleration; f = frequency; $F(f)$ = energy density; f_p = peak frequency; f_{min} = minimum frequency; $\hat{\alpha}$ = nondimensional parameter specifying the level of the high-frequency part; $\mu = f_{min}/f_p$.

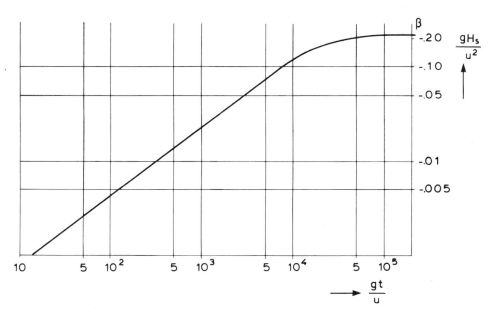

FIGURE 2. Wave growth curve for infinite fetch.

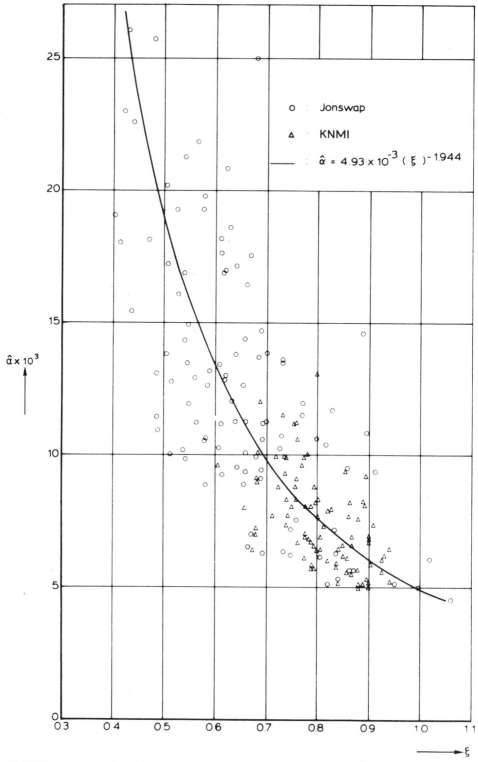

FIGURE 3. $\hat{\alpha}$ as a function of the stage of development ξ, JONSWAP data refer to orthogonal fetch. The curve is derived from Eq. (3a).

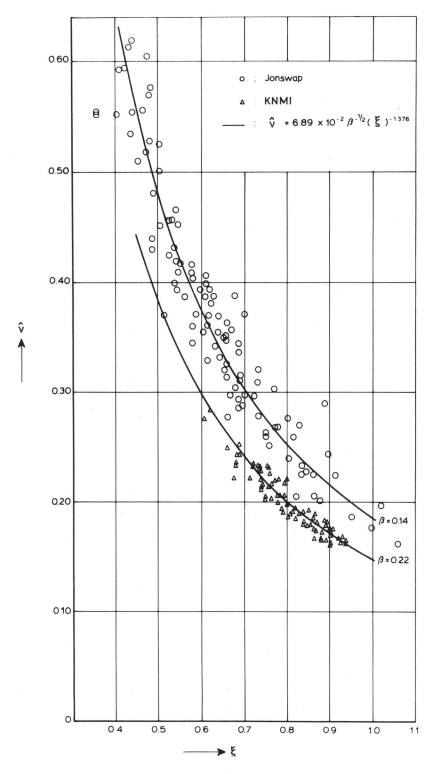

FIGURE 4. Nondimensional peak frequency as a function of the stage of development ζ. Curves derived from Eq. (3b).

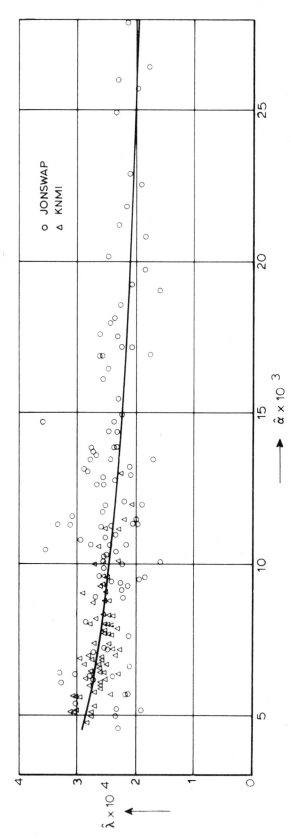

FIGURE 5. Shape factor $\hat{\lambda}$ as a function of $\hat{\alpha}$.

balance equation (see de Voogt *et al.*, 1985). The shape factor $\hat{\lambda}$ defined in terms of $\hat{\alpha}$, $\hat{v}$, and $\tilde{\varepsilon}$ by

$$\hat{\lambda} = \tilde{\varepsilon}\hat{v}^4/\hat{\alpha} \qquad \text{where } \tilde{\varepsilon} = g^2 E/u^4$$

is presented as a function of $\hat{\alpha}$ in Fig. 5.

Using (1), $\hat{\lambda}$ can also be written as

$$\hat{\lambda} = (3 - 2\mu)/4(2\pi)^4$$

From Fig. 5, it is evident that $\hat{\lambda}$, and therefore also μ, is approximately a constant.

3. WINDSEA IN SHALLOW WATER

For deep water, a model for the wind-driven sea with only one prognostic parameter, directly or indirectly assumes the existence of a specific growth curve (Sanders *et al.*, 1980). For shallow water, it is assumed that the energy input can still be inferred from the deep-water growth curve. The energy input is therefore a function of the total energy only.

The usefulness of a one-parameter model strongly depends on the relaxation time of deviations from the relations between the spectral parameters. For shallow water, as in deep water, it is tacitly assumed that the relaxation times are short enough to justify the use of a one-parameter model [For deep water, see Hasselmann *et al.* (1976)]. The α–v relation, implicitly given by Eqs. (3), is assumed to hold for shallow water. This relation differs from the relation given by Hasselmann *et al.* (1976). The shape of the growth curve and also the relaxation time of α is largely determined by the relationship between α and v. Theoretically, the α–v relation can be deduced from the prognostic equations for α and v in a two-parameter model. In the case of a uniform wind field, the parameters α and v will follow the relation

$$\alpha = Av^B \qquad \text{if } d\alpha/dt = ABv^{B-1}(dv/dt)$$

From the equations for dv/dt and $d\alpha/dt$, as given by Hasselmann *et al.* (1976), it is seen that the α–v relation is determined by the nonlinear wave–wave interactions and the atmospheric energy input. Because the atmospheric energy input is not precisely known, some uncertainty exists about the α–v relationship. On examining these equations for dv/dt and $d\alpha/dt$, one finds that the α–v relation can be closely approximated by making the right-hand side of the equation for $d\alpha/dt$ equal to zero.

In shallow water, dissipation will change the equations for dv/dt and $d\alpha/dt$. As long as the dissipation is mainly confined to the forward face of the spectrum, it will affect dv/dt but not $d\alpha/dt$ [see Hasselmann *et al.* (1976) for definition of the projection operators]. So the dissipation will not noticeably affect the α–v relationship. In extreme situations, where dissipation is also important in the high-frequency part of the spectrum, it will affect the equation for $d\alpha/dt$ so that deviation from the deep-water α–v relation can be expected. For shallow water, the assumption is made that wind-driven sea can be described by a one-parameter model. Following the description for deep water, this amounts to determining the total energy (Section 3.1). The spectral shape is then found from a stage of development defined in terms of the total energy, the local wind speed, and the depth (Section 3.2). Justification of the model comes from a comparison of the model results with measurements made during a severe storm.

3.1. Modeling the Limitation of Wave Height in Shallow Water

In this part a model for total wave energy only is constructed. Unlike the deep-water situation, the most important feature of the limited wave growth in shallow water is the inclusion of an extra source term in the energy balance equation which describes the effect of dissipation by bottom friction. The intuitive deduction of this expression is given here.

From swell measurements, it is clear that the dissipation of energy by bottom friction is almost linear. In a wind-driven sea limited by bottom friction, the wave energy in relation to the depth is much higher than encountered for swell conditions. A model has therefore been constructed that accounts for nonlinear energy dissipation. A parameter that makes it possible to tune the rate of the dissipation is also incorporated.

During the buildup of a wind-driven sea, bottom dissipation manifests itself rather sharply. One can make use of this effect by stating that the bottom dissipation starts if waves are generated with frequencies less than f_B, where f_B is a function of depth only:

$$f_B = (g/4\pi bd)^{1/2}$$

where d is depth and b is a tuning constant. The deep-water wavelength for f_B is $2bd$ so if $b = 1$, the orbital velocity at the bottom is about 1% of the orbital velocity at the sea surface.

The characteristic feature of this model is that dissipation is set proportional to the total energy E_B at frequencies smaller than f_B.

Although the transition from no bottom dissipation to appreciable bottom dissipation progresses quickly during the buildup of a wind-driven sea, the transition occurs too rapidly in the above-mentioned model. This occurs because under real conditions, bottom dissipation also takes place at frequencies greater than f_B. This means that a model designed to determine total energy only and not spectral shape must be constructed in such a way that dissipation manifests itself more gradually. The spectral shape (Fig. 6) is therefore transformed to a "Pierson–Moskowitz-like" spectral shape (Fig. 7) with the same $\dot{\alpha}$ and total energy. The transformed spectral shape has the form

$$\frac{\dot{\alpha}g^2}{(2\pi)^4} \cdot \frac{1}{f^5} \cdot e^{-a(f_0/f)^4}$$

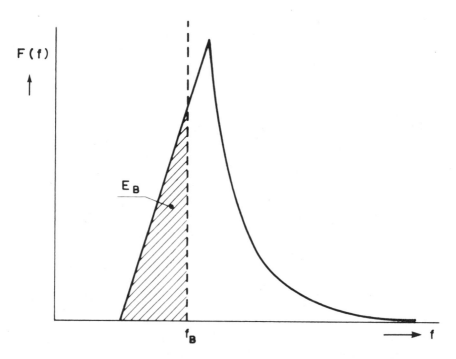

FIGURE 6. Dissipation proportional to the hatched area.

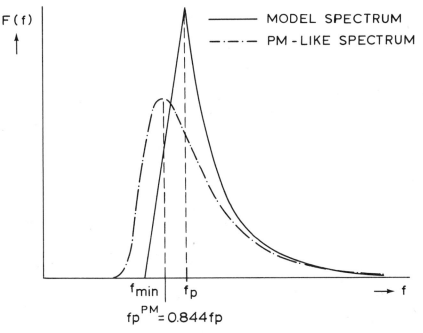

FIGURE 7. Actual and transformed spectra.

This spectral shape has its maximum at

$$f_P^{PM} = (4a/5)^{1/4} \cdot f_0 \tag{4}$$

The condition that the total energy must be the same gives

$$\frac{\hat{a}g^2}{(2\pi)^4 \cdot 4af_0^4} = \frac{H_s^2}{16} \quad (H_s = 4E^{1/2}) \tag{5}$$

In this transformation, it is required that the ratio f_P^{PM}/f_P remains constant for all stages of development ξ, where $0.5 \leqslant \xi \leqslant 1$. Using Eqs. (3), f_0 can now be solved from (4) and (5).

The condition $f_P^{PM}/f_P = \text{constant}$ is almost fulfilled if one takes

$$f_0 = (1/2\pi)(g\beta/H_s)^{1/2} \quad \text{so } a = 4\hat{a}/\beta^2 \tag{6}$$

It then follows from (3)–(5) that

$$f_P^{PM}/f_P \approx 0.84 \quad \text{for } 0.5 \leqslant \xi \leqslant 1$$

In the model, the approximate solution (6) is used, because variation over the range of ξ ($\pm 4\%$) is negligible and in addition it simplifies the calculations.

In the transformed spectral shape, energy dissipation is proportional to the total energy at frequencies $\leqslant f_B$ so that the dissipation term has the form

$$-C \int_0^{f_B} \frac{\hat{a}g^2}{(2\pi)^4 f^5} \frac{1}{e^{-a(f_0/f)^4}} df = -C \frac{H_s^2}{16} e^{-4\hat{a}(bd/\pi H_s)^2}$$

where C is the proportionality factor (the tuning constant).

Hindcast studies, with the GONO model set at grid size of 75 km and a time step of $1\frac{1}{2}$ h, were used to establish values for the constant b and c. These were found to be 4.25 and $1.85 \times 10^{-4} s^{-1}$, respectively.

In terms of the total energy E, the energy balance equation reads (see also de Voogt et al., 1985)

$$\frac{dE}{dt} = \left(\frac{dE}{dt}\right)^{g.c.} - C\frac{H_s^2}{16}e^{-4\alpha(bd/\pi H_s)^2} \tag{7}$$

with $b = 4.25$, $C = 1.85 \times 10^{-4} s^{-1}$, and $(dE/dt)^{g.c.}$ is the source term inferred from the deep-water growth curve.

In the case of a uniform depth and wind field with an infinite fetch, (7) can be evaluated using the growth curve given in Fig. 2. The results calculated with $u = 30$ m/s for a number of depths are presented in Fig. 8.

For a given wind velocity u and depth d, the maximum significant wave height can be calculated by assigning the right-hand side of (7) to zero. The resulting curves are given in Fig. 9.

In practice, water depths are not constant over the effective fetch range of the wind field. If wind is blowing in the direction of decreasing water depth, as is often the situation around the Dutch coast, then by advection of the nearby higher windsea, H_s can reach higher values than those given in Fig. 9.

Some features of the dissipation term in this model merit further discussion.

Comparison of model results with measurements of swell dissipation is not directly possible because the actual spectral shape is transformed to a "Pierson–Moskowitz-like" spectral shape. This transformation shifts the energy to the lower frequencies (see Fig. 7). As a result and because f_B applies to the transformed spectral shape, a rather large value $b (= 4.25)$ for f_B is found when tuning the model. In the case of the real spectrum, a f_B^* would be found that had a b^* smaller than the b obtained from the transformation.

By assuming that the ratio of actual spectrum peak to transformed spectrum peak frequencies is the same as the ratio of f_B^* to f_B, one can estimate b^* for actual frequencies. Using this approach, b^* is found to have a value of 3.0.

For the wavenumber k associated with f_B^*, kd is found to equal 1.24. In this model, therefore, dissipation effectively starts if waves with a $kd < 1.24$ are generated (see also Bouws,

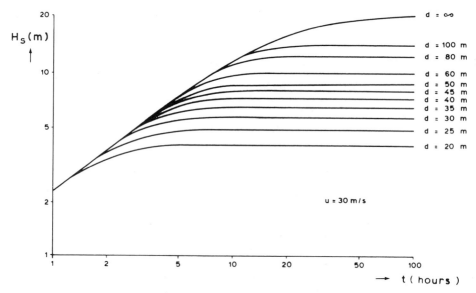

FIGURE 8. Growth curves derived from Eq. (7) with $\dot{\alpha} = 7.7 \times 10^{-3}$.

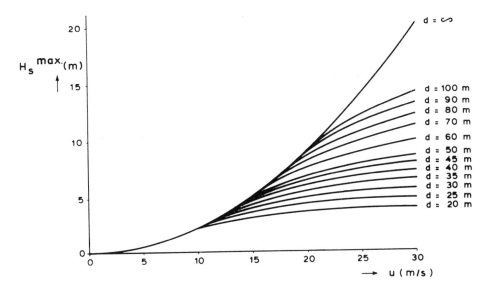

FIGURE 9. Maximum H_s for uniform wind field and depth for $\dot{\alpha} = 7.7 \times 10^{-3}$.

1980). The dissipation rate of this model can now be compared to the JONSWAP swell dissipation results. Following the JONSWAP notation (Hasselmann *et al.*, 1973) one has

$$d\ln I/d\zeta = \Gamma \tag{8}$$

where $I = $ energy density flux and

$$\zeta = \int_0^x \frac{k^2}{\omega^2 V_g \cosh^2 kd} dx$$

$k = $ wavenumber, $V_g = $ group velocity, $x = $ distance, and $\omega = 2\pi f$. For uniform depth, (8) can be written as

$$dF(f)/dt = g(f,d)F(f)$$

From Fig. 10, it can be seen that Γ is constant when plotted against frequency. If the dissipation is linear and taking $\Gamma = 0.027\,\text{m}^2/\text{s}^3$, one finds

$$g(f,d) = 0.00028\omega^2/\sin\text{h}^2 kd \tag{9}$$

From the sketch given in Fig. 11, it is clear that the constant C lies above function $g(f,d)$ given by (9).

In a situation where dissipation is mainly confined to the forward face of the spectrum, as suggested by the sketch, high energy has a relatively higher dissipation rate than low energy when compared with result (9) from JONSWAP. In the model, the dissipation is therefore nonlinear. This effect can also be seen in the JONSWAP results where the attenuation factor Γ was also plotted against I (Fig. 12).

A comparison of $g(f,d)$ with C can be made at a characteristic frequency f_D that lies in the range where bottom friction effectively limits wave growth. In this model, the dissipation takes place at frequencies smaller than $f_B{}^*$. A sensible choice for f_D in this frequency range is $0.9 f_B{}^*$, so

$$f_D = 0.9(g/4\pi b^* d)^{1/2} \qquad \text{where } b^* = 3$$

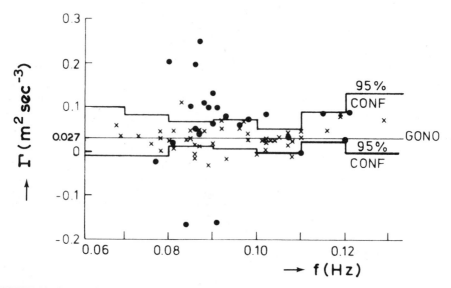

FIGURE 10. Attenuation factor Γ as a function of frequency. The value of Γ for swell dissipation in the GONO model is also indicated. (From Hasselmann *et al.*, 1973).

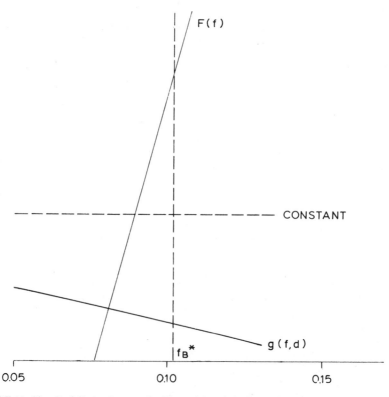

FIGURE 11. Sketch of dissipation rate in this model and the linear dissipation rate $g(f,d)$ of Eq. (9).

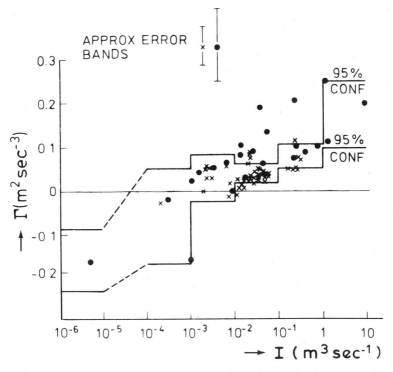

FIGURE 12. Attenuation factor Γ vs. energy density flux I. (From Hasselmann *et al.*, 1973).

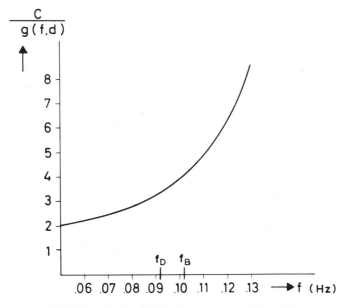

FIGURE 13. Ratio of dissipation rates at depth $d = 25\,\text{m}$.

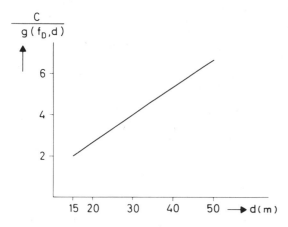

FIGURE 14. Ratio of dissipation rates at frequency f_D.

For a given depth, one can plot the ratio of C to $g(f, d)$ against frequency. This is done in Fig. 13 for $d = 25$ m. In order to specify the range of frequencies where the dissipation is important, $f_B{}^*$ and f_D are also indicated. The ratio $C/g(f_D, d)$ at the characteristic frequency f_D as a function of depth is shown in Fig. 14. One easily finds $C/g(f_D, d) = 0.133d$, which explains the straight line in Fig. 14.

In limiting situations, if depth $d1$ is greater than depth $d2$, the energy density around $f_D(d1)$ will be greater than the energy density around $f_D(d2)$. In these circumstances, the water velocity at the bottom for depth $d1$ is also greater than for depth $d2$.

The fact that the relative dissipation rate (Fig. 14) increases as d becomes larger, is in agreement with the earlier-mentioned nonlinear character of the model.

In the storm of January 3, 1976, a beautiful set of hourly Waverider measurements was recorded by the Netherlands Department of Public Works just west of the island Texel (33° 02′N, 4° 17′E, depth 30 m). Based on wind recordings made at three surrounding positions, the wind speed was estimated to be between 25 and 30 m/s for most of the day. The recordings show that the wave height was restricted to about 6 m and the spectral shape came very close to that of a pure wind-driven sea. Results produced by the model for this extreme situation are presented in Fig. 15.

This model for calculating limited wave growth in shallow water was evaluated using data from many storms recorded at several locations off the Dutch coast and in the central part of the North Sea. This evaluation for depths ranging from 20 to 60 m also covered extreme situations. The accuracy found is comparable to the results shown here (see, e.g., Hasselmann et al., 1976). In Fig. 11a of Bouws et al. (this volume), the results of the model are presented for a depth of about 60 m in a situation where bottom dissipation is very important despite the great depth (see Fig. 9 of this chapter).

3.2. Model for the Spectral Shape

In discussing the spectral shape, it is helpful to introduce a three-part classification of the effect of bottom dissipation on a wind-driven sea in shallow water. When $f_{min} = f_B{}^*$, dissipation starts and as long as $f_{min} < f_B{}^* < \frac{1}{2}(f_p + f_{min})$, limitation of wave height can be described as weak. The intermediate dissipation range can be chosen to end when $f_p = f_B{}^*$ so that strong limitation applies to situations where $f_p < f_B{}^*$. This classification can be given in terms of the wavenumber, k_p, associated with the peak frequency f_p. Writing $f_{min} = \mu f_p$, with $\mu = 0.75$, the limitation ranges are defined as

weak	$2.0 \geqslant k_p d > 1.5$
intermediate	$1.5 \geqslant k_p d > 1.24$
strong	$1.24 \geqslant k_p d$

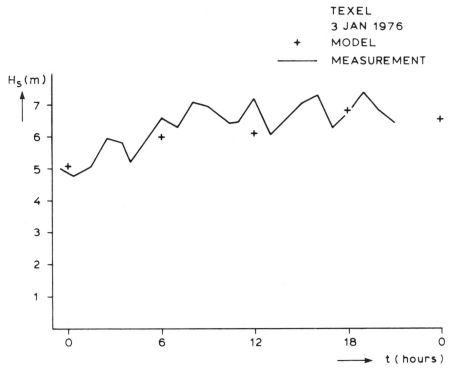

FIGURE 15. Model results and measurements for an extreme situation at a water depth of 30 m.

If the spectral shape is calculated from the stage of development ξ, defined by (2), and put into Eq. (3), the results show good agreement with measurements in the range of weak limitation and even in the first half of the intermediate range. As H_s in the intermediate range is already significantly limited by depth, the values of ξ are smaller than for the same wind field in deep water. The calculated spectra will therefore be steeper with the peak frequency shifted towards shorter waves. Seamen often report this effect when comparing the North Sea with the Atlantic Ocean.

This method for determining spectral shape in shallow water is the same as that used for deep water. Depth only indirectly determines the spectral shape because H_s is limited by the model described in Section 3.1.

Deviations from the description start when dissipation becomes very important. In the model, the beginning of this deviation is not marked by a fixed value of $k_p d$. The reason for this is given below.

Most of the measurements used to calibrate the model were made at locations with water depths of between 20 and 30 m. From these measurements, it was found that the method given above is applicable for values of $k_p d$ up to about 1.35. For smaller $k_p d$ values, deviations from the $\hat{a}-\hat{v}$ relation can be expected (see beginning of Section 3). In the model, the total energy, which is the prognostic parameter, is calculated with every time step and the associated spectral shape is determined afterwards. It follows therefore that if the spectral shape must be calculated in a different way for values of $k_p d$ smaller than 1.35, then the decision to do so in this model must be based on a critical value of total energy. From Eq. (1) with $\mu = 0.75$, it is seen that a $k_p d$ value of 1.35 is associated with a significant wave height $H_s{}^0$ given by

$$H_s{}^0 = 2.08\,\hat{a}^{1/2}d \tag{10}$$

which also involves the value of $\hat{a}$. In extreme situations where $k_p d$ is smaller than 1.35 a better description can be expected from a two-parameter model.

Instead of using a two-parameter model, the one-parameter model can be extended to determine the spectral shape for extreme situations using the dependence of H_s^0 on d only. This extension applies to values of $H_s > H_s^0$, where

$$H_s^0 = [(4 - d^{1/2})0.013 + 0.18]d \qquad \text{for } 15\,\text{m} \leqslant d \leqslant 100\,\text{m} \tag{11}$$

If in (10) $\hat{\alpha}$ is taken as a constant, H_s^0 is proportional to d. Hindcast studies indicate that for depths greater than 30 m the significant wave height that marks the beginning of deviations is better expressed by (11) than by (10) with a constant value of $\hat{\alpha}$. This would imply that for greater depths, deviations start at greater values of $k_p d$.

The distinction between wind-driven sea and swell in shallow water poses a problem. In the situation of a nonuniform wind field and/or nonuniform depth, it is possible that by advection of a nearby higher windsea, H_s will have a higher value then $H_s^{\max}$ given in Section 3.1. In these situations, the upper limit for the maximum energy $E^{\lim}$ of a wind-driven sea in the model described in this chapter is defined as

$$E^{\lim} = \int_{f_B}^{\infty} \frac{\hat{\alpha}g^2}{(2\pi)^4 f^5} \exp\left[-a(f_0/f)^4\right] df$$

where f_0 and a are given by (6) with $H_s = \beta u^2/g$. Rewriting this expression gives

$$E^{\lim}(d,u) = \left(\frac{\beta u^2}{4g}\right)^2 \left\{1 - \exp\left[-4\hat{\alpha}\left(\frac{gbd}{\pi\beta u^2}\right)^2\right]\right\} \tag{12}$$

which, for a wind-driven sea, is associated with a maximum significant wave height $H_s^{\lim}(d,u)$.

Values of f_p and $k_p d$ associated with $H_s^{\lim}$ for a number of values of d, u, and $\hat{\alpha}$ are given in Table I. It can be seen that f_p values corresponding to $k_p d$ values greater than 2.0 are

TABLE I. f_p and $k_p d$ Associated with $H_s^{\lim}$

	f_p values					$k_p d$ values				
$\alpha \times 10^3 \, u(\text{m/s})$:	10	15	20	25	30	10	15	20	25	30
$d = 20\,\text{m}, \ f_B^* = 0.114$										
5	0.140	0.113	0.108	0.107	0.106	1.70	1.23	1.16	1.14	1.13
7.5	0.154	0.117	0.110	0.107	0.107	1.98	1.29	1.17	1.14	1.13
10	0.165	0.121	0.111	0.108	0.107	2.24	1.35	1.19	1.15	1.13
12.5	0.174	0.124	0.112	0.108	0.107	2.48	1.40	1.21	1.16	1.14
$d = 25\,\text{m}, \ f_B^* = 0.102$										
5	0.139	0.105	0.098	0.096	0.095	2.01	1.30	1.18	1.14	1.13
7.5	0.153	0.110	0.100	0.097	0.096	2.40	1.38	1.20	1.15	1.14
10	0.165	0.115	0.101	0.098	0.096	2.75	1.48	1.23	1.17	1.14
12.5	0.174	0.119	0.103	0.098	0.096	3.06	1.57	1.26	1.18	1.15
$d = 30\,\text{m}, \ f_B^* = 0.093$										
5	0.138	0.100	0.091	0.088	0.087	2.36	1.37	1.20	1.15	1.14
7.5	0.153	0.106	0.093	0.089	0.088	2.85	1.50	1.24	1.17	1.14
10	0.165	0.112	0.095	0.090	0.088	3.28	1.63	1.28	1.19	1.15
12.5	0.174	0.117	0.097	0.091	0.089	3.67	1.76	1.32	1.20	1.16

independent of depth. For the maximum wave height of a wind-driven sea in shallow water, the transition from deep to shallow water is therefore concomitant with the onset of dissipation described by the model presented in Section 3.1. Given this definition of a maximum wind-driven sea, it is seen from Table I that 1.13 is the smallest $k_p d$ value one can expect. Extension of the concept of a wind-driven sea to maximum values of E given by (12) therefore gives rise to values of $k_p d$ in the range of strong limitation. For uniform depth and uniform wind field, this is not possible. Calculated values of H_s^{max} (Section 3.1) for a number of depths using $u = 30 \, \text{m/s}$ and $\alpha = 5 \times 10^{-3}$ give $k_p d$ values greater than 1.29, so they never come in the range of strong limitation.

From hindcast studies, it was found that for situations where $H_s^{\circ} \leqslant H_s \leqslant H_s^{lim}(d, u)$, the parameters $\hat{\alpha}$ and $\hat{v}$ can be calculated from Eqs. (3) using a stage of development $\tilde{\xi}$. This value of $\tilde{\xi}$ is greater than ξ defined by (2) but smaller than $\xi(d)$ defined as $\xi(d) = [H_s/H_s^{lim}(d, u)]^{1/2}$. On writing $\xi = \xi(\infty) = [H_s/H_s^{lim}(\infty, u)]^{1/2}$, the purely experimental fit for $\tilde{\xi}$ reads

$$\tilde{\xi} = A(d, u)\xi(d) + [1 - A(d, u)]\xi(\infty) \tag{13}$$

where $A(d, u) = (H_s^{lim}(d, u)/H_s^{lim}(\infty, u))^P$ with $P = 1.5$, and the parameter $\hat{\alpha}$ in $H_s^{lim}(d, u)$ [see (12)] is assigned a constant value $\hat{\alpha}_0 = 7.7 \times 10^{-3}$.

The value of $\hat{\alpha}$ in (12) has to be taken constant because this description is part of a one-parameter model. It is easily seen that for $d \to \infty$, $\xi(d) \to \xi(\infty)$ and $A(d, u) \to 1$, so $\tilde{\xi} \to \xi$.

Expression (13) can also be written as

$$\tilde{\xi} = B(d, u)\xi(\infty) \tag{14}$$

where $B(d, u) = 1 + h(d, u) - h(d, u)^{3/2}$ and $h(d, u) = \{1 - \exp[-4\hat{\alpha}_0(gbd/\pi\beta u^2)^2]\}^{1/2}$.

A consequence of transformation (14) is that the $\hat{\alpha}$–$\hat{v}$ relation changes. To see this effect, (2) can be written in the form

$$\xi = 2\left(\frac{g}{\beta}\right)^{1/2} \cdot E^{1/4} \cdot \frac{1}{u}$$

It follows from (14) that $\tilde{\xi}$ can be written as

$$\tilde{\xi} = 2\left(\frac{g}{\beta}\right)^{1/2} \cdot E^{1/4} \cdot \frac{1}{\tilde{u}}$$

where $\tilde{u} = u/B(d, u)$. Replacing ξ in (3b) with $\tilde{\xi}$ given by (14), it must be realized that a $\tilde{v}$ results from which f_p is found using $\tilde{u}$:

$$f_p = (g/\tilde{u})\tilde{v}$$

so $\hat{v} = uf_p/g = B(d, u)\tilde{v}$.

Eliminating ξ from (3a) and (3b) gives for $\beta = 0.22$ the deep-water $\hat{\alpha}$–$\hat{v}$ relation:

$$\hat{\alpha} = 0.0741\hat{v}^{1.413} \tag{15}$$

Eliminating $\tilde{\xi}$ from (3a) and (3b) gives

$$\hat{\alpha} = 0.0741\tilde{v}^{1.413}$$

which can be written as

$$\hat{\alpha} = 0.0741[B(d, u)]^{-1.413}\hat{v}^{1.413} \tag{16}$$

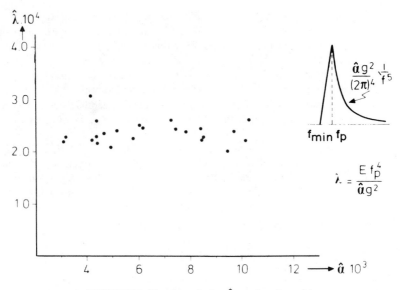

FIGURE 16. The shape factor $\hat{\lambda}$ as a function of $\hat{a}$.

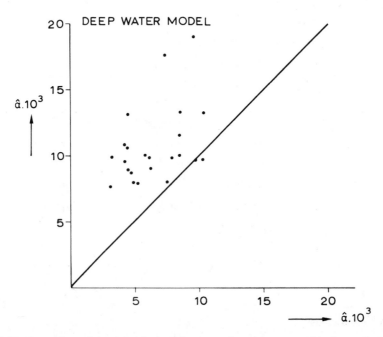

FIGURE 17. Comparison of $\hat{a}$ values obtained from actual measurements with those calculated from Eq. (3a) using $\xi(\infty)$.

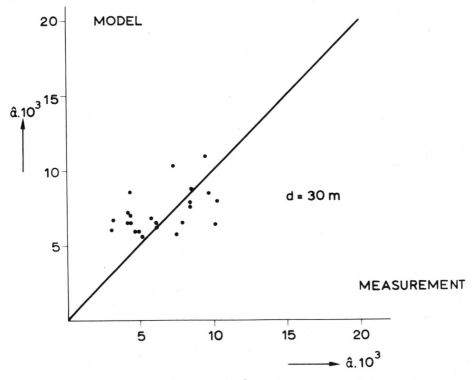

FIGURE 18. Values of $\hat{\alpha}$ calculated from (3a) using $\tilde{\xi}$ instead of ξ, plotted against measured values of $\hat{\alpha}$.

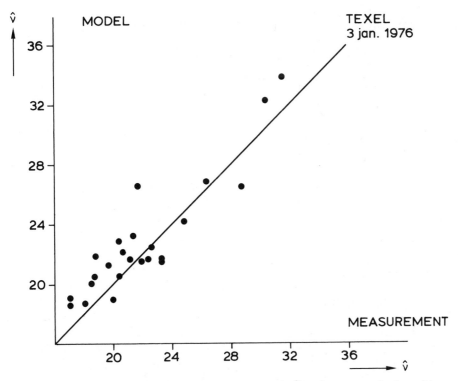

FIGURE 19. $\hat{v}$ calculated from (3b) where ξ is replaced by $\tilde{\xi}$, against measured values of $\hat{v}$.

The result of transformation (14), therefore, is that the deep-water $\hat{\alpha}$–$\hat{v}$ relation is modified by a factor $[B(d,u)]^{-1.413}$ if $H_s > H_s^\circ$.

In the storm of January 3, 1976, 23 sets of wave measurements were made. Spectra from all 23 measurements have the appearance of a wind-driven sea. Two spectra were classified as weakly limited with $1.6 > k_p d > 1.5$. Five had $1.5 \geqslant k_p d > 1.24$ and 16 were in the strong limitation range. Of the spectra in the range of strong limitation, 11 had $k_p d$ values less than 1.13. According to the definition of a maximum wind-driven sea, these 11 spectra represent a combination of sea and swell (see Table I). Spectra of this data set all have the simple form shown in Fig. 1. The parameters $\hat{\alpha}$, $\hat{v}$, and μ therefore give the same value for the shape factor $\hat{\lambda}$ (Fig. 16) as found in deep water. By calculating $\hat{\alpha}$ from (3a) using $\xi = \xi(\infty)$, the values are systematically higher than the associated measured values (Fig. 17). These measurements were made at a location where the water depth $d = 30$ m, so $H_s^\circ = 4.82$ m [Eq. (11)]. By calculating $\hat{\alpha}$ from (3a) and substituting $\tilde{\xi}$ for ξ, the calculated values of $\hat{\alpha}$ for this data set were found to be in better agreement with the $\hat{\alpha}$ values from actual measurements (see Fig. 18). The results of this procedure, applied for $\hat{v}$, are presented in Fig. 19.

In Figs. 18 and 19, the calculated values of $\hat{\alpha}$ and $\hat{v}$ are found from Eqs. (3) where ξ is replaced by $\tilde{\xi}$ given by (13). In the model described in this chapter, this procedure applies only to a wind-driven sea with $H_s^\circ \leqslant H_s \leqslant H_s^{\lim}(d, u)$. For these storm results, the hybrid computer model GONO, in which this shallow-water model is incorporated, shows that the sea state resulted from a combination of windsea and swell. The results are presented in Figs. 20–22. The quantity E_{10} appearing in Fig. 20 represents the total energy above 10 s. In Figs. 21 and 22, the quantity E represents the total energy in the period range specified on the right-hand sides of the figures.

The influence of shallow-water effects on the spectra of this data set can be seen from their $k_p d$ values. As most of these measurements have $k_p d$ values in the range of strong limitation, a deviation from the deep-water $\hat{\alpha}$–$\hat{v}$ relation can be expected. In Fig. 23, it is seen that the deep-water relation (15) represented by the drawn curve does not correspond with the data cloud.

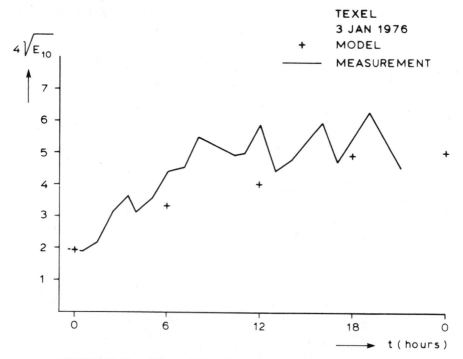

FIGURE 20. Test of the model for the low-frequency part of the spectrum.

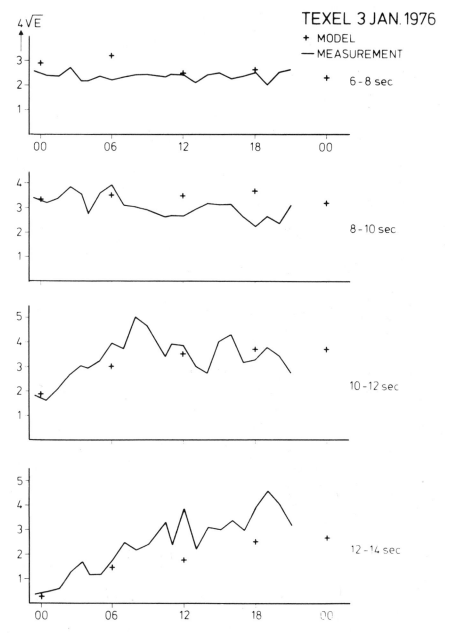

FIGURE 21. Test of the spectral shape as calculated by the model for periods above 6 s.

Relation (16), represented by the dashed line, gives an improvement although this relation applies only to the definition for a wind-driven sea given in this chapter. For this data set, the factor $[B(d, u)]^{-1.413}$ occurring in (16) is almost constant and approximately equals 0.83. Small values of $\hat{\alpha}$ deviate considerably from relation (16).

The question of whether the measured sea state should be described as that of a wind-driven sea remains open to interpretation. Results given by the model and presented here are in good agreement with actual measurements and show that the sea state resulted from a combination of sea and swell.

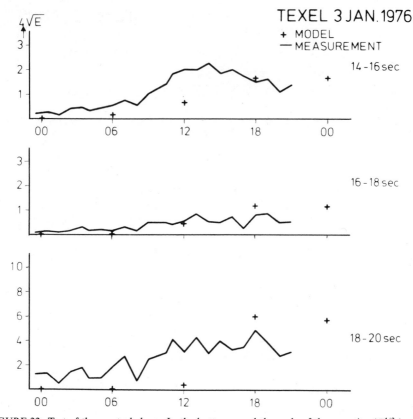

FIGURE 22. Test of the spectral shape. In the bottom panel the scale of the quantity $4E^{1/2}$ is changed.

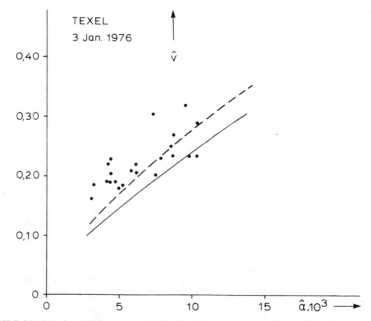

FIGURE 23. Relation (15) is represented by the drawn curve, relation (16) by the dashed line.

ACKNOWLEDGMENTS. The Waverider measurements of the January 1976 storm were performed by Rijkswaterstaat-Hoorn and processed on the computer by Evert Bouws (KNMI). We also thank David Smith for making suggestions for the concept paper.

REFERENCES

Bouws, E. (1980): Spectra of extreme wave conditions in the southern North Sea considering the influence of water depth. *Proceedings of the International Conference on Sea Climatology*, Collect. Colloq. Semin. 34, Ed. Technip, Paris, 51–71.

de Voogt, W. J. P., G. J. Komen, and J. Bruinsma (1985): The KNMI operational wave prediction model GONO. *Ocean Wave Modeling* (The SWAMP group, eds.), Plenum Press, New York, 193–200.

Hasselmann, K., T. P. Barnett, E. Bouws, H. Carlson, D. E. Cartwright, K. Enke, J. A. Ewing, H. Gienapp, D. E. Hasselmann, P. Kruseman, A. Meerburg, P. Müller, D. J. Olbers, K. Richter, W. Sell, and H. Walden (1973): Measurements of wind-wave growth and swell decay during the Joint North Sea Wave Project (JONSWAP). *Dtsch. Hydrogr. Z. Suppl. A* **8**(12).

Hasselmann, K., D. B. Ross, P. Müller, and W. Sell (1976): A parametric wave prediction model. *J. Phys. Oceanogr.* **6**, 200–228.

Sanders, J. W. (1976): A growth stage scaling model for the wind driven sea. *Dtsch. Hydrogr. Z.* **29**, 136–161.

Sanders, J. W., W. J. P. de Voogt, and J. Bruinsma (1980): Fysisch golfonderzoek Noordzee. MLTP-2 Scientific Report.

DISCUSSION

MITSUYASU: In your spectral model, the low-frequency side seems to be linearly proportional to the frequency ($\phi \propto f$). I think this form is highly different from observed spectra. The difference may not be important for the computation of total energy E, because the low-frequency side of the spectra ($f < f_m$) contains fairly small energy. However, when you compute the wave energy contained only in the low-frequency side ($f < f_d < f_m$), the difference may introduce significant error. Did you check this problem?

SANDERS: Thank you for this question because this gives me the opportunity to explain something about the spectral shape I use. It looks so simple that one gets the impression that it is no good at all. The energy contained in the low-frequency part ($f < f_m$) is about one third of the total energy, the same ratio as one finds for the JONSWAP spectrum. Moreover, if one compares this spectrum with the JONSWAP spectrum for the *same* total energy, peak frequency, and γ (corrected for the peak enhancement (factor), the differences are very small. I think that for a wind-driven sea, there is no reason to give preference to either spectrum.

CAVALERI: I refer to your transparency showing the maximum significant wave height for given wind speed and depth. For the shallower case ($d = 20$ m), you report $(H_s)_{max} = 4, 4.5$ m for $u = 30$ m/s. Our experience in northern Adriatic suggests that higher H_s are possible and happen, even in shallower depth (16 m). Do you think that the discrepancy could be explained by a different dissipation due to different bottom characteristics?

Also (suggestion by Leo Holthujsen), there could be some reason connected to different air stability conditions.

SANDERS: I think that the BORA-wind can give rise to a different growth curve. One could scale the growth curve to a higher asymptotic value β. In my model, this would be no problem as the spectral shape depends upon the stage of development and could also be used for higher atmospheric input. If that would be the case, then the same significant wave height would be reached in a shorter time, which gives the nonlinear wave–wave interactions less time to move the peak to the low-frequency part. This would give rise to steeper spectral shapes, exactly as my model would give in that situation, because the stage of development would be smaller.

42

AN EVALUATION OF OPERATIONAL
WAVE FORECASTS ON SHALLOW
WATER

E. BOUWS, G. J. KOMEN, R. A. VAN MOERKERKEN, H. H. PEECK, AND
M. J. M. SARABER

ABSTRACT. The KNMI wave model GONO (Sanders, 1976) with its shallow water extension is in operational use for the forecasting of wave conditions in the southern North Sea. The model uses a parameterized spectrum for the calculation of windsea supplemented by empirical growth laws. Swell is calculated following an idea of Haug (1968). GONO has been verified during the last two winters, against wave data from several positions. Also, during the winter 1979/80 a comparison was made with the model of Golding (1978) which is operational at the British Meteorological Office. After a brief discussion of GONO and the quality of the observational data, we will give the results of the comparison. The comparison is made for the significant wave height, H_S, and for $H_{S,10}$, which is the significant wave height derived from wave energy in waves with a period greater than 10 s. This latter quantity is important for the navigation of very large crude carriers entering Rotterdam harbor and for the construction of the Oosterschelde storm surge barrier. Also, calculated and observed winds are compared. The comparison is made for analyzed wind data as well as 12- and 24-h forecasts. First, we discuss the time series, with emphasis on interesting events, such as the extreme swell ($T \simeq 18$ s) occurring on January 15, 1980, which was generated at very northerly latitude, and the extreme situation on April 19–20, 1980, when a northerly storm of very long fetch generated waves with H_S 5–7 m on shallow water. Further, we present results of a statistical analysis, including scatter plots and monthly summary tables giving for wind direction, wind velocity, H_S, and $H_{S,10}$ the number of observations, the average of the observed values, the average error, rms error, number of cases overpredicted, and number of cases underpredicted. Also, the scatter index (ratio of standard deviation and average observed value) is given. Values range from 20 to 35%, depending mainly on the quality of wind calculations.

1. INTRODUCTION

In this chapter we will report work that we have done for the verification of our wave prediction model, GONO, described by Sanders (1976) and Bruinsma *et al.* (1980). The primary purpose was to establish the performance of the model in shallow water (depth $\gtrsim 15$ m) from an operational point of view. For a better understanding of the physics, additional studies have been made and are described elsewhere (de Voogt *et al.*, 1985; Sanders and Bruinsma, this volume).

In the present investigation we compared the calculated values of a small set of variables with observations. We also compared our results with results from the UK Met Office wave model (Golding, 1978).

E. BOUWS, G. J. KOMEN, R. A. VAN MOERKERKEN, H. H. PEECK, AND M. J. M. SARABER ● Royal Netherlands Meteorological Institute, De Bilt, The Netherlands.

This chapter consists of three parts: a brief summary of the model characteristics, a description of our verification procedure, and a discussion of some of our results.

2. MODEL CHARACTERISTICS

In order to produce operational wave forecasts for the North Sea, GONO is coupled to the KNMI atmospheric model, a four-layer quasi-geostrophic model. In addition, it is fed with information on the ice boundaries and the air–sea temperature difference. It is a hybrid model. Windsea is described with a parameterized spectrum (Fig. 1). The three parameters $\hat{\alpha}$, $u f_p/g$, and μ are determined empirically in terms of the stage of development $\xi = (H_S/H_{S,max})^{1/2}$, where H_S is the significant wave height, and $H_{S,max}$ its maximum value for the given wind speed, u. First, a rough calculation is made of the wave energy in each grid point by solving the integrated energy balance equation with an empirical source function and a best guess for the average propagation speed. For windsea the parametric model is then used, while swell is calculated by propagation along characteristics but for a few selected swell points only. This procedure follows an idea of Haug (1968). In shallow water, dissipation is taken into account, the group velocity is adjusted, and the stage of development is corrected for bottom influence (see Sanders and Bruinsma, this volume). Refraction is not included. The model is able to compute precise swell information in an economic way. Outside swell points, the possible capturing of swell by windsea is poorly treated, but this does not seem to be of great importance in North Sea conditions. GONO has a 75-km grid (Fig. 2) extending quite far to the north. It runs operationally four times a day and computes wave spectra for the swell points based on analyzed weather data and a 12- and 24-h atmospheric forecast. Spectral energies are given in seven period bands (between 6 and 20 s) and in six directional sectors. For each sector the average direction is given as well.

The UK Met Office wave model (Golding, 1978) is a spectral model coupled to a 10-layer atmospheric model. The energy balance equation is solved with a fourth-order Lax–Wendroff scheme. Source terms are empirical; relaxation to the Pierson–Moskowitz spectrum is built in. In shallow water dissipation, changes of group velocity and refraction are taken into account. The model has a coarse Atlantic and a fine (50 km) North Sea grid. It runs twice a day and gives wave spectra in every grid point. There are 11 frequency bands and 12 directions.

3. VERIFICATION PROCEDURE

We started our program December 1, 1979. It operates continuously, but we allow for a break during summer. The comparison with the UK Met Office model was during winter 1979/80 only.

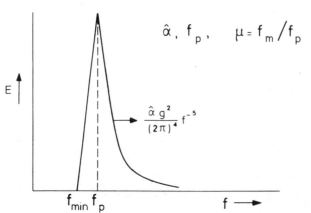

FIGURE 1. The GONO windsea spectrum is described by three parameters: the Phillips constant $\hat{\alpha}$, the peak frequency f_p, and μ, the ratio of lowest frequency to f_p.

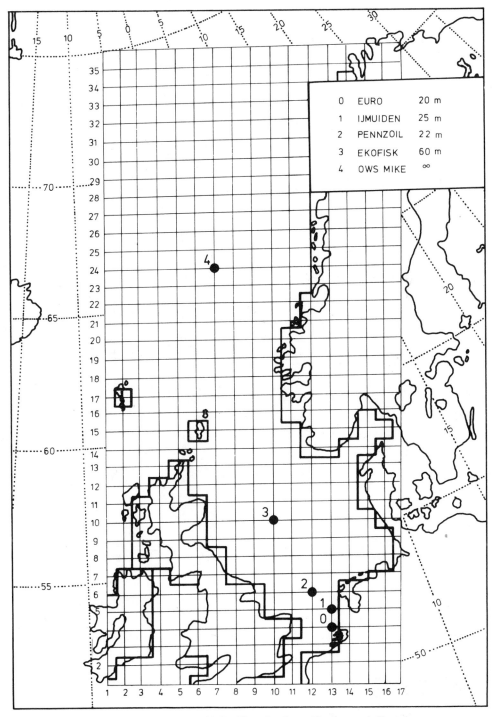

FIGURE 2. The GONO grid. Positions for the verification are indicated.

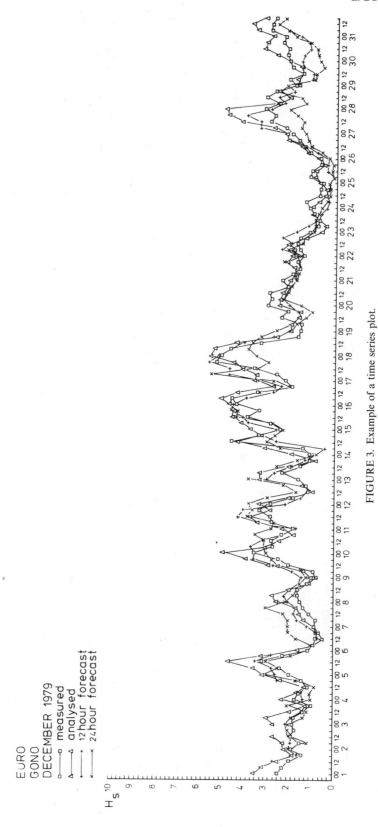

FIGURE 3. Example of a time series plot.

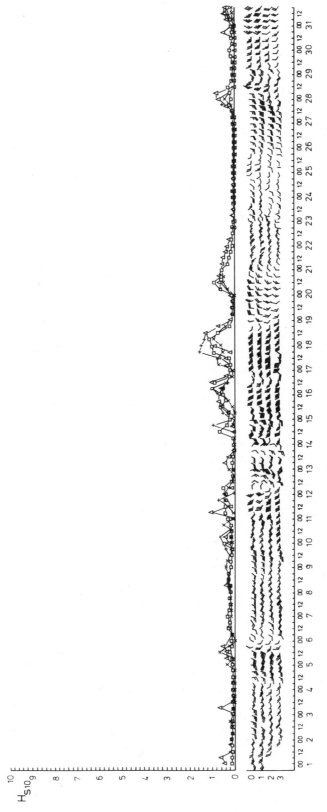

FIGURE 4. Example of a summary table.

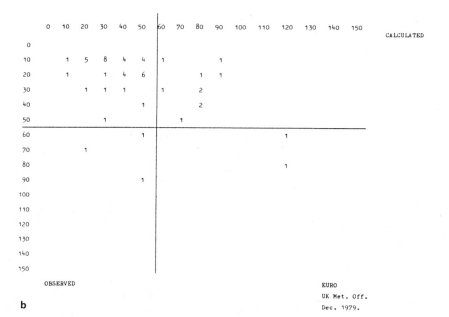

FIGURE 5. Contingency table of $H_{S,10}$ for December 1979, position Euro: (a) GONO, (b) UK Met Office model.

Four variables were selected: wind direction, wind speed, significant wave height, and a wave height parameter associated with wave components with a period exceeding 10 s:

$$H_{S,10} = 4E_{10}^{1/2}$$

where E_{10} is the variance in waves with a period greater than 10 s. This latter quantity is of importance for the navigation of very large crude carriers and for work on the storm surge barrier in the Oosterschelde estuary.

The comparison was made for the five positions indicated in Fig. 2. The quality of the observed wind and wave data is discussed at some length by Bouws et al. (1980a, b). That discussion will not be repeated here.

Since the comparison involved both the analysis and the 12- and 24-h forecasts, quite a large number of data were involved (over 10,000 numbers a month). In this section we will give an outline of the way in which we handle these data. In the next section we will discuss some of the results.

First of all, for each model and position we produce monthly plots of the time series. An example, strongly reduced in size, is given in Fig. 3. Along the abscissa date and time of date are indicated. At the top H_S is plotted, in the middle $H_{S,10}$, and at the bottom wind is indicated according to standard meteorological convention (e.g., F is northerly wind, 20 knots; 2 knots $\simeq$ 1 m/s). In each instance four values are specified: one is the observed value, the three others are analysis and 12- and 24-h forecast. These plots are very useful for the identification and analysis of individual events.

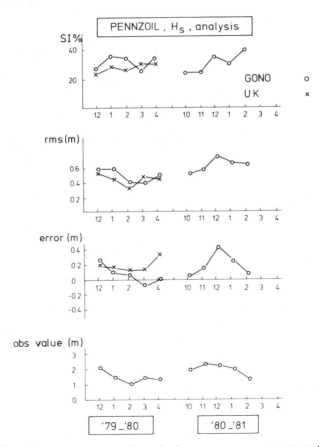

FIGURE 6. Plot of monthly averages: average observed value, average error, rms error, and scatter index.

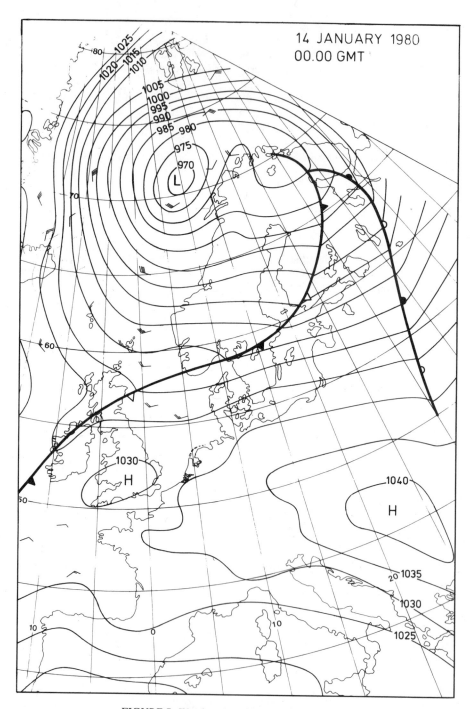

FIGURE 7. Weather chart for January 14, 1980.

On the basis of these data a monthly statistical analysis is made. Results are given in summary tables (Fig. 4). In these tables (one for each month, model, and forecast period) we indicate for each variable the number of cases, the average observed value, the average error, the rms error, and the number of cases too high and too low. For a given variable A we distinguish between

$$rms = \langle (\Delta A)^2 \rangle^{1/2} \qquad \Delta A = A_{obs} - A_{calc}$$

and

$$rms^x = \langle (\Delta A - \langle \Delta A \rangle)^2 \rangle^{1/2}$$

In practice the difference between both definitions of rms error is small, since $\langle \Delta A \rangle \sim 0$, in most

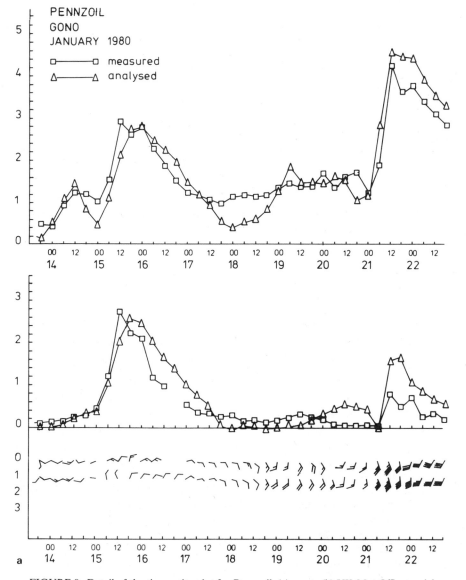

FIGURE 8. Detail of the time series plot for Pennzoil: (a) GONO, (b) UK Met Office model.

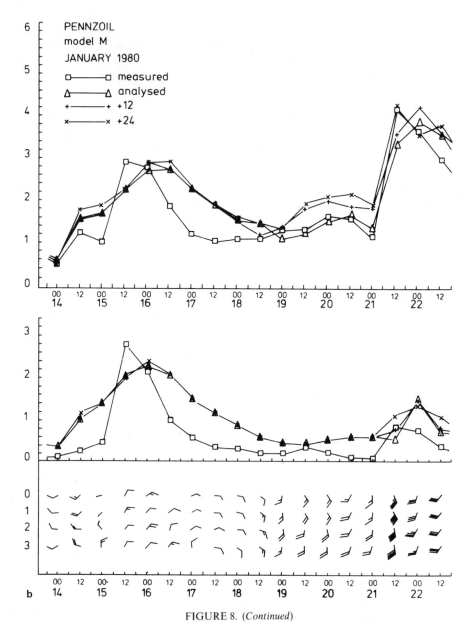

FIGURE 8. (*Continued*)

cases. We also have used the concept of scatter index, SI (Holthuijsen, 1980), the rms error divided by the average observed value. Typically (see also Ewing, 1980), wave models have a scatter index for the significant wave height which ranges between 20% (hindcasts with sophisticated models, correct wind fields) and 60% (operational runs with less accurate winds). In addition to the summary tables, we prepared contingency tables, which indicate bindwise the correspondence between observed and computed values (see Fig. 5). Finally we found it useful to produce summary plots, in which we plot the time series of monthly statistical values (Fig. 6).

4. RESULTS

In this section we give examples of the way in which we use the various tables and diagrams.

4.1. Time Series—Interesting Events

Two special events were observed in the first period of our comparison. On January 14 a severe gale in the Norwegian Sea (Fig. 7) generated swell with a period of 18 s. It reached Pennzoil on the 15th and was predicted quite well by GONO and the UK Met Office model (Figs. 8a and b). At the more southern stations, both models overpredicted (Figs. 9a and b). On April 19 a

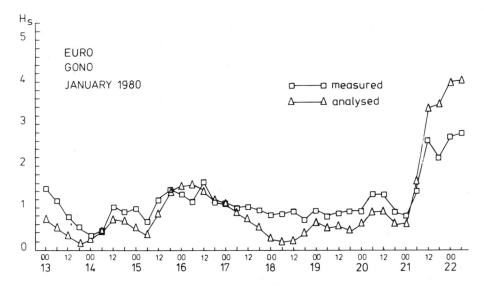

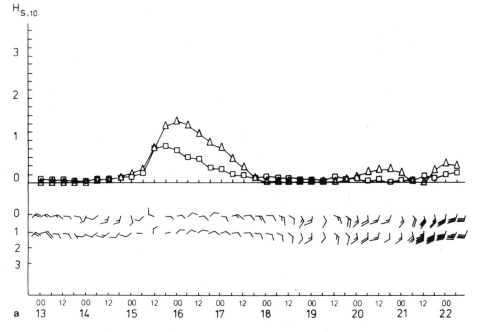

FIGURE 9. Detail of the time series plot for Euro: (a) GONO, (b) UK Met Office model.

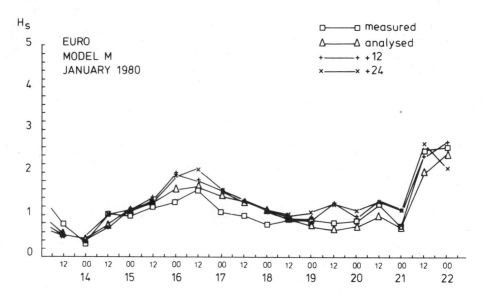

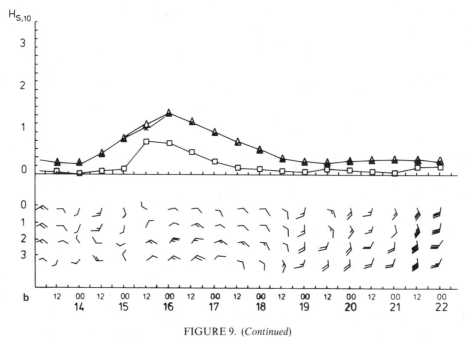

FIGURE 9. (*Continued*)

northwesterly gale, with an unusually long fetch (Fig. 10), generated waves of nearly 10 m significant wave height at Ekofisk. The time series are given in Figs. 11a and b. At Euro, H_S reached 5 m (Figs. 12a and b).

4.2. Statistical Analysis—Contingency Tables

For GONO, in December 1979 (not an exceptional month) the rms error in $H_{S,10}$ was about 22 cm at Euro (see Fig. 4). The average error was small (1 cm). Prediction of $H_{S,10}$ is very

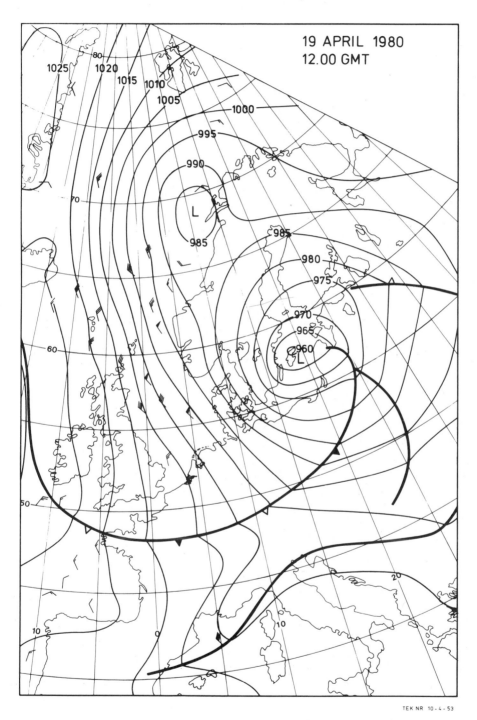

FIGURE 10. Weather chart for April 19, 1980.

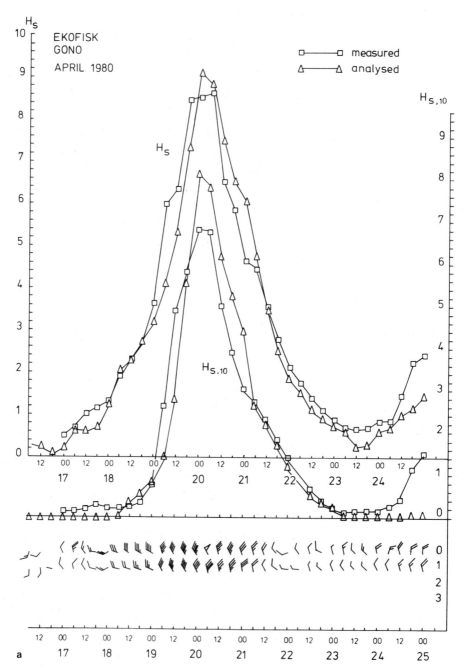

FIGURE 11. Detail of the time series plot for Ekofisk: (a) GONO, (b) UK Met Office model.

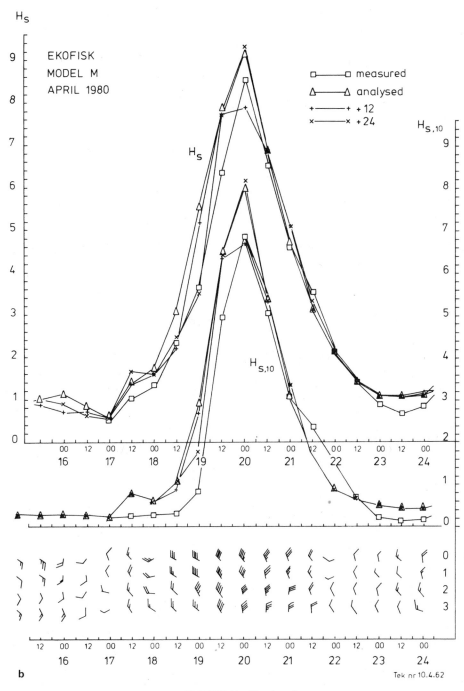

FIGURE 11. (*Continued*)

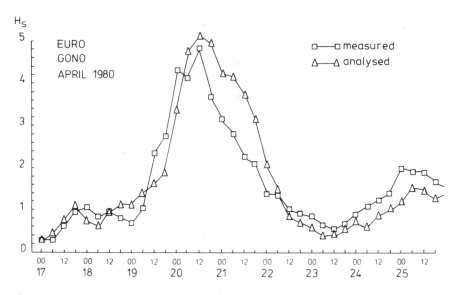

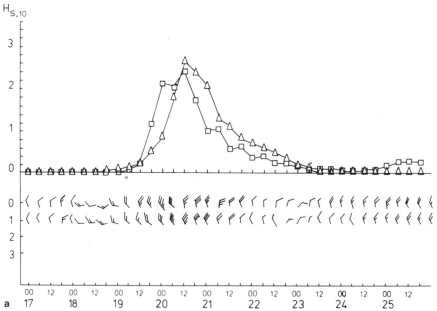

FIGURE 12. Detail of the time series plot for Euro: (a) GONO, (b) UK Met Office model.

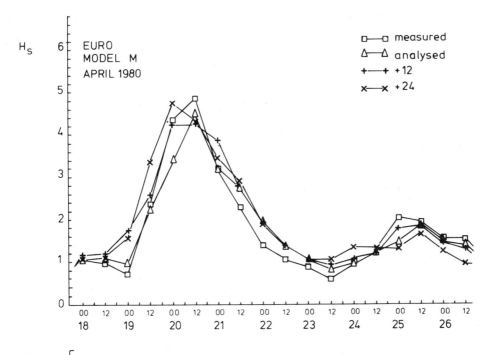

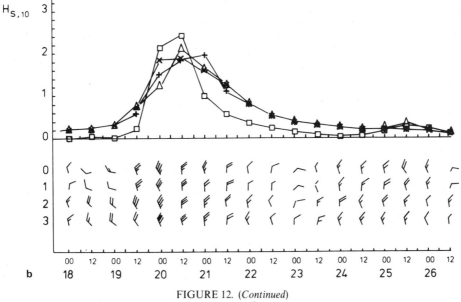

FIGURE 12. (*Continued*)

TABLE I.   Error Analysis

| LOCATION | WIND DIRECTION (DEGREES) | | | ANALYSIS GONO | | |
	NUMBER	AV. OBS	AV. ERROR	RMS	PLUS	MINUS
EURO	110	***	008	017	077	029
IJMUIDEN	104	***	014	026	081	023
PENNZOIL	107	***	− 005	016	035	070
EKOFISK	078	***	− 002	023	040	035
STATION M	098	***	− 009	025	031	063

| LOCATION | WIND SPEED (DM/SEC) | | | ANALYSIS GONO | | |
	NUMBER	AV. OBS	AV. ERROR	RMS	PLUS	MINUS
EURO	123	106	006	030	073	048
IJMUIDEN	123	106	006	028	067	054
PENNZOIL	122	099	010	027	089	033
EKOFISK	105	114	− 025	035	010	095
STATION M	118	120	− 023	041	024	093

| LOCATION | HS SIGN. WAVE HEIGHT (CM) | | | ANALYSIS GONO | | |
	NUMBER	AV. OBS	AV. ERROR	RMS	PLUS	MINUS
EURO	123	205	037	071	085	036
IJMUIDEN	088	215	040	069	067	020
PENNZOIL	122	229	028	061	084	038
EKOFISK	018	192	150	198	018	000
STATION M	117	281	088	156	092	025

| LOCATION | HS, 10 (CM) | | | ANALYSIS GONO | | |
	NUMBER	AV. OBS	AV. ERROR	RMS	PLUS	MINUS
EURO	123	023	001	022	046	076
IJMUIDEN	080	045	− 012	041	024	056
PENNZOIL	121	046	− 006	030	044	076
EKOFISK	018	104	052	116	010	008
STATION M	000					

important, because the small keel clearance of very large crude carriers near Rotterdam harbor allows a small roll angle only, which is about reached, depending on the ships' characteristics, when $H_{S,10}$ exceeds 50 cm.

Therefore, correct tuning and a scatter as small as possible are imperative. The contingency table (Fig. 5a) shows how difficult it is to give correct warnings for the exceedence of 50 cm. In nine cases correct warnings were given, eight times an unnecessary warning was given, while three times the model failed to predict the observed exceedence. The UK Met Office model did not perform better (Fig. 5b).

4.3. Conclusions of this Statistical Analysis

A careful study of the analyzed data suggested a few conclusions, some of which are rather obvious. We will list these conclusions here and give evidence in the form of a few figures.

1. The UK Met Office model has smaller average and rms errors than GONO. This is illustrated in Fig. 13, where it can be seen that GONO has a scatter index of between 30 and 40%, whereas that of the UK Met Office model is between 20 and 30%. The difference, however, is not the same for all places and all months. The + 12 forecast of Pennzoil from both models is compatible. At Euro differences are larger.

2. With respect to $H_{S,10}$ the reverse trend can be observed: the GONO calculations are

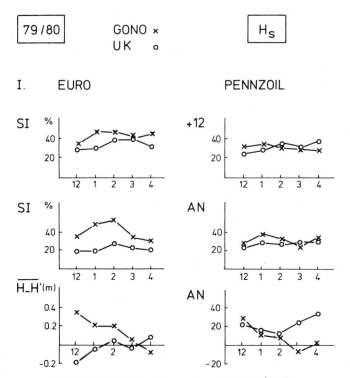

FIGURE 13. Scatter index and average error for H_S.

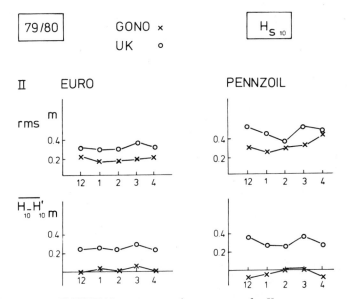

FIGURE 14. rms error and average error for $H_{S,10}$.

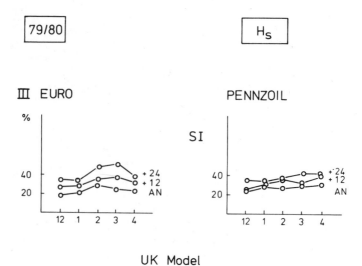

UK Model

FIGURE 15. Scatter index for different prediction times.

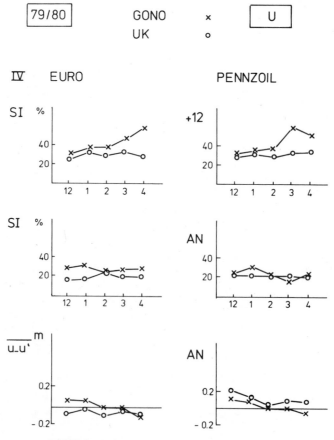

FIGURE 16. Scatter index and average error for *u*.

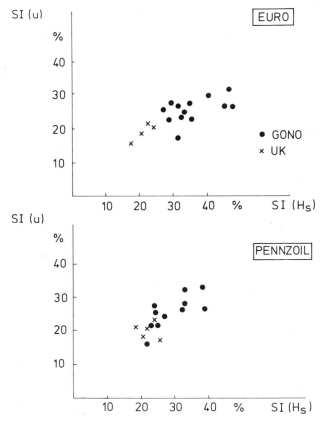

FIGURE 17. Correlation between wind speed scatter index and H_S scatter index.

better tuned and have smaller rms errors than the UK model. This can be clearly seen in Fig. 14.

3: The error in H_S increases with increasing prediction time. As an example we compare in Fig. 15 the analysis, the $+12$ and $+24$ forecast.

4. The UK Met Office model tends to have better winds than our model (Fig. 16). This is so for the analysis, where it may result from the fact that observed winds are incorporated in the analysis, and for the forecasts.

5. It is tempting to attribute errors in the calculated H_S to errors in the wind speed and direction. Figure 17, which is a scatterplot of the monthly wind speed scatter index against the H_S scatter index, shows to what extent such a relation exists. There is a correlation between wind speed error and the error in H_S, but for a given value of the wind speed scatter index, significant variation in H_S scatter index still occurs.

5. CONCLUDING REMARKS

We have shown how the verification procedure outlined above is able to quantify statements about model behavior. The program was successful in that it provided insight into the reliability of two operational North Sea wave models. Furthermore, it is useful for the determination of KNMI's future modeling strategy. Also, it gave a few suggestions (still under investigation) for possible improvements. Finally, the collected data and their analysis will be useful for testing future model modifications.

It should be stressed that the results and conclusions presented in this chapter refer (mainly) to the period December 1979–April 1980. They should not be taken to imply that the models will always behave in this manner since they have been revised since that time. This is particularly relevant for conclusion No. 2 above. After the comparison a significant programming fault was discovered in the coding of the UK Met Office model. This error led to an incorrect specification of energy dissipation in shallow water, and may be the explanation for the overprediction of long-period wave components (P. E. Francis, personal communication).

ACKNOWLEDGMENTS. We thank Brian Golding for his willingness to exchange model results. Useful discussions with Peter Janssen, Linwood Vincent, Wolfgang Rosenthal, and Jan Sanders are gratefully acknowledged. Help with data handling came from Henk Kalle and Egbert Wiggers, and last but not least we thank Toine Philippa (who wrote the operational master program for the numerical models at KNMI) and Rik de Gier for programming assistance.

REFERENCES

Bouws, E., B. W. Golding, G. J. Komen, H. H. Peeck, and M. J. M. Saraber (1980a): Preliminary results on a comparison of shallow water wave predictions. KNMI WR 80-5.

Bouws, E., G. J. Komen, R. A. van Moerkerken, H. H. Peeck, and M. J. M. Saraber (1980b): A comparison of shallow water wave predictions. KNMI V-362.

Bruinsma, J., P. A. E. M. Janssen, G. J. Komen, H. H. Peeck, M. J. M. Saraber, and W. J. P. de Voogt (1980): Description of the KNMI operational wave forecast model GONO. KNMI WR 80-8.

de Voogt, W. J. P., J. Bruinsma, and G. J. Komen (1985): The KNMI operational wave prediction model GONO. *Ocean Wave Modeling* (The SWAMP Group, eds.), Plenum Press, New York, 193–200.

Ewing, J. A. (1980): Numerical wave models and their use in hindcasting wave climate and extreme value wave heights. *Proc. Conf. Sea Climatal.*, Edition Technip, Paris, 159–178.

Golding, B. W. (1978): A depth-dependent wave model for operational forecasting. *Turbulent Fluxes Through the Sea Surface, Wave Dynamics, and Prediction* (A. Favre and K. Hasselmann, eds.), Plenum Press, New York, 593–604.

Haug, O. (1968): A numerical model for prediction of sea and swell. *Meteorol. Ann.* **5**, 139–161.

Holthuijsen, L. H. (1980): Methoden voor golfvoorspelling. Technische Adviescommissie voor de waterkeringen.

Sanders, J. W. (1976): A growth-stage scaling model for the wind-driven sea. *Dtsch. Hydrogr. Z.* **29**, 136–161.

Sanders, J. W., J. Bruinsma (1986): Modeling wind driven sea on shallow water. *Wave Dynamics and Radio Probing of the Ocean Surface* (O. M. Phillips and K. Hasselmann, eds.), Plenum Press, New York.

DISCUSSION

REECE: Since much of the measured data were obtained in relatively shallow water, I was wondering if a numerical study had been done to determine the importance of the shallow water effects predicted by each model. Since the models differ in mechanisms included, and no doubt in estimation of the strength of the bottom interaction, it seems possible that some of the scatter in the results could be explained by consideration of spatial variability of wave–bottom interaction.

KOMEN: Yes, for GONO we have done such studies. In fact, they form the subject of the next contribution and they clearly demonstrate the importance of shallow water effects. As to the second part of your question, I agree that spatial variability can contribute to the observed scatter. However, other causes are certainly important. One is the natural variability in the occurrence of particular atmospheric situations. A careful statistical analysis of our results for different directional sectors indicates that the performance of the model depends on the wind direction. This, together with the above-mentioned variability in the occurrence of specific weather systems, gives another contribution to the observed scatter.

CARDONE: In view of the general scarcity of complete wave data series in extratropical storms in deep and shallow water, it would appear that some of the cases you described may be of interest to other wave modelers. Would it be possible to obtain the wind fields and the wave measurement series for some cases?

KOMEN: Yes.

43

ANOMALOUS DISPERSION IN NUMERICAL MODELS OF WAVE SPECTRA

WILLIAM CARLISLE THACKER

ABSTRACT. When numerical models are used to compute wave spectra at relatively large distances from a storm, the results may be characterized by a patchy distribution of wave energy. This patchiness is a direct consequence of the discrete nature of the wave spectrum in numerical models. To avoid this problem by increasing resolution is computationally expensive, so it is useful to seek a numerical scheme which might minimize this effect. A finite-element analysis suggests that such a scheme might involve advection terms which are averaged over adjacent spectral contributions.

1. INTRODUCTION

Gravity waves disperse as they propagate across the surface of the ocean, longer waves outrunning shorter ones. Consequently, the continuous variety of waves that make up the chaotic sea under a storm is later manifested as a continuous and gradual increase in the length of waves encountered at increasing distance from the storm. All this is well known. What is surprising is that straightforward, reasonable methods used to compute wave propagation and dispersion in numerical models yield wave distributions that are intermittent and patchy rather than continuous and smoothly varying.

This patchiness is a direct result of the discrete nature of the wave models: space, time, wave frequency, and direction of propagation, all of which are in actuality continuous, are depicted for computational purposes by a handful of discrete, representative values. Since each discrete wave component moves at its own velocity, sooner or later the components will become so dispersed that gaps of several grid spaces separate adjacent packets of wave energy.

If the number of frequencies and directions used to specify the spectrum is increased, then adjacent spectral components will separate more slowly. Thus, by increasing spectral resolution, the patchiness can be pushed to greater and greater distances, ultimately beyond the boundaries of the spatial grid. However, wave models already saturate available computational resources, so it is useful to consider whether there might be other ways around this problem.

Smooth numerical results can be obtained by periodically averaging values computed at adjacent points on the spatial grid. Such spatial smoothing is equivalent to a diffusion process. Some numerical schemes, upwind differencing for example, can provide smooth results because

WILLIAM CARLISLE THACKER ● Atlantic Oceanographic and Meteorological Laboratories, National Oceanic and Atmospheric Administration, Miami, Florida 33149.

truncation error takes the form of numerical diffusion. Unfortunately, diffusion and wave dispersion behave differently. For the former, the width of a distribution grows like the square root of the elapsed time, while for the latter, the width increases linearly. Thus, diffusion provides too much smoothing for short times when it is not needed and too little for long times when the patches appear.

A second possibility is to average over adjacent spectral components. At first, this might appear as a change in the physics of wave dispersion, because spectral smoothing will behave like diffusion in wave-vector space. Actually, it should be viewed as an attempt to control the truncation error introduced by approximating a continuous spectrum with a discrete counterpoint. Each discrete wave component should be thought of as a band comprising waves that propagate at similar but slightly different velocities. Without some sort of spectral smoothing, the discrete wave components are more like sharp spectral spikes than like continuous bands.

Spectral smoothing is a natural consequence of the finite-element method. The reason for this is that the method enforces continuity through the use of piecewise polynomial functions to approximate continuously varying quantities. Group velocity and spectral density are required to vary continuously and thus to correspond to spectral bands rather than spectral lines. The finite-element equations take a form that is very much like finite-difference equations, the difference being that the advection terms are averaged over adjacent spectral components. The weights assigned to the various terms that contribute to the average depend on the form specified for the interpolating function.

Existing spectral models are all based upon the finite-difference method, which is most likely the better method for this problem, finite-element models being typically much more expensive for time-dependent phenomena. However, the finite-element method does suggest a simple modification to existing models: the advection terms can be replaced by spectral averages. This modification cannot totally eliminate patchy distributions because they are ultimately due to inadequate resolution, but it is inexpensive and should be tried.

2. DISPERSION AND DISCRETIZATION

The effects of dispersion on the wave spectrum are illustrated in Figs. 1 and 2. Although the spectrum is in fact a function of five variables—two spatial coordinates, wave frequency, direction of propagation, and time—in these figures it has been idealized to a function of only three variables—one spatial coordinate (in the direction of propagation), group velocity (substituted for wave frequency so that the pictures are easier to draw), and time. Figures 1 and 2 represent contour plots of the spectrum at two different times. At the earlier time all of the spectral components are simultaneously present in a small region of space. Later the energy spreads out, and only a narrow band of spectral components is present at any given location.

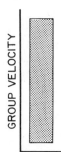

DISTANCE

FIGURE 1. On this contour plot of spectral density, shading is used to indicate the presence of wave energy. Note that waves of all frequencies are located in the same spatial region.

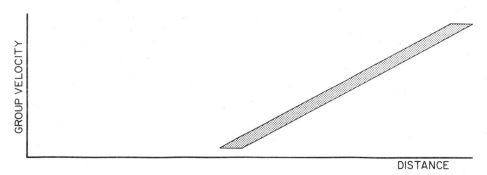

FIGURE 2. Later, after propagating and dispersing, the distribution of Fig. 1 becomes stretched out, so that only a narrow band of frequencies is present at any spatial point.

FIGURE 3. Numerical values of spectral density have been transferred to a discrete grid. The contour encloses a single region having values greater than or equal to 5.

Figures 3 and 4 illustrate the consequence of using a discrete grid to represent the spectrum. Even if *exact* solutions are transferred to the grid, contours can appear patchy. Because there is insufficient resolution to see the narrow spectral bands after the waves have propagated and dispersed, the distribution is aliased. There is no way to recognize from the discrete values on the grid the fact that the contours should enclose a narrow, oblique region. Note that the spatial resolution remains adequate at all times, which argues against spatial smoothing as a method to remove the lumpiness exhibited by the contours. On the other hand, because the lumpiness is due to inadequate resolution of narrow spectral bands, it seems reasonable to smooth the contours by averaging over adjacent spectral components.

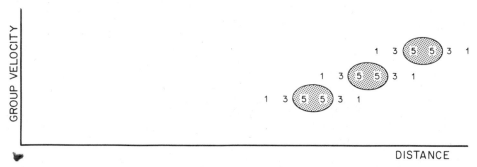

FIGURE 4. After the distribution becomes stretched out, the contour plot is lumpy. If intermediate values of group velocity were represented using a fine grid, then the patches would merge into a single region. However, better spatial resolution will not remove the lumpiness.

Although the discussion has been limited to dispersion of waves propagating in a single direction, it should be recognized that waves also disperse because they propagate in different directions. There is no essential difference in the two cases. For both cases the dispersion is due to differences in group velocity. In one case the magnitude of the velocity varies, and in the other, the direction of the velocity varies. As the waves propagate and disperse, the band of spectral components present at any given point will become narrow in direction as well as in frequency. Unfortunately, it is difficult to draw a diagram to illustrate this point. Perhaps it is sufficient to compare this to the analogous situation of all the rays of the sun incident on the earth being essentially parallel. At distances far from the storm where the waves were generated, a discrete grid of a dozen or so directions of propagation will be unable to resolve the narrow bands of directions. This aliasing of directions is just like the aliasing of frequencies illustrated in Fig. 4. Again, there is no problem with spatial resolution and no need for spatial smoothing. Spectral smoothing in both frequency and direction is more appropriate.

Spectral averaging can be carried out after every time step, or after every few time steps, during model computations. This cannot increase the resolution of the model so that the narrow spectral bands will be seen as narrow bands, but it should smooth the distributions so that they are no longer patchy. The resulting broader spectral bands can then be accepted as a consequence of limited spectral resolution.

Spectral smoothing can also be built into the numerical scheme used to approximate the equations governing propagation and dispersion. In fact, such smoothing is automatically incorporated into schemes based on the finite-element method.

3. NUMERICAL CONSIDERATIONS

If only wave propagation and dispersion are considered and all other effects, such as the generation of waves by the wind, dissipation due to breaking, and nonlinear interactions, are ignored, then the wave spectrum, $E(\mathbf{x}, \mathbf{k}, t)$, must satisfy

$$\frac{\partial E}{\partial t} + \mathbf{v} \cdot \nabla E = 0 \tag{1}$$

At first sight, this can be recognized to be the advection equation, perhaps the most studied of all partial differential equations, at least from the point of view of numerical methods. Just a few of the many papers discussing this equation are those of Courant et al. (1952), Bellman et al. (1958), Lax and Wendroff (1960), Stone and Brian (1963), Roberts and Weiss (1966), Roache (1972), Boris and Book (1973), Gadd (1978), Donéa and Giuliani (1981), and Shapiro and Pinder (1981). If the velocity $\mathbf{v}$ does not vary in space or time, then the exact solution is trivial and corresponds to uniform translation without change of shape. Unfortunately, this is not true of numerical solutions. The imperfect resolution afforded by discrete spatial and temporal grids induces a change of shape. One particularly worrisome problem is that numerical solutions, which should remain positive (e.g., the wave spectrum), might not. Another is that the solution may spread out spatially due to numerical diffusion. Although there is no way to avoid some numerical distortion, it is possible to tailor a numerical scheme to a particular application so that the most important adverse effects will be minimized.

It is important to distinguish between the numerical dispersion caused by spatial and temporal discretization of the advection equation and the anomalous dispersion that leads to patchiness in spectral models. The patchiness is a result of poor *spectral* resolution, and it would remain even if space and time were continuous so long as the spectrum is discrete. In Eq. (1), $\mathbf{v}$ represents group velocity and is a function of $\mathbf{k}$, the wave vector. Wave dispersion is a result of this functional dependence, and patchiness is a result of discretizing $\mathbf{k}$ (or, equivalently, frequency and direction of propagation). None of the references cited above are concerned with spectra, so they would consider E and $\mathbf{v}$ only as functions of $\mathbf{x}$ and t (space and time) and not as functions of $\mathbf{k}$.

Before proceeding with the discussion of anomalous dispersion, it is useful to review how spatial and temporal discretization lead to the more familiar problem of numerical dispersion. It is sufficient, for this purpose, to consider the simpler equation

$$\frac{\partial f}{\partial t} + v \frac{\partial f}{\partial x} = 0 \tag{2}$$

which describes advection in one spatial dimension. The function $f(x, t)$ might represent concentration of dye in a river and v the constant river velocity in which case Eq. (2) would describe the dye patch being washed downstream with negligible mixing. It could also describe the propagation of a *single* spectral component with group velocity given by v.

A discrete analog to (2) that is obtained when both spatial and temporal derivatives are approximated with centered finite differences is

$$\frac{f_i^+ - f_i^-}{\tau} + \frac{v}{2}\left(\frac{f_{i+1}^+ - f_{i-1}^+}{2\Delta} + \frac{f_{i+1}^- - f_{i-1}^-}{2\Delta}\right) = 0 \tag{3}$$

where $f_i^+ = f(i\Delta, (n \pm 1/2)\tau)$. The integers, i and n, index the points on the spatial and temporal grids with spacings, Δ and τ, respectively. Just as the solutions to (2) can be formed from superpositions of terms of the form $\cos(kx - \sigma t + \phi)$, with $\sigma/k = v = $ constant, solution to (3) can be formed from superpositions of terms of the form $\cos(i\Delta k - n\tau\sigma + \phi)$, with

$$\frac{\sigma}{k} = \frac{2}{k\tau}\tan^{-1}\left(\frac{v\tau}{2\Delta}\sin k\Delta\right) \tag{4}$$

Because σ/k is not constant, the numerical solution does not correspond to a simple translation without change of shape. For small $k\Delta$ and small $kv\tau$, $\sigma/k \sim v$, so the numerical dispersion is only important on the scales set by the discrete spatial and temporal grids. For any choice of Δ and τ, σ will be real, so the scheme is unconditionally stable. However, once Δ is fixed, the accuracy of the scheme depends on the value of the Courant number, $C = v\tau/\Delta$. For a fixed value of v, τ (or C) can be chosen so that σ/k differs as little as possible from v for all values of k less than π/Δ.

Another finite-difference analog to (2) is upwind differencing,

$$\frac{f_i^+ - f_i^-}{\tau} + v\left(\frac{f_i^- - f_{i-1}^-}{\Delta}\right) = 0 \tag{5}$$

which is centered neither in space nor in time. This scheme offers two advantages over the previous scheme: it is explicit, meaning that the unknown values for the advanced time are not coupled algebraically, and it preserves positivity, meaning that dye concentrations and energy densities will not become negative. However, it has the disadvantage that it is diffusive. This can be seen from a Fourier analysis leading to an equation similar to (4), which to first order in $k\Delta$ is the dispersion equation for advection with velocity, v, and diffusion with diffusivity, $\frac{1}{2}v\Delta$. This can also be seen by rewriting (5) in the form

$$\frac{f_i^+ - f_i^-}{\tau} + v\left(\frac{f_{i+1}^- - f_{i-1}^-}{2\Delta}\right) = \frac{v\Delta}{2}\left(\frac{f_{i+1}^- - 2f_i^- + f_{i-1}^-}{\Delta^2}\right) \tag{6}$$

The numerical diffusion is like spatial smoothing, and causes the solutions to spread out faster than those for the previous scheme. For this scheme, not only does the accuracy depend on the value of the Courant number, it is unstable unless $0 \leqslant C \leqslant 1$.

A third example is a scheme due to Gadd (1978), which can be written in the form

$$\frac{f_i^+ - f_i^-}{\tau} + v\left[\left(1 + \frac{2a}{3}\right)\left(\frac{f_{i+1}^- - f_{i-1}^-}{2\Delta}\right) - \frac{2a}{3}\left(\frac{f_{i+2}^- - f_{i-2}^-}{4\Delta}\right)\right]$$

$$= \frac{1}{2}v^2\tau\left[\left(1 + \frac{4a}{3}\right)\left(\frac{f_{i+1}^- - 2f_i^- + f_{i-1}^-}{\Delta^2}\right) - \frac{4a}{3}\left(\frac{f_{i+2}^- - 2f_i^- + f_{i-2}^-}{4\Delta^2}\right)\right] \tag{7}$$

where $a = \frac{3}{4}(1 - C^2)$. This scheme is also diffusive with a diffusivity, $\frac{1}{2}v^2\tau$. However, because terms computed on the subgrid with spacing 2Δ are subtracted, this scheme does not lead to as much spreading as does upwind differencing. This scheme is explicit, it is unstable unless $-1 \leqslant C \leqslant 1$, and it does not preserve positivity.

These three schemes were constructed by replacing the partial derivatives in (2) by various finite-difference counterparts, so they are examples of the finite-difference method. On the other hand, the finite-element method is based on approximations to the *function* throughout its domain rather than on approximations to the derivatives at points on a discrete mesh. For example, if $f(x,t)$ is approximated by

$$\tilde{f}(x,t) = \sum_i f_i(t)\phi_i(x) \tag{8}$$

where

$$\phi_i(x) = \begin{cases} \dfrac{x - x_{i-1}}{x_i - x_{i-1}}, & \text{if } x_{i-1} \leqslant x \leqslant x_i \\[2ex] \dfrac{x_{i+1} - x}{x_{i+1} - x_i}, & \text{if } x_i \leqslant x \leqslant x_{i+1} \\[2ex] 0, & \text{otherwise} \end{cases} \tag{9}$$

and where x_i is the coordinate of the ith point on the spatial grid, then

$$\frac{\partial \tilde{f}}{\partial t} + v\frac{\partial \tilde{f}}{\partial x} = R(x,t) \tag{10}$$

Because this piecewise linear approximation is not exact, the residual, $R(x,t)$, cannot vanish everywhere within the domain. The functions, $f_i(t)$, can be chosen so that R will be small throughout the domain by requiring that

$$\int R(x,t)\phi_i(x)dx = 0 \tag{11}$$

for each grid point, i, and for every time, t. For (11) to be satisfied, $f_i(t)$ must be a solution of

$$\frac{d}{dt}\left(\frac{1}{6}f_{i-1} + \frac{2}{3}f_i + \frac{1}{6}f_{i+1}\right) + v\left(\frac{f_{i+1} - f_{i-1}}{x_{i+1} - x_{i-1}}\right) = 0 \tag{12}$$

where $x_{i+1} - x_{i-1} = 2\Delta$. If a different choice were made for the basis functions, $\phi_i(x)$, or if a different criterion were used for judging $R(x,t)$ to be small, then some other equation would take the place of (12). In every case the spatial derivative would be replaced by some finite-difference analog, and the other term would be expressed as a spatial average. This average is a result of continuity imposed by the finite-element method.

Equation (12) is only semidiscrete, because t is still a continuous variable. Either the finite-element method or the finite-difference method can be used to construct a fully discrete computational scheme. In either case, the result will be similar to the schemes given by Eqs. (3), (5), and (7). As is the case for those schemes, finite-element solutions should not be expected to behave well when they are poorly resolved by the computational grid. The error, as before, should be a function of the Courant number. The wide choice of different finite-difference and finite-element schemes makes it possible to choose the scheme that is best for the particular problem at hand.

This digression on numerical dispersion due to spatial and temporal discretization has been relevant to the topic of modeling wave spectra, because it affects the spatial distribution of each spectral component. It has also served to illustrate how poor resolution can degrade numerical results. Now, in turning to the problem of anomalous dispersion, problems due to spectral discretization can be anticipated.

As illustrated in Figs. 1 through 4, after a sufficiently long time the wave spectrum at any point in space will become so sharp that it cannot be resolved by the spectral grid. At this time, the accuracy of the numerical solutions deteriorates and patchy results appear. Nothing can be done to force the true solution to be represented on the spectral grid, but the numerical solutions can be smoothed so that the lumpiness is removed.

It is interesting to consider what results from using the finite-element method to enforce the continuity of the spectrum. For the 1d case, the spectrum, $E(x, k, t)$, and the group velocity, $v(k)$, can be approximated by

$$\tilde{E}(x, k, t) = \sum_j E_j(x, t)\phi_j(k) \tag{13}$$

and

$$\tilde{v}(k) = \sum_j v_j \phi_j(k) \tag{14}$$

where

$$\phi_j(t) = \begin{cases} \dfrac{k - k_{j-1}}{k_j - k_{j-1}}, & k_{j-1} \leqslant k \leqslant k_j \\[2ex] \dfrac{k_{j-1} - k}{k_{j+1} - k_j}, & k_j \leqslant k \leqslant k_{j+1} \\[2ex] 0, & \text{otherwise} \end{cases} \tag{15}$$

Proceeding as before yields equations for the various spectral components that are algebraically coupled,

$$\frac{\partial}{\partial t}\left(\frac{1}{6}E_{j-1} + \frac{2}{3}E_j + \frac{1}{6}E_{j+1}\right) + \frac{1}{6}v_{j-1}\frac{\partial}{\partial x}\left(\frac{E_{j-1} + E_j}{2}\right)$$
$$+ \frac{2}{3}v_j\frac{\partial}{\partial x}\left(\frac{E_{j-1} + 6E_j + E_{j+1}}{8}\right) + \frac{1}{6}v_{j+1}\frac{\partial}{\partial x}\left(\frac{E_j + E_{j+1}}{2}\right) = 0 \tag{16}$$

This should be contrasted to the usual starting point after spectral discretization,

$$\frac{\partial E_j}{\partial t} + v_j\frac{\partial E_j}{\partial x} = 0 \tag{17}$$

Equation (17) treats the spectral components independently, even though there is the implicit assumption that k_j represents a band of wavenumbers. On the other hand, the averages in (16) reflect the fact that the waves in that band propagate with different velocities.

The particular form of the spectral averages in (16) is a result of the choice of piecewise-linear basis functions given in (15) and of the choice of criterion, similar to (11), for judging the approximation to be optimal. Varying either of these would result in differently weighted averages. Also, there is really no need to approximate the group velocity as in (14), because v is a known function of k. This assumption was made simply for ease in evaluating the weighting factors.

Another possibility is to drop the finite-element method and to retain the idea that advection should involve spectral averages. For example, (17) might be replaced by

$$\frac{\partial E_j}{\partial t} + \frac{1}{2}\left(\frac{v_{j-1} + v_j}{2}\right)\frac{\partial}{\partial x}\left(\frac{E_{j-1} + E_j}{2}\right) + \frac{1}{2}\left(\frac{v_j + v_{j+1}}{2}\right)\frac{\partial}{\partial x}\left(\frac{E_j + E_{j+1}}{2}\right) = 0 \tag{18}$$

rather than by (16). No attempt will be made here to suggest a best scheme. The variety is too great, and a lot of numerical experimentation is needed.

It is not difficult to generalize (16) or (18) to two dimensions in coordinate-space and two dimensions in wave-vector space, so that they are appropriate for the directional spectrum. For example, $E(\mathbf{x}, \omega, \theta, t)$ and $v(\omega, \theta, t)$ can be approximated as bilinear,

$$\tilde{E}(\mathbf{x}, \omega, \theta, t) = \sum_{j,k} E_{j,k}(\mathbf{x}, t)\phi_j(\omega)\phi_k(\theta) \tag{19}$$

and

$$\tilde{v}(\omega, \theta) = \sum_{j,k} v_{jk}\phi_j(\omega)\phi_k(\theta) \tag{20}$$

where $\omega^2 = g|\mathbf{k}|$ and θ is the direction of propagation, and where $\phi_j(\omega)$ and $\phi_k(\theta)$ are piecewise-linear bases functions with definitions similar to those in (9) and (15). The counterpart to (16) is straightforward to evaluate but too cumbersome to write out here. The generalization of (18) is already clumsy,

$$\frac{\partial E_{j,k}}{\partial t} + \frac{1}{4}\left(\frac{\mathbf{v}_{j-1,k-1} + \mathbf{v}_{j-1,k} + \mathbf{v}_{j,k-1} + \mathbf{v}_{j,k}}{4}\right)\cdot\nabla\left(\frac{E_{j-1,k-1} + E_{j-1,k} + E_{j,k-1} + E_{j,k}}{4}\right)$$

$$\frac{1}{4}\left(\frac{\mathbf{v}_{j,k-1} + \mathbf{v}_{j,k} + \mathbf{v}_{j+1,k-1} + \mathbf{v}_{j+1,k}}{4}\right)\cdot\nabla\left(\frac{E_{j,k-1} + E_{j,k} + E_{j+1,k-1} + E_{j+1,k}}{4}\right)$$

$$\frac{1}{4}\left(\frac{\mathbf{v}_{j-1,k} + \mathbf{v}_{j-1,k+1} + \mathbf{v}_{j,k} + \mathbf{v}_{j,k+1}}{4}\right)\cdot\nabla\left(\frac{E_{j-1,k} + E_{j-1,k+1} + E_{j,k} + E_{j,k+1}}{4}\right)$$

$$\frac{1}{4}\left(\frac{\mathbf{v}_{j,k} + \mathbf{v}_{j,k+1} + \mathbf{v}_{j+1,k} + \mathbf{v}_{j+1,k+1}}{4}\right)\cdot\nabla\left(\frac{E_{j,k} + E_{j,k+1} + E_{j+1,k} + E_{j+1,k+1}}{4}\right) = 0 \tag{21}$$

Another example could be

$$\frac{\partial E_{j,k}}{\partial t} + \mathbf{v}_{j,k}\cdot\nabla\left(\frac{4E_{j,k} + E_{j+1,k} + E_{j,k+1} + E_{j-1,k} + E_{j,k-1}}{8}\right) \tag{22}$$

The possibilities are endless. Unfortunately, there seems to be no way to choose among them without numerical experimentation.

4. CONCLUSIONS

The patchiness of computed distributions of wave energy is the result of anomalous wave dispersion caused by inadequate resolution of the wave spectrum by the discrete wave-vector grid. Because waves disperse as they propagate, the spectrum inevitably becomes sharp. The proper solution to this problem is to use an exceedingly fine mesh to represent wave frequencies and directions. If Δx and L are, respectively, the spatial resolution and the length of the spatial grid and Δu and $\bar{u}$ are, respectively, the magnitudes of the differences and averages of the group velocities of adjacent discrete wave components, then the restriction placed on the wave-vector grid via Δu is

$$\frac{\Delta u}{\bar{u}} < \frac{\Delta x}{L} \tag{23}$$

When L is large, this restriction imposes a heavy computational burden. Increasing Δx is no help because that simply degrades spatial resolution and increases numerical dispersion. The problem is essentially one of spectral resolution, not spatial resolution. If the computational expense of satisfying (23) is too great, it still may be possible to ameliorate the situation through the use of spectral smoothing. Although the spectrum should be sharp, it can be broadened until it can be resolved by the wave-vector grid. The net effect is too remove the lumpiness from the computed distributions.

Without further work it is impossible to say which of the many possible spectral smoothing schemes is best. Whatever scheme is used, there will be error because the spectrum should in fact be sharp. However, there will also be a component of error due to the details of the numerical smoothing scheme, and the choice of the scheme allows for some control over this component. Some numerical experiments are needed in order to evaluate the characteristics of a variety of schemes.

Finally, it should be recognized that, whether or not some form of spectral smoothing is used, unless (23) is satisfied, spectral models will improperly describe propagation and dispersion for times longer than $\Delta x/\Delta u$. But if after this time the contribution due to waves that have propagated long distances is small compared to those which are generated locally, then the inaccuracies of the model are relatively unimportant.

REFERENCES

Bellman, R., I. Cherry, and G. M. Wing (1958): A note on the numerical integration of a class of non-linear hyperbolic equations. *Q. Appl. Math.* **16**, 181–183.

Boris, J. P., and D. L. Book (1973): Flux-corrected transport. I. SHASTA, a fluid transport algorithm that works. *J. Comput. Phys.* **11**, 38–69.

Courant, R., E. Isaacson, and M. Rees (1952): On the solution of non-linear hyperbolic differential equations by finite differences. *Commun. Pure Appl. Math.* **5**, 243–255.

Donéa, J., and S. Giuliani (1981): A simple method to generate high-order accurate convection operators for explicit schemes based on linear finite elements. *Int. J. Num. Meth. Fluids* **1**, 63–79.

Gadd, A. J. (1978): A numerical advection scheme with small phase speed errors. *Q. J. R. Meteorol. Soc.* **104**, 583–594.

Lax, P. D., and B. Wendroff (1960): Systems of conservation laws. *Commun. Pure Appl. Math.* **13**, 217–237.

Roache, P. J. (1972): On artificial viscosity. *J. Comput. Phys.* **10**, 169–184.

Roberts, K. V., and N. O. Weiss (1966): Convective difference schemes. *Math. Comput.* **20**, 272–299.

Shapiro, A., and G. F. Pinder (1981): Analysis of an upstream weighted collocation approximation of the transport equation. *J. Comput. Phys.* **39**, 46–71.

Stone, H. L., and P. L. T. Brian (1963): Numerical solution of convection transport problems. *AIChE J.* **9**, 681–688.

44

SOME PROBLEMS IN THE DEVELOPMENT OF THE NATIONAL COASTAL WAVES PROGRAM

L. BAER, D. ESTEVA, L. HUFF, W. ISELEY, R. RIBE, AND M. EARLE

ABSTRACT. This paper describes several problems and solutions which resulted from planning the NOAA National Coastal Wave Program. Although Waverider buoys are well accepted for measurements, standard Waveriders are too limited in range and were found to have too much RF interference for a long term network. These two problems can be overcome by using buoys which have satellite communications. In earlier models of Waveriders, aging caused a low bias in height measurements. Routine calibration is thus necessary although it was not generally used in the past. Phase shifts in Waverider data limit its use in zero-crossing, wave shape, groupiness, or similar types of analyses. Uncertainties in wave measurements cause extremal wave height statistics to have a high bias and wide confidence intervals. These problems are aggravated by short record lengths.

A national Coastal Waves Program (CWP) was established in 1980 in the National Ocean Survey (NOS) of the National Oceanic and Atmospheric Administration (NOAA) primarily to provide long-term statistical information for determining environmental design criteria, information needed for government management decisions, and data to aid real-time wave forecasting in support of marine operations. In planning that program, several problems were found which affect the general wave community. This chapter describes two groups of these technical problems and their solutions: (1) wave measurements, and (2) wave statistics. Another important problem which has not yet been solved is that of developing hindcast models that use observed wave data, perhaps in a self-correcting mode.

1. MEASUREMENTS

One of the first activities of the CWP was to install a pilot network of three Waverider buoys to measure waves about 25 km off Cape Henry, Virginia; Cape May, New Jersey; and Ocean City, Maryland. These systems were conventional 0.7-m Waveriders with Warep receivers. In addition, there was a digitized interface unit at each shore station to provide serial data via telephone circuit. Real-time data were relayed on this telephone circuit to an on-line computer system at NOS Headquarters for limited real-time analysis and archiving. The first problem was sporadic loss of data in the radio link from the buoy to shore such that only about 50% of the 17-

L. BAER, D. ESTEVA, L. HUFF, W. ISELEY, AND R. RIBE ● National Ocean Service, National Oceanic and Atmospheric Administration, Rockville, Maryland 20852. M. EARLE ● MEC Systems Corporation, Manassas, Virginia 22110. *Present address for W. I.:* W&L Electronics, Capital Heights, Maryland 20743.

min records that were recorded were suitable (i.e., not over 10% of the data were missing or out of bounds and no breaks in data continuity exceeding 2 s) for straightforward spectral analysis. It was determined that these data losses were due to RF interference. There were also some further data losses in the telephone link and in the automatic processing in CWP offices. The effect of off-channel interference was reduced by adding a crystal filter and a high-gain directional antenna to the shore station. This resulted in a dramatic improvement to well above 90% data throughput in the radio link.

Another obvious problem is the limited radio communications range of Waveriders which precludes deployment in areas which are far offshore. The solution to this has been to obtain Waveriders which have been modified for satellite communications. Though satellites remove the range limitations, they impose stringent constraints on power and transmission time. These have been obviated, while maintaining data accuracy, by careful selection of duty cycles for the buoys which will transmit different messages depending on two wave height thresholds. For "low" wave heights, only spectra computed on board the buoy from 17-min records are transmitted every 3 h. For "intermediate" heights, time histories are transmitted every hour. For "high" waves 34-min time series are recorded on an on-board cassette and time series are transmitted every hour. The thresholds can be set as appropriate to each measurement site.

There are two other considerations of Waverider data which have not yet received adequate attention in the community: phase shift effects and buoy calibration.

The manufacturer's specifications describe phase shifts as shown in Fig. 1. These result from the analog double integration circuit in the buoy. CWP calibrations using a rotating arm have generally confirmed these shifts. Although such phase shifts are not important for computing spectra, they may affect many other analyses such as: "zero-crossing" analyses, studies of the distribution of wave amplitudes, studies of nonlinearities in the waveform, groupiness, and any

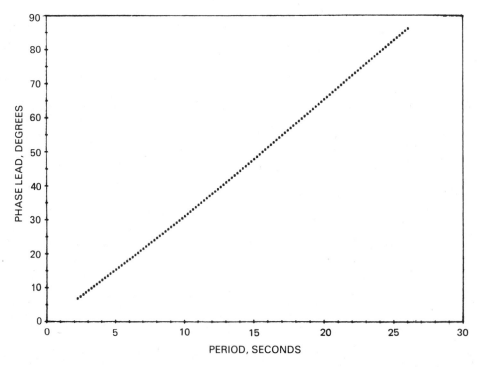

FIGURE 1. Phase shifts of Waverider as specified by the manufacturer. These shifts apply to each sinusoidal component of the waves. Although the manufacturer treats these as phase leads, they may actually be equivalent phase lags.

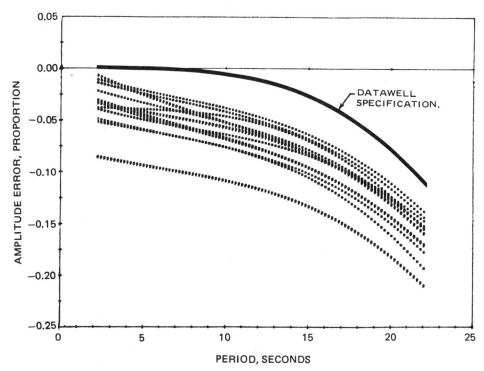

FIGURE 2. Comparison of pre- and postexperiment calibrations for amplitude measurements of Waveriders used in ARSLOE with manufacturer's specifications. Calibration curves are least-squares fit of deviations from manufacturer's specifications to approximately 10 calibration points between 4.2- and 22-s periods. Multiplicative factors to correct spectra are $1/[1\text{-(amplitude error)}]^2$.

other analysis where reconstruction of the true waveshape is important. Simulation studies have shown that if the phases of all frequency components in a wave field are uncorrelated, then the phase error in the response of the measurement system is inconsequential. However, in a real wave field with narrow crests and flattened troughs, where definite correlations exist between the phase of the primary wave components and the phase of its harmonics, the rms output of a measurement system will be affected by the phase characteristics of the measurement system. Thomas *et al.* (1982) simulated the performance of a Waverider for various assumed Pierson–Moskowitz (PM) spectra and compared the difference in data results between when phase corrections were or were not applied during data analysis. PM spectra between 10 and 50 knots (kts) were studied. It was shown that the rms of records which had not been corrected for phase shifts was in excess of the rms for the same records when phase corrections were applied. This excess approached 6% for an assumed 50-kt PM spectrum.

Figure 2 shows the amplitude calibrations of six Waveriders used in the Atlantic Remote sensing Land Ocean Experiment (ARSLOE) which were all biased to low side. Since only about half of the buoys used in ARSLOE belong to NOAA, the problem is general. Similarly low calibrations of 8 to 10% were reported by Pitt *et al.* (1978). The manufacturer attributes this bias to a change in conductivity of the fluid surrounding the accelerometer as it ages. An improved electronics circuit is now available from the manufacturer which is said to correct this problem by compensation for the conductivity changes. In addition, when calibrations are as low as those shown, the system is also temperature sensitive, requiring a further correction based on *in situ* buoy temperature which is frequently not well known. Since there are very few Waverider calibration facilities in operation and relatively few Waveriders are routinely calibrated, users of archived data should be aware of such errors.

2. STATISTICS ACCURACY

One general application of wave data and a planned major product of the CWP is for estimating the statistics of extreme waves for design purposes. The length of record and the accuracy of the measurements and/or hindcasts used to develop those statistics will impact the accuracy of the extreme statistics. Since a complete paper (Earle and Baer, 1982) on this subject is being published elsewhere, only a summary is included here.

This paper approached the question of accuracy of extreme statistics by simulating a large number of annual extreme wave records of different length (5, 10, 20, and 40 years) with different Gaussian-distributed errors (0%, $\pm 10\%$, and $\pm 30\%$ at the 90% confidence level) and for three different assumed "true" or underlying log-normal distributions of annual extremes (representing regions of very high, high, and low wave regions). Similar results would be obtained using other often-used distribution functions. All such computations assume a single distribution which may not be true when there is a mix of causative factors such as winter storms and hurricanes. In interpreting the error levels, it should be noted that a computation of significant wave height from a typical 17-min measurement of waves has a 90% confidence interval of ± 10 to 15% due to natural statistical variability besides the measurement errors. Gaussian-distributed errors at 90% confidence are 1.645 times the standard deviation.

For the usual case of relatively short records (5 to 20 years) being used to extrapolate relatively long return intervals (50 to 100 years), results of the simulation show:

1. That the equation commonly used for computing confidence intervals provides nonconservative interval widths as shown in Fig. 3. Examples of some true confidence interval widths are shown in Fig. 4.
2. There is a significant bias in computing the extreme heights by standard methods which results in overestimating the heights. An example is shown in Fig. 5.

The first of these results comes about because the equation assumes prior knowledge of the underlying distribution and no errors in the record. The prime cause of the second result (bias) is

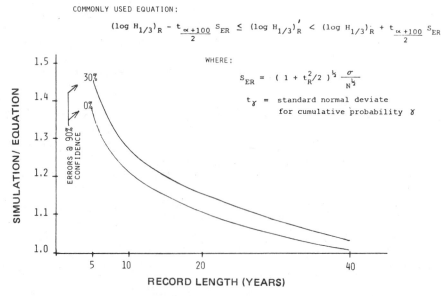

COMMONLY USED EQUATION:

$$(\log H_{1/3})_R - t_{\frac{\alpha +100}{2}} \; S_{ER} \; \leq \; (\log H_{1/3})'_R \; < \; (\log H_{1/3})_R + t_{\frac{\alpha +100}{2}} \; S_{ER}$$

WHERE:

$$S_{ER} = (1 + t_R^2/2)^{\frac{1}{2}} \frac{\sigma}{N^{\frac{1}{2}}}$$

t_γ = standard normal deviate for cumulative probability γ

FIGURE 3. Ratio of confidence intervals on significant wave height computed by simulation with those computed by the commonly used expression shown. This case is for 100-year return period in a high wave region. Mean significant wave height = 18.3 m. "Error" levels shown refer to the size of the Gaussian-distributed simulated errors in the basic time series.

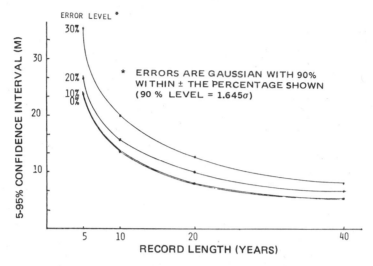

FIGURE 4. Width of the confidence interval on significant wave height for 100-year return period in a very high wave height region as determined by simulation. Mean significant wave height = 21.5 m. "Error" levels shown refer to the size of Gaussian-distributed simulated errors in the basic time series used.

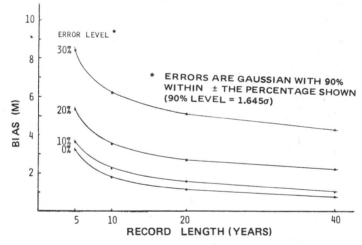

FIGURE 5. Bias in the estimated significant wave height for 100-year return period in very high wave height region as determined by simulation. Mean significant wave height = 21.5 m. All biases are positive showing that the estimates are too high. "Error" levels shown refer to the size of Gaussian-distributed simulated errors in the basic time series used.

that errors in the record increase the "noise" or standard deviation of the assumed distribution. Said another way, rank-ordering before curve fitting tends to increase the steepness of the resulting computed probability distribution curve. Since only knowledge of the accuracy of the underlying data will allow this bias to be estimated and removed and realistic confidence intervals calculated, it is important to evaluate our measurements and hindcasts critically.

3. CONCLUSIONS

Buoy systems (and other systems) are now available to measure one-dimensional waves with reasonable accuracy and reliability whenever needed (except perhaps in ice-prone areas) if care is taken in calibration, etc. Problems such as radio interference have been overcome with filters and directional antennas. Use of satellite communications removes the previous limit of having to deploy the systems relatively near the receiving station. However, without routine calibration, the quality of the data must be considered suspect. It is important to study the quality and representativeness of these measurements. Phase shifts limit the use of Waveriders for applications that need wave profile information.

Extreme wave statistics are biased and have confidence intervals that differ from those computed by the commonly used equation. These problems can only be removed by knowledge of the accuracy (including sampling representativeness) of the measurements or hindcast data used. Further, only relatively long-term data can have reasonable statistical reliability for determination of extreme waves. Accurate long-term statistics for establishing design criteria must generally be based on a new type of hindcasting model which can use the available wave measurements in a self-correcting mode. This should decrease problems with inadequate wind fields as well as incomplete theory. Similar self-correction techniques should help improve accuracy in forecasting models though the wave measurements to be used would necessarily be limited to the present and past times. However, this will require knowledge of the accuracy and representativeness of the wave measurements.

REFERENCES

Earle, M. D., and L. Baer (1982): Effects of uncertainties on extreme wave heights. *ASCE J. Waterw. Port Coastal Ocean Div.* **108**(WW4), 456–477.

Pitt, E. G., J. S. Driver, and J. A. Ewing (1978): Some intercomparisons between wave recorders. *Report No. 43*, Institute of Oceanographic Sciences, Crossway, Taunton, Somerset.

Thomas, R. W. L., E. S. Stickels, and L. C. Huff (1982): Importance of phase corrections to Waverider data. *Oceans '82 Conference Record*, 814–819.

45

MODELS FOR THE HURRICANE WAVE FIELD

D. LEE HARRIS

ABSTRACT. Comprehensive information about the wave climate in hurricanes is needed for the design of ocean structures. Several wave hindcasting models, based on the significant wave height and period concept have been proposed for satisfying this need. It is shown in this paper, by means of a brief literature survey, that the wave fields in the high wind zone of hurricanes are likely to be multiple directional and the spectra are likely to be multiple nodal. There is little reason to believe that a comprehensive investigation based on the significant wave concept will be useful.

Hurricane-generated waves are among the most destructive features of the marine environment. The need for comprehensive information about hurricane waves has been recognized for more than a century. Several algorithms for predicting the sea state in a hurricane have been proposed. In most of these models, attention is concentrated on waves of a single period. Very few data have been presented to show that the predictions made in this manner provide reliable information about the real wave field.

Early conceptual models of the hurricane wave field indicated that multinodal and multidirectional waves should be expected near the center of hurricanes. This concept gained theoretical support from a recent paper by Günther et al. (1981) who showed that low-frequency waves cannot readily change direction to conform to the local wind. Most compilations of observed wave data in hurricanes support the concept of multinodal multidirectional spectra, but few of the observations have been of sufficiently high quality to be accepted as a proof of this concept.

A demonstration that multinodal multidirectional waves are characteristic of hurricanes should be sufficient to discredit the wave prediction models based on the single dominant frequency approach for use in describing hurricane wave fields. This should encourage more concentration on the development of realistic wave fields for severe storms. The following historical review of work in this field is presented with the above end in mind.

Tannehill (1936) reviewed early work related to hurricane wave fields and reproduced a drawing credited to Reid (1849) and shown in Fig. 1. According to Reid, the waves generated within the high-wind-speed zone of a hurricane will run ahead of the storm, crossing locally generated wind waves, and producing cross seas, that is, a multidirectional wave field with at least two distinct wave trains. At a distance of several storm diameters, wave trains with different directions will be sorted and uninodal wave spectra may result.

Cline (1920) embellished Reid's sketch to indicate the increasing wave length with increasing fetch and remarked that the waves should be higher on the right side of the storm track. Tannehill

D. LEE HARRIS ● Department of Coastal Engineering, University of Florida, Gainesville, Florida 32611.

FIGURE 1. Swells and cross seas in a hurricane according to Reid.

(1936) used reports from the then recently established marine weather reporting network to show that shipboard wave reports were consistent with the concepts proposed by Reid (1849) and Cline (1920). The quality and quantity of the observations were not adequate for acceptance as a proof.

Arakawa and Suda (1953) used wave reports obtained when a typhoon passed over the Japanese Fleet in 1935 to construct a two-dimensional wave field in a typhoon. The observations obtained as a series of reports from the individual ships were replotted with reference to the position of the storm center at the time of the observation in a manner similar to the standard method for analyzing hurricane data obtained from aircraft. Arakawa and Suda documented the presence of multiple wave trains and attributed this to a local sea and one or more trains of swell generated in other regions of the storm. Arakawa (1954) presented observations from several later storms which appear to support the conclusions of the first paper.

Pore (1957) and Harris (1962) used the special data files compiled by the National Hurricane Research Project to construct additional time–space composite charts of the hurricane wave field. The quantity and quality of wave reports from merchant ships were still not adequate for the establishment of a model storm wave field based on observations.

Wilson (1955) presented a graphical technique for dealing with moving fetches so that the techniques for predicting the significant wave height and period to be expected from a uniform wind field could be applied for hurricanes. Wilson (1957) applied this approach to the prediction of maximum wave heights generated by hurricanes in the Gulf of Mexico. He acknowledged that the procedure required subjective judgement but provided no verification data.

It appears that Hamilton (1970) compiled the first set of measured wave spectra which show conclusively that multinodal spectra occur near the center of tropical cyclones. The collection of these data was sponsored by a consortium of oil companies and the data were not released for public distribution until several years later. A compilation of Hamilton's data with each spectrum plotted in its proper place relative to the storm center is shown in Fig. 2.

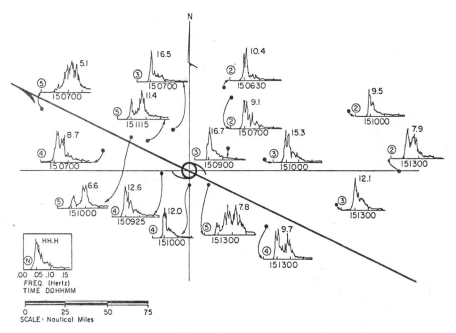

FIGURE 2. Wave spectra near the center of tropical storm Felice, September 14–16, 1970. (After Hamilton, 1970.)

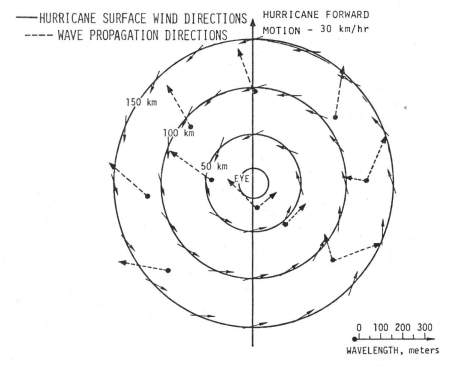

FIGURE 3. Dominant wavenumber vectors compared to wind directions in hurricane Gloria 2. (After King and Shemdin, 1978.)

Bretschneider (1972), author of many contributions to the significant height and period techniques and a widely used model spectrum, introduced a new single-frequency model for predicting waves in hurricanes.

Cardone *et al.* (1975) described an algorithm for predicting the directional spectrum of hurricane waves. This model also was sponsored by the oil companies and few details were published until several years later.

King and Shemdin (1978) presented data obtained in five hurricane penetrations by aircraft equipped with synthetic aperture radar (SAR). The sketch which they used for summarizing their data, shown here as Fig. 3, has a clear resemblance to the sketch provided by Tannehill (1936), Fig. 1.

Forristall *et al.* (1980) presented both observed and predicted directional spectra for hurricanes Carmen 1974 and Eloise 1975 which show that the observed wave spectra may be both multidirectional and multinodal. The prediction model described by Cardone *et al.* (1975) was able to predict the main features of the multidirectional wave spectrum. Further refinement of the model, however, is needed.

SUMMARY

A conceptual model of the hurricane wave field published in 1849 indicated that the wave field near the center of a hurricane will be multidirectional and multinodal. Published summaries of wave observations in hurricanes are consistent with this hypothesis but are generally of low quality.

Several schemes for predicting the wave field in hurricanes on the basis of a significant wave height and period have been proposed in spite of the contradiction between this approach and the evidence of multiple wave trains near the center of hurricanes.

Data have recently become available which show that wave fields in hurricanes do consist of overlapping wave trains. Therefore, the prediction models based on a single wave train should be discarded.

REFERENCES

Arakawa, H. (1954): On the pyramidal, mountainous, and confused sea in the right or dangerous semi-circle of typhoons. *Pap. Meteorol. Geophys.* **5**, 114–123.

Arakawa, H., and K. Suda (1953): Analysis of winds, wind waves, and swell over the sea to the east of Japan during the typhoon of September 26, 1935. *Mon. Weather Rev.* **81**, 31–37.

Bretschneider, C. L. (1972): A non-dimensional stationary hurricane wave model. *Offshore Technology Conference Proceedings* I-S1-62.

Cardone, V. J., W. J. Pierson, and E. G. Ward (1975): Hindcasting the directional spectra of hurricane generated waves. *Offshore Technology Conference Proceedings* 871–878.

Cline, I. M. (1920): Relation of changes in storm tides on the coast of the Gulf of Mexico to the center of movement of hurricanes. *Mon. Weather Rev.* **48**, 127–145.

Forristall, G. Z., E. G. Ward, and V. J. Cardone (1980): Directional wave spectra and wave kinematics in hurricanes Carmen and Eloise. *Proceedings, Coastal Engineering Conference* ASCE, **I**, 567–586.

Günther, W., W. Rosenthal, and M. Dunckel (1981): The response of gravity waves to changing wind direction. *J. Phys. Oceanogr.* **11**, 718–728.

Hamilton, R. C. (1970): Ocean Data Gathering Program Report No. 6 Covering Tropical Storm Felice. Report by Baylor Company, Houston, Texas 77036, Prepared by participants Ocean Data Gathering Program. Report is believed to be available from the National Ocean Data Center, NOAA.

Harris, D. L. (1962): Wave patterns in tropical cyclones. *Mar. Weather Log* **6**, 156–160.

King, D. B., and O. H. Shemdin (1978): Radar observations of hurricane wave directions. *Proceedings, 16th Coastal Engineering Conference* ASCE, 209–226.

Pore, N. A. (1957): Ocean surface waves produced by some recent hurricanes. *Mon. Weather Rev.* **85**, 385–392.

Reid, Lt. Col. W. (1849): *The Progress of the Development of the Law of Storms and of the Variable Winds*, London.

Tannehill, I. R. (1936): Sea swells in relation to movement and intensity of tropical storms. *Mon. Weather Rev.* **64**, 231–238.

Wilson, B. W. (1955): Graphical approach to the forecasting of waves in moving fetches. *Technical Memorandum No. 23*, Beach Erosion Board, Office of the Chief of Engineers, Washington, D. C.

Wilson, B. W. (1957): Hurricane wave statistics for the Gulf of Mexico. *Technical Memorandum No. 98*, Beach Erosion Board, Office of the Chief of Engineers, Washington, D. C.

PARTICIPANTS

WERNER ALPERS
Institut für Meereskunde
Universität Hamburg, and
Max-Planck-Institut für Meteorologie
Hamburg, West Germany
Present address:
Fachbereich Physik
Universität Bremen
Bremen, West Germany

H. M. ASSAL
Stanford Center for Radar Astronomy
Stanford University
Stanford, California 94305

L. BAER
National Ocean Service
National Oceanic and Atmospheric Administration
Rockville, Maryland 20852

M. L. BANNER
Department of Theoretical and Applied Mechanics
University of New South Wales
Sydney, Australia 2033

R. C. BEAL
Applied Physics Laboratory
The Johns Hopkins University
Laurel, Maryland 20707

A. W. BJERKAAS
Applied Physics Laboratory
The Johns Hopkins University
Laurel, Maryland 20707

LARRY F. BLIVEN
Oceanic Hydrodynamics, Inc.
Salisbury, Maryland 21801

J. BÖSENBERG
Max-Planck-Institut für Meteorologie
Hamburg, West Germany

E. BOUWS
Royal Netherlands Meteorological Institute
De Bilt, The Netherlands

J. BRUINSMA
Ministry of Transport and Public Works
Rijkswaterstaat
Directorate for Water Management and
  Hydraulic Research
The Hague, The Netherlands

WILLIAM J. CAMPBELL
Cryosphere Interactions Project
USGS
University of Puget Sound
Tacoma, Washington 98416

H. CARLSON
Deutsches Hydrographisches Institut
Hamburg, West Germany

R. D. CHAPMAN
Applied Physics Laboratory
The Johns Hopkins University
Laurel, Maryland 20707
Present address:
Department of Oceanography
Florida State University
Tallahassee, Florida 32306

ROBERT E. CHENEY
NASA Goddard Space Flight Center
Greenbelt, Maryland 20771

SIR GEORGE DEACON
Institute of Oceanographic Sciences
Wormley, England

G. P. DE LOOR
Physics Laboratory TNO
The Hague, The Netherlands

DANIELLE DE STAERKE
Institut Francais du Petrole
PP 311 92506 Rueil Malmaison
France

P. E. DEXTER
Department of Electronic and Electrical
  Engineering
University of Birmingham
Birmingham, England
Present address:
Bureau of Meteorology
Melbourne, Australia

M. DUNCKEL
Max-Planck-Institut für Meteorologie
Hamburg, West Germany

M. EARLE
MEC Systems Corporation
Manassas, Virginia 22110

DINORAH C. ESTEVA
National Ocean Service
National Oceanic and Atmospheric Administration
Rockville, Maryland 20852

ANDRÉ FONTANEL
Institut Francais du Petrole
PP 311 92506 Rueil Malmaison
France

B. L. GOTWOLS
Applied Physics Laboratory
The Johns Hopkins University
Laurel, Maryland 20707

ALBERT W. GREEN
Naval Ocean Research and Development Activity
NSTL Station, Mississippi 39529

M. GRÜNEWALD
Deutsches Hydrographisches Institut
Hamburg, West Germany

D. W. HANCOCK, III
NASA Goddard Space Flight Center
Wallops Flight Facility
Wallops Island, Virginia 23337

ROBERT O. HARGER
Department of Electrical Engineering
University of Maryland
College Park, Maryland 20742

D. LEE HARRIS
Department of Coastal Engineering
University of Florida
Gainesville, Florida 32611

D. HASSELMANN
Meteorologisches Institut
Universität Hamburg
Hamburg, West Germany

KLAUS HASSELMANN
Max-Planck-Institut für Meteorologie
Hamburg, West Germany

MITSUHIKO HATORI
Department of Geophysics
Faculty of Science
Tohoku University
Sendai, Japan
Present address:
Maritime Meteorological Division
Japan Meteorological Agency
Tokyo, Japan

D. E. HINES
NASA Goddard Space Flight Center
Wallops Flight Facility
Wallops Island, Virginia 23337

S. J. HOGAN
Department of Applied Mathematics and
  Theoretical Physics
University of Cambridge
Cambridge, England CB39EW

T. HONDA
Research Institute for Applied Mechanics
Kyushu University
Kasuga 816, Japan

P. HOOGEBOOM
Physics Laboratory TNO
The Hague, The Netherlands

NORDEN E. HUANG
NASA Goddard Space Flight Center
Greenbelt, Maryland 20771

L. HUFF
National Ocean Services
National Oceanic and Atmospheric Administration
Rockville, Maryland 20852

YUTAKA IMAI
Department of Geophysics
Faculty of Science
Tohoku University
Sendai, Japan
Present address:
Kokusai Kogyo Co., Ltd.
Hino Technical Division
Hino, Japan

G. B. IRANI
Applied Physics Laboratory
The Johns Hopkins University
Laurel, Maryland 20707

W. ISELEY
National Ocean Service
National Oceanic and Atmospheric Administration
Rockville, Maryland 20852
Present address:
W&L Electronics
Capital Heights, Maryland 20743

E. S. KASISCHKE
Environmental Research Institute of Michigan
Ann Arbor, Michigan 48107

W. C. KELLER
Space Systems and Technology Division
Naval Research Laboratory
Washington, D.C. 20375

J. E. KENNEY
Space Systems and Technology Division
Naval Research Laboratory
Washington, D.C. 20375

S. A. KITAIGORODSKII
Department of Earth and Planetary Sciences
The Johns Hopkins University
Baltimore, Maryland 21218

C. E. KNOWLES
Department of Marine, Earth, and
    Atmospheric Sciences
North Carolina State University
Raleigh, North Carolina 27695

G. J. KOMEN
Royal Netherlands Meteorological Institute
De Bilt, The Netherlands

DANIEL S. W. KWOH
TRW/Space and Technology Group
Redondo Beach, California 90278

BRUCE M. LAKE
TRW/Space and Technology Group
Redondo Beach, California 90278

M. T. LANDAHL
Department of Aeronautics and Astronautics
Massachusetts Institute of Technology
Cambridge, Massachusetts 02139

LINDA M. LAWSON
Atlantic Oceanographic and Meteorological
    Laboratories
National Oceanic and Atmospheric Administration
Miami, Florida 33149

B. LE MÉHAUTÉ
Department of Ocean Engineering
University of Miami
Miami, Florida 33149

W. T. LIU
Earth and Space Sciences Division
Jet Propulsion Laboratory
Pasadena, California 91109

ROBERT BRYAN LONG
Atlantic Oceanographic and Meteorological
    Laboratories
National Oceanic and Atmospheric Administration
Miami, Florida 33149

STEVEN R. LONG
NASA Goddard Space Flight Center
Wallops Flight Center
Wallops Island, Virginia 23337

M. S. LONGUET-HIGGINS
Department of Applied Mathematics and
    Theoretical Physics
University of Cambridge
Cambridge, England, and
Institute of Oceanographic Sciences
Wormley, England GU8 5UB

C. C. LU
Department of Ocean Engineering
University of Miami
Miami, Florida 33149

J. D. LYDEN
Environmental Research Institute of Michigan
Ann Arbor, Michigan 48107

WILLIAM MCLEISH
Atlantic Oceanographic and Meteorological
    Laboratories
National Oceanic and Atmospheric Administration
Miami, Florida 33149

JAMES G. MARSH
NASA Goddard Space Flight Center
Greenbelt, Maryland 20771

AKIRA MASUDA
Research Institute for Applied Mechanics
Kyushu University
Kasuga, 816, Japan

W. K. MELVILLE
Institute of Geophysics and Planetary Physics
University of California
San Diego, California 92037, and
Department of Civil Engineering
Massachusetts Institute of Technology
Cambridge, Massachusetts 02139

H. MITSUYASU
Research Institute for Applied Mechanics
Kyushu University
Kasuga, 816, Japan

NELLY M. MOGNARD
Centre National d'Études Spatiales
Groupe de Recherche de Geodesie Spatiale
31055 Toulouse, Cedex, France

F. M. MONALDO
Applied Physics Laboratory
The Johns Hopkins University
Laurel, Maryland 20707

B. A. NELEPO
Marine Hydrophysical Institute
Ukrainian Academy of Sciences
Sevastopol, USSR

H. H. PEECK
Royal Netherlands Meteorological Institute
De Bilt, The Netherlands

W. J. PLANT
Space Systems and Technology Division
Naval Research Laboratory
Washington, D.C. 20375

HERBERT RABIN
Deputy Assistant Secretary of the Navy
Research, Application and Space Technology
Pentagon
Washington, D.C. 20350

WILLIAM RAINEY
Deputy Administrator for Science and Applications
NASA Headquarters
Washington, D.C. 20546

R. RIBE
National Ocean Service
National Oceanic and Atmospheric Administration
Rockville, Maryland 20852

K. RICHTER
Deutsches Hydrographisches Institut
Hamburg, West Germany

WOLFGANG ROSENTHAL
Royal Netherlands Meteorological Institute
De Bilt, The Netherlands

DUNCAN ROSS
Atlantic Oceanographic and Meteorological
    Laboratories
National Oceanic and Atmospheric Administration
Miami, Florida 33149

J. W. SANDERS
Ministry of Transport and Public Works
Rijkswaterstaat
Directorate for Water Management and Hydraulic
    Research
The Hague, The Netherlands

M. J. M. SARABER
Royal Netherlands Meteorological Institute
De Bilt, The Netherlands

R. A. SHUCHMAN
Environmental Research Institute of Michigan
Ann Arbor, Michigan 48107

A. M. SHUTKO
Institute of Radioengineering and Electronics
Academy of Sciences of the USSR
Moscow, USSR

J. A. SMITH
Department of Aeronautics and Astronautics
Massachusetts Institute of Technology
Cambridge, Massachusetts 02139

N. D. SMITH
Institute of Oceanographic Sciences
Wormley, England GU8 5UB

R. SPANHOFF
Ministry of Transport and Public Works
Rijkswaterstaat
Directorate for Water Management and Hydraulic
    Research
The Hague, The Netherlands

R. H. STEWART
Scripps Institution in Oceanography
La Jolla, California 92093, and
Jet Propulsion Laboratory
Pasadena, California 91103

MING-YANG SU
Naval Ocean Research and Development Activity
NSTL Station, Mississippi 39529

WILLIAM CARLISLE THACKER
Atlantic Oceanographic and Meteorological
  Laboratories
National Oceanic and Atmospheric Administration
Miami, Florida 33149

S. THEODORIDIS
Department of Electronic and Electrical
  Engineering
University of Birmingham
Birmingham, England
Present address:
Chair of Telecommunications
School of Electrical Engineering
University of Thessaloniki
Thessaloniki, Greece

T. W. THOMPSON
Jet Propulsion Laboratory
Pasadena, California 91101

YOSHIAKI TOBA
Department of Geophysics
Faculty of Science
Tohoku University
Sendai, Japan

MASAYUKI TOKUDA
Department of Geophysics
Faculty of Science
Tohoku University
Sendai, Japan
Present address:
Institute of Coastal Oceanography
National Research Center for Disaster Prevention
Hiratsuka, Japan

DENNIS B. TRIZNA
Radar Division
Naval Research Laboratory
Washington, D.C. 20375

C. C. TUNG
Department of Civil Engineering
North Carolina State University
Raleigh, North Carolina 27650

E. W. ULMER
Department of Ocean Engineering
University of Miami
Miami, Florida 33149

G. R. VALENZUELA
Space Systems and Technology Division
Naval Research Laboratory
Washington, D.C. 20375

R. A. VAN MOERKERKEN
Royal Netherlands Meteorological Institute
De Bilt, The Netherlands

J. F. VESECKY
Stanford Center for Radar Astronomy
Stanford University
Stanford, California 94305

E. J. WALSH
NASA Goddard Space Flight Center
Wallops Flight Facility
Wallops Island, Virginia 23337

D. E. WEISSMAN
Department of Engineering Science
Hofstra University
Hempstead, New York 11550

M. A. WEISSMAN
Microscience, Inc.,
Federal Way, Washington 98003

LEWIS WETZEL
Naval Research Laboratory
Radar Division
Washington, D.C. 20375

S. E. WIDNALL
Department of Aeronautics and Astronautics
Massachusetts Institute of Technology
Cambridge, Massachusetts 02139

V. V. YEFIMOV
Marine Hydrophysical Institute
Ukrainian Academy of Sciences
Sevastopol, USSR

YELI YUAN
Department of Civil Engineering
North Carolina State University
Raleigh, North Carolina 27650
Present address:
Institute of Oceanology
Academia Sinica
Quingdao
People's Republic of China

INDEX

Ablowitz, M. J., 233
Acceptance, 47, 50
Action spectra, 50
Air flow over waves, 335
Air flow separation, 261, 323
Alber, I. E., 256
Alpers, W., 75, 76, 386, 430, 432, 509, 516,
    529, 606
Angular momentum, 226
Anomalous dispersion, 661
Apel, J. R., 465
Arakawa, H., 678
Armand, N. A., 558, 565
Atmospheric pressure spectrum, 356
Attenuation, 121
    effect of surfactant, 98
Au, B, 555
Average wave period, 183
Averaging procedures, 62
Azimuth bunching, 371
Azimuth falloff, 437, 439
Azimuth shift, 380
Azimuthal image shift, 395

Backscatter, 505
Backscattering, nonspecular, 443
Backus, G. E., 574, 583
Baker, G. R., 210
Bandlimiting phenomenon, 381, 383
Banner, M. L., 92, 96, 106, 229, 261, 263, 266,
    299, 336, 350, 352
Barger, W. R., 95
Barnett, T. P., 147, 154, 184
Barnum, J. R., 520
Barrick, D. E., 520, 542
Basharinov, A. E., 558
Bass, F. G., 375
Bathymetry, 181
Beal, R. C., 371
Beckman, P., 375
Benilov, A. Y., 36
Benjamin, T. B., 126, 138, 210
Benjamin–Feir instability, 127, 238, 241, 251, 255
Benney, D. J., 60
Bergeron, R. F., 336
Bhavanarayana, P. V., 558, 567

Bifurcation, 232
Bifurcated skew waves, 242
Bjerkaas, A. W., 165, 435
Bliven, L. F., 129
Bound harmonics, 202
Boundary conditions, 63
Boundary layer turbulence, 350
Bouws, E., 624
Bracalente, E. M., 492
Bragg scattering, 394
Breakers, surface profiles, 217
Breaking wave model, 266
Breaking waves, 31, 209, 211, 226, 255, 257,
    328, 331, 444
    in mixing, 307
Breaking waves and wind stress, 260
Bretschneider, C. L., 181
Brevig, P., 210, 223
Brown, G. S., 481
Bruinsma, J. P., 639
Brunsveld van Hulten, H. W., 506, 514
Burling, R. W., 34
Businger, J. A., 502
Bye, J., 301

Caldwell, D. R., 307
Capillary wave instability, 315
Capillary waves, 445
Cardone, V. J., 303, 483, 595, 599
Cartwright, D. E., 129, 137, 138, 479, 574
Cavaleri, L., 637
Cavanie, A., 191
Chang, P. C., 350
Chappelear, J. E., 241
Chen, B., 209, 242
Chow, S., 599
Cleaver, R. P., 218
Cline, I. M., 677
Coantic, M., 165
Cokelet, E. D., 136, 209, 255, 256, 296, 297
Collins, J. L., 181
Conditional probability
    relative wave height, 189
    relative wave period, 190
Coupling, wind-waves, 64
Coupling coefficient, 42
Cox, C. S., 321

Craik, A. D. D., 96
Crawford, D. R., 241
Critical layer, 69, 346
Crombie, D. D., 542
Cross spectra, 574

Davies, J. T., 95
Davis, R. E., 60, 68, 169, 336, 459
de Voogt, W. J. P., 639
Deacon, E. L., 104
DeRycke, R. J., 465
Dillon, T. M., 307
Directional distribution of waves, 607
Directional frequency spectra, 171, 572
Directional wavenumber spectra, 175, 449, 455
Direction wavenumber–frequency spectra, 165
Discretization, 662
Dispersion relation 193, 198, 202
Dissipation, finite depth, 145, 622
Donelan, M., 12, 23, 24, 39, 261, 264, 352, 368
Doppler shift corrections, 83, 158, 199
Doppler spectrum, 446, 519
Dorrestein, R., 95, 183
Drag coefficient, 103, 113, 261
Dungey, J. C., 42, 50

Eklund, F., 557
Energy loss
    due to wave breaking, 265, 296
    per unit surface area, 298
Energy overshoot, 145, 152, 159
Energy released from breaking, 304
Energy transfer, 41
Envelope frequency, 140
Envelope solitons, 231, 233, 236
Equilibrium
    Kolmogoroff type, 17, 29
    Phillips type, 26
Equilibrium range, 9
    constants, 35
    finite depth, 161
Esteva, D. C., 161
Evans, O. D., 85
Ewing, J. A., 138
Extrema distribution, 137

Facet motion, 395
Feir, J. E., 126, 138
Fenton, J. D., 297
Filonenko, N. N., 12, 19, 21, 23
Finite depth effects, 145
Fitzgerald, L. M., 95
Flow separation, 335, 350
Fluctuating pressure over waves, 353
Flux, wave-coherent, 326
Foam coverage, 302
Fontanel, A. N., 507
Forristall, G. Z., 12, 23, 24, 25, 26, 31, 32, 35, 544, 596

Fox, M. J. H., 41, 48, 146, 260
Frequency cross spectrum, 574
Frequency downshift, 245, 248
Frequency of whitecapping, 257
Frequency shifting, 117, 238
Fung, A. K., 492, 508
Fung, R. K., 508

Gadd, A. J., 666
Gadzhiyev, J. Z., 10, 27, 28
Garratt, J. R., 104, 481
Garrett, C., 60, 75, 96
Garwood, R. W., 306
Gelci, R., 595
Generation, finite depth, 145
Gent, P. R., 68, 72, 336
Georges, T. M., 519, 522
Gilbert, J. F., 574, 583
Giovanangeli, J. P., 165
Goda, Y., 138
Golding, B. W., 639, 640
Gonzalez, F. I., 596, 605
Goodrich, F. C., 95
Gottifredi, J. C., 96
Gotwols, B. L., 165
Graf, K. A., 393
Grant, H. L., 308, 465
Green, A. W., 231
Greenwood, J. A., 603
Groupiness, 133, 138
Growth curves, 624
Growth rate, 108, 113, 121, 122, 544
    angular dependence, 549
Gunther, E. B., 611
Gunther, W., 677
Guthart, H., 393
Guymer, L. B., 479, 486

Hamilton, J., 241
Hamilton, R. C., 678
Hamiltonian, 19
Hanratty, T. J., 336
Hasselmann, K., 10, 11, 21, 23, 35, 39, 42, 59, 60, 62, 64, 65, 75, 96, 145, 147, 150, 155, 298, 483, 509, 550, 571, 583, 606, 611, 616, 625
Hasselmann, S., 145, 155
Hatori, M., 117
Haug, O., 639
Herterich, K., 145
Hino, M., 96, 98
Holland, W. R., 306
Holtzman, J. C., 508
Honda, I., 75, 95, 299, 300
Houmb, O. G., 131
Huang, N. E., 39, 129, 230, 265
Hughes, B. A., 465
Huhnerfuss, H., 96
Hui, W. H., 42, 50, 241

Humpries, P. N., 303
Hurricane Fico, 597
Hurricane wave field, 677
Hydrodynamic contribution, 530

Image modulation transfer function, 433
Imai, Y., 117, 124
Instabilities
    of gravity-capillary waves, 315
    three-dimensional, 242, 255
Integrals, collision, 21
Interaction time function, 22
Inverse problem, 571
Inverse theory, 574
Irani, G. B., 165, 475
Israeli, M., 210

Jain, A., 378
Jameson, G. J., 96
JASIN Experiment, 517
John, F., 218
Johnson, J. W., 492
Joint probability density, wave heights and
    periods, 184
Joint probability distribution, 181
    amplitude and period, 140
Jones, L., 75, 76
Jones, W. L., 492, 508
JONSWAP spectrum, 24, 35, 43, 44, 75, 130,
    184, 616

Kahma, K. K., 12, 23, 24, 31, 32, 33, 34, 35
Kasevich, R. S., 76, 166
Kasischke, E. F., 605
Katsaros, C., 115
Kaupp, V. H., 508
Kawai, S., 350
Keller, W. C., 75
Kelly, E. J., 393
Kendall, J. M., 336, 350
Kenney, J. E., 449
Kenyon, K. E., 301
Kerman, D., 114
Keulegan, G. H., 95, 99
Khalsa, S. J. S., 502
Kinsman, B., 373
Kitaigorodskii, S. A., 9, 75, 147, 296, 306, 376
Knowles, C. E., 145
Kolmogoroff, A. N., 17
Komen, G., 516
Kondo, J., 351
Krasitskii, V. P., 10, 27, 28, 181
Kraus, E. B., 306
Kusaba, T., 106

LaFond, E. C., 465, 558, 567
Lago, B., 481
Lake, B. M., 126, 138, 140, 165, 233, 255, 256
Landahl, M. T., 59

Large, W. C., 104
Larson, T. R., 59, 432
Lau, J. C., 462
Le Marshall, J. F., 479, 486
Le Mehaute, B., 181
LeMone, M. A., 502
Level of action, 228
Levich, V. G., 95
Limiting amplitude, 297
Liu, H. -T., 115, 127, 144
Liu, H. C., 264
Liu, P. C., 150
Liu, W. T., 502
Lombardini, P. P., 95
Long, A. E., 552
Long, S. R., 129, 299
Longuet-Higgins, M. S., 41, 60, 63, 64, 65, 73,
    75, 96, 115, 129, 130, 134, 136, 137,
    138, 140, 143, 146, 181, 185, 209, 245,
    252, 255, 256, 257, 265, 295, 296, 297,
    298, 301, 307, 315, 352, 374, 461, 462,
    574
Lu, C. C., 181
Luminance spectrum, 471
Lumley, J. L., 21

Manton, M. J., 60
Marineland Experiment, 491
MARSEN, 257
Masuda, A., 41, 146
Maul, G. A., 465
McLean, J. W., 241, 255, 256
McLeish, W., 606
Meiron, D. I. 242
Melville, W., 255, 261, 263, 323, 325, 336, 350
Miche's breaking limit, 185
Microwave, passive, 555
Microwave scattering from short gravity waves, 443
Miles, J., 59, 65, 70, 71, 95, 96, 110, 209, 551
Miller, R. L., 211, 223
Miropolskii, Y. Z., 296
Mitsuyasu, H., 34, 75, 95, 142, 299, 300, 637
Mixed layer, 304
Mixing efficiency, 310
Mixing layer modeling, 307
Modeling
    inverse, 571
    shallow water, 615
Modulation of backscattered microwave signals, 529
Modulation transfer function, 91, 430, 606
Mollo-Christensen, E., 73, 502
Momentum flux, 304, 326, 328
Momentum flux to breaking waves, 321
Monahan, E. C., 260, 303
Monaldo, F. M., 76, 166
Monin, A. S., 296, 307
Moore, R. K., 492, 508
Moskowitz, L., 130
Muircheartaigh, I. O., 303

Muller, P., 299, 300
Munk, W. H., 321, 479, 595

Natural inverse, 580
Navrotskii, V. V., 296
Nelepo, B. A., 193
New, A. L., 224
Niiler, P. P., 306, 307, 309
Nonlinear capillary waves, 315
Nonlinear energy transfer, 231
Nonlinear interaction, 231
   three-dimensional, 241
Nonspecular reflection, 444
North Sea wave models, 639

Ocean wave dynamics experiment, 165
Ocean wave imaging, 394
Ochi, M. K., 191
Oil spills, 511
Okuda, K., 124, 126, 321, 329, 330, 350
Olfe, D. B., 210
Orbital velocity modulation, 432
Orbital velocity spectra, 196
Overshoot, 147, 544, 548
Overturning, 209, 215, 219

Panicker, N. N., 461
Phase speed, 176
Phase-locked components, 166
Phillips, O. M., 9, 21, 26, 30, 41, 59, 92, 96,
       146, 158, 229, 265, 266, 295, 298, 299,
       315, 332, 372, 378, 430, 447, 551
Phillips–Miles theory, 21
Phillips' constant, 11
Pierson, W. J., 39, 115, 130, 142, 181, 230, 252,
       333, 447, 595
Pierson–Moskowitz spectrum, 44, 132
Pilon, R. O., 166
Pitt, E. G., 673
Plant, W. J., 59, 73, 93, 326, 330, 378, 447
Plunging breaker, 209, 223, 258
Pond, S., 104
Pore, N. A., 678
Prediction, discrete spectral wave, 595
Pressure measurements, 324
Pressure probes, 355
Pressure spectra, 357
Pressures, surface, 326

Radar
   HF, 541
   surface contour, 449
Radar backscatter
   L-Band, 491
   measurements, 506
Radar skywave, 517
Radar wind observations, 522
Radiance image spectrum, 472
Ramamonjiarisoa, A., 165
Raney, R. K., 393, 395

Reflection, specular, 444
Regier, L. A., 169, 549
Reichardt, H., 70
Reid, W., 677
Resolution falloff, 433
Resonance, 46
Resonant interaction criteria, 157
Resonant interactions, 145
Reynolds stress modulations, 68, 71
Reynolds stresses, upper ocean, 203
Rice, S. O., 129
Richter, Karl, 75
Riedel, F. W., 435
Ripple modulation, 75
Ripple spectra, 83
Ripple spectra of encounter, 79
Roll, H. U., 502
Rosenthal, W., 75, 540
Ross, D. B., 150, 303
Rottman, J. W., 210
Rufenach, C. L., 395, 432

Saffman, P. G., 209, 242, 245
Sanders, J. W., 639
SAR azimuth impulse response, 386
SAR image, 380, 393, 491
   Gulf stream, 501
   of the sea, 377, 597
SAR imaging model, 371
SAR performance, 395
Saturation, wind-dependent, 32
Scattering, electromagnetic, 371
Scatterometer RAMSES, 531
Schooley, A. H., 315
Schroeder, L. C., 492
Schwartz, L. W., 210, 297
Scott, A. C., 233
Scott, J. C., 96
Sea surface luminance, 468
SEASAT altimeter measurements, 479
Sell, W., 41, 146
Shabat, A. B., 233
Shallow water wave statistics, 181
Shallow-water effects, 615
Shand, J. A., 470
Shearman, E. D. R., 517
Shemdin, O. H., 86, 492
Shoenfeld, P. S., 475
Shonting, D. H., 296
Short wave modulation, 75
Shuchman, A., 387, 605
Significant slope, 129, 300
Significant wave heights, Southern Ocean, 479
Sittrop, H., 508
Skew wave bifurcation, 242
Skin friction, 331
SLAR, 505
Smith, F. I. P., 96
Smith, J. A., 59, 60, 73, 74, 75, 96

Smith, N. D., 257, 574
SMMR, 597
Snodgrass, F. E., 479
Snyder, R. L., 95, 111, 353, 361, 364, 366, 549, 551, 552
Sorrell, F. Y., 133
Spectra, 13
Spectra
 demodulated, 87
 double-peaked, 55
 normalized, 23
 reduced, 14
 wave velocity, 193
Spectral window, 589
Spectrum of encounter, 76
Spilling breakers, 158, 247
Spizzichino, A., 375
Statistics of extreme waves, 674
Steele, J., 306
Steep waves, 209
Stewart, R. H., 520
Stewart, R. W., 374, 549, 550, 551, 552
Stilwell, D., 166
Stoker, J. J., 266, 297
Stokes drift, 301
Stolte, S., 75, 78
Strong, A. E., 465
Strong interactions, 124, 232, 247
Stubbs, A. R., 308
Sturm, G. V., 133
Su, M. -Y., 127, 143, 231, 256, 264
Subharmonic instabilities, 231
Suda, K., 678
Surface drift current, 301, 303
Surface jump-meter, 259
Surfactant effects, 95
Sutherland, J. A., 147, 154
Sverdrup, H. U., 595
Swell, 486
Swift, C. T., 395, 432
Symmetric wave bifurcations, 245

Takeuchi, K., 350
Tannehill, I. R., 677
Tayfun, M. A., 129, 130
Taylor, P. A., 68, 72, 336
Teague, C., 549, 550, 551, 552
Telionis, D. P., 336
Temperature, brightness, 557, 558
Thomas, R. W., 673
Thompson, R., 307
Thorpe, S. A., 303, 308
Tilt modulation, 529
Toba, Y., 11, 117
Tokuda, M., 117
Townsend, A. A., 60, 68
Townsend, W. F., 441
Transfer functions, 42, 49, 50, 52, 54
Transient wave growth, 149

Trizna, D., 38, 115, 516
Trizna, R. B., 552
Tropical storms, 596
Tsai, C. H., 191
Tung, C. C., 265
Turbulence, 17
Turbulence generation, 296
Turbulent energy dissipation rate, 308
Turbulent velocity fluctuations, 203
Two-scale electromagnetic scattering model, 374
Tyler, G. L., 431

Ulmer, E. W., 181
Uniform wave train evolution, 240

Valenzuela, G. R., 59, 62, 74, 96, 115, 375, 378
Van Dorn, W. G., 95, 104
Vanden-Broeck, J. -M., 210
Variance spectrum of encounter, 77
Velocity bunching, 393, 432
Video cameras, wave imaging, 167
Vincent, C. L., 10
Vinje, T., 210, 223
VIRR, 597
Vose, R. W., 95

Wallops spectrum, 130
Walsh, E. J., 307
Wang, J., 183
WAVDYN, 165
Wave amplitude distribution, 134
Wave breaking, 26, 136, 231, 265, 321
Wave energy, 18
Wave evolution, 120
Wave generation, 59, 263
Wave groups, three-dimensional, 231
Wave growth, 67, 71, 96, 100, 106, 111, 121, 149, 152, 541
Wave growth curve, 617
Wave interactions, 20, 42, 59
 classes of, 232
 energy flow, 46
Wave model, 603
 GONO, 639
 verification, 639
Wave modeling, shallow water, 639
Wave modulation, 126
Wave packet evolution, 233
Wave spectral development, 541
Wave statistics, 129
Wave-damping, 99
Wave-induced air flow, 340, 347
Wavenumber–frequency spectra, 172, 175, 200
Wave-pressure transfer function, 365
Waveriders, 671
Wave-tip, 213
Waves
 mechanically generated, 117
 regular, 118
 shoaling, 27

Weak interactions, 232
Webb, D. J., 41, 481
Webb, E. K., 104
Weissman, D., 516
Weissman, M., 73, 114, 368, 446
West, B. T., 21
Wetzel, L., 447
Whitecapping frequency, 257
Whitecaps, 257, 301, 321
Widnall, S. E., 59
Williams, J. M., 209, 229
Wilson, B. W., 678
Wilson, L. R., 395, 432
Wind profile, 102, 344
Wind setup, 104
Wind stress, 102
Wind wave spectrum, 100
Window effect, 124

Wishner, R. P., 393
Wright, J. W., 59, 75, 90, 91, 96, 321, 326, 329,
    375, 378, 509, 529, 542
Wu, J., 104
Wyatt, L. R., 517

Yefimov, V. V., 193
Yuan, Y., 265
Yuen, H., 126, 128, 140, 165, 233, 242, 245,
    256, 333

Zakharov, V. E., 12, 19, 21, 23, 233
Zaslavskii, M. M., 11
Zelenka, J. S., 387
Zero crossing wave period, 182
Zietlow, C. R., 260
Zilker, D. P., 336